MULTIVARIATE ANALYSIS

This is a volume in
PROBABILITY AND MATHEMATICAL STATISTICS
A Series of Monographs and Textbooks

Editors: Z. W. Birnbaum and E. Lukacs

A complete list of titles in this series is available from the publisher upon request.

MULTIVARIATE ANALYSIS

K. V. MARDIA
Department of Statistics,
University of Leeds, Leeds, U.K.

J. T. KENT
Department of Statistics,
University of Leeds, Leeds, U.K.

J. M. BIBBY
Faculty of Mathematics,
The Open University,
Milton Keynes, U.K.

ACADEMIC PRESS

An imprint of Elsevier Science

Amsterdam • Boston • Heidelberg • London • New York
Oxford • Paris • San Diego • San Francisco
Singapore • Sydney • Tokyo

This book is printed on acid-free paper.

Reprinted 2003

Academic Press
An Imprint of Elsevier Science
84 Theobald's Road, London WC1X 8RR, UK
http://www.academicpress.com

Academic Press
An Imprint of Elsevier Science
525 B Street, Suite 1900 San Diego, California 92101–4495, USA
http://www.academicpress.com

ISBN 0-12-471252-5 Pbk

A catalogue record for this book is available from the British Library

Mardia, Kantilal Varichand
Multivariate analysis. — (Probability and Mathematical statistics).
1. Multivariate analysis
I. Title II. Kent, J. T.
III. Bibby, John IV. Series
519.5'3 QA278 79–40922

Transferred to digital printing 2006

To Neeta Mardia

"Everything is related with every other thing, and this relation involves the emergence of a relational quality. The qualities cannot be known *a priori*, though a good number of them can be deduced from certain fundamental characteristics."

– Jaina philosophy

The Jaina Philosophy of Non-Absolutism by S. Mookerjee, q.v. Mahalanobis (1957).

PREFACE

Multivariate Analysis deals with observations on more than one variable where there is some inherent interdependence between the variables. With several texts already available in this area, one may very well enquire of the authors as to the need for yet another book. Most of the available books fall into two categories, either theoretical or data analytic. The present book not only combines the two approaches but it also emphasizes modern developments. The choice of material for the book has been guided by the need to give suitable matter for the beginner as well as illustrating some deeper aspects of the subject for the research worker. Practical examples are kept to the forefront and, wherever feasible, each technique is motivated by such an example.

The book is aimed at final year undergraduates and postgraduate students in Mathematics/Statistics with sections suitable for practitioners and research workers. The book assumes a basic knowledge of Mathematical Statistics at undergraduate level. An elementary course on Linear Algebra is also assumed. In particular, we assume an exposure to Matrix Algebra to the level required to read Appendix A.

Broadly speaking, Chapters 1–6 and Chapter 12 can be described as containing direct extensions of univariate ideas and techniques. The remaining chapters concentrate on specifically multivariate problems which have no meaningful analogues in the univariate case. Chapter 1 is primarily concerned with giving exploratory analyses for multivariate data and briefly introduces some of the important techniques, tools, and diagrammatic representations. Chapter 2 introduces various distributions together with some fundamental results, whereas Chapter 3 concentrates exclusively on normal distribution theory. Chapters 4–6 deal with problems in inference. Chapter 7 gives an over-view of Econometrics, whilst Principal Component Analysis, Factor Analysis, Canonical Correlation Analysis, and Discriminant Analysis are discussed from both theoretical and practical points of view in Chapters 8–11. Chapter 12 is on Multivariate Analysis of Variance, which can be better understood in terms of the techniques of previous chapters. The later chapters look into the presently developing techniques of Cluster Analysis, Multidimensional Scaling, and Directional Data.

Each chapter concludes with a set of exercises. Solving these will not only enable the reader to understand the material better but will also serve to complement the chapter itself. In general, the questions have in-built answers, but, where desirable, hints for the solution of theoretical problems are provided. Some of the numerical exercises are designed to be run on a computer, but as the main aim is on interpretation, the answers are provided. We found NAG routines and GLIM most useful, but nowadays any computer centre will have some suitable statistics and matrix algebra routines.

There are three appendices, A, B, and C, which respectively provide a sufficient background of matrix algebra, a summary of univariate statistics, and some tables of critical values. The aim of Appendix A on Matrix Algebra is not only to provide a summary of results, but also to give sufficient guidance to master these for students having little previous knowledge. Equations from Appendix A are referred to as (A.x.x) to distinguish them from (1.x.x), etc. Appendix A also includes a summary of results in n-dimensional geometry which are used liberally in the book. Appendix B gives a summary of important univariate distributions.

The reference list is by no means exhaustive. Only directly relevant articles are quoted and for a fuller bibliography we refer the reader to Anderson, Das Gupta, and Styan (1972) and Subrahmaniam and Subrahmaniam (1973). The reference list also serves as an author index. A subject index is provided.

The material in the book can be used in several different ways. For example, a one-semester elementary course of 40 lectures could cover the following topics. Appendix A; Chapter 1 (Sections 1.1–1.7); Chapter 2 (Sections 2.1–2.5); Chapter 3 (Sections 3.4.1, 3.5, 3.6.1, assuming results from previous sections, Definitions 3.7.1, 3.7.2); Chapter 4 (Section 4.2.2); Chapter 5 (Sections 5.1, 5.2.1a, 5.2.1b, 5.2.2a. 5.2.2b, 5.3.2b, 5.5); Chapter 8 (Sections 8.1, 8.2.1, 8.2.2, 8.2.5, 8.2.6, 8.4.3, 8.7); Chapter 9 (Sections 9.1–9.3, 9.4 (without details), 9.5, 9.6, 9.8); Chapter 10 (Sections 10.1, 10.2); Chapter 11 (Sections 11.1, 11.2.1–11.2.3, 11.3.1, 11.6.1). Further material which can be introduced is Chapter 12 (Sections 12.1–12.3, 12.6); Chapter 13 (Sections 13.1, 13.3.1); Chapter 14 (Sections 14.1, 14.2). This material has been covered in 40 lectures spread over two terms in different British universities. Alternatively, a one-semester course with more emphasis on foundation rather than applications could be based on Appendix A and Chapters 1–5. Two-semester courses could include all the chapters, excluding Chapters 7 and 15 on Econometrics and Directional Data, as well as the sections with asterisks. Mathematically orientated students may like to proceed to Chapter 2, omitting the data analytic ideas of Chapter 1.

Various new methods of presentation are utilized in the book. For instance the data matrix is emphasized throughout, a density-free approach is given for normal theory, the union intersection principle is used in testing as well as the likelihood ratio principle, and graphical methods are used in explanation. In view of the computer packages generally available, most of the numerical work is taken for granted and therefore, except for a few particular cases, emphasis is not placed on numerical calculations. The style of presentation is generally kept descriptive except where rigour is found to be necessary for theoretical results, which are then put in the form of theorems. If any details of the proof of a theorem are felt tedious but simple, they are then relegated to the exercises.

Several important topics not usually found in multivariate texts are discussed in detail. Examples of such material include the complete chapters on Econometrics, Cluster Analysis, Multidimensional Scaling, and Directional Data. Further material is also included in parts of other chapters: methods of graphical presentation, measures of multivariate skewness and kurtosis, the singular multinormal distribution, various non-normal distributions and families of distributions, a density-free approach to normal distribution theory, Bayesian and robust estimators, a recent solution to the Fisher–Behrens problem, a test of multinormality, a non-parametric test, discarding of variables in regression, principal component analysis and discrimination analysis, correspondence analysis, allometry, the jack-knifing method in discrimination, canonical analysis of qualitative and quantitative variables, and a test of dimensionality in MANOVA. It is hoped that coverage of these developments will be helpful for students as well as research workers.

There are various other topics which have not been touched upon partly because of lack of space as well as our own preferences, such as Control Theory, Multivariate Time Series, Latent Variable Models, Path Analysis, Growth Curves, Portfolio Analysis, and various Multivariate Designs.

In addition to various research papers, we have been influenced by particular texts in this area, especially Anderson (1958), Kendall (1975), Kshirsagar (1972), Morrison (1976), Press (1972), and Rao (1973). All these are recommended to the reader.

The authors would be most grateful to readers who draw their attention to any errors or obscurities in the book, or suggest other improvements.

January 1979

Kanti Mardia
John Kent
John Bibby

ACKNOWLEDGEMENTS

First of all we wish to express our gratitude to pioneers in this field. In particular, we should mention M. S. Bartlett, R. A. Fisher, H. Hotelling, D. G. Kendall, M. G. Kendall, P. C. Mahalanobis, C. R. Rao, S. N. Roy, W. S. Torgerson, and S. S. Wilks.

We are grateful to authors and editors who have generously granted us permission to reproduce figures and tables.

We are also grateful to many colleagues for their valuable help and comments, in particular Martin Beale, Christopher Bingham, Lesley Butler, Richard Cormack, David Cox, Ian Curry, Peter Fisk, Allan Gordon, John Gower, Peter Harris, Chunni Khatri, Conrad Leser, Eric Okell, Ross Renner, David Salmond, Cyril Smith, and Peter Zemroch. We are also indebted to Joyce Snell for making various comments on an earlier draft of the book which have led to considerable improvement. We should also express our gratitude to Rob Edwards for his help in various facets of the book, for calculations, for proof-reading, for diagrams, etc.

Some of the questions are taken from examination papers in British universities and we are grateful to various unnamed colleagues. Since the original sources of questions are difficult to trace, we apologize to any colleague who recognizes a question of his own.

The authors would like to thank their wives, Pavan Mardia, Susan Kent, and Zorina Bibby.

Finally our thanks go to Barbara Forsyth and Margaret Richardson for typing a difficult manuscript with great skill.

KVM
JTK
JMB

CONTENTS

1
Introduction

1.1 Objects and Variables

Multivariate analysis deals with data containing observations on two or more variables each measured on a set of objects. For example, we may have the set of examination marks achieved by certain students, or the cork deposit in various directions of a set of trees, or flower measurements for different species of iris (see Tables 1.2.1, 1.4.1, and 1.2.2, respectively). Each of these data has a set of "variables" (the examination marks, trunk thicknesses, and flower measurements) and a set of "objects" (the students, trees, and flowers). In general, if there are n objects, $o_1, \ldots, o_n$ and p variables, $x_1, \ldots, x_p$, the data contains np pieces of information. These may be conveniently arranged using an $(n \times p)$ "data matrix", in which each row corresponds to an object and each column corresponds to a variable. For instance, three variables on five "objects" (students) are shown as a (5×3) data matrix in Table 1.1.1.

Table 1.1.1 Data matrix with five students as objects where x_1 = age in years at entry to university, x_2 = marks out of 100 in an examination at the end of the first year, and x_3 = sex

	Variables		
Objects	x_1	x_2	x_3 †
1	18.45	70	1
2	18.41	65	0
3	18.39	71	0
4	18.70	72	0
5	18.34	94	1

† 1 indicates male; 0 indicates female.

Note that the variables need not all be of the same type: in Table 1.1.1, x_1 is a "continuous" variable, x_2 is a discrete variable, and x_3 is a dichotomous variable. Note also that attribute, characteristic, description, item, measurement, and response are synonyms for "variable", whereas individual, observation, plot, reading, and unit can be used in place of "object".

1.2 Some Multivariate Problems and Techniques

We may now illustrate various categories of multivariate technique.

1.2.1 Generalizations of univariate techniques

Most univariate questions are capable of at least one multivariate generalization. For instance, using Table 1.2.1 we may ask, as an example, "What is the appropriate underlying parent distribution of examination marks on various papers of a set of students?" "What are the summary statistics?" "Are the differences between average marks on different papers significant?", etc. These problems are direct generalizations of univariate problems and their motivation is easy to grasp. See for example Chapters 2–6 and 12.

1.2.2 Dependence and regression

Referring to Table 1.2.1, we may enquire as to the degree of dependence between performance on different papers taken by the same students. It may be useful, for counselling or other purposes, to have some idea of how final degree marks ("dependent" variables) are affected by previous examination results or by other variables such as age or sex ("explanatory" variables). This presents the so-called regression problem, which is examined in Chapter 6.

1.2.3 Linear combinations

Given examination marks on different topics (as in Table 1.2.1), the question arises of how to combine or average these marks in a suitable way. A straightforward method would use the simple arithmetic mean, but this procedure may not always be suitable. For instance, if the marks on some papers vary more than others, we may wish to weight them differently. This leads us to search for a linear combination (weighted sum) which is "optimal" in some sense. If all the examination papers fall

Table 1.2.1 Marks in open-book and closed-book examination out of 100 †

Mechanics (C)	Vectors (C)	Algebra (O)	Analysis (O)	Statistics (O)
77	82	67	67	81
63	78	80	70	81
75	73	71	66	81
55	72	63	70	68
63	63	65	70	63
53	61	72	64	73
51	67	65	65	68
59	70	68	62	56
62	60	58	62	70
64	72	60	62	45
52	64	60	63	54
55	67	59	62	44
50	50	64	55	63
65	63	58	56	37
31	55	60	57	73
60	64	56	54	40
44	69	53	53	53
42	69	61	55	45
62	46	61	57	45
31	49	62	63	62
44	61	52	62	46
49	41	61	49	64
12	58	61	63	67
49	53	49	62	47
54	49	56	47	53
54	53	46	59	44
44	56	55	61	36
18	44	50	57	81
46	52	65	50	35
32	45	49	57	64
30	69	50	52	45
46	49	53	59	37
40	27	54	61	61
31	42	48	54	68
36	59	51	45	51
56	40	56	54	35
46	56	57	49	32
45	42	55	56	40
42	60	54	49	33
40	63	53	54	25
23	55	59	53	44
48	48	49	51	37
41	63	49	46	34
46	52	53	41	40

† O indicates open-book, C indicates closed book.

Table 1.2.1 Continued

Mechanics (C)	Vectors (C)	Algebra (O)	Analysis (O)	Statistics (O)
46	61	46	38	41
40	57	51	52	31
49	49	45	48	39
22	58	53	56	41
35	60	47	54	33
48	56	49	42	32
31	57	50	54	34
17	53	57	43	51
49	57	47	39	26
59	50	47	15	46
37	56	49	28	45
40	43	48	21	61
35	35	41	51	50
38	44	54	47	24
43	43	38	34	49
39	46	46	32	43
62	44	36	22	42
48	38	41	44	33
34	42	50	47	29
18	51	40	56	30
35	36	46	48	29
59	53	37	22	19
41	41	43	30	33
31	52	37	27	40
17	51	52	35	31
34	30	50	47	36
46	40	47	29	17
10	46	36	47	39
46	37	45	15	30
30	34	43	46	18
13	51	50	25	31
49	50	38	23	9
18	32	31	45	40
8	42	48	26	40
23	38	36	48	15
30	24	43	33	25
3	9	51	47	40
7	51	43	17	22
15	40	43	23	18
15	38	39	28	17
5	30	44	36	18
12	30	32	35	21
5	26	15	20	20
0	40	21	9	14

in one group then *principal component analysis* and *factor analysis* are two techniques which can help to answer such questions (see Chapters 8 and 9). In some situations the papers may fall into more than one group—for instance, in Table 1.2.1 some examinations were "open-book" while others were "closed-book". In such situations we may wish to investigate the use of linear combinations within each group separately. This leads to the method known as *canonical correlation analysis*, which is discussed in Chapter 10.

The idea of taking linear combinations is an important one in multivariate analysis, and we shall return to it in Section 1.5.

1.2.4 Assignment and dissection

Table 1.2.2 gives three (50×4) data matrices. In each matrix the "objects" are 50 irises of species *Iris setosa*, *Iris versicolour*, and *Iris virginica*, respectively. The "variables" are

$$x_1 = \text{sepal length}, \qquad x_2 = \text{sepal width},$$
$$x_3 = \text{petal length}, \qquad x_4 = \text{petal width}.$$

If a new iris of unknown species has measurements $x_1 = 5.1$, $x_2 = 3.2$, $x_3 = 2.7$, $x_4 = 0.7$ we may ask to which species it belongs. This presents the problem of *discriminant analysis*, which is discussed in Chapter 11. However, if we were presented with the 150 observations of Table 1.2.2 in an unclassified manner (say, before the three species were established) then the aim could have been to dissect the population into homogeneous groups. This problem is handled by *cluster analysis* (see Chapter 13).

1.2.5 Building configurations

In some cases the data consists not of an $(n\times p)$ data matrix, but of $\frac{1}{2}n(n-1)$ "distances" between all pairs of points. To get an intuitive feel for the structure of such data, a configuration can be constructed of n points in a Euclidean space of low dimension (e.g. $p = 2$ or 3). Hopefully the distances between the n points of the configuration will closely match the original distances. The problems of building and interpreting such configurations are studied in Chapter 14, on *multidimensional scaling*.

1.2.6 Directional data

There are various problems which arise when the variables are directional—that is, the multivariate observations are constrained to lie on a hypersphere. For a discussion of these problems see Chapter 15.

Table 1.2.2 Measurements on three types of iris (after Fisher, 1936)

Iris setosa				*Iris versicolour*				*Iris virginica*			
Sepal length	Sepal width	Petal length	Petal width	Sepal length	Sepal width	Petal length	Petal width	Sepal length	Sepal width	Petal length	Petal width
5.1	3.5	1.4	0.2	7.0	3.2	4.7	1.4	6.3	3.3	6.0	2.5
4.9	3.0	1.4	0.2	6.4	3.2	4.5	1.5	5.8	2.7	5.1	1.9
4.7	3.2	1.3	0.2	6.9	3.1	4.9	1.5	7.1	3.0	5.9	2.1
4.6	3.1	1.5	0.2	5.5	2.3	4.0	1.3	6.3	2.9	5.6	1.8
5.0	3.6	1.4	0.2	6.5	2.8	4.6	1.5	6.5	3.0	5.8	2.2
5.4	3.9	1.7	0.4	5.7	2.8	4.5	1.3	7.6	3.0	6.6	2.1
4.6	3.4	1.4	0.3	6.3	3.3	4.7	1.6	4.9	2.5	4.5	1.7
5.0	3.4	1.5	0.2	4.9	2.4	3.3	1.0	7.3	2.9	6.3	1.8
4.4	2.9	1.4	0.2	6.6	2.9	4.6	1.3	6.7	2.5	5.8	1.8
4.9	3.1	1.5	0.1	5.2	2.7	3.9	1.4	7.2	3.6	6.1	2.5
5.4	3.7	1.5	0.2	5.0	2.0	3.5	1.0	6.5	3.2	5.1	2.0
4.8	3.4	1.6	0.2	5.9	3.0	4.2	1.5	6.4	2.7	5.3	1.9
4.8	3.0	1.4	0.1	6.0	2.2	4.0	1.0	6.8	3.0	5.5	2.1
4.3	3.0	1.1	0.1	6.1	2.9	4.7	1.4	5.7	2.5	5.0	2.0
5.8	4.0	1.2	0.2	5.6	2.9	3.6	1.3	5.8	2.8	5.1	2.4
5.7	4.4	1.5	0.4	6.7	3.1	4.4	1.4	6.4	3.2	5.3	2.3
5.4	3.9	1.3	0.4	5.6	3.0	4.5	1.5	6.5	3.0	5.5	1.8
5.1	3.5	1.4	0.3	5.8	2.7	4.1	1.0	7.7	3.8	6.7	2.2
5.7	3.8	1.7	0.3	6.2	2.2	4.5	1.5	7.7	2.6	6.9	2.3
5.1	3.8	1.5	0.3	5.6	2.5	3.9	1.1	6.0	2.2	5.0	1.5
5.4	3.4	1.7	0.2	5.9	3.2	4.8	1.8	6.9	3.2	5.7	2.3
5.1	3.7	1.5	0.4	6.1	2.8	4.0	1.3	5.6	2.8	4.9	2.0

4.6	3.6	1.0	0.2	6.3	2.5	4.9	1.5	7.7	2.8	6.7	2.0
5.1	3.3	1.7	0.5	6.1	2.8	4.7	1.2	6.3	2.7	4.9	1.8
4.8	3.4	1.9	0.2	6.4	2.9	4.3	1.3	6.7	3.3	5.7	2.1
5.0	3.0	1.6	0.2	6.6	3.0	4.4	1.4	7.2	3.2	6.0	1.8
5.0	3.4	1.6	0.4	6.8	2.8	4.8	1.4	6.2	2.8	4.8	1.8
5.2	3.5	1.5	0.2	6.7	3.0	5.0	1.7	6.1	3.0	4.9	1.8
5.2	3.4	1.4	0.2	6.0	2.9	4.5	1.5	6.4	2.8	5.6	2.1
4.7	3.2	1.6	0.2	5.7	2.6	3.5	1.0	7.2	3.0	5.8	1.6
4.8	3.1	1.6	0.2	5.5	2.4	3.8	1.1	7.4	2.8	6.1	1.9
5.4	3.4	1.5	0.4	5.5	2.4	3.7	1.0	7.9	3.8	6.4	2.0
5.2	4.1	1.5	0.1	5.8	2.7	3.9	1.2	6.4	2.8	5.6	2.2
5.5	4.2	1.4	0.2	6.0	2.7	5.1	1.6	6.3	2.8	5.1	1.5
4.9	3.1	1.5	0.2	5.4	3.0	4.5	1.5	6.1	2.6	5.6	1.4
5.0	3.2	1.2	0.2	6.0	3.4	4.5	1.6	7.7	3.0	6.1	2.3
5.5	3.5	1.3	0.2	6.7	3.1	4.7	1.5	6.3	3.4	5.6	2.4
4.9	3.6	1.4	0.1	6.3	2.3	4.4	1.3	6.4	3.1	5.5	1.8
4.4	3.0	1.3	0.2	5.6	3.0	4.1	1.3	6.0	3.0	4.8	1.8
5.1	3.4	1.5	0.2	5.5	2.5	4.0	1.3	6.9	3.1	5.4	2.1
5.0	3.5	1.3	0.3	5.5	2.6	4.4	1.2	6.7	3.1	5.6	2.4
4.5	2.3	1.3	0.3	6.1	3.0	4.6	1.4	6.9	3.1	5.1	2.3
4.4	3.2	1.3	0.2	5.8	2.6	4.0	1.2	5.8	2.7	5.1	1.9
5.0	3.5	1.6	0.6	5.0	2.3	3.3	1.0	6.8	3.2	5.9	2.3
5.1	3.8	1.9	0.4	5.6	2.7	4.2	1.3	6.7	3.3	5.7	2.5
4.8	3.0	1.4	0.3	5.7	3.0	4.2	1.2	6.7	3.0	5.2	2.3
5.1	3.8	1.6	0.2	5.7	2.9	4.2	1.3	6.3	2.5	5.0	1.9
4.6	3.2	1.4	0.2	6.2	2.9	4.3	1.3	6.5	3.0	5.2	2.0
5.3	3.7	1.5	0.2	5.1	2.5	3.0	1.1	6.2	3.4	5.4	2.3
5.0	3.3	1.4	0.2	5.7	2.8	4.1	1.3	5.9	3.0	5.1	1.8

1.3 The Data Matrix

The general $(n \times p)$ data matrix with n objects and p variables can be written as follows:

$$
\begin{array}{c c c c c c c}
 & & \multicolumn{5}{c}{\overbrace{\hspace{12em}}^{\text{Variables}}} \\
 & & x_1 & \cdots & x_j & \cdots & x_p \\
\text{Objects} & \left\{ \begin{array}{c} o_1 \\ \cdot \\ \cdot \\ \cdot \\ o_i \\ \cdot \\ \cdot \\ \cdot \\ o_n \end{array} \right. & \begin{array}{c} x_{11} \\ \cdot \\ \cdot \\ \cdot \\ x_{i1} \\ \cdot \\ \cdot \\ \cdot \\ x_{n1} \end{array} & \begin{array}{c} \cdots \\ \\ \\ \\ \cdots \\ \\ \\ \\ \cdots \end{array} & \begin{array}{c} x_{1j} \\ \cdot \\ \cdot \\ \cdot \\ x_{ij} \\ \cdot \\ \cdot \\ \cdot \\ x_{nj} \end{array} & \begin{array}{c} \cdots \\ \\ \\ \\ \cdots \\ \\ \\ \\ \cdots \end{array} & \begin{array}{c} x_{1p} \\ \cdot \\ \cdot \\ \cdot \\ x_{ip} \\ \cdot \\ \cdot \\ \cdot \\ x_{np} \end{array}
\end{array}
$$

Table 1.1.1 showed one such data matrix (5×3) with five objects and three variables. In Table 1.2.2 there were three data matrices, each having 50 rows (objects) and four columns (variables). Note that these data matrices can be considered from two alternative points of view. If we compare two columns, then we are examining the relationship between variables. On the other hand, a comparison of two rows involves examining the relationship between different objects. For example, in Table 1.1.1 we may compare the first two columns to investigate whether there is a relationship between age at entry and marks obtained. Alternatively, looking at the first two rows will give a comparison between two students ("objects"), one male and one female.

The general $(n \times p)$ data matrix will be denoted $\mathbf{X}$ or $\mathbf{X}(n \times p)$. The element in row i and column j is x_{ij}. This denotes the observation of the jth variable on the ith object. We may write the matrix $\mathbf{X} = (x_{ij})$. The rows of $\mathbf{X}$ will be written $\mathbf{x}'_1, \mathbf{x}'_2, \ldots, \mathbf{x}'_n$. Note that $\mathbf{x}_i$ denotes the ith row of $\mathbf{X}$ *written as a column*. The columns of $\mathbf{X}$ will be written with subscripts in parentheses as $\mathbf{x}_{(1)}, \mathbf{x}_{(2)}, \ldots, \mathbf{x}_{(p)}$; that is, we may write

$$\mathbf{X} = \begin{bmatrix} \mathbf{x}'_1 \\ \cdot \\ \cdot \\ \cdot \\ \mathbf{x}'_n \end{bmatrix} = [\mathbf{x}_{(1)}, \ldots, \mathbf{x}_{(p)}],$$

where

$$\mathbf{x}_i = \begin{bmatrix} x_{i1} \\ \cdot \\ \cdot \\ \cdot \\ x_{ip} \end{bmatrix} \quad (i=1,\ldots,n), \qquad \mathbf{x}_{(j)} = \begin{bmatrix} x_{1j} \\ \cdot \\ \cdot \\ \cdot \\ x_{nj} \end{bmatrix} \quad (j=1,\ldots,p).$$

Note that on the one hand $\mathbf{x}_1$ is the p-vector denoting the p observations on the first *object*, while on the other hand $\mathbf{x}_{(1)}$ is the n-vector whose elements denote the observations on the first *variable.* In multivariate analysis, the rows $\mathbf{x}_1, \ldots, \mathbf{x}_n$ usually form a random sample whereas the columns $\mathbf{x}_{(1)}, \ldots, \mathbf{x}_{(p)}$ do not. This point is emphasized in the notation by the use of parentheses.

Clearly when n and p are even moderately large, the resulting np pieces of information may prove too numerous to handle individually. Various ways of summarizing multivariate data are discussed in Sections 1.4, 1.7, 1.8.

1.4 Summary Statistics

We give here the basic summary statistics and some standard notation.

1.4.1 The mean vector and covariance matrix

An obvious extension of the univariate notion of mean and variance leads to the following definitions. The sample mean of the ith variable is

$$\bar{x}_i = \frac{1}{n}\sum_{r=1}^{n} x_{ri}, \tag{1.4.1}$$

and the sample variance of the ith variable is

$$s_{ii} = \frac{1}{n}\sum_{r=1}^{n} (x_{ri}-\bar{x}_i)^2 = s_i^2, \quad \text{say}, \quad i=1,\ldots,p. \tag{1.4.2}$$

The sample covariance between the ith and jth variables is

$$s_{ij} = \frac{1}{n}\sum_{r=1}^{n} (x_{ri}-\bar{x}_i)(x_{rj}-\bar{x}_j). \tag{1.4.3}$$

The vector of means,

$$\bar{\mathbf{x}} = \begin{bmatrix} \bar{x}_1 \\ \cdot \\ \cdot \\ \cdot \\ \bar{x}_p \end{bmatrix}, \tag{1.4.4}$$

is called the *sample mean vector*, or simply the "mean vector". It represents the centre of gravity of the sample points $\mathbf{x}_r$, $r = 1, \ldots, n$. The $p \times p$ matrix

$$\mathbf{S} = (s_{ij}), \tag{1.4.5}$$

with elements given by (1.4.2) and (1.4.3), is called the *sample covariance matrix*, or simply the "covariance matrix".

The above statistics may also be expressed in matrix notation. Corresponding to (1.4.1) and (1.4.4), we have

$$\bar{\mathbf{x}} = \frac{1}{n} \sum_{r=1}^{n} \mathbf{x}_r = \frac{1}{n} \mathbf{X}'\mathbf{1}, \tag{1.4.6}$$

where $\mathbf{1}$ is a column vector of n ones. Also

$$s_{ij} = \frac{1}{n} \sum_{r=1}^{n} x_{ri}x_{rj} - \bar{x}_i\bar{x}_j,$$

so that

$$\mathbf{S} = \frac{1}{n} \sum_{r=1}^{n} (\mathbf{x}_r - \bar{\mathbf{x}})(\mathbf{x}_r - \bar{\mathbf{x}})' = \frac{1}{n} \sum_{r=1}^{n} \mathbf{x}_r\mathbf{x}_r' - \bar{\mathbf{x}}\bar{\mathbf{x}}'. \tag{1.4.7}$$

This may also be written

$$\mathbf{S} = \frac{1}{n}\mathbf{X}'\mathbf{X} - \bar{\mathbf{x}}\bar{\mathbf{x}}' = \frac{1}{n}\left(\mathbf{X}'\mathbf{X} - \frac{1}{n}\mathbf{X}'\mathbf{1}\mathbf{1}'\mathbf{X}\right)$$

using (1.4.6). Writing

$$\mathbf{H} = \mathbf{I} - \frac{1}{n}\mathbf{1}\mathbf{1}',$$

where $\mathbf{H}$ denotes the *centring matrix*, we find that

$$\mathbf{S} = \frac{1}{n}\mathbf{X}'\mathbf{H}\mathbf{X}, \tag{1.4.8}$$

which is a convenient matrix representation of the sample covariance matrix.

Since $\mathbf{H}$ is a symmetric idempotent matrix ($\mathbf{H}=\mathbf{H}'$, $\mathbf{H}=\mathbf{H}^2$) it follows that for any p-vector $\mathbf{a}$,

$$\mathbf{a}'\mathbf{S}\mathbf{a}=\frac{1}{n}\mathbf{a}'\mathbf{X}'\mathbf{H}'\mathbf{H}\mathbf{X}\mathbf{a}=\frac{1}{n}\mathbf{y}'\mathbf{y}\geqslant 0,$$

where $\mathbf{y}=\mathbf{H}\mathbf{X}\mathbf{a}$. Hence the covariance matrix $\mathbf{S}$ is positive semi-definite ($\mathbf{S}\geqslant 0$). For continuous data we usually expect that $\mathbf{S}$ is not only positive semi-definite, but positive definite if $n\geqslant p+1$.

As in one-dimensional statistics, it is often convenient to define the covariance matrix with a divisor of $n-1$ instead of n. Set

$$\mathbf{S}_u=\frac{1}{n-1}\mathbf{X}'\mathbf{H}\mathbf{X}=\frac{n}{n-1}\mathbf{S}. \qquad (1.4.9)$$

If the data forms a random sample from a multivariate distribution, with finite second moments, then $\mathbf{S}_u$ is an *unbiased* estimate of the true covariance matrix (see Theorem 2.8.2).

The matrix

$$\mathbf{M}=\sum_{r=1}^{n}\mathbf{x}_r\mathbf{x}_r'=\mathbf{X}'\mathbf{X} \qquad (1.4.10)$$

is called the *matrix of sums of squares and products* for obvious reasons. The matrix $n\mathbf{S}$ can appropriately be labelled as the matrix of *corrected* sums of squares and products.

The *sample correlation coefficient* between the ith and the jth variables is

$$r_{ij}=s_{ij}/(s_is_j). \qquad (1.4.11)$$

Unlike s_{ij}, the correlation coefficient is invariant under both changes of scale and origin of the ith and jth variables. Clearly $|r_{ij}|\leqslant 1$. The matrix

$$\mathbf{R}=(r_{ij}) \qquad (1.4.12)$$

with $r_{ii}=1$ is called the *sample correlation matrix*. Note that $\mathbf{R}\geqslant 0$. If $\mathbf{R}=\mathbf{I}$, we say that the variables are uncorrelated. If $\mathbf{D}=\operatorname{diag}(s_i)$, then

$$\mathbf{R}=\mathbf{D}^{-1}\mathbf{S}\mathbf{D}^{-1}, \qquad \mathbf{S}=\mathbf{D}\mathbf{R}\mathbf{D}. \qquad (1.4.13)$$

Example 1.4.1 Table 1.4.1 gives a (28×4) data matrix (Rao, 1948) related to weights of bark deposits of 28 trees in the four directions north (N), east (E), south (S), and west (W).

It is found that

$$\bar{x}_1=50.536, \qquad \bar{x}_2=46.179, \qquad \bar{x}_3=49.679, \qquad \bar{x}_4=45.179.$$

Table 1.4.1 Weights of cork deposits (in centigrams) for 28 trees in the four directions (after Rao, 1948)

N	E	S	W	N	E	S	W
72	66	76	77	91	79	100	75
60	53	66	63	56	68	47	50
56	57	64	58	79	65	70	61
41	29	36	38	81	80	68	58
32	32	35	36	78	55	67	60
30	35	34	26	46	38	37	38
39	39	31	27	39	35	34	37
42	43	31	25	32	30	30	32
37	40	31	25	60	50	67	54
33	29	27	36	35	37	48	39
32	30	34	28	39	36	39	31
63	45	74	63	50	34	37	40
54	46	60	52	43	37	39	50
47	51	52	43	48	54	57	43

These means suggest *prima-facie* differences with more deposits in the N–S directions than in the E–W directions. The covariance and correlation matrices are

$$\mathbf{S}=\begin{array}{c} \\ N\\ E\\ S\\ W\end{array}\begin{array}{c}\begin{array}{cccc} N & E & S & W\end{array}\\ \begin{bmatrix} 280.034 & 215.761 & 278.136 & 218.190\\ & 212.075 & 220.879 & 165.254\\ & & 337.504 & 250.272\\ & & & 217.932\end{bmatrix}\end{array},$$

$$\mathbf{R}=\begin{array}{c} \\ N\\ E\\ S\\ W\end{array}\begin{array}{c}\begin{array}{cccc} N & E & S & W\end{array}\\ \begin{bmatrix} 1 & 0.885 & 0.905 & 0.883\\ & 1 & 0.826 & 0.769\\ & & 1 & 0.923\\ & & & 1\end{bmatrix}\end{array}.$$

Since these matrices are symmetric, only the upper half need be printed.

Comparing the diagonal terms of **S**, we note that the sample variance is largest in the south direction. Furthermore, the matrix **R** does not seem to have a "circular" pattern, e.g. the correlation between N and S is relatively high while the correlation between the other pair of opposite cardinal points W and E is lowest.

1.4.2 Measures of multivariate scatter

The matrix **S** is one possible multivariate generalization of the univariate notion of variance, measuring scatter about the mean. However, sometimes it is convenient to have a *single* number to measure multivariate scatter. Two common such measures are

(1) the *generalized variance*, $|\mathbf{S}|$; and
(2) the *total variation*, tr **S**.

A motivation for these measures is given in Section 1.5.3. For both measures, large values indicate a high degree of scatter about $\bar{\mathbf{x}}$ and low values represent concentration about $\bar{\mathbf{x}}$. However, each measure reflects different aspects of the variability in the data. The generalized variance plays an important role in maximum likelihood estimation (Chapter 5) and the total variation is a useful concept in principal component analysis (Chapter 8).

Example 1.4.2 (M. Gnanadesikan and Gupta, 1970) An experimental subject spoke 10 different words seven times each, and five speech measurements were taken on each utterance. For each word we have five variables and seven observations. The (5×5) covariance matrix was calculated for each word, and the generalized variances were as follows:

$$2.9, \quad 1.3, \quad 641.6, \quad 26828.8, \quad 262\,404.3,$$
$$169.2, \quad 3106.8, \quad 617\,671.2, \quad 6.7, \quad 3.0.$$

Ordering these generalized variances, we find that for this speaker the second word has the least variation and the eighth word has most variation. A general point of interest for identification is to find which word has least variation for a particular speaker.

1.5 Linear Combinations

Taking linear combinations of the variables is one of the most important tools of multivariate analysis. A few suitably chosen combinations may provide more information than a multiplicity of the original variables, often because the dimension of the data is reduced. Linear transformations can also simplify the structure of the covariance matrix, making interpretation of the data more straightforward.

Consider a linear combination

$$y_r = a_1 x_{r1} + \ldots + a_p x_{rp}, \qquad r = 1, \ldots, n, \tag{1.5.1}$$

where $a_1, \ldots, a_p$ are given. From (1.4.6) the mean $\bar{y}$ of the y_r is given by

$$\bar{y} = \frac{1}{n}\mathbf{a}' \sum_{r=1}^{n} \mathbf{x}_r = \mathbf{a}'\bar{\mathbf{x}}, \tag{1.5.2}$$

and the variance is given by

$$s_y^2 = \frac{1}{n}\sum_{r=1}^{n}(y_r - \bar{y})^2 = \frac{1}{n}\sum_{r=1}^{n}\mathbf{a}'(\mathbf{x}_r - \bar{\mathbf{x}})(\mathbf{x}_r - \bar{\mathbf{x}})'\mathbf{a} = \mathbf{a}'\mathbf{S}\mathbf{a}, \tag{1.5.3}$$

where we have used (1.4.7).

In general we may be interested in a q-dimensional linear transformation

$$\mathbf{y}_r = \mathbf{A}\mathbf{x}_r + \mathbf{b}, \qquad r = 1, \ldots, n, \tag{1.5.4}$$

which may be written $\mathbf{Y} = \mathbf{X}\mathbf{A}' + \mathbf{1}\mathbf{b}'$, where $\mathbf{A}$ is a $(q \times p)$ matrix and $\mathbf{b}$ is a q-vector. Usually $q \leqslant p$.

The mean vector and covariance matrix of the new objects $\mathbf{y}_r$ are given by

$$\bar{\mathbf{y}} = \mathbf{A}\bar{\mathbf{x}} + \mathbf{b}, \tag{1.5.5}$$

$$\mathbf{S}_y = \frac{1}{n}\sum_{r=1}^{n}(\mathbf{y}_r - \bar{\mathbf{y}})(\mathbf{y}_r - \bar{\mathbf{y}})' = \mathbf{A}\mathbf{S}\mathbf{A}'. \tag{1.5.6}$$

If $\mathbf{A}$ is non-singular (so, in particular, $q = p$), then

$$\mathbf{S} = \mathbf{A}^{-1}\mathbf{S}_y(\mathbf{A}')^{-1}. \tag{1.5.7}$$

Here are several important examples of linear transformations which are used later in the book. For simplicity all of the transformations are centred to have mean $\mathbf{0}$.

1.5.1 The scaling transformation

Let $\mathbf{y}_r = \mathbf{D}^{-1}(\mathbf{x}_r - \bar{\mathbf{x}})$, $r = 1, \ldots, n$, where $\mathbf{D} = \operatorname{diag}(s_i)$. This transformation scales each variable to have unit variance and thus eliminates the arbitrariness in the choice of scale. For example, if $\mathbf{x}_{(1)}$ measures lengths, then $\mathbf{y}_{(1)}$ will be the same whether $\mathbf{x}_{(1)}$ is measured in inches or metres. Note that $\mathbf{S}_y = \mathbf{R}$.

1.5.2 Mahalanobis transformation

If $\mathbf{S} > 0$ then $\mathbf{S}^{-1}$ has a unique symmetric positive definite square root $\mathbf{S}^{-1/2}$ (see equation (A.6.15)). The Mahalanobis transformation is defined by

$$\mathbf{z}_r = \mathbf{S}^{-1/2}(\mathbf{x}_r - \bar{\mathbf{x}}), \qquad r = 1, \ldots, n. \tag{1.5.8}$$

Then $\mathbf{S}_z = \mathbf{I}$, so that this transformation eliminates the correlation between the variables and standardizes the variance of each variable.

1.5.3 Principal component transformation

By the spectral decomposition theorem, the covariance matrix $\mathbf{S}$ may be written in the form

$$\mathbf{S} = \mathbf{GLG}', \tag{1.5.9}$$

where $\mathbf{G}$ is an orthogonal matrix and $\mathbf{L}$ is a diagonal matrix of the eigenvalues of $\mathbf{S}$, $l_1 \geq l_2 \geq \ldots \geq l_p \geq 0$. The principal component transformation is defined by the *rotation*

$$\mathbf{w}_r = \mathbf{G}'(\mathbf{x}_r - \bar{\mathbf{x}}), \qquad r = 1, \ldots, n. \tag{1.5.10}$$

Since $\mathbf{S}_w = \mathbf{G}'\mathbf{S}\mathbf{G} = \mathbf{L}$ is diagonal, the columns of $\mathbf{W}$, called principal components, represent *uncorrelated* linear combinations of the variables. In practice one hopes to summarize most of the variability in the data using only the principal components with the highest variances, thus reducing the dimension.

Since the principal components are uncorrelated with variances $l_1, \ldots, l_p$, it seems natural to define the "overall" spread of the data by some symmetric monotonically increasing function of $l_1, \ldots, l_p$, such as $\prod l_i$ or $\sum l_i$. From Section A.6, $|\mathbf{S}| = |\mathbf{L}| = \prod l_i$ and $\text{tr}\, \mathbf{S} = \text{tr}\, \mathbf{L} = \sum l_i$. Thus the rotation to principal components provides a motivation for the measures of multivariate scatter discussed in Section 1.4.2.

Note that alternative versions of the transformations in Sections 1.5.1–1.5.3 can be defined using $\mathbf{S}_u$ instead of $\mathbf{S}$.

Example 1.5.1 *A transformation of the cork data* If in the cork data of Table 1.4.1, the aim is to investigate whether the bark deposits are uniformly spread, our interest would be in linear combinations such as

$$y_1 = \text{N} + \text{S} - \text{E} - \text{W}, \qquad y_2 = \text{N} - \text{S}, \qquad y_3 = \text{E} - \text{W}.$$

Here

$$\mathbf{A} = \begin{bmatrix} 1 & -1 & 1 & -1 \\ 1 & 0 & -1 & 0 \\ 0 & 1 & 0 & -1 \end{bmatrix}$$

From Example 1.4.1 and (1.5.5), we find that the mean vector of the transformed variables is

$$\bar{\mathbf{y}}' = (8.857, 0.857, 1.000)$$

and the covariance matrix is

$$\begin{bmatrix} 124.12 & -20.27 & -25.96 \\ & 61.27 & 26.96 \\ & & 99.50 \end{bmatrix}.$$

Obviously, the mean of y_1 is higher than that of y_2 and y_3, indicating possibly more cork deposit along the north–south axis than along the east–west axis. However, $V(y_1)$ is larger. If we standardize the variables so that the sum of squares of the coefficients is unity, by letting

$$z_1 = (N+S-E-W)/2, \qquad z_2 = (N-S)/\sqrt{2}, \qquad z_3 = (E-W)/\sqrt{2},$$

the variances are more similar,

$$V(z_1) = 31.03, \qquad V(z_2) = 30.63, \qquad V(z_3) = 49.75.$$

1.6 Geometrical Ideas

In Section 1.3 we mentioned the two alternative perspectives which can be used to view the data matrix. On the one hand, we may be interested in comparing *columns* of the data matrix, that is the *variables*. This leads to the techniques known as *R-techniques*, so-called because the correlation matrix **R** plays an important role in this approach. *R-techniques* are important in principal component analysis, factor analysis, and canonical correlation analysis.

Alternatively, we may compare the *rows* of the data matrix, that is, the different *objects*. This leads to techniques such as discriminant analysis, cluster analysis, and multidimensional scaling, which are known as *Q-techniques*.

These two approaches correspond to different geometric ways of representing the $(n \times p)$ data matrix. First, the columns can be viewed as p points in an n-dimensional space which we shall call *R-space* or *object space*. For example, the correlation matrix has a natural interpretation in the object space of the centred matrix $\mathbf{Y} = \mathbf{HX}$. The correlation r_{ij} is just the cosine of the angle θ_{ij} subtended at the origin between the two corresponding columns,

$$\cos \theta_{ij} = \frac{\mathbf{y}'_{(i)}\mathbf{y}_{(j)}}{\|\mathbf{y}_i\| \, \|\mathbf{y}_j\|} = \frac{s_{ij}}{s_i s_j} = r_{ij}.$$

Note that the correlation coefficients are a measure of *similarity* because their values are large when variables are close to one another.

Second, the n rows may be taken as n points in p-dimensional *Q-space* or *variable space*. A natural way to compare two rows $\mathbf{x}_r$ and $\mathbf{x}_s$ is to look at the Euclidean distance between them:

$$\|\mathbf{x}_r-\mathbf{x}_s\|^2=(\mathbf{x}_r-\mathbf{x}_s)'(\mathbf{x}_r-\mathbf{x}_s).$$

An alternative procedure is to transform the data by one of the transformations of Section 1.5.1 or Section 1.5.2, and then look at the Euclidean distance between the transformed rows. Such distances play a role in cluster analysis. The most important of these distances is the *Mahalanobis distance* D_{rs} given by

$$D_{rs}^2=\|\mathbf{z}_r-\mathbf{z}_s\|^2=(\mathbf{x}_r-\mathbf{x}_s)'\mathbf{S}^{-1}(\mathbf{x}_r-\mathbf{x}_s). \tag{1.6.1}$$

The Mahalanobis distance underlies Hotelling's T^2 test and the theory of discriminant analysis. Note that the Mahalanobis distance can alternatively be defined using $\mathbf{S}_u$ instead of $\mathbf{S}$.

1.7 Graphical Representations

1.7.1 Univariate scatters

For $p=1$ and $p=2$, we can draw and interpret scatter plots for the data, but for $p=3$ the difficulties of drawing such plots can be appreciated. For $p>3$, the task becomes hopeless, although computer facilities exist which allow one to examine the projection of a multivariate scatter onto any given plane. However, the need for graphical representations when $p>3$ is greater than for the univariate case since the relationships cannot be understood by looking at a data matrix. A simple starting point is to look into univariate plots for the p variables side by side. For the cork data of Table 1.4.1, such a representation is given in Figure 1.7.1, which indicates that the distributions are somewhat skew. In this case, the variables are measured in the same units, and therefore a direct

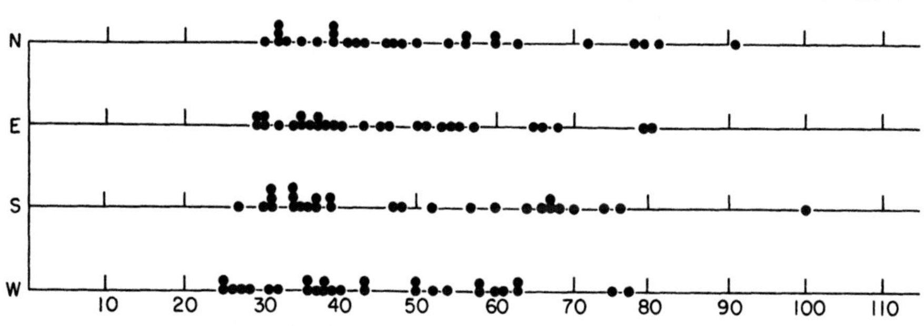

Figure 1.7.1 Univariate representation of the cork data of Table 1.4.1.

comparison of the plots is possible. In general, we should standardize the variables before using a plot.

Figure 1.7.1 does not give any idea of relationships between the variables. However, one way to exhibit the interrelationships is to plot all the observations consecutively along an axis, representing different variables by different symbols. For the cork data, a plot of this type is given in Figure 1.7.2. It shows the very noticeable tree differences as well as

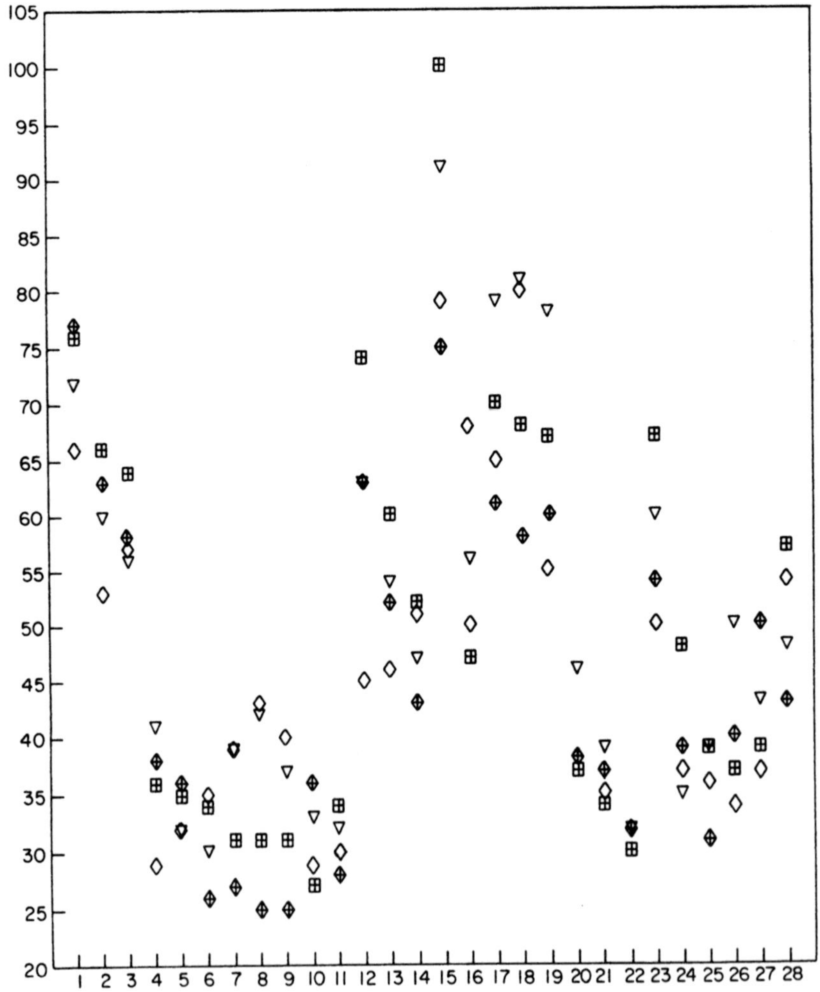

Figure 1.7.2 Consecutive univariate representation (*after* Pearson, 1956). ▽ = N, ◇ = E, ⊞ = S, ⊕ = W.

differences in pattern which are associated with the given ordering of the trees. (That is, the experimenters appear to have chosen groups of small-trunked and large-trunked trees alternately.) We also observe that the 15th tree may be an outlier. These features due to ordering would remain if the Mahalanobis residuals defined by (1.5.8) were plotted. Of course, the Mahalanobis transformation removes the main effect and adjusts every variance to unity and every covariance to zero.

1.7.2 Bivariate scatters

Another way of understanding the interrelationships is to look into all $p(p-1)/2$ bivariate scatter diagrams of the data. For the cork data, with four variables, there are six such diagrams.

A method of graphing 4 variables in two dimensions is as follows. First, draw a scatter plot for two variables, say (N, E). Then indicate the values of S and W of a point by plotting the values in two directions from

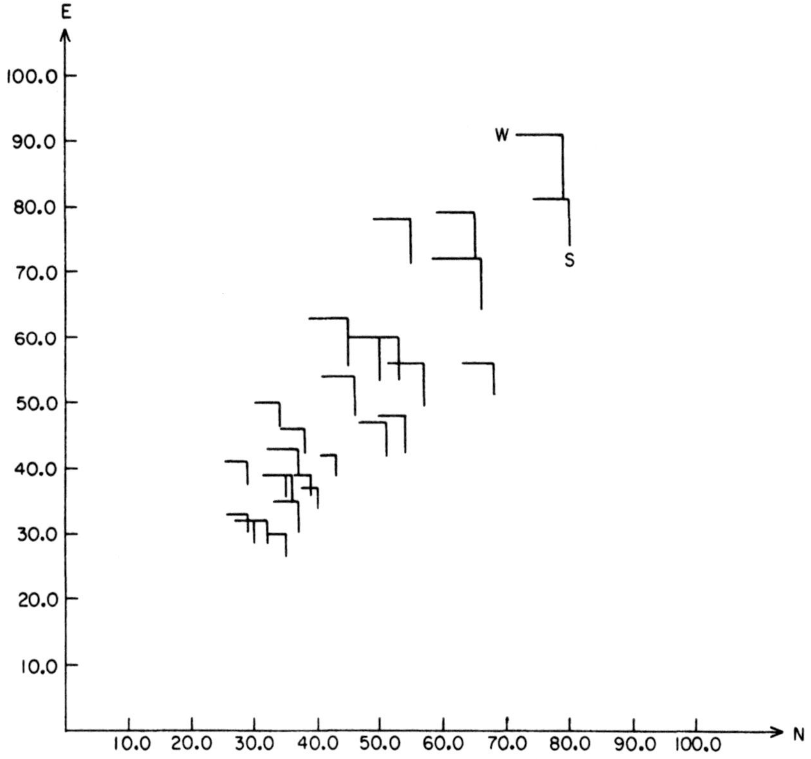

Figure 1.7.3 A glyph representation of the cork data of Table 1.4.1.

the point (N, E). Figure 1.7.3 gives such a plot where W is plotted along the negative x axis and S is plotted along the negative y axis using each point (N, E) as the origin. The markedly linear scatter of the points shows a strong dependence between N and E. A similar dependence between S and W is shown by the similar lengths of the two "legs" of each point. Dependence between N and S is reflected by the S legs being longer for large N and a similar pattern is observed between N and W, E and W, and E and S. This method of *glyphs* E. Anderson, 1960 can be extended to several variables, but it does not help in understanding the multivariate complex as a whole. A related method due to Chernoff (1973) uses similar ideas to represent the observations on each object by a human face.

1.7.3 Harmonic curves

Consider the rth data point $\mathbf{x}_r' = (x_{r1}, \ldots, x_{rp})$, $r = 1, \ldots, n$. A recent promising method (Andrews, 1972; Ball and Hall, 1970) involves plotting the curve

$$f_{\mathbf{x}_r}(t) = x_{r1}/\sqrt{2} + x_{r2}\sin t + x_{r3}\cos t + x_{r4}\sin 2t + x_{r5}\cos 2t + \ldots \quad (1.7.1)$$

for each data point $\mathbf{x}_r$, $r = 1, \ldots, n$, over the interval $-\pi \leq t \leq \pi$. Thus there will be n harmonic curves drawn in two dimensions. Two data points are compared by visually studying the curves over $[-\pi, \pi]$. Note that the square of the L_2 distance

$$\int_{-\pi}^{\pi} [f_{\mathbf{x}}(t) - f_{\mathbf{y}}(t)]^2 \, dt,$$

between two curves $f_{\mathbf{x}}(t)$ and $f_{\mathbf{y}}(t)$ simplifies to

$$\pi \|\mathbf{x} - \mathbf{y}\|^2,$$

which is proportional to the square of the Euclidean distance between $\mathbf{x}$ and $\mathbf{y}$.

An example of the use of harmonic curves applied to cluster analysis is given in Chapter 13, where clusters appear as curves bunched together for all values of t. Some practical hints for the plotting of harmonic curves with a large number of objects are given in Gnanadesikan (1977). Notice that harmonic curves depend on the order in which the variables are written down. Thus, their interpretation requires some care.

*1.8 Measures of Multivariate Skewness and Kurtosis

In Section 1.4 we have given a few basic summary statistics based on the first- and the second-order moments. In general a kth-order central moment for the variables $i_1, \ldots, i_s$ is

$$M_{i_1,\ldots,i_s}^{(j_1,\ldots,j_s)} = \frac{1}{n} \sum_{r=1}^{n} \prod_{t=1}^{s} (x_{ri_t} - \bar{x}_{i_t})^{j_t},$$

where $j_1 + \ldots + j_s = k$, $j_t \neq 0$, $t = 1, \ldots, s$. As with mean and variance, we would like extensions to the multivariate case of summary measures like $b_1 = m_3^2/s^6$ and $b_2 = m_4/s^4$, the univariate measures of skewness and kurtosis. Using the invariant functions

$$g_{rs} = (\mathbf{x}_r - \bar{\mathbf{x}})'\mathbf{S}^{-1}(\mathbf{x}_s - \bar{\mathbf{x}}),$$

Mardia (1970*a*) has defined multivariate measures of skewness and kurtosis by

$$b_{1,p} = \frac{1}{n^2} \sum_{r,s=1}^{n} g_{rs}^3 \tag{1.8.1}$$

and

$$b_{2,p} = \frac{1}{n} \sum_{r=1}^{n} g_{rr}^2. \tag{1.8.2}$$

The following properties are worth noting:

(1) $b_{1,p}$ depends only on the moments up to third order, whereas $b_{2,p}$ depends only on the moments up to fourth order excluding the third-order moments.
(2) These moments are invariant under *affine* transformations

$$\mathbf{y} = \mathbf{A}\mathbf{x} + \mathbf{b}.$$

(Similar properties hold for b_1 and b_2 under changes of scale and origin.)

(3) For $p = 1$, $b_{1,p} = b_1$ and $b_{2,p} = b_2$.
(4) Let D_r be the Mahalanobis distance between $\mathbf{x}_r$ and $\bar{\mathbf{x}}$ and let $\cos \alpha_{rs} = (\mathbf{x}_r - \bar{\mathbf{x}})'\mathbf{S}^{-1}(\mathbf{x}_s - \bar{\mathbf{x}})/D_rD_s$ denote the cosine of the Mahalanobis angle between $(\mathbf{x}_r - \bar{\mathbf{x}})$ and $(\mathbf{x}_s - \bar{\mathbf{x}})$. Then equations (1.8.1) and (1.8.2) reduce to

$$b_{1,p} = \frac{1}{n^2} \sum_{r=1}^{n} \sum_{s=1}^{n} (D_rD_s \cos \alpha_{rs})^3 \tag{1.8.3}$$

and

$$b_{2,p} = \frac{1}{n} \sum_{r=1}^{n} D_r^4. \tag{1.8.4}$$

Although $b_{1,p}$ cannot be negative, we note from (1.8.3) that if the data points are uniformly distributed on a p-dimensional hypersphere, then $b_{1,p} \simeq 0$ since $\sum \cos \alpha_{rs} \simeq 0$ and D_r and D_s are approximately equal. We should expect $b_{1,p} \gg 0$ if there is a departure from spherical symmetry.

The statistic $b_{2,p}$ will pick up extreme behaviour in the Mahalanobis distances of objects from the sample mean. The use of $b_{1,p}$ and $b_{2,p}$ to detect departure from multivariate normality is described in Chapter 5.

Exercises and Complements

1.4.1 Under the transformation

$$\mathbf{y}_r = \mathbf{A}\mathbf{x}_r + \mathbf{b}, \qquad r = 1, \ldots, n,$$

show that

$$\text{(i) } \bar{\mathbf{y}} = \mathbf{A}\bar{\mathbf{x}} + \mathbf{b}, \qquad \text{(ii) } \mathbf{S}_y = \mathbf{A}\mathbf{S}_x\mathbf{A}',$$

where $\bar{\mathbf{x}}$, $\mathbf{S}_x$ are the mean vector and the covariance matrix for the $\mathbf{x}_r$.

1.4.2 Let

$$\mathbf{S}(\mathbf{a}) = \frac{1}{n} \sum_{r=1}^{n} (\mathbf{x}_r - \mathbf{a})(\mathbf{x}_r - \mathbf{a})'$$

be the covariance matrix about $\mathbf{x} = \mathbf{a}$. Show that

$$\mathbf{S}(\mathbf{a}) = \mathbf{S} + (\bar{\mathbf{x}} - \mathbf{a})(\bar{\mathbf{x}} - \mathbf{a})'.$$

(i) Using (A.2.3k) or otherwise, show that

$$|\mathbf{S}(\mathbf{a})| = |\mathbf{S}|\{1 + (\bar{\mathbf{x}} - \mathbf{a})'\mathbf{S}^{-1}(\bar{\mathbf{x}} - \mathbf{a})\},$$

and consequently

$$\min_{\mathbf{a}} |\mathbf{S}(\mathbf{a})| = |\mathbf{S}|.$$

(ii) Show that $\min_{\mathbf{a}} \operatorname{tr} \mathbf{S}(\mathbf{a}) = \operatorname{tr} \mathbf{S}$.

*(iii) Note that $|\mathbf{S}| = \prod l_i$ and $\operatorname{tr} \mathbf{S} = \sum l_i$ are monotonically increasing symmetric functions of the eigenvalues $l_1, \ldots, l_p$ of $\mathbf{S}$. Use this observation in constructing other measures of multivariate scatter.

(iv) Using the inequality $g \leq a$, where a and g are the arithmetic and geometric means of a set of positive numbers, show that $|\mathbf{S}| \leq (n^{-1} \operatorname{tr} \mathbf{S})^n$.

1.4.3 (a) Fisher (1947) gives data relating to the body weight in kilograms (x_1) and the heart weight in grams (x_2) of 144 cats. For the 47 female cats the sums and sums of squares and products are given by

$$\mathbf{X}_1'\mathbf{1} = \begin{bmatrix} 110.9 \\ 432.5 \end{bmatrix} \quad \text{and} \quad \mathbf{X}_1'\mathbf{X}_1 = \begin{bmatrix} 265.13 & 1029.62 \\ 1029.62 & 4064.71 \end{bmatrix}.$$

Show that the mean vector and covariance matrix are $\bar{\mathbf{x}}$ and $\mathbf{S}$ given in Example 4.2.3.

(b) For the 97 male cats from Fisher's data, the statistics are

$$\mathbf{X}_2'\mathbf{1} = \begin{bmatrix} 281.3 \\ 1098.3 \end{bmatrix} \quad \text{and} \quad \mathbf{X}_2'\mathbf{X}_2 = \begin{bmatrix} 836.75 & 3275.55 \\ 3275.55 & 13056.17 \end{bmatrix}.$$

Find the mean vector and covariance matrix for the sample of male cats.

(c) Regarding the 144 male and female cats as a single sample, calculate the mean vector and covariance matrix.

(d) Calculate the correlation coefficient for each of (a), (b), and (c).

1.5.1 Let $\mathbf{M} = \mathbf{X}'\mathbf{X}$, where $\mathbf{X}$ is a data matrix. Show that

$$m_{ii} = \mathbf{x}'_{(i)}\mathbf{x}_{(i)} = n(s_{ii} + \bar{x}_i^2), \qquad m_{ij} = \mathbf{x}'_{(i)}\mathbf{x}_{(j)} = n(s_{ij} + \bar{x}_i\bar{x}_j).$$

1.5.2 Show that the scaling transformation in Section 1.5.1 can be written as

$$\mathbf{Y} = \mathbf{HXD}^{-1}, \qquad \mathbf{Y}' = (\mathbf{y}_1, \ldots, \mathbf{y}_n).$$

Use the fact that

$$\mathbf{Y}'\mathbf{1} = \mathbf{0}, \qquad \mathbf{Y}'\mathbf{HY} = \mathbf{D}^{-1}\mathbf{X}'\mathbf{HXD}^{-1}$$

to show that

$$\bar{\mathbf{y}} = \mathbf{0}, \qquad \mathbf{S}_y = \mathbf{R}.$$

1.5.3 Show that the Mahalanobis transformation in Section 1.5.2 can be written as

$$\mathbf{Z} = \mathbf{HXS}^{-1/2}, \qquad \mathbf{Z}' = (\mathbf{z}_1, \ldots, \mathbf{z}_n).$$

Hence, following the method of Exercise 1.5.2, show that $\bar{\mathbf{z}} = \mathbf{0}$, $\mathbf{S}_z = \mathbf{I}$.

1.5.4 For the mark data in Table 1.1.1 show that

$$\bar{\mathbf{x}}' = (18.458, 74.400, 0.400),$$

$$\mathbf{S} = \begin{bmatrix} 0.0159 & -0.4352 & -0.0252 \\ & 101.8400 & 3.0400 \\ & & 0.2400 \end{bmatrix},$$

$$\mathbf{S}^{-1} = \begin{bmatrix} 76.7016 & 0.1405 & 6.2742 \\ & 0.0160 & -0.1885 \\ & & 7.2132 \end{bmatrix},$$

$$\mathbf{S}^{-1/2} = \begin{bmatrix} 8.7405 & 0.0205 & 0.5521 \\ & 0.1013 & -0.0732 \\ & & 2.6274 \end{bmatrix}.$$

1.6.1 If

$$D_{rs}^2 = (\mathbf{x}_r - \mathbf{x}_s)'\mathbf{S}^{-1}(\mathbf{x}_r - \mathbf{x}_s)$$

we may write

$$D_{rs}^2 = q_{rr} + q_{ss} - 2q_{rs},$$

where

$$q_{rs} = \mathbf{x}_r'\mathbf{S}^{-1}\mathbf{x}_s.$$

Writing

$$g_{rs} = (\mathbf{x}_r - \bar{\mathbf{x}})'\mathbf{S}^{-1}(\mathbf{x}_s - \bar{\mathbf{x}}),$$

we see that

$$g_{rs} = q_{rs} + \bar{\mathbf{x}}'\mathbf{S}^{-1}\bar{\mathbf{x}} - \mathbf{x}_r'\mathbf{S}^{-1}\bar{\mathbf{x}} - \mathbf{x}_s'\mathbf{S}^{-1}\bar{\mathbf{x}}.$$

Therefore

$$D_{rs}^2 = g_{rr} + g_{ss} - 2g_{rs}.$$

(i) Show that

$$\sum_r g_{rs} = 0 \quad \text{and} \quad \sum_r g_{rr} = \sum_r \operatorname{tr} \mathbf{S}^{-1}(\mathbf{x}_r - \bar{\mathbf{x}})(\mathbf{x}_r - \bar{\mathbf{x}})' = \mathbf{n} \operatorname{tr} \mathbf{I}_p = np.$$

(ii) Therefore show that

$$\sum_r D_{rs}^2 = np + ng_{ss} \quad \text{or} \quad g_{ss} = \frac{1}{n}\sum_r D_{rs}^2 - p$$

and

$$\sum_{r,s=1}^{n} D_{rs}^2 = 2n^2 p.$$

1.8.1 (Mardia, 1970*a*) Let $u_{pq} = M_{pq}/s_1^p s_2^q$, where

$$M_{pq} = \frac{1}{n}\sum_{r=1}^{n} (x_{r1} - \bar{x}_1)^p (x_{r2} - \bar{x}_2)^q.$$

Show that

$$b_{1,2} = (1-r^2)^{-3}[u_{30}^2 + u_{03}^2 + 3(1+2r^2)(u_{12}^2 + u_{21}^2) - 2r^3 u_{30}u_{03} + 6r\{u_{30}(ru_{12} - u_{21}) + u_{03}(ru_{21} - u_{12}) - (2+r^2)u_{12}u_{21}\}]$$

and

$$b_{2,2} = (1-r^2)^{-2}[u_{40} + u_{04} + 2u_{22} + 4r(ru_{22} - u_{13} - u_{31})].$$

Hence, for $s_1 = s_2 = 1$, $r = 0$, show that

$$b_{1,2} = M_{30}^2 + M_{03}^2 + 3M_{12}^2 + 3M_{21}^2$$

and

$$b_{2,2} = M_{40} + M_{04} + 2M_{22}.$$

Thus $b_{1,2}$ accumulates the effects of M_{21}, M_{12}, M_{03}, M_{30} while $b_{2,2}$ accumulates the effects of M_{22}, M_{04}, and M_{40}.

2 Basic Properties of Random Vectors

2.1 Cumulative Distribution Functions and Probability Density Functions

Let $\mathbf{x}=(x_1, \ldots, x_p)'$ be a random vector. By analogy with univariate theory, the *cumulative distribution function* (c.d.f.) associated with $\mathbf{x}$ is the function F defined by

$$F(\mathbf{x}^\circ)=\Pr(\mathbf{x}\leq\mathbf{x}^\circ)=\Pr(x_1\leq x_1^\circ, \ldots, x_p\leq x_p^\circ). \tag{2.1.1}$$

Two important cases are absolutely continuous and discrete distributions.

A random vector $\mathbf{x}$ is *absolutely continuous* if there exists a *probability density function* (p.d.f.), $f(\mathbf{x})$, such that

$$F(\mathbf{x})=\int_{-\infty}^{\mathbf{x}} f(\mathbf{u})\, d\mathbf{u}.$$

Here $d\mathbf{u}=du_1 \ldots du_p$ represents the product of p differential elements and the integral sign denotes p-fold integration. Note that for any measurable set $D\subseteq R^p$,

$$P(\mathbf{x}\in D)=\int_D f(\mathbf{u})\, d\mathbf{u} \tag{2.1.2}$$

and

$$\int_{-\infty}^{\infty} f(\mathbf{u})\, d\mathbf{u}=1. \tag{2.1.3}$$

In this book we do not distinguish notationally between a random vector and its realization.

For a *discrete* random vector $\mathbf{x}$, the total probability is concentrated on a countable (or finite) set of points $\{\mathbf{x}_j; j=1,2,\ldots\}$. Discrete random vectors can be handled in the above framework if we call the probability function

$$\left.\begin{aligned} f(\mathbf{x}_j)&=P(\mathbf{x}=\mathbf{x}_j), \qquad j=1,2,\ldots,\\ f(\mathbf{x})&=0, \qquad \text{otherwise,}\end{aligned}\right\} \tag{2.1.4}$$

the (*discrete*) *p.d.f.* of $\mathbf{x}$ and replace the integration in (2.1.2) by the summation

$$P(\mathbf{x}\in D)=\sum_{j:\mathbf{x}_j\in D} f(\mathbf{x}_j). \tag{2.1.5}$$

However, most of the emphasis in this book is directed towards absolutely continuous random vectors.

The support S of $\mathbf{x}$ is defined as the set

$$S=\{\mathbf{x}\in R^p: f(\mathbf{x})>0\}. \tag{2.1.6}$$

In examples the p.d.f. is usually defined only on S with the value zero elsewhere being assumed.

Marginal and conditional distributions Consider the partitioned vector $\mathbf{x}'=(\mathbf{x}_1', \mathbf{x}_2')$, where $\mathbf{x}_1$ and $\mathbf{x}_2$ have k and $(p-k)$ elements, respectively $(k<p)$. The function

$$P(\mathbf{x}_1\leq\mathbf{x}_1^\circ)=F(x_1^\circ,\ldots,x_k^\circ,\infty,\ldots,\infty)$$

is called the *marginal cumulative distribution function* (marginal c.d.f.) of $\mathbf{x}_1$. In contrast $F(\mathbf{x})$ may then be described as the *joint c.d.f.* and $f(\mathbf{x})$ may be called the *joint p.d.f.*

Let $\mathbf{x}$ have joint p.d.f. $f(\mathbf{x})$. Then the *marginal p.d.f.* of $\mathbf{x}_1$ is given by the integral of $f(\mathbf{x})$ over $\mathbf{x}_2$; that is,

$$f_1(\mathbf{x}_1)=\int_{-\infty}^{\infty} f(\mathbf{x}_1,\mathbf{x}_2)\,d\mathbf{x}_2. \tag{2.1.7}$$

The marginal p.d.f. of $\mathbf{x}_2$, $f_2(\mathbf{x}_2)$, is defined similarly.

For a given value of $\mathbf{x}_1$, say, $\mathbf{x}_1=\mathbf{x}_1^\circ$, the *conditional p.d.f.* of $\mathbf{x}_2$ is proportional to $f(\mathbf{x}_1^\circ,\mathbf{x}_2)$, where the constant of proportionality can be calculated from the fact that this p.d.f. must integrate to one. In other words, the conditional p.d.f. of $\mathbf{x}_2$ given $\mathbf{x}_1=\mathbf{x}_1^\circ$ is

$$f(\mathbf{x}_2\mid\mathbf{x}_1=\mathbf{x}_1^\circ)=\frac{f(\mathbf{x}_1^\circ,\mathbf{x}_2)}{f_1(\mathbf{x}_1^\circ)}. \tag{2.1.8}$$

(It is assumed that $f_1(\mathbf{x}_1^\circ)$ is non-zero.) The conditional p.d.f. of $\mathbf{x}_1$ given $\mathbf{x}_2=\mathbf{x}_2^\circ$ can be defined similarly.

In general, two random variables can each have the same marginal distribution, even when their joint distributions are different. For instance, the marginal p.d.f.s of the following joint p.d.f.s,

$$f(x_1, x_2) = 1, \qquad 0 < x_1, x_2 < 1, \tag{2.1.9}$$

and (Morgenstern, 1956)

$$f(x_1, x_2) = 1 + \alpha(2x_1 - 1)(2x_2 - 1), \qquad 0 < x_1, x_2 < 1, \quad -1 \leqslant \alpha \leqslant 1, \tag{2.1.10}$$

are both uniform, although the two joint distributions are different.

Independence When the conditional p.d.f. $f(\mathbf{x}_2 \mid \mathbf{x}_1 = \mathbf{x}_1^\circ)$ is the same for all values of $\mathbf{x}_1^\circ$, then we say that $\mathbf{x}_1$ and $\mathbf{x}_2$ are *statistically independent* of each other. In such situations, $f(\mathbf{x}_2 \mid \mathbf{x}_1 = \mathbf{x}_1^\circ)$ must be $f_2(\mathbf{x}_2)$. Hence, the joint density must equal the product of the marginals, as stated in the following theorem.

Theorem 2.1.1 *If $\mathbf{x}_1$ and $\mathbf{x}_2$ are statistically independent then*

$$f(\mathbf{x}) = f_1(\mathbf{x}_1) f_2(\mathbf{x}_2).$$

Note that the variables x_1 and x_2 are independent for the p.d.f. given by (2.1.9), whereas the variables in (2.1.10) are dependent.

2.2 Population Moments

In this section we give the population analogues of the sample moments which were discussed in Section 1.4.

2.2.1 Expectation and correlation

If $\mathbf{x}$ is a random vector with p.d.f. $f(\mathbf{x})$ then the *expectation* or *mean* of a scalar-valued function $g(\mathbf{x})$ is defined as

$$E[g(\mathbf{x})] = \int_{-\infty}^{\infty} g(\mathbf{x}) f(\mathbf{x})\, d\mathbf{x}. \tag{2.2.1}$$

We assume that all necessary integrals converge, so that the expectations are finite. Expectations have the following properties:

(1) Linearity.

$$E[a_1 g_1(\mathbf{x}) + a_2 g_2(\mathbf{x})] = a_1 E[g_1(\mathbf{x})] + a_2 E[g_2(\mathbf{x})]. \tag{2.2.2}$$

(2) Partition, $\mathbf{x}' = (\mathbf{x}_1', \mathbf{x}_2')$. The expectation of a function of $\mathbf{x}_1$ may be

written in terms of the marginal distribution of $\mathbf{x}_1$ as follows:

$$E\{g(\mathbf{x}_1)\} = \int_{-\infty}^{\infty} g(\mathbf{x}_1)f(\mathbf{x})\,d\mathbf{x} = \int_{-\infty}^{\infty} g(\mathbf{x}_1)f_1(\mathbf{x}_1)\,d\mathbf{x}_1. \tag{2.2.3}$$

When f_1 is known, the second expression is useful for computation.

(3) If $\mathbf{x}_1$ and $\mathbf{x}_2$ are independent and $g_i(\mathbf{x}_i)$ is a function of $\mathbf{x}_i$ *only* $(i = 1, 2)$, then

$$E[g_1(\mathbf{x}_1)g_2(\mathbf{x}_2)] = E[g_1(\mathbf{x}_1)]E[g_2(\mathbf{x}_2)]. \tag{2.2.4}$$

More generally, the expectation of a *matrix*-valued (or vector-valued) function of $\mathbf{x}$, $\mathbf{G}(\mathbf{x}) = (g_{ij}(\mathbf{x}))$ is defined to be the matrix

$$E[\mathbf{G}(\mathbf{x})] = (E[g_{ij}(\mathbf{x})]).$$

2.2.2 Population mean vector and covariance matrix

The vector $E(\mathbf{x}) = \boldsymbol{\mu}$ is called the *population mean vector* of $\mathbf{x}$. Thus,

$$\mu_i = \int_{-\infty}^{\infty} x_i f(\mathbf{x})\,d\mathbf{x}, \qquad i = 1, \ldots, p.$$

The population mean possesses the linearity property

$$E(\mathbf{A}\mathbf{x} + \mathbf{b}) = \mathbf{A}E(\mathbf{x}) + \mathbf{b}, \tag{2.2.5}$$

where $\mathbf{A}(q \times p)$ and $\mathbf{b}(q \times 1)$ are constant.

The matrix

$$E\{(\mathbf{x} - \boldsymbol{\mu})(\mathbf{x} - \boldsymbol{\mu})'\} = \boldsymbol{\Sigma} = V(\mathbf{x}) \tag{2.2.6}$$

is called the *covariance matrix* of $\mathbf{x}$ also known as the variance-covariance or dispersion matrix). For conciseness, write

$$\mathbf{x} \sim (\boldsymbol{\mu}, \boldsymbol{\Sigma}) \tag{2.2.7}$$

to describe a random vector with mean vector $\boldsymbol{\mu}$ and covariance matrix $\boldsymbol{\Sigma}$.

More generally we can define the covariance between two vectors, $\mathbf{x}(p \times 1)$ and $\mathbf{y}(q \times 1)$, by the $(p \times q)$ matrix

$$C(\mathbf{x}, \mathbf{y}) = E\{(\mathbf{x} - \boldsymbol{\mu})(\mathbf{y} - \boldsymbol{\nu})'\}, \tag{2.2.8}$$

where $\boldsymbol{\mu} = E(\mathbf{x})$, $\boldsymbol{\nu} = E(\mathbf{y})$. Notice the following simple properties of covariances. Let $V(\mathbf{x}) = \boldsymbol{\Sigma} = (\sigma_{ij})$.

(1) $\sigma_{ij} = C(x_i, x_j)$, $i \neq j$; $\sigma_{ii} = V(x_i) = \sigma_i^2$, say.

(2) $\boldsymbol{\Sigma} = E(\mathbf{x}\mathbf{x}') - \boldsymbol{\mu}\boldsymbol{\mu}'$. $\quad$ (2.2.9)

(3) $V(\mathbf{a}'\mathbf{x}) = \mathbf{a}'V(\mathbf{x})\mathbf{a} = \sum a_i a_j \sigma_{ij}$, $\quad$ (2.2.10)

for all constant vectors $\mathbf{a}$. Since the left-hand side of (2.2.10) is always non-negative, we get the following result:

(4) $\mathbf{\Sigma} \geq 0$.

(5) $V(\mathbf{Ax}+\mathbf{b}) = \mathbf{A}V(\mathbf{x})\mathbf{A}'$. (2.2.11)

(6) $C(\mathbf{x}, \mathbf{x}) = V(\mathbf{x})$. (2.2.12)

(7) $C(\mathbf{x}, \mathbf{y}) = C(\mathbf{y}, \mathbf{x})'$.

(8) $C(\mathbf{x}_1+\mathbf{x}_2, \mathbf{y}) = C(\mathbf{x}_1, \mathbf{y}) + C(\mathbf{x}_2, \mathbf{y})$. (2.2.13)

(9) If $p = q$,

$$V(\mathbf{x}+\mathbf{y}) = V(\mathbf{x}) + C(\mathbf{x}, \mathbf{y}) + C(\mathbf{y}, \mathbf{x}) + V(\mathbf{y}). \tag{2.2.14}$$

(10) $C(\mathbf{Ax}, \mathbf{By}) = \mathbf{A}C(\mathbf{x}, \mathbf{y})\mathbf{B}'$. (2.2.15)

(11) If x and y are independent then $C(\mathbf{x}, \mathbf{y}) = 0$.
However the converse is *not* true. See Exercise 2.2.2.

Example 2.2.1 Let

$$f(x_1, x_2) = \begin{cases} x_1 + x_2, & 0 \leq x_1, x_2 \leq 1, \\ 0, & \text{otherwise.} \end{cases} \tag{2.2.16}$$

Then

$$\boldsymbol{\mu} = \begin{bmatrix} E(x_1) \\ E(x_2) \end{bmatrix} = \begin{bmatrix} 7/12 \\ 7/12 \end{bmatrix}, \qquad \mathbf{\Sigma} = \begin{bmatrix} \sigma_{11} & \sigma_{12} \\ \sigma_{21} & \sigma_{22} \end{bmatrix} = \begin{bmatrix} 11/144 & -1/144 \\ -1/144 & 11/144 \end{bmatrix}.$$

Correlation matrix The population correlation matrix is defined in a manner similar to its sample counterpart. Let us denote the correlation coefficient between the ith and jth variables by ρ_{ij}, so that

$$\rho_{ij} = \sigma_{ij}/\sigma_i\sigma_j, \qquad i \neq j.$$

The matrix

$$\mathbf{P} = (\rho_{ij}) \tag{2.2.17}$$

with $\rho_{ii} = 1$ is called the *population correlation matrix.* Taking $\mathbf{\Delta} = \text{diag}(\sigma_i)$, we have

$$\mathbf{P} = \mathbf{\Delta}^{-1}\mathbf{\Sigma}\mathbf{\Delta}^{-1}.$$

The matrix $\mathbf{P} \geq 0$ because $\mathbf{\Sigma} \geq 0$, and $\mathbf{\Delta}$ is symmetric.

Generalized variance By analogy with Section 1.4.2, we may also define the population *generalized variance* and *total variation* as $|\mathbf{\Sigma}|$ and $\operatorname{tr} \mathbf{\Sigma}$, respectively.

2.2.3 Mahalanobis space

We now turn to the population analogue of the Mahalanobis distance given by (1.6.1). If $\mathbf{x}$ and $\mathbf{y}$ are two points in space, then the *Mahalanobis distance* between $\mathbf{x}$ and $\mathbf{y}$, with metric $\mathbf{\Sigma}$, is the square root of

$$\Delta^2_{\mathbf{\Sigma}}(\mathbf{x}, \mathbf{y}) = (\mathbf{x}-\mathbf{y})'\mathbf{\Sigma}^{-1}(\mathbf{x}-\mathbf{y}). \tag{2.2.18}$$

(The subscript $\mathbf{\Sigma}$ may be omitted when there is no risk of confusion.) The matrix $\mathbf{\Sigma}$ is usually selected to be some convenient covariance matrix. Some examples are as follows.

(1) Let $\mathbf{x} \sim (\boldsymbol{\mu}_1, \mathbf{\Sigma})$ and let $\mathbf{y} \sim (\boldsymbol{\mu}_2, \mathbf{\Sigma})$. Then $\Delta(\boldsymbol{\mu}_1, \boldsymbol{\mu}_2)$ is a Mahalanobis distance between the parameters. It is invariant under transformations of the form

$$\mathbf{x} \to \mathbf{A}\mathbf{x}+\mathbf{b}, \qquad \mathbf{y} \to \mathbf{A}\mathbf{y}+\mathbf{b}, \quad \mathbf{\Sigma} \to \mathbf{A}\mathbf{\Sigma}\mathbf{A}',$$

where $\mathbf{A}$ is a non-singular matrix.

(2) Let $\mathbf{x} \sim (\boldsymbol{\mu}, \mathbf{\Sigma})$. The Mahalanobis distance between $\mathbf{x}$ and $\boldsymbol{\mu}$, $\Delta(\mathbf{x}, \boldsymbol{\mu})$, is here a random variable.

(3) Let $\mathbf{x} \sim (\boldsymbol{\mu}_1, \mathbf{\Sigma})$, $\mathbf{y} \sim (\boldsymbol{\mu}_2, \mathbf{\Sigma})$. The Mahalanobis distance between $\mathbf{x}$ and $\mathbf{y}$ is $\mathbf{\Delta}(\mathbf{x}, \mathbf{y})$.

2.2.4 Higher moments

Following Section 1.8, a kth-order central moment for the variables $x_{i_1}, \ldots, x_{i_s}$ is

$$\mu^{(j_1,\ldots,j_s)}_{i_1,\ldots,i_s} = E\left\{\prod_{t=1}^{s} (x_{i_t} - \mu_{i_t})^{j_t}\right\},$$

where $j_1 + \ldots + j_s = k$, $j_t \neq 0$, $t = 1, \ldots, s$. Further, suitable population counterparts of the measures of multivariate skewness and kurtosis for random vector $\mathbf{x} \sim (\boldsymbol{\mu}, \mathbf{\Sigma})$ are, respectively,

$$\beta_{1,p} = E\{(\mathbf{x}-\boldsymbol{\mu})'\mathbf{\Sigma}^{-1}(\mathbf{y}-\boldsymbol{\mu})\}^3, \tag{2.2.19}$$

$$\beta_{2,p} = E\{(\mathbf{x}-\boldsymbol{\mu})'\mathbf{\Sigma}^{-1}(\mathbf{x}-\boldsymbol{\mu})\}^2, \tag{2.2.20}$$

where $\mathbf{x}$ and $\mathbf{y}$ are identically and independently distributed (see Mardia, 1970*a*). It can be seen that these measures are invariant under linear transformations.

Example 2.2.2 Consider the p.d.f. given by (2.1.10) which has uniform marginals on (0, 1). We have

$$E(x_1)=E(x_2)=\tfrac{1}{2}, \qquad \mu_{r,s}=E(x_1-\tfrac{1}{2})^r(x_2-\tfrac{1}{2})^s=\mu_r\mu_s+4\alpha\mu_{r+1}\mu_{s+1},$$

where

$$\mu_r=2^{-r-1}\{1+(-1)^r\}/(r+1),$$

which is the rth central moment for the uniform distribution on (0, 1). Let $\gamma_{rs}=\mu_{rs}/\sigma_1^r\sigma_2^s$. Then

$$\sigma_1^2=\sigma_2^2=\tfrac{1}{12},\ \rho=\tfrac{1}{3}\alpha,\ \gamma_{12}=\gamma_{03}=0, \qquad \gamma_{22}=1,\ \gamma_{13}=\tfrac{9}{5}\rho,\ \gamma_{40}=\tfrac{9}{5}.$$

Consequently, using these in the population analogue of Exercise 1.8.1, we have

$$\beta_{1,2}=0, \qquad \beta_{2,2}=4(7-13\rho^2)/\{5(1-\rho^2)^2\}.$$

2.2.5 Conditional moments

Moments of $\mathbf{x}_1 \mid \mathbf{x}_2$ are called conditional moments. In particular, $E(\mathbf{x}_1 \mid \mathbf{x}_2)$ and $V(\mathbf{x}_1 \mid \mathbf{x}_2)$ are the conditional mean vector and the conditional variance-covariance matrix of $\mathbf{x}_1$ given $\mathbf{x}_2$. The *regression curve* of $\mathbf{x}_1$ on $\mathbf{x}_2$ is defined by the conditional expectation function

$$E(\mathbf{x}_1 \mid \mathbf{x}_2),$$

defined on the support of $\mathbf{x}_2$. If this function is linear in $\mathbf{x}_2$, then the regression is called *linear*. The conditional variance function

$$V(\mathbf{x}_1 \mid \mathbf{x}_2)$$

defines the *scedastic curve* of $\mathbf{x}_1$ on $\mathbf{x}_2$. The regression of $\mathbf{x}_1$ on $\mathbf{x}_2$ is called *homoscedastic* if $V(\mathbf{x}_1 \mid \mathbf{x}_2)$ is a constant matrix.

Example 2.2.3 Consider the p.d.f. given by (2.2.16). The marginal density of x_2 is

$$f_2(x_2)=\int_0^1 (x_1+x_2)\,dx_1=x_2+\tfrac{1}{2}, \qquad 0<x_2<1.$$

Hence the regression curve of x_1 on x_2 is

$$E(x_1 \mid x_2)=\int_0^1 x_1 f(x_1 \mid x_2)\,dx_1=\int_0^1 \frac{x_1(x_1+x_2)}{x_2+\frac{1}{2}}\,dx_1=\frac{(3x_2+2)}{3(1+2x_2)}, \qquad 0<x_2<1.$$

This is a decreasing function of x_2, so the regression is not linear. Similarly,

$$E(x_1^a \mid x_2) = \left(\frac{1}{a+2} + \frac{1}{a+1}x_2\right) \Big/ \left(\frac{1}{2} + x_2\right),$$

so that

$$V(x_1 \mid x_2) = (1 + 6x_2 + 6x_2^2)/\{18\,(1+2x_2)^2\}, \qquad 0 < x_2 < 1.$$

Hence the regression is not homoscedastic. ∎

In general, if all the specified expectations exist, then

$$E(\mathbf{x}_1) = E\{E(\mathbf{x}_1 \mid \mathbf{x}_2)\}. \tag{2.2.21}$$

However, note that the conditional expectations $E(\mathbf{x}_1 \mid \mathbf{x}_2)$ may all be finite, even when $E(\mathbf{x})$ is infinite (see Exercise 2.2.6).

2.3 Characteristic Functions

Let $\mathbf{x}$ be a random p-vector. Then the characteristic function (c.f.) of $\mathbf{x}$ is defined as the function

$$\phi_{\mathbf{x}}(\mathbf{t}) = E(e^{i\mathbf{t}'\mathbf{x}}) = \int e^{i\mathbf{t}'\mathbf{x}} f(\mathbf{x})\, d\mathbf{x}, \qquad \mathbf{t} \in R^p. \tag{2.3.1}$$

As in the univariate case, we have the following properties.

(1) The characteristic function always exists, $\phi_{\mathbf{x}}(\mathbf{0}) = 1$, and $|\phi_{\mathbf{x}}(\mathbf{t})| \leq 1$.

(2) (Uniqueness theorem.) Two random vectors have the same c.f. if and only if they have the same distribution.

(3) (Inversion theorem.) If the c.f. $\phi_{\mathbf{x}}(t)$ is absolutely integrable, then $\mathbf{x}$ has a p.d.f. given by

$$f(\mathbf{x}) = \frac{1}{(2\pi)^p} \int_{-\infty}^{\infty} e^{-i\mathbf{t}'\mathbf{x}} \phi_{\mathbf{x}}(\mathbf{t})\, d\mathbf{t}. \tag{2.3.2}$$

(4) Partition, $\mathbf{x}' = (\mathbf{x}_1', \mathbf{x}_2')$. The random vectors $\mathbf{x}_1$ and $\mathbf{x}_2$ are independent if and only if their joint c.f. factorizes into the product of their respective marginal c.f.s; that is if

$$\phi_{\mathbf{x}}(\mathbf{t}) = \phi_{\mathbf{x}_1}(\mathbf{t}_1)\phi_{\mathbf{x}_2}(\mathbf{t}_2), \tag{2.3.3}$$

where $\mathbf{t}' = (\mathbf{t}_1', \mathbf{t}_2')$.

(5)
$$E(x_1^{j_1} \dots x_p^{j_p}) = \frac{1}{i^{j_1+\dots+j_p}} \left\{ \frac{\partial^{j_1+\dots+j_p}}{\partial t_1^{j_1} \dots \partial t_p^{j_p}} \phi_{\mathbf{x}}(\mathbf{t}) \right\}_{\mathbf{t}=\mathbf{0}}$$

when this moment exists. (For a proof, differentiate both sides of (2.3.1) and put $\mathbf{t} = \mathbf{0}$.)

(6) The c.f. of the marginal distribution of $\mathbf{x}_1$ is simply $\phi_{\mathbf{x}}(\mathbf{t}_1, \mathbf{0})$.

(7) If $\mathbf{x}$ and $\mathbf{y}$ are independent random p-vectors then the c.f. of the *sum* $\mathbf{x}+\mathbf{y}$ is the *product* of the c.f.s of $\mathbf{x}$ and $\mathbf{y}$,

$$\phi_{\mathbf{x}+\mathbf{y}}(\mathbf{t}) = \phi_{\mathbf{x}}(\mathbf{t})\phi_{\mathbf{y}}(\mathbf{t}).$$

(To prove this, notice that independence implies $E(e^{i\mathbf{t}'(\mathbf{x}+\mathbf{y})}) = E(e^{i\mathbf{t}'\mathbf{x}})E(e^{i\mathbf{t}'\mathbf{y}})$.)

Example 2.3.1 Let $a(x_1, x_2)$ be a continuous positive function on an open set $S \subseteq R^2$. Let

$$f(x_1, x_2) = a(x_1, x_2)q(\theta_1, \theta_2) \exp\{x_1\theta_1 + x_2\theta_2\}, \qquad \mathbf{x} \in S,$$

be a density defined for $(\theta_1, \theta_2) \in \{(\theta_1, \theta_2) : 1/q(\theta_1, \theta_2) < \infty\}$ where

$$1/q(\theta_1, \theta_2) = \int a(x_1, x_2) \exp\{x_1\theta_1 + x_2\theta_2\}\, dx_1\, dx_2$$

is a normalization constant. Since this integral converges absolutely and uniformly over compact sets of (θ_1, θ_2), q can be extended by analytic continuation to give

$$\phi_{x_1,x_2}(t_1, t_2) = \int e^{it_1x_1+it_2x_2} f(x_1, x_2)\, dx_1\, dx_2 = q(\theta_1, \theta_2)/q(\theta_1 + it_1, \theta_2 + it_2). \quad \blacksquare$$

We end this section with an important result.

Theorem 2.3.7 (Cramér–Wold) *The distribution of a random p-vector* $\mathbf{x}$ *is completely determined by the set of all one-dimensional distributions of linear combinations* $\mathbf{t}'\mathbf{x}$, *where* $\mathbf{t} \in R^p$ *ranges through all fixed p-vectors.*

Proof Let $y = \mathbf{t}'\mathbf{x}$ and let the c.f. of y be

$$\phi_y(s) = E[e^{isy}] = E[e^{is\mathbf{t}'\mathbf{x}}].$$

Clearly for $s = 1$, $\phi_y(1) = E[e^{i\mathbf{t}'\mathbf{x}}]$, which, regarded as a function of $\mathbf{t}$, is the c.f. of $\mathbf{x}$. $\blacksquare$

The Cramér–Wold theorem implies that a multivariate probability distribution can be defined completely by specifying the distribution of all its linear combinations.

2.4 Transformations

Suppose that $f(\mathbf{x})$ is the p.d.f. of $\mathbf{x}$, and let $\mathbf{x}=\mathbf{u}(\mathbf{y})$ be a transformation from $\mathbf{y}$ to $\mathbf{x}$ which is one-to-one except possibly on sets of Lebesgue measure 0 in the supports of $\mathbf{x}$ and $\mathbf{y}$. Then the p.d.f. of $\mathbf{y}$ is

$$f\{\mathbf{u}(\mathbf{y})\}J, \tag{2.4.1}$$

where J is the Jacobian of the transformation from $\mathbf{y}$ to $\mathbf{x}$. It is defined by

$$J = \text{absolute value of } |\mathbf{J}|, \qquad \mathbf{J}=\left(\frac{\partial x_i}{\partial y_j}\right), \tag{2.4.2}$$

and we suppose that J is never zero or infinite except possibly on a set of Lebesgue measure 0. For some problems, it is easier to compute J from

$$J^{-1} = \text{absolute value of } \left|\frac{\partial y_j}{\partial x_i}\right| \tag{2.4.3}$$

using the inverse transformation $\mathbf{y}=\mathbf{u}^{-1}(\mathbf{x})$, and then substitute for $\mathbf{x}$ in terms of $\mathbf{y}$.

(1) *Linear transformation.* Let

$$\mathbf{y}=\mathbf{A}\mathbf{x}+\mathbf{b}, \tag{2.4.4}$$

where $\mathbf{A}$ is a non-singular matrix. Clearly, $\mathbf{x}=\mathbf{A}^{-1}(\mathbf{y}-\mathbf{b})$. Therefore $\partial x_i/\partial y_j = a^{ij}$, and the Jacobian of the transformation $\mathbf{y}$ to $\mathbf{x}$ is

$$\text{Abs}\,|\mathbf{A}|^{-1}. \tag{2.4.5}$$

(2) *Polar transformation.* A generalization of the two-dimensional polar transformation

$$x=r\cos\theta, \qquad y=r\sin\theta, \qquad r>0, \quad 0\leqslant\theta<2\pi,$$

to p dimensions is

$$\mathbf{x}=r\mathbf{u}(\boldsymbol{\theta}), \qquad \boldsymbol{\theta}=(\theta_1,\ldots,\theta_{p-1})', \tag{2.4.6}$$

where

$$u_i(\boldsymbol{\theta})=\cos\theta_i\prod_{j=0}^{i-1}\sin\theta_j, \qquad \sin\theta_0=\cos\theta_p=1,$$

and

$$0\leqslant\theta_j\leqslant\pi, \qquad j=1,\ldots,p-2, \qquad 0\leqslant\theta_{p-1}<2\pi, \qquad r>0.$$

Table 2.4.1 Jacobians of some transformations

Transformation Y to X	Restriction	Jacobian (absolute value)
$\mathbf{X}=\mathbf{Y}^{-1}$	$\mathbf{Y}(p\times p)$ and non-singular (all elements random)	$\lvert\mathbf{Y}\rvert^{-2p}$
$\mathbf{X}=\mathbf{Y}^{-1}$	$\mathbf{Y}$ symmetric and non-singular	$\lvert\mathbf{Y}\rvert^{-p-1}$
$\mathbf{X}=\mathbf{AY}+\mathbf{B}$	$\mathbf{Y}(p\times p)$, $\mathbf{A}(p\times p)$ non-singular, $\mathbf{B}(p\times p)$	$\lvert\mathbf{A}\rvert^{p}$
$\mathbf{X}=\mathbf{AYB}$	$\mathbf{Y}(p\times q)$, $\mathbf{A}(p\times p)$, and $\mathbf{B}(q\times q)$ non-singular	$\lvert\mathbf{A}\rvert^{q}\,\lvert\mathbf{B}\rvert^{p}$
$\mathbf{X}=\mathbf{AYA}'$	$\mathbf{Y}(p\times p)$ symmetric, $\mathbf{A}(p\times p)$ non-singular	$\lvert\mathbf{A}\rvert^{p+1}$
$\mathbf{X}=\mathbf{YY}'$	$\mathbf{Y}$ lower triangular	$2^p\prod_{i=1}^{p} y_{ii}^{p+1-i}$

The Jacobian of the transformation from $(r, \boldsymbol{\theta})$ to $\mathbf{x}$ is

$$J=r^{p-1}\prod_{i=2}^{p-1}\sin^{p-i}\theta_{i-1}. \qquad (2.4.7)$$

Note that the transformation is one to one except when $r=0$ or $\theta_i=0$ or π, for any $i=1,\ldots,p-2$.

(3) *Rosenblatt's transformation* (Rosenblatt, 1952). Suppose that $\mathbf{x}$ has p.d.f. $f(\mathbf{x})$ and denote the conditional c.d.f. of x_i given $x_1\ldots,x_{i-1}$ by $F(x_i\mid x_1,\ldots,x_{i-1})$, $i=1,2,\ldots,p$. The Jacobian of the transformation $\mathbf{x}$ to $\mathbf{y}$, where

$$y_i=F(x_i\mid x_1,\ldots,x_{i-1}), \qquad i=1,\ldots,p, \qquad (2.4.8)$$

is given by $f(x_1,\ldots,x_p)$. Hence, looking at the transformation $\mathbf{y}$ to $\mathbf{x}$, we see that $y_1,\ldots,y_p$ are independent identically distributed uniform variables on $(0,1)$.

Some other Jacobians useful in multivariate analysis are listed in Table 2.4.1. For their proof, see Deemer and Olkin (1951).

2.5 The Multinormal Distribution

2.5.1 Definition

In this section we introduce the most important multivariate probability distribution, namely the *multivariate normal* distribution. If we write the p.d.f. of $N(\mu, \sigma^2)$, the univariate normal with mean μ and variance $\sigma^2 > 0$, as

$$f(x) = \{2\pi\sigma^2\}^{-1/2} \exp\{-\tfrac{1}{2}(x-\mu)\{\sigma^2\}^{-1}(x-\mu)\},$$

then a plausible extension to p variates is

$$f(\mathbf{x}) = |2\pi\mathbf{\Sigma}|^{-1/2} \exp\{-\tfrac{1}{2}(\mathbf{x}-\boldsymbol{\mu})'\mathbf{\Sigma}^{-1}(\mathbf{x}-\boldsymbol{\mu})\}, \qquad (2.5.1)$$

where $\mathbf{\Sigma} > 0$. (Observe that the constant can be also written $\{(2\pi)^{p/2}|\mathbf{\Sigma}|^{1/2}\}^{-1}$.) Obviously, (2.5.1) is positive. It will be shown below in Theorem 2.5.1 that the total integral is unity, but first we give a formal definition.

Definition 2.5.1 *The random vector $\mathbf{x}$ is said to have a p-variate normal (or p-dimensional multinormal or multivariate normal) distribution with mean vector $\boldsymbol{\mu}$ and covariance matrix $\mathbf{\Sigma}$ if its p.d.f. is given by* (2.5.1). *We write* $\mathbf{x} \sim N_p(\boldsymbol{\mu}, \mathbf{\Sigma})$.

The quadratic form in 2.5.1 is equivalent to

$$\sum_{i=1}^{p} \sum_{j=1}^{p} \sigma^{ij}(x_i - \mu_i)(x_j - \mu_j), \qquad \mathbf{\Sigma}^{-1} = (\sigma^{ij}).$$

The p.d.f. may also be written in terms of correlations rather than covariances.

Theorem 2.5.1 *Let $\mathbf{x}$ have the p.d.f. given by* (2.5.1), *and let*

$$\mathbf{y} = \mathbf{\Sigma}^{-1/2}(\mathbf{x}-\boldsymbol{\mu}), \qquad (2.5.2)$$

where $\mathbf{\Sigma}^{-1/2}$ is the symmetric positive-definite square root of $\mathbf{\Sigma}^{-1}$. Then $y_1, \ldots, y_p$ are independent $N(0, 1)$ variables.

Proof From (2.5.2), we have

$$(\mathbf{x}-\boldsymbol{\mu})'\mathbf{\Sigma}^{-1}(\mathbf{x}-\boldsymbol{\mu}) = \mathbf{y}'\mathbf{y}. \qquad (2.5.3)$$

From (2.4.5) the Jacobian of the transformation $\mathbf{y}$ to $\mathbf{x}$ is $|\mathbf{\Sigma}|^{1/2}$. Hence, using (2.5.1), the p.d.f. of $\mathbf{y}$ is

$$g(\mathbf{y}) = \frac{1}{(2\pi)^{p/2}} e^{-\Sigma y_i^2/2}. \quad \blacksquare$$

Note that since $g(\mathbf{y})$ integrates to 1, (2.5.1) is a density.

Corollary 2.5.1.1 *If $\mathbf{x}$ has the p.d.f. given by* (2.5.1) *then*

$$E(\mathbf{x}) = \boldsymbol{\mu}, \qquad V(\mathbf{x}) = \mathbf{\Sigma}. \qquad (2.5.4)$$

Proof We have

$$E(\mathbf{y}) = \mathbf{0}, \qquad V(\mathbf{y}) = \mathbf{I}. \qquad (2.5.5)$$

From (2.5.2),

$$\mathbf{x} = \mathbf{\Sigma}^{1/2}\mathbf{y} + \boldsymbol{\mu}. \tag{2.5.6}$$

Using Theorem 2.2.1, the result follows. ■

For $p = 2$, it is usual to write ρ_{12} as ρ, $-1 < \rho < 1$. In this case the p.d.f. becomes

$$f(x_1, x_2) = \frac{1}{2\pi\sigma_1\sigma_2(1-\rho^2)^{1/2}} \times \exp\left[-\frac{1}{2(1-\rho^2)}\left\{\frac{(x_1-\mu_1)^2}{\sigma_1^2} - \frac{2\rho(x_1-\mu_1)(x_2-\mu_2)}{\sigma_1\sigma_2} + \frac{(x_2-\mu_2)^2}{\sigma_2^2}\right\}\right],$$

where $-\infty < x_1, x_2 < \infty$.

2.5.2 Geometry

We now look at some of the above ideas geometrically. The multivariate normal distribution in p dimensions has constant density on ellipses or ellipsoids of the form

$$(\mathbf{x} - \boldsymbol{\mu})'\mathbf{\Sigma}^{-1}(\mathbf{x} - \boldsymbol{\mu}) = c^2, \tag{2.5.7}$$

c being a constant. These ellipsoids are called the *contours* of the distribution or the "ellipsoids of equal concentrations". For $\boldsymbol{\mu} = \mathbf{0}$, these contours are centred at $\mathbf{x} = \mathbf{0}$, and when $\mathbf{\Sigma} = \mathbf{I}$ the contours are circles or in higher dimensions spheres or hyperspheres. Figure 2.5.1 shows a family of such contours for selected values of c for the bivariate case, and Figure 2.5.2 shows various types of contour for differing $\boldsymbol{\mu}$ and $\mathbf{\Sigma}$.

The principal component transformation facilitates interpretation of the ellipsoids of equal concentration. Using the spectral decomposition theorem (Theorem A.6.4), write $\mathbf{\Sigma} = \mathbf{\Gamma}\mathbf{\Lambda}\mathbf{\Gamma}'$, where $\mathbf{\Lambda} = \text{diag}(\lambda_1, \ldots, \lambda_p)$ is the matrix of eigenvalues of $\mathbf{\Sigma}$, and $\mathbf{\Gamma}$ is an orthogonal matrix whose columns are the corresponding eigenvectors. As in Section 1.5.3, define the *principal component transformation* by $\mathbf{y} = \mathbf{\Gamma}'(\mathbf{x} - \boldsymbol{\mu})$. In terms of $\mathbf{y}$, (2.5.7) becomes

$$\sum_{i=1}^{p} \frac{y_i^2}{\lambda_i} = c^2,$$

so that the components of $\mathbf{y}$ represent axes of the ellipsoid. This property is illustrated in Figure 2.5.1, where y_1 and y_2 represent the major and minor semi-axes of the ellipse, respectively.

2.5.3 Properties

If $\mathbf{x} \sim N_p(\boldsymbol{\mu}, \mathbf{\Sigma})$, the following results may be derived.

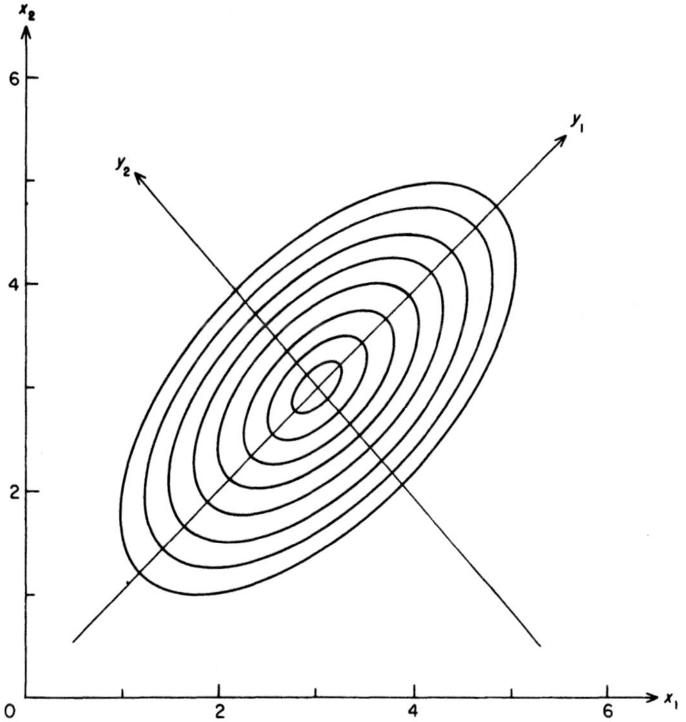

Figure 2.5.1 Ellipses of equal concentration for the bivariate normal distribution, showing the principal components y_1 *and* y_2, *where* $\boldsymbol{\mu}' = (3, 3)$, $\sigma_{11} = 3$, $\sigma_{12} = 1$, $\sigma_{22} = 3$.

Theorem 2.5.2

$$U = (\mathbf{x} - \boldsymbol{\mu})'\boldsymbol{\Sigma}^{-1}(\mathbf{x} - \boldsymbol{\mu}) \sim \chi_p^2. \tag{2.5.8}$$

Proof From (2.5.3) the left-hand side is $\sum y_i^2$, where the y_i' are independent $N(0, 1)$ by Theorem 2.5.1. Hence the result follows. ■

Using this theorem we can calculate the probability of a point $\mathbf{x}$ falling within an ellipsoid (2.5.7), from chi-square tables, since it amounts to calculating $\Pr(U \leq c^2)$.

Theorem 2.5.3 *The c.f. of* $\mathbf{x}$ *is*

$$\phi_{\mathbf{x}}(\mathbf{t}) = \exp(i\mathbf{t}'\boldsymbol{\mu} - \tfrac{1}{2}\mathbf{t}'\boldsymbol{\Sigma}\mathbf{t}). \tag{2.5.9}$$

Proof Using (2.5.6), we find that

$$\phi_{\mathbf{x}}(\mathbf{t}) = E(e^{i\mathbf{t}'\mathbf{x}}) = e^{i\mathbf{t}'\boldsymbol{\mu}}E(e^{i\boldsymbol{\alpha}'\mathbf{y}}), \tag{2.5.10}$$

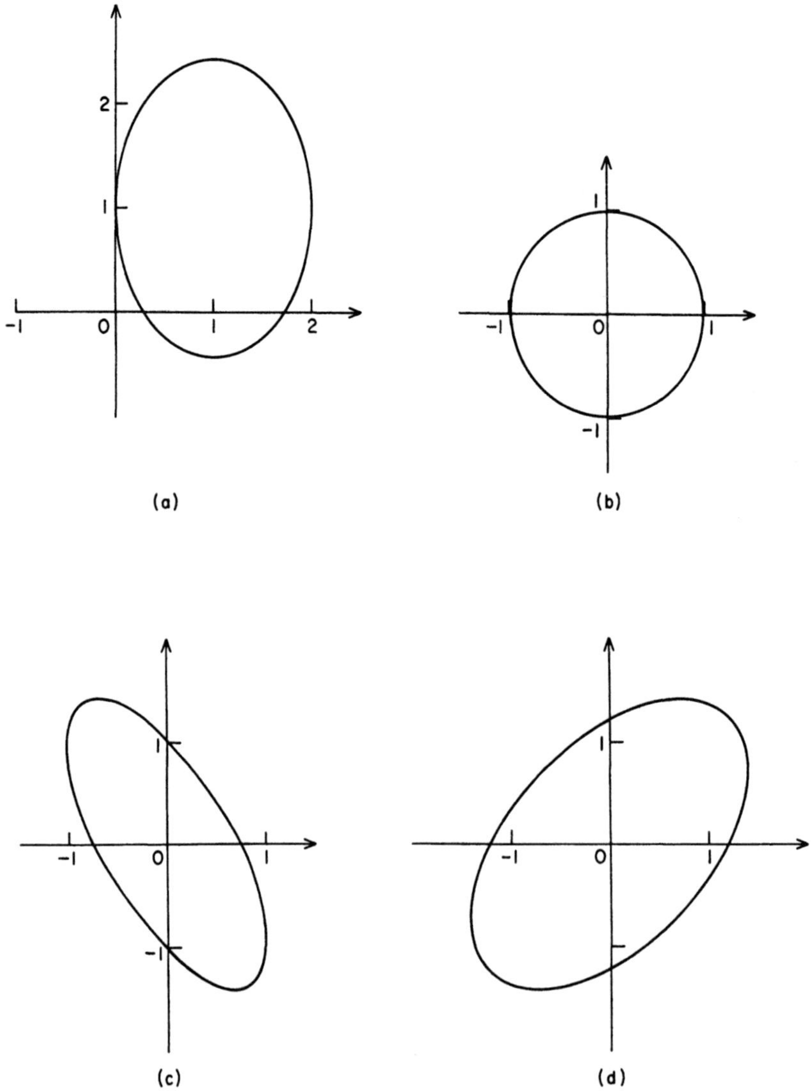

Figure 2.5.2 Ellipses of equal concentration for the bivariate normal distribution with c = 1.

(a) $\boldsymbol{\mu}' = (1, 1), \boldsymbol{\Sigma} = \begin{pmatrix} 1 & 0 \\ 0 & 2 \end{pmatrix}$. (b) $\boldsymbol{\mu}' = (0, 0), \boldsymbol{\Sigma} = \begin{pmatrix} 1 & 0 \\ 0 & 1 \end{pmatrix}$.

(c) $\boldsymbol{\mu}' = (0, 0), \boldsymbol{\Sigma} = \begin{pmatrix} 1 & -1 \\ -1 & 2 \end{pmatrix}$. (d) $\boldsymbol{\mu}' = (0, 0), \boldsymbol{\Sigma} = \begin{pmatrix} 2 & 1 \\ 1 & 2 \end{pmatrix}$.

where

$$\mathbf{u}' = \mathbf{t}'\boldsymbol{\Sigma}^{1/2}. \tag{2.5.11}$$

Since the y_i are independent $N(0, 1)$ from Theorem 2.5.1,

$$E(e^{i\mathbf{u}'\mathbf{y}}) = \prod_{i=1}^{p} \phi_{y_i}(u_i) = \prod_{i=1}^{p} e^{-u_i^2/2} = e^{-\mathbf{u}'\mathbf{u}/2}. \tag{2.5.12}$$

Substituting (2.5.12) and (2.5.11) in (2.5.10), we obtain the required result. ∎

As an example of the use of c.f.s we prove the following result.

Theorem 2.5.4 *All non-trivial linear combinations of the elements of* $\mathbf{x}$ *are univariate normal.*

Proof Let $\mathbf{a} \neq \mathbf{0}$ be a p-vector. The c.f. of $y = \mathbf{a}'\mathbf{x}$ is

$$\phi_y(t) = \phi_{\mathbf{x}}(t\mathbf{a}) = \exp\{it\mathbf{a}'\boldsymbol{\mu} - \tfrac{1}{2}t^2\mathbf{a}'\boldsymbol{\Sigma}\mathbf{a}\},$$

which is the c.f. of a normal random variable with mean $\mathbf{a}'\boldsymbol{\mu}$ and variance $\mathbf{a}'\boldsymbol{\Sigma}\mathbf{a} > 0$. ∎

Theorem 2.5.5 $\beta_{1,p} = 0$, $\beta_{2,p} = p(p+2)$.

Proof Let $V = (\mathbf{x}-\boldsymbol{\mu})'\boldsymbol{\Sigma}^{-1}(\mathbf{y}-\boldsymbol{\mu})$, where $\mathbf{x}$ and $\mathbf{y}$ are i.i.d. $N_p(\boldsymbol{\mu}, \boldsymbol{\Sigma})$. Then, from (2.2.19), $\beta_{1,p} = E(V^3)$. However, the distribution of V is symmetric about $V = 0$, and therefore $E(V^3) = 0$. From (2.2.20) and (2.5.8),

$$\beta_{2,p} = E\{(\chi_p^2)^2\} = p(p+2). \quad ∎$$

The multinormal distribution is explored in greater detail in Chapter 3 using a density-free approach.

*2.5.4 Singular multinormal distribution

The p.d.f. of $N_p(\boldsymbol{\mu}, \boldsymbol{\Sigma})$ involves $\boldsymbol{\Sigma}^{-1}$. However, if rank $(\boldsymbol{\Sigma}) = k < p$, we can define the (singular) density of $\mathbf{x}$ as

$$\frac{(2\pi)^{-k/2}}{(\lambda_1 \ldots \lambda_k)^{1/2}} \exp\{-\tfrac{1}{2}(\mathbf{x}-\boldsymbol{\mu})'\boldsymbol{\Sigma}^{-}(\mathbf{x}-\boldsymbol{\mu})\}, \tag{2.5.13}$$

where

(1) $\mathbf{x}$ lies on the hyperplane $\mathbf{N}'(\mathbf{x}-\boldsymbol{\mu}) = \mathbf{0}$, where $\mathbf{N}$ is a $p \times (p-k)$ matrix such that

$$\mathbf{N}'\boldsymbol{\Sigma} = \mathbf{0}, \quad \mathbf{N}'\mathbf{N} = \mathbf{I}_{p-k}, \tag{2.5.14}$$

(2) $\boldsymbol{\Sigma}^-$ is a g-inverse of $\boldsymbol{\Sigma}$ (see Section A.8) and $\lambda_1, \ldots, \lambda_k$ are the non-zero eigenvalues of $\boldsymbol{\Sigma}$.

There is a close connection between the singular density (2.5.13) and the non-singular multinormal distribution in k dimensions.

Theorem 2.5.6 *Let* $\mathbf{y} \sim N_k(\mathbf{0}, \boldsymbol{\Lambda}_1)$, *where* $\boldsymbol{\Lambda}_1 = \operatorname{diag}(\lambda_1, \ldots, \lambda_k)$. *Then there exists a* $(p \times k)$ *column orthonormal matrix* $\mathbf{B}$ (*that is,* $\mathbf{B}'\mathbf{B} = \mathbf{I}_k$) *such that*

$$\mathbf{x} = \mathbf{B}\mathbf{y} + \boldsymbol{\mu} \tag{2.5.15}$$

has the p.d.f. (2.5.13).

Proof The change of variables formula (2.4.1) has a generalization applicable to hypersurfaces. If $\mathbf{x} = \boldsymbol{\phi}(\mathbf{y})$, $\mathbf{y} \in R^k$, $\mathbf{x} \in R^p$, is a parametrization of a k-dimensional hypersurface in $R^p (k \leqslant p)$, and $g(\mathbf{y})$ is a p.d.f. on R^k, then $\mathbf{x}$ has a p.d.f. on the hypersurface given by

$$f(\mathbf{x}) = g(\boldsymbol{\phi}^{-1}(\mathbf{x}))\, |\mathbf{D}'\mathbf{D}|^{-1/2}, \tag{2.5.16}$$

where

$$\mathbf{D} = \mathbf{D}(\mathbf{x}) = \left(\frac{\partial \phi_i(\mathbf{y})}{\partial y_j}\right)\bigg|_{\mathbf{y}=\boldsymbol{\phi}^{-1}(\mathbf{x})}$$

is a $(p \times k)$ matrix evaluated at $\mathbf{y} = \boldsymbol{\phi}^{-1}(\mathbf{x})$, and we suppose $|\mathbf{D}'\mathbf{D}|$ is never zero.

By the spectral decomposition theorem we can write $\boldsymbol{\Sigma} = \boldsymbol{\Gamma}\boldsymbol{\Lambda}\boldsymbol{\Gamma}'$, where $\boldsymbol{\Lambda} = \operatorname{diag}(\lambda_1, \ldots, \lambda_k, 0, \ldots, 0)$ and $\boldsymbol{\Gamma} = (\mathbf{B} : \mathbf{N})$ is an orthonormal matrix partitioned so that $\mathbf{B}$ is a $(p \times k)$ matrix. Then $\boldsymbol{\Sigma} = \mathbf{B}\boldsymbol{\Lambda}_1\mathbf{B}'$, $\mathbf{B}'\mathbf{B} = \mathbf{I}_k$, $\mathbf{B}'\mathbf{N} = \mathbf{0}$, and $\mathbf{N}'\mathbf{N} = \mathbf{I}_{p-k}$. Notice that $\mathbf{N}$ can be taken to be the same as in (2.5.14).

The transformation $\mathbf{x} = \mathbf{B}\mathbf{y} + \boldsymbol{\mu}$ parametrizes the hyperplane $\mathbf{N}'(\mathbf{x} - \boldsymbol{\mu}) = 0$. The p.d.f. of $\mathbf{y}$ is given by

$$\frac{(2\pi)^{-k/2}}{(\lambda_1 \ldots \lambda_k)^{1/2}} \exp\left(-\tfrac{1}{2} \sum \frac{y_i^2}{\lambda_i}\right).$$

Now $\sum y_i^2/\lambda_i = \mathbf{y}'\boldsymbol{\Lambda}_1^{-1}\mathbf{y} = (\mathbf{x} - \boldsymbol{\mu})'\mathbf{B}\boldsymbol{\Lambda}_1^{-1}\mathbf{B}'(\mathbf{x} - \boldsymbol{\mu})$ and $\mathbf{B}\boldsymbol{\Lambda}_1^{-1}\mathbf{B}'$ is a g-inverse of $\boldsymbol{\Sigma}$. Also, for this transformation, $\mathbf{D} = \mathbf{B}$ and $|\mathbf{B}'\mathbf{B}|^{1/2} = |\mathbf{I}_k|^{1/2} = 1$. Thus, the p.d.f. of $f(\mathbf{x})$ takes the form given in (2.5.13). ■

Using Theorem 2.5.6, it is easy to show that many properties for the non-singular multinormal carry over to the singular case.

Corollary 2.5.6.1 $E(\mathbf{x}) = \boldsymbol{\mu}$, $V(\mathbf{x}) = \boldsymbol{\Sigma}$. ■

Corollary 2.5.6.2 $\phi_{\mathbf{x}}(\mathbf{t}) = \exp\{i\mathbf{t}'\boldsymbol{\mu} - \tfrac{1}{2}\mathbf{t}'\boldsymbol{\Sigma}\mathbf{t}\}$. ■ (2.5.17)

Corollary 2.5.6.3 *If* $\mathbf{a}$ *is a p-vector and* $\mathbf{a}'\mathbf{\Sigma a}>0$, *then* $\mathbf{a}'\mathbf{x}$ *has a univariate normal distribution.* ■

2.5.5 The matrix normal distribution

Let $\mathbf{X}(n\times p)$ be a matrix whose n rows, $\mathbf{x}'_1,\ldots,\mathbf{x}'_n$, are independently distributed as $N_p(\boldsymbol{\mu},\mathbf{\Sigma})$. Then $\mathbf{X}$ has the *matrix normal distribution* and represents a random sample from $N_p(\boldsymbol{\mu},\mathbf{\Sigma})$. Using (2.5.1), we find that the p.d.f. of $\mathbf{X}$ is

$$\begin{aligned} f(\mathbf{X}) &= |2\pi\mathbf{\Sigma}|^{-n/2}\exp\{-\tfrac{1}{2}\sum_{r=1}^{n}(\mathbf{x}_r-\boldsymbol{\mu})'\mathbf{\Sigma}^{-1}(\mathbf{x}_r-\boldsymbol{\mu})\} \\ &= |2\pi\mathbf{\Sigma}|^{-n/2}\exp\{-\tfrac{1}{2}\operatorname{tr}[\mathbf{\Sigma}^{-1}(\mathbf{X}-\mathbf{1}\boldsymbol{\mu}')'(\mathbf{X}-\mathbf{1}\boldsymbol{\mu}')]\}. \end{aligned}$$

2.6 Some Multivariate Generalizations of Univariate Distributions

We give below three common techniques which are used to generalize univariate distributions to higher dimensions. However, caution must be exercised because in some cases there is more than one plausible way to carry out the generalization.

2.6.1 Direct generalizations

Often a property used to derive a univariate distribution has a plausible (though not necessarily unique) extension to higher dimensions.

(1) The simplest example is the multivariate normal where the squared term $(x-\mu)^2/\sigma^2$ in the exponent of the one-dimensional p.d.f. is generalized to the quadratic form $(\mathbf{x}-\boldsymbol{\mu})'\mathbf{\Sigma}^{-1}(\mathbf{x}-\boldsymbol{\mu})$.

(2) If $\mathbf{x}\sim N_p(\boldsymbol{\mu},\mathbf{\Sigma})$ and $u_i=\exp(x_i)$, $i=1,\ldots,p$, then $\mathbf{u}$ is said to have a *multivariate log-normal distribution* with parameters $\boldsymbol{\mu}$, $\mathbf{\Sigma}$.

(3) Let $\mathbf{x}\sim N_p(\boldsymbol{\mu},\mathbf{\Sigma})$ and $y\sim\chi^2_\nu$ be independent, and set $u_i=x_i/(y/\nu)^{1/2}$, $i=1,\ldots,p$. Then $\mathbf{u}$ has a *multivariate Student's t-distribution* with parameters $\boldsymbol{\mu}$, $\mathbf{\Sigma}$, and $\boldsymbol{\nu}$ (Cornish, 1954; Dunnett and Sobel, 1954). The case $\nu=1$ is termed the *multivariate Cauchy distribution.* See Exercise 2.6.5.

(4) The *Wishart distribution* defined in Chapter 3 is a matrix generalization of the χ^2 distribution. The p.d.f. is given in (3.8.1).

(5) The multivariate Pareto distribution (Mardia, 1962, 1964*a*) has p.d.f.

$$f(\mathbf{x}) = a(a+1)\dots(a+p-1)\left(\prod_{i=1}^{p} b_i\right)^{-1}\left[1-p+\sum_{i=1}^{p}\frac{x_i}{b_i}\right]^{-(a+p)},$$

$$x_i > b_i, \quad i=1,\dots,p, \quad (2.6.1)$$

with parameters $b_i > 0$, $i=1,\dots,p$, and $a>0$. It generalizes the univariate Pareto distribution because its p.d.f. is given by a linear function of $\mathbf{x}$ raised to some power. See Exercise 2.6.4.

(6) The *Dirichlet distribution* is a generalization to p dimensions of the beta distribution (see Appendix B) with density

$$f(x_1,\dots,x_p) = \frac{\Gamma\left(\sum_{i=0}^{p}\alpha_i\right)}{\prod_{i=0}^{p}\Gamma(\alpha_i)}\left(1-\sum_{i=1}^{p}x_i\right)^{\alpha_0-1}\prod_{i=1}^{p}x_i^{\alpha_i-1},$$

$$x_i > 0, \quad i=1,\dots,p, \quad \sum_{i=1}^{p} x_i \leqslant 1, \quad (2.6.2)$$

where $\alpha_i > 0$, $i=0,1,\dots,p$, are parameters.

(7) The *multinomial distribution* is a discrete generalization to p dimensions of the binomial distribution. For parameters $a_1,\dots,a_p$ ($a_i>0$, $i=1,\dots,p$, $\sum_{i=1}^{p} a_i = 1$) and n (a positive integer), the probabilities are given by

$$P(\mathbf{x}=\mathbf{n}) = \begin{cases} \dfrac{n!}{n_1!\dots n_p!}\prod_{i=1}^{p} a_i^{n_i}, & n_i \geqslant 0, \quad i=1,\dots,p, \quad \sum_{i=1}^{p} n_i = n, \\ 0 & \text{otherwise.} \end{cases} \quad (2.6.3)$$

The mean and covariance matrix of this distribution are given in Exercise 2.6.6 and its limiting behaviour for large n is given in Example 2.9.1.

2.6.2 Common components

Let $\mathscr{F}$ be some family of distributions in one dimension and let $u_0, u_1, \dots, u_p$ denote independent members of $\mathscr{F}$. Set

$$x_i = u_i + u_0, \qquad i=1,\dots,p.$$

Then the distribution of $\mathbf{x}$ is one possible generalization of $\mathcal{F}$. This approach has been used to construct a multivariate Poisson distribution (see Exercise 2.6.2) and a multivariate gamma distribution (Ramabhandran, 1951).

2.6.3 Stochastic generalizations

Sometimes a probabilistic argument used to construct a distribution in one dimension can be generalized to higher dimensions. The simplest example is the use of the multivariate central limit theorem (Section 2.9) to justify the multivariate normal distribution. As another example, consider a multivariate exponential distribution (Weinman, 1966; Johnson and Kotz, 1972). A system has p identical components with times to failure $x_1, \ldots, x_p$. Initially, each component has an independent failure rate $\lambda_0 > 0$. (If the failure rate were constant for all time, the life time of each component would have an exponential distribution,

$$f(x, \lambda_0) = \lambda_0^{-1} \exp(-x/\lambda_0), \qquad x > 0.)$$

However, once a component fails, the failure rate of the remaining components changes. More specifically, conditional on exactly k components having failed by time $t > 0$, each of the remaining $p-k$ components has a failure rate $\lambda_k > 0$ (until the $(k+1)$th component fails). It can be shown that the joint p.d.f. is given by

$$\left(\prod_{i=0}^{p-1} \lambda_i^{-1}\right) \exp\left\{-\sum_{i=0}^{p-1} (p-i)\frac{(x_{(i+1)} - x_{(i)})}{\lambda_i}\right\}, \quad x_i > 0, \quad i = 1, \ldots p, \qquad (2.6.4)$$

where $x_{(0)} = 0$ and $x_{(1)} \leqslant x_{(2)} \leqslant \ldots \leqslant x_{(p)}$ are the order statistics.

A different multivariate exponential distribution has been described by Marshall and Olkin (1967), using a different underlying probability model (see Exercise 2.6.3).

2.7 Families of Distributions

2.7.1 The exponential family

The random vector $\mathbf{x}$ belongs to the *general exponential family* if its p.d.f. is of the form

$$f(\mathbf{x}; \boldsymbol{\theta}) = \exp\left[a_0(\boldsymbol{\theta}) + b_0(\mathbf{x}) + \sum_{i=1}^{q} a_i(\boldsymbol{\theta}) b_i(\mathbf{x})\right], \qquad \mathbf{x} \in S, \qquad (2.7.1)$$

where $\boldsymbol{\theta}' = (\theta_1, \ldots, \theta_r)$ is the vector of parameters, $\exp[a_0(\boldsymbol{\theta})]$ is the normalizing constant, and S is the support. If $r=q$ and $a_i(\boldsymbol{\theta})=\theta_i (i\neq 0)$, we say that $\mathbf{x}$ belongs to the *simple exponential family* (see (2.7.4)). The general exponential family (2.7.1) includes most of the important univariate distributions as special cases, for instance the normal, Poisson, negative binomial, and gamma distributions (Rao, 1973, p. 195).

Example 2.7.1 Putting

$$(b_1,\ldots,b_q)=(x_1,\ldots,x_p,\ x_1^2,\ldots,x_p^2,\ x_1x_2,\ x_1x_3,\ldots,x_{p-1}x_p)$$

in (2.7.1), it is seen that $N_p(\boldsymbol{\mu},\boldsymbol{\Sigma})$, whose p.d.f. is given by (2.5.1), belongs to the general exponential family.

Example 2.7.2 A discrete example is the *logit model* (Cox, 1972) defined by

$$\log P(x_1,\ldots,x_p)=a_0+\sum a_ix_i+\sum a_{ij}x_ix_j+\sum a_{ijk}x_ix_jx_k+\ldots, \tag{2.7.2}$$

for $x_i=\pm 1$, $i=1,\ldots,p$. If $a_{ij}=0$, $a_{ijk}=0,\ldots,$ the variables are independent. The logit model plays an important role in contingency tables where the variables $z_i=\frac{1}{2}(x_i+1)$, taking values 0 or 1, are of interest. The interpretation of parameters in (2.7.2) is indicated by

$$\begin{aligned}\tfrac{1}{2}\log\{P(x_1=1\mid x_2,\ldots,x_p)/P(x_1=-1\mid x_2,\ldots,x_p)\}\\ =a_1+a_{12}x_2+\ldots+a_{1p}x_p+a_{123}x_2x_3+\ldots\end{aligned} \tag{2.7.3}$$

since a_{23}, etc., do not appear in this expression.

Properties of exponential families

(1) For the simple exponential family,

$$f(\mathbf{x};\boldsymbol{\theta})=\exp\left[a_0(\boldsymbol{\theta})+b_0(\mathbf{x})+\sum_{i=1}^{q}\theta_ib_i(\mathbf{x})\right],\qquad \mathbf{x}\in S, \tag{2.7.4}$$

the vector $(b_1(\mathbf{x}),\ldots,b_q(\mathbf{x}))$ is a sufficient and complete statistic for $\boldsymbol{\theta}$.

(2) Consider the set of all p.d.f.s $g(\mathbf{x})$ with support S satisfying the constraints

$$E\{b_i(\mathbf{x})\}=c_i,\qquad i=1,\ldots,q, \tag{2.7.5}$$

where the c_i are fixed. Then the *entropy* $E\{-\log g(\mathbf{x})\}$ is maximized by the density $f(\mathbf{x};\boldsymbol{\theta})$ in (2.7.4), provided there exists $\boldsymbol{\theta}$ for which the constraints (2.7.5) are satisfied. If such a $\boldsymbol{\theta}$ exists, the maximum is unique (see Kagan *et al.*, 1973, p. 409; Mardia, 1975*a*).

The above maximum entropy property is very powerful. For example, if we fix the expectations of $x_1, \ldots, x_p$, $x_1^2, \ldots, x_p^2$, $x_1x_2, \ldots, x_{p-1}x_p$, $\mathbf{x} \in R_p$, then the maximum entropy distribution is the multinormal distribution.

Extended exponential family Let us assume that a random vector $\mathbf{y}$ is paired with an observable vector $\mathbf{x}$. Then an extended family can be defined (Dempster, 1971, 1972) as

$$f(\mathbf{y} \mid \mathbf{x}, \boldsymbol{\theta}) = \exp\left(a_0 + \sum_{i=1}^{p} \sum_{j=1}^{q} \theta_{ij} x_i y_j\right), \tag{2.7.6}$$

where a_0 depends on x_i and the parameters θ_{ij}. If $(\mathbf{x}', \mathbf{y}')'$ has a multinormal distribution, then the conditional distribution of $\mathbf{y} \mid \mathbf{x}$ is of the form (2.7.6). Obviously, the conditional distribution (2.7.3) for the logit model is also a particular case.

2.7.2 The spherical family

If the p.d.f. of $\mathbf{x}$ can be written as

$$f(\mathbf{x}) = g(\mathbf{x}'\mathbf{x}) \tag{2.7.7}$$

then the distribution of $\mathbf{x}$ is said to belong to the spherical family since it is spherically symmetric.

Note that $r = (\mathbf{x}'\mathbf{x})^{1/2}$ denotes the Euclidean distance from $\mathbf{x}$ to the origin. Hence for all vectors $\mathbf{x}$ with the same value of r the p.d.f. given by (2.7.7) has the same value. In other words, the equiprobability contours are hyperspheres.

Example 2.7.3 If $\mathbf{x} \sim N_p(\mathbf{0}, \sigma^2\mathbf{I})$, then the p.d.f. of $\mathbf{x}$ is

$$(2\pi\sigma^2)^{-p/2} \exp\left(-\tfrac{1}{2}r^2/\sigma^2\right).$$

Hence, $N_p(\mathbf{0}, \sigma^2\mathbf{I})$ is spherically symmetric.

Example 2.7.4 The multivariate Cauchy distribution with parameters $\mathbf{0}$ and $\mathbf{I}$ has p.d.f.

$$f(\mathbf{x}) = \pi^{-(p+1)/2}\Gamma\{\tfrac{1}{2}(p+1)\}(1 + \mathbf{x}'\mathbf{x})^{-(p+1)/2},$$

and belongs to the spherical family (see Exercise 2.6.5).

The following is an important property of spherically symmetric distributions.

Theorem 2.7.1 *The value of $r = (\mathbf{x}'\mathbf{x})^{1/2}$ is statistically independent of any scale-invariant function of* $\mathbf{x}$.

Proof Using the polar transformation (2.4.6) and the Jacobian (2.4.7), we see that the joint density of $(r, \boldsymbol{\theta})$,

$$r^{p-1}g(r^2)\,\mathrm{d}r\left[\prod_{i=2}^{p-1} \sin^{p-i}\theta_{i-1}\right]\mathrm{d}\theta_1 \ldots \mathrm{d}\theta_{p-1},$$

factorizes; hence r and $\boldsymbol{\theta}$ are statistically independent. Note that $\boldsymbol{\theta}$ is *uniformly* distributed on a p-dimensional hypersphere.

A function $h(\mathbf{x})$ is scale-invariant if $h(\mathbf{x}) = h(\alpha\mathbf{x})$ for all $\alpha > 0$. Setting $\alpha = 1/r$, we see that $h(\mathbf{x}) = h\{r\mathbf{u}(\boldsymbol{\theta})\} = h\{\mathbf{u}(\boldsymbol{\theta})\}$ depends only on $\boldsymbol{\theta}$. Thus, $h(\mathbf{x})$ is independent of r. ■

This theorem can also be proved in a more general setting, which will be useful later.

Theorem 2.7.2 *Let $\mathbf{X}(n \times p)$ be a random matrix which when thought of as an np-vector $\mathbf{X}^V$ is spherically symmetric. If $h(\mathbf{X})$ is any column-scale-invariant function of $\mathbf{X}$, then $h(\mathbf{X})$ is independent of $(r_1, \ldots, r_p)$, where*

$$r_j^2 = \sum_{i=1}^{n} x_{ij}^2, \qquad j = 1, \ldots, p.$$

Proof Write $R^2 = \sum r_j^2$. Then the density of $\mathbf{X}$ can be written as $g(R^2) \times \prod \mathrm{d}x_{ij}$. Transforming each column to polar coordinates, we get

$$\left[g(R^2)\prod_{j=1}^{p} r_j^{n-1}\,\mathrm{d}r_j\right]\left[\prod_{j=1}^{p}\left(\prod_{i=2}^{n-1} \sin^{n-i}\theta_{i-1,j}\right)\mathrm{d}\theta_{1j} \ldots \mathrm{d}\theta_{n-1,j}\right].$$

Thus $(r_1, \ldots, r_p)$ is independent of $\boldsymbol{\Theta} = (\theta_{ij})$.

A function $h(\mathbf{X})$ is *column-scale-invariant* if $h(\mathbf{XD}) = h(\mathbf{X})$ for all diagonal matrices $\mathbf{D} > 0$. Setting $\mathbf{D} = \operatorname{diag}(r_1^{-1}, \ldots, r_p^{-1})$, we see that

$$\begin{aligned} h(\mathbf{X}) &= h(r_1\mathbf{u}_1(\boldsymbol{\theta}_{(1)}), \ldots, r_p\mathbf{u}_p(\boldsymbol{\theta}_{(p)})) \\ &= h(\mathbf{u}_1(\boldsymbol{\theta}_{(1)}), \ldots, \mathbf{u}_p(\boldsymbol{\theta}_{(p)})) \end{aligned}$$

depends only on $\boldsymbol{\Theta}$, where each $\mathbf{u}_j(\boldsymbol{\theta}_{(j)})$ is the function of the jth column of $\boldsymbol{\Theta}$ defined as in (2.4.6). Thus, $h(\mathbf{X})$ is statistically independent of $(r_1, \ldots, r_p)$. ■

The above discussion can be extended to elliptically symmetric distributions,

$$f(\mathbf{x}) = g((\mathbf{x}-\boldsymbol{\mu})'\boldsymbol{\Sigma}^{-1}(\mathbf{x}-\boldsymbol{\mu})),$$

by noting that $\mathbf{y} = \boldsymbol{\Sigma}^{-1/2}(\mathbf{x}-\boldsymbol{\mu})$ is spherically symmetric.

Often results proved for the multinormal distribution $N_p(\mathbf{0}, \sigma^2\mathbf{I})$ are true for *all* spherically symmetric distributions. For example, if the vector

$(x_1, \ldots, x_{m+n})$ is spherically symmetric, then

$$n \sum_{j=1}^{m} x_j^2 \Big/ \left(m \sum_{j=m+1}^{m+n} x_j^2 \right) \sim F_{m,n}, \tag{2.7.8}$$

where $F_{m,n}$ is the F-variable (see Dempster, 1969; Kelker, 1972).

2.7.3 Stable distributions

A univariate random variable x with c.d.f. $F(x)$ is called *stable* if, for all $b_1 > 0$, $b_2 > 0$, there exists $b > 0$ and c real such that

$$F(x/b_1) * F(x/b_2) = F((x-c)/b),$$

where $*$ denotes convolution. This concept has been generalized to higher dimensions by Lévy (1937). A *random vector* $\mathbf{x}$ is *stable* if every linear combination of its components is univariate stable. Call $\mathbf{x}$ *symmetric* about $\mathbf{a}$ if $\mathbf{x}-\mathbf{a}$ and $-(\mathbf{x}-\mathbf{a})$ have the same distribution. Then a useful subclass of the symmetric stable laws is the set of random vectors whose c.f. is of the form

$$\log \phi_{\mathbf{x}}(\mathbf{t}) = i\mathbf{a}'\mathbf{t} - \tfrac{1}{2} \sum_{j=1}^{m} (\mathbf{t}'\boldsymbol{\Omega}_j \mathbf{t})^{\alpha/2}, \tag{2.7.9}$$

where $0 < \alpha \leq 2$, $m \geq 1$, and $\boldsymbol{\Omega}_j$ is a $(p \times p)$ positive semi-definite matrix of constants for $j = 1, \ldots, m$. Equation (2.7.9) gives the c.f. of a non-singular multivariate stable distribution if $\sum \boldsymbol{\Omega}_j > 0$ (Press, 1972, p. 155).

The Cauchy distribution and the multinormal distribution are the two most important stable distributions. For further details and applications in stable portfolio analysis see Press (1972), Chapter 12.

2.8 Random Samples

In Section 2.5.5, we met the idea of a random sample from the $N_p(\boldsymbol{\mu}, \boldsymbol{\Sigma})$ distribution. We now consider a more general situation.

Suppose that $\mathbf{x}_1, \ldots, \mathbf{x}_n$ is a random sample from a population with p.d.f. $f(\mathbf{x}; \boldsymbol{\theta})$, where $\boldsymbol{\theta}$ is a parameter vector; that is, $\mathbf{x}_1, \ldots, \mathbf{x}_n$ are independently and identically distributed (i.i.d.) where the p.d.f. of $\mathbf{x}_i$ is $f(\mathbf{x}_i; \boldsymbol{\theta})$, $i = 1, \ldots, n$. (The function f is the same for each value of i.)

We obtain now the moments of $\bar{\mathbf{x}}$, $\mathbf{S}$ and the Mahalanobis distances, under the assumption of random sampling from a population with mean $\boldsymbol{\mu}$ and covariance matrix $\boldsymbol{\Sigma}$. No assumption of normality is made in this section.

Theorem 2.8.1 *$E(\bar{\mathbf{x}})=\boldsymbol{\mu}$ and $V(\bar{\mathbf{x}})=(1/n)\boldsymbol{\Sigma}$.* (2.8.1)

Proof Since $E(\mathbf{x}_i)=\boldsymbol{\mu}$, we have

$$E(\bar{\mathbf{x}})=\frac{1}{n}\sum E(\mathbf{x}_i)=\boldsymbol{\mu}.$$

Because $V(\mathbf{x}_i)=\boldsymbol{\Sigma}$ and $C(\mathbf{x}_i, \mathbf{x}_j)=\mathbf{0}$, $i\neq j$,

$$V(\bar{\mathbf{x}})=\frac{1}{n^2}\left[\sum_{i=1}^{n} V(\mathbf{x}_i)+\sum_{i\neq j} C(\mathbf{x}_i, \mathbf{x}_j)\right]=\frac{1}{n}\boldsymbol{\Sigma}. \quad \blacksquare$$

Theorem 2.8.2 $E(\mathbf{S})=\{(n-1)/n\}\boldsymbol{\Sigma}$. (2.8.2)

Proof Since

$$\begin{aligned}\mathbf{S}&=\frac{1}{n}\sum_{i=1}^{n}\mathbf{x}_i\mathbf{x}_i'-\bar{\mathbf{x}}\bar{\mathbf{x}}'\\&=\frac{1}{n}\sum_{i=1}^{n}(\mathbf{x}_i-\boldsymbol{\mu})(\mathbf{x}_i-\boldsymbol{\mu})'-(\bar{\mathbf{x}}-\boldsymbol{\mu})(\bar{\mathbf{x}}-\boldsymbol{\mu})'\\&=\left(\frac{1}{n}-\frac{1}{n^2}\right)\sum_{i=1}^{n}(\mathbf{x}_i-\boldsymbol{\mu})(\mathbf{x}_i-\boldsymbol{\mu})'-\frac{1}{n^2}\sum_{i\neq j}(\mathbf{x}_i-\boldsymbol{\mu})(\mathbf{x}_j-\boldsymbol{\mu})'\end{aligned}$$

and $E\{(\mathbf{x}_i-\boldsymbol{\mu})(\mathbf{x}_j-\boldsymbol{\mu})'\}=\mathbf{0}$ for $i\neq j$, we see that

$$E(\mathbf{S})=n\left(\frac{1}{n}-\frac{1}{n^2}\right)\boldsymbol{\Sigma}=\frac{n-1}{n}\boldsymbol{\Sigma}. \quad \blacksquare$$

Note that

$$E(\mathbf{S}_u)=\frac{n}{n-1}E(\mathbf{S})=\boldsymbol{\Sigma},$$

so that $\mathbf{S}_u$ is an unbiased estimate of $\boldsymbol{\Sigma}$.

Theorem 2.8.3 *Let $\mathbf{G}=(g_{rs})$, where $g_{rs}=(\mathbf{x}_r-\bar{\mathbf{x}})'\mathbf{S}^{-1}(\mathbf{x}_s-\bar{\mathbf{x}})$. Then*

$$E(\mathbf{G})=\frac{np}{n-1}\mathbf{H}, \tag{2.8.3}$$

where

$$\mathbf{H}=\mathbf{I}-\frac{1}{n}\mathbf{1}\mathbf{1}'.$$

Proof We have the identities (see Exercise 1.6.1)

$$\sum_r g_{rs}=0 \quad \text{and} \quad \sum_r g_{rr}=np.$$

Under random sampling, the variables g_{rr} are identically distributed, as are the variables g_{rs} $(r \neq s)$. Let their expectations be a and b, respectively. From the above identities we see that $(n-1)b + a = 0$ and $na = np$. Therefore $a = p$ and $b = -p/(n-1)$, as stated in the theorem. ■

Note that $D_{rs}^2 = g_{rr} + g_{ss} - 2g_{rs}$. Therefore, under the assumption of random sampling,

$$E(D_{rs}^2) = a + a - 2b = 2pn/(n-1), \qquad r \neq s. \tag{2.8.4}$$

Example 2.8.1 Using the student data from Table 1.1.1, where $n = 5$ and $p = 3$, we see that, under the assumption of random sampling, $E(D_{rs}^2) = 2pn/(n-1) = 7.5$. In other words, each D_{rs} should be around 2.74. Calculations show the matrix of Mahalanobis distances D_{rs} to be

$$\begin{bmatrix} 0.00 & 2.55 & 2.92 & 3.13 & 3.07 \\ & 0.00 & 0.75 & 2.83 & 2.96 \\ & & 0.00 & 2.69 & 2.50 \\ & & & 0.00 & 3.15 \\ & & & & 0.00 \end{bmatrix}$$

It can be seen that most of the observed values of D_{rs} are indeed near 2.74, although D_{23} is substantially lower.

2.9 Limit Theorems

Although a random sample may come from a non-normal population, the *sample mean vector* $\bar{\mathbf{x}}$ will often be approximately normally distributed for large sample size n.

Theorem 2.9.1 (Central Limit Theorem) *Let* $\mathbf{x}_1, \mathbf{x}_2, \ldots$ *be an infinite sequence of independent identically distributed random vectors from a distribution with mean* $\boldsymbol{\mu}$ *and variance* $\boldsymbol{\Sigma}$. *Then*

$$n^{-1/2} \sum_{r=1}^{n} (\mathbf{x}_r - \boldsymbol{\mu}) = n^{1/2}(\bar{\mathbf{x}} - \boldsymbol{\mu}) \xrightarrow{D} N_p(\mathbf{0}, \boldsymbol{\Sigma}) \qquad \text{as } n \to \infty,$$

where $\xrightarrow{D}$ *denotes "convergence in distribution". By an abuse of notation we also write, asymptotically,*

$$\bar{\mathbf{x}} \sim N_p\left(\boldsymbol{\mu}, \frac{1}{n}\boldsymbol{\Sigma}\right).$$

The following theorem shows that a transformation of an asymptotically normal random vector with small variance is also asymptotically normal.

Theorem 2.9.2 *If* $\mathbf{t}$ *is asymptotically normal with mean* $\boldsymbol{\mu}$ *and covariance matrix* $\mathbf{V}/n$, *and if* $\mathbf{f}=(f_1,\ldots,f_q)'$ *are real-valued functions which are differentiable at* $\boldsymbol{\mu}$, *then* $\mathbf{f}(\mathbf{t})$ *is asymptotically normal with mean* $\mathbf{f}(\boldsymbol{\mu})$ *and covariance matrix* $\mathbf{D}'\mathbf{V}\mathbf{D}/n$, *where* $d_{ij}=\partial f_j/\partial t_i$ *evaluated at* $\mathbf{t}=\boldsymbol{\mu}$.

Proof First we need the following notation. Given a sequence of random variables $\{g_n\}$ and a sequence of positive numbers $\{b_n\}$, we say that $g_n=O_p(b_n)$ as $n\to\infty$ if

$$\lim_{n\to\infty}\sup_{m\geq n} P(|b_m^{-1}g_m|>c)\to 0 \qquad \text{as } c\to\infty; \tag{2.9.1}$$

that is if, for all n large enough, $|b_n^{-1}g_n|$ will with large probability not be too big. Similarly, say that $g_n=o_p(b_n)$ if

$$\lim_{n\to\infty}\sup_{m\geq n} P(|b_m^{-1}g_m|>c)=0 \qquad \text{for all } c>0; \tag{2.9.2}$$

that is, if $g_n/b_n \xrightarrow{P} 0$ as $n\to\infty$. Note that $O_p(\cdot)$ and $o_p(\cdot)$ are probabilistic versions of the $O(\cdot)$ and $o(\cdot)$ notation used for sequences of constants. Since each $\mathbf{f}_i$ is differentiable at $\boldsymbol{\mu}$ we can write

$$\mathbf{f}(\mathbf{t})-\mathbf{f}(\boldsymbol{\mu})=\mathbf{D}'(\mathbf{t}-\boldsymbol{\mu})+\|\mathbf{x}-\boldsymbol{\mu}\|\boldsymbol{\delta}(\mathbf{x}-\boldsymbol{\mu}),$$

where $\|\mathbf{x}-\boldsymbol{\mu}\|^2=\sum(x_i-\mu_i)^2$, and where $\boldsymbol{\delta}(\mathbf{x}-\boldsymbol{\mu})\to\mathbf{0}$ as $\mathbf{x}\to\boldsymbol{\mu}$.

The assumptions about $\mathbf{t}$ imply that $n^{1/2}\|\mathbf{t}-\boldsymbol{\mu}\|=O_p(1)$ and that $\|\boldsymbol{\delta}(\mathbf{t}-\boldsymbol{\mu})\|=o_p(1)$ as $n\to\infty$. Thus,

$$n^{1/2}[\mathbf{f}(\mathbf{t})-\mathbf{f}(\boldsymbol{\mu})]=n^{1/2}\mathbf{D}'(\mathbf{t}-\boldsymbol{\mu})+o_p(1)\xrightarrow{D} N_q(\mathbf{0},\mathbf{D}'\mathbf{V}\mathbf{D}). \quad \blacksquare$$

More general theorems of this type can be found in Rao (1973, p. 387).

Example 2.9.1 Let $\mathbf{x}$ be multinomially distributed with parameters n and $\mathbf{a}$, where $a_i>0$, $i=1,\ldots,p$, and set $\mathbf{z}=(1/n)\mathbf{x}$. Let $\mathbf{b}=(a_1^{1/2},\ldots,a_p^{1/2})'$ and $\mathbf{w}=(z_1^{1/2},\ldots,z_p^{1/2})'$. From Exercise 2.6.6, $\mathbf{x}$ can be written as a sum of n i.i.d. random variables with mean $\mathbf{a}$ and variance matrix $\operatorname{diag}(\mathbf{a})-\mathbf{a}\mathbf{a}'$. Thus, by the central limit theorem,

$$\mathbf{z}\sim N_p\left(\mathbf{a},\frac{1}{n}[\operatorname{diag}(\mathbf{a})-\mathbf{a}\mathbf{a}']\right)$$

asymptotically. Consider the transformation given by

$$f_i(\mathbf{z}) = z_i^{1/2}, \qquad i = 1, \ldots, p.$$

Then

$$\left.\frac{\partial f_i}{\partial z_j}\right|_{\mathbf{z}=\mathbf{a}} = \begin{cases} \frac{1}{2}b_i^{-1}, & i = j, \\ 0, & i \neq j, \end{cases}$$

so that, asymptotically,

$$\mathbf{w} \sim N_p\left(\mathbf{b}, \frac{1}{n}\mathbf{\Sigma}\right),$$

where

$$\mathbf{\Sigma} = \tfrac{1}{4}\,\text{diag}\,(b_i^{-1})\,[\text{diag}\,(\mathbf{a}) - \mathbf{aa}']\,\text{diag}\,(b_i^{-1}) = \tfrac{1}{4}[\mathbf{I} - \mathbf{bb}'].$$

Note that since $\sum b_i^2 = 1$, $\mathbf{\Sigma b} = \mathbf{0}$ and hence $\mathbf{\Sigma}$ is singular.

Exercises and Complements

2.1.1 Consider the bivariate p.d.f.

$$f(x_1, x_2) = 2, \qquad 0 < x_1 < x_2 < 1.$$

Calculate $f_1(x_1)$ and $f_2(x_2)$ to show that x_1 and x_2 are dependent.

2.1.2 As an extension of (2.2.16), let

$$f(x_1, x_2) = \begin{cases} \gamma(x_1^{\alpha} + x_2^{\beta}), & 0 < x_1, x_2 < 1, \\ 0, & \text{otherwise}, \end{cases}$$

where $\alpha > 0$, $\beta > 0$. Show that γ must equal $[1/(\alpha+1) + 1/(\beta+1)]^{-1}$ for $f(\cdot)$ to integrate to one. Calculate the corresponding c.d.f. and the probability of the following events:

$$\text{(i)}\;\; 0 < x_1, x_2 < \tfrac{1}{2}, \qquad \text{(ii)}\;\; 0 < x_1 < \tfrac{1}{2} < x_2 < 1,$$
$$\text{(iii)}\;\; \tfrac{1}{2} < x_1, x_2 < 1, \qquad \text{(iv)}\;\; 0 < x_2 < \tfrac{1}{2} < x_1 < 1.$$

2.2.1 (Mardia 1962, 1970*b*, p. 91) Consider the bivariate Pareto distribution defined by

$$f(x, y) = c(x + y - 1)^{-p-2}, \qquad x, y > 1,$$

where $p > 0$. Show that c must equal $p(p+1)$. Calculate the joint and marginal c.d.f.s of x and y. If $p > 1$ show that x and y each have mean

$p/(p-1)$, and if $p>2$ show that the covariance matrix is given by

$$\{(p-1)^2(p-2)\}^{-1}\begin{bmatrix} p & 1 \\ 1 & p \end{bmatrix}.$$

What happens to the expectations if $0<p\leqslant 1$? If $p>2$ show that $\operatorname{corr}(x, y)=1/p$, and that the generalized variance and total variation are

$$(p+1)/\{(p-1)^3(p-2)^2\} \quad \text{and} \quad 2p/\{(p-1)^2(p-2)\}.$$

Calculate the regression and scedastic curves of x on y.

2.2.2 For the random variables x and y with p.d.f.

$$(2\pi^3)^{-1/2}(x^2+y^2)^{-1/2}e^{-(x^2+y^2)/2}, \qquad -\infty<x, y<\infty,$$

show that $\rho=0$. Hence conclude that $\rho=0$ does not imply independence.

2.2.3 If $p\geqslant 2$ and $\mathbf{\Sigma}(p\times p)$ is the *equicorrelation* matrix, $\mathbf{\Sigma}=(1-\alpha)\mathbf{I}+\alpha\mathbf{11}'$, show that $\mathbf{\Sigma}\geqslant 0$ if and only if $-(p-1)^{-1}\leqslant\alpha\leqslant 1$. If $-(p-1)^{-1}<\alpha<1$, show that

$$\mathbf{\Sigma}^{-1}=(1-\alpha)^{-1}[\mathbf{I}-\alpha\{1+(p-1)\alpha\}^{-1}\mathbf{11}'].$$

If $\boldsymbol{\mu}_1'=(\delta, \mathbf{0}')$ and $\boldsymbol{\mu}_2=\mathbf{0}$ show that the Mahalanobis distance between them is given by

$$\Delta(\boldsymbol{\mu}_1, \boldsymbol{\mu}_2)=\delta\left[\frac{1+(p-2)\alpha}{(1-\alpha)\{1+(p-1)\alpha\}}\right]^{1/2}$$

2.2.4 (Fréchet inequalities: Fréchet, 1951; Mardia, 1970c) Let x and y be random variables with c.d.f. $H(x, y)$ and marginal c.d.f.s $F(x)$ and $G(y)$. Show that

$$\max(F+G-1, 0)\leqslant H\leqslant \min(F, G).$$

2.2.5 (Mardia, 1967c; Mardia and Thompson, 1972) Let x and y be random variables with c.d.f. $H(x, y)$ and corresponding marginal c.d.f.s $F(x)$ and $G(y)$. Show that

$$C(x^r, y^s)=rs\int_{-\infty}^{\infty}\int_{-\infty}^{\infty}u^{r-1}v^{s-1}\{H(u, v)-F(u)G(v)\}\,du\,dv$$

for $r, s>0$. Hence show that

$$C(x, y)=\int_{-\infty}^{\infty}\int_{-\infty}^{\infty}\{H(u, v)-F(u)G(v)\}\,du\,dv.$$

2.2.6 (Enis, 1973) Let x and y be random variables such that the p.d.f. of y is

$$g(y)=\frac{1}{\sqrt{2}}\,y^{-1/2}e^{-y/2},\qquad y>0,$$

and the conditional p.d.f. of x given y is

$$f(x\mid y)=(2\pi)^{-1/2}y^{1/2}\,e^{-yx^2/2}.$$

Show that $E[x\mid y]=0$ and hence that $E_y[E[x\mid y]]=0$, but that nevertheless the unconditional mean of x does not exist.

2.5.1 Consider the bivariate distribution of x and y defined as follows: let u and v be independent $N(0, 1)$ random variables. Set $x=u$ if $uv\geqslant 0$ while $x=-u$ if $uv<0$, and set $y=v$. Show that

(i) x and y are each $N(0, 1)$, but their joint distribution is not bivariate normal;
(ii) x^2 and y^2 are statistically independent, but x and y are not.

2.6.1 **Mixture of normals** (Mardia, 1974) Let $\phi(\mathbf{x}; \boldsymbol{\mu}, \boldsymbol{\Sigma})$ be the p.d.f. of $N_p(\boldsymbol{\mu}, \boldsymbol{\Sigma})$. Consider the mixture given by the p.d.f. $g_1(\mathbf{x})=\lambda\phi(\mathbf{x}; \boldsymbol{\mu}_1, \boldsymbol{\Sigma})+\lambda'\phi(\mathbf{x}; \boldsymbol{\mu}_2, \boldsymbol{\Sigma})$, where $0<\lambda<1$ and $\lambda'=1-\lambda$. Find a non-singular linear transformation, $\mathbf{y}=\mathbf{A}\mathbf{x}+\mathbf{b}$, for which the p.d.f. of $\mathbf{y}$ is given by

$$g_2(\mathbf{y})=\{\lambda\phi(y_1-\Delta)+\lambda'\phi(y_1)\}\prod_{r=2}^{p}\phi(y_r),\qquad -\infty<\mathbf{y}<\infty,$$

where $\phi(\cdot)$ is the p.d.f. of $N(0, 1)$, and

$$\Delta^2=(\boldsymbol{\mu}_1-\boldsymbol{\mu}_2)'\boldsymbol{\Sigma}^{-1}(\boldsymbol{\mu}_1-\boldsymbol{\mu}_2).$$

Hence show that the joint cumulant generating function $\log E\{\exp(\mathbf{t}'\mathbf{y})\}$ for $\mathbf{y}$ is

$$\tfrac{1}{2}\sum_{i=1}^{p} t_i^2+\log(\lambda'+\lambda e^{\Delta t_1}),$$

where the second term is the c.g.f. for the binomial distribution. Using the cumulant function or otherwise, show that the multivariate measures of skewness and kurtosis given by (2.2.19) and (2.2.20) for $\mathbf{y}$ (and hence also for $\mathbf{x}$) are

$$\beta_{1,p}=\{\lambda\lambda'(\lambda'-\lambda)\Delta^3\}^2/(1+\lambda\lambda'\Delta^2)^3$$

and

$$\beta_{2,p}=\{\lambda\lambda'(1-6\lambda\lambda')\Delta^4\}/(1+\lambda\lambda'\Delta^2)^2+p(p+2).$$

2.6.2 A multivariate Poisson distribution (Krishnamoorthy, 1951; Holgate, 1964) Let $u_0, u_1, \ldots, u_p$ be independent Poisson variables with parameters $\lambda_0, \lambda_1-\lambda_0, \ldots, \lambda_p-\lambda_0$, respectively. Write down the joint distribution of $x_i=u_0+u_i$, $i=1,\ldots,p$, and show that the marginal distributions of $x_1,\ldots,x_p$ are all Poisson. For $p=2$, show that the (discrete) p.d.f. is given by

$$f(x, y)=\exp(-\lambda_1-\lambda_2+\lambda_0)\frac{a^x b^y}{x!y!}\sum_{r=0}^{s}\frac{x^{(r)}}{a^r}\cdot\frac{y^{(r)}}{b^r}\cdot\frac{\lambda_0^r}{r!},$$

where $s=\min(x, y)$, $a=\lambda_1-\lambda_0$, $b=\lambda_2-\lambda_0$, $\lambda_1>\lambda_0>0$, $\lambda_2>\lambda_0>0$, and $x^{(r)}=x(x-1)\ldots(x-r+1)$. Furthermore,

$$E(y \mid x)=b+(\lambda_0/\lambda_1)x, \qquad \operatorname{var}(y \mid x)=b+\{a\lambda_0/\lambda_1^2\}x.$$

2.6.3 A multivariate exponential distribution (Marshall and Olkin, 1967) In a system of p components there are 2^p-1 types of shocks, each of which is fatal to one of the subsets $(i_1,\ldots,i_r)$ of the p components. A component dies when one of the subsets containing it receives a shock. The different types of shock have independent timing, and the shock which is fatal to the subset $(i_1,\ldots,i_r)$ has an exponential distribution with parameter $\lambda_{i_1,\ldots,i_r}^{-1}$ representing the expected value of the distribution. Show that the lifetimes $x_1,\ldots,x_p$ of the components are distributed as

$$-\log P(\mathbf{x}>\mathbf{a})=\sum_{i=1}^{p}\lambda_i a_i + \sum_{i_1<i_2}\lambda_{i_1i_2}\max(a_{i_1}, a_{i_2})+\ldots+\lambda_{1,2,\ldots,p}\max(a_1,\ldots,a_p).$$

2.6.4 A multivariate Pareto distribution (Mardia, 1962, 1964*a*) Consider the p.d.f.

$$f(\mathbf{x})=a(a+1)\ldots(a+p-1)\left(\prod_{i=1}^{p}b_i\right)^{-1}\left(\sum_{i=1}^{p}b_i^{-1}x_i-p+1\right)^{-(a+p)},$$

$$x_i>b_i, \quad i=1,\ldots,p,$$

where $a>0$, $b_i>0$.

(i) Show that any subset of the components of $\mathbf{x}$ has the same type of distribution.

(ii) Show that

$$P(\mathbf{x}>\mathbf{c})=\left(\sum_{i=1}^{p} b_i^{-1}c_i-p+1\right)^{-a}.$$

(iii) Let $\mathbf{x}_r$, $r=1,\ldots,n$, be a random sample from this population. Show that $\mathbf{u}=\min(\mathbf{x}_r)$ has the above distribution with the parameter a replaced by na.

2.6.5 Multivariate Student's *t* and Cauchy distribution (Cornish, 1954; Dunnett and Sobel, 1954) A random vector has a multivariate t distribution if its p.d.f. is of the form

$$g_\nu(\mathbf{t};\boldsymbol{\mu},\boldsymbol{\Sigma})=\frac{c_p\,|\boldsymbol{\Sigma}|^{-1/2}}{[1+\nu^{-1}(\mathbf{t}-\boldsymbol{\mu})'\boldsymbol{\Sigma}^{-1}(\mathbf{t}-\boldsymbol{\mu})]^{(\nu+p)/2}},$$

where

$$c_p=\frac{\Gamma((\nu+p)/2)}{(\pi\nu)^{p/2}\Gamma(\nu/2)}$$

and ν is known as the number of degrees of freedom of the distribution.

(i) Let $x_j=y_j/(S/\sqrt{\nu})$, $j=1,\ldots,p$, where $\mathbf{y}\sim N_p(\mathbf{0},\mathbf{I})$, $S^2\sim\chi^2_\nu$, and $\mathbf{y}$ and S are independent. Show that $\mathbf{x}$ has the p.d.f. $g_\nu(\mathbf{x};\mathbf{0},\mathbf{I})$.

(ii) For $\nu=1$, the distribution is known as a *multivariate Cauchy distribution*. Show that its c.f. is

$$\exp\{i\boldsymbol{\mu}'\mathbf{t}-(\mathbf{t}'\boldsymbol{\Sigma}\mathbf{t})^{1/2}\}.$$

2.6.6 Let $\mathbf{e}_i$ denote the unit vector with 1 in the ith place and 0 elsewhere. Let $\mathbf{y}_j$, $j=1,2,\ldots$, be a sequence of independent identically distributed random vectors such that $P(\mathbf{y}=\mathbf{e}_i)=a_i$, $i=1,\ldots,p$, where $a_i>0$, $\sum a_i=1$.

(i) Show that $E(\mathbf{y})=\mathbf{a}$, $V(\mathbf{y})=\operatorname{diag}(\mathbf{a})-\mathbf{a}\mathbf{a}'$.

(ii) Let $\mathbf{x}=\sum_{j=1}^{n}\mathbf{y}_j$. Show that $\mathbf{x}$ has the multinomial distribution given in (2.6.3).

(iii) Show that $E(\mathbf{x})=n\mathbf{a}$, $V(\mathbf{x})=n[\operatorname{diag}(\mathbf{a})-\mathbf{a}\mathbf{a}']$.

(iv) Verify that $[\operatorname{diag}(\mathbf{a})-\mathbf{a}\mathbf{a}']\mathbf{1}=\mathbf{0}$, and hence deduce that $V(\mathbf{x})$ is singular.

2.7.1 Show that the only spherical distribution for which the components of $\mathbf{x}$ are independent is the spherical multinormal distribution.

2.8.1 In the terminology of Section 2.8, suppose $E(g_{rr}^2)=c$, $E(g_{rs}^2)=d$, $r\neq s$. Using (1.8.2) and $nc=\sum_{i=1}^{n}E(g_{rr}^2)$, show that $c=E(b_{2,p})$. Further, using

$$nc+\frac{n(n-1)d}{2}=E\left\{\sum_{r=1}^{n}\sum_{s=1}^{n}g_{rs}^{2}\right\}=n,$$

show that

$$d=\frac{2}{n-1}\{1-E(b_{2,p})\}.$$

2.8.2 (Mardia, 1964*b*) Let $\mathbf{U}=(\mathbf{u}_1,\ldots,\mathbf{u}_n)'$ be an $(n\times p)$ data matrix from a p.d.f. $f(\mathbf{u})$ and set

$$x_i=\min_{r=1,\ldots,n}(u_{ri}),\qquad y_i=\max_{r=1,\ldots,n}(u_{ri}).$$

(i) Show that the joint p.d.f. of $(\mathbf{x},\mathbf{y})$ is

$$(-1)^p\frac{\partial^{2p}}{\partial x_1\ldots\partial x_p\,\partial y_1\ldots\partial y_p}\left\{\int_{\mathbf{x}}^{\mathbf{y}}f(\mathbf{u})\,d\mathbf{u}\right\}^n.$$

(ii) Let $R_i=y_i-x_i$ denote the *range* of the ith variable. Show that the joint p.d.f. of $R_1,\ldots,R_p$ is

$$\int_{-\infty}^{\infty}\left[(-1)^p\frac{\partial^{2p}}{\partial x_1\ldots\partial x_p\,\partial y_1\ldots\partial y_p}\left\{\int_{\mathbf{x}}^{\mathbf{y}}f(\mathbf{u})\,d\mathbf{u}\right\}^n\right]_{\mathbf{y}=\mathbf{x}+\mathbf{R}}d\mathbf{x}.$$

2.9.3 Karl Pearson-type inequality Let $\psi(\mathbf{x})$ be a non-negative function. Show that

$$P(\psi(\mathbf{x})<\epsilon^2)\geq 1-\frac{E\{\psi(\mathbf{x})\}^s}{\epsilon^{2s}}.$$

In particular,

$$P[(\mathbf{x}-\boldsymbol{\mu})'\boldsymbol{\Sigma}^{-1}(\mathbf{x}-\boldsymbol{\mu})<\epsilon^2]\geq 1-\beta_{2,p}/\epsilon^4,$$

where $\beta_{2,p}$ is the measure of kurtosis for $\mathbf{x}$.

3
Normal Distribution Theory

3.1 Characterization and Properties

3.1.1 The central role of multivariate normal theory

There has been a tendency in multivariate analysis to assume that all random vectors come from the multivariate normal or "multinormal" family of distributions. Among the reasons for its preponderance in the multivariate context are the following:

(1) The multinormal distribution is an easy generalization of its univariate counterpart, and the multivariate analysis runs almost parallel to the corresponding analysis based on univariate normality. The same cannot be said of other multivariate generalizations: different authors have given different extensions of the gamma, Poisson, and exponential distributions, and attempts to derive entirely suitable definitions have not yet proved entirely successful (see Section 2.6).
(2) The multivariate normal distribution is entirely defined by its first and second moments—a total of only $\frac{1}{2}p(p+3)$ parameters in all. This compares with 2^p-1 for the multivariate binary or logit distribution (2.7.2). This economy of parameters simplifies the problems of estimation.
(3) In the case of normal variables zero correlation implies independence, and pairwise independence implies total independence. Again, other distributions do not necessarily have these properties (see Exercise 2.2.2).
(4) Linear functions of a multinormal vector are themselves univariate normal. This opens the door to an extremely *simple derivation* of

multinormal theory, as developed here. Again, other distributions may not have this property, e.g. linear functions of multivariate binary variables are not themselves binary.

(5) Even when the original data is not multinormal, one can often appeal to central limit theorems which prove that certain functions such as the sample mean are normal for large samples (Section 2.9).

(6) The equiprobability contours of the multinormal distribution are simple ellipses, which by a suitable change of coordinates can be made into circles (or, in the general case, hyperspheres). This geometric simplicity, together with the associated invariance properties, allows us to derive many crucial properties through intuitively appealing arguments.

In this chapter, unlike Section 2.5, we shall use a density-free approach, and try to emphasize the interrelationships between different distributions without using their actual p.d.f.s.

3.1.2 A definition by characterization

In this chapter we shall define the multinormal distribution with the help of the Cramér–Wold theorem (Theorem 2.3.7). This states that the multivariate distribution of any random p-vector $\mathbf{x}$ is completely determined by the univariate distributions of linear functions such as $\mathbf{a}'\mathbf{x}$, where $\mathbf{a}$ may be any non-random p-vector.

Definition 3.1.1 *We say that* $\mathbf{x}$ *has a* p-variate normal distribution *if and only if* $\mathbf{a}'\mathbf{x}$ *is univariate normal for all fixed* p*-vectors* $\mathbf{a}$. ■

To allow for the case $\mathbf{a} = \mathbf{0}$, we regard constants as degenerate forms of the normal distribution.

The above definition of multinormality has a useful geometric interpretation. If $\mathbf{x}$ is visualized as a random point in p-dimensional space, then linear combinations such as $\mathbf{a}'\mathbf{x}$ can be regarded as projections of $\mathbf{x}$ onto a one-dimensional subspace. Definition 3.1.1 therefore implies that the projection of $\mathbf{x}$ onto all one-dimensional subspaces has a univariate normal distribution. This geometric interpretation makes it clear that even after $\mathbf{x}$ is transformed by any arbitrary shift, rotation, or projection, it will still have the property of normality. In coordinate-dependent terms, this may be stated more precisely as follows. (In this theorem and others that follow we will assume that matrices and vectors such as $\mathbf{A}$, $\mathbf{b}$, and $\mathbf{c}$ are non-random unless otherwise stated.)

Theorem 3.1.1 *If* $\mathbf{x}$ *has a* p*-variate normal distribution, and if* $\mathbf{y}=\mathbf{A}\mathbf{x}+\mathbf{c}$, *where* $\mathbf{A}$ *is any* $(q\times p)$ *matrix and* $\mathbf{c}$ *is any* q*-vector, then* $\mathbf{y}$ *has a* q*-variate normal distribution.* ■

Proof Let $\mathbf{b}$ be any fixed q-vector. Then $\mathbf{b}'\mathbf{y}=\mathbf{a}'\mathbf{x}+\mathbf{b}'\mathbf{c}$, where $\mathbf{a}=\mathbf{A}'\mathbf{b}$. Since $\mathbf{x}$ is multinormal, $\mathbf{a}'\mathbf{x}$ is univariate normal by Definition 3.1.1. Therefore $\mathbf{b}'\mathbf{y}$ is also univariate normal for all fixed vectors $\mathbf{b}$, and therefore $\mathbf{y}$ is multinormal by virtue of Definition 3.1.1. ■

Corollary 3.1.1.1 *Any subset of elements of a multinormal vector itself has a multinormal distribution. In particular the individual elements each have univariate normal distributions.* ■

Note that the above theorem and corollary need not assume that the covariance matrix $\mathbf{\Sigma}$ is of full rank. Therefore these results apply also to the singular multinormal distribution (Section 2.5.4). Also, the proofs do not use any intrinsic property of normality. Therefore similar results hold in principle for any other multivariate distribution defined in a similar way. That is, if we were to say that $\mathbf{x}$ has a p-variate "M" distribution whenever $\mathbf{a}'\mathbf{x}$ is univariate "M" for all fixed $\mathbf{a}$ ("M" could be "Cauchy") then results analogous to Theorem 3.1.1 and Corollary 3.1.1.1 could be derived.

However, before proceeding further, we must prove the *existence* of the multinormal distribution. This is done by showing that Definition 3.1.1 leads to the c.f. which has already been referred to in (2.5.9) and (2.5.17).

Theorem 3.1.2 *If* $\mathbf{x}$ *is multinormal with mean vector* $\boldsymbol{\mu}$ *and covariance matrix* $\mathbf{\Sigma}(\mathbf{\Sigma}\geqslant 0)$, *then its c.f. is given by*

$$\phi_{\mathbf{x}}(\mathbf{t})=\exp(i\mathbf{t}'\boldsymbol{\mu}-\tfrac{1}{2}\mathbf{t}'\mathbf{\Sigma}\mathbf{t}). \tag{3.1.1}$$

Proof We follow the lines of the Cramér–Wold theorem (Theorem 2.3.7), and note that if $y=\mathbf{t}'\mathbf{x}$ then y has mean $\mathbf{t}'\boldsymbol{\mu}$ and variance $\mathbf{t}'\mathbf{\Sigma}\mathbf{t}$. Since y is univariate normal, $y\sim N(\mathbf{t}'\boldsymbol{\mu},\mathbf{t}'\mathbf{\Sigma}\mathbf{t})$. Therefore from the c.f. of the univariate normal distribution, the c.f. of y is

$$\phi_y(s)=E(\exp isy)=\exp(is\mathbf{t}'\boldsymbol{\mu}-\tfrac{1}{2}s^2\mathbf{t}'\mathbf{\Sigma}\mathbf{t}).$$

Hence the c.f. of $\mathbf{x}$ must be given by

$$\phi_{\mathbf{x}}(\mathbf{t})=E(\exp i\mathbf{t}'\mathbf{x})=E(\exp iy)=\phi_y(1)=\exp(i\mathbf{t}'\boldsymbol{\mu}-\tfrac{1}{2}\mathbf{t}'\mathbf{\Sigma}\mathbf{t}).$$

From Section 2.5, we see that (3.1.1) is indeed the c.f. of a multivariate distribution. Hence the multinormal distribution with mean $\boldsymbol{\mu}$ and covariance matrix $\boldsymbol{\Sigma}$ exists and its c.f. has the stated form. ■

As in Section 2.5, we may summarize the statement, "$\mathbf{x}$ is p-variate normal with mean $\boldsymbol{\mu}$ and covariance matrix $\boldsymbol{\Sigma}$", by writing $\mathbf{x} \sim N_p(\boldsymbol{\mu}, \boldsymbol{\Sigma})$. When the dimension is clear we may omit the subscript p. We can obtain the p.d.f. when $\boldsymbol{\Sigma} > 0$ using the inversion formula (2.3.2), and it is given by (2.5.1).

Theorem 3.1.3 (a) *Two jointly multinormal vectors are independent if and only if they are uncorrelated*

(b) *For two jointly multinormal vectors, pair-wise independence of their components implies complete independence.* ■

Proof The c.f. given in Theorem 3.1.2 factorizes as required only when the corresponding submatrix of $\boldsymbol{\Sigma}$ is zero. This happens only when the vectors are uncorrelated. ■

3.2 Linear Forms

Theorem 3.1.1 proved that if $\mathbf{x} \sim N_p(\boldsymbol{\mu}, \boldsymbol{\Sigma})$ and $\mathbf{y} = \mathbf{A}\mathbf{x} + \mathbf{c}$, where $\mathbf{A}$ is any $(q \times p)$ matrix, then $\mathbf{y}$ has a q-variate normal distribution. Now from Section 2.2.2 we know that the moments of $\mathbf{y}$ are $\mathbf{A}\boldsymbol{\mu} + \mathbf{c}$ and $\mathbf{A}\boldsymbol{\Sigma}\mathbf{A}'$. Hence we deduce immediately the following results.

Theorem 3.2.1 *If* $\mathbf{x} \sim N_p(\boldsymbol{\mu}, \boldsymbol{\Sigma})$ *and* $\mathbf{y} = \mathbf{A}\mathbf{x} + \mathbf{c}$, *then* $\mathbf{y} \sim N_q(\mathbf{A}\boldsymbol{\mu} + \mathbf{c}, \mathbf{A}\boldsymbol{\Sigma}\mathbf{A}')$. ■

Corollary 3.2.1.1 *If* $\mathbf{x} \sim N_p(\boldsymbol{\mu}, \boldsymbol{\Sigma})$ *and* $\boldsymbol{\Sigma} > 0$, *then* $\mathbf{y} = \boldsymbol{\Sigma}^{-1/2}(\mathbf{x} - \boldsymbol{\mu}) \sim N_p(\mathbf{0}, \mathbf{I})$ *and* $(\mathbf{x} - \boldsymbol{\mu})'\boldsymbol{\Sigma}^{-1}(\mathbf{x} - \boldsymbol{\mu}) = \sum y_i^2 \sim \chi_p^2$. ■

Corollary 3.2.1.2 *If* $\mathbf{x} \sim N_p(\boldsymbol{\mu}, \sigma^2\mathbf{I})$ *and* $\mathbf{G}(q \times p)$ *is any row-orthonormal matrix, i.e. satisfying* $\mathbf{G}\mathbf{G}' = \mathbf{I}_q$, *then* $\mathbf{G}\mathbf{x} \sim N_q(\mathbf{G}\boldsymbol{\mu}, \sigma^2\mathbf{I})$. ■

Corollary 3.2.1.3 *If* $\mathbf{x} \sim N_p(\mathbf{0}, \mathbf{I})$ *and* $\mathbf{a}$ *is any non-zero* p*-vector, then* $\mathbf{a}'\mathbf{x}/\sqrt{\mathbf{a}'\mathbf{a}}$ *has the standard univariate normal distribution.* ■

Corollary 3.2.1.1 shows that any normal vector can easily be converted into standard form. It also gives an important quadratic expression which has a chi-squared distribution. From Corollary 3.2.1.2 we note that the standard multinormal distribution has a certain invariance under orthogonal transformations. Note that Corollary 3.2.1.3 also applies if $\mathbf{a}$ is a random vector independent of $\mathbf{x}$ (see Exercise 3.2.4). A further direct result of Theorem 3.1.3 is the following.

Theorem 3.2.2 *If* $\mathbf{x} \sim N_p(\boldsymbol{\mu}, \boldsymbol{\Sigma})$, *then* $\mathbf{Ax}$ *and* $\mathbf{Bx}$ *are independent if and only if* $\mathbf{A\Sigma B'} = \mathbf{0}$. ∎

Corollary 3.2.2.1 *If* $\mathbf{x} \sim N_p(\boldsymbol{\mu}, \sigma^2\mathbf{I})$ *and* $\mathbf{G}$ *is any row-orthonormal matrix, then* $\mathbf{Gx}$ *is independent of* $(\mathbf{I} - \mathbf{G'G})\mathbf{x}$. ∎

If $\mathbf{x}$ is partitioned into two subvectors, with r and s elements, respectively, then by noting two particular matrices which satisfy the conditions of Theorem 3.2.2, we may prove the following.

Theorem 3.2.3 *If* $\mathbf{x} = (\mathbf{x}_1', \mathbf{x}_2')' \sim N_p(\boldsymbol{\mu}, \boldsymbol{\Sigma})$, *then* $\mathbf{x}_1$ *and* $\mathbf{x}_{2.1} = \mathbf{x}_2 - \boldsymbol{\Sigma}_{21}\boldsymbol{\Sigma}_{11}^{-1}\mathbf{x}_1$ *have the following distributions and are statistically independent:*

$$\mathbf{x}_1 \sim N_r(\boldsymbol{\mu}_1, \boldsymbol{\Sigma}_{11}), \qquad \mathbf{x}_{2.1} \sim N_s(\boldsymbol{\mu}_{2.1}, \boldsymbol{\Sigma}_{22.1})$$

where

$$\boldsymbol{\mu}_{2.1} = \boldsymbol{\mu}_2 - \boldsymbol{\Sigma}_{21}\boldsymbol{\Sigma}_{11}^{-1}\boldsymbol{\mu}_1, \qquad \boldsymbol{\Sigma}_{22.1} = \boldsymbol{\Sigma}_{22} - \boldsymbol{\Sigma}_{21}\boldsymbol{\Sigma}_{11}^{-1}\boldsymbol{\Sigma}_{12}. \qquad (3.2.1)$$

Proof We may write $\mathbf{x}_1 = \mathbf{Ax}$ where $\mathbf{A} = [\mathbf{I}, \mathbf{0}]$, and $\mathbf{x}_{2.1} = \mathbf{Bx}$ where $\mathbf{B} = [-\boldsymbol{\Sigma}_{21}\boldsymbol{\Sigma}_{11}^{-1}, \mathbf{I}]$. Therefore, by Theorem 3.2.1, $\mathbf{x}_1$ and $\mathbf{x}_{2.1}$ are normal. Their moments are $\mathbf{A}\boldsymbol{\mu}$, $\mathbf{A\Sigma A'}$, $\mathbf{B}\boldsymbol{\mu}$, and $\mathbf{B\Sigma B'}$, which simplify to the given expressions. To prove independence note that $\mathbf{A\Sigma B'} = \mathbf{0}$, and use Theorem 3.2.2. ∎

Similar results hold (using g-inverses) for the case of singular distributions. The above theorem can now be used to find the conditional distribution of $\mathbf{x}_2$ when $\mathbf{x}_1$ is known.

Theorem 3.2.4 *Using the assumptions and notation of Theorem* 3.2.3, *the conditional distribution of* $\mathbf{x}_2$ *for a given value of* $\mathbf{x}_1$ *is*

$$\mathbf{x}_2 \mid \mathbf{x}_1 \sim N_s(\boldsymbol{\mu}_2 + \boldsymbol{\Sigma}_{21}\boldsymbol{\Sigma}_{11}^{-1}(\mathbf{x}_1 - \boldsymbol{\mu}_1), \boldsymbol{\Sigma}_{22.1}).$$

Proof Since $\mathbf{x}_{2.1}$ is independent of $\mathbf{x}_1$, its conditional distribution for a given value of $\mathbf{x}_1$ is the same as its marginal distribution, which was stated in Theorem 3.2.3. Now $\mathbf{x}_2$ is simply $\mathbf{x}_{2.1}$ plus $\boldsymbol{\Sigma}_{21}\boldsymbol{\Sigma}_{11}^{-1}\mathbf{x}_1$, and this term is constant when $\mathbf{x}_1$ is given. Therefore the conditional distribution of $\mathbf{x}_2 \mid \mathbf{x}_1$ is normal, and its conditional mean is

$$E[\mathbf{x}_2 \mid \mathbf{x}_1] = \boldsymbol{\mu}_{2.1} + \boldsymbol{\Sigma}_{21}\boldsymbol{\Sigma}_{11}^{-1}\mathbf{x}_1 = \boldsymbol{\mu}_2 + \boldsymbol{\Sigma}_{21}\boldsymbol{\Sigma}_{11}^{-1}(\mathbf{x}_1 - \boldsymbol{\mu}_1). \qquad (3.2.2)$$

The conditional covariance matrix of $\mathbf{x}_2$ is the same as that of $\mathbf{x}_{2.1}$, namely $\boldsymbol{\Sigma}_{22.1}$. ∎

If the assumption of normality is dropped from Theorem 3.2.3, then $\mathbf{x}_1$ and $\mathbf{x}_{2.1}$ still have the means and covariances stated. Instead of being independent of each other, however, all that can be said in general is that $\mathbf{x}_1$ and $\mathbf{x}_{2.1}$ are uncorrelated.

When $p=2$ and x_1 and x_2 are both scalars, then the expressions given above simplify. Putting σ_1^2, σ_2^2, and $\rho\sigma_1\sigma_2$ in place of $\mathbf{\Sigma}_{11}$, $\mathbf{\Sigma}_{22}$, and $\mathbf{\Sigma}_{12}$ we find that

$$\mathbf{\Sigma}_{22.1}=\sigma_2^2(1-\rho^2), \tag{3.2.3}$$

so that the conditional distribution of x_2 given x_1 is

$$x_2 \mid x_1 \sim N_1\{\mu_2+\rho\sigma_2\sigma_1^{-1}(x_1-\mu_1), \sigma_2^2(1-\rho^2)\}.$$

Example 3.2.1 If $\mathbf{\Sigma}$ is the equicorrelation matrix $\mathbf{\Sigma}=(1-\rho)\mathbf{I}+\rho\mathbf{1}\mathbf{1}'$, then the conditional distributions take a special form. Note that $\mathbf{\Sigma}_{11}$ and $\mathbf{\Sigma}_{22}$ are equicorrelation matrices of order $(r\times r)$ and $(s\times s)$, respectively. Also $\mathbf{\Sigma}_{12}=\rho\mathbf{1}_r\mathbf{1}_s'$ and $\mathbf{\Sigma}_{21}=\rho\mathbf{1}_s\mathbf{1}_r'$. Furthermore, we know from (A.3.2b) that

$$\mathbf{\Sigma}_{11}^{-1}=(1-\rho)^{-1}[\mathbf{I}-\alpha\mathbf{1}\mathbf{1}'], \qquad \alpha=\frac{\rho}{1+(r-1)\rho}.$$

Therefore, $\mathbf{\Sigma}_{21}\mathbf{\Sigma}_{11}^{-1}=\rho\mathbf{1}\mathbf{1}'\mathbf{\Sigma}_{11}^{-1}=\rho(1-\rho)^{-1}(1-\alpha r)\mathbf{1}\mathbf{1}'=\alpha\mathbf{1}\mathbf{1}'$. Substituting in (3.2.2) we find that the conditional mean is

$$E[\mathbf{x}_2 \mid \mathbf{x}_1]=\boldsymbol{\mu}_2+\alpha\mathbf{1}_r'(\mathbf{x}_1-\boldsymbol{\mu}_1)\mathbf{1}_s$$

and the conditional covariance matrix is

$$\mathbf{\Sigma}_{22.1}=(1-\rho)\mathbf{I}+\rho\mathbf{1}_s\mathbf{1}_s'-\rho\alpha\mathbf{1}_s\mathbf{1}_r'\mathbf{1}_r\mathbf{1}_s'=(1-\rho)\mathbf{I}+\rho(1-r\alpha)\mathbf{1}\mathbf{1}'.$$

Note that the conditional mean is just the original mean $\boldsymbol{\mu}_2$ with each element altered by the same amount. Moreover, this amount is proportional to $\mathbf{1}'(\mathbf{x}_1-\boldsymbol{\mu}_1)$, the sum of the deviations of the elements of $\mathbf{x}_1$ from their respective means. If $r=1$, then the conditional mean is

$$E(\mathbf{x}_2 \mid x_1)=\boldsymbol{\mu}_2+\rho(x_1-\mu_1)\mathbf{1}.$$

3.3 Transformations of Normal Data Matrices

Let $\mathbf{x}_1,\ldots,\mathbf{x}_n$ be a random sample from $N(\boldsymbol{\mu},\mathbf{\Sigma})$. We call $\mathbf{X}=(\mathbf{x}_1,\ldots,\mathbf{x}_n)'$ a data matrix from $N(\boldsymbol{\mu},\mathbf{\Sigma})$, or simply a "normal data matrix". In this section we shall consider linear functions such as $\mathbf{Y}=\mathbf{AXB}$, where $\mathbf{A}(m\times n)$ and $\mathbf{B}(p\times q)$ are fixed matrices of real numbers. The most important linear function is the sample mean $\bar{\mathbf{x}}'=n^{-1}\mathbf{1}'\mathbf{X}$, where $\mathbf{A}=n^{-1}\mathbf{1}'$ and $\mathbf{B}=\mathbf{I}_p$. The following result is immediate from Theorem 2.8.1.

Theorem 3.3.1 *If $\mathbf{X}(n\times p)$ is a data matrix from $N_p(\boldsymbol{\mu},\boldsymbol{\Sigma})$, and if $n\bar{\mathbf{x}}=\mathbf{X}'\mathbf{1}$, then $\bar{\mathbf{x}}$ has the $N_p(\boldsymbol{\mu}, n^{-1}\boldsymbol{\Sigma})$ distribution.* ■

We may ask under what conditions $\mathbf{Y}=\mathbf{AXB}$ is itself a normal data matrix. Since $y_{ij}=\sum_{\alpha,\beta} a_{i\alpha}x_{\alpha\beta}b_{\beta j}$, clearly each element of $\mathbf{Y}$ is univariate normal. However, for $\mathbf{Y}$ to be a normal data matrix we also require (a) that the rows of $\mathbf{Y}$ should be independent of each other and (b) that each row should have the same distribution. The following theorem gives necessary and sufficient conditions on $\mathbf{A}$ and $\mathbf{B}$.

Theorem 3.3.2 *If $\mathbf{X}(n\times p)$ is a normal data matrix from $N_p(\boldsymbol{\mu},\boldsymbol{\Sigma})$ and if $\mathbf{Y}(m\times q)=\mathbf{AXB}$, then $\mathbf{Y}$ is a normal data matrix if and only if*

(a) $\mathbf{A1}=\alpha\mathbf{1}$ *for some scalar* α, *or* $\mathbf{B}'\boldsymbol{\mu}=\mathbf{0}$, *and*
(b) $\mathbf{AA}'=\beta\mathbf{I}$ *for some scalar* β, *or* $\mathbf{B}'\boldsymbol{\Sigma}\mathbf{B}=\mathbf{0}$.

When both these conditions are satisfied then $\mathbf{Y}$ is a normal data matrix from $N_q(\alpha\mathbf{B}'\boldsymbol{\mu}, \beta\mathbf{B}'\boldsymbol{\Sigma}\mathbf{B})$. ■

Proof See Exercise 3.3.4. ■

To understand this theorem, note that post-multiplication of $\mathbf{X}$ involves adding weighted *variables*, while pre-multiplication of $\mathbf{X}$ adds weighted *objects*. Since the original objects (rows of $\mathbf{X}$) are independent, the transformed objects (rows of $\mathbf{Y}$) are also independent unless the pre-multiplication by $\mathbf{A}$ has introduced some interdependence. This clearly cannot happen when $\mathbf{A}$ is $k\mathbf{I}$, since then $\alpha=k$ and $\beta=k^2$, and both conditions of the theorem are satisfied. Similarly, all permutation matrices satisfy the conditions required on $\mathbf{A}$.

We may also investigate the correlation structure between two linear transformations of $\mathbf{X}$. Conditions for independence are stated in the following theorem.

Theorem 3.3.3 *If $\mathbf{X}$ is a data matrix from $N(\boldsymbol{\mu},\boldsymbol{\Sigma})$, and if $\mathbf{Y}=\mathbf{AXB}$ and $\mathbf{Z}=\mathbf{CXD}$, then the elements of $\mathbf{Y}$ are independent of the elements of $\mathbf{Z}$ if and only if either* (a) $\mathbf{B}'\boldsymbol{\Sigma}\mathbf{D}=\mathbf{0}$ *or* (b) $\mathbf{AC}'=\mathbf{0}$. ■

Proof See Exercise 3.3.5. ■

This result is also valid in the situation where the rows of $\mathbf{X}$ do not have the same mean; see Exercise 3.4.20.

Corollary 3.3.3.1 *Under the conditions of the theorem, if* $\mathbf{X}=(\mathbf{X}_1,\mathbf{X}_2)$ *then* $\mathbf{X}_1$ *is independent of* $\mathbf{X}_{2.1}=\mathbf{X}_2-\mathbf{X}_1\mathbf{\Sigma}_{11}^{-1}\mathbf{\Sigma}_{12}$. *Also* $\mathbf{X}_1$ *is a data matrix from* $N(\boldsymbol{\mu}_1,\mathbf{\Sigma}_{11})$ *and* $\mathbf{X}_{2.1}$ *is a data matrix from* $N(\boldsymbol{\mu}_{2.1},\mathbf{\Sigma}_{22.1})$, *where* $\boldsymbol{\mu}_{2.1}$ *and* $\mathbf{\Sigma}_{22.1}$ *are defined in* (3.2.1). ■

Proof We have $\mathbf{X}_1=\mathbf{XB}$ where $\mathbf{B}'=(\mathbf{I},\mathbf{0})$, and $\mathbf{X}_{2.1}=\mathbf{XD}$ where $\mathbf{D}'=(-\mathbf{\Sigma}_{21}\mathbf{\Sigma}_{11}^{-1},\mathbf{I})$. Since $\mathbf{B}'\mathbf{\Sigma}\mathbf{D}=\mathbf{0}$ the result follows from part (a) of the theorem. ■

Corollary 3.3.3.2 *Under the conditions of the theorem,* $\bar{\mathbf{x}}=n^{-1}\mathbf{X}'\mathbf{1}$ *is independent of* $\mathbf{HX}$, *and therefore* $\bar{\mathbf{x}}$ *is independent of* $\mathbf{S}=n^{-1}\mathbf{X}'\mathbf{HX}$. ■

Proof Put $\mathbf{A}=n^{-1}\mathbf{1}'$ and $\mathbf{C}=\mathbf{H}=\mathbf{I}-n^{-1}\mathbf{11}'$ in the theorem. Since $\mathbf{AC}'=\mathbf{0}$, the result follows. ■

3.4. The Wishart Distribution

3.4.1 Introduction

We now turn from linear functions to consider matrix-valued quadratic functions of the form $\mathbf{X}'\mathbf{CX}$, where $\mathbf{C}$ is a symmetric matrix. Among such functions the most important special case is the sample covariance matrix $\mathbf{S}$ obtained by putting $\mathbf{C}=n^{-1}\mathbf{H}$, where $\mathbf{H}$ is the centring matrix. (However, other quadratic functions can also be used, for instance, in permuting the rows of $\mathbf{X}$, or in finding within-group and between group covariance matrices in regression analysis). These quadratic forms often lead to the Wishart distribution, which constitutes a matrix generalization of the univariate chi squared distribution, and has many similar properties.

Definition 3.4.1 *If* $\mathbf{M}(p\times p)$ *can be written* $\mathbf{M}=\mathbf{X}'\mathbf{X}$, *where* $\mathbf{X}(m\times p)$ *is a data matrix from* $N_p(\mathbf{0},\mathbf{\Sigma})$, *then* $\mathbf{M}$ *is said to have a* Wishart distribution *with scale matrix* $\mathbf{\Sigma}$ *and degrees of freedom parameter* m. *We write* $\mathbf{M}\sim W_p(\mathbf{\Sigma},m)$. *When* $\mathbf{\Sigma}=\mathbf{I}_p$, *the distribution is said to be in standard form.* ■

Note when $p=1$ that the $W_1(\sigma^2,m)$ distribution is given by $\mathbf{x}'\mathbf{x}$, where the elements of $\mathbf{x}(m\times 1)$ are i.i.d. $N_1(0,\sigma^2)$ variables; that is the $W_1(\sigma^2,m)$ distribution is the same as the $\sigma^2\chi^2_m$ distribution.

The scale matrix $\mathbf{\Sigma}$ plays the same role in the Wishart distribution as σ^2 does in the $\sigma^2\chi^2_m$ distribution. We shall usually suppose $\mathbf{\Sigma}>0$.

Note that the first moment of $\mathbf{M}$ is given by

$$E[\mathbf{M}]=\sum_{i=1}^{m}E[\mathbf{x}_i\mathbf{x}_i']=m\mathbf{\Sigma}.$$

3.4.2 Properties of Wishart matrices

Theorem 3.4.1 *If* $\mathbf{M} \sim W_p(\mathbf{\Sigma}, m)$ *and* $\mathbf{B}$ *is a* $(p \times q)$ *matrix, then* $\mathbf{B}'\mathbf{M}\mathbf{B} \sim W_q(\mathbf{B}'\mathbf{\Sigma}\mathbf{B}, m)$. ■

Proof The theorem follows directly from Definition 3.4.1, since $\mathbf{B}'\mathbf{M}\mathbf{B} = \mathbf{B}'\mathbf{X}'\mathbf{X}\mathbf{B} = \mathbf{Y}'\mathbf{Y}$ where $\mathbf{Y} = \mathbf{X}\mathbf{B}$ and the rows of $\mathbf{X}$ are i.i.d. $N_p(\mathbf{0}, \mathbf{\Sigma})$. From Theorem 3.3.2, the rows of $\mathbf{Y}$ are i.i.d. $N_q(\mathbf{0}, \mathbf{B}'\mathbf{\Sigma}\mathbf{B})$. Therefore, using Definition 3.4.1 again, $\mathbf{Y}'\mathbf{Y}$ has the stated distribution. ■

Corollary 3.4.1.1 *Diagonal submatrices of* $\mathbf{M}$ *themselves have a Wishart distribution.* ■

Corollary 3.4.1.2 $\mathbf{\Sigma}^{-1/2}\mathbf{M}\mathbf{\Sigma}^{-1/2} \sim W_p(\mathbf{I}, m)$. ■

Corollary 3.4.1.3 *If* $\mathbf{M} \sim W_p(\mathbf{I}, m)$ *and* $\mathbf{B}(p \times q)$ *satisfies* $\mathbf{B}'\mathbf{B} = \mathbf{I}_q$, *then* $\mathbf{B}'\mathbf{M}\mathbf{B} \sim W_q(\mathbf{I}, m)$. ■

The corollaries follow by inserting particular values of $\mathbf{B}$ in the theorem. Note from Corollary 3.4.1.1 that each diagonal element of $\mathbf{M}$ has a $\sigma_i^2 \chi_m^2$ distribution. The following theorem generalizes and emphasizes this important relationship between the chi-squared and Wishart distributions.

Theorem 3.4.2 *If* $\mathbf{M} \sim W_p(\mathbf{\Sigma}, m)$, *and* $\mathbf{a}$ *is any fixed p-vector such that* $\mathbf{a}'\mathbf{\Sigma}\mathbf{a} \neq 0$, *then*

$$\mathbf{a}'\mathbf{M}\mathbf{a}/\mathbf{a}'\mathbf{\Sigma}\mathbf{a} \sim \chi_m^2.$$

Proof From Theorem 3.4.1 we see that $\mathbf{a}'\mathbf{M}\mathbf{a} \sim W_1(\mathbf{a}'\mathbf{\Sigma}\mathbf{a}, m)$, which is equivalent to the stated result. ■

Note that the converse of Theorem 3.4.2 is untrue—see Exercise 3.4.4.

Corollary 3.4.2.1 $m_{ii} \sim \sigma_i^2 \chi_m^2$. ■

Theorem 3.4.1 showed that the class of Wishart matrices is closed under the transformation $\mathbf{M} \rightarrow \mathbf{B}'\mathbf{M}\mathbf{B}$. The Wishart family is also closed under addition.

Theorem 3.4.3 *If* $\mathbf{M}_1 \sim W_p(\mathbf{\Sigma}, m_1)$ *and* $\mathbf{M}_2 \sim W_p(\mathbf{\Sigma}, m_2)$, *and if* $\mathbf{M}_1$ *and* $\mathbf{M}_2$ *are independent, then* $\mathbf{M}_1 + \mathbf{M}_2 \sim W_p(\mathbf{\Sigma}, m_1 + m_2)$. ■

Proof We may write $\mathbf{M}_i$ as $\mathbf{X}_i'\mathbf{X}_i$, where $\mathbf{X}_i$ has m_i independent rows taken from $N_p(\mathbf{0}, \mathbf{\Sigma})$, $i = 1, 2$. But

$$\mathbf{M}_1 + \mathbf{M}_2 = \mathbf{X}_1'\mathbf{X}_1 + \mathbf{X}_2'\mathbf{X}_2 = \mathbf{X}'\mathbf{X}.$$

Now $\mathbf{X}_1$ and $\mathbf{X}_2$ may be chosen so as to be independent, in which case all the (m_1+m_2) rows of $\mathbf{X}$ are i.i.d. $N_p(\mathbf{0}, \mathbf{\Sigma})$ variables. The result then follows from Definition 3.4.1. ■

So far we have taken a normal data matrix $\mathbf{X}$ with zero mean, and derived a Wishart distribution based on $\mathbf{X}'\mathbf{X}$. However, it is possible that other functions of $\mathbf{X}$ apart from $\mathbf{X}'\mathbf{X}$ also have a Wishart distribution. Clearly any matrix containing the sums of squares and cross-products from a subset of the rows of $\mathbf{X}$ also has a Wishart distribution. Such a matrix equals $\mathbf{X}'\mathbf{C}\mathbf{X}$, where $c_{ii}=1$ whenever the ith row of $\mathbf{X}$ is in the subset, and all other elements of $\mathbf{C}$ are zero. This matrix is symmetric and idempotent, which suggests the following theorem.

Theorem 3.4.4 (Cochran, 1934) *If* $\mathbf{X}(n\times p)$ *is a data matrix from* $N_p(\mathbf{0}, \mathbf{\Sigma})$, *and if* $\mathbf{C}(n\times n)$ *is a symmetric matrix, then*

(a) $\mathbf{X}'\mathbf{C}\mathbf{X}$ *has the same distribution as a weighted sum of independent* $W_p(\mathbf{\Sigma}, 1)$ *matrices, where the weights are eigenvalues of* $\mathbf{C}$;
(b) $\mathbf{X}'\mathbf{C}\mathbf{X}$ *has a Wishart distribution if and only if* $\mathbf{C}$ *is idempotent, in which case* $\mathbf{X}'\mathbf{C}\mathbf{X} \sim W_p(\mathbf{\Sigma}, r)$ *where* $r = \operatorname{tr} \mathbf{C} = \operatorname{rank} \mathbf{C}$;
(c) *If* $\mathbf{S}=n^{-1}\mathbf{X}'\mathbf{H}\mathbf{X}$ *is the sample covariance matrix, then* $n\mathbf{S}\sim W_p(\mathbf{\Sigma}, n-1)$. ■

Proof Using the spectral decomposition theorem (Theorem A.6.4), write

$$\mathbf{C} = \sum_{i=1}^{n} \lambda_i \boldsymbol{\gamma}_{(i)}\boldsymbol{\gamma}'_{(i)} \quad \text{where} \quad \boldsymbol{\gamma}'_{(i)}\boldsymbol{\gamma}_{(j)} = \delta_{ij} \tag{3.4.1}$$

and λ_i and $\boldsymbol{\gamma}_{(i)}$ are the ith eigenvalue and eigenvector of $\mathbf{C}$, respectively. Using (3.4.1) we see that

$$\mathbf{X}'\mathbf{C}\mathbf{X} = \sum \lambda_i \mathbf{y}_i \mathbf{y}'_i \quad \text{where} \quad \mathbf{y}_i = \mathbf{X}'\boldsymbol{\gamma}_{(i)}. \tag{3.4.2}$$

Writing $\mathbf{Y}=\mathbf{\Gamma}'\mathbf{X}$ where $\mathbf{\Gamma}$ is orthogonal, it is easily seen from Theorem 3.3.2 that $\mathbf{Y}$ is a data matrix from $N_p(\mathbf{0}, \mathbf{\Sigma})$. Therefore $\mathbf{y}_1\mathbf{y}'_1, \ldots, \mathbf{y}_n\mathbf{y}'_n$ in (3.4.2) are a set of independent Wishart matrices each having rank one. This proves part (a) of the theorem. For part (b) note that if $\mathbf{C}$ is idempotent and of rank r, then exactly r of the λ_i are non-zero, and each non-zero λ_i equals 1. Also $r=\operatorname{tr} \mathbf{C}$. Hence $\mathbf{X}'\mathbf{C}\mathbf{X}\sim W_p(\mathbf{\Sigma}, r)$ as required.

To prove (c) we note that $\mathbf{H}$ is idempotent and of rank $(n-1)$.

For the proof in (b) that idempotence of $\mathbf{C}$ is in fact a *necessary* condition for $\mathbf{X}'\mathbf{C}\mathbf{X}$ to have a Wishart distribution we refer the reader to Anderson (1958, p. 164). ■

The above theorem is also valid when $\boldsymbol{\mu} \neq \mathbf{0}$ if $\mathbf{C}$ is a symmetric idempotent matrix whose row sums are 0; see Exercise 3.4.5. For an extension to the case where the rows of $\mathbf{X}$ do not have the same mean, see Exercise 3.4.20.

Using Exercise 3.4.5, and also results from Theorem 3.3.1 and Corollary 3.3.3.2, we emphasize the following important results concerning the sample mean and covariance matrix of a random sample $\mathbf{x}_1, \ldots, \mathbf{x}_n$ from $N_p(\boldsymbol{\mu}, \boldsymbol{\Sigma})$:

$$\bar{\mathbf{x}} \sim N_p(\boldsymbol{\mu}, n^{-1}\boldsymbol{\Sigma}), \qquad n\mathbf{S} \sim W_p(\boldsymbol{\Sigma}, n-1), \qquad \bar{\mathbf{x}} \text{ and } \mathbf{S} \text{ are independent.} \tag{3.4.3}$$

For an alternative proof of (3.4.3), see Exercise 3.4.12.

We turn now to consider pairs of functions such as $\mathbf{X}'\mathbf{C}\mathbf{X}$ and $\mathbf{X}'\mathbf{D}\mathbf{X}$, and investigate the conditions under which they are independent.

Theorem 3.4.5 (Craig, 1943; Lancaster, 1969, p. 23) *If the rows of* $\mathbf{X}$ *are i.i.d.* $N_p(\boldsymbol{\mu}, \boldsymbol{\Sigma})$, *and if* $\mathbf{C}_1, \ldots, \mathbf{C}_k$ *are symmetric matrices, then* $\mathbf{X}'\mathbf{C}_1\mathbf{X}, \ldots, \mathbf{X}'\mathbf{C}_k\mathbf{X}$ *are jointly independent if* $\mathbf{C}_r\mathbf{C}_s = \mathbf{0}$ *for all* $r \neq s$. ■

Proof First consider the case $k = 2$ and let $\mathbf{C}_1 = \mathbf{C}$, $\mathbf{C}_2 = \mathbf{D}$. As in (3.4.2) we can write

$$\mathbf{X}'\mathbf{C}\mathbf{X} = \sum \lambda_i \mathbf{y}_i \mathbf{y}_i', \qquad \mathbf{X}'\mathbf{D}\mathbf{X} = \sum \psi_j \mathbf{z}_j \mathbf{z}_j',$$

where $\mathbf{y}_i = \mathbf{X}'\boldsymbol{\gamma}_{(i)}$ and $\mathbf{z}_j = \mathbf{X}'\boldsymbol{\delta}_{(j)}$, $\boldsymbol{\gamma}_{(i)}$ and $\boldsymbol{\delta}_{(j)}$ being eigenvectors of $\mathbf{C}$ and $\mathbf{D}$, respectively, with λ_i and ψ_j the corresponding eigenvalues. From Theorem 3.3.3 we note that $\mathbf{y}_i$ and $\mathbf{z}_j$ are independent if and only if $\boldsymbol{\gamma}_{(i)}'\boldsymbol{\delta}_{(j)} = 0$. Thus, the np-dimensional normal random vectors $(\lambda_1^{1/2}\mathbf{y}_1', \ldots, \lambda_n^{1/2}\mathbf{y}_n')'$ and $(\psi_1^{1/2}\mathbf{z}_1', \ldots, \psi_n^{1/2}\mathbf{z}_n')'$ will be independent if $\boldsymbol{\gamma}_{(i)}'\boldsymbol{\delta}_{(j)} = 0$ whenever $\lambda_i\psi_j$ is non-zero; that is, if

$$\lambda_i \psi_j \boldsymbol{\gamma}_{(i)}'\boldsymbol{\delta}_{(j)} = 0, \qquad \text{for all } i, j. \tag{3.4.4}$$

But

$$\mathbf{C}\mathbf{D} = \sum \lambda_i \psi_j \boldsymbol{\gamma}_{(i)} \boldsymbol{\gamma}_{(i)}' \boldsymbol{\delta}_{(j)} \boldsymbol{\delta}_{(j)}'.$$

If $\mathbf{C}\mathbf{D} = \mathbf{0}$, then pre-multiplying by $\boldsymbol{\gamma}_{(u)}'$ and post-multiplying by $\boldsymbol{\delta}_{(v)}$ gives $\lambda_u \psi_v \boldsymbol{\gamma}_{(u)}'\boldsymbol{\delta}_{(v)} = 0$. This holds for all u and v, and therefore (3.4.4) holds. Thus, since $\mathbf{X}'\mathbf{C}\mathbf{X}$ and $\mathbf{X}'\mathbf{D}\mathbf{X}$ are functions of independent normal np-vectors, they are independent.

To deal with the case $k>2$ notice that normal np-vectors which are pairwise independent are also jointly independent. (This may be easily proved in the same way as Theorem 3.1.3(b).) Hence, the matrices $\mathbf{X}'\mathbf{C}_r\mathbf{X}$, $r=1,\ldots,k$, are jointly independent. ■

The converse of the theorem also holds. For a proof, the reader is referred to Ogawa (1949). A similar theorem gives the condition for $\mathbf{X}'\mathbf{C}\mathbf{X}$ to be independent of a linear function like $\mathbf{AXB}$ (see Exercise 3.4.7). This theorem is also valid when the rows of $\mathbf{X}$ have different means; see Exercise 3.4.20.

Note that Craig's theorem does not require the $\mathbf{C}_r$ to be idempotent, although if they are, and if $\boldsymbol{\mu}=\mathbf{0}$, then by Cochran's theorem the quadratic forms to which they lead are not only independent but also each have the Wishart distribution. An important decomposition of $\mathbf{X}'\mathbf{X}$ when $\boldsymbol{\mu}=\mathbf{0}$ (and of $\mathbf{X}'\mathbf{H}\mathbf{X}$ for general $\boldsymbol{\mu}$) into a sum of independent Wishart matrices is described in Exercise 3.4.6. This decomposition forms the basis for the multivariate analysis of variance (see Chapter 12.)

These theorems can be easily extended to cover $(n\times n)$ matrices such as $\mathbf{XCX}'$ (in contrast to $(p\times p)$ matrices such as $\mathbf{X}'\mathbf{CX}$). This is done by noting that if the rows of $\mathbf{X}$ are i.i.d $N_p(\boldsymbol{\mu},\boldsymbol{\Sigma})$, then in general the rows of $\mathbf{X}'$ (that is the columns of $\mathbf{X}$) are *not* i.i.d. However, when $\boldsymbol{\mu}=\mathbf{0}$ and $\boldsymbol{\Sigma}=\mathbf{I}_p$, the rows of $\mathbf{X}'$ are i.i.d. $N_n(\mathbf{0},\mathbf{I})$ vectors, since in this case all the np elements of $\mathbf{X}$ are i.i.d. Hence we can get the necessary extensions for the standard case (see Exercise 3.4.9), and can thereby derive the relevant properties for general $\boldsymbol{\Sigma}$ (Exercise 3.4.10). The special case where $n=1$ leads to quadratic forms proper and is discussed in Exercise 3.4.11.

*3.4.3 Partitioned Wishart matrices

If $\mathbf{M}\sim W_p(\boldsymbol{\Sigma}, m)$, it is often useful to partition $\mathbf{M}$ into submatrices in the usual way. For instance, we may want to divide the p original variables into two subgroups, consisting of say a and b variables, respectively, where $a+b=p$. Suppose then that $\mathbf{M}_{11}$ is $(a\times a)$, and $\mathbf{M}_{22}$ is $(b\times b)$, where $a+b=p$.

We have already noted (Corollary 3.4.1.1) that $\mathbf{M}_{11}$ and $\mathbf{M}_{22}$ have Wishart distributions, although these distributions are not in general independent. However, $\mathbf{M}_{11}$ *is* independent of

$$\mathbf{M}_{22.1}=\mathbf{M}_{22}-\mathbf{M}_{21}\mathbf{M}_{11}^{-1}\mathbf{M}_{12}. \tag{3.4.5}$$

Here $\mathbf{M}_{22.1}$ is in fact just m times the sample analogue of $\boldsymbol{\Sigma}_{22.1}$ defined in (3.2.1).

Note that when p equals 2, the matrix $\mathbf{M}$ may be written as

$$\mathbf{M}=m\begin{bmatrix} s_1^2 & rs_1s_2 \\ rs_1s_2 & s_2^2 \end{bmatrix}. \tag{3.4.6}$$

The matrix $\mathbf{M}_{22.1}$ simplifies in this context to

$$\mathbf{M}_{22.1}=ms_2^2(1-r^2). \tag{3.4.7}$$

Various properties concerning the joint distribution of $\mathbf{M}_{11}$, $\mathbf{M}_{12}$, and $\mathbf{M}_{22.1}$ are proved in the following theorem.

Theorem 3.4.6 *Let* $\mathbf{M}\sim W_p(\mathbf{\Sigma}, m)$, $m>a$. *Then*

(a) $\mathbf{M}_{22.1}$ *has the* $W_b(\mathbf{\Sigma}_{22.1}, m-a)$ *distribution and is independent of* $(\mathbf{M}_{11}, \mathbf{M}_{12})$, *and*
(b) *if* $\mathbf{\Sigma}_{12}=\mathbf{0}$, *then* $\mathbf{M}_{22}-\mathbf{M}_{22.1}=\mathbf{M}_{21}\mathbf{M}_{11}^{-1}\mathbf{M}_{12}$ *has the* $W_b(\mathbf{\Sigma}_{22}, a)$ *distribution, and* $\mathbf{M}_{21}\mathbf{M}_{11}^{-1}\mathbf{M}_{12}$, $\mathbf{M}_{11}$, *and* $\mathbf{M}_{22.1}$ *are jointly independent.*

Proof Write $\mathbf{M}=\mathbf{X}'\mathbf{X}$, where the rows of $\mathbf{X}(m\times p)$ are i.i.d. $N_p(\mathbf{0},\mathbf{\Sigma})$ random vectors. Then $\mathbf{M}_{22.1}$ may be written

$$\mathbf{M}_{22.1}=\mathbf{X}_2'\mathbf{X}_2-\mathbf{X}_2'\mathbf{X}_1\mathbf{M}_{11}^{-1}\mathbf{X}_1'\mathbf{X}_2=\mathbf{X}_2'\mathbf{P}\mathbf{X}_2, \tag{3.4.8}$$

where $\mathbf{P}$ is the symmetric, idempotent matrix defined by $\mathbf{P}=\mathbf{I}-\mathbf{X}_1\mathbf{M}_{11}^{-1}\mathbf{X}_1'$. Note that $\mathbf{P}$ has rank $(m-a)$ and is a function of $\mathbf{X}_1$ alone. Since $\mathbf{X}_1\mathbf{P}=\mathbf{0}$ we see that $\mathbf{X}_2'\mathbf{P}\mathbf{X}_2=\mathbf{X}_{2.1}'\mathbf{P}\mathbf{X}_{2.1}$, where $\mathbf{X}_{2.1}=\mathbf{X}_2-\mathbf{X}_1\mathbf{\Sigma}_{11}^{-1}\mathbf{\Sigma}_{12}$. By Corollary 3.3.3.1, $\mathbf{X}_{2.1}\mid\mathbf{X}_1$ is distributed as a data matrix from $N_p(\mathbf{0},\mathbf{\Sigma}_{22.1})$. Therefore, using Theorem 3.4.4(b), for any given value of $\mathbf{X}_1$, the conditional distribution of $\mathbf{M}_{22.1}\mid\mathbf{X}_1$ is $W_b(\mathbf{\Sigma}_{22.1}, m-a)$. But this conditional distribution is free of $\mathbf{X}_1$. Therefore it is the unconditional (marginal) distribution, and moreover $\mathbf{M}_{22.1}$ is independent of $\mathbf{X}_1$.

Since $\mathbf{P}(\mathbf{I}-\mathbf{P})=\mathbf{0}$, we see from Theorem 3.3.3 that, given $\mathbf{X}_1$, the matrices $\mathbf{P}\mathbf{X}_{2.1}$ and $(\mathbf{I}-\mathbf{P})\mathbf{X}_{2.1}=\mathbf{X}_1\mathbf{M}_{11}^{-1}\mathbf{M}_{12}-\mathbf{X}_1\mathbf{\Sigma}_{11}^{-1}\mathbf{\Sigma}_{12}$ are independent. Hence, given $\mathbf{X}_1$, the matrices $\mathbf{M}_{22.1}=(\mathbf{P}\mathbf{X}_{2.1})'(\mathbf{P}\mathbf{X}_{2.1})$ and $(\mathbf{I}-\mathbf{P})\mathbf{X}_{2.1}$ are independent. But from the above paragraph, $\mathbf{M}_{22.1}$ is independent of $\mathbf{X}_1$. Hence $\mathbf{M}_{22.1}$ is independent of $(\mathbf{X}_1, (\mathbf{I}-\mathbf{P})\mathbf{X}_{2.1})$. Since $\mathbf{M}_{11}$ and $\mathbf{M}_{12}$ can be expressed in terms of $\mathbf{X}_1$ and $(\mathbf{I}-\mathbf{P})\mathbf{X}_{2.1}$, we see that $\mathbf{M}_{22.1}$ is independent of $(\mathbf{M}_{11}, \mathbf{M}_{12})$. Thus, the proof of part (a) of the theorem is completed.

For part (b) note that

$$\begin{aligned}\mathbf{M}_{22}-\mathbf{M}_{22.1}&=\mathbf{X}_2'(\mathbf{I}-\mathbf{P})\mathbf{X}_2\\ &=\mathbf{X}_{2.1}'(\mathbf{I}-\mathbf{P})\mathbf{X}_{2.1}+\mathbf{\Sigma}_{21}\mathbf{\Sigma}_{11}^{-1}\mathbf{M}_{12}+\mathbf{M}_{21}\mathbf{\Sigma}_{11}^{-1}\mathbf{\Sigma}_{12}-\mathbf{\Sigma}_{21}\mathbf{\Sigma}_{11}^{-1}\mathbf{M}_{11}\mathbf{\Sigma}_{11}^{-1}\mathbf{\Sigma}_{12}\end{aligned}$$

(see Exercise 3.3.2). Now when $\boldsymbol{\Sigma}_{12}$ is zero, the last three terms of this expression disappear, leaving just $\mathbf{X}_{2.1}'(\mathbf{I}-\mathbf{P})\mathbf{X}_{2.1}$. Because $(\mathbf{I}-\mathbf{P})$ is symmetric, idempotent, and has rank a, the distribution of $\mathbf{M}_{22}-\mathbf{M}_{22.1}$ conditional upon a given value of $\mathbf{X}_1$ is $W_b(\boldsymbol{\Sigma}_{22}, a)$. Moreover, Craig's theorem (Theorem 3.4.5) implies that $\mathbf{M}_{22.1}$ and $\mathbf{M}_{22}-\mathbf{M}_{22.1}$ are independent for any given value of $\mathbf{X}_1$, since $\mathbf{P}(\mathbf{I}-\mathbf{P})=\mathbf{0}$. As the conditional distributions of $\mathbf{M}_{22.1}$ and $\mathbf{M}_{22}-\mathbf{M}_{22.1}$ do not involve $\mathbf{X}_1$, we see that $\mathbf{M}_{22.1}$ and $\mathbf{M}_{22}-\mathbf{M}_{22.1}$ are (unconditionally) independent, and further, $(\mathbf{M}_{22.1}, \mathbf{M}_{22}-\mathbf{M}_{22.1})$ is independent of $\mathbf{X}_1$ (and hence independent of $\mathbf{M}_{11}$). Therefore in their unconditional joint distribution, the three matrices $\mathbf{M}_{11}, \mathbf{M}_{22.1}$, and $\mathbf{M}_{22}-\mathbf{M}_{22.1}$ are independent of one another. ∎

Recall from (A.2.4g) that $\mathbf{M}^{22}=\mathbf{M}_{22.1}^{-1}$. Hence parts (a) and (b) of the theorem may also be written as follows.

Corollary 3.4.6.1 (a) $(\mathbf{M}^{22})^{-1} \sim W_b((\boldsymbol{\Sigma}^{22})^{-1}, m-a)$ *and is independent of* $(\mathbf{M}_{11}, \mathbf{M}_{12})$.

(b) *If* $\boldsymbol{\Sigma}_{12}=\mathbf{0}$ *then* $\mathbf{M}_{22}-(\mathbf{M}^{22})^{-1} \sim W_b(\boldsymbol{\Sigma}_{22}, a)$, *and* $\mathbf{M}_{22}-(\mathbf{M}^{22})^{-1}$, $\mathbf{M}_{11}$, *and* $(\mathbf{M}^{22})^{-1}$ *are jointly independent.* ∎

Hence, when $\boldsymbol{\Sigma}_{12}=\mathbf{0}$, the Wishart matrix $\mathbf{M}_{22}$ can be decomposed into the sum $(\mathbf{M}_{22}-\mathbf{M}_{22.1})+(\mathbf{M}_{22.1})$, where the two components of the sum have independent Wishart distributions. Moreover, the degrees of freedom are additive in a similar manner.

If the population correlation coefficient is zero then, for the bivariate case, part (b) of Theorem 3.4.6 leads to the following results with the help of (3.4.6) and (3.4.7):

$$m_{11} \sim \sigma_1^2\chi_m^2, \qquad m_{22.1}=ms_2^2(1-r^2) \sim \sigma_2^2\chi_{m-1}^2,$$
$$m_{22}-m_{22.1}=ms_2^2r^2 \sim \sigma_2^2\chi_1^2.$$

Moreover, these three chi-squared variables are jointly statistically independent.

Theorem 3.4.7 *If* $\mathbf{M} \sim W_p(\boldsymbol{\Sigma}, m)$, $m>p$, *then*

(a) *the ratio* $\mathbf{a}'\boldsymbol{\Sigma}^{-1}\mathbf{a}/\mathbf{a}'\mathbf{M}^{-1}\mathbf{a}$ *has the* χ_{m-p+1}^2 *distribution for any fixed* p-*vector* $\mathbf{a}$ *and in particular,* $\sigma^{ii}/m^{ii} \sim \chi_{m-p+1}^2$ *for* $i=1,\ldots,p$;

(b) m^{ii} *is independent of all the elements of* $\mathbf{M}$ *except* m_{ii}.

Proof From Corollary 3.4.6.1, putting $b=1$ and $a=p-1$, we get

$$(m^{pp})^{-1} \sim (\sigma^{pp})^{-1}\chi_{m-p+1}^2. \qquad (3.4.9)$$

This proves part (a) for the special case where $\mathbf{a}=(0,\ldots,0,1)'$.

For general $\mathbf{a}$, let $\mathbf{A}$ be a non-singular matrix whose last column equals

$\mathbf{a}$, and set $\mathbf{N} = \mathbf{A}^{-1}\mathbf{M}(\mathbf{A}^{-1})'$. Then $\mathbf{N} \sim W_p(\mathbf{A}^{-1}\mathbf{\Sigma}(\mathbf{A}^{-1})', m)$. Since $\mathbf{N}^{-1} = \mathbf{A}'\mathbf{M}^{-1}\mathbf{A}$, we see from (3.4.9) that

$$(n^{pp})^{-1} = (\mathbf{a}'\mathbf{M}^{-1}\mathbf{a})^{-1} \sim [(\mathbf{A}'\mathbf{\Sigma}^{-1}\mathbf{A})_{pp}]^{-1}\chi^2_{m-p+1}$$
$$= (\mathbf{a}'\mathbf{\Sigma}^{-1}\mathbf{a})^{-1}\chi^2_{m-p+1}.$$

For part (b), note that $(\mathbf{M}^{22})^{-1}$ in Theorem 3.4.6 is independent of $(\mathbf{M}_{11}, \mathbf{M}_{12})$. Hence for this particular case, m^{pp} is independent of all the elements of $\mathbf{M}$ except m_{pp}. By a suitable permutation of the rows and columns a similar result can be proved for all i. ■

For a generalization of this theorem, see Exercise 3.4.19.

The following theorem describes the distribution of $|\mathbf{M}|$.

Theorem 3.4.8 *If $\mathbf{M} \sim W_p(\mathbf{\Sigma}, m)$ and $m \geq p$, then $|\mathbf{M}|$ is $|\mathbf{\Sigma}|$ times p independent chi-squared random variables with degrees of freedom m, $m-1, \ldots, m-p+1$.*

Proof We proceed by induction on p. Clearly the theorem is true if $p = 1$. For $p > 1$ partition $\mathbf{M}$ with $a = p-1$ and $b = 1$. Suppose by the induction hypothesis that $|\mathbf{M}_{11}|$ can be written as $|\mathbf{\Sigma}_{11}|$ times $p-1$ independent chi-squared random variables with degrees of freedom m, $m-1, \ldots, m-p+2$. By Theorem 3.4.6, $\mathbf{M}_{11}$ is independent of the scalar $\mathbf{M}_{22.1} \sim \mathbf{\Sigma}_{22.1}\chi^2_{m-p+1}$. Since $|\mathbf{M}| = |\mathbf{M}_{11}|\,|\mathbf{M}_{22.1}|$ and $|\mathbf{\Sigma}| = |\mathbf{\Sigma}_{11}|\,|\mathbf{\Sigma}_{22.1}|$ (see equation (A.2.3j)), the theorem follows. ■

Corollary 3.4.8.1 *If $\mathbf{M} \sim W_p(\mathbf{\Sigma}, m)$, $\mathbf{\Sigma} > 0$, and $m > p$, then $\mathbf{M} > 0$ with probability one.*

Proof Since a chi-squared variate is strictly positive with probability one and $|\mathbf{\Sigma}| > 0$, it follows that $|\mathbf{M}| > 0$. Hence, since by construction $\mathbf{M}$ is p.s.d., all of the eigenvalues of $\mathbf{M}$ are strictly positive with probability one. ■

3.5 The Hotelling T^2 Distribution

We now turn to functions such as $\mathbf{d}'\mathbf{M}^{-1}\mathbf{d}$, where $\mathbf{d}$ is normal, $\mathbf{M}$ is Wishart, and $\mathbf{d}$ and $\mathbf{M}$ are independent. For instance, $\mathbf{d}$ may be the sample mean, and $\mathbf{M}$ proportional to the sample covariance matrix (see equation (3.4.3)). This important special case is examined in Corollary 3.5.1.1.

We shall now derive the general distribution of quadratic forms such as the above. This work was initiated by Hotelling (1931).

Definition 3.5.1 *If α can be written as $m\mathbf{d}'\mathbf{M}^{-1}\mathbf{d}$ where $\mathbf{d}$ and $\mathbf{M}$ are independently distributed as $N_p(\mathbf{0}, \mathbf{I})$ and $W_p(\mathbf{I}, m)$, respectively, then we say that α has the* Hotelling T^2 distribution *with parameters p and m. We write $\alpha \sim T^2(p, m)$.*

Theorem 3.5.1 *If $\mathbf{x}$ and $\mathbf{M}$ are independently distributed as $N_p(\boldsymbol{\mu}, \boldsymbol{\Sigma})$ and $W_p(\boldsymbol{\Sigma}, m)$, respectively, then*

$$m(\mathbf{x}-\boldsymbol{\mu})'\mathbf{M}^{-1}(\mathbf{x}-\boldsymbol{\mu}) \sim T^2(p, m). \qquad (3.5.1)$$

Proof If $\mathbf{d}^* = \boldsymbol{\Sigma}^{-1/2}(\mathbf{x}-\boldsymbol{\mu})$ and $\mathbf{M}^* = \boldsymbol{\Sigma}^{-1/2}\mathbf{M}\boldsymbol{\Sigma}^{-1/2}$, we see that $\mathbf{d}^*$ and $\mathbf{M}^*$ satisfy the requirements of Definition 3.5.1. Therefore $\alpha \sim T^2(p, m)$ where

$$\alpha = m\mathbf{d}^{*\prime}\mathbf{M}^{*-1}\mathbf{d}^* = m(\mathbf{x}-\boldsymbol{\mu})'\mathbf{M}^{-1}(\mathbf{x}-\boldsymbol{\mu}).$$

Hence the theorem is proved. ■

Corollary 3.5.1.1 *If $\bar{\mathbf{x}}$ and $\mathbf{S}$ are the mean vector and covariance matrix of a sample of size n* from *$N_p(\boldsymbol{\mu}, \boldsymbol{\Sigma})$, and $\mathbf{S}_u = (n/(n-1))\mathbf{S}$, then*

$$(n-1)(\bar{\mathbf{x}}-\boldsymbol{\mu})'\mathbf{S}^{-1}(\bar{\mathbf{x}}-\boldsymbol{\mu}) = n(\bar{\mathbf{x}}-\boldsymbol{\mu})'\mathbf{S}_u^{-1}(\bar{\mathbf{x}}-\boldsymbol{\mu}) \sim T^2(p, n-1). \quad (3.5.2)$$

Proof Substituting $\mathbf{M} = n\mathbf{S}$, $m = n-1$, and $\mathbf{x}-\boldsymbol{\mu}$ for $n^{1/2}(\bar{\mathbf{x}}-\boldsymbol{\mu})$ in the theorem, the result follows immediately. ■

Corollary 3.5.1.2. *The T^2 statistic is invariant under any non-singular linear transformation $\mathbf{x} \to \mathbf{A}\mathbf{x}+\mathbf{b}$.* ■

Of course the univariate t statistic also has this property of invariance mentioned above. Indeed the square of the univariate t_m variable has the $T^2(1, m)$ distribution (see Exercise 3.5.1). In other words, the $F_{1,m}$ distribution and the $T^2(1, m)$ distribution are the same. The following theorem extends the result.

Theorem 3.5.2

$$T^2(p, m) = \{mp/(m-p+1)\}F_{p,m-p+1}. \qquad (3.5.3)$$

Proof We use the characterization $\alpha = m\mathbf{d}'\mathbf{M}^{-1}\mathbf{d}$ given in Definition 3.5.1. Write

$$\alpha = m(\mathbf{d}'\mathbf{M}^{-1}\mathbf{d}/\mathbf{d}'\mathbf{d})\mathbf{d}'\mathbf{d}.$$

Since $\mathbf{M}$ is independent of $\mathbf{d}$ we see from Theorem 3.4.7(a) that the conditional distribution of $\beta = \mathbf{d}'\mathbf{d}/\mathbf{d}'\mathbf{M}^{-1}\mathbf{d}$ given $\mathbf{d}$ is χ^2_{m-p+1}. Since this conditional distribution does not depend on $\mathbf{d}$, it is also the marginal distribution of β, and furthermore β is independent of $\mathbf{d}$. By Theorem 2.5.2, $\mathbf{d}'\mathbf{d} \sim \chi^2_p$, so we can express α as a ratio of independent χ^2 variables:

$$\alpha = m\chi^2_p/\chi^2_{m-p+1} = \{mp/(m-p+1)\}F_{p,m-p+1}. \quad \blacksquare$$

Corollary 3.5.2.1 *If $\bar{\mathbf{x}}$ and $\mathbf{S}$ are the mean and covariance of a sample of size n from $N_p(\boldsymbol{\mu}, \boldsymbol{\Sigma})$, then*

$$\{(n-p)/p\}(\bar{\mathbf{x}}-\boldsymbol{\mu})'\mathbf{S}^{-1}(\bar{\mathbf{x}}-\boldsymbol{\mu}) \sim F_{p,n-p}. \tag{3.5.4}$$

Proof The result follows immediately from (3.5.2). ■

Corollary 3.5.2.2 $|\mathbf{M}|/|\mathbf{M}+\mathbf{d}\mathbf{d}'| \sim B(\frac{1}{2}(m-p+1), \frac{1}{2}p)$ *where $B(\cdot,\cdot)$ is a beta variable.*

Proof From equation (A.2.3n) the given ratio equals $m/(m+\alpha)$. Using the F distribution of α, and the univariate relationship between F and beta distributions, the result follows. (See also Exercise 3.5.2.) ■

Since $\mathbf{d}'\mathbf{d}$ and β are independent chi-squared statistics, as shown in the proof of Theorem 3.5.2, their ratio is independent of their sum, which also has a chi-squared distribution. (See Theorem B. 4.1.) The sum is

$$\mathbf{d}'\mathbf{d} + \beta = \mathbf{d}'\mathbf{d}\left(1 + \frac{1}{\mathbf{d}'\mathbf{M}^{-1}\mathbf{d}}\right).$$

Hence we have the following:

Theorem 3.5.3 *If $\mathbf{d}$ and $\mathbf{M}$ are independently distributed as $N_p(\mathbf{0}, \mathbf{I})$ and $W_p(\mathbf{I}, m)$, respectively, then*

$$\mathbf{d}'\mathbf{d}\left(1 + \frac{1}{\mathbf{d}'\mathbf{M}^{-1}\mathbf{d}}\right) \sim \chi^2_{m+1}$$

and is distributed independently of $\mathbf{d}'\mathbf{M}^{-1}\mathbf{d}$. ■

Theorem 3.5.3 is one extension of the univariate result that if $d \sim N(0, 1)$ and $u^2 \sim \chi^2$, then $d^2 + u^2$ is independent of d^2/u^2. Another generalization of the same result is given in Theorem 3.5.4, which requires the following lemma.

Lemma *Let $\mathbf{W}$ be a square symmetric $(p \times p)$ random matrix and let $\mathbf{x}$ be a random p-vector. If $\mathbf{x}$ is independent of $(\mathbf{g}_1'\mathbf{W}\mathbf{g}_1, \ldots, \mathbf{g}_p'\mathbf{W}\mathbf{g}_p)$ for all non-random orthogonal matrices $\mathbf{G} = (\mathbf{g}_1, \ldots, \mathbf{g}_p)'$, then $\mathbf{x}$ is independent of $\mathbf{W}$.*

Proof Using the joint characteristic function of $\mathbf{W}$ and $\mathbf{x}$ it is seen that if $\text{tr}\,\mathbf{AW}$ is independent of $\mathbf{x}$ for every square symmetric matrix $\mathbf{A}$, then $\mathbf{W}$ is independent of $\mathbf{x}$. Now $\mathbf{A}$ may be written in canonical form as $\sum \lambda_i \mathbf{g}_i \mathbf{g}_i'$ say, where $\mathbf{G} = (\mathbf{g}_1, \ldots, \mathbf{g}_p)'$ is orthogonal. Then

$$\text{tr}\,\mathbf{AW} = \text{tr}\left(\sum \lambda_i \mathbf{g}_i \mathbf{g}_i'\right)\mathbf{W} = \sum \lambda_i \mathbf{g}_i' \mathbf{W} \mathbf{g}_i.$$

If the conditions of the lemma are satisfied, then all of the terms in the summation, and the summation itself, are independent of $\mathbf{x}$. Thus, $\text{tr}\,\mathbf{AW}$ is independent of $\mathbf{x}$ for all $\mathbf{A}$, and therefore $\mathbf{W}$ is independent of $\mathbf{x}$. ■

Theorem 3.5.4 *If* $\mathbf{d}$ *and* $\mathbf{M}$ *are independently distributed as* $N_p(\mathbf{0}, \mathbf{I})$ *and* $W_p(\mathbf{I}, m)$, *respectively, then* $\mathbf{d}'\mathbf{M}^{-1}\mathbf{d}$ *is independent of* $\mathbf{M} + \mathbf{dd}'$.

Proof Let $\mathbf{G} = (\mathbf{g}_1, \ldots, \mathbf{g}_p)'$ be an orthogonal matrix and consider the quadratic forms $q_j = \mathbf{g}_j'(\mathbf{M} + \mathbf{dd}')\mathbf{g}_j$, $j = 1, \ldots, p$. Write

$$\mathbf{M} = \mathbf{X}_1'\mathbf{X}_1 = \sum_{i=1}^{m} \mathbf{x}_i \mathbf{x}_i' \quad \text{and} \quad \mathbf{d} = \mathbf{x}_{m+1},$$

where $\mathbf{X} = (\mathbf{X}_1', \mathbf{x}_{m+1})'$ is a data matrix from $N_p(\mathbf{0}, \mathbf{I})$. Set $\mathbf{Y} = \mathbf{XG}'$ so that $\mathbf{Y}_1 = \mathbf{X}_1\mathbf{G}'$ and $\mathbf{y}_{m+1} = \mathbf{G}\mathbf{x}_{m+1}$. Then

$$q_j = \mathbf{g}_j'(\mathbf{M} + \mathbf{dd}')\mathbf{g}_j = \sum_{i=1}^{m+1} (\mathbf{g}_j'\mathbf{x}_i)^2 = \sum_{i=1}^{m+1} y_{ij}^2.$$

Now $\mathbf{Y}$ is also a data matrix from $N_p(\mathbf{0}, \mathbf{I})$, so, thought of as a $p(m+1)$-vector $\mathbf{Y}^V$, it has the $N_{p(m+1)}(\mathbf{0}, \mathbf{I})$ distribution, and hence is spherically symmetric. Therefore, by Theorem 2.7.2, $(q_1, \ldots, q_p)$ is statistically independent of any column-scale-invariant function of $\mathbf{Y}$. In particular, $(q_1, \ldots, q_p)$ is independent of

$$\mathbf{d}'\mathbf{M}^{-1}\mathbf{d} = \mathbf{y}_{m+1}'(\mathbf{Y}_1'\mathbf{Y}_1)^{-1}\mathbf{y}_{m+1}.$$

Since this result holds for all orthogonal matrices $\mathbf{G}$, we see from the above lemma that the theorem is proved. ■

3.6 Mahalanobis Distance

3.6.1 The two-sample Hotelling T^2 statistic

The so-called Mahalanobis distance between two populations with means $\boldsymbol{\mu}_1$ and $\boldsymbol{\mu}_2$, and common covariance matrix $\boldsymbol{\Sigma}$ has already been defined in

Section 2.2.3. It is given by Δ, where

$$\Delta^2 = (\boldsymbol{\mu}_1 - \boldsymbol{\mu}_2)'\boldsymbol{\Sigma}^{-1}(\boldsymbol{\mu}_1 - \boldsymbol{\mu}_2). \tag{3.6.1}$$

Of course, only rarely are the population parameters known, and it is usual for them to be estimated by the corresponding sample values. Suppose we have two samples of size n_1 and n_2, where $n_1 + n_2 = n$. Then the sample Mahalanobis distance, D, can be defined by

$$D^2 = (\bar{\mathbf{x}}_1 - \bar{\mathbf{x}}_2)'\mathbf{S}_u^{-1}(\bar{\mathbf{x}}_1 - \bar{\mathbf{x}}_2), \tag{3.6.2}$$

where $\mathbf{S}_u = (n_1\mathbf{S}_1 + n_2\mathbf{S}_2)/(n-2)$ is an unbiased estimate of $\boldsymbol{\Sigma}$. (The sample mean and covariance matrix for sample i, $i = 1, 2$, are denoted $\bar{\mathbf{x}}_i$ and $\mathbf{S}_i$.) The statistical distribution of D^2 under one particular set of assumptions is given by the following theorem.

Theorem 3.6.1 *If $\mathbf{X}_1$ and $\mathbf{X}_2$ are independent data matrices, and if the n_i rows of $\mathbf{X}_i$ are i.i.d. $N_p(\boldsymbol{\mu}_i, \boldsymbol{\Sigma}_i)$, $i = 1, 2$, then when $\boldsymbol{\mu}_1 = \boldsymbol{\mu}_2$ and $\boldsymbol{\Sigma}_1 = \boldsymbol{\Sigma}_2$, $(n_1 n_2/n)D^2$ is a $T^2(p, n-2)$ variable.* ■

Proof Since $\bar{\mathbf{x}}_i \sim N_p(\boldsymbol{\mu}_i, n_i^{-1}\boldsymbol{\Sigma}_i)$, $i = 1, 2$, the general distribution of $\mathbf{d} = \bar{\mathbf{x}}_1 - \bar{\mathbf{x}}_2$ is normal with mean $\boldsymbol{\mu}_1 - \boldsymbol{\mu}_2$, and covariance matrix $n_1^{-1}\boldsymbol{\Sigma}_1 + n_2^{-1}\boldsymbol{\Sigma}_2$. When $\boldsymbol{\mu}_1 = \boldsymbol{\mu}_2$ and $\boldsymbol{\Sigma}_1 = \boldsymbol{\Sigma}_2 = \boldsymbol{\Sigma}$, $\mathbf{d} \sim N_p(\mathbf{0}, c\boldsymbol{\Sigma})$, where $c = n/n_1 n_2$.

If $\mathbf{M}_i = n_i\mathbf{S}_i$ then $\mathbf{M}_i \sim W_p(\boldsymbol{\Sigma}_i, n_i - 1)$. Thus, when $\boldsymbol{\Sigma}_1 = \boldsymbol{\Sigma}_2 = \boldsymbol{\Sigma}$,

$$\mathbf{M} = (n-2)\mathbf{S}_u = \mathbf{M}_1 + \mathbf{M}_2 \sim W_p(\boldsymbol{\Sigma}, n-2).$$

So $c\mathbf{M} \sim W_p(c\boldsymbol{\Sigma}, n-2)$. Moreover, $\mathbf{M}$ is independent of $\mathbf{d}$ since $\bar{\mathbf{x}}_i$ is independent of $\mathbf{S}_i$ for $i = 1, 2$, and the two samples are independent of one another. Therefore

$$(n-2)\mathbf{d}'(c\mathbf{M})^{-1}\mathbf{d} \sim T^2(p, n-2). \tag{3.6.3}$$

Simplifying the left-hand side gives the required result. ■

The quantity

$$(n_1 n_2/n)D^2 = (n_1 n_2/n)(\bar{\mathbf{x}}_1 - \bar{\mathbf{x}}_2)'\mathbf{S}_u^{-1}(\bar{\mathbf{x}}_1 - \bar{\mathbf{x}}_2) \tag{3.6.4}$$

is known as *Hotelling's two-sample T^2 statistic*. Using the relationship (Theorem 3.5.2) between T^2 and F statistics, we may also deduce that, under the stated conditions,

$$\frac{n_1 n_2(n-p-1)}{n(n-2)p} D^2 \sim F_{p,n-p-1}. \tag{3.6.5}$$

*3.6.2 A decomposition of Mahalanobis distance

The Mahalanobis distance of $\boldsymbol{\mu}$ from $\mathbf{0}$ is

$$\Delta_p^2 = \boldsymbol{\mu}'\boldsymbol{\Sigma}^{-1}\boldsymbol{\mu}. \tag{3.6.6}$$

Partition $\boldsymbol{\mu}' = (\boldsymbol{\mu}_1', \boldsymbol{\mu}_2')$, where $\boldsymbol{\mu}_1$ contains the first k variables. Then, using (A.2.4f) and (A.2.4g) to partition $\boldsymbol{\Sigma}^{-1}$, we can write

$$\Delta_p^2 = \boldsymbol{\mu}_1'\boldsymbol{\Sigma}_{11}^{-1}\boldsymbol{\mu}_1 + \boldsymbol{\mu}_{2.1}'\boldsymbol{\Sigma}_{22.1}^{-1}\boldsymbol{\mu}_{2.1}, \tag{3.6.7}$$

where $\boldsymbol{\mu}_{2.1} = \boldsymbol{\mu}_2 - \boldsymbol{\Sigma}_{21}\boldsymbol{\Sigma}_{11}^{-1}\boldsymbol{\mu}_1$ and $\boldsymbol{\Sigma}_{22.1} = \boldsymbol{\Sigma}_{22} - \boldsymbol{\Sigma}_{21}\boldsymbol{\Sigma}_{11}^{-1}\boldsymbol{\Sigma}_{12}$. Note that $\Delta_k^2 = \boldsymbol{\mu}_1'\boldsymbol{\Sigma}_{11}^{-1}\boldsymbol{\mu}_1$ represents the Mahalanobis distance based on the first k variables, so the condition $\boldsymbol{\mu}_{2.1} = \mathbf{0}$ is equivalent to $\Delta_k^2 = \Delta_p^2$.

Similarly, if $\mathbf{u} \sim N_p(\boldsymbol{\mu}, \boldsymbol{\Sigma})$ and $\mathbf{M} \sim W_p(\boldsymbol{\Sigma}, m)$ then the sample Mahalanobis distance

$$D_p^2 = m\mathbf{u}'\mathbf{M}^{-1}\mathbf{u} \tag{3.6.8}$$

can be partitioned as

$$D_p^2 = D_k^2 + m\mathbf{z}'\mathbf{M}_{22.1}^{-1}\mathbf{z}, \tag{3.6.9}$$

where $D_k^2 = m\mathbf{u}_1'\mathbf{M}_{11}^{-1}\mathbf{u}_1$, $\mathbf{M}_{22.1} = \mathbf{M}_{22} - \mathbf{M}_{21}\mathbf{M}_{11}^{-1}\mathbf{M}_{12}$, and

$$\mathbf{z} = \mathbf{u}_2 - \mathbf{M}_{21}\mathbf{M}_{11}^{-1}\mathbf{u}_1 \tag{3.6.10}$$

(see Exercise 3.6.1).

The following theorem gives the distribution of $D_p^2 - D_k^2$ when $\boldsymbol{\mu}_{2.1} = \mathbf{0}$.

Theorem 3.6.2 *If D_p^2 and D_k^2 are as defined above and $\boldsymbol{\mu}_{2.1} = \mathbf{0}$, then*

$$\frac{D_p^2 - D_k^2}{m + D_k^2} \sim \frac{p-k}{m-p+1} F_{p-k, m-p+1},$$

and is independent of D_k^2.

Proof Suppose $\mathbf{M} = \mathbf{X}'\mathbf{X}$, where $\mathbf{X}$ is a data matrix from $N_p(\mathbf{0}, \boldsymbol{\Sigma})$ independent of $\mathbf{u}$.

By Theorem 3.4.6, $\mathbf{M}_{22.1} \sim W_{p-k}(\boldsymbol{\Sigma}_{22.1}, m-k)$ and is independent of $(\mathbf{M}_{21}, \mathbf{M}_{11})$. By hypothesis $\mathbf{u}$ is independent of $\mathbf{M}$. Thus, $\mathbf{M}_{22.1}$, $(\mathbf{M}_{21}, \mathbf{M}_{11})$, and $\mathbf{u}$ are jointly independent and hence

$$\mathbf{M}_{22.1} \text{ is independent of } (\mathbf{M}_{21}, \mathbf{M}_{11}, \mathbf{u}). \tag{3.6.11}$$

Let us examine the distribution of $\mathbf{z}$ in (3.6.10) conditional on $\mathbf{u}_1$ and

$\mathbf{X}_1$. First write

$$\mathbf{u}_2 = \mathbf{u}_{2.1} + \boldsymbol{\Sigma}_{21}\boldsymbol{\Sigma}_{11}^{-1}\mathbf{u}_1, \tag{3.6.12}$$

where $\mathbf{u}_{2.1} = \mathbf{u}_2 - \boldsymbol{\Sigma}_{21}\boldsymbol{\Sigma}_{11}^{-1}\mathbf{u}_1$ is normally distributed with mean $\boldsymbol{\mu}_{2.1} = \mathbf{0}$ and covariance matrix $\boldsymbol{\Sigma}_{22.1}$. By Theorem 3.2.3, $\mathbf{u}_{2.1}$ is independent of $(\mathbf{u}_1, \mathbf{X})$ and hence

$$\mathbf{u}_{2.1} \text{ is independent of } (\mathbf{M}_{21}\mathbf{M}_{11}^{-1}\mathbf{u}_1, \mathbf{u}_1, \mathbf{X}_1). \tag{3.6.13}$$

The second term in (3.6.12), $\boldsymbol{\Sigma}_{21}\boldsymbol{\Sigma}_{11}^{-1}\mathbf{u}_1$, is a constant given $\mathbf{u}_1, \mathbf{X}_1$.

To study $\mathbf{M}_{21}\mathbf{M}_{11}^{-1}\mathbf{u}_1 \mid \mathbf{u}_1, \mathbf{X}_1$, write

$$\mathbf{X}_2 = \mathbf{X}_{2.1} + \mathbf{X}_1\boldsymbol{\Sigma}_{11}^{-1}\boldsymbol{\Sigma}_{12},$$

where $\mathbf{X}_{2.1}$ is a data matrix from $N_{p-k}(\mathbf{0}, \boldsymbol{\Sigma}_{22.1})$ independent of $(\mathbf{X}_1, \mathbf{u}_1)$. Then

$$\mathbf{M}_{21}\mathbf{M}_{11}^{-1}\mathbf{u}_1 = \mathbf{X}_2'\mathbf{X}_1\mathbf{M}_{11}^{-1}\mathbf{u}_1 = \mathbf{X}_{2.1}'\mathbf{b} + \boldsymbol{\Sigma}_{21}\boldsymbol{\Sigma}_{11}^{-1}\mathbf{X}_1'\mathbf{b},$$

where $\mathbf{b} = \mathbf{X}_1\mathbf{M}_{11}^{-1}\mathbf{u}_1$ and $\boldsymbol{\Sigma}_{21}\boldsymbol{\Sigma}_{11}^{-1}\mathbf{X}_1'\mathbf{b} = \boldsymbol{\Sigma}_{21}\boldsymbol{\Sigma}_{11}^{-1}\mathbf{u}_1$ are constants given $(\mathbf{X}_1, \mathbf{u}_1)$. Given $(\mathbf{u}_1, \mathbf{X}_1)$, $\mathbf{X}_{2.1}'\mathbf{b}$ is a linear combination of the (statistically independent) rows of $\mathbf{X}_{2.1}$ so that $\mathbf{X}_{2.1}'\mathbf{b} \mid \mathbf{u}_1, \mathbf{X}_1$ is normally distributed with mean $\mathbf{0}$ and covariance matrix $(\mathbf{b}'\mathbf{b})\boldsymbol{\Sigma}_{22.1} = (\mathbf{u}_1'\mathbf{M}_{11}^{-1}\mathbf{u}_1)\boldsymbol{\Sigma}_{22.1}$.

From (3.6.13), $\mathbf{u}_{2.1} \mid \mathbf{u}_1, \mathbf{X}_1$ is independent of $\mathbf{X}_{2.1}'\mathbf{b}$. Thus, adding the two terms of $\mathbf{z}$ together, we see that

$$\mathbf{z} \mid \mathbf{u}_1, \mathbf{X}_1 \sim N_{p-k}(\mathbf{0}, (1 + \mathbf{u}_1'\mathbf{M}_{11}^{-1}\mathbf{u}_1)\boldsymbol{\Sigma}_{22.1}).$$

Let $\mathbf{y} = (1 + D_k^2/m)^{-1/2}\mathbf{z}$. Since $D_k^2 = m\mathbf{u}_1'\mathbf{M}_{11}^{-1}\mathbf{u}_1$ is a function of $\mathbf{u}_1$ and $\mathbf{X}_1$ only,

$$\mathbf{y} \mid \mathbf{u}_1, \mathbf{X}_1 \sim N_{p-k}(\mathbf{0}, \boldsymbol{\Sigma}_{22.1}).$$

As this conditional distribution does not depend on $(\mathbf{u}_1, \mathbf{X}_1)$ it is also the marginal distribution. Furthermore, $\mathbf{y}$ is independent of $(\mathbf{u}_1, \mathbf{X}_1)$ and hence independent of D_k^2.

Now $\mathbf{y}$ and D_k^2 are functions of $(\mathbf{M}_{11}, \mathbf{M}_{12}, \mathbf{u})$, so, by (3.6.11), $\mathbf{M}_{22.1}, \mathbf{y}$, and D_k^2 are jointly independent. Thus, from Theorem 3.5.1,

$$\mathbf{y}'\mathbf{M}_{22.1}^{-1}\mathbf{y} = \frac{D_p^2 - D_k^2}{m + D_k^2} \sim (m-k)^{-1}T^2(p-k, m-k) = \frac{p-k}{m-p+1}F_{p-k, m-p+1}$$

and further, this quantity is independent of D_k^2. ■

3.7 Statistics Based on the Wishart Distribution

In univariate analysis many tests are based on statistics having independent chi-squared distributions. A particular hypothesis may imply, say, that $a \sim \sigma^2\chi^2_\alpha$ and $b \sim \sigma^2\chi^2_\beta$, where a and b are statistics based on the data. If a and b are independent then, as is well known, a/b is α/β times an $F_{\alpha,\beta}$ variable, and $a/(a+b)$ has a beta distribution with parameters $\frac{1}{2}\alpha$ and $\frac{1}{2}\beta$. Neither of these functions involves the parameter σ, which in general is unknown.

In the multivariate case many statistics are based on independent Wishart distributions. Let $\mathbf{A} \sim W_p(\mathbf{\Sigma}, m)$ be independent of $\mathbf{B} \sim W_p(\mathbf{\Sigma}, n)$ where $m \geq p$. Since $m \geq p$, $\mathbf{A}^{-1}$ exists, and the non-zero eigenvalues of the matrix $\mathbf{A}^{-1}\mathbf{B}$ are the quantities of interest. Note that since $\mathbf{A}^{-1}\mathbf{B}$ is similar to the p.s.d. matrix $\mathbf{A}^{-1/2}\mathbf{B}\mathbf{A}^{-1/2}$, all of the non-zero eigenvalues will be positive. Also, with probability 1, the number of non-zero eigenvalues equals $\min(n, p)$. Further, the scale matrix $\mathbf{\Sigma}$ has no effect on the distribution of these eigenvalues, so without loss of generality we may suppose $\mathbf{\Sigma} = \mathbf{I}$ (see Exercise 3.7.1).

For convenience denote the joint distribution of the $\min(n, p)$ non-zero roots of $\mathbf{A}^{-1}\mathbf{B}$ by $\Psi(p, m, n)$. Then the following theorem gives an important relationship between the Ψ distributions for different values of the parameters. See also Exercise 3.7.3.

Theorem 3.7.1 *For $m \geq p$ and $n, p \geq 1$, the $\Psi(p, m, n)$ distribution is identical to the $\Psi(n, m+n-p, p)$ distribution.* ■

Proof First, note that the number of non-zero eigenvalues is the same for each distribution. Note also that $m \geq p$ implies $m+n-p \geq n$ so the latter distribution makes sense.

Suppose $n \leq p$. The $\Psi(p, m, n)$ distribution is the joint distribution of the non-zero eigenvalues of $\mathbf{A}^{-1}\mathbf{B}$, where $\mathbf{A} \sim W_p(\mathbf{I}, m)$ independently of $\mathbf{B} \sim W_p(\mathbf{I}, n)$. Write $\mathbf{B} = \mathbf{X}'\mathbf{X}$, where $\mathbf{X}(n \times p)$ is a data matrix from $N_p(\mathbf{0}, \mathbf{I})$ independent of $\mathbf{A}$. Because $n \leq p$, $\mathbf{X}\mathbf{X}'$ is a non-singular $(n \times n)$ matrix (and so $\mathbf{X}\mathbf{X}' > 0$) with probability 1. Define

$$\mathbf{G} = (\mathbf{X}\mathbf{X}')^{-1/2}\mathbf{X}. \tag{3.7.1}$$

Then $\mathbf{G}$ is a row orthonormal matrix ($\mathbf{G}\mathbf{G}' = \mathbf{I}_n$) and also $\mathbf{X}\mathbf{G}'\mathbf{G} = \mathbf{X}$. Thus,

$$\mathbf{A}^{-1}\mathbf{B} = \mathbf{A}^{-1}\mathbf{X}'\mathbf{X} = \mathbf{A}^{-1}\mathbf{G}'\mathbf{G}\mathbf{X}'\mathbf{X}\mathbf{G}'\mathbf{G},$$

which has the same eigenvalues as

$$(\mathbf{G}\mathbf{A}^{-1}\mathbf{G}')(\mathbf{G}\mathbf{X}'\mathbf{X}\mathbf{G}') = \mathbf{C}^{-1}\mathbf{D}, \text{ say.} \tag{3.7.2}$$

We shall now show that $\mathbf{C} \sim W_n(\mathbf{I}, m+n-p)$ independently of $\mathbf{D} \sim W_n(\mathbf{I}, p)$. The theorem will then follow.

Since $\mathbf{G}$ is a function of $\mathbf{X}$, and $\mathbf{A}$ is independent of $\mathbf{X}$ we find (see Exercise 3.4.19) that

$$(\mathbf{G}\mathbf{A}^{-1}\mathbf{G}')^{-1} \mid \mathbf{X} = \mathbf{C} \mid \mathbf{X} \sim W_n(\mathbf{I}, m+n-p).$$

Since this distribution does not depend on $\mathbf{X}$, it is also the unconditional distribution, and $\mathbf{C}$ is independent of $\mathbf{X}$. Hence $\mathbf{C}$ is also independent of $\mathbf{D} = \mathbf{G}\mathbf{X}'\mathbf{X}\mathbf{G}'$. Finally, since all np elements of $\mathbf{X}$ are independent $N(0, 1)$ random variables, $\mathbf{X}'(p \times n)$ can be considered as a data matrix from $N_n(0, \mathbf{I})$, so that

$$\mathbf{D} = \mathbf{G}\mathbf{X}'\mathbf{X}\mathbf{G}' = \mathbf{X}\mathbf{X}' \sim W_n(\mathbf{I}, \mathrm{p}).$$

Thus the result is proved when $n \leq p$. If $n > p$, then start with the $\Psi(n, m+n-p, p)$ distribution instead of the $\Psi(p, m, n)$ distribution in the above discussion. ■

The following result describes the distribution of a generalization of the F statistic.

Theorem 3.7.2 *If* $\mathbf{A} \sim W_p(\mathbf{\Sigma}, m)$ *and* $\mathbf{B} \sim W_p(\mathbf{\Sigma}, n)$ *are independent and if* $m \geq p$ *and* $n \geq p$, *then*

$$\phi = |\mathbf{A}^{-1}\mathbf{B}| = |\mathbf{B}|/|\mathbf{A}| \tag{3.7.3}$$

is proportional to the product of p *independent* F *variables, of which the* ith *has degrees of freedom* $(n-i+1)$ *and* $(m-i+1)$. ■

Proof From Theorem 3.4.8, $|\mathbf{A}|$ and $|\mathbf{B}|$ are each $|\mathbf{\Sigma}|$ times the product of p independent chi-squared variables. Therefore ϕ is the product of p ratios of independent chi-squared statistics. The ith ratio is $\chi^2_{n-i+1}/\chi^2_{m-i+1}$ i.e. $(n-i+1)/(m-i+1)$ times an $F_{n-i+1, m-i+1}$ statistic. Allowing i to vary from 1 to p the result follows. ■

The multivariate extension of the beta variable will now be defined.

Definition 3.7.1 *When* $\mathbf{A} \sim W_p(\mathbf{I}, m)$ *and* $\mathbf{B} \sim W_p(\mathbf{I}, n)$ *are independent,* $m \geq p$, *we say that*

$$\Lambda = |\mathbf{A}|/|\mathbf{A}+\mathbf{B}| = |\mathbf{I}+\mathbf{A}^{-1}\mathbf{B}|^{-1} \sim \Lambda(p, m, n) \tag{3.7.4}$$

has a Wilks' lambda distribution *with parameters* p, m, *and* n. ■

The Λ family of distributions occurs frequently in the context of likelihood ratio tests. The parameter m usually represents the "error" degrees of freedom and n the "hypothesis" degrees of freedom. Thus $m+n$ represents the "total" degrees of freedom. Unfortunately the notation for this statistic is far from standard. Like the T^2 statistic, Wilks' lambda distribution is invariant under changes of the scale parameters of $\mathbf{A}$ and $\mathbf{B}$ (see Exercise 3.7.4). The distribution of Λ is given in the following theorem.

Theorem 3.7.3 *We have*

$$\Lambda(p, m, n) \sim \prod_{i=1}^{n} u_i, \tag{3.7.5}$$

where $u_1, \ldots, u_n$ are n independent variables and $u_i \sim B(\frac{1}{2}(m+i-p), \frac{1}{2}p)$, $i=1, \ldots, n$. ■

Proof Write $\mathbf{B}=\mathbf{X}'\mathbf{X}$, where the n rows of $\mathbf{X}$ are i.i.d. $N_p(\mathbf{0}, \mathbf{I})$ variables. Let $\mathbf{X}_i$ be the $(i \times p)$ matrix consisting of the first i rows of $\mathbf{X}$, and let

$$\mathbf{M}_i = \mathbf{A}+\mathbf{X}_i'\mathbf{X}_i, \qquad i=1, \ldots, n.$$

Note that $\mathbf{M}_0=\mathbf{A}$, $\mathbf{M}_n=\mathbf{A}+\mathbf{B}$, and $\mathbf{M}_i=\mathbf{M}_{i-1}+\mathbf{x}_i\mathbf{x}_i'$. Now write

$$\Lambda(p, m, n) = \frac{|\mathbf{A}|}{|\mathbf{A}+\mathbf{B}|} = \frac{|\mathbf{M}_0|}{|\mathbf{M}_n|} = \frac{|\mathbf{M}_0|}{|\mathbf{M}_1|}\frac{|\mathbf{M}_1|}{|\mathbf{M}_2|}\cdots\frac{|\mathbf{M}_{n-1}|}{|\mathbf{M}_n|}.$$

This product may be written as $u_1u_2 \ldots u_n$, where

$$u_i = \frac{|\mathbf{M}_{i-1}|}{|\mathbf{M}_i|}, \qquad i=1, \ldots, n.$$

Now $\mathbf{M}_i=\mathbf{M}_{i-1}+\mathbf{x}_i\mathbf{x}_i'$, and therefore, by Corollary 3.5.2.2, with $\mathbf{M}_{i-1}$ corresponding to $\mathbf{M}$ and $\mathbf{x}_i$ corresponding to $\mathbf{d}$,

$$u_i \sim B\{\tfrac{1}{2}(m+i-p), \tfrac{1}{2}p\}, \qquad i=1, \ldots, n.$$

It remains to be shown that the u_i are statistically independent. From Theorem 3.5.4, $\mathbf{M}_i$ is independent of

$$1+\mathbf{x}_i'\mathbf{M}_{i-1}^{-1}\mathbf{x}_i = |\mathbf{M}_i|/|\mathbf{M}_{i-1}| = u_i^{-1}.$$

Since u_i is independent of $\mathbf{x}_{i+1}, \ldots, \mathbf{x}_n$, and

$$\mathbf{M}_{i+j} = \mathbf{M}_i + \sum_{k=1}^{j} \mathbf{x}_{i+k}\mathbf{x}_{i+k}',$$

it follows that u_i is also independent of $\mathbf{M}_{i+1}, \mathbf{M}_{i+2}, \ldots, \mathbf{M}_n$, and hence independent of $u_{i+1}, \ldots, u_n$. The result follows. ■

Theorem 3.7.4 *The $\Lambda(p, m, n)$ and $\Lambda(n, m+n-p, p)$ distributions are the same.*

Proof Let $\lambda_1 \geq \ldots \geq \lambda_p$ denote the eigenvalues of $\mathbf{A}^{-1}\mathbf{B}$ in Definition 3.7.1 and let $k = \min(n, p)$ denote the number of non-zero such eigenvalues. Then, using Section A.6, we can write

$$\Lambda(p, m, n) = |\mathbf{I} + \mathbf{A}^{-1}\mathbf{B}|^{-1} = \prod_{i=1}^{p} (1+\lambda_i)^{-1} = \prod_{i=1}^{k} (1+\lambda_i)^{-1}. \quad (3.7.6)$$

Thus Λ is a function of the non-zero eigenvalues of $\mathbf{A}^{-1}\mathbf{B}$, so the result follows from Theorem 3.7.1 ■

Special cases of these results are as follows:

(a) The statistics $\Lambda(p, m, 1)$ and $\Lambda(1, m+1-p, p)$ are equivalent, and each corresponds to a single $B\{\frac{1}{2}(m-p+1), \frac{1}{2}p\}$ statistic.

(b) $\Lambda(p, m, 2)$ and $\Lambda(2, m+2-p, p)$ are equivalent, and correspond to the product of a $B\{\frac{1}{2}(m-p+1), \frac{1}{2}p\}$ statistic with an independent $B\{\frac{1}{2}(m-p+2), \frac{1}{2}p\}$ statistic.

From the relationship between β and F variables, functions of $\Lambda(p, m, 1)$ and $\Lambda(p, m, 2)$ statistics can also be expressed in terms of the F distribution as follows:

$$\frac{1-\Lambda(p, m, 1)}{\Lambda(p, m, 1)} \sim \frac{p}{m-p+1} F_{p, m-p+1}; \quad (3.7.7)$$

$$\frac{1-\Lambda(1, m, n)}{\Lambda(1, m, n)} \sim \frac{n}{m} F_{n,m}; \quad (3.7.8)$$

$$\frac{1-\sqrt{\Lambda(p, m, 2)}}{\sqrt{\Lambda(p, m, 2)}} \sim \frac{p}{m-p+1} F_{2p, 2(m-p+1)}; \quad (3.7.9)$$

$$\frac{1-\sqrt{\Lambda(2, m, n)}}{\sqrt{\Lambda(2, m, n)}} \sim \frac{n}{m-1} F_{2n, 2(m-1)}. \quad (3.7.10)$$

Formulae (3.7.7) and (3.7.8) are easy to verify (see Exercise 3.7.5), but (3.7.9) and (3.7.10) are more complicated (see Anderson, 1958, pp. 195–196).

For other values of n and p, provided m is large, we may use *Bartlett's approximation*:

$$-\{m-\tfrac{1}{2}(p-n+1)\}\log \Lambda(p, m, n) \sim \chi^2_{np} \tag{3.7.11}$$

asymptotically as $m \to \infty$. Pearson and Hartley (1972, p. 333) tabulate values of a constant factor $C(p, n, m-p+1)$ which improves the approximation.

An approximation which uses the F distribution with non-integer degrees of freedom is discussed in Mardia and Zemroch (1978). Another approximation based on Theorem 3.7.3 is given in Exercise 3.7.6.

The Λ statistic arises naturally in likelihood ratio tests, and this property explains the limiting χ^2 distribution in (3.7.11). Another important statistic in hypothesis testing is the greatest root statistic defined below:

Definition 3.7.2 *Let* $\mathbf{A} \sim W_p(\mathbf{I}, m)$ *be independent of* $\mathbf{B} \sim W_p(\mathbf{I}, n)$, *where* $m \geq p$. *Then the largest eigenvalue* θ *of* $(\mathbf{A}+\mathbf{B})^{-1}\mathbf{B}$ *is called the* greatest root statistic *and its distribution is denoted* $\theta(p, m, n)$.

Note that $\theta(p, m, n)$ can also be defined as the largest root of the determinental equation

$$|\mathbf{B} - \theta(\mathbf{A}+\mathbf{B})| = 0.$$

If λ is an eigenvalue of $\mathbf{A}^{-1}\mathbf{B}$, then $\lambda/(1+\lambda)$ is an eigenvalue of $(\mathbf{A}+\mathbf{B})^{-1}\mathbf{B}$ (Exercise 3.7.8). Since this is a monotone function of λ, θ is given by

$$\theta = \lambda_1/(1+\lambda_1), \tag{3.7.12}$$

where λ_1 denotes the largest eigenvalue of $\mathbf{A}^{-1}\mathbf{B}$. Since $\lambda_1 > 0$ we see that $0 < \theta < 1$.

Using Theorem 3.7.1 and (3.7.6) we easily get the following properties:

(1) $\theta(p, m, n)$ and $\theta(n, m+n-p, p)$ have the same distribution; (3.7.13)

(2) $$\frac{\theta(1, m, n)}{1-\theta(1, m, n)} = \frac{1-\Lambda(1, m, n)}{\Lambda(1, m, n)} \sim \frac{n}{m} F_{n,m}; \tag{3.7.14}$$

(3) $$\frac{\theta(p, m, 1)}{1-\theta(p, m, 1)} = \frac{1-\Lambda(p, m, 1)}{\Lambda(p, m, 1)} \sim \frac{p}{m-p+1} F_{p,m-p+1}. \tag{3.7.15}$$

For $p \geq 2$, critical values of the θ statistic must be found from tables.

Table C.4 in Appendix C gives upper percentage points for $p=2$. Pearson and Hartley (1972, pp. 98–104, 336–350) give critical values for general values of p. Of course the relation (3.7.13) can be used to extend the tables.

As above, note that p = dimension, m = "error" degrees of freedom, and n = "hypothesis" degrees of freedom.

3.8 Other Distributions Related to the Multinormal

Wishart distribution

The Wishart distribution $W_p(\mathbf{\Sigma}, m)$ has already been described. For reference, its p.d.f. (when $\mathbf{\Sigma}>0$ and $m \geq p$) is given by

$$f(\mathbf{M}) = \frac{|\mathbf{M}|^{(m-p-1)/2} \exp(-\frac{1}{2} \operatorname{tr} \mathbf{\Sigma}^{-1}\mathbf{M})}{2^{mp/2}\pi^{p(p-1)/4} |\mathbf{\Sigma}|^{m/2} \prod_{i=1}^{p} \Gamma(\frac{1}{2}(m+1-i))} \tag{3.8.1}$$

with respect to Lebesgue measure $\prod_{i \leq j} dm_{ij}$ in $R^{p(p+1)/2}$, restricted to the set where $\mathbf{M}>0$ (see Anderson, 1958, p. 154.)

Inverted Wishart distribution

(See Siskind, 1972.) If $\mathbf{M} \sim W_p(\mathbf{\Sigma}, m)$ where $\mathbf{\Sigma}>0$ and $m \geq p$, then $\mathbf{U}=\mathbf{M}^{-1}$ is said to have an inverted Wishart distribution $W_p^{-1}(\mathbf{\Sigma}, m)$. Using the Jacobian from Table 2.5.1 we see that its p.d.f. is

$$g(\mathbf{U}) = \frac{|\mathbf{U}|^{-(m+p+1)/2} \exp(-\frac{1}{2} \operatorname{tr} \mathbf{\Sigma}^{-1}\mathbf{U}^{-1})}{2^{mp/2}\pi^{p(p-1)/4} |\mathbf{\Sigma}|^{m/2} \prod_{i=1}^{p} \Gamma(\frac{1}{2}(m+1-i))}. \tag{3.8.2}$$

The expected value of $\mathbf{U}$ is given (see Exercise 3.4.13) by

$$E(\mathbf{U}) = \mathbf{\Sigma}^{-1}/(m-p-1). \tag{3.8.3}$$

Complex multinormal distribution

(See Wooding, 1956; Goodman, 1963; Khatri, 1965.) Let $\mathbf{z}=(\mathbf{x}', \mathbf{y}')' \sim N_{2p}((\boldsymbol{\mu}_1', \boldsymbol{\mu}_2')', \mathbf{\Sigma})$ where $\mathbf{\Sigma}_{12}(p \times p)$ is a skew-symmetric matrix ($\mathbf{\Sigma}_{12} = -\mathbf{\Sigma}_{21}$). Then the distribution of $\mathbf{w}=\mathbf{x}+i\mathbf{y}$ is known as complex mutinormal.

Non-central Wishart distribution

(See James, 1964.) Let $\mathbf{X}$ be a data matrix from $N_p(\boldsymbol{\mu}, \mathbf{\Sigma}), \boldsymbol{\mu} \neq \mathbf{0}$. Then $\mathbf{M}=\mathbf{X}'\mathbf{X}$ has a non-central Wishart distribution.

Matrix T distribution
(See Kshirsagar, 1960; Dickey, 1967.) Let $\mathbf{X}(n \times p)$ be a data matrix from $N_p(\mathbf{0}, \mathbf{Q})$ which is independent of $\mathbf{M} \sim W_n(\mathbf{P}, \nu)$. Then $\mathbf{T} = \mathbf{X}'\mathbf{M}^{-1/2}$ has the matrix T distribution.

Matrix beta type I distribution
(See Kshirsagar, 1961; Mitra, 1969; Khatri and Pillai, 1965.) Let $\mathbf{M}_i \sim W_p(\mathbf{\Sigma}, \nu_i)$, $i = 1, 2$. Then $(\mathbf{M}_1 + \mathbf{M}_2)^{-1/2}\mathbf{M}_1(\mathbf{M}_1 + \mathbf{M}_2)^{-1/2}$ has the matrix beta type I distribution.

Matrix beta type II distribution
(See Kshirsagar, 1960; Khatri and Pillai, 1965.) Let $\mathbf{M}_i \sim W_p(\mathbf{\Sigma}, \nu_i)$, $i = 1, 2$. Then $\mathbf{M}_2^{-1/2}\mathbf{M}_1\mathbf{M}_2^{-1/2}$ has the matrix beta type II distribution.

Exercises and Complements

3.2.1 If the rows of $\mathbf{X}$ are i.i.d. $N_p(\mathbf{0}, \mathbf{I})$, then using Theorem 2.7.1 show that $\operatorname{tr} \mathbf{X}'\mathbf{X}$ is independent of any scale-invariant function of $\mathbf{X}$.

3.2.2 If $\mathbf{x} \sim N_p(\boldsymbol{\mu}, \mathbf{\Sigma})$, show that $\mathbf{x}$ and $\mathbf{Gx}$ have the same distribution for all orthogonal matrices $\mathbf{G}$ if and only if $\boldsymbol{\mu} = \mathbf{0}$ and $\mathbf{\Sigma} = \sigma^2\mathbf{I}$. How does this relate to the property of spherical symmetry?

3.2.3 If $\mathbf{x} \sim N(\mathbf{0}, \sigma^2\mathbf{I})$, show that $\mathbf{Ax}$ and $(\mathbf{I} - \mathbf{A}^-\mathbf{A})\mathbf{x}$, where $\mathbf{A}^-$ is a generalized inverse satisfying $\mathbf{AA}^-\mathbf{A} = \mathbf{A}$, are independent and each has a normal distribution.

3.2.4 (a) If $\mathbf{x} \sim N_p(\boldsymbol{\mu}, \mathbf{\Sigma})$ and $\mathbf{a}$ is any fixed vector, show that

$$f = \frac{\mathbf{a}'(\mathbf{x} - \boldsymbol{\mu})}{\sqrt{\mathbf{a}'\mathbf{\Sigma}\mathbf{a}}} \sim N(0, 1).$$

(b) If $\mathbf{a}$ is now a random vector independent of $\mathbf{x}$ for which $P(\mathbf{a}'\mathbf{\Sigma}\mathbf{a} = 0) = 0$, show that $f \sim N(0, 1)$ and is independent of $\mathbf{a}$.

(c) Hence show that if $\mathbf{x} \sim N_3(\mathbf{0}, \mathbf{I})$ then

$$\frac{x_1 e^{x_3} + x_2 \log |x_3|}{[e^{2x_3} + (\log |x_3|)^2]^{1/2}} \sim N(0, 1).$$

3.2.5 (a) The ordinary least squares coefficient for a line passing through the origin is given by $b = \sum x_i y_i / \sum x_i^2 = \mathbf{x}'\mathbf{y}/\mathbf{x}'\mathbf{x}$. If $\mathbf{y} \sim N_n(\mathbf{0}, \mathbf{I})$ and if

$\mathbf{x}$ is statistically independent of $\mathbf{y}$ then show that

$$\alpha = \mathbf{x}'\mathbf{y}/(\mathbf{x}'\mathbf{x})^{1/2} \sim N(0, 1)$$

and is independent of $\mathbf{x}$.

(b) Suppose $\mathbf{x} \sim N_n(\mathbf{0}, \mathbf{I})$. Since $b = \alpha/(\mathbf{x}'\mathbf{x})^{1/2}$ where $\alpha \sim N(0, 1)$ and is independent of $\mathbf{x}$, and since $\mathbf{x}'\mathbf{x} \sim \chi_n^2$ we deduce that $n^{1/2}b \sim t_n$. This is the null distribution of the ordinary least squares coefficient for the bivariate normal model.

3.2.6 Using the covariance matrix

$$\mathbf{\Sigma} = \begin{bmatrix} 1 & \rho & \rho^2 \\ & 1 & 0 \\ & & 1 \end{bmatrix},$$

show that the conditional distribution of (x_1, x_2) given x_3 has mean vector $[\mu_1 + \rho^2(x_3 - \mu_3), \mu_2]'$ and covariance matrix

$$\begin{bmatrix} 1-\rho^4 & \rho \\ \rho & 1 \end{bmatrix}.$$

3.2.7 If $\mathbf{x}_1, \mathbf{x}_2, \mathbf{x}_3$ are i.i.d. $N_p(\boldsymbol{\mu}, \mathbf{\Sigma})$ random variables, and if $\mathbf{y}_1 = \mathbf{x}_1 + \mathbf{x}_2$, $\mathbf{y}_2 = \mathbf{x}_2 + \mathbf{x}_3$, $\mathbf{y}_3 = \mathbf{x}_1 + \mathbf{x}_3$, then obtain the conditional distribution of $\mathbf{y}_1$ given $\mathbf{y}_2$, and of $\mathbf{y}_1$ given $\mathbf{y}_2$ and $\mathbf{y}_3$

3.3.1 If $\mathbf{x} \sim N_p(\boldsymbol{\mu}, \mathbf{\Sigma})$ and $\mathbf{Q\Sigma Q}'(q \times q)$ is non-singular, then, given that $\mathbf{Qx} = \mathbf{q}$, show that the conditional distribution of $\mathbf{x}$ is normal with mean $\boldsymbol{\mu} + \mathbf{\Sigma Q}'(\mathbf{Q\Sigma Q}')^{-1}(\mathbf{q} - \mathbf{Q}\boldsymbol{\mu})$ and (singular) covariance matrix $\mathbf{\Sigma} - \mathbf{\Sigma Q}'(\mathbf{Q\Sigma Q}')^{-1}\mathbf{Q\Sigma}$.

3.3.2 If $\mathbf{X}_{2.1}$ is as defined in Corollary 3.3.3.1, and if $\mathbf{Q} = \mathbf{X}_1(\mathbf{X}_1'\mathbf{X}_1)^{-1}\mathbf{X}_1'$, show that

(i) $\mathbf{QX}_{2.1} = \mathbf{QX}_2 - \mathbf{X}_1\mathbf{\Sigma}_{11}^{-1}\mathbf{\Sigma}_{12}$, (ii) $(\mathbf{I} - \mathbf{Q})\mathbf{X}_{2.1} = (\mathbf{I} - \mathbf{Q})\mathbf{X}_2$.

Hence prove that

(iii) $\mathbf{X}_{2.1}'(\mathbf{I} - \mathbf{Q})\mathbf{X}_{2.1} = \mathbf{X}_2'(\mathbf{I} - \mathbf{Q})\mathbf{X}_2$,

(iv) $\mathbf{X}_{2.1}'\mathbf{QX}_{2.1} = \mathbf{X}_2'\mathbf{QX}_2 - \mathbf{\Sigma}_{21}\mathbf{\Sigma}_{11}^{-1}\mathbf{M}_{12} - \mathbf{M}_{21}\mathbf{\Sigma}_{11}^{-1}\mathbf{\Sigma}_{12} + \mathbf{\Sigma}_{21}\mathbf{\Sigma}_{11}^{-1}\mathbf{M}_{11}\mathbf{\Sigma}_{11}^{-1}\mathbf{\Sigma}_{12}$,

where $\mathbf{M}_{ij} = \mathbf{X}_i'\mathbf{X}_j$.

(See also Exercise 3.4.15.)

3.3.3 Suppose that $\mathbf{x} \sim (\boldsymbol{\mu}, \boldsymbol{\Sigma})$ and $\mathbf{a}$ is a fixed vector. If r_i is the correlation between x_i and $\mathbf{a}'\mathbf{x}$, show that $\mathbf{r} = (c\boldsymbol{D})^{-1/2}\boldsymbol{\Sigma}\mathbf{a}$ where $c = \mathbf{a}'\boldsymbol{\Sigma}\mathbf{a}$ and $\mathbf{D} = \text{Diag}\,(\boldsymbol{\Sigma})$. When does $\mathbf{r} = \boldsymbol{\Sigma}\mathbf{a}$?

3.3.4 (Proof of Theorem 3.3.2) (a) Let $\mathbf{X}^V$ be the np-vector obtained by stacking the columns of $\mathbf{X}$ on top of one another (Section A.2.5). Then $\mathbf{X}(n \times p)$ is a random data matrix from $N_p(\boldsymbol{\mu}, \boldsymbol{\Sigma})$ if and only if

$$\mathbf{X}^V \sim N_{np}(\boldsymbol{\mu} \otimes \mathbf{1}, \boldsymbol{\Sigma} \otimes \mathbf{I}),$$

where $\otimes$ denotes Kronecker multiplication (see Section A.2.5).

(b) Using the fact that

$$(\mathbf{AXB})^V = (\mathbf{B}' \otimes \mathbf{A})\mathbf{X}^V, \qquad (*)$$

we deduce that $(\mathbf{AXB})^V$ is multivariate normal with mean

$$(\mathbf{B}' \otimes \mathbf{A})(\boldsymbol{\mu} \otimes \mathbf{1}) = \mathbf{B}'\boldsymbol{\mu} \otimes \mathbf{A1}$$

and with covariance matrix

$$(\mathbf{B}' \otimes \mathbf{A})(\boldsymbol{\Sigma} \otimes \mathbf{I})(\mathbf{B}' \otimes \mathbf{A})' = \mathbf{B}'\boldsymbol{\Sigma}\mathbf{B} \otimes \mathbf{AA}'.$$

(c) Therefore $\mathbf{AXB}$ is a normal data matrix if and only if

(i) $\mathbf{A1} = \alpha\mathbf{1}$ for some scalar α or $\mathbf{B}'\boldsymbol{\mu} = \mathbf{0}$, and
(ii) $\mathbf{AA}' = \beta\mathbf{I}$ for some scalar β or $\mathbf{B}'\boldsymbol{\Sigma}\mathbf{B} = \mathbf{0}$.

(d) If both the above conditions are satisfied then $\mathbf{AXB}$ is a random data matrix from $N_q(\alpha\mathbf{B}'\boldsymbol{\mu}, \beta\mathbf{B}'\boldsymbol{\Sigma}\mathbf{B})$.

3.3.5 (Proof of Theorem 3.3.3) Suppose that $\mathbf{X}(n \times p)$ is a data matrix from $N_p(\boldsymbol{\mu}, \boldsymbol{\Sigma})$, and that $\mathbf{Y} = \mathbf{AXB}$ and $\mathbf{Z} = \mathbf{CXD}$. Then, using $(*)$ from Exercise 3.3.4, show that

$$\mathbf{Y}^V\mathbf{Z}^{V\prime} = (\mathbf{B}' \otimes \mathbf{A})\mathbf{X}^V\mathbf{X}^{V\prime}(\mathbf{D}' \otimes \mathbf{C})'.$$

But using Exercise 3.3.4 (a) we know that $V(X^V) = \boldsymbol{\Sigma} \otimes \mathbf{I}$. Therefore

$$C[\mathbf{Y}^V, \mathbf{Z}^{V\prime}] = (\mathbf{B}' \otimes \mathbf{A})(\boldsymbol{\Sigma} \otimes \mathbf{I})(\mathbf{D} \otimes \mathbf{C}') = \mathbf{B}'\boldsymbol{\Sigma}\mathbf{D} \otimes \mathbf{AC}'.$$

The elements of $\mathbf{Y}$ and $\mathbf{Z}$ are uncorrelated if and only if the above matrix is the zero matrix, i.e. if and only if either $\mathbf{B}'\boldsymbol{\Sigma}\mathbf{D} = \mathbf{0}$ or $\mathbf{AC}' = \mathbf{0}$.

3.4.1 If $\mathbf{M} \sim W_p(\boldsymbol{\Sigma}, m)$ and $\mathbf{a}$ is any random p-vector which satisfies $\mathbf{a}'\boldsymbol{\Sigma}\mathbf{a} \neq 0$ with probability one, and is independent of $\mathbf{M}$, then $\mathbf{a}'\mathbf{M}\mathbf{a}/\mathbf{a}'\boldsymbol{\Sigma}\mathbf{a}$ has the χ^2_m distribution, and is independent of $\mathbf{a}$.

3.4.2 If $\mathbf{M} \sim W_p(\mathbf{\Sigma}, m)$, show that $\mathbf{b}'\mathbf{M}\mathbf{b}$ and $\mathbf{d}'\mathbf{M}\mathbf{d}$ are statistically independent if $\mathbf{b}'\mathbf{\Sigma}\mathbf{d} = 0$. (Hint: use Theorem 3.3.3.) Hence show that m_{ii} and m_{jj} are independent if $\sigma_{ij} = 0$, and that when $\mathbf{\Sigma} = \mathbf{I}$, $\operatorname{tr} \mathbf{M}$ has the χ^2_{mp} distribution. Give an alternative proof of this result which follows directly from the representation of $\mathbf{M}$ as $\mathbf{X}'\mathbf{X}$.

3.4.3 (Mitra, 1969) Show that the following conditions taken together are necessary (and sufficient) for $\mathbf{M}$ to have the $W_p(\mathbf{\Sigma}, m)$ distribution:

(a) $\mathbf{M}$ is symmetric, and if $\mathbf{a}'\mathbf{\Sigma}\mathbf{a} = 0$ then $\mathbf{a}'\mathbf{M}\mathbf{a} = 0$ with probability one;
(b) for every $(q \times p)$ matrix $\mathbf{L}$ which satisfies $\mathbf{L}\mathbf{\Sigma}\mathbf{L}' = \mathbf{I}$, the diagonal elements of $\mathbf{L}\mathbf{M}\mathbf{L}'$ are independent χ^2_m variables.

3.4.4 (Mitra, 1969) The converse of Theorem 3.4.2 does not hold. That is, if $\mathbf{a}'\mathbf{T}\mathbf{a}/\mathbf{a}'\mathbf{\Sigma}\mathbf{a} \sim \chi^2_f$ for all $\mathbf{a}$, then $\mathbf{T}$ does not necessarily have a Wishart distribution.

(a) *Construction* Consider $\mathbf{T} = \alpha\mathbf{M}$ where $\mathbf{M} \sim W_p(\mathbf{\Sigma}, n)$, $\alpha \sim B(\frac{1}{2}f, \frac{1}{2}(n-f))$, and α and $\mathbf{M}$ are independent. From Theorem 3.4.2, $\mathbf{a}'\mathbf{T}\mathbf{a}/\mathbf{a}'\mathbf{\Sigma}\mathbf{a}$ is the product of independent $B(\frac{1}{2}f, \frac{1}{2}(n-f))$ and χ^2_n variables. Hence, using the hint, show that $\mathbf{a}'\mathbf{T}\mathbf{a}/\mathbf{a}'\mathbf{\Sigma}\mathbf{a} \sim \chi^2_f$. Thus $\mathbf{T} = \alpha\mathbf{M}$ satisfies the required property. (Hint: If $x_1, \ldots, x_n$ are i.i.d. $N(0, 1)$ variables then, using the fact that $\mathbf{x}'\mathbf{x}$ is independent of any scale-invariant function, see Exercise 3.2.1, note that

$$\sum_{i=1}^{f} x_i^2 \Big/ \sum_{i=1}^{n} x_i^2 \quad \text{and} \quad \sum_{i=1}^{n} x_i^2$$

are independent variables with $B(\frac{1}{2}f, \frac{1}{2}(n-f))$ and χ^2_n distributions, respectively. Hence the product of the above variables is $\sum_{i=1}^{f} x_i^2$, which has a χ^2_f distribution.)

(b) *Contradiction* But $\mathbf{T}$ cannot have a Wishart distribution. For if it does, it must have f degrees of freedom. In that case $r_{ij} = t_{ij}/(t_{ii}t_{jj})^{1/2}$ would have the distribution of a sample correlation coefficient based on a normal sample of size $(f+1)$. But r_{ij} is also $m_{ij}/(m_{ii}m_{jj})^{1/2}$, which has the distribution of a sample correlation coefficient based on a normal sample of size $(n+1)$. Hence we have a contradiction, and $\mathbf{T}$ cannot have a Wishart distribution.

3.4.5 (a) If the rows of $\mathbf{X}$ are i.i.d. $N_p(\boldsymbol{\mu}, \mathbf{\Sigma})$, and $\mathbf{Y} = \mathbf{X} - \mathbf{1}\boldsymbol{\mu}'$, then the rows of $\mathbf{Y}$ are i.i.d. $N_p(\mathbf{0}, \mathbf{\Sigma})$. Now

$$\mathbf{X}'\mathbf{C}\mathbf{X} = \mathbf{Y}'\mathbf{C}\mathbf{Y} + \mathbf{Y}'\mathbf{C}\mathbf{1}\boldsymbol{\mu}' + \boldsymbol{\mu}\mathbf{1}'\mathbf{C}\mathbf{Y} + (\mathbf{1}'\mathbf{C}\mathbf{1})\boldsymbol{\mu}\boldsymbol{\mu}'.$$

If $\mathbf{C}$ is symmetric and idempotent, then $\mathbf{Y'CY}$ has a Wishart distribution by virtue of Theorem 3.4.4, and $\mathbf{X'CX}$ is the sum of a Wishart distribution, two non-independent normal distributions, and a constant. However, a special case arises when the rows of $\mathbf{C}$ sum to zero, since then $\mathbf{C1}=\mathbf{0}$ and the final three terms above are all zero. Thus we have the following result which generalizes Theorem 3.4.4.

(b) If the rows of $\mathbf{X}$ are i.i.d. $N_p(\boldsymbol{\mu}, \boldsymbol{\Sigma})$ and $\mathbf{C}$ is a symmetric matrix, then $\mathbf{X'CX}$ has a $W_p(\boldsymbol{\Sigma}, r)$ distribution if (and only if) (i) $\mathbf{C}$ is idempotent and either (ii) $\boldsymbol{\mu}=\mathbf{0}$ or (iii) $\mathbf{C1}=\mathbf{0}$. In either case $r=\operatorname{tr}\mathbf{C}$.

(c) (Proof of $\mathbf{X'HX} \sim W_p(\boldsymbol{\Sigma}, n-1)$) As a corollary to the above we deduce that if the rows of $\mathbf{X}(n\times p)$ are i.i.d. $N_p(\boldsymbol{\mu}, \boldsymbol{\Sigma})$, then $\mathbf{X'HX}\sim W_p(\boldsymbol{\Sigma}, n-1)$, and $\mathbf{S}=n^{-1}\mathbf{X'HX}\sim W_p(n^{-1}\boldsymbol{\Sigma}, n-1)$ because the centring matrix $\mathbf{H}$ is idempotent of rank $(n-1)$, and all its row sums are zero.

3.4.6 (a) Let $\mathbf{C}_1, \ldots, \mathbf{C}_k$ be $(n\times n)$ symmetric idempotent matrices such that $\mathbf{C}_1+\ldots+\mathbf{C}_k=\mathbf{I}$ and let $\mathbf{X}(n\times p)$ be a data matrix from $N_p(\boldsymbol{\mu}, \boldsymbol{\Sigma})$. Show that $\mathbf{C}_i\mathbf{C}_j=\mathbf{0}$ for $i\neq j$. Hence if $\mathbf{M}_i=\mathbf{X}'\mathbf{C}_i\mathbf{X}$, deduce that $\mathbf{X'X}=\mathbf{M}_1+\ldots+\mathbf{M}_k$ is a decomposition of $\mathbf{X'X}$ into a sum of independent matrices.

(b) If $\boldsymbol{\mu}=\mathbf{0}$, show that $\mathbf{M}_i\sim W_p(\boldsymbol{\Sigma}, r_i)$ for $i=1,\ldots,k$, where $r_i=\operatorname{rank}(\mathbf{C}_i)$.

(c) For general $\boldsymbol{\mu}$, if $\mathbf{C}_1=n^{-1}\mathbf{11}'$, use Exercise 3.4.5 to show that $\mathbf{M}_i\sim W_p(\boldsymbol{\Sigma}, r_i)$ for $i=2,\ldots,k$.

3.4.7 If the rows of $\mathbf{X}$ are i.i.d. $N_p(\boldsymbol{\mu}, \boldsymbol{\Sigma})$ and if $\mathbf{C}$ is a symmetric $(n\times n)$ matrix, then $\mathbf{X'CX}$ and $\mathbf{AXB}$ are independent if either $\mathbf{B}'\boldsymbol{\Sigma}=\mathbf{0}$ or $\mathbf{AC}=\mathbf{0}$.

(Hint: As in (3.4.2) we have $\mathbf{X'CX}=\sum \lambda_i\mathbf{y}_i\mathbf{y}_i'$, where $\mathbf{y}_i=\mathbf{X}'\boldsymbol{\gamma}_{(i)}$ and $\mathbf{C}=\sum \lambda_i\boldsymbol{\gamma}_{(i)}\boldsymbol{\gamma}_{(i)}'$. Now by Theorem 3.3.3, $\mathbf{AXB}$ is independent of $\lambda_i^{1/2}\boldsymbol{\gamma}_{(i)}'\mathbf{X}$ if and only if either $\mathbf{B}'\boldsymbol{\Sigma}=\mathbf{0}$ or $\lambda_i^{1/2}A\boldsymbol{\gamma}_{(i)}=\mathbf{0}$. The second condition holds for all i if and only if $\mathbf{AC}=\mathbf{0}$. Hence the result follows.)

3.4.8 If the rows of $\mathbf{X}$ are i.i.d. and $n\mathbf{S}=\mathbf{X'HX}$ is statistically independent of $n\bar{\mathbf{x}}=\mathbf{X}'\mathbf{1}$, then the rows of $\mathbf{X}$ must have a multivariate normal distribution (Kagan *et al.*, 1973). However, note that $\bar{\mathbf{x}}$ is not independent of $\mathbf{X'X}$.

3.4.9 If the rows of $\mathbf{X}(n\times p)$ are i.i.d. $N_p(\mathbf{0}, \mathbf{I})$, and if $\mathbf{C}$ and $\mathbf{D}$ are symmetric $(p\times p)$ matrices, then the following results may be derived from Theorems 3.4.4 and 3.4.5:

(a) $\mathbf{XCX}'\sim W_n(\mathbf{I}, r)$ if and only if $\mathbf{C}$ is idempotent, and $r=\operatorname{tr}\mathbf{C}$;
(b) $\mathbf{XCX}'$ and $\mathbf{XDX}'$ are independent if and only if $\mathbf{CD}=\mathbf{0}$;
(c) $\mathbf{XCX}'$ and $\mathbf{AX'B}$ are independent if (and only if) $\mathbf{AC}=\mathbf{0}$ or $\mathbf{B}=\mathbf{0}$.

(Hint: the columns of $\mathbf{X}$ are the rows of $\mathbf{X}'$, and these also are i.i.d.)

3.4.10 Extend the results of Exercise 3.4.9 to the case of general $\mathbf{\Sigma}>0$, and show that if the rows of $\mathbf{X}$ are i.i.d. $N_p(\mathbf{0}, \mathbf{\Sigma})$ then

(a) $\mathbf{XCX}' \sim W_n(\mathbf{I}, r)$ if and only if $\mathbf{C\Sigma C}=\mathbf{C}$, in which case $r=\operatorname{tr}\mathbf{C\Sigma}$;
(b) $\mathbf{XCX}'$ and $\mathbf{XDX}'$ are independent if and only if $\mathbf{C\Sigma D}=\mathbf{0}$;
(c) $\mathbf{XCX}'$ and $\mathbf{AX'B}$ are independent if (and only if) $\mathbf{A\Sigma C}=\mathbf{0}$ or $\mathbf{b}=\mathbf{0}$.

(Hint: note that the rows of $\mathbf{X\Sigma}^{-1/2}$ are i.i.d. $N_p(\mathbf{0}, \mathbf{I})$, and use Exercise 3.4.9.)

3.4.11 From Exercise 3.4.9 show that if $\mathbf{x} \sim N_p(\boldsymbol{\mu}, \mathbf{\Sigma})$ then

(a) $\mathbf{x}'\mathbf{Cx} \sim \chi_r^2$ if and only if $\boldsymbol{\mu}=\mathbf{0}$ and $\mathbf{C\Sigma C}=\mathbf{C}$, in which case $r=\operatorname{tr}\mathbf{C\Sigma}$ (as a special case, $(\mathbf{x}-\boldsymbol{\mu})'\mathbf{\Sigma}^{-1}(\mathbf{x}-\boldsymbol{\mu}) \sim \chi_p^2$);
(b) $\mathbf{x}'\mathbf{Cx}$ and $\mathbf{x}'\mathbf{Dx}$ are independent if and only if $\mathbf{C\Sigma D}=\mathbf{0}$;
(c) $\mathbf{x}'\mathbf{Cx}$ and $\mathbf{Ax}$ are independent if (and only if) $\mathbf{A\Sigma C}=\mathbf{0}$.

3.4.12 (Alternative proof that $\bar{\mathbf{x}}$ and $\mathbf{S}$ are independent) Let $\mathbf{X}(n \times p)$ be a data matrix from $N_p(\boldsymbol{\mu}, \mathbf{\Sigma})$ and let $\mathbf{A}(n \times n)$ be an orthogonal matrix whose last row is given by $\mathbf{a}_n = n^{-1/2}\mathbf{1}$. If $\mathbf{Y}=\mathbf{AX}$ show that

(a) the rows of $\mathbf{Y}$ are independent;
(b) $\mathbf{y}_n = n^{1/2}\bar{\mathbf{x}} \sim N_p(n^{1/2}\boldsymbol{\mu}, \mathbf{\Sigma})$;
(c) $\mathbf{y}_i \sim N_p(\mathbf{0}, \mathbf{\Sigma})$ for $i=1, \ldots, n-1$;
(d) $n\mathbf{S}=\mathbf{X}'\mathbf{HX}=\sum_{i=1}^{n-1} \mathbf{y}_i\mathbf{y}_i' \sim W_p(\mathbf{\Sigma}, n-1)$ independently of $\bar{\mathbf{x}}$.

3.4.13 (Expectation of the inverted Wishart distribution) If $\mathbf{M} \sim W_p(\mathbf{\Sigma}, m)$, $m \geq p+2$, show that $E(\mathbf{M}^{-1})=\mathbf{\Sigma}^{-1}/(m-p-1)$.

(Hint: If $x \sim \chi_n^2$, $n \geq 3$, then use its p.d.f. to show that $E(x^{-1})=1/(n-2)$. Also note that $\mathbf{a}'\mathbf{\Sigma}^{-1}\mathbf{a}/\mathbf{a}'\mathbf{M}^{-1}\mathbf{a} \sim \chi_{m-p+1}^2$ for all constant vectors $\mathbf{a}$.)

3.4.14 If $c \sim \chi_m^2$ then, using the central limit theorem (Theorem 2.9.1) and the transformation theorem (Theorem 2.9.2), it is easy to see that $\log c$ has an asymptotic $N(\log m, 2/m)$ distribution as $m \to \infty$. Deduce a corresponding result for the asymptotic distribution of $\log |\mathbf{M}|$, where $\mathbf{M} \sim W(\mathbf{\Sigma}, m)$.

3.4.15 (a) Let $\mathbf{X}$ be a data matrix from $N_p(\mathbf{0}, \mathbf{\Sigma})$. If $\mathbf{X}_{2.1}=\mathbf{X}_2-\mathbf{X}_1\mathbf{\Sigma}_{11}^{-1}\mathbf{\Sigma}_{12}$ then

$$\mathbf{X}_{2.1}'\mathbf{X}_{2.1}=\mathbf{F}=\mathbf{M}_{22}+\mathbf{\Sigma}_{21}\mathbf{\Sigma}_{11}^{-1}\mathbf{M}_{11}\mathbf{\Sigma}_{11}^{-1}\mathbf{\Sigma}_{12}-\mathbf{\Sigma}_{21}\mathbf{\Sigma}_{11}^{-1}\mathbf{M}_{12}-\mathbf{M}_{21}\mathbf{\Sigma}_{11}^{-1}\mathbf{\Sigma}_{12},$$

and $\mathbf{F} \sim W(\mathbf{\Sigma}_{22.1}, m)$.

(Hint: use Theorem 3.4.1; see also Exercise 3.3.2.)

(b) $\mathbf{F}$ is independent of $\mathbf{X}_1$ and of $\mathbf{M}_{11}$.

(c) If $\hat{\mathbf{F}}$ is the matrix obtained from $\mathbf{F}$ by substituting $n^{-1}\mathbf{M}$ for $\mathbf{\Sigma}$, then $\hat{\mathbf{F}}=\mathbf{M}_{22.1}$ defined in (3.4.5).

3.4.16 If $\mathbf{A}$ and $\mathbf{B}$ are $(p\times p)$ symmetric idempotent matrices of rank r and s and if $\mathbf{AB}=\mathbf{0}$, show that, for $\mathbf{x}\sim N_p(\mathbf{0}, \sigma^2\mathbf{I})$,

$$\frac{\mathbf{x}'\mathbf{A}\mathbf{x}/r}{\mathbf{x}'\mathbf{B}\mathbf{x}/s}\sim F_{r,s}, \qquad \frac{\mathbf{x}'\mathbf{A}\mathbf{x}}{\mathbf{x}'(\mathbf{A}+\mathbf{B})\mathbf{x}}\sim B(\tfrac{1}{2}r, \tfrac{1}{2}s),$$

and

$$\frac{(p-r)\mathbf{x}'\mathbf{A}\mathbf{x}}{r\mathbf{x}'(\mathbf{I}-\mathbf{A})\mathbf{x}}\sim F_{r,p-r}.$$

3.4.17 (a) Suppose that the elements of $\mathbf{x}$ are i.i.d. with mean 0, variance 1, and third and fourth moments μ_3 and μ_4. Consider the matrix $\mathbf{M}=\mathbf{x}\mathbf{x}'$, and show that

$$E(m_{ij})=\delta_{ij}, \qquad C(m_{ij}, m_{kl})=(\mu_4-3)\delta_{ijkl}+(\delta_{ik}\delta_{jl}+\delta_{il}\delta_{jk}),$$

$$V(m_{ij})=(\mu_4-2)\delta_{ij}+1,$$

where δ_{ij} is the Kronecker delta and $\delta_{ijkl}=1$ if and only if $i=j=k=l$, and is 0 otherwise.

(b) Suppose that the elements of $\mathbf{X}(n\times p)$ are i.i.d. with the moments given above. If $\mathbf{M}=\mathbf{X}'\mathbf{X}$ show that

$$C(m_{ij}, m_{kl})=n[(\mu_4-3)\delta_{ijkl}+\delta_{ik}\delta_{jl}+\delta_{il}\delta_{jk}].$$

(c) Using the fact that $\mu_4=3$ for $N(0, 1)$, show that if $\mathbf{M}\sim W_p(\mathbf{\Sigma}, n)$ then

$$V(m_{ij})=n(\sigma_{ij}^2+\sigma_{ii}\sigma_{jj}), \qquad C(m_{ij}, m_{kl})=n(\sigma_{ik}\sigma_{jl}+\sigma_{il}\sigma_{jk}).$$

3.4.18 (Alternative proof of Corollary 3.4.8.1) Let $f(\mathbf{x})$ be a p.d.f. on R^p and let $\mathbf{X}=(\mathbf{x}_1,\ldots,\mathbf{x}_n)'$ be a random sample from $f(\mathbf{x})$. Show that rank $(\mathbf{X})=\min(n, p)$ with probability one, and hence, using (A.4.2e), show that if $n\geq p$, then $\mathbf{X}'\mathbf{X}>0$ with probability one.

3.4.19 (Eaton, 1972, §8.26) (a) Note the following generalization of Theorem 3.4.7(a): if $\mathbf{M}\sim W_p(\mathbf{\Sigma}, m)$ and $\mathbf{A}(k\times p)$ has rank k, then

$$(\mathbf{A}\mathbf{M}^{-1}\mathbf{A}')^{-1}\sim W_k((\mathbf{A}\mathbf{\Sigma}^{-1}\mathbf{A}')^{-1}, m-p+k).$$

(Hint: If $\mathbf{A}=[\mathbf{I}, \mathbf{0}]$, then this theorem is simply a statement that $(\mathbf{M}^{11})^{-1}\sim W_k((\mathbf{\Sigma}^{11})^{-1}, m-p+k)$, which has already been proved in Corollary

3.4.6.1. Now any $(k \times p)$ matrix of rank k can be written $\mathbf{A} = \mathbf{B}[\mathbf{I}_k, \mathbf{0}]\boldsymbol{\Gamma}\boldsymbol{\Sigma}^{1/2}$, where $\mathbf{B}(k \times k)$ is non-singular and $\boldsymbol{\Gamma} = (\boldsymbol{\Gamma}_1', \boldsymbol{\Gamma}_2')'$ is orthogonal. (The first k rows, $\boldsymbol{\Gamma}_1$, of $\boldsymbol{\Gamma}$ form an orthonormal basis for the rows of $\mathbf{A}\boldsymbol{\Sigma}^{-1/2}$. The rows of $\boldsymbol{\Gamma}_2$ are orthogonal to the rows of $\mathbf{A}\boldsymbol{\Sigma}^{-1/2}$, so $\mathbf{A}\boldsymbol{\Sigma}^{-1/2}\boldsymbol{\Gamma}_2' = \mathbf{0}$.) Let $\mathbf{V} = \boldsymbol{\Gamma}\boldsymbol{\Sigma}^{-1/2}\mathbf{M}\boldsymbol{\Sigma}^{-1/2}\boldsymbol{\Gamma}'$. Clearly $\mathbf{V} \sim W_p(\mathbf{I}, m)$. Now $\mathbf{A}\mathbf{M}^{-1}\mathbf{A}' = \mathbf{B}[\mathbf{I}, \mathbf{0}]\mathbf{V}^{-1}[\mathbf{I}, \mathbf{0}]'\mathbf{B}' = \mathbf{B}\mathbf{V}^{11}\mathbf{B}'$ where $\mathbf{V}^{11}$ is the upper left-hand submatrix of $\mathbf{V}^{-1}$. Now $(\mathbf{A}\mathbf{M}^{-1}\mathbf{A}')^{-1} = (\mathbf{B}')^{-1}(\mathbf{V}^{11})^{-1}\mathbf{B}^{-1}$. But $(\mathbf{V}^{11})^{-1} \sim W_k(\mathbf{I}, m-p+k)$. Therefore $(\mathbf{A}\mathbf{M}^{-1}\mathbf{A}')^{-1} \sim W_k((\mathbf{B}\mathbf{B}')^{-1}, m-p+k)$. But $\mathbf{B} = \mathbf{A}\boldsymbol{\Sigma}^{-1/2}\boldsymbol{\Gamma}'[\mathbf{I}, \mathbf{0}]' = \mathbf{A}\boldsymbol{\Sigma}^{-1/2}\boldsymbol{\Gamma}_1'$. Thus, since $\mathbf{A}\boldsymbol{\Sigma}^{-1/2}\boldsymbol{\Gamma}_2' = \mathbf{0}$ and $\mathbf{I} = \boldsymbol{\Gamma}_1'\boldsymbol{\Gamma}_1 + \boldsymbol{\Gamma}_2'\boldsymbol{\Gamma}_2$, $(\mathbf{B}\mathbf{B}')^{-1} = (\mathbf{A}\boldsymbol{\Sigma}^{-1}\mathbf{A}')^{-1}$.)

(b) Hence deduce Theorem 3.4.7(a) as a special case.

3.4.20 Let the rows of $\mathbf{X}(n \times p)$ be independently distributed $\mathbf{x}_i \sim N_p(\boldsymbol{\mu}_i, \boldsymbol{\Sigma})$ for $i = 1, \ldots, n$, where the $\boldsymbol{\mu}_i$ are not necessarily equal.

(a) Show that Theorem 3.3.3 remains valid with this assumption on $\mathbf{X}$.

(b) Show that Craig's theorem (Theorem 3.4.5) remains valid with this assumption on $\mathbf{X}$.

(c) If $E(\mathbf{C}\mathbf{X}) = \mathbf{0}$, show that parts (a) and (b) of Cochran's theorem (Theorem 3.4.4) remain valid with this assumption on $\mathbf{X}$.

3.5.1 (a) Examine the case $p = 1$ in Theorem 3.5.1. Putting $d \sim N_1(0, \sigma^2)$ and $ms^2 \sim \sigma^2\chi_m^2$ to correspond to $\mathbf{d}$ and $\mathbf{M}$, respectively, show that α corresponds to $(d/s)^2$, which is the square of a t_m variable.

(b) If $\mathbf{d} \sim N_p(\boldsymbol{\mu}, \boldsymbol{\Sigma})$ and $\mathbf{M} \sim W_p(\boldsymbol{\Sigma}, m)$ then we say that $\alpha = m\mathbf{d}'\mathbf{M}^{-1}\mathbf{d}$ has a *non-central* T^2 distribution. Show that

$$\alpha = \beta + 2m\boldsymbol{\mu}'\mathbf{M}^{-1}\mathbf{x} + m\boldsymbol{\mu}'\mathbf{M}^{-1}\boldsymbol{\mu},$$

where β has a central T^2 distribution and $\mathbf{x} \sim N_p(\mathbf{0}, \boldsymbol{\Sigma})$.

(c) Under the conditions stated in part (b), show that α is proportional to a non-central F statistic, where the non-centrality parameter depends on $\boldsymbol{\mu}'\boldsymbol{\Sigma}^{-1}\boldsymbol{\mu}$, the Mahalanobis distance from $\boldsymbol{\mu}$ to the origin.

3.5.2 Using the assumptions of Theorem 3.5.1, show that if $\alpha = m\mathbf{d}'\mathbf{M}^{-1}\mathbf{d}$ then

(a)
$$\frac{\alpha}{\alpha+m} = \frac{\mathbf{d}'\mathbf{M}^{-1}\mathbf{d}}{1+\mathbf{d}'\mathbf{M}^{-1}\mathbf{d}} \sim B(\tfrac{1}{2}p, \tfrac{1}{2}(m-p+1));$$

(b)
$$\frac{m}{\alpha+m} = \frac{1}{1+\mathbf{d}'\mathbf{M}^{-1}\mathbf{d}} \sim B(\tfrac{1}{2}(m-p+1), \tfrac{1}{2}p);$$

(c)
$$\frac{|\mathbf{M}|}{|\mathbf{M}+\mathbf{d}\mathbf{d}'|} = \frac{m}{\alpha+m} \sim B(\tfrac{1}{2}(m-p+1), \tfrac{1}{2}p).$$

(Hint: use equation (A.2.3n).)

3.5.3 Show that if $m\alpha \sim T^2(p, m)$ then $(m+1)\log(1+\alpha)$ has an asymptotic χ_p^2 distribution as $m \to \infty$.

3.6.1 Partition $\boldsymbol{\mu} = (\boldsymbol{\mu}_1', \boldsymbol{\mu}_2')'$ and suppose $\boldsymbol{\Sigma}$ is also partitioned. Using equations (A.2.4g) and (A.2.4f), show that

$$\boldsymbol{\mu}'\boldsymbol{\Sigma}^{-1}\boldsymbol{\mu} = \boldsymbol{\mu}_1'\boldsymbol{\Sigma}_{11}^{-1}\boldsymbol{\mu}_1 + \boldsymbol{\mu}_{2.1}'\boldsymbol{\Sigma}_{22.1}^{-1}\boldsymbol{\mu}_{2.1},$$

where

$$\boldsymbol{\mu}_{2.1} = \boldsymbol{\mu}_2 - \boldsymbol{\Sigma}_{21}\boldsymbol{\Sigma}_{11}^{-1}\boldsymbol{\mu}_1 \quad \text{and} \quad \boldsymbol{\Sigma}_{22.1} = \boldsymbol{\Sigma}_{22} - \boldsymbol{\Sigma}_{21}\boldsymbol{\Sigma}_{11}^{-1}\boldsymbol{\Sigma}_{12}.$$

3.7.1 If $\mathbf{A} \sim W_p(\boldsymbol{\Sigma}, m)$ independently of $\mathbf{B} \sim W_p(\boldsymbol{\Sigma}, n)$ and $m \geq p$, show that $\mathbf{A}^{-1}\mathbf{B}$ has the same eigenvalues as $\mathbf{A}^{*-1}\mathbf{B}^*$, where $\mathbf{A}^* = \boldsymbol{\Sigma}^{-1/2}\mathbf{A}\boldsymbol{\Sigma}^{-1/2} \sim W_p(\mathbf{I}, m)$ independently of $\mathbf{B}^* = \boldsymbol{\Sigma}^{-1/2}\mathbf{B}\boldsymbol{\Sigma}^{-1/2} \sim W_p(\mathbf{I}, n)$. Hence deduce that the joint distribution of the eigenvalues of $\mathbf{A}^{-1}\mathbf{B}$ does not depend on $\boldsymbol{\Sigma}$.

3.7.2 In Theorem 3.7.1, show that $\mathbf{G} = (\mathbf{X}\mathbf{X}')^{-1/2}\mathbf{X}$ is row orthonormal. Also show that $\mathbf{X}\mathbf{G}'\mathbf{G} = \mathbf{X}$ and that $\mathbf{G}\mathbf{X}'\mathbf{X}\mathbf{G}' = \mathbf{X}\mathbf{X}'$.

3.7.3 (Alternative proof of Theorem 3.7.1) Let $\mathbf{X}((m+n)\times(n+p))$, where $m \geq p$, be a data matrix from $N_{n+p}(\mathbf{0}, \mathbf{I})$ and partition

$$\mathbf{X}'\mathbf{X} = \mathbf{M} = \begin{pmatrix} \mathbf{M}_{11} & \mathbf{M}_{12} \\ \mathbf{M}_{21} & \mathbf{M}_{22} \end{pmatrix} \begin{matrix} n \\ p \end{matrix}.$$

Using Theorem 3.4.6, show that $(\mathbf{M}^{22})^{-1} = \mathbf{M}_{22.1} = \mathbf{M}_{22} - \mathbf{M}_{21}\mathbf{M}_{11}^{-1}\mathbf{M}_{12} \sim W_p(\mathbf{I}, m)$ independently of $\mathbf{M}_{22} - \mathbf{M}_{22.1} = \mathbf{M}_{21}\mathbf{M}_{11}^{-1}\mathbf{M}_{12} \sim W_p(\mathbf{I}, n)$. Thus, using equation (A.2.4g), the non-zero roots of $\mathbf{M}^{22}\mathbf{M}_{21}\mathbf{M}_{11}^{-1}\mathbf{M}_{12} = -\mathbf{M}^{21}\mathbf{M}_{12}$ have the $\Psi(p, m, n)$ distribution. Since $\mathbf{M}$ and $\mathbf{M}^{-1}$ are symmetric, $-\mathbf{M}^{21}\mathbf{M}_{12} = -\mathbf{M}^{12\prime}\mathbf{M}_{21}' = -(\mathbf{M}_{21}\mathbf{M}^{12})'$, which has the same non-zero eigenvalues as $-\mathbf{M}^{12}\mathbf{M}_{21}$(since $\mathbf{AB}$, $(\mathbf{BA})'$, and $\mathbf{BA}$ all have the same non-zero eigenvalues). Interchanging the roles of 1 and 2 above, and applying Theorem 3.4.6 again, show that the roots of $-\mathbf{M}^{12}\mathbf{M}_{21}$ have the $\Psi(n, m+n-p, p)$ distribution. Thus the $\Psi(p, m, n)$ and $\Psi(n, m+n-p, p)$ distributions are the same.

3.7.4 If $\mathbf{A} \sim W_p(\boldsymbol{\Sigma}, m)$ and $\mathbf{B} \sim W_p(\boldsymbol{\Sigma}, n)$ are independent Wishart matrices, show that $|\mathbf{A}|/|\mathbf{A}+\mathbf{B}|$ has the $\Lambda(p, m, n)$ distribution.

3.7.5 Verify the F distribution given in (3.7.7) and (3.7.8).

3.7.6 (a) (Bartlett, 1947) Show that $-\{t - \frac{1}{2}(p+q+1)\}\log \Lambda(p, t-q, q)$ has approximately a χ_{pq}^2 distribution for large t.

(b) (Rao, 1951, 1973, p. 556) If $\Lambda \sim \Lambda(p, t-q, q)$, then an alternative approximation uses

$$R = \frac{ms - 2\lambda}{pq} \frac{1 - \Lambda^{1/s}}{\Lambda^{1/s}},$$

where $m = t - \frac{1}{2}(p+q+1)$, $\lambda = \frac{1}{4}(pq-2)$, and $s^2 = (p^2q^2-4)/(p^2+q^2-5)$. Then R has an asymptotic F distribution with pq and $(ms-2\lambda)$ degrees of freedom, even though $(ms-2\lambda)$ need not be an integer.

3.7.7 Show that if m or n is less than p then the ϕ statistic defined in Theorem 3.7.2 can not be used.

3.7.8 If $\mathbf{A} > 0$ and $\mathbf{B} \geq 0$ and $\mathbf{Bx} = \lambda \mathbf{Ax}$ for some $\mathbf{x} \neq \mathbf{0}$, then show that $\mathbf{Bx} = (\lambda/(1+\lambda))(\mathbf{A}+\mathbf{B})\mathbf{x}$. Hence deduce that if λ is an eigenvalue of $\mathbf{A}^{-1}\mathbf{B}$, then $\lambda/(1+\lambda)$ is an eigenvalue of $(\mathbf{A}+\mathbf{B})^{-1}\mathbf{B}$.

Added in proof

3.4.21 Let $\mathbf{x} \sim N_p(\mathbf{0}, \mathbf{\Sigma})$. Using the characteristic function of $\mathbf{x}$, show that

$$E(x_i x_j x_k x_l) = \sigma_{ij}\sigma_{kl} + \sigma_{ik}\sigma_{jl} + \sigma_{il}\sigma_{jk},$$

for $i, j, k, l = 1, \ldots, p$. Hence prove that

$$C(\mathbf{x}'\mathbf{A}\mathbf{x}, \mathbf{x}'\mathbf{B}\mathbf{x}) = 2 \operatorname{tr}(\mathbf{A\Sigma B\Sigma}),$$

where $\mathbf{A}$ and $\mathbf{B}$ are $p \times p$ symmetric matrices. Hint: prove the result for the transformed variables $\mathbf{y} = \mathbf{\Sigma}^{-1/2}\mathbf{x}$.

4
Estimation

4.1 Likelihood and Sufficiency

4.1.1 The likelihood function

Suppose that $\mathbf{x}_1,\ldots,\mathbf{x}_n$ is a random sample from a population with p.d.f. $f(\mathbf{x};\boldsymbol{\theta})$, where $\boldsymbol{\theta}$ is a parameter vector. The likelihood function of the whole sample is

$$L(\mathbf{X};\boldsymbol{\theta})=\prod_{i=1}^{n} f(\mathbf{x}_i;\boldsymbol{\theta}). \tag{4.1.1}$$

The log likelihood function is

$$l(\mathbf{X};\boldsymbol{\theta})=\log L(\mathbf{X};\boldsymbol{\theta})=\sum_{i=1}^{n}\log f(\mathbf{x}_i;\boldsymbol{\theta}). \tag{4.1.2}$$

Given a sample $\mathbf{X}$ both $L(\mathbf{X};\boldsymbol{\theta})$ and $l(\mathbf{X};\boldsymbol{\theta})$ are considered as functions of the parameter $\boldsymbol{\theta}$. For the special case $n=1$, $L(\mathbf{x};\boldsymbol{\theta})=f(\mathbf{x};\boldsymbol{\theta})$ and the distinction between the p.d.f. and the likelihood function is to be noted: $f(\mathbf{x};\boldsymbol{\theta})$ is interpreted as a p.d.f. when $\boldsymbol{\theta}$ is fixed and $\mathbf{x}$ is allowed to vary and it is interpreted as the likelihood function when $\mathbf{x}$ is fixed and $\boldsymbol{\theta}$ is allowed to vary. As we know, the p.d.f. plays a key role in probability theory whereas the likelihood is central to the theory of statistical inference.

Example 4.1.1. Suppose $\mathbf{x}_1,\ldots,\mathbf{x}_n$ is a random sample from $N_p(\boldsymbol{\mu},\boldsymbol{\Sigma})$. Then (4.1.1) and (4.1.2) become, respectively,

$$L(\mathbf{X};\boldsymbol{\mu},\boldsymbol{\Sigma})=|2\pi\boldsymbol{\Sigma}|^{-n/2}\exp\left\{-\tfrac{1}{2}\sum_{i=1}^{n}(\mathbf{x}_i-\boldsymbol{\mu})'\boldsymbol{\Sigma}^{-1}(\mathbf{x}_i-\boldsymbol{\mu})\right\}, \tag{4.1.3}$$

and

$$l(\mathbf{X}; \boldsymbol{\mu}, \boldsymbol{\Sigma}) = \log L(\mathbf{X}; \boldsymbol{\theta}) = -\frac{n}{2}\log|2\pi\boldsymbol{\Sigma}| - \tfrac{1}{2}\sum_{i=1}^{n}(\mathbf{x}_i - \boldsymbol{\mu})'\boldsymbol{\Sigma}^{-1}(\mathbf{x}_i - \boldsymbol{\mu}). \tag{4.1.4}$$

Equation (4.1.4) can be simplified as follows. When the identity

$$(\mathbf{x}_i - \boldsymbol{\mu})'\boldsymbol{\Sigma}^{-1}(\mathbf{x}_i - \boldsymbol{\mu}) = (\mathbf{x}_i - \bar{\mathbf{x}})'\boldsymbol{\Sigma}^{-1}(\mathbf{x}_i - \bar{\mathbf{x}}) + (\bar{\mathbf{x}} - \boldsymbol{\mu})'\boldsymbol{\Sigma}^{-1}(\bar{\mathbf{x}} - \boldsymbol{\mu}) + 2(\bar{\mathbf{x}} - \boldsymbol{\mu})'\boldsymbol{\Sigma}^{-1}(\mathbf{x}_i - \bar{\mathbf{x}}), \tag{4.1.5}$$

is summed over the index $i = 1, \ldots, n$, the final term on the right-hand side vanishes, yielding

$$\sum_{i=1}^{n}(\mathbf{x}_i - \boldsymbol{\mu})'\boldsymbol{\Sigma}^{-1}(\mathbf{x}_i - \boldsymbol{\mu}) = \sum_{i=1}^{n}(\mathbf{x}_i - \bar{\mathbf{x}})'\boldsymbol{\Sigma}^{-1}(\mathbf{x}_i - \bar{\mathbf{x}}) + n(\bar{\mathbf{x}} - \boldsymbol{\mu})'\boldsymbol{\Sigma}^{-1}(\bar{\mathbf{x}} - \boldsymbol{\mu}). \tag{4.1.6}$$

Since each term $(\mathbf{x}_i - \bar{\mathbf{x}})'\boldsymbol{\Sigma}^{-1}(\mathbf{x}_i - \bar{\mathbf{x}})$ is a scalar, it equals the trace of itself. Hence (see Section A.2.2)

$$(\mathbf{x}_i - \bar{\mathbf{x}})'\boldsymbol{\Sigma}^{-1}(\mathbf{x}_i - \bar{\mathbf{x}}) = \operatorname{tr} \boldsymbol{\Sigma}^{-1}(\mathbf{x}_i - \bar{\mathbf{x}})(\mathbf{x}_i - \bar{\mathbf{x}})'. \tag{4.1.7}$$

Summing (4.1.7) over the index i and substituting in (4.1.6) yields

$$\sum_{i=1}^{n}(\mathbf{x}_i - \boldsymbol{\mu})'\boldsymbol{\Sigma}^{-1}(\mathbf{x}_i - \boldsymbol{\mu}) = \operatorname{tr} \boldsymbol{\Sigma}^{-1}\left\{\sum_{i=1}^{n}(\mathbf{x}_i - \bar{\mathbf{x}})(\mathbf{x}_i - \bar{\mathbf{x}})'\right\} + n(\bar{\mathbf{x}} - \boldsymbol{\mu})'\boldsymbol{\Sigma}^{-1}(\bar{\mathbf{x}} - \boldsymbol{\mu}). \tag{4.1.8}$$

Writing

$$\sum_{i=1}^{n}(\mathbf{x}_i - \bar{\mathbf{x}})(\mathbf{x}_i - \bar{\mathbf{x}})' = n\mathbf{S}$$

and using (4.1.8) in (4.1.4) gives

$$l(\mathbf{X}; \boldsymbol{\mu}, \boldsymbol{\Sigma}) = -\frac{n}{2}\log|2\pi\boldsymbol{\Sigma}| - \frac{n}{2}\operatorname{tr} \boldsymbol{\Sigma}^{-1}\mathbf{S} - \frac{n}{2}(\bar{\mathbf{x}} - \boldsymbol{\mu})'\boldsymbol{\Sigma}^{-1}(\bar{\mathbf{x}} - \boldsymbol{\mu}). \tag{4.1.9}$$

For the special case when $\boldsymbol{\Sigma} = \mathbf{I}$ and $\boldsymbol{\mu} = \boldsymbol{\theta}$ then (4.1.9) becomes

$$l(\mathbf{X}; \boldsymbol{\theta}) = -\frac{np}{2}\log 2\pi - \frac{n}{2}\operatorname{tr} \mathbf{S} - \frac{n}{2}(\bar{\mathbf{x}} - \boldsymbol{\theta})'(\bar{\mathbf{x}} - \boldsymbol{\theta}). \tag{4.1.10}$$

4.1.2 Efficient scores and Fisher's information

The *score function* or efficient score $\mathbf{s} = \mathbf{s}(\mathbf{X}; \boldsymbol{\theta})$ is defined as

$$\mathbf{s}(\mathbf{X}; \boldsymbol{\theta}) = \frac{\partial}{\partial\boldsymbol{\theta}} l(\mathbf{X}; \boldsymbol{\theta}) = \frac{1}{L(\mathbf{X}; \boldsymbol{\theta})}\frac{\partial}{\partial\boldsymbol{\theta}} L(\mathbf{X}; \boldsymbol{\theta}). \tag{4.1.11}$$

Of course, $\mathbf{s}(\mathbf{X};\boldsymbol{\theta})$ is a random vector which we may write simply as $\mathbf{s}$. The covariance matrix of $\mathbf{s}$ is called *Fisher's information matrix*, which we denote by $\mathbf{F}$.

Theorem 4.1.1 *If $\mathbf{s}(\mathbf{X};\boldsymbol{\theta})$ is the score of a likelihood function and if $\mathbf{t}$ is any function of $\mathbf{X}$ and $\boldsymbol{\theta}$, then, under certain regularity conditions,*

$$E(\mathbf{s}\mathbf{t}')=\frac{\partial}{\partial\boldsymbol{\theta}}E(\mathbf{t}')-E\left(\frac{\partial\mathbf{t}'}{\partial\boldsymbol{\theta}}\right). \tag{4.1.12}$$

Proof We have

$$E(\mathbf{t}')=\int \mathbf{t}'L\,\mathrm{d}\mathbf{X}.$$

On differentiating both sides with respect to $\boldsymbol{\theta}$ and taking the differentiation under the integral sign in the right-hand side (assuming this operation is valid), we obtain

$$\frac{\partial}{\partial\boldsymbol{\theta}}E(\mathbf{t}')=\int\frac{\partial \log L}{\partial\boldsymbol{\theta}}\mathbf{t}'L\,\mathrm{d}\mathbf{X}+\int\frac{\partial\mathbf{t}'}{\partial\boldsymbol{\theta}}L\,\mathrm{d}\mathbf{X}.$$

The result follows on simplifying and rearranging this expression. ■

Corollary 4.1.1.1 *If $\mathbf{s}(\mathbf{X};\boldsymbol{\theta})$ is the score corresponding to a regular likelihood function then $E(\mathbf{s})=\mathbf{0}$.*

Proof Choose $\mathbf{t}$ as any constant vector. Then $E(\mathbf{s})\mathbf{t}'=\mathbf{0}$ for all $\mathbf{t}$. Thus,

$$E(\mathbf{s})=\mathbf{0}. \quad ■ \tag{4.1.13}$$

Corollary 4.1.1.2 *If $\mathbf{s}(\mathbf{X};\boldsymbol{\theta})$ is the score corresponding to a regular likelihood function and if $\mathbf{t}$ is any unbiased estimator of $\boldsymbol{\theta}$ then*

$$E(\mathbf{s}\mathbf{t}')=\mathbf{I}. \tag{4.1.14}$$

Proof We have $E(\mathbf{t})=\boldsymbol{\theta}$ and, since $\mathbf{t}$ does not involve $\boldsymbol{\theta}$, $\dfrac{\partial\mathbf{t}'}{\partial\boldsymbol{\theta}}=\mathbf{0}$. ■

Corollary 4.1.1.3 *If $\mathbf{s}(\mathbf{X};\boldsymbol{\theta})$ is the score corresponding to a regular likelihood function and if $\mathbf{t}$ is an estimator such that $E(\mathbf{t})=\boldsymbol{\theta}+\mathbf{b}(\boldsymbol{\theta})$, then $E(\mathbf{s}\mathbf{t}')=\mathbf{I}+\mathbf{B}$ where $(\mathbf{B})_{ij}=\partial b_j/\partial\theta_i$.*

Proof Use $\partial\mathbf{t}'/\partial\boldsymbol{\theta}=\mathbf{I}+\mathbf{B}$. ■

From Corollary 4.1.1.1 we see that $\mathbf{F}=V(\mathbf{s})=E(\mathbf{s}\mathbf{s}')$. Applying Theorem 4.1.1 with $\mathbf{t}=\mathbf{s}$ gives

$$\mathbf{F}=E(\mathbf{s}\mathbf{s}')=-E\left(\frac{\partial\mathbf{s}'}{\partial\boldsymbol{\theta}}\right)=-E\left(\frac{\partial^2 l}{\partial\boldsymbol{\theta}\,\partial\boldsymbol{\theta}'}\right). \tag{4.1.15}$$

Example 4.1.2 Suppose $\mathbf{x}_1, \ldots, \mathbf{x}_n$ is a random sample from $N_p(\boldsymbol{\theta}, \mathbf{I})$. From (4.1.10),

$$\mathbf{s}(\mathbf{X}; \boldsymbol{\theta}) = \frac{\partial}{\partial \boldsymbol{\theta}} l(\mathbf{X}; \boldsymbol{\theta}) = n(\bar{\mathbf{x}} - \boldsymbol{\theta}), \qquad \frac{\partial \mathbf{s}'}{\partial \boldsymbol{\theta}} = -n\mathbf{I}.$$

Hence, from (4.1.15), $\mathbf{F} = n\mathbf{I}$. Alternatively, $\mathbf{F}$ can be obtained as the covariance matrix of $n(\bar{\mathbf{x}} - \boldsymbol{\theta})$.

Example 4.1.3 For $n = 1$, the simple exponential family defined in Section 2.7.1 has likelihood function

$$f(\mathbf{x}; \boldsymbol{\theta}) = \exp\left[a_0(\boldsymbol{\theta}) + b_0(\mathbf{x}) + \sum_{i=1}^{q} \theta_i b_i(\mathbf{x})\right].$$

Hence

$$l(\mathbf{x}; \boldsymbol{\theta}) = a_0(\boldsymbol{\theta}) + b_0(\mathbf{x}) + \sum_{i=1}^{q} \theta_i b_i(\mathbf{x}),$$

$$s(\mathbf{x}; \boldsymbol{\theta}) = \frac{\partial a_0(\boldsymbol{\theta})}{\partial \boldsymbol{\theta}} + \mathbf{b},$$

where $\mathbf{b}' = (b_1(\mathbf{x}), \ldots, b_q(\mathbf{x}))$. Using (4.1.15),

$$\mathbf{F} = -E\left(\frac{\partial \mathbf{s}}{\partial \boldsymbol{\theta}'}\right) = -\frac{\partial^2 a_0(\boldsymbol{\theta})}{\partial \boldsymbol{\theta} \partial \boldsymbol{\theta}'}. \tag{4.1.16}$$

4.1.3 The Cramér–Rao lower bound

Theorem 4.1.2 *If $\mathbf{t} = t(\mathbf{X})$ is an unbiased estimator of $\boldsymbol{\theta}$ based on a regular likelihood function, then*

$$V(\mathbf{t}) \geq \mathbf{F}^{-1}, \tag{4.1.17}$$

where $\mathbf{F}$ is the Fisher information matrix.

Proof Consider $\text{corr}^2[\alpha, \gamma]$, where $\alpha = \mathbf{a}'\mathbf{t}$, $\gamma = \mathbf{c}'\mathbf{s}$, and $\mathbf{s}$ is the score function. From (4.1.13) and (4.1.14), we have

$$C[\alpha, \gamma] = \mathbf{a}' C[\mathbf{t}, \mathbf{s}]\mathbf{c} = \mathbf{a}'\mathbf{c}, \qquad V(\gamma) = \mathbf{c}' V(\mathbf{s})\mathbf{c} = \mathbf{c}'\mathbf{F}\mathbf{c}.$$

Hence

$$\text{corr}^2[\alpha, \gamma] = (\mathbf{a}'\mathbf{c})^2 / \{\mathbf{a}' V(\mathbf{t})\mathbf{a}\mathbf{c}'\mathbf{F}\mathbf{c}\} \leq 1. \tag{4.1.18}$$

Maximizing the left-hand side of (4.1.18) with respect to $\mathbf{c}$ with the help of (A.9.13) gives, for all $\mathbf{a}$,

$$\mathbf{a}'\mathbf{F}^{-1}\mathbf{a} / \mathbf{a}' V(\mathbf{t})\mathbf{a} \leq 1;$$

that is $\mathbf{a}'\{V(\mathbf{t}) - \mathbf{F}^{-1}\}\mathbf{a} \geq 0$ for all $\mathbf{a}$, which is equivalent to (4.1.17). ■

Note that the Cramér–Rao lower bound is attained if and only if the estimator is a linear function of the score vector.

Example 4.1.4 A genetic example presented by Fisher (1970, p. 305) has four outcomes with probabilities $(2+\theta)/4$, $(1-\theta)/4$, $(1-\theta)/4$, and $\theta/4$, respectively. If in n trials one observes x_i results for outcome i, where $\sum x_i = n$, then $\mathbf{x}=(x_1, x_2, x_3, x_4)'$ has a multinomial distribution and the likelihood function is

$$L(\mathbf{x};\theta)=c(2+\theta)^{x_1}(1-\theta)^{x_2+x_3}\theta^{x_4}/4^n,$$

where $c=n!/(x_1!\,x_2!\,x_3!\,x_4!)$. Then

$$l(\mathbf{x};\theta)=\log c-n\log 4+x_1\log(2+\theta)+(x_2+x_3)\log(1-\theta)+x_4\log\theta.$$

The score function is given by

$$s(\mathbf{x};\theta)=\frac{\partial l}{\partial\theta}=\frac{x_1}{2+\theta}-\frac{x_2+x_3}{1-\theta}+\frac{x_4}{\theta}.$$

From (4.1.15),

$$F=-E\left(\frac{-x_1}{(2+\theta)^2}-\frac{(x_2+x_3)}{(1-\theta)^2}-\frac{x_4}{\theta^2}\right)=\frac{n(1+2\theta)}{\{2\theta(1-\theta)(2+\theta)\}}.$$

The Cramér–Rao lower bound is F^{-1}. Here $4x_4/n$ is an unbiased estimator of θ with variance $\theta(4-\theta)/n$ which exceeds F^{-1}. Hence $4x_4/n$ does not attain the Cramér–Rao lower bound. Since the lower bound can only be obtained by a linear function of the score, any such function (which will involve θ) cannot be a statistic. Hence no unbiased estimator can attain the lower bound in this example.

4.1.4 Sufficiency

Suppose $\mathbf{X}=(\mathbf{x}_1,\ldots,\mathbf{x}_n)'$ is a sequence of independent identically distributed random vectors whose distribution depends on the parameter $\boldsymbol{\theta}$, and let $\mathbf{t}(\mathbf{X})$ be any statistic. The statistic $\mathbf{t}$ is said to be sufficient for $\boldsymbol{\theta}$ if $L(\mathbf{X};\boldsymbol{\theta})$ can be factorized as

$$L(\mathbf{X};\boldsymbol{\theta})=g(\mathbf{t};\boldsymbol{\theta})h(\mathbf{X}), \tag{4.1.19}$$

where h is a non-negative function not involving $\boldsymbol{\theta}$ and g is a function of $\boldsymbol{\theta}$ and $\mathbf{t}$.

Note that (4.1.19) implies that the efficient score $\mathbf{s}$ depends on the data only through the sufficient statistic.

Example 4.1.5 For the multinormal case we have from (4.1.9) taking exponentials,

$$L(\mathbf{X}; \boldsymbol{\mu}, \boldsymbol{\Sigma}) = |2\pi\boldsymbol{\Sigma}|^{-n/2} \exp\left\{-\frac{n}{2} \operatorname{tr} \boldsymbol{\Sigma}^{-1}\mathbf{S} - \frac{n}{2} (\bar{\mathbf{x}} - \boldsymbol{\mu})' \boldsymbol{\Sigma}^{-1} (\bar{\mathbf{x}} - \boldsymbol{\mu})\right\}. \tag{4.1.20}$$

In (4.1.19), taking $h(\mathbf{X}) = 1$, we see that $\bar{\mathbf{x}}$ and $\mathbf{S}$ are sufficient for $\boldsymbol{\mu}$ and $\boldsymbol{\Sigma}$.

Example 4.1.6 From Example 4.1.4,

$$L(\mathbf{x}; \theta) = c(2+\theta)^{x_1}(1-\theta)^{x_2+x_3}\theta^{x_4}/4^{x_1+x_2+x_3+x_4}.$$

Taking $g(\mathbf{t}; \theta) = (2+\theta)^{x_1}(1-\theta)^{x_2+x_3}\theta^{x_4}$ and $\mathbf{t}' = (x_1, x_2+x_3, x_4)$ in (4.1.19), we see that $\mathbf{t}$ is sufficient for θ. Since $n = x_1+x_2+x_3+x_4$ is fixed, $\mathbf{t}^+ = (x_1, x_4)'$ is also sufficient, as x_2+x_3 can be written in terms of $\mathbf{t}^+$.

Example 4.1.7 For a sample of size n from the simple exponential family as in Example 4.1.3 the likelihood can be written as

$$f(\mathbf{X}; \boldsymbol{\theta}) = \exp\left[na_0(\boldsymbol{\theta}) + \sum_{i=1}^{q} \theta_i \sum_{j=1}^{n} b_i(\mathbf{x}_j)\right] \exp\left[\sum_{j=1}^{n} b_0(\mathbf{x}_j)\right].$$

In (4.1.19), take $g(\mathbf{t}; \boldsymbol{\theta})$ as the first factor in this expression and $h(\mathbf{X})$ as the second. The vector $\mathbf{t}$, where

$$t_i = \sum_{j=1}^{n} b_i(\mathbf{x}_j),$$

is sufficient for $\boldsymbol{\theta}$.

A sufficient statistic is said to be minimal sufficient if it is a function of every other sufficient statistic. The Rao–Blackwell theorem states that if the minimal sufficient statistic is *complete*, then any unbiased estimator which is a function of the minimal sufficient statistic must necessarily be the unique minimum variance unbiased estimator (MVUE); that is, it will have a smaller covariance matrix than any other unbiased estimator.

Example 4.1.8 Suppose that the n rows of $\mathbf{X}$ are i.i.d. $N_p(\boldsymbol{\mu}, \mathbf{I})$ vectors and we seek the best unbiased estimate of the quadratic function $\theta = \boldsymbol{\mu}'\boldsymbol{\mu} + \mathbf{1}'\boldsymbol{\mu}$. The mean $\bar{\mathbf{x}}$ is minimal sufficient for $\boldsymbol{\mu}$, and we consider a quadratic function of $\bar{\mathbf{x}}$, $t = a\bar{\mathbf{x}}'\bar{\mathbf{x}} + \mathbf{b}'\bar{\mathbf{x}} + c$. Then

$$E(t) = a\left(\boldsymbol{\mu}'\boldsymbol{\mu} + \frac{p}{n}\right) + \mathbf{b}'\boldsymbol{\mu} + c.$$

If t is to be an unbiased estimator of θ, then $a = 1$, $b = 1$, and $c = -ap/n$.

Since this unbiased estimator is a function of the minimal sufficient statistic $\bar{\mathbf{x}}$, which is complete, then, from the Rao–Blackwell theorem, t is the minimum variance unbiased estimator of θ.

4.2 Maximum Likelihood Estimation

4.2.1 General case

The maximum likelihood estimate (m.l.e.) of an unknown parameter is that value of the parameter which maximizes the likelihood of the given observations. In regular cases this maximum may be found by differentiation and, since $l(\mathbf{X};\boldsymbol{\theta})$ is at a maximum when $L(\mathbf{X};\boldsymbol{\theta})$ is at a maximum, the equation $(\partial l/\partial\boldsymbol{\theta})=\mathbf{s}=\mathbf{0}$ is solved for $\boldsymbol{\theta}$. The m.l.e. $\hat{\boldsymbol{\theta}}$ of $\boldsymbol{\theta}$ is that value which provides an overall maximum. Since $\mathbf{s}$ is a function of a sufficient statistic, then so is the m.l.e. Also, if the density $f(\mathbf{x};\boldsymbol{\theta})$ satisfies certain regularity conditions and if $\hat{\boldsymbol{\theta}}_n$ is the m.l.e. of $\boldsymbol{\theta}$ for a random sample of size n, then $\hat{\boldsymbol{\theta}}_n$ is asymptotically normally distributed with mean $\boldsymbol{\theta}$ and covariance matrix $\mathbf{F}_n^{-1}=(n\mathbf{F})^{-1}$, where $\mathbf{F}$ is the Fisher information matrix for a *single* observation (see, for example, Rao, 1973, p. 416). In particular, since $V(\hat{\boldsymbol{\theta}}_n)\to\mathbf{0}$, $\hat{\boldsymbol{\theta}}_n$ is a *consistent* estimate of $\boldsymbol{\theta}$, i.e. plim $\hat{\boldsymbol{\theta}}_n=\boldsymbol{\theta}$, where plim denotes limit in probability.

Example 4.2.1 The m.l.e. of θ in Example 4.1.4 is found by solving the equation $\mathbf{s}=\mathbf{0}$, which becomes

$$n\theta^2+(2x_2+2x_3-x_1+x_4)\theta-2x_4=0.$$

This quadratic equation gives two roots, one positive and one negative. Only the positive root is admissible. Writing $p_i=x_i/n$, $i=1,2,3,4$, the quadratic equation becomes

$$\theta^2+\theta(2-3p_1-p_4)-2p_4=0.$$

Discarding the negative root of this quadratic equation, $\hat{\theta}=\alpha+(\alpha^2+2p_4)^{1/2}$, where $\alpha=\frac{1}{2}(3p_1+p_4)-1$.

Example 4.2.2 If $(\mathbf{x}_1,\ldots,\mathbf{x}_n)$ is a random sample from Weinman's p-variate exponential distribution defined in Section 2.6.3, then the log likelihood is given by

$$n^{-1}l=-\sum_{j=0}^{p-1}\left(\log\lambda_j+\frac{\delta_j}{\lambda_j}\right), \tag{4.2.1}$$

where

$$n\delta_j=(p-j)\sum_{i=1}^{n}\{x_{i(j+1)}-x_{i(j)}\}$$

with $x_{i(j)}$ the jth smallest element of $\mathbf{x}_i$ $(i=1,\ldots,n;\ j=1,\ldots,p-1)$, and $x_{i(0)}=0$. Note that although the elements of $\mathbf{x}_i$ are not independent, the elements $\delta_1,\ldots,\delta_{p-1}$ are independent. Differentiating the log likelihood gives the elements s_j of the score $\mathbf{s}$ as

$$s_j = \partial l/\partial\lambda_j = -n(\lambda_j - \delta_j)/\lambda_j^2.$$

The m.l.e.s are therefore $\hat{\lambda}_j = \delta_j$. Since $E(s_j) = 0$ we see that $\hat{\lambda}_j$ is an unbiased estimate of λ_j. In the special case when all the parameters are equal, say λ, then (4.2.1) becomes

$$n^{-1}l = -p\log\lambda - \left(\sum_{j=0}^{p-1}\delta_j\right)\bigg/\lambda. \tag{4.2.2}$$

Differentiating with respect to λ gives

$$s = \frac{\partial l}{\partial\lambda} = -\frac{np}{\lambda} + n\left(\sum_{j=0}^{p-1}\delta_j\right)\bigg/\lambda^2 = n\left(\sum_{j=0}^{p-1}\delta_j - p\lambda\right)\bigg/\lambda^2.$$

Solving $s = 0$ gives

$$\hat{\lambda} = \left(\sum_{j=0}^{p-1}\delta_j\right)\bigg/p.$$

Again, since $E(s) = 0$ we see that $\hat{\lambda}$ is unbiased for λ.

4.2.2 Multivariate normal case

4.2.2.1 Unconstrained case

For the multivariate normal distribution we have the log likelihood function from (4.1.9):

$$l(\mathbf{X}; \boldsymbol{\mu}, \boldsymbol{\Sigma}) = -\frac{n}{2}\log|2\pi\boldsymbol{\Sigma}| - \frac{n}{2}\operatorname{tr}\boldsymbol{\Sigma}^{-1}\mathbf{S} - \frac{n}{2}\operatorname{tr}\boldsymbol{\Sigma}^{-1}(\bar{\mathbf{x}}-\boldsymbol{\mu})(\bar{\mathbf{x}}-\boldsymbol{\mu})'.$$

We shall now show that

$$\hat{\boldsymbol{\mu}} = \bar{\mathbf{x}}, \qquad \hat{\boldsymbol{\Sigma}} = \mathbf{S} \tag{4.2.3}$$

if $n \geq p+1$. Note that from Corollary 3.4.8.1 that $\mathbf{S} > 0$ with probability 1.

The parameters here are $\boldsymbol{\mu}$ and $\boldsymbol{\Sigma}$. Using new parameters $\boldsymbol{\mu}$ and $\mathbf{V}$ with $\mathbf{V} = \boldsymbol{\Sigma}^{-1}$, we first calculate $\partial l/\partial\boldsymbol{\mu}$ and $\partial l/\partial\mathbf{V}$. Equation (4.1.9) becomes

$$l(\mathbf{X}; \boldsymbol{\mu}, \mathbf{V}) = -\frac{np}{2}\log 2\pi + \frac{n}{2}\log|\mathbf{V}| - \frac{n}{2}\operatorname{tr}\mathbf{VS} - \frac{n}{2}\operatorname{tr}\mathbf{V}(\bar{\mathbf{x}}-\boldsymbol{\mu})(\bar{\mathbf{x}}-\boldsymbol{\mu})'. \tag{4.2.4}$$

Using Section A.9,

$$\frac{\partial l}{\partial\boldsymbol{\mu}} = n\mathbf{V}(\bar{\mathbf{x}}-\boldsymbol{\mu}). \tag{4.2.5}$$

To calculate $\partial l/\partial \mathbf{V}$, consider separately each term on the right-hand side of (4.2.4). From (A.9.3),

$$\frac{\partial}{\partial v_{ij}} \log |\mathbf{V}| = \begin{cases} 2V_{ij}/|\mathbf{V}|, & i \neq j, \\ V_{ii}/|\mathbf{V}|, & i = j, \end{cases}$$

where V_{ij} is the ijth cofactor of $\mathbf{V}$. Since $\mathbf{V}$ is symmetric, the matrix with elements $V_{ij}/|\mathbf{V}|$ equals $\mathbf{V}^{-1} = \mathbf{\Sigma}$. Thus

$$\frac{\partial \log |\mathbf{V}|}{\partial \mathbf{V}} = 2\mathbf{\Sigma} - \text{Diag}\,\mathbf{\Sigma}.$$

From (A.9.4),

$$\frac{\partial \operatorname{tr} \mathbf{VS}}{\partial \mathbf{V}} = 2\mathbf{S} - \text{Diag}\,\mathbf{S}$$

and

$$\frac{\partial \operatorname{tr} \mathbf{V}(\bar{\mathbf{x}} - \boldsymbol{\mu})(\bar{\mathbf{x}} - \boldsymbol{\mu})'}{\partial \mathbf{V}} = 2(\bar{\mathbf{x}} - \boldsymbol{\mu})(\bar{\mathbf{x}} - \boldsymbol{\mu})' - \text{Diag}\,(\bar{\mathbf{x}} - \boldsymbol{\mu})(\bar{\mathbf{x}} - \boldsymbol{\mu})'.$$

Combining these equations we see that

$$\frac{\partial l}{\partial \mathbf{V}} = \frac{n}{2}(2\mathbf{M} - \text{Diag}\,\mathbf{M}), \tag{4.2.6}$$

where $\mathbf{M} = \mathbf{\Sigma} - \mathbf{S} - (\bar{\mathbf{x}} - \boldsymbol{\mu})(\bar{\mathbf{x}} - \boldsymbol{\mu})'$.

To find the m.l.e.s of $\boldsymbol{\mu}$ and $\mathbf{\Sigma}$ we must solve

$$\frac{\partial l}{\partial \boldsymbol{\mu}} = \mathbf{0} \quad \text{and} \quad \frac{\partial l}{\partial \mathbf{V}} = \mathbf{0}.$$

From (4.2.5) we see that $\mathbf{V}(\bar{\mathbf{x}} - \boldsymbol{\mu}) = \mathbf{0}$. Hence the m.l.e. of $\boldsymbol{\mu}$ is $\hat{\boldsymbol{\mu}} = \bar{\mathbf{x}}$ and from (4.2.6) $\partial l/\partial \mathbf{V} = \mathbf{0}$ gives $2\mathbf{M} - \text{Diag}\,\mathbf{M} = \mathbf{0}$. This implies $\mathbf{M} = \mathbf{0}$, i.e. the m.l.e. of $\mathbf{\Sigma}$ is given by

$$\hat{\mathbf{\Sigma}} = \mathbf{S} + (\bar{\mathbf{x}} - \boldsymbol{\mu})(\bar{\mathbf{x}} - \boldsymbol{\mu})'. \tag{4.2.7}$$

Since $\hat{\boldsymbol{\mu}} = \bar{\mathbf{x}}$, (4.2.7) gives $\hat{\mathbf{\Sigma}} = \mathbf{S}$.

Strictly speaking, the above argument only tells us that $\bar{\mathbf{x}}$ and $\mathbf{S}$ give a stationary point of the likelihood. In order to show that $\bar{\mathbf{x}}$ and $\mathbf{S}$ give the overall maximum value, consider the following theorem.

Theorem 4.2.1 *For any fixed* $(p \times p)$ *matrix* $\mathbf{A} > 0$,

$$f(\mathbf{\Sigma}) = |\mathbf{\Sigma}|^{-n/2} \exp\left(-\tfrac{1}{2} \operatorname{tr} \mathbf{\Sigma}^{-1}\mathbf{A}\right)$$

is maximized over $\mathbf{\Sigma} > 0$ *by* $\mathbf{\Sigma} = n^{-1}\mathbf{A}$, *and* $f(n^{-1}\mathbf{A}) = |n^{-1}\mathbf{A}|^{-n/2} e^{-np/2}$.

Proof It is easily seen that $f(n^{-1}\mathbf{A})$ takes the form given above. Then we can write

$$\log f(n^{-1}\mathbf{A}) - \log f(\mathbf{\Sigma}) = \tfrac{1}{2}np(a - 1 - \log g),$$

where $a = \operatorname{tr} \mathbf{\Sigma}^{-1}\mathbf{A}/np$ and $g = |n^{-1}\mathbf{\Sigma}^{-1}\mathbf{A}|^{1/p}$ are the arithmetic and geometric means of the eigenvalues of $n^{-1}\mathbf{\Sigma}^{-1}\mathbf{A}$. Note that all of these eigenvalues are positive from Corollary A.7.3.1. From Exercise 4.2.3, $a - 1 - \log g \geq 0$, and hence $f(n^{-1}\mathbf{A}) \geq f(\mathbf{\Sigma})$ for all $\mathbf{\Sigma} > 0$. ■

If we maximize the likelihood (4.1.9) over $\mathbf{\Sigma}$ for fixed $\boldsymbol{\mu}$, we find from Theorem 4.2.1 that $\hat{\mathbf{\Sigma}}$ is given by (4.2.7). Then since from (A.2.3m)

$$\begin{aligned} l(\mathbf{X}; \boldsymbol{\mu}, \mathbf{S} + (\bar{\mathbf{x}} - \boldsymbol{\mu})(\bar{\mathbf{x}} - \boldsymbol{\mu})') &= -\frac{n}{2} |\mathbf{S} + (\bar{\mathbf{x}} - \boldsymbol{\mu})(\bar{\mathbf{x}} - \boldsymbol{\mu})'| \\ &= -\frac{n}{2} |\mathbf{S}|\{1 + (\bar{\mathbf{x}} - \boldsymbol{\mu})'\mathbf{S}^{-1}(\bar{\mathbf{x}} - \boldsymbol{\mu})\} \\ &\leq -\frac{n}{2} |\mathbf{S}|, \end{aligned}$$

we see that the log likelihood is maximized by $\hat{\boldsymbol{\mu}} = \bar{\mathbf{x}}$. Alternatively, we could maximize over $\boldsymbol{\mu}$ first (Exercise 4.2.11). (In later examples we shall in general leave it to the reader to verify that the stationary points obtained are in fact overall maxima.)

Note that the m.l.e. of $\boldsymbol{\mu}$ could have been deduced from (4.1.9) directly, whether $\mathbf{\Sigma}$ is known or not. Since $\mathbf{\Sigma}^{-1} > 0$, $-(\bar{\mathbf{x}} - \boldsymbol{\mu})'\mathbf{\Sigma}^{-1}(\bar{\mathbf{x}} - \boldsymbol{\mu}) \leq 0$ and is maximized when $\boldsymbol{\mu} = \bar{\mathbf{x}}$. Thus $l(\mathbf{X}; \boldsymbol{\mu}, \mathbf{\Sigma})$ in (4.1.9) is maximized at $\boldsymbol{\mu} = \bar{\mathbf{x}}$ whether $\mathbf{\Sigma}$ is constrained or not. Consequently, $\hat{\boldsymbol{\mu}} = \bar{\mathbf{x}}$ when $\mathbf{\Sigma}$ is known. However, in cases where $\boldsymbol{\mu}$ is constrained the m.l.e. of $\mathbf{\Sigma}$ will be affected, e.g. when $\boldsymbol{\mu}$ is known *a priori*, the m.l.e. of $\mathbf{\Sigma}$ is found by solving $\partial l/\partial \mathbf{V} = \mathbf{0}$ and is given by (4.2.7).

In the case where $n = 1$ and $\mathbf{\Sigma} = \mathbf{I}$ is known, Stein (1956) showed that when $p \geq 3$, m.l.e. $\hat{\boldsymbol{\mu}} = \bar{\mathbf{x}} = \mathbf{x}$ is an *inadmissible* estimator of $\boldsymbol{\mu}$ under quadratic loss. For further discussion of this remarkable fact, see Cox and Hinkley (1974, p. 447).

4.2.2.2 Constraints on the mean vector $\boldsymbol{\mu}$

Consider the case where $\boldsymbol{\mu}$ is known to be proportional to a known vector, so $\boldsymbol{\mu} = k\boldsymbol{\mu}_0$. For example, the elements of $\mathbf{x}$ could represent a sample of repeated measurements, in which case $\boldsymbol{\mu} = k\mathbf{1}$. With $\mathbf{\Sigma}$ also known (4.1.9) becomes

$$l(\mathbf{X}; k) = -\frac{n}{2}\log|2\pi\mathbf{\Sigma}| - \frac{n}{2}\operatorname{tr}\mathbf{\Sigma}^{-1}\mathbf{S} - \frac{n}{2}(\bar{\mathbf{x}} - k\boldsymbol{\mu}_0)'\mathbf{\Sigma}^{-1}(\bar{\mathbf{x}} - k\boldsymbol{\mu}_0).$$

To find the m.l.e. of k, we solve $\partial l/\partial k = 0$. This gives

$$n\boldsymbol{\mu}_0'\mathbf{\Sigma}^{-1}(\bar{\mathbf{x}} - k\boldsymbol{\mu}_0) = 0, \tag{4.2.8}$$

i.e. the m.l.e. $\hat{k}$ of k is

$$\hat{k} = \boldsymbol{\mu}_0'\mathbf{\Sigma}^{-1}\bar{\mathbf{x}}/\boldsymbol{\mu}_0'\mathbf{\Sigma}^{-1}\boldsymbol{\mu}_0. \tag{4.2.9}$$

Then $\hat{k}$ is unbiased with variance $(n\boldsymbol{\mu}_0'\mathbf{\Sigma}^{-1}\boldsymbol{\mu}_0)^{-1}$; see Exercise 4.2.1.

When $\mathbf{\Sigma}$ is unknown the two equations to solve for $\hat{\mathbf{\Sigma}}$ and $\hat{k}$ are (4.2.7) and (4.2.9). Pre- and post-multiplying (4.2.7) by $\hat{\mathbf{\Sigma}}^{-1}$ and $\mathbf{S}^{-1}$, respectively, gives

$$\mathbf{S}^{-1} = \hat{\mathbf{\Sigma}}^{-1} + \hat{\mathbf{\Sigma}}^{-1}(\bar{\mathbf{x}} - \boldsymbol{\mu})(\bar{\mathbf{x}} - \boldsymbol{\mu})'\mathbf{S}^{-1}. \tag{4.2.10}$$

Pre-multiplying (4.2.10) by $\boldsymbol{\mu}_0'$ and using (4.2.8) gives

$$\boldsymbol{\mu}_0'\mathbf{S}^{-1} = \boldsymbol{\mu}_0'\hat{\mathbf{\Sigma}}^{-1}.$$

Thus, from (4.2.9),

$$\hat{k} = \boldsymbol{\mu}_0'\mathbf{S}^{-1}\bar{\mathbf{x}}/\boldsymbol{\mu}_0'\mathbf{S}^{-1}\boldsymbol{\mu}_0. \tag{4.2.11}$$

A further type of constraint is $\mathbf{R}\boldsymbol{\mu} = \mathbf{r}$, where $\mathbf{R}$ and $\mathbf{r}$ are pre-specified. Maximizing the log likelihood subject to this constraint may be achieved by augmenting the log likelihood with a Lagrangian expression; thus we maximize

$$l^+ = l - n\boldsymbol{\lambda}'(\mathbf{R}\boldsymbol{\mu} - \mathbf{r}),$$

where $\boldsymbol{\lambda}$ is a vector of Lagrangian multipliers and l is as given in (4.1.9). With $\mathbf{\Sigma}$ assumed known, to find the m.l.e. of $\boldsymbol{\mu}$ we are required to find a $\boldsymbol{\lambda}$ for which the solution to

$$\frac{\partial l^+}{\partial \boldsymbol{\mu}} = \mathbf{0} \tag{4.2.12}$$

satisfies the constraint $\mathbf{R}\boldsymbol{\mu} = \mathbf{r}$. From (4.2.5)

$$\frac{\partial l^+}{\partial \boldsymbol{\mu}} = n\mathbf{\Sigma}^{-1}(\bar{\mathbf{x}} - \boldsymbol{\mu}) - n\mathbf{R}'\boldsymbol{\lambda}.$$

Thus

$$\bar{\mathbf{x}} - \boldsymbol{\mu} = \mathbf{\Sigma}\mathbf{R}'\boldsymbol{\lambda}.$$

Pre-multiplying by $\mathbf{R}$ gives $\mathbf{R}\bar{\mathbf{x}} - \mathbf{r} = (\mathbf{R}\mathbf{\Sigma}\mathbf{R}')\boldsymbol{\lambda}$ if the constraint is to be satisfied. Thus, we take $\boldsymbol{\lambda}=(\mathbf{R\Sigma R})^{-1}(\mathbf{R}\bar{\mathbf{x}}-\mathbf{r})$, so

$$\hat{\boldsymbol{\mu}}=\bar{\mathbf{x}}-\boldsymbol{\Sigma}\mathbf{R}'\boldsymbol{\lambda}=\bar{\mathbf{x}}-\boldsymbol{\Sigma}\mathbf{R}'(\mathbf{R}\boldsymbol{\Sigma}\mathbf{R}')^{-1}(\mathbf{R}\bar{\mathbf{x}}-\mathbf{r}). \tag{4.2.13}$$

When $\boldsymbol{\Sigma}$ is unknown the m.l.e. of $\boldsymbol{\mu}$ becomes

$$\hat{\boldsymbol{\mu}}=\bar{\mathbf{x}}-\mathbf{S}\mathbf{R}'(\mathbf{R}\mathbf{S}\mathbf{R}')^{-1}(\mathbf{R}\bar{\mathbf{x}}-\mathbf{r}). \tag{4.2.14}$$

See Exercise 4.2.8.

4.2.2.3 Constraints on $\boldsymbol{\Sigma}$

First we consider the case where $\boldsymbol{\Sigma}=k\boldsymbol{\Sigma}_0$, where $\boldsymbol{\Sigma}_0$ is known. From (4.1.9),

$$2n^{-1}l(\mathbf{X};\boldsymbol{\mu},k)=-p\log k-\log|2\pi\boldsymbol{\Sigma}_0|-k^{-1}\alpha \tag{4.2.15}$$

where $\alpha=\operatorname{tr}\boldsymbol{\Sigma}_0^{-1}\mathbf{S}+(\bar{\mathbf{x}}-\boldsymbol{\mu})'\boldsymbol{\Sigma}_0^{-1}(\bar{\mathbf{x}}-\boldsymbol{\mu})$ is independent of k. If $\boldsymbol{\mu}$ is known, then to obtain the m.l.e. of k we solve $\partial l/\partial k=0$. This gives

$$-p/k+\alpha/k^2=0.$$

Thus the m.l.e. of k is $\hat{k}=\alpha/p$. If $\boldsymbol{\mu}$ is unconstrained, then we solve $\partial l/\partial\boldsymbol{\mu}=\mathbf{0}$ and $\partial l/\partial k=0$. These give $k=\alpha/p$ and $\hat{\boldsymbol{\mu}}=\bar{\mathbf{x}}$. Together these give

$$\hat{k}=\operatorname{tr}\boldsymbol{\Sigma}_0^{-1}\mathbf{S}/p. \tag{4.2.16}$$

The constraint $\boldsymbol{\Sigma}_{12}=\mathbf{0}$, where $\boldsymbol{\Sigma}_{12}$ is an off-diagonal submatrix of $\boldsymbol{\Sigma}$, implies that the two corresponding groups of variables are independent. The m.l.e. of $\boldsymbol{\Sigma}$ can be found by considering each subgroup separately and is

$$\hat{\boldsymbol{\Sigma}}=\begin{bmatrix}\mathbf{S}_{11} & \mathbf{0}\\ \mathbf{0} & \mathbf{S}_{22}\end{bmatrix}.$$

Example 4.2.3 For 47 female cats the body weight (kgs) and heart weight (gms) were recorded; see Fisher (1947) and Exercise 1.4.3. The sample mean vector and covariance matrix are

$$\bar{\mathbf{x}}=(2.36, 9.20)', \qquad \mathbf{S}=\begin{bmatrix}0.0735 & 0.1937\\ 0.1937 & 1.8040\end{bmatrix}.$$

Thus $\bar{\mathbf{x}}$ and $\mathbf{S}$ are the unconstrained m.l.e's for $\boldsymbol{\mu}$ and $\boldsymbol{\Sigma}$. However, if from other information we know that $\boldsymbol{\mu}=(2.5, 10)'$, then $\hat{\boldsymbol{\Sigma}}$ is given by (4.2.7),

$$\bar{\mathbf{x}}-\boldsymbol{\mu}=(-0.14, -0.80)',$$

$$\hat{\boldsymbol{\Sigma}}=\mathbf{S}+(\bar{\mathbf{x}}-\boldsymbol{\mu})(\bar{\mathbf{x}}-\boldsymbol{\mu})'=\begin{bmatrix}0.0931 & 0.3057\\ 0.3057 & 2.4440\end{bmatrix}.$$

If instead we assume $\boldsymbol{\mu} = k(2.5, 10)'$, then from (4.2.11),

$$\hat{k} = (2.5, 10)\mathbf{S}^{-1}(2.36, 9.20)'/(2.5, 10)\mathbf{S}^{-1}(2.5, 10)'.$$

This gives $\hat{k} = 0.937$, so the m.l.e. of $\boldsymbol{\mu}$ is $(2.34, 9.37)'$ and the m.l.e. for $\boldsymbol{\Sigma}$ is now $\hat{\boldsymbol{\Sigma}} = \mathbf{S} + (0.02, -0.17)'(0.02, -0.17)$ which is quite close to $\mathbf{S}$.

If the covariance matrix was known to be proportional to

$$\boldsymbol{\Sigma}_0 = \begin{bmatrix} 0.1 & 0.2 \\ 0.2 & 2.0 \end{bmatrix}$$

with $\boldsymbol{\mu}$ unconstrained, then from (4.2.16), $\hat{k} = \operatorname{tr} \boldsymbol{\Sigma}_0^{-1}\mathbf{S}/p = 0.781$. Hence

$$\hat{\boldsymbol{\Sigma}} = \begin{bmatrix} 0.078 & 0.156 \\ 0.156 & 1.562 \end{bmatrix}.$$

4.2.2.4 *Samples with linked parameters*

We turn now to consider situations with several normal samples, where we know *a priori* that some relationship exists between their parameters. For instance we may have independent data matrices $\mathbf{X}_1, \ldots, \mathbf{X}_k$, where the rows of $\mathbf{X}_i(n_i \times p)$ are i.i.d. $N_p(\boldsymbol{\mu}_i, \boldsymbol{\Sigma}_i)$, $i = 1, \ldots, k$. The most common constraints are

(a) $\boldsymbol{\Sigma}_1 = \ldots = \boldsymbol{\Sigma}_k$ or
(b) $\boldsymbol{\Sigma}_1 = \ldots = \boldsymbol{\Sigma}_k$ and $\boldsymbol{\mu}_1 = \ldots = \boldsymbol{\mu}_k$.

Of course if (b) holds we can treat all the data matrices as constituting one sample from a single population.

To calculate the m.l.e.s if (a) holds, note that, from (4.1.9), the log likelihood function is given by

$$l = -\tfrac{1}{2} \sum_i [n_i \log |2\pi\boldsymbol{\Sigma}| + n_i \operatorname{tr} \boldsymbol{\Sigma}^{-1}(\mathbf{S}_i + \mathbf{d}_i\mathbf{d}_i')], \qquad (4.2.17)$$

where $\mathbf{S}_i$ is the covariance matrix of the ith sample, $i = 1, \ldots, k$, and $\mathbf{d}_i = \bar{\mathbf{x}}_i - \boldsymbol{\mu}_i$. Since there is no restriction on the population means, the m.l.e. of $\boldsymbol{\mu}_i$ is $\bar{\mathbf{x}}_i$ and setting $n = \sum n_i$, (4.2.17) becomes

$$l = -\tfrac{1}{2}n \log |2\pi\boldsymbol{\Sigma}| - \tfrac{1}{2} \operatorname{tr} \boldsymbol{\Sigma}^{-1}\mathbf{W}, \quad \text{where} \quad \mathbf{W} = \sum_i n_i\mathbf{S}_i. \qquad (4.2.18)$$

Differentiating (4.2.18) with respect to $\boldsymbol{\Sigma}$ and equating to zero gives $\boldsymbol{\Sigma} = n^{-1}\mathbf{W}$. Therefore $\hat{\boldsymbol{\Sigma}} = n^{-1}\mathbf{W}$ is the m.l.e. of $\boldsymbol{\Sigma}$ under the conditions stated.

4.3 Other Techniques and Concepts

4.3.1 Bayesian inference

Until now the vector $\boldsymbol{\theta}$ has been regarded as fixed but unknown. However, a Bayesian approach would regard $\boldsymbol{\theta}$ as a random variable whose distribution reflects subjective beliefs in what the value of $\boldsymbol{\theta}$ is likely to be. A fairly complete exposition of the multivariate aspects of Bayesian inference appears in Chapter 8 of Box and Tiao (1973).

Suppose that before observing the vector $\mathbf{x}$, our beliefs about the value of $\boldsymbol{\theta}$ could be represented by a prior distribution $\pi(\boldsymbol{\theta})$. For instance, if θ is a scalar which is believed to lie between 1 and 5 with a 95% probability, then we may take the prior distribution to be normal with mean 3 and variance 1, since about 95% of this distribution falls between 1 and 5. This would lead to

$$\pi(\theta)=\frac{1}{\sqrt{2\pi}}\exp\{-\tfrac{1}{2}(\theta-3)^2\}.$$

Alternatively, if our beliefs were not symmetric about 3, some other prior distribution might be more suitable. For instance if θ is non-negative, then a gamma distribution could be used.

The prior density, given by $\pi(\boldsymbol{\theta})$, and the likelihood of the observed data $f(\mathbf{x};\boldsymbol{\theta})$ together define the so-called posterior distribution, which is the conditional distribution of $\boldsymbol{\theta}$ given $\mathbf{x}$. By Bayes' theorem, the posterior density is given by

$$\pi(\boldsymbol{\theta}\mid\mathbf{x})=\{\pi(\boldsymbol{\theta})f(\mathbf{x};\boldsymbol{\theta})\}\bigg/\int\pi(\boldsymbol{\theta})f(\mathbf{x};\boldsymbol{\theta})\,d\boldsymbol{\theta}\propto\pi(\boldsymbol{\theta})f(\mathbf{x};\boldsymbol{\theta}). \qquad (4.3.1)$$

All Bayesian inference is based upon the posterior probability function. In particular, the Bayesian estimate of $\boldsymbol{\theta}$ is given by the mean of the posterior density $\pi(\boldsymbol{\theta}\mid\mathbf{x})$.

Sometimes it is convenient to use an improper prior density ($\int\pi(\boldsymbol{\theta})\,d\boldsymbol{\theta}=\infty$). Such priors are allowable as long as $\int\pi(\boldsymbol{\theta})f(\mathbf{x};\boldsymbol{\theta})\,d\boldsymbol{\theta}<\infty$, so that the posterior density is defined.

Example 4.3.1 Suppose that $\mathbf{X}=(\mathbf{x}_1,\ldots,\mathbf{x}_n)'$ is a random sample from $N_p(\boldsymbol{\mu},\boldsymbol{\Sigma})$ and that there is no prior knowledge concerning $\boldsymbol{\mu}$ and $\boldsymbol{\Sigma}$. Then a natural choice of prior distribution is the non-informative (or vague) prior based on Jeffreys' principle of invariance (Jeffreys, 1961, p. 179). This principle assumes first that the location parameter $\boldsymbol{\mu}$ and scale parameter $\boldsymbol{\Sigma}$ are independent, so $\pi(\boldsymbol{\mu},\boldsymbol{\Sigma})=\pi(\boldsymbol{\mu})\pi(\boldsymbol{\Sigma})$, and second that

$$\pi(\boldsymbol{\mu})\propto|F(\boldsymbol{\mu})|^{1/2},\qquad \pi(\boldsymbol{\Sigma})\propto|F(\boldsymbol{\Sigma})|^{1/2},$$

where $F(\boldsymbol{\mu})$ and $F(\boldsymbol{\Sigma})$ are the information matrices for $\boldsymbol{\mu}$ and $\boldsymbol{\Sigma}$. Hence, using Exercise 4.1.12, we see that

$$\pi(\boldsymbol{\mu}, \boldsymbol{\Sigma}) \propto |\boldsymbol{\Sigma}|^{-(p+1)/2}. \tag{4.3.2}$$

The marginal posterior of $\boldsymbol{\mu}$ can now be obtained by substituting for $\pi(\boldsymbol{\mu}, \boldsymbol{\Sigma})$ in (4.3.1) and integrating $\pi(\boldsymbol{\mu}, \boldsymbol{\Sigma} \mid \mathbf{X})$ with respect to $\boldsymbol{\Sigma}$, to give

$$\pi(\boldsymbol{\mu} \mid \mathbf{X}) \propto \int |\boldsymbol{\Sigma}|^{-(n+p+1)/2} \exp\left(-\tfrac{1}{2} \operatorname{tr} \boldsymbol{\Sigma}^{-1}\mathbf{V}\right) d\boldsymbol{\Sigma}, \tag{4.3.3}$$

where $\mathbf{V} = n\mathbf{S} + n(\bar{\mathbf{x}} - \boldsymbol{\mu})(\bar{\mathbf{x}} - \boldsymbol{\mu})'$.

The function to be integrated here is similar to the p.d.f. of the inverted Wishart distribution given in (3.8.2). Using the normalizing constant of that expression, it is clear from (4.3.3) that

$$\pi(\boldsymbol{\mu} \mid \mathbf{X}) \propto |\mathbf{V}|^{-n/2}.$$

But

$$|\mathbf{V}| = n^p |\mathbf{S}|[1 + (\bar{\mathbf{x}} - \boldsymbol{\mu})'\mathbf{S}^{-1}(\bar{\mathbf{x}} - \boldsymbol{\mu})].$$

Therefore

$$\pi(\boldsymbol{\mu} \mid \mathbf{X}) \propto [1 + (\bar{\mathbf{x}} - \boldsymbol{\mu})'\mathbf{S}^{-1}(\bar{\mathbf{x}} - \boldsymbol{\mu})]^{-n/2}.$$

Thus the posterior distribution of $\boldsymbol{\mu}$ is a multivariate t distribution with $(n-p)$ degrees of freedom and parameters $\bar{\mathbf{x}}$ and $\mathbf{S}$ (see Exercise 2.6.5). The mean of this posterior distribution is the sample mean $\bar{\mathbf{x}}$, so in this case the Bayes' estimate of $\boldsymbol{\mu}$ is the same as the maximum likelihood estimate.

The posterior distribution of $\boldsymbol{\Sigma}$ is obtained by integrating $\pi(\boldsymbol{\mu}, \boldsymbol{\Sigma} \mid \mathbf{X})$ with respect to $\boldsymbol{\mu}$. This leads to an inverted Wishart posterior distribution, with a mean of $n\mathbf{S}/(n-p-2)$ (see Exercises 3.4.13 and 4.3.1). Note that in this case the Bayes' estimate differs from the maximum likelihood estimate.

Example 4.3.2 When one wishes to incorporate *informative* prior information into the analysis then a useful prior distribution is the "natural conjugate prior" (see, for example, Press, 1972, p. 76). For example, in the case of a random sample from the $N_p(\boldsymbol{\mu}, \boldsymbol{\Sigma})$ distribution, the density of $\bar{\mathbf{x}}$ and $\mathbf{S}$ from (3.4.3) and (3.8.1) is

$$f(\bar{\mathbf{x}}, \mathbf{S}; \boldsymbol{\mu}, \boldsymbol{\Sigma}) \propto \{|\boldsymbol{\Sigma}|^{-1/2} \exp[-\tfrac{1}{2}n(\bar{\mathbf{x}} - \boldsymbol{\mu})'\boldsymbol{\Sigma}^{-1}(\bar{\mathbf{x}} - \boldsymbol{\mu})]\} \times \{|\boldsymbol{\Sigma}|^{-(n-1)/2} \exp(-\tfrac{1}{2}n \operatorname{tr} \boldsymbol{\Sigma}^{-1}\mathbf{S})\}. \tag{4.3.4}$$

Then the conjugate prior for $\boldsymbol{\mu}$ and $\boldsymbol{\Sigma}$ is obtained by thinking of (4.3.4) as a density in $\boldsymbol{\mu}$ and $\boldsymbol{\Sigma}$ after interpreting the quantities $\mathbf{x}$, $n\mathbf{S}$, and $n-1$ as parameters of the prior distribution $\boldsymbol{\phi}$, $\mathbf{G} > 0$, and $m > 2p - 1$, and adjusting the proportionality constant so that the density integrates to 1. Thus,

$$\pi(\boldsymbol{\mu}, \boldsymbol{\Sigma}) \propto \{|\boldsymbol{\Sigma}|^{-m/2} \exp(-\tfrac{1}{2} \operatorname{tr} \boldsymbol{\Sigma}^{-1}\mathbf{G})\}\{|\boldsymbol{\Sigma}|^{-1/2} \exp[-\tfrac{1}{2}(\boldsymbol{\mu} - \boldsymbol{\phi})'\boldsymbol{\Sigma}^{-1}(\boldsymbol{\mu} - \boldsymbol{\phi})]\}. \tag{4.3.5}$$

The Bayes' estimates of $\boldsymbol{\mu}$ and $\boldsymbol{\Sigma}$ are, respectively (see Exercise 4.3.2),

$$(\boldsymbol{\phi}+n\bar{\mathbf{x}})/(1+n), \qquad \left\{n\mathbf{S}+\mathbf{G}+\frac{n}{1+n}(\bar{\mathbf{x}}-\boldsymbol{\phi})(\bar{\mathbf{x}}-\boldsymbol{\phi})'\right\}\Big/(n+m-2p-2). \tag{4.3.6}$$

Neither of these estimates is the same as the corresponding maximum likelihood estimate, although they do indicate that the Bayesian procedure leads to a weighted average of prior estimates and sample-based estimates.

4.3.2 Robust estimation of location and dispersion for multivariate distributions

Estimators of location

One possible way to estimate the location of a multivariate sample is to treat each variable separately. For example, one can use the α-trimmed mean for each variable (i.e. the mean after omitting a proportion α of the smallest and a proportion α of the largest observations on each variable). An alternative estimator based on each variable separately is the M-estimator (maximum likelihood type estimator). For a univariate sample $x_1, \ldots, x_n$ it is defined implicitly by

$$\sum_{i=1}^{n} \psi(x_i - T_n) = 0,$$

where ψ is a given function. If the sample comes from a symmetric distribution F, and if one takes $\psi(x) = \max[-k, \min(x, k)]$ where $F(-k) = \alpha$, then this estimator has the same asymptotic behaviour as the α-trimmed mean (Huber, 1972). (Note that if $\psi(x) = x$, then $T_n = \bar{x}$).

An inherently multivariate technique to estimate the mean vector has been given by Gentleman (1965) (also see, Barnard, 1976). For fixed k, $1 \leq k \leq 2$, he proposed an estimator $\mathbf{x}_k^*$ which minimizes

$$\sum_{i=1}^{n} \|\mathbf{x}_i - \mathbf{x}_k^*\|^k,$$

where $\|\cdot\|$ denotes the usual Euclidean norm. He recommends $k = \frac{3}{2}$ in practice to minimize the effect of outliers and to retain as much precision as possible. Note that one gets the usual mean vector when $k = 2$.

Tukey has proposed a multivariate analogue to trimming called "peeling" (see Barnett, 1976). This consists of deleting extreme points of the convex hull of the sample and either repeating this a fixed number of times or until a fixed percentage of the points has been removed. Very little is known about the behaviour of this procedure.

Example 4.3.3 We illustrate the above techniques on the iris data of Table 1.2.2 for *Iris versicolour* with x_1 = petal length and x_2 = sepal length. Figure 4.3.1 shows the set of convex hulls for this data. The mean vector after "peeling once", i.e. excluding the extreme points (nine in all) of the convex hull, is $\bar{\mathbf{x}}_p^* = (4.26, 5.91)'$. This is not far distant from the sample mean vector $\bar{\mathbf{x}} = (4.26, 5.94)'$. The excluded points are listed in Table 4.3.1.

The α-trimmed values for $\alpha = 0.04$ are also shown in Fig. 4.3.1, i.e. the largest two and smallest two extreme values are trimmed for each variable. The trimmed mean vector is $\bar{\mathbf{x}}_{(\alpha)}^* = (4.27, 5.93)'$. This is very close to the untrimmed mean $\bar{\mathbf{x}}$. Hence on this data the values of the three estimates are similar. However, the properties of the estimators are vastly different, especially on samples which are contaminated in some way.

Estimates of dispersion

There are various univariate techniques for producing robust estimates of variance which can be extended to covariance, but in general a covariance matrix produced in this way is not necessarily positive definite.

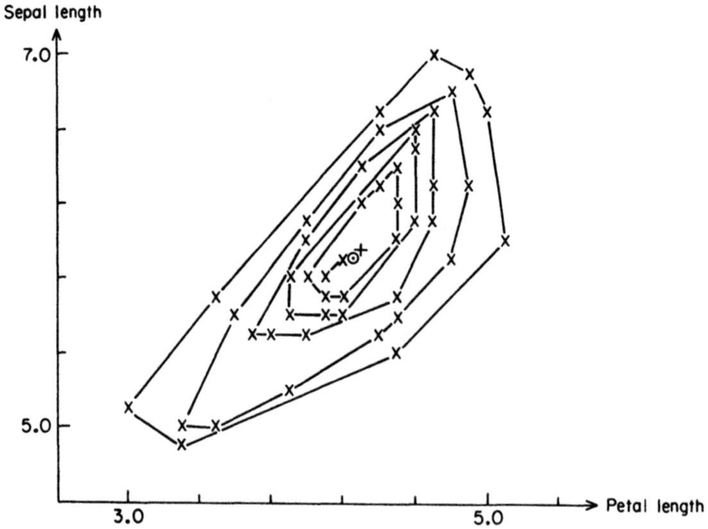

Figure 4.3.1 Convex hull for the iris data (I. versicolour variety) with ⊙ = mean after "peeling" once, + = α-trimmed mean (α = 0.04).

Table 4.3.1 Extreme values to be excluded from robust estimates in the iris data of Table 1.2.2

Values excluded after peeling once (x_1, x_2)	α-trimming, $\alpha = 0.04$
(4.7, 7.0), (4.9, 6.9), (3.3, 4.9), (4.4, 6.7) (5.0, 6.7), (3.5, 5.7), (5.1, 6.0), (4.5, 5.4), (3.0, 5.1)	x_1: 3.3, 5.0, 5.1, 3.0 x_2: 7.0, 6.9, 4.9, 5.0

The following procedure looks promising (see Wilk *et al.*, 1962).

(1) Rank the observations $\mathbf{x}_i (i = 1, \ldots, n)$ in terms of Euclidean distance $\|\mathbf{x}_i - \mathbf{x}^*\|$, where $\mathbf{x}^*$ is a robust estimator of location.

(2) Take a subset R whose ranks are the smallest $100(1-\alpha)\%$ and compute the sum of squares and products matrix:

$$\mathbf{A}_0 = \sum_{\mathbf{x}_r \in R} (\mathbf{x}_r - \mathbf{x}^*)(\mathbf{x}_r - \mathbf{x}^*)',$$

where α should be small to ensure that $\mathbf{A}_0$ is non-singular.

(3) Rank all n observations in terms of $(\mathbf{x}_i - \mathbf{x}^*)'\mathbf{A}_0^{-1}(\mathbf{x}_i - \mathbf{x}^*)$.

(4) A subset T of the n observations whose ranks are the smallest $100(1-\beta)\%$ are chosen and the robust estimator of the dispersion matrix is

$$\mathbf{S}^* = \frac{k}{n(1-\beta)} \sum_{\mathbf{x}_r \in T} (\mathbf{x}_r - \mathbf{x}^*)(\mathbf{x}_r - \mathbf{x}^*)'.$$

Here the value of k is chosen to make $\mathbf{S}^*$ sufficiently unbiased and β is a small positive number. If β is small and $n(1-\beta) \geq p$, then $\mathbf{S}^*$ is non-singular with probability one. More work needs to be done, however, to recommend k, α, and β in practice.

Gnanadesikan and Kettenring (1972) and Huber (1972) have made a comparison of the many robust estimators available, but the subject is still in a developing state.

Exercises and Complements

4.1.1 Let $\mathbf{x}$ come from the general exponential family with p.d.f.

$$f(\mathbf{x}; \boldsymbol{\theta}) = \exp\left\{a_0(\boldsymbol{\theta}) + b_0(\mathbf{x}) + \sum_{i=1}^{q} a_i(\boldsymbol{\theta}) b_i(\mathbf{x})\right\}.$$

(a) Show that the score function is

$$\mathbf{s}=\frac{\partial a_0}{\partial \boldsymbol{\theta}}+\sum_{i=1}^{q}\frac{\partial a_i}{\partial \boldsymbol{\theta}}b_i(\mathbf{x})=\frac{\partial a_0}{\partial \boldsymbol{\theta}}+\mathbf{Ab},$$

where $\mathbf{A}$ is the matrix whose ith column is $\partial a_i/\partial \boldsymbol{\theta}$, and $\mathbf{b}'$ is the vector $(b_1(\mathbf{x}),\dots, b_q(\mathbf{x}))$.

(b) Note that $\partial a_0/\partial \boldsymbol{\theta}$ and $\mathbf{A}$ are both free of $\mathbf{x}$. Therefore, since the score has mean zero, prove that

$$\mathbf{A}E[\mathbf{b}]=-\frac{\partial a_0}{\partial \boldsymbol{\theta}}.$$

If $\mathbf{A}$ is a non-singular matrix, then

$$E[\mathbf{b}]=-\mathbf{A}^{-1}\frac{\partial a_0}{\partial \boldsymbol{\theta}}.$$

(c) Comparing (a) and (b), prove that the value of $\boldsymbol{\theta}$ for which $\mathbf{s}$ is zero (i.e. the maximum likelihood estimator) is also the value which equates $\mathbf{b}$ to its expected value.

(d) Show that the information matrix is given by

$$\mathbf{F}=-E\left(\frac{\partial \mathbf{s}'}{\partial \boldsymbol{\theta}}\right)=-\frac{\partial^2 a_0}{\partial \boldsymbol{\theta}\,\partial \boldsymbol{\theta}'}-\sum_{i=1}^{q}\frac{\partial^2 a_i}{\partial \boldsymbol{\theta}\,\partial \boldsymbol{\theta}'}E[b_i(\mathbf{x})].$$

4.1.2 (a) If $\mathbf{x}$ has the multinomial distribution, so that $V(\mathbf{x})=\operatorname{diag}(\mathbf{a})-\mathbf{aa}'$, as shown in Section 2.6.1, then show that

$$V(\boldsymbol{\beta}'\mathbf{x})=\boldsymbol{\beta}'V(\mathbf{x})\boldsymbol{\beta}=\sum \beta_i^2 a_i-\left(\sum \beta_i a_i\right)^2.$$

(b) Note that the score s defined in Example 4.1.4 may be written as $s=\boldsymbol{\beta}'\mathbf{x}$, where $a_1=\frac{1}{4}(2+\theta)$, $a_2=a_3=\frac{1}{4}(1-\theta)$, and $a_4=\theta/4$, and $\beta_1=1/(4a_1)$, $\beta_2=\beta_3=-1/(4a_2)$ and $\beta_4=1/(4a_4)$. Hence deduce that $V(s)=\frac{1}{16}\sum a_i^{-1}$.

4.1.3 Show from (4.1.2) that the score of a set of independent random variables equals the sum of their individual scores.

4.1.4 If $l(\mathbf{X};\boldsymbol{\theta})$ is as defined in (4.1.10) with $\mathbf{t}=\bar{\mathbf{x}}$, verify Corollary 4.1.1.2.

4.1.5 If $\mathbf{x}_1,\dots,\mathbf{x}_n$ are independent variables whose distributions depend upon the parameter $\boldsymbol{\theta}$, and if $\mathbf{F}_i=E(\mathbf{s}_i\mathbf{s}_i')$ is the Fisher information matrix corresponding to $\mathbf{x}_i$, then show that $E(\mathbf{s}_i\mathbf{s}_j')=\mathbf{0}$ and hence that $\mathbf{F}=\mathbf{F}_1+\dots+\mathbf{F}_n$ is the Fisher information matrix corresponding to the matrix $\mathbf{X}=(\mathbf{x}_1,\dots,\mathbf{x}_n)'$.

4.1.6 Obtain the Fisher information corresponding to the genetic experiment of Example 4.1.4 by squaring the score and evaluating $E(s^2)$. Check that the result is equal to that obtained in Example 4.1.4.

4.1.7 Show that if $E(\mathbf{t})=\boldsymbol{\theta}+\mathbf{b}(\boldsymbol{\theta})$, where $\mathbf{t}$ does not depend on $\boldsymbol{\theta}$, then equation (4.1.18) becomes

$$\text{corr}^2(\alpha,\gamma)=\{\mathbf{c}'(\mathbf{I}+\mathbf{B})\mathbf{a}\}^2/(\mathbf{a}'\mathbf{V}\mathbf{a}\mathbf{c}'\mathbf{F}\mathbf{c}), \qquad \mathbf{B}=(\partial b_j(\boldsymbol{\theta})/\partial\theta_i),$$

so

$$\max_{\mathbf{c}} \text{corr}^2(\alpha,\gamma)=\mathbf{a}'(\mathbf{I}+\mathbf{B})'\mathbf{F}^{-1}(\mathbf{I}+\mathbf{B})\mathbf{a}/\mathbf{a}'\mathbf{V}\mathbf{a}.$$

Hence show that the Cramér–Rao lower bound can be generalized.

4.1.8 If $\mathbf{A}\sim W_p(\boldsymbol{\Sigma}, m)$ show that its likelihood is given by

$$L(\mathbf{A};\boldsymbol{\Sigma})=c|\boldsymbol{\Sigma}|^{-m/2}|\mathbf{A}|^{(m-p-1)/2}\exp(-\tfrac{1}{2}\operatorname{tr}\boldsymbol{\Sigma}^{-1}\mathbf{A}),$$

where c includes all the terms which depend only on m and p. Taking $\mathbf{V}=\boldsymbol{\Sigma}^{-1}$ as the parameter as in Section 4.2.2.1, the log likelihood is therefore

$$l(\mathbf{A};\mathbf{V})=\log c+\tfrac{1}{2}m\log|\mathbf{V}|+\tfrac{1}{2}(m-p-1)\log|\mathbf{A}|-\tfrac{1}{2}\operatorname{tr}\mathbf{V}\mathbf{A}.$$

Writing $\mathbf{K}=\tfrac{1}{2}m\boldsymbol{\Sigma}-\tfrac{1}{2}\mathbf{A}$, use the results of Section 4.2.2.1 to show that the score is the matrix

$$\mathbf{U}=\frac{\partial l}{\partial \mathbf{V}}=2\mathbf{K}-\operatorname{Diag}\mathbf{K}.$$

Note that, since $E(\mathbf{A})=m\boldsymbol{\Sigma}$, it is clear that $\mathbf{U}$ has mean zero.

4.1.9 If $\mathbf{x}\sim N_p(\mu\mathbf{a},\sigma^2\mathbf{I})$, where $\mathbf{a}$ is known, then show that $\mathbf{x}$ enters the likelihood function only through the bivariate vector $\mathbf{t}=(\mathbf{a}'\mathbf{x},\mathbf{x}'\mathbf{x})$. Hence show that $\mathbf{t}$ is sufficient for (μ,σ^2). Hence show that when the elements of $\mathbf{x}$ are i.i.d. $N(\mu,\sigma^2)$, then $(\bar{x},\mathbf{x}'\mathbf{x})$ is sufficient for (μ,σ^2).

4.1.10 (Generalization of Example 4.1.4 and Exercise 4.1.2) If $\mathbf{x}=(x_1,\dots,x_p)'$ has a multinomial distribution with parameters $a_1(\boldsymbol{\theta}),\dots,a_p(\boldsymbol{\theta})$, and $N=\sum x_i$,

(a) show that the score function is

$$\mathbf{s}=\sum\frac{x_i}{a_i}\frac{\partial a_i}{\partial\boldsymbol{\theta}};$$

(b) show that the condition $E[\mathbf{s}]=\mathbf{0}$ is equivalent to the condition $\sum(\partial a_i/\partial\boldsymbol{\theta})=\mathbf{0}$, which is implied by the condition $\sum a_i=1$;

(c) show that

$$\frac{\partial\mathbf{s}'}{\partial\boldsymbol{\theta}}=\sum\frac{x_i}{a_i^2}\left(a_i\frac{\partial^2 a_i}{\partial\boldsymbol{\theta}\,\partial\boldsymbol{\theta}'}-\frac{\partial a_i}{\partial\boldsymbol{\theta}}\frac{\partial a_i}{\partial\boldsymbol{\theta}'}\right),$$

and that the information matrix is

$$\mathbf{F} = -E\left(\frac{\partial \mathbf{s}'}{\partial \boldsymbol{\theta}}\right) = N\sum \left(\frac{1}{a_i}\frac{\partial a_i}{\partial \boldsymbol{\theta}}\frac{\partial a_i}{\partial \boldsymbol{\theta}'} - \frac{\partial^2 a_i}{\partial \boldsymbol{\theta}\partial \boldsymbol{\theta}'}\right)$$

—when $\boldsymbol{\theta}$ is a scalar this simplifies to

$$f = N\sum \left(\frac{1}{a_i}\left(\frac{\partial a_i}{\partial \theta}\right)^2 - \frac{\partial^2 a_i}{\partial \theta^2}\right).$$

4.1.11 Suppose that $\mathbf{s}(\mathbf{x}; \boldsymbol{\theta})$ and $\mathbf{F}(\boldsymbol{\theta})$ are the score vector and information matrix with respect to a particular parameter vector $\boldsymbol{\theta}$. Let $\boldsymbol{\phi} = \boldsymbol{\phi}(\boldsymbol{\theta})$, where $\boldsymbol{\phi}(\cdot)$ is a one-to-one differentiable transformation. Then show that the new score vector and information matrix are, respectively,

$$\mathbf{s}(\mathbf{x}; \boldsymbol{\phi}) = \frac{\partial \boldsymbol{\theta}'}{\partial \boldsymbol{\phi}}\mathbf{s}(\mathbf{x}; \boldsymbol{\theta}), \qquad \mathbf{F}(\boldsymbol{\phi}) = \frac{\partial \boldsymbol{\theta}'}{\partial \boldsymbol{\phi}}\mathbf{F}(\boldsymbol{\theta})\frac{\partial \boldsymbol{\theta}}{\partial \boldsymbol{\phi}'}.$$

4.1.12 Let $\mathbf{x}_1, \ldots, \mathbf{x}_n$ be a sample from $N_p(\boldsymbol{\mu}, \boldsymbol{\Sigma})$ with $\boldsymbol{\mu}$ known and $p = 2$. For simplicity take $\mathbf{V} = \boldsymbol{\Sigma}^{-1}$ as the parameter matrix. Put $\mathbf{v}' = (v_{11}, v_{22}, v_{12})$. Then, using (4.2.4), show that

$$\mathbf{F}(\mathbf{V}) = -E\left(\frac{\partial^2 l}{\partial \mathbf{v}\partial \mathbf{v}'}\right)$$

$$= \frac{n}{2(v_{11}v_{22} - v_{12}^2)^2}\begin{pmatrix} v_{22}^2 & v_{12}^2 & -2v_{22}v_{12} \\ v_{12}^2 & v_{11}^2 & -2v_{11}v_{12} \\ -2v_{22}v_{12} & -2v_{11}v_{12} & 2(v_{11}v_{22} + v_{12}^2) \end{pmatrix}.$$

Hence show that

$$|\mathbf{F}(\mathbf{V})| = \tfrac{1}{4}n^3|\mathbf{V}|^{-3} = \tfrac{1}{4}n^3|\boldsymbol{\Sigma}|^3.$$

This result can be generalized to the general case to give

$$|\mathbf{F}(\mathbf{V})| \propto |\mathbf{V}|^{-(p+1)} \propto |\boldsymbol{\Sigma}|^{p+1}$$

(see, for example, Box and Tiao, 1973). Hence, using Exercise 4.1.11 and the Jacobian from Table 2.5.1, show that

$$|\mathbf{F}(\boldsymbol{\Sigma})| \propto |\boldsymbol{\Sigma}|^{-(p+1)}.$$

4.1.13 Show that the information matrix derived in Exercise 4.1.12 for $p = 2$ can also be written

$$F(\mathbf{V}) = \frac{n}{2}\begin{pmatrix} \sigma_{11}^2 & \sigma_{12}^2 & -2\sigma_{11}\sigma_{12} \\ \sigma_{12}^2 & \sigma_{22}^2 & -2\sigma_{22}\sigma_{12} \\ -2\sigma_{11}\sigma_{12} & -2\sigma_{22}\sigma_{12} & 2\ \sigma_{22}\sigma_{11} + \sigma_{12}^2 \end{pmatrix}.$$

4.2.1 If $\boldsymbol{\mu}=k\boldsymbol{\mu}_0$ and $\hat{k}$ is the m.l.e. of k given by (4.2.9), show that $E(\hat{k})=k$ and $V(\hat{k})=(n\boldsymbol{\mu}_0'\boldsymbol{\Sigma}^{-1}\boldsymbol{\mu}_0)^{-1}$, where n is the sample size.

4.2.2 Since $\bar{\mathbf{x}}$ is independent of $\mathbf{S}$ and has expectation $k\boldsymbol{\mu}_0$, show that the estimator $\hat{k}$ given in (4.2.11) is unbiased, and that its variance is

$$n^{-1}E\{\boldsymbol{\mu}_0'\mathbf{S}^{-1}\boldsymbol{\Sigma}\mathbf{S}^{-1}\boldsymbol{\mu}_0/(\boldsymbol{\mu}_0'\mathbf{S}^{-1}\boldsymbol{\mu}_0)^2\}.$$

4.2.3 If $x\geqslant 0$ then show that

$$f(x)=x-1-\log x$$

has a minimum of 0 when $x=1$. (The first derivative of x is positive when $x>1$, negative when $x<1$, and zero when $x=1$). Therefore show that

$$x\geqslant \log x+1.$$

If a and g are the arithmetic and geometric means of a set of positive numbers, then establish that $a\geqslant 1+\log g$ and that equality holds only when each positive number equals unity.

4.2.4 Using the data in Example 4.2.3, show that if $\boldsymbol{\mu}'$ was known to be proportional to (2, 8), then the constant of proportionality would be $k=1.17$.

4.2.5 If $\mathbf{x}\sim N_p(\boldsymbol{\mu}, \sigma^2\mathbf{I})$, where $\boldsymbol{\mu}$ is known to lie on the unit sphere (i.e. $\boldsymbol{\mu}'\boldsymbol{\mu}=1$), show that the m.l.e. of $\boldsymbol{\mu}$ is $\mathbf{x}/(\mathbf{x}'\mathbf{x})^{1/2}$.

4.2.6 If $\mathbf{R}\boldsymbol{\mu}=\mathbf{r}$ and $\boldsymbol{\mu}$ is given by (4.2.13), show that $E(\hat{\boldsymbol{\mu}})=\boldsymbol{\mu}$ and $V(\hat{\boldsymbol{\mu}})=n^{-1}(\boldsymbol{\Sigma}-\boldsymbol{\Sigma}\mathbf{R}'(\mathbf{R}\boldsymbol{\Sigma}\mathbf{R}')^{-1}\mathbf{R}\boldsymbol{\Sigma})$. Note that $V(\hat{\boldsymbol{\mu}})$ is less than $V(\bar{\mathbf{x}})$ by a non-negative definite matrix.

4.2.7 If $\boldsymbol{\Sigma}=k\boldsymbol{\Sigma}_0$ and $\boldsymbol{\mu}$ is unconstrained, so that $\hat{k}$ is given by $p^{-1}\operatorname{tr}\boldsymbol{\Sigma}_0^{-1}\mathbf{S}$, as in (4.2.16), show that $E[\hat{k}]=(n-1)k/n$.

4.2.8 (Proof of (4.2.14)) If $\mathbf{R}\boldsymbol{\mu}=\mathbf{r}$ and $\boldsymbol{\Sigma}$ is unknown, then the m.l.e. of $\boldsymbol{\mu}$ is obtained by solving simultaneously

$$\frac{\partial l^+}{\partial \boldsymbol{\mu}}=\mathbf{0} \quad \text{and} \quad \mathbf{R}\boldsymbol{\mu}=\mathbf{r}.$$

Note that $\hat{\boldsymbol{\Sigma}}=\mathbf{S}+\mathbf{d}\mathbf{d}'$ where $\mathbf{d}=\hat{\boldsymbol{\Sigma}}\mathbf{R}'(\mathbf{R}\hat{\boldsymbol{\Sigma}}\mathbf{R}')^{-1}\mathbf{a}$ and $\mathbf{a}=\mathbf{R}\bar{\mathbf{x}}-\mathbf{r}$. Also, note that $\mathbf{R}\mathbf{d}=\mathbf{a}$ and therefore $\hat{\boldsymbol{\Sigma}}\mathbf{R}'=\mathbf{S}\mathbf{R}'+\mathbf{d}\mathbf{a}'$ and $\mathbf{R}\hat{\boldsymbol{\Sigma}}\mathbf{R}'=\mathbf{R}\mathbf{S}\mathbf{R}'+\mathbf{a}\mathbf{a}'$. Thus,

$$\mathbf{d}=(\mathbf{S}\mathbf{R}'+\mathbf{d}\mathbf{a}')(\mathbf{a}\mathbf{a}'+\mathbf{R}\mathbf{S}\mathbf{R}')^{-1}\mathbf{a}=\mathbf{S}\mathbf{R}'(\mathbf{R}\mathbf{S}\mathbf{R}')^{-1}\mathbf{a},$$

where we have used (A.2.4f) with $\mathbf{A}=\mathbf{R}\mathbf{S}\mathbf{R}'$, $\mathbf{B}=\mathbf{D}'=\mathbf{a}$, and $\mathbf{C}=1$ for the second part, plus some simplification. But $\mathbf{d}=\bar{\mathbf{x}}-\hat{\boldsymbol{\mu}}$. Therefore $\hat{\boldsymbol{\mu}}=\bar{\mathbf{x}}-\mathbf{S}\mathbf{R}'(\mathbf{R}\mathbf{S}\mathbf{R}')^{-1}(\mathbf{R}\bar{\mathbf{x}}-\mathbf{r})$. This is the same as the expression obtained in (4.2.13) when $\boldsymbol{\Sigma}$ is known, with the substitution of $\mathbf{S}$ for $\boldsymbol{\Sigma}$.

4.2.9 Use (4.2.13) to show that when $\boldsymbol{\mu}$ is known *a priori* to be of the form $(\boldsymbol{\mu}_1', \mathbf{0}')'$, then the m.l.e. of $\boldsymbol{\mu}_1$ is $\bar{\mathbf{x}}_1 - \boldsymbol{\Sigma}_{12}\boldsymbol{\Sigma}_{22}^{-1}\bar{\mathbf{x}}_2$.

4.2.10 (Mardia, 1962, 1964*a*) (a) Show that the log likelihood of a random sample $(x_1, y_1)', \ldots, (x_n, y_n)'$, from the bivariate Pareto distribution of Exercise 2.2.1 is given by

$$n^{-1}l = \log p(p+1) - (p+2)\bar{a},$$

where

$$n\bar{a} = \sum_{i=1}^{n} \log(x_i + y_i - 1).$$

(b) Show that maximum likelihood estimator of p is given by

$$\hat{p} = \bar{a}^{-1} - \tfrac{1}{2} + (\bar{a}^{-2} + \tfrac{1}{4})^{1/2}.$$

(c) By evaluating $\partial^2 l/\partial p^2$, show that the Fisher information matrix is $n(2p^2+2p+1)/\{p(p+1)\}^2$, and hence show that the asymptotic variance of $\hat{p}$ is $n^{-1}\{p(p+1)\}^2/(2p^2+2p+1)$.

4.2.11 Show that for any value of $\boldsymbol{\Sigma} > 0$, the multinormal log likelihood function (4.1.9) is maximized when $\boldsymbol{\mu} = \bar{\mathbf{x}}$. Then, using Theorem 4.2.1, show that if $\boldsymbol{\mu} = \bar{\mathbf{x}}$, then (4.1.9) is maximized when $\boldsymbol{\Sigma} = \mathbf{S}$.

4.2.12 If $\mathbf{X}$ is a data matrix from $N_p(\boldsymbol{\mu}, \boldsymbol{\Sigma})$, where $\boldsymbol{\Sigma} = \sigma^2[(1-\rho)\mathbf{I} + \rho\mathbf{1}\mathbf{1}']$ is proportional to the equicorrelation matrix, show that the m.l.e.s are given by

$$\hat{\mu} = \bar{x}, \qquad \hat{\sigma}^2 = p^{-1}\operatorname{tr}\mathbf{S}, \qquad \hat{\rho} = \frac{2}{p(p-1)\hat{\sigma}^2}\sum_{i<j} s_{ij}.$$

(Hint: using (A.3.2b), write $\boldsymbol{\Sigma}^{-1} = a\mathbf{I} + b\mathbf{1}\mathbf{1}'$ and differentiate the log likelihood with respect to a and b.)

4.3.1 (Press, 1972, p. 167) Show that if $\mathbf{x}_1, \ldots, \mathbf{x}_n$ are i.i.d. $N_p(\boldsymbol{\mu}, \boldsymbol{\Sigma})$, then $\bar{\mathbf{x}}$ and $\mathbf{M} = n\mathbf{S}$ have the joint density

$$f(\bar{\mathbf{x}}, \mathbf{S}; \boldsymbol{\mu}, \boldsymbol{\Sigma}) \propto \frac{|\mathbf{S}|^{(n-p-2)/2}}{|\boldsymbol{\Sigma}|^{n/2}} \exp\{-\tfrac{1}{2}n \operatorname{tr}\boldsymbol{\Sigma}^{-1}[\mathbf{S} + (\bar{\mathbf{x}}-\boldsymbol{\mu})(\bar{\mathbf{x}}-\boldsymbol{\mu})'\}.$$

If the vague prior $\pi(\boldsymbol{\mu}, \boldsymbol{\Sigma}) \propto |\boldsymbol{\Sigma}|^{-(p+1)/2}$ is taken, show that the posterior density can be written as

$$\pi(\boldsymbol{\mu}, \boldsymbol{\Sigma} \mid \mathbf{X}) \propto |\boldsymbol{\Sigma}|^{-(\nu+p+1)/2} \exp\{-\tfrac{1}{2}\operatorname{tr}\mathbf{V}^{-1}\boldsymbol{\Sigma}^{-1}\},$$

where $\nu = n$ and $\mathbf{V} = [\mathbf{S} + (\bar{\mathbf{x}}-\boldsymbol{\mu})(\bar{\mathbf{x}}-\boldsymbol{\mu})']^{-1}n^{-1}$. By integrating $\pi(\boldsymbol{\mu}, \boldsymbol{\Sigma} \mid \mathbf{X})$ with respect to $\boldsymbol{\Sigma}$ and using the normalizing constant of the inverse

Wishart distribution, show that the posterior density of $\boldsymbol{\mu}$ is

$$\pi(\boldsymbol{\mu} \mid \mathbf{X}) \propto \{1+(\bar{\mathbf{x}}-\boldsymbol{\mu})'\mathbf{S}^{-1}(\bar{\mathbf{x}}-\boldsymbol{\mu})\}^{-n/2},$$

i.e. $\boldsymbol{\mu}$ has a multivariate t distribution with parameters $\bar{\mathbf{x}}$, $\mathbf{S}$, and $\nu = n-p$. By integrating $\pi(\boldsymbol{\mu}, \boldsymbol{\Sigma} \mid \mathbf{X})$ with respect to $\boldsymbol{\mu}$, show that the posterior density of $\boldsymbol{\Sigma}$ is inverted Wishart, $W_p^{-1}(n\mathbf{S}, n-1)$ (see (3.8.2)). Hence show that the posterior mean of $\boldsymbol{\mu}$ is $\bar{\mathbf{x}}$, and if $n > p-2$ show that the posterior mean of $\boldsymbol{\Sigma}$ is $n\mathbf{S}/(n-p-2)$ (see Exercise 3.4.13).

4.3.2 (Press, 1972, p. 168) (a) If the informative prior from (4.3.5) for $\boldsymbol{\mu}$ and $\boldsymbol{\Sigma}$ is taken, then show that the posterior density is

$$\pi(\boldsymbol{\mu}, \boldsymbol{\Sigma} \mid \mathbf{X}) = |\boldsymbol{\Sigma}|^{-(n+m+1)/2} \exp\{-\tfrac{1}{2} \operatorname{tr} \boldsymbol{\Sigma}^{-1}[n\mathbf{S}+\mathbf{G} + (\boldsymbol{\mu}-\boldsymbol{\phi})(\boldsymbol{\mu}-\boldsymbol{\phi})' + n(\boldsymbol{\mu}-\bar{\mathbf{x}})(\boldsymbol{\mu}-\bar{\mathbf{x}})']\}$$

(b) Show that the posterior density of $\boldsymbol{\mu}$ is multivariate t with parameters

$$\mathbf{a} = \frac{\boldsymbol{\phi}+n\bar{\mathbf{x}}}{1+n}, \qquad \mathbf{C} = \frac{n\mathbf{S}+\mathbf{G}}{1+n} + \frac{n}{(1+n)^2}(\boldsymbol{\phi}-\bar{\mathbf{x}})(\boldsymbol{\phi}-\bar{\mathbf{x}})', \qquad \nu = n+m-p.$$

(c) Show that the posterior density of $\boldsymbol{\Sigma}$ is inverted Wishart, $W_p^{-1}((n+1)\mathbf{C}, n+m-p-1)$ (see equation (4.8.2)).

(d) Hence show that the posterior mean of $\boldsymbol{\mu}$ is $\mathbf{a}$, and if $n+m-2p-2>0$ show that the posterior mean of $\boldsymbol{\Sigma}$ is $(1+n)\mathbf{C}/(n+m+2p-2)$.
(see Exercise 3.4.13).

5
Hypothesis Testing

5.1 Introduction

The fundamental problems of multivariate hypothesis testing may be attributed to two sources—the sheer number of different hypotheses that exist, and the difficulty in choosing between various plausible test statistics. In this chapter we shall present two general approaches based on the likelihood ratio rest (LRT) and union intersection test (UIT), respectively. On some occasions, the LRT and UIT both lead to the same test statistics, but on other occasions they lead to different statistics.

The sheer number of possible hypotheses is well illustrated by the p-dimensional normal distribution. This has $\frac{1}{2}p(p+3)$ parameters, and therefore quite apart from other hypotheses, there are $2^{p(p+3)/2}$ hypotheses which only specify values for a subset of these parameters. This function of p is large even for p as small as 3. Other hypotheses could specify the values of ratios between parameters or other functions, and further hypotheses could test linear or non-linear restrictions. Of course, in practice the number of hypotheses one is interested in is much smaller than this.

The second source of fundamental problems mentioned in the opening paragraph concerns the difficulty of choosing between certain plausible test statistics, and the key question of whether to use a sequence of univariate tests or whether some new multivariate test would be better. These problems are apparent even in the bivariate case, as shown in the following example. (See also Exercise 5.3.6.)

Example 5.1.1 To illustrate the difference between multivariate hypothesis-testing and the corresponding univariate alternatives, consider only the variables x_1 and x_3, where

$$x_1 = \text{head length of first son} \quad \text{and} \quad x_3 = \text{head length of second son},$$

from Frets' head data given in Table 5.1.1. For the complete data of Table 5.1.1, $n = 25$,

$$\bar{\mathbf{x}} = (185.72, 151.12, 183.84, 149.24)'$$

and

$$\mathbf{S} = \begin{bmatrix} 91.481 & 50.753 & 66.875 & 44.267 \\ & 52.186 & 49.259 & 33.651 \\ & & 96.775 & 54.278 \\ & & & 43.222 \end{bmatrix}.$$

We assume initially that x_1 and x_3 are independent, and that each is normally distributed with a variance of 100. (The assumption of independence is woefully unrealistic, and is advanced here purely for pedagogic

Table 5.1.1 The measurements on the first and second adult sons in a sample of 25 families. (Data from Frets, 1921.)

First son		Second son	
Head length	Head breadth	Head length	Head breadth
191	155	179	145
195	149	201	152
181	148	185	149
183	153	188	149
176	144	171	142
208	157	192	152
189	150	190	149
197	159	189	152
188	152	197	159
192	150	187	151
179	158	186	148
183	147	174	147
174	150	185	152
190	159	195	157
188	151	187	158
163	137	161	130
195	155	183	158
186	153	173	148
181	145	182	146
175	140	165	137
192	154	185	152
174	143	178	147
176	139	176	143
197	167	200	158
190	163	187	150

reasons.) Suppose that we wish to test the univariate hypotheses (simultaneously) that both means are 182, i.e.

$$H_1: x_1 \sim N(182, 100), \qquad H_2: x_3 \sim N(182, 100).$$

They may be tested using the following z statistics:

$$z_1 = \frac{\bar{x}_1 - 182}{10/\sqrt{25}} = 1.86, \qquad z_2 = \frac{\bar{x}_3 - 182}{10/\sqrt{25}} = 0.92.$$

Since $|z_i| < 1.96$ for $i = 1, 2$, neither of the univariate hypotheses H_1 nor H_2 would be rejected at the 5% level of significance.

Now let us consider the bivariate hypothesis

$$H_3: x_1 \sim N(185, 100) \quad \text{and} \quad x_3 \sim N(185, 100).$$

This hypothesis is true if and only if both H_1 and H_2 are true. There are various ways of testing H_3.

One way would be to accept H_3 if and only if the respective z tests led to acceptance of both H_1 and H_2. Another approach would be to note that if both H_1 and H_2 are true then

$$z_3 = \frac{1}{\sqrt{2}}(z_1 + z_2) \sim N(0, 1).$$

An acceptance region based on z_3 would lie between the two lines $z_1 + z_2 = \pm c$. Note that the observed value of z_3 equals 1.966, which is just significant at the 5% level, even though neither z_1 nor z_2 were significant. Hence a multivariate hypothesis may be rejected even when each of its univariate components is accepted.

A third approach is based on the observation that if H_1 and H_2 are true, then

$$z_1^2 + z_2^2 \sim \chi_2^2.$$

This would lead to a circular acceptance region. In the above example, $z_1^2 + z_2^2 = 4.306$, which is below the 95th percentile of the χ_2^2 distribution.

A final possibility, to be considered in detail below, would use information on the correlation between z_1 and z_2 to derive a test statistic based on a quadratic form in z_1 and z_2. This leads to an elliptical acceptance region. Thus, depending upon which test statistic is chosen, our acceptance region can be rectangular, linear, circular, or elliptical. These different acceptance regions can easily lead to conflicting results.

We shall concentrate on developing general strategies and particular test statistics which can be used in many frequently occurring situations. Section 5.2 gives two systematic strategies, based on the likelihood ratio test (LRT) and union intersection test (UIT), respectively. The tests are introduced briefly in Section 5.2, with several examples of each. We find that for some hypotheses the LRT and UIT lead to identical test statistics, while for other hypotheses they lead to different statistics. Sections 5.3 and 5.4 discuss more detailed applications of the LRT and UIT. Section 5.5 considers the construction of simultaneous confidence intervals, while Section 5.6 makes some points of comparison between different types of hypothesis testing. Throughout the chapter we shall use the tables in Appendix C and also those in Pearson and Hartley (1972, pp. 98–116, 333–358). Sections 5.7 and 5.8 consider non-normal situations and non-parametric tests.

5.2 The Techniques Introduced

This section introduces the likelihood ratio test (LRT) and union intersection test (UIT) and gives a few applications.

5.2.1 The likelihood ratio test (LRT)

We assume that the reader is already familiar with the LR procedure from his knowledge of univariate statistics. The general strategy of the LRT is to maximize the likelihood under the null hypothesis H_0, and also to maximize the likelihood under the alternative hypothesis H_1. These main results are given in the following definitions and theorem.

Definition 5.2.1 *If the distribution of the random sample* $\mathbf{X}=(\mathbf{x}_1, \ldots, \mathbf{x}_n)'$ *depends upon a parameter vector* $\boldsymbol{\theta}$, *and if* $H_0:\boldsymbol{\theta}\in\Omega_0$ *and* $H_1:\boldsymbol{\theta}\in\Omega_1$ *are any two hypotheses, then the likelihood ratio (LR) statistic for testing* H_0 *against* H_1 *is defined as*

$$\lambda(\mathbf{x}) = L_0^*/L_1^*, \tag{5.2.1}$$

where L_i^* *is the largest value which the likelihood function takes in region* Ω_i, $i=0,1$.

Equivalently, we may use the statistic

$$-2\log\lambda = 2(l_1^* - l_0^*), \tag{5.2.2}$$

where $l_1^* = \log L_1^*$ and $l_0^* = \log L_0^*$.

In the case of simple hypotheses, where Ω_0 and Ω_1 each contain only a single point, the optimal properties of the LR statistic are proved in the well-known Neyman–Pearson lemma. For LR properties when H_0 and H_1 are composite hypotheses, see Kendall and Stuart (1973, p. 195). In general one tends to favour H_1 when the LR statistic is low, and H_0 when it is high. A test procedure based on the LR statistic may be defined as follows:

Definition 5.2.2 *The likelihood ratio test (LRT) of size α for testing H_0 against H_1 has as its rejection region*

$$R = \{\mathbf{x} \mid \lambda(\mathbf{x}) < c\}$$

where c is determined so that

$$\sup_{\theta \in \Omega_0} P_\theta(\mathbf{x} \in R) = \alpha.$$

For the hypotheses we are interested in, the distribution of λ does not, in fact, depend on the particular value of $\theta \in \Omega_0$, so the above supremum is unnecessary. The LRT has the following very important asymptotic property.

Theorem 5.2.1 *In the notation of* (5.2.1), *if Ω_1 is a region in R^q, and if Ω_0 is an r-dimensional subregion of Ω_1, then under suitable regularity conditions, for each $\boldsymbol{\theta} \in \Omega_0$, $-2 \log \lambda$ has an asymptotic χ^2_{q-r} distribution as $n \to \infty$.*

Proof See, for example, Silvey (1970, p. 113). ■

To illustrate the LRT, we examine three hypotheses assuming that $\mathbf{X} = (\mathbf{x}_1, \ldots, \mathbf{x}_n)'$ is a random sample from $N_p(\boldsymbol{\mu}, \boldsymbol{\Sigma})$.

5.2.1a The hypothesis $H_0: \boldsymbol{\mu} = \boldsymbol{\mu}_0$, $\boldsymbol{\Sigma}$ known

When $\boldsymbol{\Sigma}$ is known, then H_0 is a simple hypothesis and, from (4.1.9), the maximized log likelihood under H_0 is

$$l_0^* = l(\boldsymbol{\mu}_0, \boldsymbol{\Sigma}) = -\tfrac{1}{2}n \log|2\pi\boldsymbol{\Sigma}| - \tfrac{1}{2}n \operatorname{tr} \boldsymbol{\Sigma}^{-1}\mathbf{S} - \tfrac{1}{2}n(\bar{\mathbf{x}} - \boldsymbol{\mu}_0)'\boldsymbol{\Sigma}^{-1}(\bar{\mathbf{x}} - \boldsymbol{\mu}_0).$$

Since H_1 places no constraints on $\boldsymbol{\mu}$, the m.l.e. of $\boldsymbol{\mu}$ is $\bar{\mathbf{x}}$ and

$$l_1^* = l(\bar{\mathbf{x}}, \boldsymbol{\Sigma}) = -\tfrac{1}{2}n \log|2\pi\boldsymbol{\Sigma}| - \tfrac{1}{2}n \operatorname{tr} \boldsymbol{\Sigma}^{-1}\mathbf{S}. \tag{5.2.3}$$

Therefore, using (5.2.2), we get

$$-2 \log \lambda = 2(l_1^* - l_0^*) = n(\bar{\mathbf{x}} - \boldsymbol{\mu}_0)'\boldsymbol{\Sigma}^{-1}(\bar{\mathbf{x}} - \boldsymbol{\mu}_0). \tag{5.2.4}$$

Now from Theorem 2.5.2 we know that this function has an exact χ^2_p distribution under H_0. Hence we have a statistic whose distribution is known. It can therefore be used to test the null hypothesis. (Note that in this case the asymptotic distribution given by Theorem 5.2.1. is also the small sample distribution.)

Example 5.2.1 Consider the first and third variables of Frets' head measurement data from Table 5.1.1 and the hypothesis considered in Section 5.1.1, namely that $(x_1, x_3)' \sim N_2(\boldsymbol{\mu}_0, \boldsymbol{\Sigma})$, where

$$\boldsymbol{\mu}_0 = \begin{bmatrix} 182 \\ 182 \end{bmatrix} \quad \text{and} \quad \boldsymbol{\Sigma} = \begin{bmatrix} 100 & 0 \\ 0 & 100 \end{bmatrix}.$$

Using (5.2.4) we deduce that

$$-2 \log \lambda = 25(3.72, 1.84) \begin{bmatrix} 0.01 & 0 \\ 0 & 0.01 \end{bmatrix} \begin{pmatrix} 3.72 \\ 1.84 \end{pmatrix} = 4.31.$$

Since this is below 5.99, the 95th percentile of χ_2^2, we accept the hypothesis that $\boldsymbol{\mu} = \boldsymbol{\mu}_0$.

In this case, we can find a confidence region for μ_1 and μ_3 using Theorem 2.5.2. A 95% confidence region for the means of x_1 and x_3 can be given using the inequality

$$25(185.72 - \mu_1, 183.84 - \mu_3) \operatorname{diag}(0.01, 0.01) \begin{pmatrix} 185.72 - \mu_1 \\ 183.84 - \mu_3 \end{pmatrix} < 5.99,$$

i.e.

$$(185.72 - \mu_1)^2 + (183.84 - \mu_3)^2 < 23.96.$$

Because $\boldsymbol{\Sigma}$ is assumed to be proportional to $\mathbf{I}$, this gives a confidence region which is circular in the parameter space.

A more useful way to express this confidence region in terms of simultaneous confidence intervals is given in Exercise 5.5.2. See also Section 5.5.

5.2.1b *The hypothesis $H_0 : \boldsymbol{\mu} = \boldsymbol{\mu}_0$, $\boldsymbol{\Sigma}$ unknown (Hotelling one-sample T^2 test)*

In this case $\boldsymbol{\Sigma}$ must be estimated under H_0 and also under H_1. Therefore both hypotheses are composite. Using results from Section 4.2.2.1, we know that the m.l.e.s are as follows:

$$\text{under } H_0, \quad \hat{\boldsymbol{\mu}} = \boldsymbol{\mu}_0 \quad \text{and} \quad \hat{\boldsymbol{\Sigma}} = \mathbf{S} + \mathbf{dd}' \quad \text{where} \quad \mathbf{d} = \bar{\mathbf{x}} - \boldsymbol{\mu}_0;$$

whereas

$$\text{under } H_1, \quad \hat{\boldsymbol{\mu}} = \bar{\mathbf{x}} \quad \text{and} \quad \hat{\boldsymbol{\Sigma}} = \mathbf{S}.$$

Now from (4.1.9) we deduce that

$$l_0^* = l(\boldsymbol{\mu}_0, \mathbf{S} + \mathbf{dd}') = -\tfrac{1}{2}n\{p \log 2\pi + \log |\mathbf{S}| + \log(1 + \mathbf{d}'\mathbf{S}^{-1}\mathbf{d}) + p\}, \tag{5.2.5}$$

whereas $l_1^* = l(\bar{\mathbf{x}}, \mathbf{S})$ is obtained by putting $\mathbf{d} = \mathbf{0}$ in this expression. Thus,

$$-2 \log \lambda = 2(l_1^* - l_0^*) = n \log (1 + \mathbf{d}'\mathbf{S}^{-1}\mathbf{d}). \tag{5.2.6}$$

This statistic depends upon $(n-1)\mathbf{d}'\mathbf{S}^{-1}\mathbf{d}$, which is known from Corollary 3.5.1.1 to be a $T^2(p, n-1)$ statistic, often known as the Hotelling one-sample T^2 statistic. Further, $\{(n-p)/p\}\mathbf{d}'\mathbf{S}^{-1}\mathbf{d}$ has an $F_{p,n-p}$ distribution.

Example 5.2.2 Consider again the first and third variables of Frets' head data and let us test the hypothesis $H_0: (x_1, x_3)' \sim N_2(\boldsymbol{\mu}_0, \boldsymbol{\Sigma})$ where $\boldsymbol{\mu}_0 = (182, 182)'$ and $\boldsymbol{\Sigma}$ is now assumed unknown. Using the numbers given in Example 5.1.1, we test

$$\frac{n-p}{p}\mathbf{d}'\mathbf{S}^{-1}\mathbf{d} = \frac{23}{3}(3.72, 1.84)\begin{bmatrix} 91.481 & 66.875 \\ 66.875 & 96.775 \end{bmatrix}^{-1}\begin{pmatrix} 3.72 \\ 1.84 \end{pmatrix} = 1.28$$

against the $F_{2,23}$ distribution. Since $F_{2,23;0.05} = 3.44$, we accept the null hypothesis.

5.2.1c The hypothesis $H_0: \boldsymbol{\Sigma} = \boldsymbol{\Sigma}_0$, $\boldsymbol{\mu}$ unknown

The m.l.e.s for this example under H_0 are $\hat{\boldsymbol{\mu}} = \bar{\mathbf{x}}$ and $\boldsymbol{\Sigma} = \boldsymbol{\Sigma}_0$ and, under H_1, $\hat{\boldsymbol{\mu}} = \bar{\mathbf{x}}$ and $\hat{\boldsymbol{\Sigma}} = \mathbf{S}$. Therefore

$$l_0^* = l(\bar{\mathbf{x}}, \boldsymbol{\Sigma}_0) = -\tfrac{1}{2} n \log |2\pi\boldsymbol{\Sigma}_0| - \tfrac{1}{2} n \operatorname{tr} \boldsymbol{\Sigma}_0^{-1} \mathbf{S},$$

and

$$l_1^* = l(\bar{\mathbf{x}}, \mathbf{S}) = -\tfrac{1}{2} n \log |2\pi\mathbf{S}| - \tfrac{1}{2} np.$$

Therefore

$$-2 \log \lambda = 2(l_1^* - l_0^*) = n \operatorname{tr} \boldsymbol{\Sigma}_0^{-1} \mathbf{S} - n \log |\boldsymbol{\Sigma}_0^{-1} \mathbf{S}| - np. \tag{5.2.7}$$

Note that this statistic is a function of the eigenvalues of $\boldsymbol{\Sigma}_0^{-1}\mathbf{S}$, and also that, as $\mathbf{S}$ approaches $\boldsymbol{\Sigma}_0$, $-2 \log \lambda$ approaches zero. In fact, if we write a for the arithmetic mean of the eigenvalues of $\boldsymbol{\Sigma}_0^{-1}\mathbf{S}$ and g for the geometric mean, so that $\operatorname{tr} \boldsymbol{\Sigma}_0^{-1}\mathbf{S} = pa$ and $|\boldsymbol{\Sigma}_0^{-1}\mathbf{S}| = g^p$, then (5.2.7) becomes

$$-2 \log \lambda = np(a - \log g - 1). \tag{5.2.8}$$

One problem with the statistic given by (5.2.7) is that its distribution is far from simple. Anderson (1958, p. 265) finds a formula for the moments of λ, and also derives the characteristic function of $-2 \log \lambda$, under both H_0 and H_1. Korin (1968) has expressed (5.2.8) as an infinite sum of chi-squared variables, for certain small values of n and p (see Pearson and Hartley, 1972, pp. 111, 358). However, these results are not easy to use. The general result cited in Theorem 5.2.1 indicates, however, that $-2 \log \lambda$ given by (5.2.8) has an asymptotic χ^2_m distribution under H_0, where m equals the number of independent parameters in $\mathbf{\Sigma}$, i.e. $m = \frac{1}{2}p(p+1)$.

Example 5.2.3 Using the first and third variables of Frets' data from Table 5.1.1, we may test the hypothesis $\mathbf{\Sigma} = \text{diag}\,(100, 100)$, which was assumed in Example 5.1.1. Here

$$\mathbf{\Sigma}_0^{-1}\,\mathbf{S} = \begin{bmatrix} 0.01 & 0 \\ 0 & 0.01 \end{bmatrix} \begin{bmatrix} 91.481 & 66.875 \\ 66.875 & 96.775 \end{bmatrix} = \begin{bmatrix} 0.9148 & 0.6688 \\ 0.6688 & 0.9678 \end{bmatrix}.$$

The eigenvalues of this matrix are given by $\lambda_1 = 1.611$ and $\lambda_2 = 0.272$. Therefore, $a = 0.9413$ and $g = 0.6619$. Hence from (5.2.8) we find that

$$-2 \log \lambda = 17.70.$$

This must be compared asymptotically with a χ^2_3 distribution, and we see quite clearly that the hypothesis is rejected; that is, our original assumption regarding the covariance matrix was false. This might perhaps have been expected from a cursory look at the sample covariance matrix, which shows the presence of a strong correlation.

Using the same data, and having rejected the hypothesis that $\mathbf{\Sigma} = \text{diag}\,(100, 100)$, we might now examine the hypothesis that

$$\mathbf{\Sigma} = \mathbf{\Sigma}_0 = \begin{bmatrix} 100 & 50 \\ 50 & 100 \end{bmatrix}.$$

Certainly judging from the sample covariance matrix $\mathbf{S}$, this hypothesis seems distinctly more plausible. It is found that

$$\mathbf{\Sigma}_0^{-1}\,\mathbf{S} = \begin{bmatrix} 0.7739 & 0.2465 \\ 0.2818 & 0.8445 \end{bmatrix}.$$

The arithmetic and geometric means of the eigenvalues are $a = 0.8092$ and $g = 0.7642$. Note that these are far closer together than the values obtained previously, reflecting the fact that $\mathbf{\Sigma}^{-1}\,\mathbf{S}$ is closer to the form $k\mathbf{I}$. Inserting the values of a and g in (5.2.8) we find that $-2 \log \lambda = 3.9065$, which is well below the 95% percentile of the χ^2_3 distribution. Hence we deduce that this revised value of $\mathbf{\Sigma}$ is quite plausible under the given assumptions.

5.2.2 The union intersection test (UIT)

Consider a random vector $\mathbf{x}$, which has the $N_p(\boldsymbol{\mu}, \mathbf{I})$ distribution and a non-random p-vector $\mathbf{a}$. Then if $y_{\mathbf{a}} = \mathbf{a}'\mathbf{x}$, we know that $y_{\mathbf{a}} \sim N_1(\mathbf{a}'\boldsymbol{\mu}, \mathbf{a}'\mathbf{a})$. Moreover, this is true for all p-vectors $\mathbf{a}$.

Now suppose that we wish to test the hypothesis $H_0: \boldsymbol{\mu} = \mathbf{0}$. Then under H_0 we know that $y_{\mathbf{a}} \sim N(0, \mathbf{a}'\mathbf{a})$, a hypothesis which we may call $H_{0\mathbf{a}}$. Moreover, $H_{0\mathbf{a}}$ is true for all p-vectors $\mathbf{a}$. In other words, the multivariate hypothesis H_0 can be written as the intersection of the set of univariate hypothesis $H_{0\mathbf{a}}$; that is

$$H_0 = \cap H_{0\mathbf{a}}. \tag{5.2.9}$$

The intersection sign is used here because *all* the $H_{0\mathbf{a}}$ must be true in order for H_0 to be true. We call $H_{0\mathbf{a}}$ a *component* of H_0.

Now let us consider how we would test the univariate hypothesis $H_{0\mathbf{a}}$. One obvious way is to use $z_{\mathbf{a}} = y_{\mathbf{a}}/\sqrt{\mathbf{a}'\mathbf{a}}$, which has the standard normal distribution. A rejection region for $H_{0\mathbf{a}}$ based on $z_{\mathbf{a}}$ would be of the form

$$R_{\mathbf{a}} = \{z_{\mathbf{a}} : |z_{\mathbf{a}}| > c\}.$$

where c is some arbitrary critical value, say 1.96. This rejection region for $H_{0\mathbf{a}}$ could also be written

$$R_{\mathbf{a}} = \{z_{\mathbf{a}} : z_{\mathbf{a}}^2 > c^2\}. \tag{5.2.10}$$

Hence we have a rejection region for each of the univariate hypotheses which together imply H_0 in (5.2.9).

We turn now to consider a sensible rejection region for the composite hypothesis H_0. Since H_0 is true if and only if *every* component hypothesis $H_{0\mathbf{a}}$ is true, it seems sensible to accept H_0 if and only if *every* component hypothesis $H_{0\mathbf{a}}$ is accepted; that is, we reject H_0 if *any* component hypothesis is rejected. This leads to a rejection region for H_0 which is the union of the rejection regions for the component hypotheses; that is

$$R = \cup R_{\mathbf{a}}. \tag{5.2.11}$$

The *union* of the rejection regions given by (5.2.11) and the *intersection* of component hypotheses formulated in (5.2.9) provide the basis of the union intersection strategy, which was due initially to Roy (1957).

Definition 5.2.2 *A union intersection test (UIT) for the hypothesis H_0 is a test whose rejection region R can be written as in* (5.2.11), *where $R_{\mathbf{a}}$ is the rejection region corresponding to a component hypothesis $H_{0\mathbf{a}}$, and where H_0 can be written as in* (5.2.9).

Applied to the above example, the union intersection strategy based on (5.2.10) leads to a rejection of H_0 if and only if any $z_{\mathbf{a}}^2$ exceeds c^2; that is, H_0 is *accepted* if and only if $z_{\mathbf{a}}^2 \leq c^2$ for *all* $z_{\mathbf{a}}$. This is the same as saying that H_0 is accepted if and only if

$$\max_{\mathbf{a}} z_{\mathbf{a}}^2 \leq c^2.$$

In general the union intersection test often leads to the maximization or minimization of some composite test statistic such as $z_{\mathbf{a}}^2$. In this example,

$$z_{\mathbf{a}}^2 = y_{\mathbf{a}}^2/\mathbf{a}'\mathbf{a} = \mathbf{a}'\mathbf{x}\mathbf{x}'\mathbf{a}/\mathbf{a}'\mathbf{a}.$$

This is maximized when $\mathbf{a} = \mathbf{x}$, so that $\max z_{\mathbf{a}}^2 = \mathbf{x}'\mathbf{x}$. Hence in this example the UIT statistic would be $\mathbf{x}'\mathbf{x}$, whose distribution under H_0 is known to be χ_p^2.

The method of constructing UITs has important practical consequences. If the null hypothesis is rejected, then one can ask which of the component rejection regions $R_{\mathbf{a}}$ were responsible, thus getting a clearer idea about the nature of the deviation from the null hypothesis. In the above example, if one rejects H_0, then one can ask which linear combinations $\mathbf{a}'\mathbf{x}$ were responsible. In particular one can look at the variables $x_i = \mathbf{e}_i'\mathbf{x}$ on their own. For example it might be the case that some of the variables x_i lie in $R_{\mathbf{e}_i}$ whereas the others are all acceptable. See Section 5.5 for further details.

Unfortunately, the LRT does not have this property. If one rejects the null hypothesis using a LRT, then one cannot ask for more details about the reasons for rejection.

We shall now take the hypotheses used to illustrate the LRT procedure in Sections 5.2.1a–5.2.1c above and study them using UITs. As in those examples, we take $\mathbf{X}$ to be a matrix whose n rows are i.i.d. $N_p(\boldsymbol{\mu}, \boldsymbol{\Sigma})$ vectors. Note that this implies that $\mathbf{y} = \mathbf{X}\mathbf{a}$ is a vector whose n elements are i.i.d. $N(\mathbf{a}'\boldsymbol{\mu}, \mathbf{a}'\boldsymbol{\Sigma}\mathbf{a})$ variables.

5.2.2a The hypothesis $H_0: \boldsymbol{\mu} = \boldsymbol{\mu}_0$, $\boldsymbol{\Sigma}$ known (union intersection approach)

Under H_0 the elements of $\mathbf{y}$ are i.i.d. $N(\mu_y, \sigma_y^2)$, where $\mathbf{y} = \mathbf{X}\mathbf{a}$, $\mu_y = \mathbf{a}'\boldsymbol{\mu}_0$ and $\sigma_y^2 = \mathbf{a}'\boldsymbol{\Sigma}\mathbf{a}$. An obvious test statistic based on $\bar{y} = \mathbf{a}'\bar{\mathbf{x}}$ is

$$z_{\mathbf{a}}^2 = n\mathbf{a}'(\bar{\mathbf{x}} - \boldsymbol{\mu}_0)(\bar{\mathbf{x}} - \boldsymbol{\mu}_0)'\mathbf{a}/\mathbf{a}'\boldsymbol{\Sigma}\mathbf{a}.$$

Using (5.2.11) we wish to reject H_0 for large values of

$$\max_{\mathbf{a}} z_{\mathbf{a}}^2 = n(\bar{\mathbf{x}} - \boldsymbol{\mu}_0)'\boldsymbol{\Sigma}^{-1}(\bar{\mathbf{x}} - \boldsymbol{\mu}_0). \tag{5.2.12}$$

This chi-squared statistic for the UIT has already been derived as the LR statistic in Section 5.2.1a.

Thus, for this hypothesis, the UIT and LRT both lead to the same test statistic. However, this property is not true in general, as we shall see below in Section 5.2.2c. Note that the critical values for the UI test statistic should be calculated on the basis of the distribution of the UI statistic, in this case a χ_p^2 distribution, and not from the value 1.96 or

$(1.96)^2$ which relates to one of the component hypotheses of H_0 rather than H_0 itself.

5.2.2b The hypothesis $H_0: \boldsymbol{\mu} = \boldsymbol{\mu}_0, \boldsymbol{\Sigma}$ unknown (Hotelling one-sample T^2 test, union intersection approach)

Once more we note that under H_0 the elements of $\mathbf{y}$ are i.i.d. $N(\mu_y, \sigma_y^2)$. However, in this example σ_y^2 is unknown and therefore must be estimated. An obvious estimator is

$$s_y^2 = \frac{1}{n} \sum (y_i - \bar{y})^2 = \frac{1}{n} \sum (\mathbf{x}_i'\mathbf{a} - \bar{\mathbf{x}}'\mathbf{a})^2 = \mathbf{a}'\mathbf{S}\mathbf{a}. \qquad (5.2.13)$$

This leads to the test statistic for $H_{0\mathbf{a}}$, namely

$$t_{\mathbf{a}} = \frac{\bar{y} - \mu_y}{\sqrt{s_y^2/(n-1)}}.$$

We note that

$$t_{\mathbf{a}}^2 = (n-1) \frac{\mathbf{a}'(\bar{\mathbf{x}} - \boldsymbol{\mu}_0)(\bar{\mathbf{x}} - \boldsymbol{\mu}_0)'\mathbf{a}}{\mathbf{a}'\mathbf{S}\mathbf{a}},$$

once more a ratio of quadratic forms. This time the UIT statistic is

$$\max_{\mathbf{a}} t_{\mathbf{a}}^2 = (n-1)(\bar{\mathbf{x}} - \boldsymbol{\mu}_0)'\mathbf{S}^{-1}(\bar{\mathbf{x}} - \boldsymbol{\mu}_0). \qquad (5.2.14)$$

Note that (5.2.14) is Hotelling's one-sample T^2 statistic, which has already been met in Section 5.2.1b. Hence once more the UIT and LRT procedures have led to the same test statistic.

5.2.2c The hypothesis $H_0: \boldsymbol{\Sigma} = \boldsymbol{\Sigma}_0, \boldsymbol{\mu}$ unknown (union intersection test)

Once more the elements of $\mathbf{y}$ are i.i.d. $N(\mu_y, \sigma_y^2)$. However, this time we wish to examine σ_y^2 and see whether it equals

$$\sigma_{0y}^2 = \mathbf{a}'\boldsymbol{\Sigma}_0\mathbf{a}.$$

An obvious estimator of σ_y^2 is s_y^2 defined in (5.2.12). This leads to the test statistic

$$U_{\mathbf{a}} = ns_y^2/\sigma_{0y}^2 = n\mathbf{a}'\mathbf{S}\mathbf{a}/\mathbf{a}'\boldsymbol{\Sigma}_0\mathbf{a}$$

and we reject $H_{0\mathbf{a}}$ if $U_{\mathbf{a}} \leqslant c_{1\mathbf{a}}$ or $U_{\mathbf{a}} \geqslant c_{2\mathbf{a}}$, where $c_{1\mathbf{a}}$ and $c_{2\mathbf{a}}$ are chosen to make the size of the test equal to α. Since

$$\max_{\mathbf{a}} \frac{n\mathbf{a}'\mathbf{S}\mathbf{a}}{\mathbf{a}'\boldsymbol{\Sigma}_0\mathbf{a}} = n\lambda_1(\boldsymbol{\Sigma}_0^{-1}\mathbf{S}), \qquad \min_{\mathbf{a}} \frac{n\mathbf{a}'\mathbf{S}\mathbf{a}}{\mathbf{a}'\boldsymbol{\Sigma}_0\mathbf{a}} = n\lambda_p(\boldsymbol{\Sigma}_0^{-1}\mathbf{S}), \qquad (5.2.15)$$

where λ_i denotes the ith largest eigenvalue, we see that the critical region for the UIT takes the form

$$\lambda_p(\mathbf{\Sigma}_0^{-1}\mathbf{S})<c_1 \quad \text{or} \quad \lambda_1(\mathbf{\Sigma}_0^{-1}\mathbf{S})>c_2,$$

where c_1 and c_2 are chosen to make the size of the test equal to α. However, the joint distribution of the roots of $\mathbf{\Sigma}^{-1}\mathbf{S}$ is quite complicated and its critical values have not yet been tabulated. Note that this UIT is *not* the same as the corresponding LRT which was obtained in (5.2.7), although both statistics depend only on the eigenvalues of $\mathbf{\Sigma}_0^{-1}\mathbf{S}$.

5.3 The Techniques Further Illustrated

In this section we take various further hypotheses, and derive the corresponding LRTs and UITs.

5.3.1 One-sample hypotheses on $\boldsymbol{\mu}$

We have already seen above that the hypothesis $\boldsymbol{\mu}=\boldsymbol{\mu}_0$ leads to the same test statistic using both the LRT and UIT principle, whether or not $\mathbf{\Sigma}$ is known. The following general result allows us to deduce the LRT for a wide variety of hypotheses on $\boldsymbol{\mu}$. As before, we assume that $\mathbf{x}_1, \ldots, \mathbf{x}_n$ is a random sample from $N_p(\boldsymbol{\mu}, \mathbf{\Sigma})$.

Theorem 5.3.1 *If H_0 and H_1 are hypotheses which lead to m.l.e.s $\hat{\boldsymbol{\mu}}$ and $\bar{\mathbf{x}}$, respectively, and if there are no constraints on $\mathbf{\Sigma}$, then the m.l.e.s of $\mathbf{\Sigma}$ are $\mathbf{S}+\mathbf{dd}'$ and $\mathbf{S}$, respectively, where $\mathbf{d}=\bar{\mathbf{x}}-\hat{\boldsymbol{\mu}}$. The LR test for testing H_0 against H_1 is given by*

$$-2\log\lambda = n\mathbf{d}'\mathbf{\Sigma}^{-1}\mathbf{d} \qquad \text{if } \mathbf{\Sigma} \text{ is known} \tag{5.3.1}$$

and

$$-2\log\lambda = n\log(1+\mathbf{d}'\mathbf{S}^{-1}\mathbf{d}) \qquad \text{if } \mathbf{\Sigma} \text{ is unknown.} \tag{5.3.2}$$

Proof The proof is identical to that followed in deriving (5.2.4) and (5.2.6). ■

Unfortunately, no result of corresponding generality exists concerning the UI test. Therefore, UITs will have to be derived separately for each of the following examples.

5.3.1a *The hypothesis* $H_0: \mathbf{R}\boldsymbol{\mu} = \mathbf{r}, \boldsymbol{\Sigma}$ *known (hypothesis of linear constraints)*

Likelihood ratio test Here $\mathbf{R}$ and $\mathbf{r}$ are pre-specified. The m.l.e. of $\boldsymbol{\mu}$ under $H_0: \mathbf{R}\boldsymbol{\mu} = \mathbf{r}$ is given by (4.2.13). The corresponding LR test is given by (5.3.1), where

$$\mathbf{d} = \bar{\mathbf{x}} - \hat{\boldsymbol{\mu}} = \boldsymbol{\Sigma}\mathbf{R}'(\mathbf{R}\boldsymbol{\Sigma}\mathbf{R}')^{-1}(\mathbf{R}\bar{\mathbf{x}} - \mathbf{r}).$$

The LR test is therefore given by

$$-2 \log \lambda = n(\mathbf{R}\bar{\mathbf{x}} - \mathbf{r})'(\mathbf{R}\boldsymbol{\Sigma}\mathbf{R}')^{-1}(\mathbf{R}\bar{\mathbf{x}} - \mathbf{r}). \quad (5.3.3)$$

Under H_0 the rows of $\mathbf{XR}'$ are i.i.d. $N_q(\mathbf{r}, \mathbf{R}\boldsymbol{\Sigma}\mathbf{R}')$ random vectors, where $q < p$ is the number of elements in r. Therefore, by Theorem 2.5.2, (5.3.3) has an exact χ_q^2 distribution under H_0.

An important special case occurs when $\boldsymbol{\mu}$ is partitioned, $\boldsymbol{\mu} = (\boldsymbol{\mu}_1', \boldsymbol{\mu}_2')'$, and we wish to test whether $\boldsymbol{\mu}_1 = \mathbf{0}$. This hypothesis can be expressed in the form $\mathbf{R}\boldsymbol{\mu} = \mathbf{r}$ on setting $\mathbf{R} = (\mathbf{I}, \mathbf{0})$ and $\mathbf{r} = \mathbf{0}$.

In an obvious notation (5.3.3) becomes

$$-2 \log \lambda = n\bar{\mathbf{x}}_1\boldsymbol{\Sigma}_{11}^{-1}\bar{\mathbf{x}}_1.$$

Another special case occurs when we wish to test the hypothesis

$$H_0: \boldsymbol{\mu} = k\boldsymbol{\mu}_0 \quad \text{for some } k, \quad (5.3.4)$$

where $\boldsymbol{\mu}_0$ is a given vector. This hypothesis may be expressed in the form $\mathbf{R}\boldsymbol{\mu} = \mathbf{0}$, if we take $\mathbf{R}$ to be a $((p-1) \times p)$ matrix of rank $p-1$, whose rows are all orthogonal to $\boldsymbol{\mu}_0$. Under H_0, the m.l.e. $\hat{k}$ of k is given by (4.2.19) and $\hat{k}$ may be used to express the LR statistic without explicit use of the matrix $\mathbf{R}$. See Exercise 5.3.3.

An alternative method of deriving the LRT in (5.3.3) can be given by using the methods of Section 5.2.1a to test the hypothesis $\mathbf{R}\boldsymbol{\mu} = \mathbf{r}$ for the transformed data matrix $\mathbf{Y} = \mathbf{XR}'$. This approach also leads to (5.3.3).

Union intersection test A sensible union intersection test for this hypothesis can be obtained by applying the methods of Section 5.2.2a to the transformed variables $\mathbf{XR}'$. Since the methods of 5.2.1a and 5.2.2a lead to the same test, the UIT is the same as the LRT for this hypothesis.

5.3.1b *The hypothesis* $H_0: \mathbf{R}\boldsymbol{\mu} = \mathbf{r}, \boldsymbol{\Sigma}$ *unknown (hypothesis of linear constraints)*

Likelihood ratio test Using (4.2.14) and (5.3.2) we know that

$$-2 \log \lambda = n \log (1 + \mathbf{d}'\mathbf{S}^{-1}\mathbf{d}), \quad (5.3.5)$$

where

$$\mathbf{d} = \mathbf{SR}'(\mathbf{RSR}')^{-1}(\mathbf{R\bar{x}} - \mathbf{r}).$$

Note that this test is based on the statistic

$$(n-1)\mathbf{d}'\mathbf{S}^{-1}\mathbf{d} = (n-1)(\mathbf{R\bar{x}} - \mathbf{r})'(\mathbf{RSR}')^{-1}(\mathbf{R\bar{x}} - \mathbf{r}), \tag{5.3.6}$$

which has a $T^2(q, n-1)$ distribution under H_0 since $\mathbf{R\bar{x}} \sim N_q(\mathbf{r}, n^{-1}\mathbf{R\Sigma R}')$ independently of $n\mathbf{RSR}' \sim W_q(\mathbf{R\Sigma R}', n-1)$. For the hypothesis that $\boldsymbol{\mu}_1 = \mathbf{0}$, this T^2 statistic becomes

$$(n-1)\mathbf{\bar{x}}_1\mathbf{S}_{11}^{-1}\mathbf{\bar{x}}_1. \tag{5.3.7}$$

Example 5.3.1 We may use the cork tree data from Table 1.4.1 to examine the hypothesis H_0 that the depth of bark deposit on $n = 28$ trees is the same in all directions. Let $\mathbf{x}$ denote the four directions (N, E, S, W)$'$. One way to represent the null hypothesis is $\mathbf{R}\boldsymbol{\mu} = \mathbf{0}$, where

$$\mathbf{R} = \begin{bmatrix} 1 & -1 & 1 & -1 \\ 1 & 0 & -1 & 0 \\ 0 & 1 & 0 & -1 \end{bmatrix}.$$

Here $q = 3$. To test H_0 we use the statistic

$$(n-1)\mathbf{\bar{y}}'\mathbf{S}_y^{-1}\mathbf{\bar{y}} = 20.74 \sim T^2(3, 27),$$

where $\mathbf{\bar{y}}$ and $\mathbf{S}_y$ are the sample mean and covariance matrix for the transformed variables $\mathbf{Y} = \mathbf{XR}'$, and were calculated in Example 1.5.1. The corresponding F statistic is given by

$$\frac{n-q}{q}\mathbf{\bar{y}}'\mathbf{S}_y^{-1}\mathbf{\bar{y}} = 6.40 \sim F_{3,25}.$$

Since $F_{3,25;0.01} = 4.68$, we conclude that H_0 must be rejected.

Note that the null hypothesis for this example may also be expressed in the form $H_0 : \boldsymbol{\mu} = k\mathbf{1}$, as in (5.3.4).

Union intersection test As in the previous section, the UIT gives the same test here as the LRT.

5.3.2 One-sample hypotheses on Σ

Theorem 5.3.2 *Let $\mathbf{x}_1, \ldots, \mathbf{x}_n$ be a random sample from $N_p(\boldsymbol{\mu}, \boldsymbol{\Sigma})$. If H_0 and H_1 are hypotheses which lead to $\hat{\boldsymbol{\Sigma}}$ and $\mathbf{S}$ as the m.l.e.s. for $\boldsymbol{\Sigma}$, and if $\mathbf{\bar{x}}$ is the m.l.e. of $\boldsymbol{\mu}$ under both hypotheses, then the LRT for testing H_0 against H_1 is given by*

$$-2\log\lambda = np(a - \log g - 1), \tag{5.3.8}$$

where a and g are the arithmetic and geometric means of the eigenvalues of $\hat{\mathbf{\Sigma}}^{-1}\mathbf{S}$.

Proof The proof is identical to that followed in deriving (5.2.7) and (5.2.8). ∎

A similar result holds if $\boldsymbol{\mu}$ is known. Unfortunately, however, no corresponding results are known for the union intersection test.

5.3.2a The hypothesis $H_0: \mathbf{\Sigma} = k\mathbf{\Sigma}_0$, $\boldsymbol{\mu}$ *unknown*

Likelihood ratio test The m.l.e. $\hat{k}$ was found in (4.2.16). The LRT is therefore given by (5.3.8), where a and g relate to the eigenvalues of $\mathbf{\Sigma}_0^{-1}\mathbf{S}/\hat{k}$. But in this particular case (5.3.8) conveniently simplifies. We may write $a = a_0/\hat{k}$ and $g = g_0/\hat{k}$, where a_0 and g_0 are the arithmetic and geometric means of the eigenvalues of $\mathbf{\Sigma}_0^{-1}\mathbf{S}$. But from (4.2.16), $\hat{k} = p^{-1}\operatorname{tr}\mathbf{\Sigma}_0^{-1}\mathbf{S} = a_0$. Therefore, in fact, $a = 1$ and $g = g_0/a_0$. Hence, from (5.3.8), we get

$$-2\log\lambda = np\log(a_0/g_0). \tag{5.3.9}$$

In other words, the likelihood ratio criterion for this hypothesis depends simply upon the ratio between the arithmetic and geometric means of eigenvalues of $\mathbf{\Sigma}_0^{-1}\mathbf{S}$. This result is intuitively appealing, since, as $\mathbf{S}$ tends to $k\mathbf{\Sigma}_0$, the values of a_0 and g_0 both tend to k, and the expression given by (5.3.9) tends to zero, just as one would expect when H_0 is satisfied. From Theorem 5.2.1, we know that (5.3.9) has an asymptotic χ^2 distribution with $\frac{1}{2}p(p+1)-1 = \frac{1}{2}(p-1)(p+2)$ degrees of freedom. Korin (1968) has used the technique of Box (1949) to express $-2\log\lambda$ as a sum of chi-squared variates, and thereby found appropriate χ^2 and F approximations.

Union intersection test There seems to be no straightforward UIT for this hypothesis, but see Olkin and Tomsky (1975).

Example 5.3.2 In the so-called "test for sphericity" we wish to test whether $\mathbf{\Sigma} = k\mathbf{I}$. The LRT is given by (5.3.9), where a_0 and g_0 now relate to the eigenvalues of $\mathbf{S}$ in the following way:

$$a_0 = \frac{1}{p}\sum s_{ii}, \qquad g_0^p = |\mathbf{S}|.$$

For the first and third variables of Frets' data, which has already been used in Example 5.2.3, $n = 25$, $p = 2$, $a_0 = 94.13$, $|\mathbf{S}| = 4380.81$, and $g_0 = 66.19$. Therefore, from (5.3.9), $-2\log\lambda = 17.6$. This has an asymptotic χ^2 distribution with $\frac{1}{2}(p-1)(p+2) = 2$ degrees of freedom. Since the upper 5% critical value is exceeded, we conclude that $\mathbf{\Sigma}$ does not have the stated form.

5.3.2b The hypothesis $H_0: \mathbf{\Sigma}_{12} = \mathbf{0}, \boldsymbol{\mu}$ unknown

Likelihood ratio test Partition the variables into two sets with dimensions p_1 and p_2, respectively, $p_1 + p_2 = p$. If $\mathbf{\Sigma}_{12} = 0$, the likelihood splits into two factors. Each factor can be maximized separately over $\mathbf{\Sigma}_{11}$, $\boldsymbol{\mu}_1$ and $\mathbf{\Sigma}_{22}$, $\boldsymbol{\mu}_2$, respectively, giving $\hat{\boldsymbol{\mu}} = (\hat{\boldsymbol{\mu}}_1', \hat{\boldsymbol{\mu}}_2')' = \bar{\mathbf{x}}$ and

$$\hat{\mathbf{\Sigma}} = \begin{pmatrix} \mathbf{S}_{11} & \mathbf{0} \\ \mathbf{0} & \mathbf{S}_{22} \end{pmatrix}.$$

Since the m.l.e. of $\boldsymbol{\mu}$ under H_0 and H_1 is $\bar{\mathbf{x}}$, we know from (5.3.8) that the LRT for this hypothesis depends upon the eigenvalues of

$$\hat{\mathbf{\Sigma}}^{-1}\mathbf{S} = \begin{bmatrix} \mathbf{S}_{11} & \mathbf{0} \\ \mathbf{0} & \mathbf{S}_{22} \end{bmatrix}^{-1} \begin{bmatrix} \mathbf{S}_{11} & \mathbf{S}_{12} \\ \mathbf{S}_{21} & \mathbf{S}_{22} \end{bmatrix} = \begin{bmatrix} \mathbf{I} & \mathbf{S}_{11}^{-1}\mathbf{S}_{12} \\ \mathbf{S}_{22}^{-1}\mathbf{S}_{21} & \mathbf{I} \end{bmatrix}.$$

Clearly $\operatorname{tr} \hat{\mathbf{\Sigma}}^{-1}\mathbf{S} = p$, and therefore the arithmetic mean of the eigenvalues, a, is one. Also, using (A.2.3j),

$$g^p = |\hat{\mathbf{\Sigma}}^{-1}\mathbf{S}| = |\mathbf{S}|/(|\mathbf{S}_{11}|\,|\mathbf{S}_{22}|) = |\mathbf{S}_{22} - \mathbf{S}_{21}\mathbf{S}_{11}^{-1}\mathbf{S}_{12}|/|\mathbf{S}_{22}|. \tag{5.3.10}$$

Hence

$$\begin{aligned} -2\log\lambda &= -n \log(|\mathbf{S}_{22} - \mathbf{S}_{21}\mathbf{S}_{11}^{-1}\mathbf{S}_{12}|/|\mathbf{S}_{22}|) \\ &= -n \log|\mathbf{I} - \mathbf{S}_{22}^{-1}\mathbf{S}_{21}\mathbf{S}_{11}^{-1}\mathbf{S}_{12}| \\ &= -n \log \prod_{i=1}^{k} (1-\lambda_i), \end{aligned} \tag{5.3.11}$$

where the λ_i are the non-zero eigenvalues of $\mathbf{S}_{22}^{-1}\mathbf{S}_{21}\mathbf{S}_{11}^{-1}\mathbf{S}_{12}$, and $k = \min(p_1, p_2)$.

This result is intuitively appealing since if $\mathbf{S}_{12}$ is close to $\mathbf{0}$, as it should be if H_0 is true, then $-2\log\lambda$ also takes a small value. Note that $-2\log\lambda$ can also be written in terms of the correlation matrix as $|\mathbf{I} - \mathbf{R}_{22}^{-1}\mathbf{R}_{21}\mathbf{R}_{11}^{-1}\mathbf{R}_{12}|$.

To find the distribution of the LR statistic, write $\mathbf{M}_{11} = n\mathbf{S}_{11}$, $\mathbf{M}_{22} = n\mathbf{S}_{22}$, and $\mathbf{M}_{22.1} = n(\mathbf{S}_{22} - \mathbf{S}_{21}\mathbf{S}_{11}^{-1}\mathbf{S}_{12})$. Then, from Theorem 3.4.6(b), we see that under H_0, $\mathbf{M}_{22.1}$ and $\mathbf{M}_{22} - \mathbf{M}_{22.1}$ are independently distributed Wishart matrices, $\mathbf{M}_{22.1} \sim W_{p_2}(\mathbf{\Sigma}_{22}, n-1-p_1)$ and $\mathbf{M}_{22} - \mathbf{M}_{22.1} \sim W_{p_2}(\mathbf{\Sigma}_{22}, p_1)$. Hence, provided $n - 1 \geq p_1 + p_2$,

$$\lambda^{2/n} = |\mathbf{M}_{22.1}|/|\mathbf{M}_{22.1} + (\mathbf{M}_{22} - \mathbf{M}_{22.1})| \sim \Lambda(p_2, n-1-p_1, p_1). \tag{5.3.12}$$

The null hypothesis is rejected for small $\lambda^{2/n}$ and the test can be conducted using Wilks' Λ. For the cases $p_1=1,2$ (or $p_2=1,2$) the exact distribution of Λ from (3.7.7)–(3.7.10) can be used. For general values of p_1, p_2, we may use Bartlett's approximation (3.7.11):

$$-(n-\tfrac{1}{2}(p_1+p_2+3))\log|\mathbf{I}-\mathbf{S}_{22}^{-1}\mathbf{S}_{21}\mathbf{S}_{11}^{-1}\mathbf{S}_{12}|\sim\chi^2_{p_1p_2}, \qquad (5.3.13)$$

asymptotically, for large n.

As we shall see in Chapter 10 the LRT can be used in canonical correlation analysis. The exact distribution of (5.3.11) was investigated by Hotelling (1936), Girshick (1939), and Anderson (1958, p. 237). Narain (1950) has shown that the LRT has the desirable property of being unbiased.

When one set of variables has just a single member, so that $p_1=1$ and $p_2=p-1$, these formulae simplify: in that case (5.3.11) is just $-n\log|\mathbf{I}-\mathbf{R}_{22}^{-1}\boldsymbol{\alpha}\boldsymbol{\alpha}'|$, where $\boldsymbol{\alpha}=\mathbf{R}_{21}$ is a vector. Now, using (A.2.3k), this equals

$$-n\log(1-\boldsymbol{\alpha}'\mathbf{R}_{22}^{-1}\boldsymbol{\alpha})=-n\log(1-\mathbf{R}_{12}\mathbf{R}_{22}^{-1}\mathbf{R}_{21})$$
$$=-n\log(1-R^2),$$

where R is the multiple correlation coefficient between the first variable and the others (see Section 6.5).

The union intersection test If $\mathbf{x}_1$ and $\mathbf{x}_2$ are the two subvectors of $\mathbf{x}$, then independence between $\mathbf{x}_1$ and $\mathbf{x}_2$ implies that the scalars $\mathbf{a}'\mathbf{x}_1$ and $\mathbf{b}'\mathbf{x}_2$ are also independent, whatever the vectors $\mathbf{a}$ and $\mathbf{b}$. Now an obvious way of testing whether or not the two scalars are independent is to look at their sample correlation coefficient, r. Clearly

$$r^2=\frac{[C(\mathbf{a}'\mathbf{x}_1,\mathbf{b}'\mathbf{x}_2)]^2}{V(\mathbf{a}'\mathbf{x}_1)V(\mathbf{b}'\mathbf{x}_2)}=\frac{(\mathbf{a}'\mathbf{S}_{12}\mathbf{b})^2}{\mathbf{a}'\mathbf{S}_{11}\mathbf{a}\mathbf{b}'\mathbf{S}_{22}\mathbf{b}}. \qquad (5.3.14)$$

The UIT based on this decomposition uses the statistic $\max_{\mathbf{a},\mathbf{b}} r^2(\mathbf{a},\mathbf{b})$, which equals the largest eigenvalue of $\mathbf{S}_{11}^{-1}\mathbf{S}_{12}\mathbf{S}_{22}^{-1}\mathbf{S}_{21}$, and also equals the largest eigenvalue of $\mathbf{S}_{22}^{-1}\mathbf{S}_{21}\mathbf{S}_{11}^{-1}\mathbf{S}_{12}$. (This result is proved in Theorem 10.2.1.)

Since $\mathbf{S}_{22}^{-1}\mathbf{S}_{21}\mathbf{S}_{11}^{-1}\mathbf{S}_{12}$ can be written as $[\mathbf{M}_{22.1}+(\mathbf{M}_{22}-\mathbf{M}_{22.1})]^{-1}(\mathbf{M}_{22}-\mathbf{M}_{22.1})$, where $\mathbf{M}_{22}=n\mathbf{S}_{22}$ and $\mathbf{M}_{22}-\mathbf{M}_{22.1}=n\mathbf{S}_{21}\mathbf{S}_{11}^{-1}\mathbf{S}_{12}$ have the above independent Wishart distributions, we see that the largest eigenvalue has the greatest root distribution, $\theta(p_2, n-1-p_1, p_1)$, described in Section 3.7.

Writing the non-zero eigenvalues of $\mathbf{S}_{22}^{-1}\mathbf{S}_{21}\mathbf{S}_{11}^{-1}\mathbf{S}_{12}$ as $\lambda_1, \lambda_2, \ldots, \lambda_k$, note that the UIT is based on λ_1, while the LRT derived in (5.3.11) is based on $\prod(1-\lambda_i)$.

Example 5.3.3 The test statistics derived above may be used to examine whether there is a correlation between head length and breadth measurements of first sons and those of second sons. The matrices $\mathbf{S}_{11}$, $\mathbf{S}_{22}$, and $\mathbf{S}_{12}$ are the relevant submatrices of the matrix $\mathbf{S}$ given in Example 5.1.1, and $n=25$, $p_1=2$, $p_2=2$. The LRT and UIT both require evaluation of the eigenvalues of

$$\mathbf{S}_{22}^{-1}\mathbf{S}_{21}\mathbf{S}_{11}^{-1}\mathbf{S}_{12}=\begin{bmatrix}0.3014 & 0.2006\\ 0.4766 & 0.3232\end{bmatrix}.$$

Here $\lambda_1=0.6218$ and $\lambda_2=0.0029$.

For the likelihood ratio test (5.3.12) we require

$$\lambda^{2/n}=(1-\lambda_1)(1-\lambda_2)=0.377.$$

Using the asymptotic distribution (5.3.13), $-(25-\frac{7}{2})\log(0.377)=21.0$ is to be tested against the 5% critical value of a $\chi^2_{p_1p_2}=\chi^2_4$ distribution. The observed value of the statistic is clearly significant, and hence the null hypothesis is rejected. To use an exact distribution, note that $\lambda^{2/n}\sim\Lambda(p_2, n-1-p_1, p_1)=\Lambda(2, 22, 2)$, and from (3.7.10),

$$21(1-\sqrt{\Lambda})/2\sqrt{\Lambda}=6.60\sim F_{4,42}.$$

This value is significant at the 5% level and hence we still reject H_0.

For the union intersection test we require the largest eigenvalue of $\mathbf{S}_{22}^{-1}\mathbf{S}_{21}\mathbf{S}_{11}^{-1}\mathbf{S}_{12}$, which is $\lambda_1=0.6218$. From Table C.4 in Appendix C with $\nu_1=n-p_2-1=22$ and $\nu_2=p_2=2$, and $\alpha=0.05$, the critical value is 0.330. Hence the null hypothesis is rejected for this test also.

5.3.2c *The hypothesis H_0:$\mathbf{P}=\mathbf{I}$, $\boldsymbol{\mu}$ unknown*

The hypothesis of the last section may be generalized to consider the hypothesis that all the variables are uncorrelated with one another; that is $\mathbf{P}=\mathbf{I}$ or, equivalently, $\boldsymbol{\Sigma}$ is diagonal. Under H_0, the mean and variance of each variable are estimated separately, so that $\hat{\boldsymbol{\mu}}=\bar{\mathbf{x}}$ and $\hat{\boldsymbol{\Sigma}}=\operatorname{diag}(s_{11},\ldots,s_{pp})$.

Hence, using (5.3.8), it is easily seen that the LRT is given in terms of the correlation matrix $\mathbf{R}$ by the statistic

$$-2\log\lambda=-n\log|\mathbf{R}|, \tag{5.3.15}$$

which, by Theorem 5.2.1, has an asymptotic χ^2 distribution under H_0 with $\frac{1}{2}p(p+1)-p=\frac{1}{2}p(p-1)$ degrees of freedom. Box (1949) showed that the χ^2 approximation is improved if n is replaced by

$$n'=n-1-\tfrac{1}{6}(2p+5)=n-\tfrac{1}{6}(2p+11). \tag{5.3.16}$$

Thus, we shall use the test statistic

$$-n' \log |\mathbf{R}| \sim \chi^2_{p(p-1)/2} \tag{5.3.17}$$

asymptotically.

5.3.3 Multi-sample hypotheses

We now consider the situation of k independent normal samples, whose likelihood was given by (4.2.17). We shall use some results from Section 4.2.2.4 in deriving LRTs.

5.3.3a The hypothesis $H_b: \boldsymbol{\mu}_1 = \ldots = \boldsymbol{\mu}_k$ *given that* $\boldsymbol{\Sigma}_1 = \ldots = \boldsymbol{\Sigma}_k$ *(one-way multivariate analysis of variance)*

Likelihood ratio test (Wilks' Λ test) The LRT of H_b is easily derived from the result of Section 4.2.2.4. The m.l.e.s under H_b are $\bar{\mathbf{x}}$ and $\mathbf{S}$, since the observations can be viewed under H_b as constituting a single random sample. The m.l.e. of $\boldsymbol{\mu}_i$ under the alternative hypothesis is $\bar{\mathbf{x}}_i$, the ith sample mean, and the m.l.e. of the common variance matrix is $n^{-1}\mathbf{W}$, where $\mathbf{W} = \sum n_i \mathbf{S}_i$ is the "within-groups" sum of squares and products (SSP) matrix and $n = \sum n_i$. Using (4.2.17), the LRT is given by

$$\lambda_b = \{|\mathbf{W}|/|n\mathbf{S}|\}^{n/2} = |\mathbf{T}^{-1}\mathbf{W}|^{n/2}. \tag{5.3.19}$$

Here $\mathbf{T} = n\mathbf{S}$ is the "total" SSP matrix, derived by regarding all the data matrices as if they constituted a single sample. In contrast, the matrix $\mathbf{W}$ is the "within-groups" SSP matrix and

$$\mathbf{B} = \mathbf{T} - \mathbf{W} = \sum n_i (\bar{\mathbf{x}}_i - \bar{\mathbf{x}})(\bar{\mathbf{x}}_i - \bar{\mathbf{x}})' \tag{5.3.20}$$

may be regarded as the "between-groups" SSP matrix. Hence, from (5.3.19),

$$\lambda_b^{2/n} = |\mathbf{W}|/|\mathbf{B} + \mathbf{W}| = |\mathbf{I} + \mathbf{W}^{-1}\mathbf{B}|^{-1}. \tag{5.3.21}$$

The matrix $\mathbf{W}^{-1}\mathbf{B}$ is an obvious generalization of the univariate variance ratio. It will tend to zero if H_0 is true.

We now find the distribution of (5.3.21). Write the k samples as a single data matrix,

$$\mathbf{X}(n \times p) = \begin{bmatrix} \mathbf{X}_1 \\ \cdot \\ \cdot \\ \cdot \\ \mathbf{X}_k \end{bmatrix},$$

where $\mathbf{X}_i(n_i \times p)$ represents the observations from the ith sample, $i = 1, \ldots, k$. Let $\mathbf{1}_i$ denote the n-vector with 1 in the places corresponding to the ith sample and 0 elsewhere, and set $\mathbf{I}_i = \text{diag}(\mathbf{1}_i)$. Then $\mathbf{I} = \sum \mathbf{I}_i$ and $\mathbf{1} = \sum \mathbf{1}_i$. Let $\mathbf{H}_i = \mathbf{I}_i - n_i^{-1}\mathbf{1}_i\mathbf{1}_i'$ represent the centring matrix for the ith sample, so that $n_i\mathbf{S}_i = \mathbf{X}'\mathbf{H}_i\mathbf{X}$, and set

$$\mathbf{C}_1 = \sum \mathbf{H}_i, \qquad \mathbf{C}_2 = \sum n_i^{-1}\mathbf{1}_i\mathbf{1}_i' - n^{-1}\mathbf{1}\mathbf{1}'.$$

It is easily verified that $\mathbf{W} = \mathbf{X}'\mathbf{C}_1\mathbf{X}$ and $\mathbf{B} = \mathbf{X}'\mathbf{C}_2\mathbf{X}$. Further, $\mathbf{C}_1$ and $\mathbf{C}_2$ are idempotent matrices of ranks $n-k$ and $k-1$, respectively, and $\mathbf{C}_1\mathbf{C}_2 = \mathbf{0}$.

Now under H_0, $\mathbf{X}$ is a data matrix from $N_p(\boldsymbol{\mu}, \boldsymbol{\Sigma})$. Thus by Cochran's theorem (Theorem 3.4.4) and Craig's theorem (Theorem 3.4.5),

$$\mathbf{W} = \mathbf{X}'\mathbf{C}_1\mathbf{X} \sim W_p(\boldsymbol{\Sigma}, n-k),$$
$$\mathbf{B} = \mathbf{X}'\mathbf{C}_2\mathbf{X} \sim W_p(\boldsymbol{\Sigma}, k-1),$$

and, furthermore, $\mathbf{W}$ and $\mathbf{B}$ are independent. Therefore, provided $n \geq p+k$,

$$|\mathbf{I} + \mathbf{W}^{-1}\mathbf{B}|^{-1} \sim \Lambda(p, n-k, k-1)$$

under H_0, where Wilks' Λ statistic is described in Section 3.7.

Union intersection test The univariate analogue of H_0 would be tested using the analysis of variance statistics

$$\sum n_i(\bar{x}_i - \bar{x})^2 \Big/ \sum n_i s_i^2,$$

where $\bar{x}$ is the overall mean. The corresponding formula for a linear combination $\mathbf{X}\mathbf{a}$ is $\sum n_i(\mathbf{a}'(\bar{\mathbf{x}}_i - \bar{\mathbf{x}}))^2 / \sum n_i \mathbf{a}'\mathbf{S}_i\mathbf{a}$. This expression's maximum value is the largest eigenvalue of

$$\left(\sum n_i\mathbf{S}_i\right)^{-1} \sum n_i(\bar{\mathbf{x}}_i - \bar{\mathbf{x}})(\bar{\mathbf{x}}_i - \bar{\mathbf{x}})' = \mathbf{W}^{-1}\mathbf{B}, \tag{5.3.22}$$

where $\mathbf{W}$ and $\mathbf{B}$ were defined above. Thus the UI statistic for this hypothesis is the greatest root of $\mathbf{W}^{-1}\mathbf{B}$, which is not the same as the LR statistic. However, note from (5.3.21) that the LRT is in fact based on $\prod (1+\lambda_i)^{-1}$, where $\lambda_1, \ldots, \lambda_p$ are the eigenvalues of $\mathbf{W}^{-1}\mathbf{B}$. Hence again we see the familiar pattern that the LRT and UIT are both functions of the eigenvalues of the same matrix, but they lead to different test statistics.

Two-sample Hotelling T^2 test $(k=2)$ When $k=2$, the LRT can be simplified in terms of the two-sample Hotelling T^2 statistic given in Section 3.6. In this case,

$$\mathbf{B} = n_1(\bar{\mathbf{x}}_1 - \bar{\mathbf{x}})(\bar{\mathbf{x}}_1 - \bar{\mathbf{x}})' + n_2(\bar{\mathbf{x}}_2 - \bar{\mathbf{x}})(\bar{\mathbf{x}}_2 - \bar{\mathbf{x}})'.$$

But $\bar{\mathbf{x}}_1-\bar{\mathbf{x}}=(n_2/n)\mathbf{d}$, where $\mathbf{d}=\bar{\mathbf{x}}_1-\bar{\mathbf{x}}_2$. Also $\bar{\mathbf{x}}_2-\bar{\mathbf{x}}=-n_1\mathbf{d}/n$. Therefore,

$$\mathbf{B}=(n_1n_2/n)\,\mathbf{d}\mathbf{d}', \tag{5.3.23}$$

and

$$|\mathbf{I}+\mathbf{W}^{-1}\mathbf{B}|=|\mathbf{I}+(n_1n_2/n)\mathbf{W}^{-1}\,\mathbf{d}\mathbf{d}'|=1+(n_1n_2/n)\mathbf{d}'\,\mathbf{W}^{-1}\,\mathbf{d}. \tag{5.3.24}$$

The second term is, of course, proportional to the Hotelling two-sample T^2 statistic, and we reject H_0 for large values of this statistic.

Furthermore, $\mathbf{W}^{-1}\mathbf{B}$ is of rank one, and its largest eigenvalue is

$$(n_1n_2/n)\mathbf{d}'\mathbf{W}^{-1}\mathbf{d} \tag{5.3.25}$$

(Corollary A.6.2.1). Therefore the UIT is also given in terms of the two-sample Hotelling T^2 statistic. Hence, although in general the LRT and UIT are different for this hypothesis, they are the same in the two-sample case. For numerical examples of these tests see Examples 12.3.1 and 12.6.1.

*5.3.3*b *The hypothesis* $H_a:\mathbf{\Sigma}_1=\ldots=\mathbf{\Sigma}_k$ *(test of homogeneity of covariances)*

Likelihood ratio test (Box's M test) The m.l.e. of $\boldsymbol{\mu}_i$ is $\bar{\mathbf{x}}_i$ under both H_a and the alternative. The m.l.e. of $\mathbf{\Sigma}_i$ is $\mathbf{S}=n^{-1}\mathbf{W}$ under H_a, and $\mathbf{S}_i$ under the alternative. Therefore

$$-2\log\lambda_a=n\log|\mathbf{S}|-\sum n_i\log|\mathbf{S}_i|=\sum n_i\log|\mathbf{S}_i^{-1}\mathbf{S}|. \tag{5.3.26}$$

Using Theorem 5.2.1, this has an asymptotic chi-squared distribution with $\frac{1}{2}p(p+1)(k-1)$ degrees of freedom. It may be argued that if n_i is small then (5.3.26) gives too much weight to the contribution of $\mathbf{S}$. This consideration led Box (1949) to propose the test statistic M in place of that given by (5.3.26). Box's M is given by

$$M=\gamma\sum(n_i-1)\log|\mathbf{S}_{ui}^{-1}\mathbf{S}_u|, \tag{5.3.27}$$

where

$$\gamma=1-\frac{2p^2+3p-1}{6(p+1)(k-1)}\left(\sum\frac{1}{n_i-1}-\frac{1}{n-k}\right),$$

and $\mathbf{S}_u$ and $\mathbf{S}_{ui}$ are the unbiased estimators

$$\mathbf{S}_u=\frac{n}{n-k}\mathbf{S},\qquad \mathbf{S}_{ui}=\frac{n_i}{n_i-1}\mathbf{S}_i.$$

Box's M also has an asymptotic $\chi^2_{p(p+1)(k-1)/2}$ distribution. Box's approximation seems to be good if each n_i exceeds 20, and if k and p do not exceed 5. Box also gave an asymptotic F distribution (see Pearson and

Hartley, 1972, p. 108). For $p=1$, (5.3.26) reduces to the test statistic for Bartlett's test of homogeneity, viz.

$$n \log s^2 - \sum n_i \log s_i^2, \qquad (5.3.28)$$

where s_i^2 is the variance of sample i and s^2 is the pooled estimate of the variance.

No simple UIT seems to be available for this hypothesis.

Example 5.3.4 Jolicoeur and Mosimann (1960) measured the shell length, width, and height of 48 painted turtles, 24 male and 24 female. The resulting mean vectors and covariance matrices for males and females, respectively, were as follows:

$$\bar{\mathbf{x}}_1 = \begin{bmatrix} 113.38 \\ 88.29 \\ 40.71 \end{bmatrix}, \qquad \mathbf{S}_1 = \begin{bmatrix} 132.99 & 75.85 & 35.82 \\ & 47.96 & 20.75 \\ & & 10.79 \end{bmatrix},$$

$$\bar{\mathbf{x}}_2 = \begin{bmatrix} 136.00 \\ 102.58 \\ 51.96 \end{bmatrix}, \qquad \mathbf{S}_2 = \begin{bmatrix} 432.58 & 259.87 & 161.67 \\ & 164.57 & 98.99 \\ & & 63.87 \end{bmatrix}.$$

Here $n_1 = n_2 = 24$, $n = 48$. The overall covariance matrix $\mathbf{S} = (n_1\mathbf{S}_1 + n_2\mathbf{S}_2)/n$ is

$$\begin{bmatrix} 282.79 & 167.86 & 98.75 \\ & 106.24 & 59.87 \\ & & 37.34 \end{bmatrix}.$$

We have $|\mathbf{S}_1| = 698.87$, $|\mathbf{S}_2| = 11\,124.82$, $|\mathbf{S}| = 4899.74$, while the corresponding determinants of the unbiased estimators are

$$|\mathbf{S}_{u1}| = 794.05, \qquad |\mathbf{S}_{u2}| = 12\,639.89, \qquad |\mathbf{S}_u| = 5567.03.$$

Further, $k=2$, $p=3$, and therefore $\gamma = 0.9293$. Hence, from (5.3.27), $M = 24.099$, which is to be tested against the 5% critical value 12.6 provided by the χ_6^2 distribution. Since this is highly significant we conclude that the male and female population covariance matrices are not equal.

5.3.3c The hypothesis $H_c: \boldsymbol{\mu}_1 = \ldots = \boldsymbol{\mu}_k$ and $\boldsymbol{\Sigma}_1 = \ldots = \boldsymbol{\Sigma}_k$ (test of complete homogeneity)

Likelihood ratio test Let us first discuss the relationship between this hypothesis, H_c, and the hypotheses H_a and H_b which were considered in the last two sections. Note that H_b is a "conditional" hypothesis; that is, if L_a^*, L_b^*, and L_c^* are the maximized likelihoods under the three hypotheses and if L^* is the unconstrained maximum likelihood, then

$$\lambda_a = L_a^*/L^*, \qquad \lambda_b = L_b^*/L_a^*, \qquad \lambda_c = L_c^*/L^*.$$

But L_c^* and L_b^* are equal, and therefore a relationship exists between the λ. In fact $\lambda_c = \lambda_a \lambda_b$, and this observation enables us to obtain λ_c directly from (5.3.19) and (5.3.26). We must have

$$-2 \log \lambda_c = -2 \log \lambda_a - 2 \log \lambda_b = n \log |\tfrac{1}{n}\mathbf{W}| - \sum n_i \log |\mathbf{S}_i|.$$

This statistic has an asymptotic chi-squared distribution with $\frac{1}{2}p(k-1)\times(p+3)$ degrees of freedom.

No simple UIT seems to be available for this hypothesis.

*5.4 The Behrens–Fisher Problem

The problem of testing equality of means with no assumption about the corresponding covariance matrices is an extension of the Behrens–Fisher problem. The essential points appear in the univariate situation, where the confidence interval solution differs from those reached by the fiducial and the Bayesian approaches. An excellent discussion of the univariate case appears in Kendall and Stuart (1973, pp. 146–161). The multivariate problem is as follows. Suppose that $\mathbf{X}_1$ and $\mathbf{X}_2$ are independent random data matrices, where $\mathbf{X}_i(n_i \times p)$ is drawn from $N_p(\boldsymbol{\mu}_i, \boldsymbol{\Sigma}_i)$, $i = 1, 2$; how can we test the hypothesis that $\boldsymbol{\mu}_1 = \boldsymbol{\mu}_2$? Equivalently, we may write $\boldsymbol{\delta} = \boldsymbol{\mu}_1 - \boldsymbol{\mu}_2$ and test the hypothesis $H_0: \boldsymbol{\delta} = \mathbf{0}$ against the hypothesis $H_1: \boldsymbol{\delta} \neq \mathbf{0}$.

5.4.1 The likelihood ratio approach

The m.l.e.s of $(\boldsymbol{\mu}_1, \boldsymbol{\mu}_2, \boldsymbol{\Sigma}_1, \boldsymbol{\Sigma}_2)$ are as follows. Under H_1 the likelihood is maximized when $\boldsymbol{\mu}_i = \bar{\mathbf{x}}_i$ and $\boldsymbol{\Sigma}_i = \mathbf{S}_i$. From (4.2.17) the maximized log likelihood is

$$l_1^* = -\tfrac{1}{2}(n_1 \log |\mathbf{S}_1| + n_2 \log |\mathbf{S}_2|) - \tfrac{1}{2}np \log 2\pi - \tfrac{1}{2}np.$$

Under H_0 the m.l.e. of $\boldsymbol{\mu}$, the common mean, may be shown by differentiation to satisfy

$$\hat{\boldsymbol{\mu}} = (n_1 \hat{\boldsymbol{\Sigma}}_1^{-1} + n_2 \hat{\boldsymbol{\Sigma}}_2^{-1})^{-1}(n_1 \hat{\boldsymbol{\Sigma}}_1^{-1}\bar{\mathbf{x}}_1 + n_2 \hat{\boldsymbol{\Sigma}}_2^{-1}\bar{\mathbf{x}}_2) \tag{5.4.1}$$

(see (A.9.2)), and the m.l.e.s of $\boldsymbol{\Sigma}_i$ satisfy

$$\hat{\boldsymbol{\Sigma}}_i = \mathbf{S}_i + \mathbf{d}_i \mathbf{d}_i', \tag{5.4.2}$$

where

$$\mathbf{d}_i = \bar{\mathbf{x}}_i - \hat{\boldsymbol{\mu}},\ i = 1, 2, \text{ as in Section 4.2.2.4.}$$

A degree of simplification of these equations is possible, but their main value is in suggesting the following iterative algorithm:

(1) Start with the initial estimates $\hat{\boldsymbol{\Sigma}}_i = \mathbf{S}_i$, $i = 1, 2$.
(2) Find the corresponding estimate of $\boldsymbol{\mu}$ from (5.4.1). Call this $\hat{\boldsymbol{\mu}}$.
(3) Using $\hat{\boldsymbol{\mu}}$ in (5.4.2), calculate new $\hat{\boldsymbol{\Sigma}}_i$ by

$$\hat{\boldsymbol{\Sigma}}_i = \mathbf{S}_i + (\bar{\mathbf{x}}_i - \hat{\boldsymbol{\mu}})(\bar{\mathbf{x}}_i - \hat{\boldsymbol{\mu}})'.$$

(4) Return to (2) using the new $\hat{\boldsymbol{\Sigma}}_i$.

This iterative procedure may be repeated until covergence is reached.

5.4.2 Union intersection principle

One approach to the problem consists in using a solution to the univariate Fisher–Behrens problem as proposed by Yao (1965). First, we note that

$$\bar{\mathbf{x}}_i \sim N_p(\boldsymbol{\mu}_i, \boldsymbol{\Gamma}_i), \qquad \mathbf{S}_i \sim W_p(\boldsymbol{\Gamma}_i, f_i), \qquad i = 1, 2, \tag{5.4.3}$$

where

$$f_i = n_i - 1, \qquad \boldsymbol{\Gamma}_i = n_i^{-1}\boldsymbol{\Sigma}_i, \qquad i = 1, 2. \tag{5.4.4}$$

Let

$$\left.\begin{array}{lll} \mathbf{d} = \bar{\mathbf{x}}_1 - \bar{\mathbf{x}}_2, & \mathbf{U}_i = \mathbf{S}_i/f_i, & i = 1, 2, \\ \mathbf{U} = \mathbf{U}_1 + \mathbf{U}_2, & \boldsymbol{\Gamma} = \boldsymbol{\Gamma}_1 + \boldsymbol{\Gamma}_2. & \end{array}\right\} \tag{5.4.5}$$

Note that $\mathbf{U}$ is an unbiased estimator of $\boldsymbol{\Gamma}$, and, when the null hypothesis is true,

$$\mathbf{d} \sim N_p(\mathbf{0}, \boldsymbol{\Gamma}).$$

Let us assume that the following statement is true:

$$f\mathbf{U} \sim W_p(\boldsymbol{\Gamma}, f) \tag{5.4.6}$$

where f is to be selected, independently of $\mathbf{d}$, so that for all p-vectors $\mathbf{a}$,

$$f\mathbf{a}'\mathbf{U}\mathbf{a} \sim (\mathbf{a}'\boldsymbol{\Gamma}\mathbf{a})\chi_f^2.$$

Then, as in Section 5.2.2b, it can be shown that

$$w_{\mathbf{a}} = (\mathbf{a}'\mathbf{d})^2/(\mathbf{a}'\mathbf{U}\mathbf{a}) \sim t_{\mathbf{a}}^2(f), \tag{5.4.7}$$

where $t_{\mathbf{a}}(f)$ denotes a t statistic with f degrees of freedom. Further,

$$w_{\mathbf{a}}^* = \max_{\mathbf{a}} w_{\mathbf{a}} = \mathbf{d}'\mathbf{U}^{-1}\mathbf{d} \sim T^2(p, f), \tag{5.4.8}$$

where the maximizing $\mathbf{a}$ is given by

$$\mathbf{a}^* = \mathbf{U}^{-1}\mathbf{d}. \tag{5.4.9}$$

Welch (1947) has shown that although (5.4.6) is not in general valid, we have, approximately,

$$w_{\mathbf{a}} \sim t_{\mathbf{a}}^2(f_{\mathbf{a}}), \tag{5.4.10}$$

where

$$\frac{1}{f_{\mathbf{a}}} = \frac{1}{f_1}\left(\frac{\mathbf{a}'\mathbf{U}_1\mathbf{a}}{\mathbf{a}'\mathbf{U}\mathbf{a}}\right)^2 + \frac{1}{f_{\mathbf{a}}}\left(\frac{\mathbf{a}'\mathbf{U}_2\mathbf{a}}{\mathbf{a}'\mathbf{U}\mathbf{a}}\right)^2. \tag{5.4.11}$$

On substituting $\mathbf{a}^*$ from (5.4.9) in (5.4.11), we expect from (5.4.8) that, approximately,

$$\mathbf{d}'\mathbf{U}^{-1}\mathbf{d} \sim T^2(p, f^*),$$

where

$$\frac{1}{f^*} = \frac{1}{f_1}\left(\frac{\mathbf{d}'\mathbf{U}^{-1}\mathbf{U}_1\mathbf{U}^{-1}\mathbf{d}}{\mathbf{d}'\mathbf{U}^{-1}\mathbf{d}}\right)^2 + \frac{1}{f_2}\left(\frac{\mathbf{d}'\mathbf{U}^{-1}\mathbf{U}_2\mathbf{U}^{-1}\mathbf{d}}{\mathbf{d}'\mathbf{U}^{-1}\mathbf{d}}\right)^2.$$

Thus a test can be carried out. For other solutions to this problem, see Bennett (1951) and G. S. James (1954).

5.5 Simultaneous Confidence Intervals

We now show through examples how the union intersection principle leads to the construction of so-called "simultaneous confidence intervals" for parameters.

5.5.1 The one-sample Hotelling T^2 case

From (5.2.14) we have

$$t_{\mathbf{a}}^2 \leq T^2 \qquad \text{for all } p\text{-vectors } \mathbf{a}, \tag{5.5.1}$$

where

$$t_{\mathbf{a}} = \mathbf{a}'(\bar{\mathbf{x}} - \boldsymbol{\mu})/[\mathbf{a}'\mathbf{S}\mathbf{a}/(n-1)]^{1/2} \tag{5.5.2}$$

and T^2 is the one-sample Hotelling T^2 statistic. Since, from Theorem 3.5.2, T^2 is proportional to an F distribution, the upper 100α percentage point of T^2 is

$$\Pr(T^2 \geq T_\alpha^2) = \alpha, \tag{5.5.3}$$

where

$$T_\alpha^2 = \{(n-1)p/(n-p)\}F_{p,n-p;\alpha} \tag{5.5.4}$$

and $F_{p,n-p;\alpha}$ is the upper 100α percentage point of $F_{p,n-p}$. Now if $T^2 \leq T_\alpha^2$, then (5.5.1) implies

$$t_{\mathbf{a}}^2 \leq T_\alpha^2 \qquad \text{for all } p\text{-vectors } \mathbf{a}. \tag{5.5.5}$$

On substituting (5.5.2) and (5.5.4) in (5.5.5), and inverting the inequality, we can rewrite (5.5.3) as

$$\Pr\{\mathbf{a}'\boldsymbol{\mu} \in (\mathbf{a}'\bar{\mathbf{x}} - b, \mathbf{a}'\bar{\mathbf{x}} + b) \text{ for all } \mathbf{a}\} = 1 - \alpha, \tag{5.5.6}$$

where

$$b = [\{\mathbf{a}'\mathbf{S}\mathbf{a}p/(n-p)\}F_{p,n-p;\alpha}]^{1/2}.$$

Thus

$$(\mathbf{a}'\bar{\mathbf{x}} - b, \mathbf{a}'\bar{\mathbf{x}} + b) \tag{5.5.7}$$

is the $100(1-\alpha)\%$ confidence interval for $\mathbf{a}'\boldsymbol{\mu}$. Note that *there is a probability of* $(1-\alpha)$ *that all confidence intervals for* $\mathbf{a}'\boldsymbol{\mu}$ *obtained by varying* $\mathbf{a}$ *will be true*. Hence, they are called the simultaneous confidence intervals for $\mathbf{a}'\boldsymbol{\mu}$.

(a) It can be seen that, for fixed $\mathbf{a}$, we could obtain $100(1-\alpha)\%$ confidence intervals for $\mathbf{a}'\boldsymbol{\mu}$ from $t_{\mathbf{a}}$ given by (5.5.2) since it is distributed as Student's t with $(n-1)$ d.f. Namely, in (5.5.7), take $b = [\{\mathbf{a}'\mathbf{S}\mathbf{a}/(n-1)\}F_{1,n-1;\alpha}]^{1/2}$ since $F_{1,n-1} = (t_{n-1})^2$. (Of course, the confidence coefficient will change if $\mathbf{a}$ is varied.)

(b) The simultaneous confidence intervals may be used to study more specific hypotheses concerning $\boldsymbol{\mu}$. For example, the hypothesis $H_0: \boldsymbol{\mu} = \mathbf{0}$ is accepted at $100\alpha\%$ level of significance if and only if every $100(1-\alpha)\%$ simultaneous confidence interval contains the zero value. *Thus if H_0 is rejected, there must be at least one vector* $\mathbf{a}$ *for which the corresponding confidence interval does not contain zero.* Hence this method allows us to examine which particular linear combinations lead H_0 to be rejected.

Example 5.5.1 In Example 5.3.1 the differences in cork depth in the four directions were summarized by the transformed variables $\mathbf{Y} = \mathbf{X}\mathbf{R}'$ and $\mathbf{Y}$ was assumed to be a data matrix from $N_3(\mathbf{R}\boldsymbol{\mu}, \mathbf{R}\boldsymbol{\Sigma}\mathbf{R}')$, where $\mathbf{R}\boldsymbol{\mu} = \boldsymbol{\nu}$, say. Hotelling's T^2 test for the cork data led to the rejection of

$H_0: \boldsymbol{\nu} = \mathbf{0}$. We now examime which mean might have led to the rejection of H_0. If we put $\mathbf{a} = (1, 0, 0)'$, then $\mathbf{a}'\mathbf{S}\mathbf{a} = s_{11}$. Using Example 1.5.1,

$$n = 28, \qquad p = 3, \qquad \mathbf{y}_1 = 8.8571, \qquad F_{3,25;0.01} = 4.68, \qquad s_{11} = 124.123,$$

so (5.5.7) gives the 99% confidence interval for ν_1 as

$$0.51 < \nu_1 < 17.21.$$

Similarly, using the values $\mathbf{a} = (0, 1, 0)'$ and $\mathbf{a} = (0, 0, 1)'$, respectively, gives the intervals

$$-5.01 < \nu_2 < 6.72 \quad \text{and} \quad -6.48 < \nu_3 < 8.48.$$

Hence only the interval for ν_1 does not contain zero, and thus it is this particular mean which led to the rejection of the null hypothesis.

In general, when the hypothesis of zero mean is rejected, there must exist at least one $\mathbf{a}$ for which the associated confidence interval excludes zero. However, it is not necessary that $\mathbf{a}$ be one of the vectors $\mathbf{e}_i = (0, 0, \ldots, 1, 0, \ldots, 0)'$ as in the above example.

5.5.2 The two-sample Hotelling T^2 case

As in Section 5.5.1, we can construct a simultaneous confidence interval for $\mathbf{a}'\boldsymbol{\delta} = \mathbf{a}'(\boldsymbol{\mu}_1 - \boldsymbol{\mu}_2)$ using the two sample means $\bar{\mathbf{x}}_1$ and $\bar{\mathbf{x}}_2$ based on sample sizes n_1 and n_2. A UIT test for $\boldsymbol{\delta} = \mathbf{0}$ was developed in (5.3.25). Using the same argument as in Section 5.5.1, it is found that the $100(1-\alpha)\%$ simultaneous confidence intervals for $\mathbf{a}'\boldsymbol{\delta}$ are given by

$$\mathbf{a}'(\bar{\mathbf{x}}_1 - \bar{\mathbf{x}}_2) \pm \{\mathbf{a}'(n_1\mathbf{S}_1 + n_2\mathbf{S}_2)\mathbf{a}p(n_1 + n_2)F_{p,m;\alpha}/n_1 n_2 m\}^{1/2}, \qquad (5.5.7)$$

where $m = n_1 + n_2 - p - 1$.

5.5.3 Other examples

The union intersection approach can be used to give simultaneous confidence intervals in more complicated situations. For example, SCIs can be evaluated for the one-way multivariate analysis of variance of Section 5.3.3a to examine which group differences on which variables are important in the rejection of the null hypothesis. These SCIs are discussed in Section 12.6.

More generally, simultaneous confidence intervals can be calculated for the parameters of a multivariate regression. See Section 6.3.3.

5.6 Multivariate Hypothesis Testing: Some General Points

Having discussed the LRT and UIT, what conclusions can one draw? They are different procedures, and in general they may lead to different test statistics.

An important statistical property of LRT is the general asymptotic chi-squared result stated in Theorem 5.2.1. No such result exists for UITs. However, the importance of this difference should not be overemphasized, since no corresponding small-sample result is known. In addition, the LRT depends upon the assumption of a specific distributional form. This is not crucial to the UI procedure, which merely requires a "sensible" way of testing each of the component hypotheses. For example, the t statistic may be sensible for many non-normal distributions.

A more important advantage of the UI approach is that, if H_0 is rejected, it is a simple matter to calculate which of the component hypotheses led to the rejection of H_0. This can suggest a particular way of amending the hypothesis in order to accord more with the data. For instance, if the T^2 statistic devised under both the LR and UI approaches were "significant", then the implication using the LR approach would be simply "reject H_0". However, the UI approach would indicate more than this. As well as knowing that H_0 should be rejected, one could enquire which specific linear combinations $\mathbf{a}'\mathbf{x}$ led to its rejection. We saw this property illustrated in the discussion of simultaneous confidence intervals in Section 5.5.

Another nice property of the LRT is that it sometimes leads to convenient factorizations. For example, in Theorem 3.7.3 it was shown that Wilks' Λ can be written as the product of independent beta variates.

From extensive studies it has become clear that the UI criterion is insensitive to alternative hypotheses involving more than one non-zero eigenvalue. On the other hand, Wilks' criterion provides good protection over a wide range of alternatives (Schatzoff, 1966; Pillai and Jayachandran, 1967). However, Wilks' criterion in itself does not provide simultaneous confidence intervals, a feature which might in practice outweigh its theoretical power properties.

Gabriel (1970) has examined theoretically the relationship between LR and UI tests. Other criteria have also been proposed, including the following functions of the eigenvalues:

$$\text{Pillai (1955)} \qquad \sum \lambda_i;$$

$$\text{Giri (1968)} \qquad \sum \frac{\lambda_i}{1+\lambda_i};$$

$$\text{Kiefer and Schwartz (1965)} \qquad \prod \left(\frac{1+\lambda_i}{\lambda_i}\right).$$

Mardia (1971) has investigated the effect of non-normality upon Pillai's criterion. He found that it is fairly robust when used for testing hypotheses on means, but not when used for hypotheses on covariance matrices. These properties are reminiscent of similar results from robustness studies on univariate statistics.

Of course there are further criteria which can be used for assessing multivariate tests—the criterion of maximal invariance is one (Cox and Hinkley, 1974, pp. 157–170).

*5.7 Non-normal Data

We have so far assumed that the parent population is multinormal. For non-normal populations when the form of the density is known, we can proceed to obtain the likelihood ratio test for a given testing problem. The asymptotic distribution of the likelihood ratio criterion is known through Wilks' theorem (see for example Exercise 5.7.1). However, the exact distribution is not known in general.

Since multinormal theory is so developed, it is worthwhile to enquire which tests of multinormal theory can safely be applied to non-normal data. Initially, of course, a test of multinormality should be applied. We shall now describe two such tests. For a review of various tests of multinormality, see Mardia (1978).

In Theorem 2.5.5 it was shown that, for the multinormal distribution, the multivariate measures of skewness and kurtosis have values $\beta_{1,p}=0$ and $\beta_{2,p}=p(p+2)$. Mardia (1970*a*) has shown that the sample counterparts of the measures given in Section 1.8, $b_{1,p}$ and $b_{2,p}$, have the following asymptotic distributions as $n\to\infty$ when the underlying population is multinormal:

$$\tfrac{1}{6}nb_{1,p} \sim \chi_f^2, \quad \text{where} \quad f=\tfrac{1}{6}p(p+1)(p+2), \tag{5.7.1}$$

and

$$\{b_{2,p}-p(p+2)\}/\{8p(p+2)/n\}^{1/2} \sim N(0, 1). \tag{5.7.2}$$

These statistics can be used to test the null hypothesis of multinormality. One rejects the null hypothesis for large $b_{1,p}$ and for both large and small values of $b_{2,p}$. Critical values of these statistics for small samples are given in Mardia (1974).

Example 5.7.1 Consider the cork data of Table 1.4.1, where $n=28$ and $p=4$. From (1.8.1) and (1.8.2) it is found that

$$b_{1,4}=4.476 \quad \text{and} \quad b_{2,4}=22.957.$$

Then $\frac{1}{6}nb_{1,4}=20.9$ is clearly not significant when compared to the upper 5% critical value of the χ^2_{20} distribution. From (5.7.2) we find that the observed value of the standard normal variate is -0.40, which again is not significant.

The main advantage of these tests is that if the hypotheses of $\beta_{1,p}=0$ and $\beta_{2,p}=p(p+2)$ are accepted then we can use the normal theory tests on mean vectors and covariance matrices because Mardia (1970*a*, 1971, 1974, 1975*a*) has shown the following:

(a) The size of Hotelling's T^2 test is overall robust to non-normality, although both Hotelling's one-sample T^2 test and the two-sample test for $n_1 \neq n_2$ are more sensitive to $\beta_{1,p}$ than to $\beta_{2,p}$. If $n_1=n_2$ the two-sample test is hardly influenced by $\beta_{1,p}$ or $\beta_{2,p}$.
(b) Tests on covariances are sensitive, and sensitivity is measured by $\beta_{2,p}$.

Thus, broadly speaking, in the presence of non-normality, the normal theory tests on means are influenced by $\beta_{1,p}$ whereas tests on covariances are influenced by $\beta_{2,p}$.

If the hypothesis of multinormality is rejected, one can sometimes take remedial steps to transform the data to normality, e.g. using the log transformation. Another possibility is to use a non-parametric test, but the field of multivariate non-parametric methods is not a rich one (see Puri and Sen, 1971). This fact partly results from the lack of a natural way to order multivariate data. However, we give a bivariate non-parametric test in the next section.

5.8 A Non-parametric Test for the Bivariate Two-sample Problem

Suppose $\mathbf{x}_{1i}$, $i=1,\ldots,n_1$, and $\mathbf{x}_{2j}$, $j=1,\ldots,n_2$, are independent random samples from populations with c.d.f.s $F(\mathbf{x})$ and $G(\mathbf{x})$, respectively. Suppose $\bar{\mathbf{x}}$ is the mean vector of the combined sample. Let the angles made with the x_1 axis in the positive direction by the vectors $(\mathbf{x}_{1i}-\bar{\mathbf{x}})$, $i=1,\ldots,n_1$, and $(\mathbf{x}_{2j}-\bar{\mathbf{x}})$, $j=1,\ldots,n_2$, be ranked in a single sequence. Let $(r_1,\ldots,r_{n_1})$ and $(r'_1,\ldots,r'_{n_2})$ be the ranks of the angles in the first and the second samples, respectively. Let $N=n_1+n_2$. We replace the angles of the first and second samples by

$$\mathbf{y}_{1i}=\{\cos(2\pi r_i/N), \sin(2\pi r_i/N)\}', \qquad i=1,\ldots,n_1, \tag{5.8.1}$$

and

$$\mathbf{y}_{2j}=\{\cos(2\pi r'_j/N), \sin(2\pi r'_j/N)\}', \qquad j=1,\ldots,n_2. \tag{5.8.2}$$

This means that the angles are replaced by the "uniform scores" on a circle. For these observations it can be shown that

$$\bar{\mathbf{y}}=\mathbf{0}, \qquad \mathbf{S}_y=\tfrac{1}{2}\mathbf{I}. \tag{5.8.3}$$

Let T^2 be Hotelling's two-sample T^2 statistic for the $\mathbf{y}$ given by (3.6.4), and define

$$U=\frac{(N-1)T^2}{(N-2)}=\frac{n_1(N-1)}{n_2}(\bar{\mathbf{y}}_1-\bar{\mathbf{y}})'\mathbf{S}_y^{-1}(\bar{\mathbf{y}}_1-\bar{\mathbf{y}}). \tag{5.8.4}$$

(The second equality follows because $\bar{\mathbf{y}}_1-\bar{\mathbf{y}}_2=\bar{\mathbf{y}}_1-\bar{\mathbf{y}}+\bar{\mathbf{y}}-\bar{\mathbf{y}}_2=(N/n_2)\times(\bar{\mathbf{y}}_1-\bar{\mathbf{y}})$.) Consider the hypotheses $H_0: F(\mathbf{x})=G(\mathbf{x})$ and $H_1: F(\mathbf{x})=G(\mathbf{x}+\boldsymbol{\delta})$, $\boldsymbol{\delta}\neq 0$. Since we expect $\bar{\mathbf{y}}_1-\bar{\mathbf{y}}$ to be shifted in the direction $\boldsymbol{\delta}$ when the alternative hypothesis is true, U is a sensible statistic for this test. On substituting (5.8.3) in (5.8.4), we get

$$U=\frac{2(N-1)}{n_1n_2}\left\{\left(\sum_{i=1}^{n_1}\cos\frac{2\pi r_i}{N}\right)^2+\left(\sum_{i=1}^{n_1}\sin\frac{2\pi r_i}{N}\right)^2\right\}. \tag{5.8.5}$$

We reject H_0 for large values of U.

As n_1, $n_2\to\infty$, $n_1/n_2\to\beta$, $0<\beta<1$, it is found under H_0 that, asymptotically,

$$U\sim\chi_2^2. \tag{5.8.6}$$

Example 5.8.1 The distance between the shoulders of the larger left valve (x) and the length of specimens (y) of *Bairda oklahomaensis* from two geological levels are given in Table 5.8.1. Here $n_1=8$, $n_2=11$, $N=19$. It is found that the ranks r_i for level 1 are

5, 13, 14, 15, 16, 17, 18, 19,

and

$$\sum\cos\frac{2\pi r_i}{N}=2.9604, \qquad \sum\sin\frac{2\pi r_i}{N}=-3.6613.$$

Table 5.8.1 Data for the geological problem (Shaver, 1960)

Level 1		Level 2	
x	y	x	y
631	1167	682	1257
606	1222	631	1227
682	1278	631	1237
480	1045	707	1368
606	1151	631	1227
556	1172	682	1262
429	970	707	1313
454	1166	656	1283
		682	1298
		656	1283
		672	1278

Values of x and y are in micrometres.

Consequently U given by (5.8.4) is 9.07, which is significant since the 5% value of χ_2^2 is 5.99.

We note the following points. Like T^2, U is invariant under non-singular linear transformations of the variables $\mathbf{x}_{1i}$ and $\mathbf{x}_{2i}$. The Pitman's asymptotic efficiency of this test relative to Hotelling's T^2 test for the normal model is $\pi/4$. The U test is strictly non-parametric in the sense that the null distribution does not depend on the underlying distribution of the variables. This test was proposed by Mardia (1967*b*). For critical values and extensions see Mardia (1967*b*, 1968, 1969, 1972*a*,*b*) and Mardia and Spurr (1977).

Another non-parametric test can be constructed by extending the Mann–Whitney test as follows. First combine the two groups and rank the observations for each variable separately. Then use T^2 for the resulting ranks. This test is *not* invariant under all linear transformations, but it is invariant under monotone transformations of each variable separately. For further non-parametric tests, see Puri and Sen (1971).

Exercises and Complements

5.2.1 Derive a likelihood ratio test for the hypothesis $\boldsymbol{\mu}'\boldsymbol{\mu}=1$, based on one observation $\mathbf{x}$ from $N_p(\boldsymbol{\mu}, \sigma^2\mathbf{I})$, where σ^2 is known.

5.2.2 Show that $\boldsymbol{\Sigma}_0^{-1}\mathbf{S}$, a, g, and $-2\log\lambda$ take the values indicated in Example 5.2.3.

5.2.3 If $\mathbf{M}$ is a $(p \times p)$ matrix, distributed as $W_p(\mathbf{\Sigma}, n)$, where

$$\mathbf{\Sigma} = \begin{bmatrix} \mathbf{\Sigma}_{11} & \mathbf{0} \\ \mathbf{0} & \mathbf{\Sigma}_{22} \end{bmatrix} \begin{matrix} p_1 \\ p_2 \end{matrix},$$

and $\mathbf{M}$ is partitioned conformably, show that $\mathbf{F} = \mathbf{M}_{21}\mathbf{M}_{11}^{-1}\mathbf{M}_{12}$ and $\mathbf{G} = \mathbf{M}_{22} - \mathbf{M}_{21}\mathbf{M}_{11}^{-1}\mathbf{M}_{12}$ are distributed independently as $W_{p_2}(\mathbf{\Sigma}_{22}, p_1)$ and $W_{p_2}(\mathbf{\Sigma}_{22}, n-p_1)$.

Using the fact that $\mathbf{a}'\mathbf{F}\mathbf{a}$ and $\mathbf{a}'\mathbf{G}\mathbf{a}$ are independent χ^2 variables for all $\mathbf{a}$, derive a two-sided union intersection test, based on the F-ratio, of the hypothesis that $\mathbf{\Sigma}$ is in fact of the form given above.

5.2.4 (a) Show that if Example 5.2.1 is amended so that

$$\mathbf{\Sigma} = \begin{bmatrix} 100 & 50 \\ 50 & 100 \end{bmatrix},$$

then $-2 \log \lambda = 3.46$, and H_0 is still not rejected.

(b) Show that with $\mathbf{\Sigma}$ given above, the 95% confidence region for μ_1 and μ_3 is elliptical, and takes the form

$$(185.72-\mu_1)^2+(183.84-\mu_3)^2-(185.72-\mu_1)(183.84-\mu_3) \leqslant 17.97.$$

5.2.5 Many of the tests in this chapter have depended upon the statistic $\mathbf{d}'\mathbf{S}^{-1}\mathbf{d}$, where $\mathbf{S}$ is a $(p \times p)$ matrix. Show that in certain special cases this statistic simplifies as follows:

(a) If $\mathbf{S}$ is diagonal, then $\mathbf{d}'\mathbf{S}^{-1}\mathbf{d} = t_1^2 + \ldots + t_p^2$, where $t_i = d_i/\sqrt{s_{ii}}$.

(b) When $p = 2$,

$$\mathbf{d}'\mathbf{S}^{-1}\mathbf{d} = \frac{(t_1^2 + t_2^2 - 2t_1t_2r_{12})}{1 - r_{12}^2}.$$

(c) When $p = 3$. $\mathbf{d}'\mathbf{S}^{-1}\mathbf{d}$ is given by

$$\frac{t_1^2(1-r_{23}^2)+t_2^2(1-r_{13}^2)+t_3^2(1-r_{12}^2)-2t_1t_2u_{12.3}-2t_1t_3u_{13.2}-2t_2t_3u_{23.1}}{1+2r_{12}r_{13}r_{23}-r_{23}^2-r_{13}^2-r_{12}^2}$$

where $u_{12.3} = r_{12} - r_{13}r_{23}$, etc.

5.3.1 (Rao, 1948, p. 63) (a) Consider the cork-tree data from Example 5.3.1 and the contrasts

$$z_1 = (\mathrm{N}+\mathrm{S})-(\mathrm{E}+\mathrm{W}), \qquad z_2 = \mathrm{S}-\mathrm{W}, \qquad z_3 = \mathrm{N}-\mathrm{S}.$$

Show that $z_1 = y_1$, $z_3 = y_2$, and $z_2 = \frac{1}{2}(y_1 - y_2 + y_3)$, where y_1, y_2, y_3 were defined in Example 5.3.1.

(b) Show that the sample mean and covariance matrix for $\mathbf{z} = (z_1, z_2, z_3)'$ are

$$\bar{\mathbf{z}} = (8.86, 4.50, 0.86)' \quad \text{and} \quad \mathbf{S} = \begin{bmatrix} 124.12 & 59.22 & -20.27 \\ & 54.89 & -27.29 \\ & & 61.26 \end{bmatrix}.$$

(c) Hence show that the value of T^2 for testing the hypothesis $E(\mathbf{z}) = \mathbf{0}$ is $T^2 = 20.736$ and the value of F is 6.40. Show that this is significant at the 1% level. (Note that the value of T^2 obtained in this exercise is the same as that obtained in Example 5.3.1, thus illustrating that T^2 is invariant under linear transformations.)

5.3.2 (Seal, 1964, p. 106) (a) Measurements of cranial length (x_1) and cranial breadth (x_2) on a sample of 35 mature female frogs led to the following statistics:

$$\bar{\mathbf{x}} = \begin{bmatrix} 22.860 \\ 24.397 \end{bmatrix}, \qquad \mathbf{S}_1 = \begin{bmatrix} 17.178 & 19.710 \\ & 23.710 \end{bmatrix}.$$

Test the hypothesis that $\mu_1 = \mu_2$.

(b) Similar measurements on 14 male frogs led to the statistics

$$\bar{\mathbf{x}} = \begin{bmatrix} 21.821 \\ 22.843 \end{bmatrix}, \qquad \mathbf{S}_2 = \begin{bmatrix} 17.159 & 17.731 \\ & 19.273 \end{bmatrix}.$$

Again test the hypothesis that $\mu_1 = \mu_2$.

(c) Show that the pooled covariance matrix obtained from the above data is

$$\mathbf{S} = \frac{n_1\mathbf{S}_1 + n_2\mathbf{S}_2}{n_1 + n_2} = \begin{bmatrix} 17.173 & 19.145 \\ & 22.442 \end{bmatrix}.$$

Use this to test the hypothesis that the covariance matrices of male and female populations are the same.

(d) Test the hypothesis that $\boldsymbol{\mu}_1 = \boldsymbol{\mu}_2$, stating your assumptions carefully.

5.3.3 (a) Consider the hypothesis $H_0: \boldsymbol{\mu} = k\boldsymbol{\mu}_0$, where $\boldsymbol{\Sigma}$ is known. From (4.2.9), the m.l.e. of k under H_0 is given by $\hat{k} = \boldsymbol{\mu}_0'\boldsymbol{\Sigma}^{-1}\bar{\mathbf{x}}/\boldsymbol{\mu}_0\boldsymbol{\Sigma}^{-1}\boldsymbol{\mu}_0$. Using (5.3.1), show that the LRT is given by

$$-2\log\lambda = n\bar{\mathbf{x}}'\boldsymbol{\Sigma}^{-1}\{\boldsymbol{\Sigma} - (\boldsymbol{\mu}_0'\boldsymbol{\Sigma}^{-1}\boldsymbol{\mu}_0)^{-1}\boldsymbol{\mu}_0\boldsymbol{\mu}_0'\}\boldsymbol{\Sigma}^{-1}\bar{\mathbf{x}}.$$

Using Section 5.3.1a, deduce that $-2\log\lambda \sim \chi^2_{p-1}$ exactly.

(b) Consider now the hypothesis $H_0: \boldsymbol{\mu} = k\boldsymbol{\mu}_0$, where $\boldsymbol{\Sigma}$ is *unknown*. From (4.2.11) the m.l.e. of k under H_0 is given by $\hat{k} = \boldsymbol{\mu}_0'\mathbf{S}^{-1}\bar{\mathbf{x}}/\boldsymbol{\mu}_0'\mathbf{S}^{-1}\boldsymbol{\mu}_0$. Using (5.3.2), show that

$$-2 \log \lambda = n \log (1 + \mathbf{d}'\mathbf{S}^{-1}\mathbf{d}),$$

where

$$\mathbf{d}'\mathbf{S}^{-1}\mathbf{d} = \bar{\mathbf{x}}\mathbf{S}^{-1}\{\mathbf{S} - (\boldsymbol{\mu}_0'\mathbf{S}^{-1}\boldsymbol{\mu}_0)^{-1}\boldsymbol{\mu}_0\boldsymbol{\mu}_0'\}\mathbf{S}^{-1}\bar{\mathbf{x}}.$$

Using Section 5.3.1b, show that $(n-1)\mathbf{d}'\mathbf{S}^{-1}\mathbf{d} \sim T^2(p-1, n-1)$.

5.3.4 If λ_a, λ_b, and λ_c are as defined as in Section 5.3.3c and if m_a, m_b, and m_c are the respective asymptotic degrees of freedom, show that $m_c = m_a + m_b$. Explain.

5.3.5 Let $\mathbf{X}$ be a data matrix from $N_p(\boldsymbol{\mu}, \boldsymbol{\Sigma})$.

(a) Show that the LRT for the hypothesis that x_1 is uncorrelated with $x_2, \ldots, x_p$ is given by

$$\lambda = (1 - R^2)^{n/2},$$

where $R^2 = \mathbf{R}_{12}\mathbf{R}_{22}^{-1}\mathbf{R}_{21}$ is the squared multiple correlation coefficient between x_1 and $x_2, \ldots, x_p$.

(b) Show that the LRT for the hypothesis that $\boldsymbol{\Sigma} = \sigma^2\{(1-\rho)\mathbf{I} + \rho\mathbf{1}\mathbf{1}'\}$ is given by

$$\lambda = \{|\mathbf{S}|/[v^p(1-r)^{p-1}(1+(p-1)r)]\}^{n/2},$$

where v is the average of the diagonal elements of $\mathbf{S}$ and vr is the average of the off-diagonal elements. (See Exercise 4.2.12 and also Wilks, 1962, p. 477.)

5.3.6 (a) Rao (1966, p. 91) gives an anthropometric example in which each of two variables taken separately gives a "significant" difference between two communities, whereas the T^2 or F test utilizing both variables together does not indicate a significant difference.

Community	Sample size	Mean length (in millimetres) of Femur	Mean length (in millimetres) of Humerus
1	27	460.4	335.1
2	20	444.3	323.1
t statistics for each character		2.302	2.214

The pooled unbiased estimate of the common covariance matrix can be constructed from the information in Rao (1966) as

$$\mathbf{S}_u = \begin{bmatrix} 561.8 & 380.7 \\ & 343.2 \end{bmatrix}.$$

Show that each of the individual t statistics is significant at the 5% level (using 45 degrees of freedom). However, if one tests whether the population means of femur and humerus are the same in both communities, show that the resulting F statistic takes the value 2.685, which is not significant at the 5% level (using two and 44 degrees of freedom).

(b) Rao shows that the power of the T^2 test depends upon p, the number of variables, and Δ_p, the Mahalanobis distance in p dimensions between the two populations. He gives charts which illustrate the power functions for values of p between 2 and 9. It can be seen from these charts that, for a given sample size, the power can decrease when the number of variables increases from p to $(p+q)$, unless the increase in Mahalanobis distance $\Delta^2_{p+q}-\Delta^2_p$ is of a certain order of magnitude. In the example given in part (a), the D^2 for femur alone was $D_1^2=0.4614$, while that for femur and humerus together was 0.4777. Such a small increase led to a loss of power with samples of the given size. Comment on this behaviour.

5.5.1 Following Section 5.5.1, show that the simultaneous confidence intervals for $\mathbf{a}'(\boldsymbol{\mu}_1-\boldsymbol{\mu}_2)$ in the two-sample case are given by (5.5.7).

5.5.2 (a) In the one-sample case when $\boldsymbol{\Sigma}$ is known, use the approach of Section 5.5.1 to show that simultaneous $100(1-\alpha)\%$ confidence intervals for $\mathbf{a}'\boldsymbol{\mu}$ are given by

$$\mathbf{a}'\boldsymbol{\mu}\in\mathbf{a}'\bar{\mathbf{x}}\pm\left\{\frac{1}{n}(\mathbf{a}'\boldsymbol{\Sigma}\mathbf{a})\chi^2_{p;\alpha}\right\}^{1/2}\qquad\text{for all } \mathbf{a}.$$

(b) Show that $\mathbf{a}'\boldsymbol{\mu}$ lies in the above confidence interval for all $\mathbf{a}$ if and only if $\boldsymbol{\mu}$ lies in the $100(1-\alpha)\%$ confidence ellipsoid of Example 5.2.1,

$$n(\boldsymbol{\mu}-\bar{\mathbf{x}})'\boldsymbol{\Sigma}^{-1}(\boldsymbol{\mu}-\bar{\mathbf{x}})<\chi^2_{p;\alpha}.$$

(Hint: see Corollary A.9.2.2).

(c) Construct a corresponding confidence ellipsoid for the one-sample case when $\boldsymbol{\Sigma}$ is unknown.

5.7.1 Consider a random sample of size n from Weinman's p-variate exponential distribution (Section 2.6.3) and consider the null hypothesis $H_0:\lambda_0=\lambda_1=\ldots=\lambda_{p-1}$. The maximum log likelihood may be found from Example 4.2.2. Show that

$$-2\log\lambda=2np\log(\bar{a}/\bar{g}),$$

where $\bar{a}$ and $\bar{g}$ are arithmetic and geometric means, and give the asymptotic distribution of $-2\log\lambda$.

5.7.2 For the cork data of Table 1.4.1, show that the measures of skewness and kurtosis of Section 1.8, taking one variable at a time, are

	N	E	S	W
$b_{1,1}$	0.6542	0.6925	0.5704	0.1951
$b_{2,1}$	2.6465	2.7320	2.8070	2.2200

whereas taking two variables at a time, they are

	NE	NS	NW	ES	EW	SW
$b_{1,2}$	1.214	1.019	1.138	2.192	0.965	0.935
$b_{2,2}$	7.512	6.686	7.186	9.359	6.896	7.835

What can be concluded regarding normality of variables in the lower dimensions? Can the large values of $b_{1,2}$ and $b_{2,2}$ for ES be partially ascribed to the larger values of $b_{1,1}$ and $b_{2,1}$ for E and S, respectively?

5.7.3 For the cork data (Table 1.4.1) let $y_1 = \mathrm{N}+\mathrm{S}-\mathrm{E}-\mathrm{W}$, $y_2 = \mathrm{N}-\mathrm{S}$, $y_3 = \mathrm{E}-\mathrm{W}$. Show that

$$b_{1,3} = 1.777, \qquad b_{2,3} = 13.558.$$

Test the hypothesis of normality.

5.8.1 Show that

$$\sum_{i=1}^{N} \cos\frac{2\pi i}{N} = \sum_{i=1}^{N} \sin\frac{2\pi i}{N} = 0,$$

and

$$\sum_{i=1}^{N} \cos^2\frac{2\pi i}{N} = \sum_{i=1}^{N} \sin^2\frac{2\pi i}{N} = \frac{1}{2}N, \qquad \sum_{i=1}^{N} \cos\frac{2\pi i}{N}\sin\frac{2\pi i}{N} = 0.$$

Hence prove (5.8.3). Verify the second equality in (5.8.4) and prove (5.8.5).

5.8.2 Apply Hotelling's T^2 test to data in Table 5.8.1 and compare the conclusions with those in Example 5.8.1.

6
Multivariate Regression Analysis

6.1 Introduction

Consider the model defined by

$$\mathbf{Y} = \mathbf{X}\mathbf{B} + \mathbf{U}, \tag{6.1.1}$$

where $\mathbf{Y}(n \times p)$ is an observed matrix of p response variables on each of n individuals, $\mathbf{X}(n \times q)$ is a known matrix, $\mathbf{B}(q \times p)$ is a matrix of unknown regression parameters, and $\mathbf{U}$ is a matrix of unobserved random disturbances whose rows for given $\mathbf{X}$ are uncorrelated, each with mean $\mathbf{0}$ and common covariance matrix $\mathbf{\Sigma}$.

When $\mathbf{X}$ represents a matrix of q "independent" variables observed on each of the n individuals, (6.1.1) is called the *multivariate regression model*. (If $\mathbf{X}$ is a random matrix then the distribution of $\mathbf{U}$ is assumed to be uncorrelated with $\mathbf{X}$.) When $\mathbf{X}$ is a design matrix (usually 0s and 1s), (6.1.1) is called the *general linear model*. Since the mathematics is the same in both cases, we shall not emphasize the difference, and we shall usually suppose that the first column of $\mathbf{X}$ equals $\mathbf{1}$ (namely $\mathbf{X} = (\mathbf{1}, \mathbf{X}_1)$) to allow for an overall mean effect. For simplicity, we shall treat $\mathbf{X}$ as a fixed matrix throughout the chapter. If $\mathbf{X}$ is random, then all likelihoods and expectations are to be interpreted as "conditional on $\mathbf{X}$".

The columns of $\mathbf{Y}$ represent "dependent" variables which are to be explained in terms of the "independent" variables given by the columns of $\mathbf{X}$. Note that

$$E(y_{ij}) = \mathbf{x}_i'\boldsymbol{\beta}_{(j)}$$

so that the expected value of y_{ij} depends on the ith row of $\mathbf{X}$ and the jth column of the matrix of regression coefficients. (The case $p = 1$, where there is only one dependent variable, is the familiar *multiple regression model* which we will write as $\mathbf{y}=\mathbf{X}\boldsymbol{\beta}+\mathbf{u}$.)

In most applications we shall suppose that $\mathbf{U}$ is normally distributed, so that

$$\mathbf{U} \text{ is a data matrix from } N_p(\mathbf{0}, \boldsymbol{\Sigma}), \tag{6.1.2}$$

where $\mathbf{U}$ is independent of $\mathbf{X}$. Under the assumption of normal errors, the log likelihood for the data $\mathbf{Y}$ in terms of the parameters $\mathbf{B}$ and $\boldsymbol{\Sigma}$ is given from (4.1.4) by

$$l(\mathbf{B}, \boldsymbol{\Sigma}) = -\tfrac{1}{2}n \log|2\pi\boldsymbol{\Sigma}| - \tfrac{1}{2}\operatorname{tr}(\mathbf{Y}-\mathbf{X}\mathbf{B})\boldsymbol{\Sigma}^{-1}(\mathbf{Y}-\mathbf{X}\mathbf{B})'. \tag{6.1.3}$$

6.2 Maximum Likelihood Estimation

6.2.1 Maximum likelihood estimators for B and Σ

We shall now give the m.l.e.s for $\mathbf{B}$ and $\boldsymbol{\Sigma}$ in the multivariate regression when $n \geq p+q$. In order for the estimates of $\mathbf{B}$ to be unique we shall suppose that $\mathbf{X}$ has full rank q, so that the inverse $(\mathbf{X}'\mathbf{X})^{-1}$ exists. The situation where $\operatorname{rank}(\mathbf{X})<q$ is discussed in Section 6.4.

Let

$$\mathbf{P} = \mathbf{I} - \mathbf{X}(\mathbf{X}'\mathbf{X})^{-1}\mathbf{X}'. \tag{6.2.1}$$

Then $\mathbf{P}(n \times n)$ is a symmetric idempotent matrix of rank $n-q$ which projects onto the subspace of R^n orthogonal to the columns of $\mathbf{X}$. (See Exercise 6.2.1.) In particular, note that $\mathbf{P}\mathbf{X}=\mathbf{0}$.

Theorem 6.2.1 *For the log likelihood function* (6.1.3), *the m.l.e.s of* $\mathbf{B}$ *and* $\boldsymbol{\Sigma}$ *are*

$$\hat{\mathbf{B}} = (\mathbf{X}'\mathbf{X})^{-1}\mathbf{X}'\mathbf{Y} \tag{6.2.2}$$

and

$$\hat{\boldsymbol{\Sigma}} = n^{-1}\mathbf{Y}'\mathbf{P}\mathbf{Y}. \tag{6.2.3}$$

Proof Let

$$\hat{\mathbf{Y}} = \mathbf{X}\hat{\mathbf{B}} = \mathbf{X}(\mathbf{X}'\mathbf{X})^{-1}\mathbf{X}'\mathbf{Y} \tag{6.2.4}$$

and

$$\hat{\mathbf{U}} = \mathbf{Y} - \mathbf{X}\hat{\mathbf{B}} = \mathbf{P}\mathbf{Y} \tag{6.2.5}$$

denote the "fitted" value of $\mathbf{Y}$ and the "fitted" error matrix, respectively. Then $\mathbf{Y}-\mathbf{X}\mathbf{B} = \hat{\mathbf{U}} + \mathbf{X}\hat{\mathbf{B}} - \mathbf{X}\mathbf{B}$, so the second term in the right-hand side of (6.1.3) can be written

$$-\tfrac{1}{2}\operatorname{tr}\boldsymbol{\Sigma}^{-1}(\mathbf{Y}-\mathbf{X}\mathbf{B})'(\mathbf{Y}-\mathbf{X}\mathbf{B}) = -\tfrac{1}{2}\operatorname{tr}\boldsymbol{\Sigma}^{-1}[\hat{\mathbf{U}}'\hat{\mathbf{U}} + (\hat{\mathbf{B}}-\mathbf{B})'\mathbf{X}'\mathbf{X}(\hat{\mathbf{B}}-\mathbf{B})].$$

Substituting in (6.1.3) we get

$$l(\mathbf{B}, \boldsymbol{\Sigma}) = -\tfrac{1}{2}n \log |2\pi\boldsymbol{\Sigma}| - \tfrac{1}{2}n \operatorname{tr} \boldsymbol{\Sigma}^{-1}\hat{\boldsymbol{\Sigma}} - \tfrac{1}{2} \operatorname{tr} \boldsymbol{\Sigma}^{-1}(\hat{\mathbf{B}} - \mathbf{B})'\mathbf{X}'\mathbf{X}(\hat{\mathbf{B}} - \mathbf{B}). \tag{6.2.6}$$

Only the final term here involves $\mathbf{B}$, and this is maximized (at the value zero) when $\mathbf{B} = \hat{\mathbf{B}}$. Then the "reduced" log likelihood function is given by

$$l(\hat{\mathbf{B}}, \boldsymbol{\Sigma}) = -\tfrac{1}{2}np \log 2\pi - \tfrac{1}{2}n(\log |\boldsymbol{\Sigma}| + \operatorname{tr} \boldsymbol{\Sigma}^{-1}\hat{\boldsymbol{\Sigma}}),$$

which by Theorem 4.2.1 is maximized when $\boldsymbol{\Sigma} = \hat{\boldsymbol{\Sigma}}$.

The maximum value of the likelihood is given from (6.2.6) by

$$l(\hat{\mathbf{B}}\ \hat{\boldsymbol{\Sigma}}) = -\tfrac{1}{2}n \log |2\pi\hat{\boldsymbol{\Sigma}}| - \tfrac{1}{2}np. \quad \blacksquare \tag{6.2.7}$$

The statistics $\hat{\mathbf{B}}$ and $\hat{\boldsymbol{\Sigma}}$ are sufficient for $\mathbf{B}$ and $\boldsymbol{\Sigma}$ (Exercise 6.2.2). For an efficient method of calculating $\hat{\mathbf{B}}$ and $\hat{\boldsymbol{\Sigma}}$, see Anderson (1958, pp. 184–187).

Note that

$$\hat{\mathbf{U}}'\hat{\mathbf{U}} = \mathbf{Y}'\mathbf{Y} - \hat{\mathbf{Y}}'\hat{\mathbf{Y}} = \mathbf{Y}'\mathbf{P}\mathbf{Y}. \tag{6.2.8}$$

In other words, the residual sum of squares and products (SSP) matrix $\hat{\mathbf{U}}'\hat{\mathbf{U}}$ equals the total SSP matrix $\mathbf{Y}'\mathbf{Y}$ minus the fitted SSP matrix $\hat{\mathbf{Y}}'\hat{\mathbf{Y}}$. Geometrically, each column of $\mathbf{Y}$ can be split into two orthogonal parts. First, $\hat{\mathbf{y}}_{(i)}$ is the projection of $\mathbf{y}_{(i)}$ onto the space spanned by the columns of $\mathbf{X}$. Second, $\hat{\mathbf{u}}_{(i)} = \mathbf{P}\mathbf{y}_{(i)}$ is the projection of $\mathbf{y}_{(i)}$ onto the space orthogonal to the columns of $\mathbf{X}$. This property is illustrated for $p = 1$ in Figure 6.2.1.

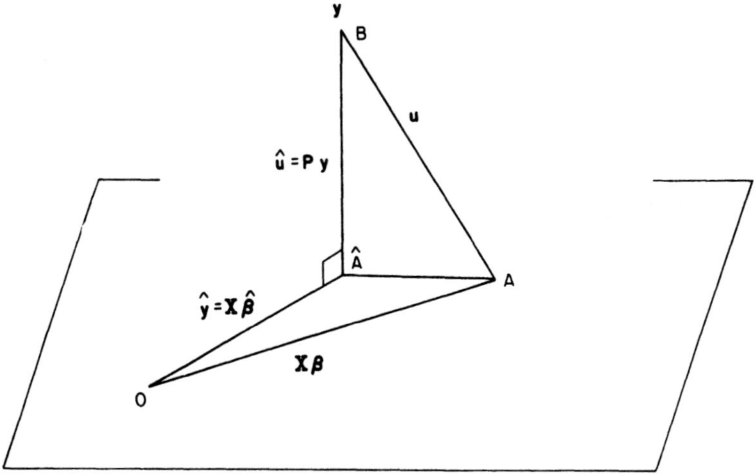

Figure 6.2.1 Geometry of multiple regression model, $\mathbf{y} = \mathbf{X}\boldsymbol{\beta} + \mathbf{u}$.

The true value of $\mathbf{X}\boldsymbol{\beta}=E(\mathbf{y})$ is denoted by the point A. The observed value of $\mathbf{y}$ (the point B) is projected onto the plane spanned by the columns of $\mathbf{X}$ to give the point $\hat{\text{A}}$ as an estimate of A.

If we let $\mathbf{M}_{11}=\mathbf{X}'\mathbf{X}, \mathbf{M}_{22}=\mathbf{Y}'\mathbf{Y}$, and $\mathbf{M}_{12}=\mathbf{X}'\mathbf{Y}$, then we can write

$$n\hat{\mathbf{\Sigma}}=\hat{\mathbf{U}}'\hat{\mathbf{U}}=\mathbf{M}_{22}-\mathbf{M}_{21}\mathbf{M}_{11}^{-1}\mathbf{M}_{12}.$$

Thus the residual SSP matrix can be viewed as a residual Wishart matrix $\mathbf{M}_{22.1}$ conditioned on $\mathbf{X}$, as in (3.4.5).

6.2.2 The distribution of $\hat{\mathbf{B}}$ and $\hat{\mathbf{\Sigma}}$

We now consider the joint distribution of the m.l.e.s derived in Theorem 6.2.1 under the multivariate linear model with normal errors.

Theorem 6.2.2 *Under the model defined by* (6.1.1) and (6.1.2)

(a) $\hat{\mathbf{B}}$ is unbiased for $\mathbf{B}$,

(b) $E(\hat{\mathbf{U}})=\mathbf{0}$,

(c) $\hat{\mathbf{B}}$ *and* $\hat{\mathbf{U}}$ *are multivariate normal,*

(d) $\hat{\mathbf{B}}$ *is statistically independent of* $\hat{\mathbf{U}}$, *and hence of* $\hat{\mathbf{\Sigma}}$.

Proof (a) Inserting $\mathbf{Y}=\mathbf{X}\mathbf{B}+\mathbf{U}$ in the formula for $\hat{\mathbf{B}}$ defined in Theorem 6.2.1,

$$\hat{\mathbf{B}}=(\mathbf{X}'\mathbf{X})^{-1}\mathbf{X}'(\mathbf{X}\mathbf{B}+\mathbf{U})=\mathbf{B}+(\mathbf{X}'\mathbf{X})^{-1}\mathbf{X}'\mathbf{U}. \qquad (6.2.9)$$

Therefore, $E(\hat{\mathbf{B}})=\mathbf{B}$ since $E(\mathbf{U})=\mathbf{0}$. Hence (a) is proved.

(b) We have $\hat{\mathbf{U}}=\mathbf{Y}-\mathbf{X}\hat{\mathbf{B}}$. Therefore

$$E(\hat{\mathbf{U}})=E(\mathbf{Y})-\mathbf{X}E(\hat{\mathbf{B}})=\mathbf{X}\mathbf{B}-\mathbf{X}\mathbf{B}=\mathbf{0}.$$

(c) We have

$$\hat{\mathbf{U}}=\mathbf{P}\mathbf{Y}=\mathbf{P}\mathbf{U}, \qquad (6.2.10)$$

where $\mathbf{P}$ was defined in (6.2.1). Also $\hat{\mathbf{B}}=\mathbf{B}+(\mathbf{X}'\mathbf{X})^{-1}\mathbf{X}'\mathbf{U}$ from (6.2.9). Hence both $\hat{\mathbf{U}}$ and $\hat{\mathbf{B}}$ are linear functions of $\mathbf{U}$. Since $\mathbf{U}$ has a multivariate normal distribution, the normality of $\hat{\mathbf{U}}$ and $\hat{\mathbf{B}}$ follows immediately.

(d) This follows by applying Theorem 3.3.3 to the linear functions defined in part (c) above. ■

Of particular importance in the above theorem is the fact the $\hat{\mathbf{B}}$ and $\hat{\mathbf{\Sigma}}$ are independent for normal data. More precise results concerning the distribution of $\hat{\mathbf{B}}$ and $\hat{\mathbf{\Sigma}}$ are given in the following theorem.

Theorem 6.2.3 (a) *The covariance between* $\hat{\beta}_{ij}$ *and* $\hat{\beta}_{kl}$ *is* $\sigma_{jl}g_{ik}$, *where* $\mathbf{G}=(\mathbf{X}'\mathbf{X})^{-1}$.

(b) $n\hat{\mathbf{\Sigma}}\sim \mathbf{W}_p(\mathbf{\Sigma}, n-q)$.

Proof (a) From (6.2.9) we see that $\hat{\beta}_{ij}-\beta_{ij}=\mathbf{a}_i'\mathbf{u}_{(j)}$, where $\mathbf{A}=(\mathbf{X}'\mathbf{X})^{-1}\mathbf{X}'$. Therefore

$$C(\hat{\beta}_{ij}, \hat{\beta}_{kl})=\mathbf{a}_i'E[\mathbf{u}_{(j)}\mathbf{u}_{(l)}']\mathbf{a}_k=\mathbf{a}_i'[\sigma_{jl}\mathbf{I}]\mathbf{a}_k=\sigma_{jl}(\mathbf{A}\mathbf{A}')_{ik}=\sigma_{jl}g_{ik}$$

since $\mathbf{A}\mathbf{A}'=\mathbf{G}$. Therefore part (a) is proved.

(b) From (6.2.10), $n\hat{\mathbf{\Sigma}}=\hat{\mathbf{U}}'\hat{\mathbf{U}}=\mathbf{U}'\mathbf{P}\mathbf{U}$, where $\mathbf{P}$ is an idempotent matrix of rank $(n-q)$. Since each row of $\mathbf{U}$ has mean $\mathbf{0}$, the result follows from Theorem 3.4.4. ∎

Note that when $\mathbf{X}'\mathbf{X}=\mathbf{I}$, or nearly so, the elements of a column $\hat{\boldsymbol{\beta}}_{(i)}$ will tend to have a low correlation with one another; this property is considered desirable.

6.3 The General Linear Hypothesis

We wish to consider hypotheses of the form $\mathbf{C}_1\mathbf{B}\mathbf{M}_1=\mathbf{D}$, where $\mathbf{C}_1(g\times q)$, $\mathbf{M}_1(p\times r)$, and $\mathbf{D}(g\times r)$ are given matrices and $\mathbf{C}_1$ and $\mathbf{M}_1$ have ranks g and r, respectively. In many cases $\mathbf{D}=\mathbf{0}$ and $\mathbf{M}_1=\mathbf{I}$, so the hypothesis takes the form $\mathbf{C}_1\mathbf{B}=\mathbf{0}$. Note that the rows of $\mathbf{C}_1$ make assertions about the effect on the regression from linear combinations of the *independent* variables, whereas the columns of $\mathbf{M}_1$ focus attention on particular linear combinations of the *dependent* variables.

We shall follow the two-pronged approach begun in Chapter 5; that is, we shall consider the hypothesis from the point of view of the likelihood ratio test (LRT) and the union intersection test (UIT).

6.3.1 The likelihood ratio test (LRT).

Consider first the hypothesis $\mathbf{C}_1\mathbf{B}=\mathbf{D}$ (so $\mathbf{M}_1=\mathbf{I}$). It is convenient to define some additional matrices before constructing the LRT. Let $\mathbf{C}_2((q-g)\times q)$ be a matrix such that $\mathbf{C}'=(\mathbf{C}_1', \mathbf{C}_2')$ is a non-singular matrix of order $q\times q$, and let $\mathbf{B}_0(q\times p)$ be any matrix satisfying $\mathbf{C}_1\mathbf{B}_0=\mathbf{D}$.

Then the model (6.1.1) can be rewritten as

$$\mathbf{Y}_+=\mathbf{Z}\mathbf{\Delta}+\mathbf{U}, \tag{6.3.1}$$

where $\mathbf{Y}_+=\mathbf{Y}-\mathbf{X}\mathbf{B}_0$, $\mathbf{Z}=\mathbf{X}\mathbf{C}^{-1}$, and $\mathbf{\Delta}=(\mathbf{\Delta}_1', \mathbf{\Delta}_2')'=\mathbf{C}(\mathbf{B}-\mathbf{B}_0)$. The hypothesis $\mathbf{C}_1\mathbf{B}=\mathbf{D}$ becomes $\mathbf{\Delta}_1=\mathbf{0}$. Partition $\mathbf{C}^{-1}=(\mathbf{C}^{(1)}, \mathbf{C}^{(2)})$ and let

$$\mathbf{P}_1=\mathbf{I}-\mathbf{X}\mathbf{C}^{(2)}(\mathbf{C}^{(2)\prime}\mathbf{X}'\mathbf{X}\mathbf{C}^{(2)})^{-1}\mathbf{C}^{(2)\prime}\mathbf{X}' \tag{6.3.2}$$

represent the projection onto the subspace orthogonal to the columns of $\mathbf{X}\mathbf{C}^{(2)}$.

From (6.2.7) we see that the maximized likelihoods under the null and alternative hypotheses are given by

$$|2\pi n^{-1}\mathbf{Y}_+'\mathbf{P}_1\mathbf{Y}_+|^{-n/2}\exp(-\tfrac{1}{2}np) \quad \text{and} \quad |2\pi n^{-1}\mathbf{Y}_+'\mathbf{P}_+\mathbf{Y}_+|^{-n/2}\exp(-\tfrac{1}{2}np),$$

respectively. Thus the LR statistic is given by

$$\lambda^{2/n} = |\mathbf{Y}_+'\mathbf{P}\mathbf{Y}_+|/|\mathbf{Y}_+'\mathbf{P}_1\mathbf{Y}_+|. \tag{6.3.3}$$

Note that $\mathbf{P}$ is a projection onto a subspace of $\mathbf{P}_1$ (a vector orthogonal to all the columns of $\mathbf{X}$ is orthogonal to the columns of $\mathbf{XC}^{(2)}$), so that $\mathbf{P}_2 = \mathbf{P}_1 - \mathbf{P}$ is also a projection. It can be shown that

$$\mathbf{P}_2 = \mathbf{X}(\mathbf{X}'\mathbf{X})^{-1}\mathbf{C}_1'[\mathbf{C}_1(\mathbf{X}'\mathbf{X})^{-1}\mathbf{C}_1']^{-1}\mathbf{C}_1(\mathbf{X}'\mathbf{X})^{-1}\mathbf{X}' \tag{6.3.4}$$

(see Exercise 6.3.1). Also note that $\mathbf{PY} = \mathbf{PY}_+$ because $\mathbf{PX} = \mathbf{0}$. This observation is useful in proving the following result.

Theorem 6.3.1 *The LRT of the hypothesis* $\mathbf{C}_1\mathbf{B} = \mathbf{D}$ *for the model* (6.3.1) *has test statistic*

$$\lambda^{2/n} = |\mathbf{Y}'\mathbf{P}\mathbf{Y}|/|\mathbf{Y}'\mathbf{P}\mathbf{Y} + \mathbf{Y}_+'\mathbf{P}_2\mathbf{Y}_+|, \tag{6.3.5}$$

which has $\Lambda(p, n-q, g)$ *distribution under the null hypothesis.*

Proof The formula for $\lambda^{2/n}$ was derived above. For its distribution, note from (6.2.10) that $\mathbf{Y}'\mathbf{P}\mathbf{Y} = \mathbf{U}'\mathbf{P}\mathbf{U}$ under the null and alternative hypotheses. Also, under the null hypothesis, $\mathbf{Y}_+'\mathbf{P}_2\mathbf{Y}_+ = \mathbf{U}'\mathbf{P}_2\mathbf{U}$ (Exercise 6.3.2). Since $\mathbf{P}$ and $\mathbf{P}_2$ are independent projections ($\mathbf{PP}_2 = \mathbf{0}$) of ranks $n-q$ and g, respectively, it follows from Theorems 3.4.3 and 3.4.4 that $\mathbf{Y}'\mathbf{P}\mathbf{Y}$ and $\mathbf{Y}_+'\mathbf{P}_2\mathbf{Y}_+$ have independent $W_p(\mathbf{\Sigma}, n-q)$ and $W_p(\mathbf{\Sigma}, g)$ distributions. The distribution of $\lambda^{2/n}$ follows. ∎

Now let us consider the hypothesis $\mathbf{C}_1\mathbf{B}\mathbf{M}_1 = \mathbf{D}$ for the model (6.1.1). From Exercise 6.3.3, we see that it is equivalent to study the hypothesis $\mathbf{C}_1\mathbf{B}\mathbf{M}_1 = \mathbf{D}$ for the transformed model

$$\mathbf{Y}\mathbf{M}_1 = \mathbf{X}\mathbf{B}\mathbf{M}_1 + \mathbf{U}\mathbf{M}_1, \tag{6.3.6}$$

where $\mathbf{U}\mathbf{M}_1$ is now a data matrix from $N_p(\mathbf{0}, \mathbf{M}_1'\mathbf{\Sigma}\mathbf{M}_1)$ and $\mathbf{B}\mathbf{M}_1$ contains qr parameters.

Define matrices

$$\begin{aligned}\mathbf{H} &= \mathbf{M}_1'\mathbf{Y}_+'\mathbf{P}_2\mathbf{Y}_+\mathbf{M}_1\\ &= \mathbf{M}_1'\mathbf{Y}_+'\mathbf{X}(\mathbf{X}'\mathbf{X})^{-1}\mathbf{C}_1'[\mathbf{C}_1(\mathbf{X}'\mathbf{X})^{-1}\mathbf{C}_1']^{-1}\mathbf{C}_1(\mathbf{X}'\mathbf{X})^{-1}\mathbf{X}'\mathbf{Y}_+\mathbf{M}_1,\end{aligned} \tag{6.3.7}$$

and

$$\begin{aligned}\mathbf{E} &= \mathbf{M}_1'\mathbf{Y}'\mathbf{P}\mathbf{Y}\mathbf{M}_1\\ &= \mathbf{M}_1'\mathbf{Y}'[\mathbf{I} - \mathbf{X}(\mathbf{X}'\mathbf{X})^{-1}\mathbf{X}']\mathbf{Y}\mathbf{M}_1\end{aligned} \tag{6.3.8}$$

to be the SSP matrix due to the regression and the residual SSP matrix, respectively. Then the LRT is given by the following theorem.

Theorem 6.3.2 *The LRT of the hypothesis* $\mathbf{C}_1\mathbf{B}\mathbf{M}_1=\mathbf{D}$ *for the model* (6.1.1) *is given by the test statistic*

$$\lambda^{2/n}=|\mathbf{E}|/|\mathbf{H}+\mathbf{E}|\sim\Lambda(r, n-q, g). \tag{6.3.9}$$

Proof Apply Theorem 6.3.1 to the transformed model (6.3.6). ■

Note that the LR statistic can also be written

$$|\mathbf{E}|/|\mathbf{H}+\mathbf{E}|=\prod_{i=1}^{r}(1+\lambda_i)^{-1}, \tag{6.3.10}$$

where $\lambda_1,\ldots,\lambda_r$ are the eigenvalues of $\mathbf{H}\mathbf{E}^{-1}$.

When $p=1$ the matrices in (6.3.9) are scalars and it is more usual to use the equivalent F statistic

$$\mathbf{H}/\mathbf{E}\sim[g/(n-q)]F_{g,n-q}. \tag{6.3.11}$$

The null hypothesis is rejected for large values of this statistic.

6.3.2 The union intersection test (UIT)

The multivariate hypothesis $\mathbf{C}_1\mathbf{B}\mathbf{M}_1=\mathbf{D}$ is true if and only if $\mathbf{b}'\mathbf{C}_1\mathbf{B}\mathbf{M}_1\mathbf{a}=\mathbf{b}'\mathbf{D}\mathbf{a}$ for all $\mathbf{a}$ and $\mathbf{b}$. Replacing $\mathbf{C}_1$ and $\mathbf{M}_1$ by $\mathbf{b}'\mathbf{C}_1$ and $\mathbf{M}_1\mathbf{a}$ in (6.3.7)–(6.3.8), we see from (6.3.11) that each of these univariate hypotheses is tested by the statistic

$$\frac{\{\mathbf{b}'\mathbf{C}_1(\mathbf{X}'\mathbf{X})^{-1}\mathbf{X}'\mathbf{Y}_+\mathbf{M}_1\mathbf{a}\}^2}{\{\mathbf{b}'\mathbf{C}_1(\mathbf{X}'\mathbf{X})^{-1}\mathbf{C}_1'\mathbf{b}\}\{\mathbf{a}'\mathbf{M}_1'\mathbf{Y}'\mathbf{P}\mathbf{Y}\mathbf{M}_1\mathbf{a}\}}, \tag{6.3.12}$$

which has the $(n-q)^{-1}F_{1,n-q}$ distribution under the null hypothesis for *fixed* $\mathbf{a}$ and $\mathbf{b}$. Maximizing (6.3.12) over $\mathbf{b}$ gives

$$\mathbf{a}'\mathbf{H}\mathbf{a}/\mathbf{a}'\mathbf{E}\mathbf{a}, \tag{6.3.13}$$

which has the $[g/(n-q)]F_{g,n-q}$ distribution for *fixed* $\mathbf{a}$ (because $\mathbf{H}$ and $\mathbf{E}$ have independent Wishart distributions). Finally, maximizing (6.3.13) over $\mathbf{a}$ shows the UI statistic to be λ_1, the largest eigenvalue of $\mathbf{H}\mathbf{E}^{-1}$. Let $\theta=\lambda_1/(1+\lambda_1)$ denote the greatest root of $\mathbf{H}(\mathbf{H}+\mathbf{E})^{-1}$.

The distribution of $\theta=\theta(r, n-q, g)$ was described in Chapter 3. The null hypothesis is rejected for large values of θ.

If $\text{rank}(\mathbf{M}_1)=r=1$, then maximizing over $\mathbf{a}$ is unnecessary and (6.3.13) becomes simply the ratio of scalars (6.3.11). Thus the UIT is equivalent to the LRT in this case.

If $\text{rank}(\mathbf{C}_1) = g = 1$, then maximizing over $\mathbf{b}$ is unnecessary. Since $\mathbf{H}$ and $\mathbf{HE}^{-1}$ have rank 1 in this case, we see from (6.3.10) that the UIT and LRT are equivalent here also. The only non-zero eigenvalue of $\mathbf{HE}^{-1}$ equals its trace, so from (6.3.7)–(6.3.8),

$$\lambda_1 = \text{tr}\,\mathbf{HE}^{-1} = \{\mathbf{C}_1(\mathbf{X}'\mathbf{X})^{-1}\mathbf{C}_1'\}^{-1}\mathbf{C}_1(\mathbf{X}'\mathbf{X})^{-1}\mathbf{X}'\mathbf{Y}_+\mathbf{M}_1\mathbf{E}^{-1}\mathbf{M}_1'\mathbf{Y}_+'\mathbf{X}(\mathbf{X}'\mathbf{X})^{-1}\mathbf{C}_1'.$$

Let $\mathbf{d} = \mathbf{X}(\mathbf{X}'\mathbf{X})^{-1}\mathbf{C}_1'$. Under the null hypothesis $\mathbf{M}_1'\mathbf{Y}_+'\mathbf{d} \sim N_p(\mathbf{0}, (\mathbf{d}'\mathbf{d})\mathbf{M}_1'\mathbf{\Sigma}\mathbf{M}_1)$ (Exercise 6.3.4) and under both the null and alternative hypotheses $\mathbf{E} \sim W_r(\mathbf{M}_1'\mathbf{\Sigma}\mathbf{M}_1, n-q)$ independently. Thus under the null hypothesis λ_1 has the

$$(n-q)^{-1}T^2_{r,n-q} = [r/(n-q-r+1)]F_{r,n-q-r+1}$$

distribution.

6.3.3 Simultaneous confidence intervals

Suppose $\mathbf{B}$ is the true value of the regression parameter matrix and let $\mathbf{Y}_+ = \mathbf{Y} - \mathbf{XB}$. Then the UI principle states that the probability that the ratio in (6.3.12) will be less than or equal to $\theta_\alpha/(1-\theta_\alpha)$ simultaneously for all $\mathbf{a}$ and $\mathbf{b}$ for which the denominator does not vanish, equals $1-\alpha$. Here θ_α denotes the upper α critical value of the $\theta(r, n-q, g)$ distribution. Rearranging gives the simultaneous confidence intervals (SCIs)

$$P\Big(\mathbf{b}'\mathbf{C}_1\mathbf{B}\mathbf{M}_1\mathbf{a} \in \mathbf{b}'\mathbf{C}_1(\mathbf{X}'\mathbf{X})^{-1}\mathbf{X}'\mathbf{Y}\mathbf{M}_1\mathbf{a} \pm \left\{\frac{\theta_\alpha}{1-\theta_\alpha}(\mathbf{a}'\mathbf{E}\mathbf{a})[\mathbf{b}'\mathbf{C}_1(\mathbf{X}'\mathbf{X})^{-1}\mathbf{C}_1'\mathbf{b}]\right\}^{1/2}, \text{ for all } \mathbf{a}, \mathbf{b}\Big) = 1-\alpha. \qquad (6.3.14)$$

Note that $\mathbf{b}'\mathbf{C}_1(\mathbf{X}'\mathbf{X})^{-1}\mathbf{X}'\mathbf{Y}\mathbf{M}_1\mathbf{a} = \mathbf{b}'\mathbf{C}_1\hat{\mathbf{B}}\mathbf{M}_1\mathbf{a}$ is an unbiased estimator of $\mathbf{b}'\mathbf{C}_1\mathbf{B}\mathbf{M}_1\mathbf{a}$.

In some applications $\mathbf{a}$ and/or $\mathbf{b}$ are given *a priori* and the confidence limits in (6.3.14) can be narrowed. If $\mathbf{a}$ is fixed then SCIs for $\mathbf{b}$ can be obtained by replacing $\theta_\alpha/(1-\theta_\alpha)$ by $[g/(n-q)]F_{g,n-q;\alpha}$. If $\mathbf{b}$ is fixed then SCIs for $\mathbf{a}$ can be found by replacing $\theta_\alpha/(1-\theta_\alpha)$ by

$$(n-q)^{-1}T^2_{r,n-q;\alpha} = [r/(n-q-r+1)]\,F_{r,n-q-r+1;\alpha}.$$

Finally, if both $\mathbf{a}$ and $\mathbf{b}$ are known, a single confidence interval is obtained by replacing $\theta_\alpha/(1-\theta_\alpha)$ by $(n-q)^{-1}F_{1,n-q;\alpha}$.

6.4 Design Matrices of Degenerate Rank

In the design of experiments it is often convenient for symmetry reasons to consider design matrices $\mathbf{X}$ which are not of full rank. Unfortunately the parameter matrix $\mathbf{B}$ in (6.1.1) is not well-defined in this case. The simplest way to resolve this indeterminacy is to restrict the regression to a subset of the columns of $\mathbf{X}$.

Suppose $\mathbf{X}(n \times q)$ has rank $k<q$. Rearrange the columns of $\mathbf{X}$ and partition $\mathbf{X}=(\mathbf{X}_1, \mathbf{X}_2)$ so that $\mathbf{X}_1(n \times k)$ has full rank. Then $\mathbf{X}_2$ can be written in terms of $\mathbf{X}_1$ as $\mathbf{X}_2=\mathbf{X}_1\mathbf{A}$ for some matrix $\mathbf{A}(k \times (q-k))$. Partition $\mathbf{B}'=(\mathbf{B}_1', \mathbf{B}_2')$. If the regression model (6.1.1) is valid, then it can be reformulated in terms of $\mathbf{X}_1$ alone as

$$\mathbf{Y}=\mathbf{X}_1\mathbf{B}_1^*+\mathbf{U}, \tag{6.4.1}$$

where $\mathbf{B}_1^*=\mathbf{B}_1+\mathbf{A}\mathbf{B}_2$, *is* uniquely determined.

Consider a hypothesis $\mathbf{C}_1\mathbf{B}\mathbf{M}_1=\mathbf{D}$ and partition $\mathbf{C}_1=(\mathbf{C}_{11}, \mathbf{C}_{12})$. The hypothesis matrix $\mathbf{C}_1$ is called *testable* if, as a function of $\mathbf{B}$,

$$\mathbf{X}\mathbf{B}=\mathbf{0} \quad \text{implies} \quad \mathbf{C}_1\mathbf{B}=\mathbf{0}. \tag{6.4.2}$$

Then although $\mathbf{B}$ is indeterminate in the model (6.1.1), $\mathbf{C}_1\mathbf{B}=\mathbf{C}_{11}\mathbf{B}_1^*$ is uniquely determined. Also, note that the condition (6.4.2) does not involve $\mathbf{M}_1$. With a similar motivation, a linear combination $\mathbf{d}'\mathbf{B}$ is called *estimable if* $\mathbf{X}\mathbf{B}=\mathbf{0}$ implies $\mathbf{d}'\mathbf{B}=\mathbf{0}$.

An alternative way to describe testability was given by Roy (1957). He showed that (6.4.2) holds if and only if

$$\mathbf{C}_{12}=\mathbf{C}_{11}(\mathbf{X}_1'\mathbf{X}_1)^{-1}\mathbf{X}_1'\mathbf{X}_2. \tag{6.4.3}$$

(See Exercise 6.4.1.) This criterion is convenient in practice.

If $\hat{\mathbf{B}}_1^*$ is the m.l.e. of $\mathbf{B}_1^*$ in the model (6.4.1) and if $\mathbf{C}_1$ is testable, then the (unique) m.l.e. of $\mathbf{C}_1\mathbf{B}$ is given by

$$\mathbf{C}_{11}\hat{\mathbf{B}}_1^*=\mathbf{C}_{11}(\mathbf{X}_1'\mathbf{X}_1)^{-1}\mathbf{X}_1'\mathbf{Y}.$$

Thus, for a design matrix of non-full rank, the estimation of estimable linear combinations and the testing of testable hypotheses is most conveniently carried out using the model (6.4.1) instead of (6.1.1).

Example 6.4.1 (One-way multivariate analysis of variance). In Section 5.3.3a we considered k multinormal samples, $\mathbf{y}_1, \ldots, \mathbf{y}_{n_1}$, $\mathbf{y}_{n_1+1}, \ldots, \mathbf{y}_{n_1+n_2}$, $\mathbf{y}_{n_1+n_2+1}, \ldots, \mathbf{y}_n$, where the ith sample consists of n_i observations from $N_p(\boldsymbol{\mu}+\boldsymbol{\tau}_i, \boldsymbol{\Sigma})$, $i=1,\ldots, k$, and $n=n_1+\cdots+n_k$. We can write this model in the form (6.1.1) using the design and parameter matrices

$$\mathbf{X}=\begin{bmatrix}1 & 0 & \cdots & 0 & 1\\ \cdot & \cdot & & \cdot & \cdot\\ \cdot & \cdot & & \cdot & \cdot\\ \cdot & \cdot & & \cdot & \cdot\\ 1 & 0 & \cdots & 0 & 1\\ 0 & 1 & \cdots & 0 & 1\\ \cdot & \cdot & & \cdot & \cdot\\ \cdot & \cdot & & \cdot & \cdot\\ \cdot & \cdot & & \cdot & \cdot\\ 0 & 1 & \cdots & 0 & 1\\ \cdot & \cdot & & \cdot & \cdot\\ \cdot & \cdot & & \cdot & \cdot\\ \cdot & \cdot & & \cdot & \cdot\\ 0 & 0 & \cdots & 1 & 1\\ \cdot & \cdot & & \cdot & \cdot\\ \cdot & \cdot & & \cdot & \cdot\\ \cdot & \cdot & & \cdot & \cdot\\ 0 & 0 & \cdots & 1 & 1\end{bmatrix} \quad \text{and} \quad \mathbf{B}'=(\boldsymbol{\tau}_1,\ldots,\boldsymbol{\tau}_k,\boldsymbol{\mu}).$$

The matrix $\mathbf{X}(n\times(k+1))$ has rank k. The indeterminacy of this parametrization is usually resolved by requiring $\sum_1^k \boldsymbol{\tau}_i=\mathbf{0}$. However, in this section it is more convenient to drop $\boldsymbol{\mu}$ from the regression; that is, partition $\mathbf{X}$ so that $\mathbf{X}_1$ contains the first k columns, and use the model (6.4.1). Since

$$(\mathbf{X}_1'\mathbf{X}_1)^{-1}=\operatorname{diag}(n_i^{-1})$$

we see that the m.l.e.s of $\boldsymbol{\tau}_1^*,\ldots,\boldsymbol{\tau}_k^*$ in the model (6.4.1) are given by

$$(\hat{\boldsymbol{\tau}}_1^*,\ldots,\hat{\boldsymbol{\tau}}_k^*)'=(\mathbf{X}_1'\mathbf{X}_1)^{-1}\mathbf{X}_1'\mathbf{Y}=(\bar{\mathbf{y}}_1,\ldots,\bar{\mathbf{y}}_k)'.$$

One way to express the hypothesis that $\boldsymbol{\tau}_1=\cdots=\boldsymbol{\tau}_k$ for the model (6.1.1) is to use the $((k-1)\times(k+1))$ matrix

$$\mathbf{C}_1=\begin{pmatrix}1 & 0 & 0 & \cdots & 0 & -1 & 0\\ 0 & 1 & 0 & \cdots & 0 & -1 & 0\\ \cdot & \cdot & \cdot & & \cdot & \cdot & \cdot\\ \cdot & \cdot & \cdot & & \cdot & \cdot & \cdot\\ \cdot & \cdot & \cdot & & \cdot & \cdot & \cdot\\ 0 & 0 & 0 & \cdots & 1 & -1 & 0\end{pmatrix}.$$

Partition $\mathbf{C}_1=(\mathbf{C}_{11},\mathbf{C}_{12})$ so that $\mathbf{C}_{11}$ contains the first k columns. It is easily checked that the hypothesis $\mathbf{C}_1\mathbf{B}=\mathbf{0}$ is testable (and is equivalent to the hypothesis $\boldsymbol{\tau}_1^*=\cdots=\boldsymbol{\tau}_k^*$ for the model (6.4.1). The matrices $\mathbf{H}$ and $\mathbf{E}$

defined by (6.3.7) and (6.3.8), with $\mathbf{X}$ and $\mathbf{C}_1$ replaced by $\mathbf{X}_1$ and $\mathbf{C}_{11}$, can be shown to equal $\mathbf{B}$ and $\mathbf{W}$, respectively, the "between-groups" and "within-groups" SSP matrices given in Section 5.3.3a. (See Exercise 6.4.2.) Thus, as expected, the LRT and UIT take the same forms as given there.

As $\mathbf{b}$ varies through all $(k-1)$-vectors, $\mathbf{C}'_{11}\mathbf{b}=\mathbf{c}$ varies through all k-vectors satisfying $\sum_1^k c_i = 0$. Such vectors are called *contrasts* and $\sum c_i\boldsymbol{\tau}_i = \sum c_i\boldsymbol{\tau}_i^*$ for all $\boldsymbol{\mu}$. Thus, from (6.3.14), simultaneous confidence intervals for linear combinations $\mathbf{a}$ of contrasts $\mathbf{c}$ between group means are given by

$$\mathbf{c}'\boldsymbol{\tau}\mathbf{a} \in \mathbf{c}'\bar{\mathbf{Y}}\mathbf{a} \pm \left\{\frac{\theta_\alpha}{1-\theta_\alpha}\mathbf{a}'\mathbf{W}\mathbf{a}\sum_{i=1}^{k}\frac{c_i^2}{n_i}\right\}^{1/2},$$

where we have written $\boldsymbol{\tau}' = (\boldsymbol{\tau}_1, \ldots, \boldsymbol{\tau}_k)$ and $\bar{\mathbf{Y}}' = (\bar{\mathbf{y}}_1, \ldots, \bar{\mathbf{y}}_k)$, and θ_α is the upper α critical value of $\theta(p, n-k, k-1)$. If $\mathbf{a}$ and/or $\mathbf{c}$ are known *a priori*, the adjustments described at the end of Section 6.3.3 are appropriate.

6.5 Multiple Correlation

6.5.1 The effect of the mean

In the model (6.1.1) it is sometimes convenient to separate the effect of the mean from the other "independent" variables. Rewrite the regression model as

$$\mathbf{Y} = \mathbf{1}\boldsymbol{\mu}' + \mathbf{X}\mathbf{B} + \mathbf{U}, \tag{6.5.1}$$

where $(\mathbf{1}, \mathbf{X})$ now represents an $(n\times(1+q))$ matrix of full rank. Without loss of generality we shall suppose the columns of $\mathbf{X}$ are each centred to have mean $\mathbf{0}$. (If $\mathbf{x}_{(i)}$ is replaced by $\mathbf{x}_{(i)} - \bar{x}_i\mathbf{1}$, then (6.5.1) remains valid if $\boldsymbol{\mu}$ is replaced by $\boldsymbol{\mu} + \sum \bar{x}_i\boldsymbol{\beta}_{i\cdot}$.) Then

$$\left[\begin{pmatrix}\mathbf{1}'\\ \mathbf{X}'\end{pmatrix}(\mathbf{1}, \mathbf{X})\right]^{-1} = \begin{pmatrix} n^{-1} & \mathbf{0} \\ \mathbf{0} & (\mathbf{X}'\mathbf{X})^{-1}\end{pmatrix} \tag{6.5.2}$$

because $\mathbf{X}'\mathbf{1} = \mathbf{0}$. Thus, by Theorem 6.2.3(a), $\hat{\boldsymbol{\mu}} = \bar{\mathbf{y}}$ is independent of $\hat{\mathbf{B}} = (\mathbf{X}'\mathbf{X})^{-1}\mathbf{X}'\mathbf{Y} = (\mathbf{X}'\mathbf{X})^{-1}\mathbf{X}'(\mathbf{Y} - \mathbf{1}\bar{\mathbf{y}}')$. If the mean of $\mathbf{Y}$ is estimated by $\bar{\mathbf{y}}$, then (6.5.1) can be written

$$\mathbf{Y} - \mathbf{1}\bar{\mathbf{y}}' = \mathbf{X}\mathbf{B} + \mathbf{W}, \tag{6.5.3}$$

where $\mathbf{W} = \mathbf{U} - \mathbf{1}\bar{\mathbf{u}}'$ denotes the centred error matrix which, thought of as a column vector, $\mathbf{W}^V$, has distribution $N_{np}(\mathbf{0}, \boldsymbol{\Sigma}\otimes\mathbf{H})$. Here $\mathbf{H} = \mathbf{I} - n^{-1}\mathbf{1}\mathbf{1}'$. (See Exercise 6.5.1.)

6.5.2 Multiple correlation coefficient

Suppose $p=1$ so that (6.5.1) becomes

$$\mathbf{y}-\bar{y}\mathbf{1}=\mathbf{X}\boldsymbol{\beta}+\mathbf{v}, \tag{6.5.4}$$

where $\mathbf{v}\sim N_n(\mathbf{0},\sigma^2\mathbf{H})$ and again we suppose the columns of $\mathbf{X}$ are each centred to have zero mean. Define

$$\mathbf{S}=\begin{pmatrix} s_{11} & \mathbf{S}_{12} \\ \mathbf{S}_{21} & \mathbf{S}_{22}\end{pmatrix}=n^{-1}\begin{pmatrix}\mathbf{y}'-\bar{y}\mathbf{1}' \\ \mathbf{X}'\end{pmatrix}(\mathbf{y}-\bar{y}\mathbf{1},\mathbf{X}) \tag{6.5.5}$$

to be the sample covariance matrix for $\mathbf{y}$ and the columns of $\mathbf{X}$. Let $\mathbf{R}$ denote the corresponding correlation matrix. Since

$$\hat{\boldsymbol{\beta}}=(\mathbf{X}'\mathbf{X})^{-1}\mathbf{X}'\mathbf{y}=(\mathbf{X}'\mathbf{X})^{-1}\mathbf{X}'(\mathbf{y}-\bar{y}\mathbf{1})=\mathbf{S}_{22}^{-1}\mathbf{S}_{21},$$

we see that the (sample) correlation between $\mathbf{y}$ and $\mathbf{X}\hat{\boldsymbol{\beta}}$ is given by

$$\begin{aligned}R_{y\cdot\mathbf{x}}=\operatorname{corr}(\mathbf{y},\mathbf{X}\hat{\boldsymbol{\beta}})&=\mathbf{S}_{12}\mathbf{S}_{22}^{-1}\mathbf{S}_{21}/\{s_{11}\mathbf{S}_{12}\mathbf{S}_{22}^{-1}\mathbf{S}_{21}\}^{1/2}\\ &=\{\mathbf{S}_{12}\mathbf{S}_{22}^{-1}\mathbf{S}_{21}/s_{11}\}^{1/2}.\end{aligned} \tag{6.5.6}$$

Then $R_{y\cdot\mathbf{x}}$ is called the *multiple correlation coefficient of* $\mathbf{y}$ with $\mathbf{X}$. It can also be defined in terms of the correlations as

$$R^2_{y\cdot\mathbf{x}}=\mathbf{R}_{12}\mathbf{R}_{22}^{-1}\mathbf{R}_{21}. \tag{6.5.7}$$

An important property of the multiple correlation is given by the following theorem.

Theorem 6.5.1 *The largest sample correlation between* $\mathbf{y}$ *and a linear combination of the columns of* $\mathbf{X}$ *is given by* $R_{y\cdot\mathbf{x}}$, *and is attained by the linear combination* $\mathbf{X}\hat{\boldsymbol{\beta}}$, *where* $\hat{\boldsymbol{\beta}}=(\mathbf{X}'\mathbf{X})^{-1}\mathbf{X}'\mathbf{y}$ *is the regression coefficient of* $\mathbf{y}$ *on* $\mathbf{X}$.

Proof Let $\mathbf{X}\boldsymbol{\beta}$ denote a linear combination of the columns of $\mathbf{X}$. Then

$$\operatorname{corr}^2(\mathbf{y},\mathbf{X}\boldsymbol{\beta})=(\boldsymbol{\beta}'\mathbf{S}_{21})^2/(s_{11}\boldsymbol{\beta}'\mathbf{S}_{22}\boldsymbol{\beta}).$$

This function is a ratio of quadratic forms in $\boldsymbol{\beta}$ and, by Corollary A.9.2.2, its maximum is given by $\mathbf{S}_{12}\mathbf{S}_{22}^{-1}\mathbf{S}_{21}/s_{11}$, and is attained when $\hat{\boldsymbol{\beta}}=\mathbf{S}_{22}^{-1}\mathbf{S}_{21}=(\mathbf{X}'\mathbf{X})^{-1}\mathbf{X}'\mathbf{y}$. ■

Another common notation, which is used for a single matrix $\mathbf{X}(n\times q)$ to label the multiple correlation of $\mathbf{x}_{(1)}$ with $\mathbf{x}_{(2)},\ldots,\mathbf{x}_{(q)}$, is $R_{1\cdot2\ldots q}$. It is necessary to exercise some caution when interpreting the multiple correlation, as the following example shows.

Example 6.5.1 Consider the case when

$$r_{13}=0,\qquad r_{12}=\cos\theta,\qquad r_{23}=\sin\theta.$$

These correlations will arise when $\mathbf{x}_{(1)}$, $\mathbf{x}_{(2)}$, and $\mathbf{x}_{(3)}$ lie in the same plane in R^n and the angle between $\mathbf{x}_{(1)}$ and $\mathbf{x}_{(2)}$ is θ. If θ is small then r_{13} and r_{12} are both small. However, $R_{3\cdot 12}=1$. This example illustrates the seemingly paradoxical result that, for small θ, $\mathbf{x}_{(3)}$ has a low correlation with both $\mathbf{x}_{(1)}$ and $\mathbf{x}_{(2)}$, but the multiple correlation coefficient $R_{3\cdot 12}$ is at its maximum.

The multiple correlation coefficient can be used to partition the total sum of squares as in (6.2.8). If $\hat{\mathbf{y}}=\mathbf{X}\hat{\boldsymbol{\beta}}$ denotes the fitted value of $\mathbf{y}$, and $\hat{\mathbf{u}}=\mathbf{y}-\bar{y}\mathbf{1}-\mathbf{X}\hat{\boldsymbol{\beta}}$ denotes the fitted residual, then the total sum of squares (about the mean) can be expressed as

$$\mathbf{y}'\mathbf{y}-n\bar{y}^2=\hat{\mathbf{y}}'\hat{\mathbf{y}}+\hat{\mathbf{u}}'\hat{\mathbf{u}},$$

which can also be written

$$ns_y^2=ns_y^2R_{y\cdot\mathbf{x}}^2+ns_y^2(1-R_{y\cdot\mathbf{x}}^2). \tag{6.5.8}$$

Thus, the squared multiple correlation coefficient represents the proportion of the total sum of squares explained by the regression on $\mathbf{X}$.

6.5.3 Partial correlation coefficient

It is sometimes of interest to study the correlation between two variables after allowing for the effects of other variables. Consider the multivariate regression (6.5.3) with $p=2$ and centre the columns of $\mathbf{Y}$ so that $\bar{y}_1=\bar{y}_2=0$. We can write each column of $\mathbf{Y}$ as

$$\mathbf{y}_{(i)}=\hat{\mathbf{y}}_i+\hat{\mathbf{u}}_{(i)}, \qquad i=1,2,$$

where $\hat{\mathbf{y}}_i=\mathbf{X}(\mathbf{X}'\mathbf{X})^{-1}\mathbf{X}'\mathbf{y}_{(i)}$ is the "fitted value" of $\mathbf{y}_{(i)}$ and $\hat{\mathbf{u}}_{(i)}$ is the "residual value" of $\mathbf{y}_{(i)}$. Since $\hat{\mathbf{u}}_{(1)}$ and $\hat{\mathbf{u}}_{(2)}$ are both residuals after fitting $\mathbf{X}$, a correlation between $\mathbf{y}_{(1)}$ and $\mathbf{y}_{(2)}$ after "eliminating" the effect of $\mathbf{X}$ can be defined by

$$r_{12\cdot\mathbf{x}}=\hat{\mathbf{u}}_{(1)}'\hat{\mathbf{u}}_{(2)}/\{\|\hat{\mathbf{u}}_{(1)}\|\,\|\hat{\mathbf{u}}_{(2)}\|\}. \tag{6.5.9}$$

The coefficient $r_{12\cdot\mathbf{x}}$ is called the (*sample*) *partial correlation coefficient between* $\mathbf{y}_{(1)}$ *and* $\mathbf{y}_{(2)}$ *given* $\mathbf{X}$.

To calculate the partial correlation coefficient, partition the covariance matrix of $\mathbf{Y}$ and $\mathbf{X}$ as

$$\mathbf{S}=\begin{pmatrix} s_{11} & s_{12} & \mathbf{s}_1' \\ s_{21} & s_{22} & \mathbf{s}_2' \\ \mathbf{s}_1 & \mathbf{s}_2 & \mathbf{S}_{22}\end{pmatrix}=n^{-1}\begin{pmatrix}\mathbf{y}_{(1)}' \\ \mathbf{y}_{(2)}' \\ \mathbf{X}\end{pmatrix}(\mathbf{y}_{(1)},\mathbf{y}_{(2)},\mathbf{X}).$$

If we set

$$s_{ij}^*=s_{ij}-\mathbf{s}_i'\mathbf{S}_{22}^{-1}\mathbf{s}_j, \qquad i,j=1,2,$$

then

$$r_{12\cdot\mathbf{x}} = s_{12}^*/\{s_{11}^* s_{22}^*\}^{1/2}. \tag{6.5.10}$$

(See Exercise 6.5.5.)

If we partition the correlation matrix $\mathbf{R}$ similarly, and set $r_{ij}^* = r_{ij} - \mathbf{r}_i'\mathbf{R}_{22}^{-1}\mathbf{r}_j$, $i, j = 1, 2$, then $r_{12\cdot\mathbf{x}}$ can also be written

$$r_{12\cdot\mathbf{x}} = r_{12}^*/\{r_{11}^* r_{22}^*\}^{1/2}. \tag{6.5.11}$$

Another notation, used for a single matrix $\mathbf{X}(n \times q)$ to describe the partial correlation between $\mathbf{x}_{(1)}$ and $\mathbf{x}_{(2)}$ given $\mathbf{x}_{(3)}, \ldots, \mathbf{x}_{(q)}$, is $r_{12\cdot 3\ldots q}$.

In general if $\mathbf{X} = (\mathbf{X}_1, \mathbf{X}_2)$ is a partitioned data matrix with sample covariance matrix $\mathbf{S}$, we can define the "partial" or "residual" covariance matrix of $\mathbf{X}_1$ after allowing for the effect of $\mathbf{X}_2$ by

$$\mathbf{S}_{11\cdot 2} = \mathbf{S}_{11} - \mathbf{S}_{12}\mathbf{S}_{22}^{-1}\mathbf{S}_{21}. \tag{6.5.12}$$

Note that $\mathbf{S}_{11\cdot 2}$ can also be interpreted as the sample conditional covariance matrix of $\mathbf{X}_1$ given $\mathbf{X}_2$. (A population version of (6.5.12) for the multinormal distribution was given in (3.2.1).)

Example 6.5.2 Observations on the intelligence (x_1), weight (x_2), and age (x_3) of school children produced the following correlation matrix

$$\mathbf{R} = \begin{bmatrix} 1.0000 & 0.6162 & 0.8267 \\ & 1.0000 & 0.7321 \\ & & 1.0000 \end{bmatrix}.$$

This in fact suggests a high dependence (0.6162) between weight and intelligence. However, calculating the partial correlation coefficient $r_{12\cdot 3} = 0.0286$ shows that the correlation between weight and intelligence is very much lower, in fact almost zero, after the effect of age has been taken into account.

The multiple correlation coefficient between intelligence and the other two variables is evaluated from (6.5.7) as

$$R_{1\cdot 23} = 0.8269.$$

However, the simple correlation r_{13} is 0.8267, so weight obviously plays little further part in explaining intelligence. ■

6.5.4 Measures of correlation between vectors

In the univariate regression model defined by (6.5.1), that is when $p = 1$, we have seen that the squared multiple correlation coefficient R^2 represents the "proportion of variance explained" by the model, and, from

(6.5.8),

$$1-R^2=\hat{\mathbf{u}}'\hat{\mathbf{u}}/\mathbf{y}'\mathbf{y}.$$

(We assume $\mathbf{y}$ is centred so that $\bar{y}=0$.)

It seems reasonable to seek a similar measure for the multivariate correlation between matrices $\mathbf{X}$ and $\mathbf{Y}$ in the model $\mathbf{Y}=\mathbf{XB}+\mathbf{U}$.

Let us start by writing $\mathbf{D}=(\mathbf{Y}'\mathbf{Y})^{-1}\hat{\mathbf{U}}'\hat{\mathbf{U}}$. (Again, we assume that $\mathbf{Y}$ is centred so that the columns of $\mathbf{Y}$ have zero mean.) The matrix $\mathbf{D}$ is a simple generalization of $1-R^2$ in the univariate case. Note that $\hat{\mathbf{U}}'\hat{\mathbf{U}}=\mathbf{Y}'\mathbf{PY}$ ranges between zero, when all the variation in $\mathbf{Y}$ is "explained" by the model, and $\mathbf{Y}'\mathbf{Y}$ at the other extreme, when no part of the variation in $\mathbf{Y}$ is explained. Therefore $\mathbf{I}-\mathbf{D}$ varies between the identity matrix and the zero matrix. (See Exercise 6.5.7.) Any sensible measure of multivariate correlation should range between one and zero at these extremes and this property is satisfied by at least two oft-used coefficients, the trace correlation r_T, and the determinant correlation r_D, defined as follows (see Hooper, 1959, pp. 249–250),

$$r_T^2=p^{-1}\operatorname{tr}(\mathbf{I}-\mathbf{D}),\qquad r_D^2=\det(\mathbf{I}-\mathbf{D}). \tag{6.5.13}$$

For their population counterparts, see Exercise 6.5.8. Hotelling (1936) suggested the "vector alienation coefficient" (r_A), which in our notation is $\det\mathbf{D}$, and ranges from zero when $\hat{\mathbf{U}}=\mathbf{0}$ to one when $\hat{\mathbf{U}}=\mathbf{Y}$.

If $d_1,\ldots,d_p$ are the eigenvalues of $\mathbf{I}-\mathbf{D}$ then

$$r_T^2=p^{-1}\sum d_i,\qquad r_D^2=\prod d_i,\qquad r_A=\prod(1-d_i).$$

From this formulation, it is clear that r_D or r_A is zero if *just one* d_i is zero or one, respectively, but r_T is zero only if *all* the d_i are zero. Note that these measures of vector correlations are invariant under commutation, viz. they would remain the same if we were to define $\mathbf{D}$ as equal to $(\hat{\mathbf{U}}'\hat{\mathbf{U}})(\mathbf{Y}'\mathbf{Y})^{-1}$.

6.6 Least Squares Estimation

6.6.1 Ordinary least squares (OLS) estimation

Consider the multiple regression model

$$\mathbf{y}=\mathbf{X}\boldsymbol{\beta}+\mathbf{u}, \tag{6.6.1}$$

where $\mathbf{u}(n\times 1)$ is the vector of disturbances. Suppose that the matrix $\mathbf{X}(n\times q)$ is a known matrix of rank q, and let us relax the assumptions about the disturbances, assuming merely that

$$E(\mathbf{u}) = \mathbf{0}, \qquad V(\mathbf{u}) = \mathbf{\Omega}. \tag{6.6.2}$$

When $\mathbf{u}$ is not normally distributed then the estimation of $\boldsymbol{\beta}$ can be approached by the method of least squares.

Definition 6.6.1 *The* ordinary least squares (OLS) estimator *of* $\boldsymbol{\beta}$ *is given by*

$$\hat{\boldsymbol{\beta}} = (\mathbf{X}'\mathbf{X})^{-1}\mathbf{X}'\mathbf{y}. \tag{6.6.3}$$

Differentiation shows that $\hat{\boldsymbol{\beta}}$ minimizes the residual sum of squares

$$(\mathbf{y} - \mathbf{X}\boldsymbol{\beta})'(\mathbf{y} - \mathbf{X}\boldsymbol{\beta}), \tag{6.6.4}$$

hence the terminology. Note that $\hat{\boldsymbol{\beta}}$ is exactly the same as the m.l.e. of $\boldsymbol{\beta}$ given in Theorem 6.2.1 (with $p = 1$) under the assumption of normality, and the assumption $\mathbf{\Omega} = \sigma^2\mathbf{I}_n$.

Note that

$$E(\hat{\boldsymbol{\beta}}) = (\mathbf{X}'\mathbf{X})^{-1}\mathbf{X}'(\mathbf{X}\boldsymbol{\beta} + E\mathbf{u}) = \boldsymbol{\beta}$$

so that $\hat{\boldsymbol{\beta}}$ is an *unbiased* estimate of $\boldsymbol{\beta}$. Also

$$V(\hat{\boldsymbol{\beta}}) = (\mathbf{X}'\mathbf{X})^{-1}(\mathbf{X}'\mathbf{\Omega}\mathbf{X})(\mathbf{X}'\mathbf{X})^{-1}. \tag{6.6.5}$$

In particular, if $\mathbf{\Omega} = \sigma^2\mathbf{I}_n$, this formula simplifies to

$$V(\hat{\boldsymbol{\beta}}) = \sigma^2(\mathbf{X}'\mathbf{X})^{-1}, \tag{6.6.6}$$

and in this case $\hat{\boldsymbol{\beta}}$ has the following optimal property.

Theorem 6.6.1 (Gauss–Markov) *Consider the multiple regression model* (6.6.1) *and suppose the disturbance terms are uncorrelated with one another,* $V(\mathbf{u}) = \sigma^2\mathbf{I}_n$. *Then the OLS estimator* (6.6.3) *has a covariance matrix which is smaller than that of any other linear estimator.* ■

In other words the OLS estimator is the *best* linear unbiased estimator (BLUE). The proof of this theorem is outlined in Exercise 6.6.2.

6.6.2 Generalized least squares estimation

When $\mathbf{u}$ does not have a covariance matrix equal to $\sigma^2\mathbf{I}$, then, in general, the OLS estimator is *not* the BLUE. However, a straightforward transformation adjusts the covariance matrix so that OLS can be used.

Let

$$E(\mathbf{u}) = \mathbf{0}, \qquad V(\mathbf{u}) = \mathbf{\Omega}.$$

Suppose that $\mathbf{\Omega}$ is known and consider the transformed model

$$\mathbf{z}=\mathbf{\Omega}^{-1/2}\mathbf{X}\boldsymbol{\beta}+\mathbf{v}, \tag{6.6.7}$$

where

$$\mathbf{z}=\mathbf{\Omega}^{-1/2}\mathbf{y}, \qquad \mathbf{v}=\mathbf{\Omega}^{-1/2}\mathbf{u}.$$

Since $V(\mathbf{v})=\mathbf{I}$, this transformed model satisfies the assumptions of the Gauss–Markov theorem (Theorem 6.6.1). Thus, the best linear unbiased estimate of $\boldsymbol{\beta}$ is given by

$$\tilde{\boldsymbol{\beta}}=(\mathbf{X}'\mathbf{\Omega}^{-1}\mathbf{X})^{-1}\mathbf{X}'\mathbf{\Omega}^{-1/2}\mathbf{z}=(\mathbf{X}'\mathbf{\Omega}^{-1}\mathbf{X})^{-1}\mathbf{X}'\mathbf{\Omega}^{-1}\mathbf{y} \tag{6.6.8}$$

and the covariance matrix of $\tilde{\boldsymbol{\beta}}$ is

$$V(\tilde{\boldsymbol{\beta}})=(\mathbf{X}'\mathbf{\Omega}^{-1}\mathbf{X})^{-1}. \tag{6.6.9}$$

The estimator defined by (6.6.8) is called the *generalized least squares* (GLS) estimator and is an obvious generalization of the OLS estimator. The fact that the GLS estimator is the BLUE can also be proved directly (Exercise 6.6.3).

Note that (6.6.8) is not an estimator if $\mathbf{\Omega}$ is not known. However, in some applications, although $\mathbf{\Omega}$ is unknown, there exists a consistent estimator $\hat{\mathbf{\Omega}}$ of $\mathbf{\Omega}$. Then $\hat{\mathbf{\Omega}}$ may be used in place of $\mathbf{\Omega}$ in (6.6.8).

6.6.3 Application to multivariate regression

Multivariate regression is an example where OLS and GLS lead to the same estimators. Write the multivariate regression model (6.1.1) in vector form,

$$\mathbf{Y}^V=\mathbf{X}^*\mathbf{B}^V+\mathbf{U}^V, \tag{6.6.10}$$

where $\mathbf{Y}^V=(\mathbf{y}'_{(1)},\ldots,\mathbf{y}'_{(p)})'$ is obtained by stacking the columns of $\mathbf{Y}$ on top of one another, and $\mathbf{X}^*=\mathbf{I}_p\otimes\mathbf{X}$. The disturbance $\mathbf{U}^V$ is assumed to have mean $\mathbf{0}$ and covariance matrix

$$V(\mathbf{U}^V)=\mathbf{\Omega}=\mathbf{\Sigma}\otimes\mathbf{I}_n.$$

Then the GLS estimator of $\mathbf{B}^V$ is given by

$$\begin{aligned}\hat{\mathbf{B}}^V&=(\mathbf{X}^{*\prime}\mathbf{\Omega}^{-1}\mathbf{X}^*)^{-1}\mathbf{X}^{*\prime}\mathbf{\Omega}^{-1}\mathbf{Y}^V\\&=(\mathbf{\Sigma}\otimes\mathbf{X}'\mathbf{X})^{-1}(\mathbf{\Sigma}^{-1}\otimes\mathbf{X}')\mathbf{Y}^V\\&=[\mathbf{I}\otimes(\mathbf{X}'\mathbf{X})^{-1}\mathbf{X}']\mathbf{Y}^V.\end{aligned} \tag{6.6.11}$$

Note that (6.6.10) does not depend upon $\mathbf{\Sigma}$ and hence defines an estimator whether $\mathbf{\Sigma}$ is known or not. In particular, the OLS estimator is obtained when $\mathbf{\Sigma}=\mathbf{I}$, so the OLS and GLS estimators are the same in this case. Further (6.6.11) also coincides with the m.l.e. of $\mathbf{B}$ obtained in Theorem 6.2.1 under the assumption of normality.

6.6.4 Asymptotic consistency of least squares estimators

Let us examine the asymptotic behaviour of the GLS estimator as the sample size n tends to ∞. Let $\mathbf{x}_1, \mathbf{x}_2, \ldots$ be a sequence of (fixed) "independent" variables and let $u_1, u_2, \ldots$ be a sequence of random disturbances. Writing $\mathbf{X} = \mathbf{X}_n = (\mathbf{x}_1', \ldots, \mathbf{x}_n')'$, $\mathbf{\Omega} = \mathbf{\Omega}_n$, and $\boldsymbol{\beta} = \boldsymbol{\beta}_n$, to emphasize the dependence on n, suppose that

$$\lim (\mathbf{X}_n'\mathbf{\Omega}_n^{-1}\mathbf{X}_n)^{-1} = \mathbf{0}. \tag{6.6.12}$$

Then $E(\tilde{\boldsymbol{\beta}}_n) = \boldsymbol{\beta}$ for all n and $V(\hat{\boldsymbol{\beta}}_n) = (\mathbf{X}_n'\mathbf{\Omega}_n^{-1}\mathbf{X}_n)^{-1} \to \mathbf{0}$ as $n \to \infty$. Hence $\tilde{\boldsymbol{\beta}}$ is *consistent*.

If the "independent" variables $\mathbf{x}_i$ are *random* rather than fixed, but still uncorrelated with the random disturbances, then the results of Sections 6.6.1–6.6.2 remain valid provided all expectations and variances are interpreted as "conditional on $\mathbf{X}$". In particular, suppose that

$$E(\mathbf{u} \mid \mathbf{X}) = \mathbf{0} \quad \text{and} \quad V(\mathbf{u} \mid \mathbf{X}) = \mathbf{\Omega} \tag{6.6.13}$$

do not depend on $\mathbf{X}$.

If we replace condition (6.6.12) by

$$\operatorname{plim} (\mathbf{X}_n'\mathbf{\Omega}_n^{-1}\mathbf{X}_n)^{-1} = \mathbf{0}, \tag{6.6.14}$$

then consistency holds in this situation also. (Here plim denotes limit in probability; that is, if $\mathbf{A}_n$ is a sequence of random matrices and $\mathbf{A}_0$ is a constant matrix, then $\operatorname{plim} \mathbf{A}_n = \mathbf{A}_0$ or $\mathbf{A}_n \xrightarrow{p} \mathbf{A}_0$ if and only if

$$P(\|\mathbf{A}_n - \mathbf{A}_0\| > \varepsilon) \to 0 \quad \text{as} \quad n \to \infty, \tag{6.6.15}$$

for each $\varepsilon > 0$, where $\|\mathbf{A}\| = \sum\sum a_{ij}^2$.) The proof is outlined in Exercise 6.6.5.

If

$$\lim n^{-1}\mathbf{X}_n'\mathbf{\Omega}_n^{-1}\mathbf{X}_n = \mathbf{\Psi}, \tag{6.6.16}$$

or

$$\operatorname{plim} n^{-1}\mathbf{X}_n'\mathbf{\Omega}_n^{-1}\mathbf{X}_n = \mathbf{\Psi}, \tag{6.6.17}$$

where $\mathbf{\Psi}(q \times q)$ is non-singular, then $(\mathbf{X}_n'\mathbf{\Omega}_n^{-1}\mathbf{X}_n)^{-1}$ tends to zero at rate n^{-1} so that (6.6.12) or (6.6.14) holds. In this situation, with some further regularity assumptions on $\mathbf{X}_n$ and/or $\mathbf{u}_n$, it can be shown that in fact $\hat{\boldsymbol{\beta}}_n$ is asymptotically *normally* distributed, with mean $\boldsymbol{\beta}$ and covariance matrix

$n^{-1}\mathbf{\Psi}^{-1}$. Note that (6.6.16) will hold, for example, if $\mathbf{\Omega}_n = \sigma^2\mathbf{I}_n$ and $\mathbf{X}_n$ is a random sample from some distribution with mean $\boldsymbol{\mu}$ and covariance matrix $\mathbf{\Sigma}$, letting $\mathbf{\Psi} = \mathbf{\Sigma} + \boldsymbol{\mu}\boldsymbol{\mu}'$.

6.7 Discarding of Variables

6.7.1. Dependence analysis

Consider the multiple regression

$$\mathbf{y} = \mathbf{X}\boldsymbol{\beta} + \mu\mathbf{1} + \mathbf{u}$$

and suppose that some of the columns of $\mathbf{X}(n \times q)$ are "nearly" collinear. Since rank ($\mathbf{X}$) equals the number of linearly independent columns of $\mathbf{X}$ and also equals the number of non-zero eigenvalues of $\mathbf{X}'\mathbf{X}$, we see that in this situation some of the eigenvalues of $\mathbf{X}'\mathbf{X}$ will be "nearly" zero. Hence some of the eigenvalues of $(\mathbf{X}'\mathbf{X})^{-1}$ will be very large. The variance of the OLS estimator $\hat{\boldsymbol{\beta}}$ equals $\sigma^2(\mathbf{X}'\mathbf{X})^{-1}$, and so at least some of these regression estimates will have large variances.

Clearly, because of the high collinearity, some of the independent variables are contributing little to the regression. Therefore, it is of interest to ask how well the regression can be explained using a smaller number of independent variables. There are two important reasons for discarding variables here; namely

(1) to increase the precision of the regression estimates for the retained variables, and

(2) to reduce the number of measurements needed on similar data in the future.

Suppose it has been decided to retain k variables ($k < q$). Then a natural choice of retained variables is the subset $\mathbf{x}_{i_1}, \ldots, \mathbf{x}_{i_k}$ which maximizes the squared multiple correlation coefficient

$$R^2_{\mathbf{y}\cdot i_1,\ldots,i_k}. \tag{6.7.1}$$

This choice will explain more of the variation in $\mathbf{y}$ than any other set of k variables.

Computationally, the search for the best subset of k variables is not simple, but an efficient algorithm has been developed by Beale *et al.* (1967). See also Beale (1970).

The choice of k, the number of retained variables, is somewhat arbitrary. One rule of thumb is to retain enough variables so that the squared multiple correlation with $\mathbf{y}$ using k variables is at least 90 or 95% of the squared multiple correlation using all q variables (but see Thompson, 1978, for some test criteria).

Further, although there is usually just one set of k variables maximizing (6.7.1), there are often other k-sets which are "nearly optimal", so that for practical purposes the choice of variables is not uniquely determined.

An alternative method of dealing with multicollinearity in a multiple regression can be developed using principal components (see Section 8.8).

Example 6.7.1 In a study of $n = 180$ pitprops cut from the Corsican pine tree, Jeffers (1967) studied the dependence of maximum compressive strength ($\mathbf{y}$) on 13 other variables ($\mathbf{X}$) measured on each prop. The strength of pitprops is of interest because they are used to support mines.

The physical variables measured on each prop were

x_1 = the top diameter of the prop (in inches);
x_2 = the length of the prop (in inches);
x_3 = the moisture content of the prop, expressed as a percentage of the dry weight;
x_4 = the specific gravity of the timber at the time of the test;
x_5 = the oven-dry specific gravity of the timber;
x_6 = the number of annual rings at the top of the prop;
x_7 = the number of annual rings at the base of the prop;
x_8 = the maximum bow (in inches);
x_9 = the distance of the point of maximum bow from the top of the prop (in inches);
x_{10} = the number of knot whorls;
x_{11} = the length of clear prop from the top of the prop (in inches);
x_{12} = the average number of knots per whorl;
x_{13} = the average diameter of the knots (in inches).

The columns of $\mathbf{X}$ have each been standardized to have mean 0 and variance 1 so that $\mathbf{X}'\mathbf{X}$, given in Table 6.7.1, represents the correlation matrix for the independent variables. The correlation between each of the independent variables and $\mathbf{y}$ is given in Table 6.7.2. The eigenvalues of $\mathbf{X}'\mathbf{X}$ are 4.22, 2.38, 1.88, 1.11, 0.91, 0.82, 0.58, 0.44, 0.35, 0.19, 0.05, 0.04, 0.04, and clearly the smallest ones are nearly zero.

For each value of k, the subset of k variables which maximizes (6.7.1) is given in Table 6.7.3. Note that if all the variables are retained then 73.1% of the variance of $\mathbf{y}$ can be explained by the regression. If only $k = 6$ variables are retained, then we can still explain 71.6% of the variance of $\mathbf{y}$, which is $(0.716/0.731) \times 100\%$ or over 95% of the "explainable" variance of $\mathbf{y}$.

Table 6.7.1 Correlation matrix for physical properties of props

x_1												
0.954	x_2											
0.364	0.297	x_3										
0.342	0.284	0.882	x_4									
−0.129	−0.118	−0.148	0.220	x_5								
0.313	0.291	0.153	0.381	0.364	x_6							
0.496	0.503	−0.029	0.174	0.296	0.813	x_7						
0.424	0.419	−0.054	−0.059	0.004	0.090	0.372	x_8					
0.592	0.648	0.125	0.137	−0.039	0.211	0.465	0.482	x_9				
0.545	0.569	−0.081	−0.014	0.037	0.274	0.679	0.557	0.526	x_{10}			
0.084	0.076	0.162	0.097	−0.091	−0.036	−0.113	0.061	0.085	−0.319	x_{11}		
−0.019	−0.036	0.220	0.169	−0.145	0.024	−0.232	−0.357	−0.127	−0.368	0.029	x_{12}	
0.134	0.144	0.126	0.015	−0.208	−0.329	−0.424	−0.202	−0.076	−0.291	0.007	0.184	x_{13}

Table 6.7.2 Correlation between the independent variables and **y** for pitprop data

Variable	Correlation
1	−0.419
2	−0.338
3	−0.728
4	−0.543
5	0.247
6	0.117
7	0.110
8	−0.253
9	−0.235
10	−0.101
11	−0.055
12	−0.117
13	−0.153

Note that for this data, the optimal $(k+1)$-set of variables equals the optimal k-set of variables plus one new variable, for $k=1,\dots, 12$. Unfortunately, this property does not hold in general.

Further discussion of this data appears in Example 8.8.1.

For the discussion of recent methods of discarding variables and related tests see Thompson (1978).

Table 6.7.3 Variables selected in multiple regression for pitprop data

k	Variables selected	R^2
1	3	0.530
2	3, 8	0.616
3	3, 6, 8	0.684
4	3, 6, 8, 11	0.695
5	1, 3, 6, 8, 11	0.706
6	1, 2, 3, 6, 8, 11	0.716
7	1, 2, 3, 4, 6, 8, 11	0.721
8	1, 2, 3, 4, 6, 8, 11, 12	0.725
9	1, 2, 3, 4, 5, 6, 8, 11, 12	0.727
10	1, 2, 3, 4, 5, 6, 8, 9, 11, 12	0.729
11	1, 2, 3, 4, 5, 6, 7, 8, 9, 11, 12	0.729
12	1, 2, 3, 4, 5, 6, 7, 8, 9, 10, 11, 12	0.731
13	1, 2, 3, 4, 5, 6, 7, 8, 9, 10, 11, 12, 13	0.731

6.7.2 Interdependence analysis

Consider now a situation where q explanatory variables $\mathbf{X}$ are considered on their own, rather than as independent variables in a regression. When the columns of $\mathbf{X}$ are nearly collinear, it is desirable to discard some of the variables in order to reduce the number of measurements needed to describe the data effectively. In this context we need a measure of how well a retained set of k variables, $\mathbf{x}_{i_1},\ldots,\mathbf{x}_{i_k}$, explains the whole data set $\mathbf{X}$.

A measure of how well one rejected variable $\mathbf{x}_{j_1}$ is explained by the retained variables $\mathbf{x}_{i_1},\ldots,\mathbf{x}_{i_k}$ is given by the squared multiple correlation coefficient $R^2_{j_1\cdot i_1,\ldots,i_k}$. Thus, an overall measure of the ability of the retained variables to explain the data is obtained by looking at the worst possible case; namely

$$\min_{j_l} R^2_{j_l\cdot i_1,\ldots,i_k}, \tag{6.7.2}$$

where j_l runs through all the rejected variables. Then the best choice for the retained variables is obtained by maximizing (6.7.2) over all k-sets of variables.

Computationally this problem is similar to the problem of Section 6.7.1. See Beale *et al.* (1967) and Beale (1970).

As in Section 6.7.1 the choice of k is somewhat arbitrary. A possible rule of thumb is to retain enough variables so that the minimum R^2 with any rejected variable is at least 0.50, say. Further, although the best choice of k variables is usually analytically well-defined, there are often several "nearly optimal" choices of k variables making the selection non-unique for practical purposes.

Another method of rejecting variables in interdependence analysis based on principal components is discussed in Section 8.7.

Example 6.7.2 Let us carry out an interdependence analysis for the 13 explanatory variables of Jeffers' pitprop data of Example 6.7.1. For each value of $k = 1,\ldots, 13$ the optimal k-set of variables is given in Table 6.7.4, together with the minimum value of the squared multiple correlation of any of the rejected variables with the retained variables. For example, eight variables, namely the variables 3, 5, 7, 8, 9, 11, 12, 13, are needed to explain at least 50% of the variation in each of the rejected variables.

Note from Table 6.7.4 that the best k-set of variables is *not* in general a subset of the best $(k+1)$-set.

Further discussion of this data is given in Example 8.7.1.

Table 6.7.4 Variables selected in interdependence analysis for pitprop data

k	Variables selected	min R^2 with any rejected variable
1	9	0.002
2	4, 10	0.050
3	1, 5, 10	0.191
4	1, 5, 8, 10	0.237
5	1, 4, 7, 11, 12	0.277
6	3, 4, 10, 11, 12, 13	0.378
7	2, 3, 4, 8, 11, 12, 13	0.441
8	3, 5, 7, 8, 9, 11, 12, 13	0.658
9	2, 4, 5, 7, 8, 9, 11, 12, 13	0.751
10	1, 3, 5, 6, 8, 9, 10, 11, 12, 13	0.917
11	2, 3, 5, 6, 7, 8, 9, 10, 11, 12, 13	0.919
12	1, 3, 4, 5, 6, 7, 8, 9, 10, 11, 12, 13	0.927
13	1, 2, 3, 4, 5, 6, 7, 8, 9, 10, 11, 12, 13	—

Exercises and Complements

6.1.1 (Box and Tiao, 1973, p. 439) For the multivariate regression model (6.1.1) with normal errors, show that Bayesian analysis with a non-informative prior leads to a posterior p.d.f. for **B** which is proportional to

$$|n\hat{\mathbf{\Sigma}}+(\hat{\mathbf{B}}-\mathbf{B})'\mathbf{X}'\mathbf{X}(\hat{\mathbf{B}}-\mathbf{B})|^{-n/2},$$

where $\hat{\mathbf{\Sigma}}$ and $\hat{\mathbf{B}}$ are defined in (6.2.2)–(6.2.3). Hence the posterior distribution of **B** is a matrix T distribution.

6.2.1 Show that the $(n\times n)$ matrix **P** defined in (6.2.1) is symmetric and idempotent. Show that $\mathbf{Pw}=\mathbf{w}$ if and only if **w** is orthogonal to all columns of **X** and that $\mathbf{Pw}=\mathbf{0}$ if and only if **w** is a linear combination of the columns of **X**. Hence **P** is a projection onto the subspace of R^n orthogonal to the columns of **X**.

6.2.2 (a) From (6.2.6) show that $l(\mathbf{B},\mathbf{\Sigma})$ depends on **Y** only through $\hat{\mathbf{B}}$ and $\hat{\mathbf{\Sigma}}$, and hence show that $(\hat{\mathbf{B}},\hat{\mathbf{\Sigma}})$ is sufficient for $(\mathbf{B},\mathbf{\Sigma})$.

(b) Is $\hat{\mathbf{B}}$ sufficient for **B** when $\mathbf{\Sigma}$ is known?

(c) Is $\hat{\mathbf{\Sigma}}$ sufficient for $\mathbf{\Sigma}$ when **B** is known?

6.2.3 Show the following from Theorem 6.2.3(a):

(a) The correlation between $\hat{\beta}_{ij}$ and $\hat{\beta}_{kl}$ is $\rho_{jl}g_{ik}/(g_{ii}g_{kk})^{1/2}$, where $\rho_{jl}=\sigma_{jl}/(\sigma_{jj}\sigma_{ll})^{1/2}$.

(b) The covariance between two rows of $\hat{\mathbf{B}}$ is $C(\hat{\boldsymbol{\beta}}_i, \hat{\boldsymbol{\beta}}_k) = g_{ik}\boldsymbol{\Sigma}$.
(c) The covariance between two columns of $\hat{\boldsymbol{\beta}}$ is $C(\hat{\boldsymbol{\beta}}_{(j)}, \hat{\boldsymbol{\beta}}_{(l)}) = \sigma_{jl}\mathbf{G}$.

6.2.4 Show from (6.2.9) that

$$\hat{\mathbf{B}} - \mathbf{B} = \mathbf{AU},$$

where $\mathbf{A} = (\mathbf{X}'\mathbf{X})^{-1}\mathbf{X}'$ and $\mathbf{AA}' = (\mathbf{X}'\mathbf{X})^{-1} = \mathbf{G}$. Hence, following the approach adopted in Exercise 3.3.4, show that $\hat{\mathbf{B}} - \mathbf{B}$, thought of as a column vector, has the distribution

$$(\hat{\mathbf{B}} - \mathbf{B})^V \sim \mathbf{N}(\mathbf{0}, \boldsymbol{\Sigma} \otimes \mathbf{G}),$$

thus confirming Theorem 6.2.3(a) and Exercise 6.2.3.

6.3.1 Verify formula (6.3.4) for the projection $\mathbf{P}_2$. Hint: let $\mathbf{Z} = (\mathbf{Z}_1, \mathbf{Z}_2) = \mathbf{XC}^{-1} = (\mathbf{XC}^{(1)}, \mathbf{XC}^{(2)})$ and let $\mathbf{A} = \mathbf{Z}'\mathbf{Z}$. Note that $\mathbf{CC}^{-1} = \mathbf{I}$ implies $\mathbf{C}_1\mathbf{C}^{(1)} = \mathbf{I}$ and $\mathbf{C}_1\mathbf{C}^{(2)} = \mathbf{0}$. Show that $\mathbf{P}_1 - \mathbf{P} = \mathbf{ZA}^{-1}\mathbf{Z}' - \mathbf{Z}_2\mathbf{A}_{22}^{-1}\mathbf{Z}_2'$ and that $\mathbf{P}_2 = \mathbf{ZA}^{-1}[\mathbf{I}, \mathbf{0}]'(\mathbf{A}^{11})^{-1}[\mathbf{I}, \mathbf{0}]\mathbf{A}^{-1}\mathbf{Z}'$. Hence using (A.2.4g), show that $\mathbf{P}_1 - \mathbf{P}$ and $\mathbf{P}_2$ can both be written in the form

$$[\mathbf{Z}_1, \mathbf{Z}_2]\begin{bmatrix} \mathbf{A}^{11} & \mathbf{A}^{12} \\ \mathbf{A}^{21} & \mathbf{A}_{21}(\mathbf{A}^{11})^{-1}\mathbf{A}_{12} \end{bmatrix}\begin{bmatrix} \mathbf{Z}_1' \\ \mathbf{Z}_2' \end{bmatrix}.$$

6.3.2 In Theorem 6.3.1 verify that $\mathbf{Y}_+'\mathbf{P}_2\mathbf{Y}_+ = \mathbf{U}'\mathbf{P}_2\mathbf{U}$.

6.3.3 (a) Partition $\mathbf{X} = (\mathbf{X}_1, \mathbf{X}_2)$, $\mathbf{Y} = (\mathbf{Y}_1, \mathbf{Y}_2)$ and

$$\mathbf{B} = \begin{pmatrix} \mathbf{B}_{11} & \mathbf{B}_{12} \\ \mathbf{B}_{21} & \mathbf{B}_{22} \end{pmatrix}, \qquad \boldsymbol{\Sigma} = \begin{pmatrix} \boldsymbol{\Sigma}_{11} & \boldsymbol{\Sigma}_{12} \\ \boldsymbol{\Sigma}_{21} & \boldsymbol{\Sigma}_{22} \end{pmatrix},$$

and consider the hypothesis $H_0 : \mathbf{B}_{11} = \mathbf{0}$. Let $L_1(\mathbf{Y}; \mathbf{B}, \boldsymbol{\Sigma})$ denote the likelihood of the whole data set $\mathbf{Y}$ under the model (6.1.1), which is given by (6.1.3), and let $L_2(\mathbf{Y}_1; \mathbf{B}_{11}, \mathbf{B}_{21}, \boldsymbol{\Sigma}_{11})$ denote the likelihood of the data $\mathbf{Y}_1$ under the model (6.3.6). Show that

$$\frac{\max L_1 \text{ over } (\boldsymbol{\Sigma}, \mathbf{B}) \text{ such that } \mathbf{B}_{11} = \mathbf{0}}{\max L_1 \text{ over } (\boldsymbol{\Sigma}, \mathbf{B})} = \frac{\max L_2 \text{ over } (\boldsymbol{\Sigma}_{11}, \mathbf{B}_{21}) \text{ such that } \mathbf{B}_{11} = \mathbf{0}}{\max L_2 \text{ over } (\boldsymbol{\Sigma}_{11}, \mathbf{B}_{11}, \mathbf{B}_{21})}$$

and hence the LRT for H_0 is the same whether one uses the model (6.1.1) or (6.3.6). (Hint: Split the maximization of L_1 into two parts by factorizing the joint density of $(\mathbf{Y}_1, \mathbf{Y}_2)$ as the product of the marginal density of $\mathbf{Y}_1$ (which depends on $\mathbf{B}_{11}, \mathbf{B}_{21}$, and $\boldsymbol{\Sigma}_{11}$ alone) times the conditional density of $\mathbf{Y}_2 \mid \mathbf{Y}_1$ (which when maximized over $\mathbf{B}_{12}, \mathbf{B}_{22}, \boldsymbol{\Sigma}_{12}$, and $\boldsymbol{\Sigma}_{22}$, takes the same value, whatever the value of $\mathbf{B}_{11}$, $\mathbf{B}_{21}$, and $\boldsymbol{\Sigma}_{11}$.)
(b) Carry out a similar analysis for the hypothesis $\mathbf{C}_1\mathbf{BM}_1 = \mathbf{D}$.

6.3.4 If $\mathbf{Z}(n \times p)$ is a data matrix from $N_p(\mathbf{0}, \mathbf{\Sigma})$, and $\mathbf{a}$ is a fixed n-vector, show that $\mathbf{Z}'\mathbf{a} \sim N_p(\mathbf{0}, (\mathbf{a}'\mathbf{a})\mathbf{\Sigma})$.

6.4.1 If a hypothesis matrix $\mathbf{C}_1$ is testable in the sense of (6.4.2), show that the parameter matrix $\mathbf{C}_1\mathbf{B}$ is uniquely determined in the model (6.1.1) because $E\mathbf{Y} = \mathbf{X}\mathbf{B}$. Show that $\mathbf{C}_1$ is testable if and only if Roy's criterion (6.4.3) holds.

6.4.2 In Example 6.4.1 show that $\mathbf{H} = \mathbf{Y}'\mathbf{P}_2\mathbf{Y}$ and $\mathbf{E} = \mathbf{Y}'\mathbf{P}\mathbf{Y}$ are the "between-groups" and "within-groups" SSP matrices, respectively, which were defined in Section 5.3.3a. (Hint: Set $N_0 = 0$ and $N_i = n_1 + \cdots + n_i$ for $i = 1, \ldots, k$. Consider the following orthogonal basis in R^n. Let $\mathbf{e} = \mathbf{1}$. Let $\mathbf{f}_{ij}$ have $+1/(j-1)$ in the places $N_{i-1}+1, \ldots, N_{i-1}+j-1$, and -1 in the $(N_{i-1}+j)$th place and 0 elsewhere for $j = 2, \ldots, n_i$; $i = 1, \ldots, k$. Let $\mathbf{g}_i$ have $1/N_{i-1}$ in the places $1, \ldots, 1/N_{i-1}$ and $-1/n_i$ in the places $N_{i-1}+1, \ldots, N_i$, with 0 elsewhere, for $i = 2, \ldots, k$. Then verify that $\mathbf{P}_2\mathbf{e} = \mathbf{P}_2\mathbf{f}_{ij} = \mathbf{0}$ and $\mathbf{P}_2\mathbf{g}_i = \mathbf{g}_i$, so that

$$\mathbf{P}_2 = \sum_{i=2}^{k} (\mathbf{g}_i'\mathbf{g}_i)^{-1}\mathbf{g}_i\mathbf{g}_i'.$$

Similarly verify that $\mathbf{P}\mathbf{e} = \mathbf{P}\mathbf{g}_i = \mathbf{0}$ and $\mathbf{P}\mathbf{f}_{ij} = \mathbf{f}_{ij}$ so that $\mathbf{P} = \mathbf{I} - n^{-1}\mathbf{1}\mathbf{1}' - \mathbf{P}_2$. Thus, verify that $\mathbf{H}$ and $\mathbf{E}$ are the "between-groups" and "within-groups" SSP matrices from Section 5.5.3a.)

6.4.3 Give the form for the simultaneous confidence intervals in Example 6.4.1 when the contrast $\mathbf{c}$ and/or the linear combination of variables $\mathbf{a}$ is given *a priori*.

6.5.1 Let $\mathbf{U}$ be a data matrix from $N_p(\boldsymbol{\mu}, \mathbf{\Sigma})$ and set $\mathbf{W} = \mathbf{H}\mathbf{U} = \mathbf{U} - \mathbf{1}\bar{\mathbf{u}}'$. Show that $E(\mathbf{w}_i) = \mathbf{0}$ for $i = 1, \ldots, n$ and that

$$C(\mathbf{w}_i, \mathbf{w}_j) = E(\mathbf{w}_i\mathbf{w}_j') = \begin{cases} (1-n^{-1})\mathbf{\Sigma}, & i = j, \\ -n^{-1}\mathbf{\Sigma}, & i \neq j. \end{cases}$$

Hence, writing $\mathbf{W}$ as a vector (see Section A.2.5), deduce $\mathbf{W}^V \sim N_{np}(\mathbf{0}, \mathbf{\Sigma} \otimes \mathbf{H})$.

6.5.2 (Population multiple correlation coefficient) If $\mathbf{x}$ is a random p-vector with covariance matrix $\mathbf{\Sigma}$ partitioned as (6.5.5), show that the largest correlation between x_1 and a linear combination of $x_2, \ldots, x_p$ equals $\sigma_{11}^{-1}\mathbf{\Sigma}_{12}\mathbf{\Sigma}_{22}^{-1}\mathbf{\Sigma}_{21}$.

6.5.3 (a) If $\mathbf{R}$ is a (2×2) correlation matrix show that $R_{1\cdot 2} = |r_{12}|$.

(b) If $\mathbf{R}$ is a (3×3) correlation matrix show that

$$R_{1\cdot 23}^2 = (r_{12}^2 + r_{13}^2 - 2r_{12}r_{13}r_{23})/(1 - r_{23}^2).$$

6.5.4 Let $\mathbf{R}$ be the correlation matrix of $\mathbf{X}(n \times p)$. Show that

(a) $0 \leq R_{1\cdot 2\ldots p} \leq 1$;
(b) $R_{1\cdot 2\ldots p} = 0$ if and only if $\mathbf{x}_{(1)}$ is uncorrelated with each of $\mathbf{x}_{(2)}, \ldots, \mathbf{x}_{(p)}$;
(c) $R_{1\cdot 2\ldots p} = 1$ if and only if $\mathbf{x}_{(1)}$ is a linear combination of $\mathbf{x}_{(2)}, \ldots, \mathbf{x}_{(p)}$.

6.5.5 Verify formula (6.5.10) for the partial correlation coefficient and show that it can be expressed in the form (6.5.11).

6.5.6 Let $\mathbf{R}$ be a (3×3) correlation matrix. Show that

(a) $r_{12\cdot 3} = (r_{12} - r_{13}r_{23})/\{(1-r_{13}^2)(1-r_{23}^2)\}^{1/2}$;
(b) $r_{12\cdot 3}$ may be non-zero when $r_{12} = 0$ and vice versa;
(c) $r_{12\cdot 3}$ and r_{12} have different signs when $r_{13}r_{23} > r_{12} > 0$ or $r_{13}r_{23} < r_{12} < 0$.

6.5.7 (a) Show that the eigenvalues $d_1, \ldots, d_p$ of $\mathbf{I} - \mathbf{D}$ in (6.5.13) all lie between 0 and 1.

(b) If a and g are arithmetic and geometric means of $d_1, \ldots, d_p$, show that

$$r_T^2 = a, \qquad r_D^2 = g^p.$$

(c) If a_1 and g_1 are arithmetic and geometric means of $1-d_1, \ldots, 1-d_p$, show that

$$a_1 = 1 - a \quad \text{and} \quad r_A = g_1^p.$$

(d) Hence show that $r_D^2 \leq r_T^{2p}$ and $r_A \leq (1 - r_T^2)^p$.

6.5.8 Let $\mathbf{x} = (\mathbf{x}_1, \mathbf{x}_2)$ be a random $(p+q)$-vector with known mean $\boldsymbol{\mu}$ and covariance matrix $\boldsymbol{\Sigma}$. If $\mathbf{x}_2$ is regressed on $\mathbf{x}_1$ using

$$E\{(\mathbf{x}_2 - \boldsymbol{\mu}_2) \mid \mathbf{x}_1\} = \boldsymbol{\Sigma}_{21}\boldsymbol{\Sigma}_{11}^{-1}(\mathbf{x}_1 - \boldsymbol{\mu}_1),$$

then show that

$$\boldsymbol{\Delta} = \boldsymbol{\Sigma}_{22}^{-1}\boldsymbol{\Sigma}_{22\cdot 1} = \mathbf{I} - \boldsymbol{\Sigma}_{22}^{-1}\boldsymbol{\Sigma}_{21}\boldsymbol{\Sigma}_{11}^{-1}\boldsymbol{\Sigma}_{12}.$$

The corresponding population values of ρ_T, the trace correlation, and ρ_D, the determinant correlation, are given by

$$\rho_T^2 = \frac{1}{q}\operatorname{tr}(\mathbf{I} - \boldsymbol{\Delta}) = \frac{1}{q}\operatorname{tr}\boldsymbol{\Sigma}_{22}^{-1}\boldsymbol{\Sigma}_{21}\boldsymbol{\Sigma}_{11}^{-1}\boldsymbol{\Sigma}_{12}$$

and

$$\rho_D^2 = \det(\mathbf{I} - \boldsymbol{\Delta}) = |\boldsymbol{\Sigma}_{21}\boldsymbol{\Sigma}_{11}^{-1}\boldsymbol{\Sigma}_{12}|/|\boldsymbol{\Sigma}_{22}|.$$

6.6.1 If (6.6.1) holds, show that $E(\mathbf{y}) = \mathbf{X}\boldsymbol{\beta}$. Show that the residual sum of squares $(\mathbf{y} - \mathbf{X}\boldsymbol{\beta})'(\mathbf{y} - \mathbf{X}\boldsymbol{\beta})$ is minimized when $\boldsymbol{\beta} = \hat{\boldsymbol{\beta}}$, given by (6.6.3).

6.6.2 If $\mathbf{t}=\mathbf{By}$ is any linear estimator of $\boldsymbol{\beta}$ and $\mathbf{y}$ satisfies (6.6.1), show that

(a) $\mathbf{t}$ is unbiased if $\mathbf{BX}=\mathbf{I}$ and hence confirm that the OLS estimator is unbiased;
(b) the covariance of $\mathbf{t}$ is $\mathbf{B\Omega B}'$, and hence confirm (6.6.5)–(6.6.6);
(c) $\mathbf{t}=\hat{\boldsymbol{\beta}}+\mathbf{Cy}$, where $\mathbf{C}=\mathbf{B}-(\mathbf{X}'\mathbf{X})^{-1}\mathbf{X}'$, and hence show that $\mathbf{t}$ is unbiased if and only if $\mathbf{CX}=\mathbf{0}$.

If now $\mathbf{t}$ is an unbiased estimator and $\boldsymbol{\Omega}=\sigma^2\mathbf{I}_n$, show that the covariance matrix of $\mathbf{t}$ is

$$E[(\mathbf{t}-\boldsymbol{\beta})(\mathbf{t}-\boldsymbol{\beta})']=E[(\mathbf{t}-\hat{\boldsymbol{\beta}})(\mathbf{t}-\hat{\boldsymbol{\beta}})']+E[(\hat{\boldsymbol{\beta}}-\boldsymbol{\beta})(\hat{\boldsymbol{\beta}}-\boldsymbol{\beta})'],$$

and hence that $\hat{\boldsymbol{\beta}}$ has smaller covariance matrix than any other linear unbiased estimator. (This proves the Gauss–Markov result, Theorem 6.6.1.)

6.6.3 If $\mathbf{t}=\mathbf{By}$ is a linear unbiased estimator of $\boldsymbol{\beta}$ where $\mathbf{y}$ satisfies (6.6.1) and $\tilde{\boldsymbol{\beta}}$ is given by (6.6.8), show that $\mathbf{t}=\tilde{\boldsymbol{\beta}}+\tilde{\mathbf{C}}\mathbf{y}$ where

$$\tilde{\mathbf{C}}=\mathbf{B}-(\mathbf{X}'\boldsymbol{\Omega}^{-1}\mathbf{X})^{-1}\mathbf{X}'\boldsymbol{\Omega}^{-1}.$$

Hence show that $\mathbf{t}$ is unbiased if and only if $\tilde{\mathbf{C}}\mathbf{X}=\mathbf{0}$ and that, if $\mathbf{t}$ is unbiased, then its covariance matrix is not less than that of $\tilde{\boldsymbol{\beta}}$.

6.6.4 If $\mathbf{y}\sim N_n(\mathbf{X}\boldsymbol{\beta},\boldsymbol{\Omega})$ with known covariance matrix $\boldsymbol{\Omega}$, then show that the GLS estimator is the maximum likelihood estimator of $\boldsymbol{\beta}$ and that it is normal with mean $\boldsymbol{\beta}$ and variance $(\mathbf{X}'\boldsymbol{\Omega}^{-1}\mathbf{X})^{-1}$.

6.6.5 To prove the consistency of the GLS estimator $\tilde{\boldsymbol{\beta}}_n$ under assumptions (6.6.13) and (6.6.14) it is necessary and sufficient to prove that $\operatorname{plim}\tilde{\beta}_{n,i}=\beta_i$ for each component, $i=1,\ldots,n$.

(a) Show that

$$E(\tilde{\boldsymbol{\beta}}_n\mid\mathbf{X}_n)=\boldsymbol{\beta},\qquad V(\tilde{\boldsymbol{\beta}}_n\mid\mathbf{X}_n)=(\mathbf{X}_n'\boldsymbol{\Omega}_n^{-1}\mathbf{X}_n)^{-1}.$$

(b) Let σ_n^{ii} denote the (i,i)th element of $(\mathbf{X}_n'\boldsymbol{\Omega}_n^{-1}\mathbf{X}_n)^{-1}$, which is a function of the random matrix $\mathbf{X}_n$. Then, by (6.6.14), $\operatorname{plim}\sigma_n^{ii}=0$. Using Chebyshev's inequality, show that

$$P(|\tilde{\beta}_{n,i}-\beta_i|>\varepsilon\mid\mathbf{X}_n)<\sigma_n^{ii}/\varepsilon^2.$$

(c) Show that for all $\varepsilon>0$, $\delta>0$

$$P(|\tilde{\beta}_{n,i}-\beta_i|>\varepsilon)\leqslant P(|\tilde{\beta}_{n,i}-\beta_i|>\varepsilon\,|\,\sigma_n^{ii}<\delta)P(\sigma_n^{ii}<\delta)+P(\sigma_n^{ii}\geqslant\delta).$$

Using (b) and (6.6.14) deduce that for any fixed $\varepsilon>0$, the right-hand side of the above equation can be made arbitrarily small by choosing δ small enough and restricting n to be sufficiently large. Hence $P(|\tilde{\beta}_{n,i}-\beta_i|>\varepsilon)\to 0$ as $n\to\infty$ so that $\operatorname{plim}\tilde{\beta}_{n,i}=\beta_i$.

7
Econometrics

7.1 Introduction

Economic theory postulates relationships between economic variables. It is the purpose of econometrics to quantify these relationships. Typically, an economic model might predict that one economic variable is approximately determined as a linear combination of other economic variables. The econometrician is interested in estimating the coefficients in this relationship and testing hypotheses about it.

At first sight this framework appears similar to the regression model of Chapter 6. However, several complications can arise in econometrics (and also in other fields) which require special treatment.

(1) *Heteroscedasticity* The variance of the error term might not be constant. This problem can sometimes be resolved by transforming the data.
(2) *Autocorrelation* If the data represents a series of observations taken over time then the error terms might be correlated with one another, in which case time series techniques are appropriate.
(3) *Dependence* In the usual multiple regression model,

$$\mathbf{y} = \mathbf{X}\boldsymbol{\beta} + \mathbf{u},$$

it is assumed that $\mathbf{X}$ is independent of the disturbance term $\mathbf{u}$. However, in econometrics this assumption is often not justifiable.

In this chapter we shall only be concerned with complications arising from (3). For this purpose, it is convenient to distinguish two sorts of economic variables. Suppose we are interested in modelling an *economic system*. *Endogenous variables* are variables measured *within* the system. Their values are affected both by other variables in the system and also by variables outside the system. On the other hand, *exogenous variables* are variables which are measured *outside* the system. They can still affect the behaviour of the system, but are not themselves affected by fluctuations in the system. (Economists sometimes also consider the effect of *lagged endogenous* variables, that is endogenous variables observed at an earlier time. These variables usually play the same role as exogenous variables, but for simplicity we shall not discuss them here.)

A typical equation in econometrics tries to explain one of the endogenous variables in terms of some of the other endogenous variables and some of the exogenous variables, plus a disturbance term. An essential assumption in this model states that because the exogenous variables are determined outside the system, they are *uncorrelated* with the disturbance term. This assumption is *not* made for the endogenous variables; one usually assumes that all the endogenous variables are correlated with the disturbance term.

Example 7.1.1 (Kmenta, 1971, p. 563) The supply equation in a simplified model of food consumption and prices is

$$Q_t = b_1 + b_2 P_t + b_3 F_t + b_4 A_t + u_t, \tag{7.1.1}$$

where Q_t = food consumption per head, P_t = ratio of food prices to general consumer prices, F_t = ratio of farm prices to general consumer prices, A_t = time in years (a trend term), and u_t is a disturbance term. The subscript $t = 1, \ldots, 20$ represents observations over 20 years. Here Q_t and P_t are assumed to be endogenous; their values represent the state of the food market. On the other hand F_t and A_t are assumed to be exogenous variables; their values can affect but are not affected by the food market. Of course, the constant term 1 allowing for the mean effect is always taken to be exogenous.

This example is discussed further throughout the chapter.

One of the purposes of econometrics in the above regression-like equation is to estimate the parameters. However, because P_t is assumed to be correlated with u_t the usual OLS regression estimates are not appropriate, and other techniques must be sought.

7.2 Instrumental Variables and Two-stage Least Squares

7.2.1 Instrumental variables (IV) estimation

Unfortunately, when the independent variables are correlated with the random disturbances, the OLS estimator is in general *not consistent.*

Consider the regression-like equation

$$\mathbf{y} = \mathbf{X}\boldsymbol{\beta} + \mathbf{u}, \qquad (7.2.1)$$

where $\mathbf{y}$ and $\mathbf{u}$ are $(n \times 1)$, $\mathbf{X}$ is $(n \times q)$, $\boldsymbol{\beta}$ is $(q \times 1)$, and $\mathbf{X}$ is correlated with $\mathbf{u}$. An example in which the OLS estimator $\hat{\boldsymbol{\beta}}$ is *not* consistent is outlined in Exercise 7.2.1.

One technique which can be used to overcome this inconsistency is the method of *instrumental variables* (IV). Suppose we can find a *new* set of variables $\mathbf{Z}(n \times q)$ with the same dimension as $\mathbf{X}$, such that $\mathbf{Z}$ is uncorrelated with the disturbances $\mathbf{u}$. Then $\mathbf{Z}$ can be used to construct a consistent estimate of $\boldsymbol{\beta}$.

More specifically, suppose the following assumptions are satisfied for $\mathbf{X}$, $\mathbf{u}$, and the IVs $\mathbf{Z}$:

$$E(\mathbf{u} \mid \mathbf{Z}) = \mathbf{0}, \qquad V(\mathbf{u} \mid \mathbf{Z}) = \sigma^2\mathbf{I}, \qquad (7.2.2)$$

$$\text{plim } n^{-1}\mathbf{Z}'\mathbf{X} = \boldsymbol{\Psi}, \text{ non-singular as } n \to \infty, \qquad (7.2.3)$$

$$\text{plim } n^{-1}\mathbf{Z}'\mathbf{Z} = \boldsymbol{\Theta}, \text{ non-singular as } n \to \infty. \qquad (7.2.4)$$

Thus, the random disturbances are uncorrelated with one another and with the instrumental variables $\mathbf{Z}$; however, (7.2.3) states that $\mathbf{Z}$ *is* correlated with $\mathbf{X}$.

Tables 7.1.1 and 7.1.2 summarize the types of estimation to be considered in this chapter.

Definition 7.2.1 *The* instrumental variable (IV) estimator *of* $\boldsymbol{\beta}$, *using the instruments* $\mathbf{Z}$, *is defined by*

$$\boldsymbol{\beta}^* = (\mathbf{Z}'\mathbf{X})^{-1}\mathbf{Z}'\mathbf{y}. \qquad (7.2.5)$$

Theorem 7.2.1 *For the model* (7.2.1) *satisfying assumptions* (7.2.2)–7.2.4), *the instrumental variables estimator* (7.2.5) *is consistent, i.e.* plim $\boldsymbol{\beta}^* = \boldsymbol{\beta}$.

Table 7.1.1 General methods of estimation for a single equation

Method	Estimator	Location in text
(1) Ordinary least squares (OLS)	$\hat{\boldsymbol{\beta}} = (\mathbf{X}'\mathbf{X})^{-1}\mathbf{X}'\mathbf{y}$	(6.6.3)
(2) Instrumental variables (IV)	$\boldsymbol{\beta}^* = (\mathbf{Z}'\mathbf{X})^{-1}\mathbf{Z}'\mathbf{y}$	(7.2.5)
(3) Two-stage least squares	$\boldsymbol{\beta}^{**} = (\hat{\mathbf{X}}'\hat{\mathbf{X}})^{-1}\hat{\mathbf{X}}'\mathbf{y}$	(7.2.10)

Table 7.1.2 Methods of estimation for a simultaneous system of equations

Method	Location in text
(A) Single-equation methods	
(1) Indirect least squares (ILS)	(7.3.19)
(2) Two-stage least squares (2SLS)	(7.4.3)
(3) Limited information maximum likelihood (LIML)	Section 7.4.2
(B) System methods	
(1) Generalized least squares (GLS) and Zellner's two-stage estimator for seemingly unrelated regressions	(7.5.3) and (7.5.7)
(2) Three-stage least squares (3SLS)	(7.5.9)
(3) Full information maximum likelihood (FIML)	Section 7.5.3

Proof The IV estimator $\boldsymbol{\beta}^*$ can be expressed as

$$\boldsymbol{\beta}^* = (\mathbf{Z}'\mathbf{X})^{-1}\mathbf{Z}'(\mathbf{X}\boldsymbol{\beta}+\mathbf{u})$$
$$= \boldsymbol{\beta} + (\mathbf{Z}'\mathbf{X})^{-1}\mathbf{Z}'\mathbf{u}.$$

Thus to show consistency we must show that the second term converges in probability to $\mathbf{0}$ as $n \to \infty$. Since by (7.2.3) plim $(n^{-1}\mathbf{Z}'\mathbf{X})^{-1} = \boldsymbol{\Psi}^{-1}$, it is sufficient to show that

$$\text{plim } n^{-1}\mathbf{Z}'\mathbf{u} = \mathbf{0}. \tag{7.2.6}$$

Now from (7.2.2), $E(n^{-1}\mathbf{Z}'\mathbf{u} \mid \mathbf{Z}) = \mathbf{0}$ and

$$V(n^{-1}\mathbf{Z}'\mathbf{u} \mid \mathbf{Z}) = n^{-2}\mathbf{Z}'V(\mathbf{u} \mid \mathbf{Z})\mathbf{Z} = n^{-2}\sigma^2\mathbf{Z}'\mathbf{Z}.$$

Since, from (7.2.4), plim $n^{-1}\mathbf{Z}'\mathbf{Z} = \boldsymbol{\Theta}$, this conditional variance tends to $\mathbf{0}$ at rate $1/n$; hence (7.2.6) follows (see also Exercise 6.6.5). ■

Remarks (1) Note that although $\boldsymbol{\beta}^*$ is consistent, it is not necessarily an unbiased estimator of $\boldsymbol{\beta}$. We cannot evaluate the term $E((\mathbf{Z}'\mathbf{X})^{-1}\mathbf{Z}'\mathbf{u})$ because $\mathbf{X}$ is correlated with $\mathbf{u}$.

(2) With some further regularity assumptions on $\mathbf{X}$, $\mathbf{Z}$, and $\mathbf{u}$, it can be shown that $\boldsymbol{\beta}^*$ is asymptotically unbiased and normally distributed,

$$\boldsymbol{\beta}^* \sim N_q(\boldsymbol{\beta}, n^{-1}\sigma^2\boldsymbol{\Psi}^{-1}\boldsymbol{\Theta}(\boldsymbol{\Psi}')^{-1}).$$

In practice $\boldsymbol{\Psi}$ and $\boldsymbol{\Theta}$ can be estimated by $n^{-1}\mathbf{Z}'\mathbf{X}$ and $n^{-1}\mathbf{Z}'\mathbf{Z}$, respectively, and σ^2 can be estimated from the residual vector by

$$\hat{\sigma}^2 = (\mathbf{y}-\mathbf{X}\boldsymbol{\beta}^*)'(\mathbf{y}-\mathbf{X}\boldsymbol{\beta}^*)/(n-q). \tag{7.2.7}$$

Example 7.2.1 (Errors in variables) Consider the simple regression

$$\tilde{\mathbf{y}} = \alpha + \beta\tilde{\mathbf{x}} + \mathbf{u},$$

where $\tilde{\mathbf{x}}$ is a random sample from $N(\mu, \sigma_{\tilde{x}}^2)$ independent of $\mathbf{u}$ which is a random sample from $N(0, \sigma_u^2)$. Suppose, however, that we can only observe $\tilde{\mathbf{x}}$ and $\tilde{\mathbf{y}}$ subject to measurement errors, i.e. we observe

$$\mathbf{y} = \tilde{\mathbf{y}} + \mathbf{v} \quad \text{and} \quad \mathbf{x} = \tilde{\mathbf{x}} + \mathbf{w},$$

where $\mathbf{v}$ and $\mathbf{w}$ are random samples from $N(0, \sigma_v^2)$ and $N(0, \sigma_w^2)$, and $\tilde{\mathbf{x}}$, $\mathbf{u}$, $\mathbf{v}$, and $\mathbf{w}$ are all independent. Then we are trying to fit the regression line using the equation

$$\mathbf{y} = \alpha + \beta\mathbf{x} + \boldsymbol{\epsilon}, \qquad \boldsymbol{\epsilon} = \mathbf{u} + \mathbf{v} - \beta\mathbf{w}.$$

However, since $C(\epsilon_i, x_i) = -\beta\sigma_w^2$, $i = 1, \ldots, n$, the OLS estimator of β will not be consistent. Note that $\mathbf{v}$ and $\mathbf{w}$ can be considered "errors in variables" whereas $\mathbf{u}$ is an "error in equation".

Suppose we can find a variable $\mathbf{z}$ which is correlated with $\tilde{\mathbf{x}}$, but uncorrelated with $\mathbf{u}$, $\mathbf{v}$, and $\mathbf{w}$. Then using $(\mathbf{1}, \mathbf{z})$ as instrumental variables for $(\mathbf{1}, \mathbf{x})$, the IV estimators of β and α are found from (7.2.5) to be

$$\beta^* = s_{zy}/s_{zx}, \qquad \alpha^* = \bar{y} - \beta^*\bar{x}, \tag{7.2.8}$$

where s_{zy} is the sample covariance between $\mathbf{z}$ and $\mathbf{y}$. (See Exercise 7.2.2.) This estimate of $\boldsymbol{\beta}$ is also found in factor analysis (Chapter 9) in slightly disguised form when estimating the parameters for $p = 3$ variables with $k = 1$ factor (see Example 9.2.1).

Example 7.2.2 In a "cross-sectional" study of capital–labour substitution in the furniture industry (Arrow *et al.*, 1961; Kmenta, 1971, p. 313), the authors examine the relationship

$$y_i = \alpha + \beta x_i + u_i,$$

where $y = \log$ (value added ÷ labour input), $x = \log$ (wage rate ÷ price of product), and i runs through different countries. The data is given in Table 7.2.1. The OLS estimates for α and β lead to the estimated regression equation (with standard errors in parentheses underneath):

$$y_i = \underset{(0.10)}{-2.28} + \underset{(0.03)}{0.84}\, x_i.$$

However, as the explanatory variables may contain errors of measurement, it seems preferable to use instrumental variable estimation. The value of $z = \log$ (wage rate ÷ price of product) for knitting-mill products is an exogenous variable for the furniture industry and seems a plausible choice of instrumental variable, since it is unlikely to be correlated with measurement errors for x or with the disturbance term in the regression.

Table 7.2.1 Capital–labour substitution data (Kmenta, 1971, p. 313)

Country	y	x	z
United States	0.7680	3.5459	3.4241
Canada	0.4330	3.2367	3.1748
New Zealand	0.4575	3.2865	3.1686
Australia	0.5002	3.3202	3.2989
Denmark	0.3462	3.1585	3.1742
Norway	0.3068	3.1529	3.0492
United Kingdom	0.3787	3.2101	3.1175
Colombia	−0.1188	2.6066	2.5681
Brazil	−0.1379	2.4872	2.5682
Mexico	−0.2001	2.4280	2.6364
Argentina	−0.3845	2.3182	2.5703

The values of z are given in Table 7.2.1. The instrumental variable estimates from (7.2.7) and (7.2.8) (with estimated standard errors) lead to the equation

$$y_i = \underset{(0.10)}{-2.30} + \underset{(0.03)}{0.84}\, x_i.$$

It will be noted that IV estimates and OLS estimates are very similar. Thus, in this example, the measurement errors do not seem to be severe.

7.2.2 Two-stage least squares (2SLS) estimation

The instrumental variable matrix $\mathbf{Z}(n \times k)$ in Section 7.2.1 is assumed to have the same dimension as the "independent" variable matrix $\mathbf{X}(n \times q)$, i.e. $k = q$. However, if $k > q$ then an extension of IV estimation may be given using the method of *two-stage least squares* (2SLS). This method is defined as follows.

First, regress $\mathbf{X}$ on $\mathbf{Z}$ using the usual OLS multivariate regression estimates to get a fitted value of $\mathbf{X}$,

$$\hat{\mathbf{X}} = \mathbf{Z}(\mathbf{Z}'\mathbf{Z})^{-1}\mathbf{Z}'\mathbf{X}. \tag{7.2.9}$$

Note that $\hat{\mathbf{X}}(n \times q)$ is a linear combination of the columns of $\mathbf{Z}$.

Second, substitute $\hat{\mathbf{X}}$ for $\mathbf{X}$ in the original equation (7.2.1) and use OLS to estimate $\boldsymbol{\beta}$ by

$$\begin{aligned}\boldsymbol{\beta}^{**} &= (\hat{\mathbf{X}}'\hat{\mathbf{X}})^{-1}\hat{\mathbf{X}}'\mathbf{y} \\ &= [\mathbf{X}'\mathbf{Z}(\mathbf{Z}'\mathbf{Z})^{-1}\mathbf{Z}'\mathbf{X}]^{-1}(\mathbf{X}'\mathbf{Z})(\mathbf{Z}'\mathbf{Z})^{-1}\mathbf{Z}'\mathbf{y}. \qquad (7.2.10)\end{aligned}$$

Then $\boldsymbol{\beta}^{**}$ is called the 2SLS estimator of $\boldsymbol{\beta}$.

Note that since $\hat{\mathbf{X}}'\hat{\mathbf{X}} = \hat{\mathbf{X}}'\mathbf{X}$, we can write

$$\boldsymbol{\beta}^{**} = (\hat{\mathbf{X}}'\mathbf{X})^{-1}\hat{\mathbf{X}}'\mathbf{y},$$

so that $\boldsymbol{\beta}^{**}$ can be interpreted as an IV estimator using $\hat{\mathbf{X}}$ as the instrumental variables for $\mathbf{X}$. In fact if $k = q$ then $\boldsymbol{\beta}^{**}$ reduces to the IV estimator $\boldsymbol{\beta}^{*}$ in (7.2.5) (see Exercise 7.2.3). For general $k \geq q$ it can be seen that $\boldsymbol{\beta}^{**}$ is consistent, provided that $\mathbf{Z}$ is uncorrelated with $\mathbf{u}$, plim $n^{-1}\mathbf{Z}'\mathbf{Z} = \boldsymbol{\Theta}(k \times k)$ non-singular, and plim $n^{-1}\mathbf{Z}'\mathbf{X} = \boldsymbol{\Psi}(k \times q)$ of rank q. (See Exercise 7.2.3.)

We have not yet commented on the choice of instrumental variables in practice. Intuitively, it is clear that we should choose $\mathbf{Z}$ to be as highly correlated with $\mathbf{X}$ as possible, subject to $\mathbf{Z}$ being uncorrelated with $\mathbf{u}$. However, in practice it is often difficult to determine the correlation between $\mathbf{Z}$ and $\mathbf{u}$ because the random disturbances $\mathbf{u}$ are unobserved.

In the context of measurement errors (Examples 7.2.1 and 7.2.2), the problem of choosing instrumental variables is often avoided by assuming that the measurement errors are "negligible" compared to the equation errors, so that OLS may be safely used. However, we shall see in the next section that there are situations in which the use of instrumental variables estimation does play an important practical role.

7.3 Simultaneous Equation Systems

7.3.1 Structural form

Imagine an economic system which can be described by a collection of endogenous variables $y_1, \ldots, y_p$. Suppose this system can also be affected by a number of exogenous variables $x_1, \ldots, x_q$, but these exogenous variables are not affected by changes in the endogenous variables.

An *econometric model* of this economic system consists of a system of p simultaneous equations in which certain linear combinations of the endogenous variables are approximately explained by linear combinations of some of the exogenous variables. More specifically, consider matrices $\mathbf{Y}(n \times p)$ and $\mathbf{X}(n \times q)$. The n components of $\mathbf{Y}$ and $\mathbf{X}$ might represent observations on the same unit over different periods of time (longitudinal studies) or observations on different units at the same period in time (cross-sectional studies). Suppose that $\mathbf{Y}$ and $\mathbf{X}$ satisfy p simultaneous equations

$$\mathbf{YB}+\mathbf{X\Gamma}=\mathbf{U}. \tag{7.3.1}$$

Equation (7.3.1) is called the *structural form* of the model under study. Here $\mathbf{B}(p\times p)=(\beta_{ij})$ and $\mathbf{\Gamma}(q\times p)=(\gamma_{ij})$ are parameter matrices and $\mathbf{U}(n\times p)$ is a matrix of disturbances. The jth column of (7.3.1) represents the jth structural equation on n observational units, $j=1,\ldots,p$.

The parameters in each equation are defined only up to a scaling factor, and this indeterminacy is resolved by taking one element in each column $\boldsymbol{\beta}_{(j)}$ equal to 1. It is usually convenient to take $\beta_{jj}=1$ so that the jth equation can be rewritten as a regression expressing the jth endogenous variable in terms of some of the other endogenous and exogenous variables. For example, the first equation can then be written

$$\mathbf{y}_{(1)}=-\mathbf{y}_{(2)}\beta_{21}-\cdots-\mathbf{y}_{(p)}\beta_{p1}-\mathbf{X}\boldsymbol{\gamma}_{(1)}+\mathbf{u}_{(1)}. \tag{7.3.2}$$

We shall suppose one column of $\mathbf{X}$ is set equal to $\mathbf{1}$ in order to allow for an overall mean effect.

In the specification of the model some of the parameters in $\mathbf{B}$ and $\mathbf{\Gamma}$ will be known to be zero. Thus, not all of the variables appear in every equation. Also, some of the equations may be *exact identities*; that is, the parameters are known exactly and the error terms vanish identically for these equations.

For the error term $\mathbf{U}=(\mathbf{u}_1,\ldots,\mathbf{u}_n)'$ we make the following assumptions for $s, t=1,\ldots,n$:

$$E(\mathbf{u}_s)=\mathbf{0}, \qquad V(\mathbf{u}_s)=\mathbf{\Sigma}, \qquad C(\mathbf{u}_s,\mathbf{u}_t)=\mathbf{0}.$$

Thus the errors may be correlated between equations, but there is no correlation between different observational units. If r of the structural equations are exact identities, then the corresponding entries in $\mathbf{\Sigma}$ are known to be 0, and we suppose $\operatorname{rank}(\mathbf{\Sigma})=p-r$. When convenient we shall further suppose that $\mathbf{U}$ is multinormally distributed.

An essential assumption of the model states that *the disturbance* $\mathbf{U}$ *is uncorrelated with the exogenous variables* $\mathbf{X}$. However, in general, the disturbance $\mathbf{U}$ *will* be correlated with the endogenous variables. Thus in any equation containing more than one endogenous variable, the OLS estimates of the parameters will not be appropriate and other estimates must be sought.

Example 7.3.1 Consider a simple two-equation model consisting of an income identity

$$\mathbf{w}=\mathbf{c}+\mathbf{z} \tag{7.3.3}$$

and a consumption function

$$\mathbf{c}=\alpha+\beta\mathbf{w}+\mathbf{u}. \tag{7.3.4}$$

Here $\mathbf{w}$ = income, $\mathbf{c}$ = consumption expenditure, and $\mathbf{z}$ = non-consumption expenditure. In this model $\mathbf{w}$ and $\mathbf{c}$ are considered as endogenous variables (determined inside the model), whereas $\mathbf{z}$ is considered as exogenous (perhaps determined by government policy). All the vectors are of length n, representing observations on n units. This model can be easily put into the form (7.3.1) by setting $\mathbf{Y}=(\mathbf{w},\mathbf{c})$, $\mathbf{Z}=(\mathbf{1},\mathbf{z})$, $\mathbf{U}=(\mathbf{0},\mathbf{u})$ and

$$\mathbf{B}=\begin{bmatrix}1 & -\beta\\ -1 & 1\end{bmatrix},\qquad \mathbf{\Gamma}=\begin{bmatrix}0 & -\alpha\\ -1 & 0\end{bmatrix},\qquad \mathbf{\Sigma}=\begin{bmatrix}0 & 0\\ 0 & \sigma_{22}\end{bmatrix}.$$

7.3.2 Reduced form

If the structural model (7.3.1) is well-defined, the endogenous variables will be determined by the values of the exogenous variables and the random disturbances. In particular when $\mathbf{U}$ vanishes then $\mathbf{Y}$ has a unique value (as a function of $\mathbf{X}$) representing the equilibrium behaviour of the economic model, and any non-zero random disturbances reflect perturbations away from this equilibrium.

To solve (7.3.1) for $\mathbf{Y}$ we assume $\mathbf{B}$ is non-singular. Then (7.3.1) becomes

$$\mathbf{Y}=\mathbf{X\Pi}+\mathbf{V}, \tag{7.3.5}$$

where we set

$$\mathbf{\Pi}=-\mathbf{\Gamma B}^{-1} \tag{7.3.6}$$

and $\mathbf{V}=\mathbf{UB}^{-1}$. Equation (7.3.5) is called the *reduced form* of the model and $\mathbf{\Pi}(q\times p)$ is called the matrix of *reduced parameters.* Note that the rows of $\mathbf{V}$ are uncorrelated each with mean $\mathbf{0}$ and covariance matrix $(\mathbf{B}^{-1})'\mathbf{\Sigma B}^{-1}=\mathbf{\Psi}$, say. Furthermore, the disturbances $\mathbf{V}$ are uncorrelated with $\mathbf{X}$, so that (7.3.5) satisfies the assumptions of the usual multivariate regression model (6.1.1). Hence we may estimate the reduced parameters $\mathbf{\Pi}$ using, for example, the usual OLS estimator

$$\hat{\mathbf{\Pi}}=(\mathbf{X}'\mathbf{X})^{-1}\mathbf{X}'\mathbf{Y}. \tag{7.3.7}$$

(We shall suppose throughout that $\mathbf{X}'\mathbf{X}$ is non-singular so that the reduced parameter matrix $\mathbf{\Pi}$ is uniquely defined.)

However, the interesting problem is how to transform an estimate of $\mathbf{\Pi}$ into estimates of $\mathbf{B}$ and $\mathbf{\Gamma}$. We shall discuss this problem in the next section.

Example 7.3.2 In Example 7.3.1, we can solve (7.3.3) and (7.3.4) for $\mathbf{w}$ and $\mathbf{c}$ to get

$$\mathbf{w}=\frac{\alpha}{1-\beta}+\frac{1}{1-\beta}\mathbf{z}+\frac{1}{1-\beta}\mathbf{u}, \tag{7.3.8}$$

$$\mathbf{c}=\frac{\alpha}{1-\beta}+\frac{\beta}{1-\beta}\mathbf{z}+\frac{1}{1-\beta}\mathbf{u}. \tag{7.3.9}$$

Thus for this model

$$\mathbf{\Pi}=\begin{bmatrix}\pi_{11} & \pi_{12}\\ \pi_{21} & \pi_{22}\end{bmatrix}=\begin{bmatrix}\alpha/(1-\beta) & \alpha/(1-\beta)\\ 1/(1-\beta) & \beta/(1-\beta)\end{bmatrix}. \tag{7.3.10}$$

Note that it is clear from (7.3.8) that $\mathbf{w}$ and $\mathbf{u}$ are correlated, so that OLS estimation applied directly to (7.3.4) is not appropriate.

7.3.3 The identification problem

Since we are supposing $\mathbf{X}'\mathbf{X}$ is non-singular, the parameter matrix $\mathbf{\Pi}$ for the reduced model (7.3.5) is uniquely defined. However, the parameters of interest are $\mathbf{B}$ and $\mathbf{\Gamma}$, so the question here is whether we can solve (7.3.6) for $\mathbf{B}$ and $\mathbf{\Gamma}$ in terms of $\mathbf{\Pi}$. If so, the structural parameters $\mathbf{B}$ and $\mathbf{\Gamma}$ are well-defined.

Equation (7.3.6) can be rewritten

$$\mathbf{\Pi B}=-\mathbf{\Gamma} \quad \text{or} \quad \mathbf{\Pi}\boldsymbol{\beta}_{(j)}=-\boldsymbol{\gamma}_{(j)}, \qquad j=1,\ldots,p. \tag{7.3.11}$$

We shall attempt to solve (7.3.11) for $\mathbf{B}$ and $\mathbf{\Gamma}$ in terms of $\mathbf{\Pi}$ one column at a time. Note that the jth columns $\boldsymbol{\beta}_{(j)}$ and $\boldsymbol{\gamma}_{(j)}$ correspond to the parameters for the jth structural equation in (7.3.1). For notational simplicity we shall look at the *first* equation; solving for $\boldsymbol{\beta}_{(1)}$ and $\boldsymbol{\gamma}_{(1)}$ and we suppose that this equation is not an exact identity.

Write $\boldsymbol{\beta}=\boldsymbol{\beta}_{(1)}$ and $\boldsymbol{\gamma}=\boldsymbol{\gamma}_{(1)}$. Some of the components of $\boldsymbol{\beta}(p\times 1)$ and $\boldsymbol{\gamma}(q\times 1)$ will be known to be 0. Without loss of generality let the endogenous variables $y_1,\ldots,y_p$ and the exogenous variables $x_1,\ldots,x_q$ be ordered in such a way that

$$\boldsymbol{\beta}=\begin{bmatrix}\boldsymbol{\beta}_0\\ \mathbf{0}\end{bmatrix}, \qquad \boldsymbol{\gamma}=\begin{bmatrix}\boldsymbol{\gamma}_0\\ \mathbf{0}\end{bmatrix}. \tag{7.3.12}$$

Then $\boldsymbol{\beta}_0$ and $\boldsymbol{\gamma}_0$ include all the components of $\boldsymbol{\beta}$ and $\boldsymbol{\gamma}$ which have not been explicitly set to zero. Let p_1 and q_1 denote the lengths of $\boldsymbol{\beta}_0(p_1\times 1)$ and $\boldsymbol{\gamma}_0(q_1\times 1)$, respectively.

Partitioning $\mathbf{\Pi}(q\times p)$ as

$$\mathbf{\Pi}=\begin{bmatrix}\mathbf{\Pi}_{00} & \mathbf{\Pi}_{01}\\ \mathbf{\Pi}_{10} & \mathbf{\Pi}_{11}\end{bmatrix}, \tag{7.3.13}$$

where $\mathbf{\Pi}_{00}$ is $(q_1\times p_1)$, we see that the equation $\mathbf{\Pi}\boldsymbol{\beta}=-\boldsymbol{\gamma}$ becomes

$$\mathbf{\Pi}_{00}\boldsymbol{\beta}_0=-\boldsymbol{\gamma}_0, \tag{7.3.14}$$

$$\mathbf{\Pi}_{10}\boldsymbol{\beta}_0=\mathbf{0}. \tag{7.3.15}$$

Note that if we can solve (7.3.15) for $\boldsymbol{\beta}_0$, we can immediately get $\boldsymbol{\gamma}_0$ from (7.3.14).

Now (7.3.15) consists of $q-q_1$ equations, stating that $\boldsymbol{\beta}_0(p_1\times 1)$ is orthogonal to each of the rows of $\mathbf{\Pi}_{00}((q-q_1)\times p_1)$. Thus, (7.3.15) can be solved uniquely for $\boldsymbol{\beta}_0$ if and only if

$$\operatorname{rank}(\mathbf{\Pi}_{10})=p_1-1, \tag{7.3.16}$$

in which case $\boldsymbol{\beta}_0$ lies in the one-dimensional subspace orthogonal to the rows of $\mathbf{\Pi}_{10}$, usually normalized by $\beta_{11}=1$.

Since we are supposing that $\mathbf{\Pi}$ is the true reduced parameter matrix for the structural parameters $\mathbf{B}$ and $\mathbf{\Gamma}$, (7.3.15) holds; that is, the rows of $\mathbf{\Pi}_{10}$ are p_1-vectors lying in the subspace of R^{p_1} orthogonal to $\boldsymbol{\beta}_0$. Hence, it is always true that

$$\operatorname{rank}(\mathbf{\Pi}_{10})\leqslant p_1-1. \tag{7.3.17}$$

Also, since $\mathbf{\Pi}_{10}$ is a $((q-q_1)\times p_1)$ matrix, it is always the case that $q-q_1\geqslant\operatorname{rank}(\mathbf{\Pi}_{10})$. Thus it is convenient to distinguish the following three mutually exclusive and exhaustive possibilities:

(1) If $\operatorname{rank}(\mathbf{\Pi}_{10})<p_1-1$, the first equation is said to be *under-identified.*

(2) If $\operatorname{rank}(\mathbf{\Pi}_{10})=p_1-1$ and $q-q_1=p_1-1$, the first equation is said to be *exactly identified.*

(3) If $\operatorname{rank}(\mathbf{\Pi}_{10})=p_1-1$ and $q-q_1>p_1-1$, the first equation is said to be *over-identified.*

An equation is said to be *identified* if it is either exactly identified or over-identified.

7.3.4 Under-identified equations

To say the first equation is under-identified means that there is more than one set of values for the parameters $\boldsymbol{\beta}_0$ and $\boldsymbol{\gamma}_0$ which correspond to a given value of $\mathbf{\Pi}$. Thus, the parameters of the structural model are not uniquely defined; different values for the structural parameters define the same model. Hence no estimation of the structural parameters is possible here. In this case the equation defining the structural model are incomplete and the econometrician must rethink his model. The problem is similar to the lack of uniqueness for the parameters in a regression when the design matrix is singular.

We shall assume for the rest of the chapter that none of the equations in the system is under-identified.

7.3.5 Exactly identified equations and indirect least squares (ILS) estimation

If the first equation is exactly identified then the p_1-1 equations of (7.3.15) may be solved for the p_1-1 unknown elements of $\boldsymbol{\beta}_0$. Note that in this case $\boldsymbol{\Pi}_{10}$ is a $((p_1-1)\times p_1)$ matrix of full rank p_1-1. Also, $\boldsymbol{\gamma}_0$ may be found from (7.3.14).

Suppose now that $\boldsymbol{\Pi}$ is unknown. A natural estimate of $\boldsymbol{\Pi}$ is the OLS estimate

$$\hat{\boldsymbol{\Pi}}=(\mathbf{X}'\mathbf{X})^{-1}\mathbf{X}'\mathbf{Y}. \tag{7.3.18}$$

Then, with probability 1, $\hat{\boldsymbol{\Pi}}_{10}$ will have full rank p_1-1 (the same rank as $\boldsymbol{\Pi}_{10}$), so that the equations

$$\hat{\boldsymbol{\Pi}}_{10}\hat{\boldsymbol{\beta}}_0=\mathbf{0},\qquad \hat{\boldsymbol{\Pi}}_{00}\hat{\boldsymbol{\beta}}_0=-\hat{\boldsymbol{\gamma}}_0,\qquad \hat{\beta}_{11}=1 \tag{7.3.19}$$

can be solved for $\hat{\boldsymbol{\beta}}_0$ and $\hat{\boldsymbol{\gamma}}_0$. These estimates are called the *indirect least squares* (ILS) estimates of $\boldsymbol{\beta}_0$ and $\boldsymbol{\gamma}_0$. If regularity conditions are satisfied so that $\hat{\boldsymbol{\Pi}}$ is consistent, then $\boldsymbol{\beta}_0$ and $\boldsymbol{\gamma}_0$ will also be consistent.

7.3.6 Over-identified equations

Suppose the first equation is over-identified. Then (7.3.15) consists of $q-q_1$ equations to be solved for the p_1-1 unknown parameters of $\boldsymbol{\beta}_0$, where $q-q_1>p_1-1$. When $\boldsymbol{\Pi}_{10}$ is known exactly, this excess of equations is irrelevant.

However, if $\boldsymbol{\Pi}$ is unknown and is estimated by $\hat{\boldsymbol{\Pi}}$ in (7.3.6), then a problem arises. For in general $\hat{\boldsymbol{\Pi}}_{10}$ is a $((q-q_1)\times p_1)$ matrix of full rank p_1, whereas $\boldsymbol{\Pi}_{10}$ has rank p_1-1. Thus the equations

$$\hat{\boldsymbol{\Pi}}_{10}\hat{\boldsymbol{\beta}}_0=\mathbf{0} \tag{7.3.20}$$

are inconsistent and an exact solution for $\hat{\boldsymbol{\beta}}_0$ does not exist. Hence the method of ILS does not work here and an approximate solution to (7.3.20) must be sought. Techniques to deal with this situation will be considered in Sections 7.4–7.5.

7.3.7 Conditions for identifiability

Conditions for the identifiability of the first structural equation have been given above in terms of the reduced parameter matrix $\mathbf{\Pi}_{10}$. In practice one is usually more interested in the structural parameters $\mathbf{B}$ and $\mathbf{\Gamma}$, so it is useful to describe the conditions for identifiability in terms of these parameters. Partition

$$\mathbf{B}=\begin{bmatrix}\boldsymbol{\beta}_0 & \mathbf{B}_0\\ \mathbf{0} & \mathbf{B}_1\end{bmatrix}, \qquad \mathbf{\Gamma}=\begin{bmatrix}\boldsymbol{\gamma}_0 & \mathbf{\Gamma}_0\\ \mathbf{0} & \mathbf{\Gamma}_1\end{bmatrix} \tag{7.3.21}$$

and define $\mathbf{D}((p-p_1+q-q_1)\times(p-1))$ by

$$\mathbf{D}=\begin{bmatrix}\mathbf{B}_1\\ \mathbf{\Gamma}_1\end{bmatrix}. \tag{7.3.22}$$

Then it can be shown that

$$\operatorname{rank}(\mathbf{\Pi}_{10})=\operatorname{rank}(\mathbf{D})-(p-p_1). \tag{7.3.23}$$

(See Exercise 7.3.2.)

Thus, the identifiability conditions can be rephrased as follows. For completeness, we give the conditions for any structural equation which is not an exact identity. (The identification problem does not arise for exact identities in the system because there are no parameters to be estimated.)

Consider the jth structural equation of (7.3.1). Let p_j and q_j denote the number of components of $\boldsymbol{\beta}_{(j)}$ and $\boldsymbol{\gamma}_{(j)}$, respectively, which are not explicitly set to zero. Let $\mathbf{B}_1^{(j)}((p-p_j)\times(p-1))$ and $\mathbf{\Gamma}_1^{(j)}((q-q_j)\times(p-1))$ denote the matrices obtained from $\mathbf{B}$ and $\mathbf{\Gamma}$ by *deleting* the jth column and by *deleting* all the rows $\boldsymbol{\beta}_i$ for which $\beta_{ij}\neq 0$, and all the rows $\boldsymbol{\gamma}_i$ for which $\gamma_{ij}\neq 0$, respectively. Set $\mathbf{D}^{(j)}=(\mathbf{B}_1^{(j)\prime}, \mathbf{\Gamma}_1^{(j)\prime})'((p-p_j+q-q_j)\times(p-1))$.

Theorem 7.3.1 *Suppose the jth structural equation of* (7.3.1) *is not an exact identity. Then this equation is*

(a) *under-identified if* $\operatorname{rank}(\mathbf{D}^{(j)})<p-1$,
(b) *exactly identified if* $\operatorname{rank}(\mathbf{D}^{(j)})=p-1$ *and* $q-q_j=p_j-1$,
(c) *over-identified if* $\operatorname{rank}(\mathbf{D}^{(j)})=p-1$ *and* $q-q_j>p_j-1$. ∎

Note that $q-q_j=p_j-1$ if and only if $\mathbf{D}^{(j)}$ is a square matrix. Also $q-q_j<p_j-1$ implies that $\operatorname{rank}(\mathbf{D}^{(j)})<p-1$, so that the jth equation is under-identified in this case.

Example 7.3.3 Let us illustrate this theorem on the equations of Example 7.3.1. Only the consumption equation (7.3.4) is not an exact identity here. Checking the second column of $\mathbf{B}$ and $\mathbf{\Gamma}$ we see that $p_2=2$, $q_2=1$ so that $\mathbf{B}_1^{(2)}$ is empty and $\mathbf{\Gamma}_1^{(2)}=(-1)$. Thus $\mathbf{D}^{(2)}=(-1)$ is a (1×1) square matrix of full rank, so this equation is exactly identified.

Example 7.3.4 (Kmenta, 1971, p. 563) Consider the following model of food consumption and prices:

$$Q_t = a_1 + a_2P_t + a_3D_t + u_{t1} \qquad \text{(demand)},$$
$$Q_t = b_1 + b_2P_t + b_3F_t + b_4A_t + u_{t2} \qquad \text{(supply)},$$

where D_t is disposable income at constant prices and t ranges from 1 to 20 years. The supply equation has already been discussed in Example 7.1.1 and the other variables are defined there. The variables Q_t and P_t are endogenous, whereas 1, D_t, F_t, and A_t are exogenous. This model can be put into the form (7.3.1) by setting

$$\mathbf{Y}=\begin{bmatrix} Q_1 & P_1 \\ \vdots & \vdots \\ Q_{20} & P_{20} \end{bmatrix}, \qquad \mathbf{X}=\begin{bmatrix} 1 & D_1 & F_1 & A_1 \\ \vdots & \vdots & \vdots & \vdots \\ 1 & D_{20} & F_{20} & A_{20} \end{bmatrix},$$

$$\mathbf{B}=\begin{bmatrix} 1 & 1 \\ -a_2 & -b_2 \end{bmatrix}, \qquad \mathbf{\Gamma}=\begin{bmatrix} -a_1 & -b_1 \\ -a_3 & 0 \\ 0 & -b_3 \\ 0 & -b_4 \end{bmatrix}.$$

Note that Q_t has been interpreted as the "dependent" variable in both the demand and supply equations.

It is easily checked that, in the notation of Theorem 7.3.1,

$$\mathbf{D}^{(1)}=\begin{pmatrix} -b_3 \\ -b_4 \end{pmatrix}, \qquad \mathbf{D}^{(2)}=(-a_3),$$

so that the demand equation is over-identified and the supply equation is exactly identified.

In most practical situations the structural equations will be over-identified. We now turn to methods of estimation for such cases, and shall give two classes of estimators—*single-equation methods* and *system methods*. Single-equation methods give estimates for one equation at a time, making only limited use of the other equations in the system, whereas system methods give simultaneous estimates for all the parameters in the system.

7.4 Single-equation Estimators

7.4.1 Two-stage least squares (2SLS)

A structural equation can be written as a regression-like equation by putting one endogenous variable on the left-hand side and all the other variables on the right-hand side, as in (7.3.2). Unfortunately, OLS is not usually applicable here because some of the endogenous variables appear amongst the explanatory variables, and the endogenous variables are correlated with the disturbance term. One way to deal with this problem is the method of two-stage least squares (2SLS) given in Section 7.2.2, where we choose our instrumental variables from *all* the exogenous variables.

For notational convenience let us work on the first structural equation and take the first endogenous variable as the "dependent" variable, so $\beta_{11}=1$. This equation can be written in the form of a regression as

$$\mathbf{y}_{(1)}=-\mathbf{Y}_*\boldsymbol{\beta}_*-\mathbf{X}_0\boldsymbol{\gamma}_0+\mathbf{u}_{(1)}. \tag{7.4.1}$$

Here $\mathbf{Y}_*$ and $\mathbf{X}_0$ denote those endogenous and exogenous variables, respectively, appearing in the first structural equation (other than $\mathbf{y}_{(1)}$). In the notation of Section 7.3, $\boldsymbol{\beta}'=(\boldsymbol{\beta}_0',\mathbf{0}')=(1,\boldsymbol{\beta}_*',\mathbf{0}')$ and $\mathbf{Y}_0=(\mathbf{y}_{(1)},\mathbf{Y}_*)$.

Since $\mathbf{X}_0$ is exogenous and hence uncorrelated with the random disturbance, it can serve as its own instrument. On the other hand, $\mathbf{Y}_*$ should be replaced in (7.4.1) by its fitted value after regressing $\mathbf{Y}_*$ on $\mathbf{X}$; that is,

$$\hat{\mathbf{Y}}_*=\mathbf{X}(\mathbf{X}'\mathbf{X})^{-1}\mathbf{X}'\mathbf{Y}_*.$$

After inserting $\hat{\mathbf{Y}}_*$ in (7.4.1) we get the new equation

$$\mathbf{y}_{(1)}=-\hat{\mathbf{Y}}_*\boldsymbol{\beta}_*-\mathbf{X}_0\boldsymbol{\gamma}_0+\tilde{\mathbf{u}}_{(1)}, \tag{7.4.2}$$

where $\tilde{\mathbf{u}}_{(1)}=\mathbf{u}_{(1)}-(\mathbf{Y}_*-\hat{\mathbf{Y}}_*)\boldsymbol{\beta}_*$. Using OLS on (7.4.2) we get the estimator

$$\begin{bmatrix}-\hat{\boldsymbol{\beta}}_*\\-\hat{\boldsymbol{\gamma}}_0\end{bmatrix}=\begin{bmatrix}\hat{\mathbf{Y}}_*'\hat{\mathbf{Y}}_* & \hat{\mathbf{Y}}_*'\mathbf{X}_0\\ \mathbf{X}_0'\hat{\mathbf{Y}}_* & \mathbf{X}_0'\mathbf{X}_0\end{bmatrix}^{-1}\begin{bmatrix}\hat{\mathbf{Y}}_*'\mathbf{y}_{(1)}\\ \mathbf{X}_0'\mathbf{y}_{(1)}\end{bmatrix}, \tag{7.4.3}$$

which is called the *two-stage least squares* estimator of $\boldsymbol{\beta}_*$, $\boldsymbol{\gamma}_0$.

Note that (7.4.3) can be written

$$\begin{bmatrix}-\hat{\boldsymbol{\beta}}_*\\-\hat{\boldsymbol{\gamma}}_0\end{bmatrix}=(\hat{\mathbf{Z}}'\hat{\mathbf{Z}})^{-1}\hat{\mathbf{Z}}'\mathbf{y}_{(1)},$$

where $\hat{\mathbf{Z}}=(\hat{\mathbf{Y}}_*,\mathbf{X}_0)$. Then the variance of $(\hat{\boldsymbol{\beta}}_*',\hat{\boldsymbol{\gamma}}_0')'$ given $\mathbf{X}$ can be estimated by

$$\hat{\sigma}_{11}(\hat{\mathbf{Z}}'\hat{\mathbf{Z}})^{-1},$$

where $\hat{\sigma}_{11}$ is an estimate of σ_{11} obtained from the residual vector by

$$\hat{\sigma}_{11}=\frac{(\mathbf{y}_{(1)}+\mathbf{Y}_*\hat{\boldsymbol{\beta}}_*+\mathbf{X}_0\hat{\boldsymbol{\gamma}}_0)'(\mathbf{y}_{(1)}+\mathbf{Y}_*\hat{\boldsymbol{\beta}}_*+\mathbf{X}_0\hat{\boldsymbol{\gamma}}_0)}{n-p_1+1-q_1}.$$

The denominator here of course represents the degrees of freedom of the chi-squared distribution that the residual sum of squares would have if $\mathbf{Y}_*$ were equal to $\hat{\mathbf{Y}}_*$, that is if $\mathbf{Y}_*$ were non-random given $\mathbf{X}$.

If the first structural equation is exactly identified then the 2SLS estimator of $\boldsymbol{\beta}_*$ and $\boldsymbol{\gamma}_0$ reduces to the ILS estimate (Exercise 7.4.1).

7.4.2 Limited information maximum likelihood (LIML)

Another method, which is applicable when the errors are normally distributed, is maximum likelihood estimation. In this section we look not at the likelihood of the whole system, but only at the likelihood of those endogenous variables, $y_1, \ldots, y_{p_1}$, occurring in the first structural equation. We wish to obtain m.l.e.s for the structural parameters $\boldsymbol{\beta}_{(1)}$, $\boldsymbol{\gamma}_{(1)}$, and σ_{11}.

For simplicity suppose that none of the first p_1 equations is an identity. Then the disturbance terms $\mathbf{v}_{(1)}, \ldots, \mathbf{v}_{(p_1)}$ in the reduced equations are normally distributed with likelihood

$$|2\pi\boldsymbol{\Psi}_{00}|^{-n/2}\exp\left(-\tfrac{1}{2}\operatorname{tr}\boldsymbol{\Psi}_{00}^{-1}\mathbf{V}_0'\mathbf{V}_0\right),$$

where $\mathbf{V}_0=(\mathbf{v}_{(1)}, \ldots, \mathbf{v}_{(p_1)})$ and $\boldsymbol{\Psi}_{00}>0$. Writing $\mathbf{V}_0$ in terms of the observed quantities $\mathbf{Y}_0=(\mathbf{y}_{(1)}, \ldots, \mathbf{y}_{(p_1)})$ and $\mathbf{X}=(\mathbf{X}_0, \mathbf{X}_1)$, $n\times[q_1+(q-q_1)]$, the likelihood becomes

$$|2\pi\boldsymbol{\Psi}_{00}|^{-n/2}\exp\left[-\tfrac{1}{2}\operatorname{tr}\boldsymbol{\Psi}_{00}^{-1}(\mathbf{Y}_0-\mathbf{X}_0\boldsymbol{\Pi}_{00}-\mathbf{X}_1\boldsymbol{\Pi}_{10})'(\mathbf{Y}_0-\mathbf{X}_0\boldsymbol{\Pi}_{00}-\mathbf{X}_1\boldsymbol{\Pi}_{10})\right]. \tag{7.4.4}$$

This quantity is to be maximized with respect to $\boldsymbol{\Psi}_{00}(p_1\times p_1)$, $\boldsymbol{\Pi}_{00}(q_1\times p_1)$, and $\boldsymbol{\Pi}_{10}((q-q_1)\times p_1)$ subject to the constraint that $\boldsymbol{\Pi}_{10}$ has rank p_1-1. The constraints on $\boldsymbol{\Pi}$ imposed by the other structural equations in the system are ignored in this maximization.

Denote the resulting m.l.e.s by $\hat{\boldsymbol{\Psi}}_{00}$, $\hat{\boldsymbol{\Pi}}_{00}$, $\hat{\boldsymbol{\Pi}}_{10}$. Then the m.l.e.s of the structural parameters of interest are given by solving

$$\hat{\boldsymbol{\Pi}}_{10}\hat{\boldsymbol{\beta}}_0=\mathbf{0},$$

normalized by $\hat{\beta}_{11}=1$, say, and setting

$$\hat{\boldsymbol{\gamma}}_0=-\hat{\boldsymbol{\Pi}}_{00}\hat{\boldsymbol{\beta}}_0, \qquad \hat{\sigma}_{11}=\hat{\boldsymbol{\beta}}_0'\hat{\boldsymbol{\Psi}}_{00}\hat{\boldsymbol{\beta}}_0.$$

The details of this maximization are a bit intricate and will not be given here. (See, for example, Dhrymes, 1970, pp. 344–348.) In particular it can be shown that $\hat{\boldsymbol{\beta}}_0$ minimizes the ratio

$$\frac{\boldsymbol{\beta}_0'\mathbf{Y}_0'(\mathbf{I}-\mathbf{X}_0(\mathbf{X}_0'\mathbf{X}_0)^{-1}\mathbf{X}_0')\mathbf{Y}_0\boldsymbol{\beta}_0}{\boldsymbol{\beta}_0'\mathbf{Y}_0'(\mathbf{I}-\mathbf{X}(\mathbf{X}'\mathbf{X})^{-1}\mathbf{X}')\mathbf{Y}_0\boldsymbol{\beta}_0} \tag{7.4.5}$$

and that $\hat{\boldsymbol{\gamma}}_0$ and $\hat{\sigma}_{11}$ are then obtained by regressing the linear combination $\mathbf{Y}_0\hat{\boldsymbol{\beta}}_0(n \times 1)$ on $\mathbf{X}_0$.

Note that the numerator of (7.4.5) is the residual sum of squares for the regression of the linear combination $\mathbf{Y}_0\boldsymbol{\beta}$ on $\mathbf{X}_0$, and the denominator for the regression on $\mathbf{X}$. For this reason the LIML estimate of $\boldsymbol{\beta}$ is also called the *least variance ratio* (LVR) estimate. Since $\mathbf{X}_0$ is a submatrix of $\mathbf{X}$, the ratio (7.4.5) is always greater than 1. If we write (7.4.5) as $\boldsymbol{\beta}_0'\mathbf{A}\boldsymbol{\beta}_0/\boldsymbol{\beta}_0'\mathbf{C}\boldsymbol{\beta}_0$, it is easily seen that $\hat{\boldsymbol{\beta}}_0$ is the eigenvector of $\mathbf{C}^{-1}\mathbf{A}$ corresponding to the smallest eigenvalue, which for convenience we have standardized by $\hat{\beta}_{11}=1$.

An explicit estimate of the covariance matrix of $(\hat{\boldsymbol{\beta}}_0', \hat{\boldsymbol{\gamma}}_0')'$ can be given, but we shall not consider further details here. (See, for example, Kmenta, 1971, pp. 567–570.)

It is clear that aside from the normalization constraint $\beta_{11}=1$, the LIML estimator $\hat{\boldsymbol{\beta}}_0$ is a symmetric function of $\mathbf{y}_{(1)}, \ldots, \mathbf{y}_{(p_1)}$. Hence it does not matter which endogenous variable is chosen as the "dependent" variable in the first structural equation. This invariance property is not shared by the 2SLS estimator.

However, it can be shown (for example, Dhrymes, 1970, p. 355) that, under mild regularity conditions, the 2SLS and LIML estimators are asymptotically equivalent to one another; thus they are both asymptotically unbiased with the same covariance matrix. When the first structural equation is exactly identified, the LIML estimator is identical to the 2SLS estimator (and hence the same as the ILS estimator). See Exercise 7.4.2.

Example 7.4.1 (Kmenta, 1971, pp. 564, 572–573) To illustrate the 2SLS and LIML estimators let us return to the food consumption and prices model of Example 7.3.4. Twenty observations were simulated from the multinormal distribution using the true parameters

$$Q_t = 96.5 - 0.25P_t + 0.30D_t + u_{t1} \qquad \text{(demand)},$$

$$Q_t = 62.5 + 0.15P_t + 0.20F_t + 0.36A_t + u_{t2} \qquad \text{(supply)},$$

with covariance matrix for u_{t1} and u_{t2},

$$\boldsymbol{\Sigma} = \begin{bmatrix} 3.125 & 3.725 \\ 3.725 & 4.645 \end{bmatrix}.$$

Table 7.4.1 Data for food consumption and prices model

Q_t	P_t	D_t	F_t	A_t
98.485	100.323	87.4	98.0	1
99.187	104.264	97.6	99.1	2
102.163	103.435	96.7	99.1	3
101.504	104.506	98.2	98.1	4
104.240	98.001	99.8	110.8	5
103.243	99.456	100.5	108.2	6
103.993	101.066	103.2	105.6	7
99.900	104.763	107.8	109.8	8
100.350	96.446	96.6	108.7	9
102.820	91.228	88.9	100.6	10
95.435	93.085	75.1	81.0	11
92.424	98.801	76.9	68.6	12
94.535	102.908	84.6	70.9	13
98.757	98.756	90.6	81.4	14
105.797	95.119	103.1	102.3	15
100.225	98.451	105.1	105.0	16
103.522	86.498	96.4	110.5	17
99.929	104.016	104.4	92.5	18
105.223	105.769	110.7	89.3	19
106.232	113.490	127.1	93.0	20

Table 7.4.2 Estimators (with standard errors) for food consumption and prices model

	True coefficient	OLS	2SLS	LIML	3SLS	FIML
Demand equation						
Constant	96.5	99.90	94.63 (7.9)	93.62 (8.0)	Same as 2SLS	Same as LIML
P	−0.25	−0.32	−0.24 (0.10)	−0.23 (0.10)		
D	0.30	0.33	0.31 (0.05)	0.31 (0.05)		
Supply equation						
Constant	62.5	58.28	49.53 (12.01)	Same as 2SLS	52.11 (11.89)	51.94 (12.75)
D	0.15	0.16	0.24 (0.10)		0.23 (0.10)	0.24 (0.11)
F	0.20	0.25	0.26 (0.05)		0.23 (0.04)	0.22 (0.05)
A	0.36	0.25	0.25 (0.10)		0.36 (0.07)	0.37 (0.08)

Values for D_t and F_t are taken from Girschick and Haavelmo (1947) and the data is summarized in Table 7.4.1.

The estimated coefficients using various procedures are summarized in Table 7.4.2. Note that 2SLS was carried out using Q_t as the dependent variable in each equation.

Standard errors are not given for the (inconsistent) OLS estimators and it can be seen how inaccurate these estimates are compared with the other procedures. Note that 2SLS and LIML estimates for the supply equation are identical because this equation is exactly identified. (See Exercises 7.4.1 and 7.4.2.) The 3SLS and FIML estimators will be described in the next section.

7.5 System Estimators

7.5.1 Seemingly unrelated regressions

Before we turn to general system estimators we first consider the special case when only *one* endogenous variable appears in each equation. With a slightly different notation from the previous sections we can write this model as

$$\mathbf{y}_{(j)} = \mathbf{X}_j \boldsymbol{\delta}_{(j)} + \mathbf{u}_{(j)}, \qquad j = 1, \ldots, p, \tag{7.5.1}$$

where $\mathbf{X}_j(n \times q_j)$ denotes those exogenous variables occurring in the jth equation. Note that since only exogenous variables appear on the right-hand side of each equation, this model is already in reduced form. Hence the assumptions of the multiple regression model are satisfied and OLS applied to each equation separately will give consistent estimates of the parameters. However, since the disturbances are assumed correlated between equations, greater efficiency can be attained by treating the system as a whole.

Write the model (7.5.1) as a single multiple regression

$$\mathbf{Y}^V = \mathbf{X}\boldsymbol{\Delta}^V + \mathbf{U}^V \tag{7.5.2}$$

where $\mathbf{Y}^V = (\mathbf{y}'_{(1)}, \ldots, \mathbf{y}'_{(p)})'$ denotes stacking the columns of $\mathbf{Y}$ on top of one another and set

$$\mathbf{X} = \begin{bmatrix} \mathbf{X}_1 & \mathbf{0} & \ldots & \mathbf{0} \\ \mathbf{0} & \mathbf{X}_2 & \ldots & \mathbf{0} \\ \cdot & \cdot & & \cdot \\ \cdot & \cdot & & \cdot \\ \cdot & \cdot & & \cdot \\ \mathbf{0} & \mathbf{0} & \ldots & \mathbf{X}_p \end{bmatrix}.$$

The covariance matrix of $\mathbf{U}^V$ is $C(\mathbf{U}^V) = \mathbf{\Sigma} \otimes \mathbf{I} = \mathbf{\Omega}$, say. If $\mathbf{\Sigma}$ is known, then estimation of $\mathbf{\Delta}$ in (7.5.2) can be carried out by generalized least squares to give the estimate

$$\mathbf{\Delta}^{*V} = (\mathbf{X}'\mathbf{\Omega}^{-1}\mathbf{X})^{-1}\mathbf{X}'\mathbf{\Omega}^{-1}\mathbf{Y}^V. \qquad (7.5.3)$$

(See Section 6.6.2.) Note that $\mathbf{\Delta}^*$ is also the m.l.e. of $\mathbf{\Delta}$ when the data is normally distributed.

The OLS estimate $\hat{\mathbf{\Delta}}$ of $\mathbf{\Delta}$ is given by

$$\hat{\mathbf{\Delta}}^V = (\mathbf{X}'\mathbf{X})^{-1}\mathbf{X}'\mathbf{Y}^V; \qquad (7.5.4)$$

that is,

$$\hat{\mathbf{\delta}}_{(j)} = (\mathbf{X}_j'\mathbf{X}_j)^{-1}\mathbf{X}_j'\mathbf{y}_{(j)}, \qquad j = 1, \ldots, p. \qquad (7.5.5)$$

There are two special cases in which the GLS estimate $\mathbf{\Delta}^*$ and the OLS estimate $\hat{\mathbf{\Delta}}$ are the same:

(a) $\mathbf{X}_1 = \ldots = \mathbf{X}_p$. In this case (7.5.1) reduces to the *multivariate* regression (6.6.1). See Section 6.6.3.

(b) $\mathbf{\Sigma}$ is diagonal. See Exercise 7.5.1.

In general both the GLS and OLS estimators are consistent, but by the Gauss–Markov theorem $\mathbf{\Delta}^*$ has a smaller covariance matrix and hence is preferred.

In practice $\mathbf{\Sigma}$ is not usually known, so that the GLS estimate (7.5.3) would appear to be of limited use. However, it is possible to estimate $\mathbf{\Sigma}$ and to use this estimate of $\mathbf{\Sigma}$ in (7.5.3) to give a modified GLS estimate.

First, we estimate $\mathbf{\Delta}$ by OLS, and use the resulting residuals to estimate $\mathbf{\Sigma}$ by $\hat{\mathbf{\Sigma}}$, where

$$\hat{\sigma}_{ij} = \hat{\mathbf{u}}_{(i)}'\hat{\mathbf{u}}_{(j)}/[(n-q_i)(n-q_j)]^{1/2} \qquad (7.5.6)$$

and

$$\hat{\mathbf{u}}_{(i)} = \mathbf{y}_{(i)} - \mathbf{X}_i\hat{\mathbf{\delta}}_{(i)}.$$

Using this estimate, $\hat{\mathbf{\Omega}} = \hat{\mathbf{\Sigma}} \otimes \mathbf{I}$, in (7.5.3), we get the new estimate of $\mathbf{\Delta}$,

$$\hat{\hat{\mathbf{\Delta}}}^V = (\mathbf{X}'\hat{\mathbf{\Omega}}^{-1}\mathbf{X})^{-1}\mathbf{X}'\hat{\mathbf{\Omega}}^{-1}\mathbf{Y}^V, \qquad (7.5.7)$$

which is known as *Zellner's two-stage* estimator. Under mild regularity conditions the variance matrix of $\hat{\hat{\mathbf{\Delta}}}^V$ can be consistently estimated by

$$(\mathbf{X}'\hat{\mathbf{\Omega}}^{-1}\mathbf{X})^{-1}.$$

Example 7.5.1 (Kmenta, 1971, p. 527) Consider data on the investment performance of two firms, General Electric Company and Westinghouse Electric Company, over the period 1935–1954. Each firm's investment (I) is related to the value of its capital stock (C), and the value of its shares (F). The assumed relationship is

$$I_t = \alpha C_t + \beta F_t + \gamma + u_t, \qquad t = 1935, \ldots, 1954.$$

The results for General Electric are as follows (standard errors in parentheses):

(a) Using ordinary least squares,

$$I_t = \underset{(0.026)}{0.152C} + \underset{(0.016)}{0.027F_t} - \underset{(31.37)}{9.956}.$$

(b) Using Zellner's two-stage method,

$$I_t = \underset{(0.025)}{0.139C_t} + \underset{(0.015)}{0.038F_t} - \underset{(29.32)}{27.72}.$$

The results for Westinghouse were as follows:

(a) Using ordinary least squares,

$$I_t = \underset{(0.056)}{0.092C_t} + \underset{(0.016)}{0.053F_t} - \underset{(8.02)}{0.509}.$$

(b) Using Zellner's two-stage method,

$$I_t = \underset{(0.053)}{0.058C_t} + \underset{(0.015)}{0.064F_t} - \underset{(7.55)}{1.25}.$$

It can be seen that in the case of each of the six coefficients, Zellner's estimate has a lower estimated standard error than does the ordinary least squares estimate.

7.5.2 Three-stage least squares (3SLS)

The method of three-stage least squares involves an application of Zellner's estimator to the general system of structural equations.

As in Section 7.4.1, write each of the structural equations as a regression-like equation,

$$\mathbf{y}_{(j)} = \mathbf{Z}_j \boldsymbol{\delta}_{(j)} + \mathbf{u}_{(j)}, \qquad j = 1, \ldots, p - r.$$

Here $\mathbf{Z}_j = (\mathbf{Y}_{*,j}, \mathbf{X}_{0,j})$ denotes those endogenous and exogenous variables (other than $\mathbf{y}_{(j)}$) appearing in the jth equation and $\boldsymbol{\delta}_{(j)} = (-\boldsymbol{\beta}'_{*,(j)}, -\boldsymbol{\gamma}'_{0,(j)})'$ represents the corresponding structural coefficients. Also, r denotes the number of exact identities in the system which we omit from consideration.

This system can be written more compactly as

$$\mathbf{Y}^V = \mathbf{Z}\boldsymbol{\Delta}^V + \mathbf{U}^V,$$

where

$$\mathbf{Z} = \begin{bmatrix} \mathbf{Z}_1 & & \mathbf{0} \\ & \ddots & \\ \mathbf{0} & & \mathbf{Z}_{p-r} \end{bmatrix}, \qquad \mathbf{Y}^V = \begin{bmatrix} \mathbf{y}_{(1)} \\ \vdots \\ \mathbf{y}_{(p-r)} \end{bmatrix},$$

and similarly for $\boldsymbol{\Delta}^V$ and $\mathbf{U}^V$.

The covariance matrix of $\mathbf{U}^V$ is given by

$$C(\mathbf{U}^V) = \boldsymbol{\Sigma} \otimes \mathbf{I} = \boldsymbol{\Omega},$$

say, which suggests using the method of generalized least squares. The correlation between the disturbance terms can be eliminated by making the transformation

$$\boldsymbol{\Omega}^{-1/2}\mathbf{Y}^V = \boldsymbol{\Omega}^{-1/2}\mathbf{Z}\boldsymbol{\Delta}^V + \boldsymbol{\Omega}^{-1/2}\mathbf{U}^V. \tag{7.5.8}$$

Since $\mathbf{Z}$ is correlated with $\mathbf{U}^V$, OLS is not appropriate here. However 2SLS can be used to replace $\mathbf{Z}$ in (7.5.8) by its fitted value after regressing on $\mathbf{X}$; that is, let

$$\hat{\mathbf{Z}}_j = (\hat{\mathbf{Y}}_{*,j}, \mathbf{X}_{0,j}), \quad \text{where} \quad \hat{\mathbf{Y}}_{*,j} = \mathbf{X}(\mathbf{X}'\mathbf{X})^{-1}\mathbf{X}'\mathbf{Y}_{*,j}$$

and set

$$\hat{\mathbf{Z}} = \begin{bmatrix} \hat{\mathbf{Z}}_1 & & \mathbf{0} \\ & \ddots & \\ \mathbf{0} & & \mathbf{Z}_{p-r} \end{bmatrix}.$$

This substitution leads to the estimate of $\boldsymbol{\Delta}^V$

$$(\hat{\mathbf{Z}}'\boldsymbol{\Omega}^{-1}\hat{\mathbf{Z}})\hat{\mathbf{Z}}'\boldsymbol{\Omega}^{-1}\mathbf{Y}^V, \quad \text{where} \quad \boldsymbol{\Omega} = \boldsymbol{\Sigma} \otimes \mathbf{I}.$$

Note that this application of 2SLS, which assumes a knowledge of $\boldsymbol{\Sigma}$, is different from the single-equation application of 2SLS in Section 7.4.1.

Of course in practice $\boldsymbol{\Omega}$ is unknown, but it can be estimated by $\hat{\boldsymbol{\Omega}} = \hat{\boldsymbol{\Sigma}} \otimes \mathbf{I}$, where

$$\hat{\sigma}_{ij} = (\mathbf{y}_{(i)} - \mathbf{Z}_i\hat{\boldsymbol{\delta}}_{(i)})'(\mathbf{y}_{(j)} - \mathbf{Z}_j\hat{\boldsymbol{\delta}}_{(j)})/[(n - p_i + 1 - q_i)(n - p_j + 1 - q_j)]^{1/2}$$

and $\hat{\boldsymbol{\delta}}_{(i)}$ and $\hat{\boldsymbol{\delta}}_{(j)}$ are the corresponding single equation 2SLS estimators from Section 7.4.1. Then the estimator

$$\tilde{\boldsymbol{\Delta}}^V = (\hat{\mathbf{Z}}'\hat{\boldsymbol{\Omega}}^{-1}\hat{\mathbf{Z}})^{-1}\hat{\mathbf{Z}}'\hat{\boldsymbol{\Omega}}^{-1}\mathbf{Y}^V \tag{7.5.9}$$

is called the *three-stage least squares* estimator. Under mild regularity conditions it is consistent and the variance can be consistently estimated by

$$(\hat{\mathbf{Z}}'\hat{\boldsymbol{\Omega}}^{-1}\hat{\mathbf{Z}})^{-1}.$$

For notational convenience we have used the jth endogenous variable as the "dependent" variable in the jth structural equation. In fact any of the endogenous variables appearing in the jth equation can be chosen as the "dependent" variable. However, different choices will in general lead to different estimates of the structural parameters (beyond the differences of scaling).

Example 7.5.2 The application of 3SLS to the food consumption and prices model of Example 7.4.1 is summarized in Table 7.4.2 with Q_t chosen as the "dependent" variable for both equations. Note that for the demand equation 3SLS gives the same estimates as 2SLS. This occurs because the supply equation is exactly identified and hence adds no information when estimating the over-identified demand equations. (See Exercise 7.5.2.)

7.5.3 Full information maximum likelihood (FIML)

Maximum likelihood estimation can also be applied to the whole system of equations. For simplicity suppose there are no exact identities in the system. Then the likelihood of the reduced form residuals is given by

$$|2\pi\boldsymbol{\Psi}|^{-n/2}\exp[-\tfrac{1}{2}\operatorname{tr}(\boldsymbol{\Psi}^{-1}\mathbf{V}'\mathbf{V})] = |2\pi\boldsymbol{\Psi}|^{-n/2}\exp[-\tfrac{1}{2}\operatorname{tr}\{\boldsymbol{\Psi}^{-1}(\mathbf{Y}-\mathbf{X\Pi})'(\mathbf{Y}-\mathbf{X\Pi})\}]. \tag{7.5.10}$$

This likelihood is then maximized with respect to $\boldsymbol{\Psi}>0$ and $\boldsymbol{\Pi}(q\times p)$, satisfying the constraint

$$\boldsymbol{\Pi}\mathbf{B} = -\boldsymbol{\Gamma} \tag{7.5.11}$$

for some pair of matrices $\mathbf{B}(p\times p)$ and $\boldsymbol{\Gamma}(q\times p)$. Each column of $\mathbf{B}$ usually normalized by $\beta_{ii}=1$, $i=1,\ldots,p$ and some of the elements of $\mathbf{B}$ and $\boldsymbol{\Gamma}$ are known to be 0 from the structural form of the model. The resulting maximizing value of $\mathbf{B}$ and $\boldsymbol{\Gamma}$ is called the *full information maximum likelihood* (FIML) estimator of the structural parameters.

Unfortunately this maximization cannot be carried out analytically, and an iterative procedure must be used. Note that (7.5.10) is the likelihood of all the equations in the system, whereas (7.4.4) included only some of the equations, and that (7.5.11) is a more restrictive set of conditions on Π than those given in Section 7.4.2.

It can be shown that, under mild regularity conditions, for large n the FIML estimator is asymptotically the same as the 3SLS estimator; hence they are both asymptotically unbiased with the same covariance matrix. (See Dhrymes, 1970, p. 371.)

Example 7.5.3 Consider again the food supply and demand model of Example 7.4.1. The FIML estimates are given in Table 7.4.1. Note that the FIML and LIML estimates of the demand equation are identical because the supply equation is exactly identified and system methods offer no advantage over single equation methods in this situation. (See Exercise 7.5.3.) Also, the FIML and 3SLS estimates for the supply equation are similar to one another.

7.6 Comparison of Estimates

To summarize this chapter, we compare the different estimators used in a simultaneous system of equations. First, OLS and GLS applied directly to the system generally lead to biased and inconsistent estimates. Single-equation methods such as 2SLS and LIML are consistent but ignore the information provided by the system as a whole. On the other hand, system methods such as 3SLS and FIML take this information into account and have smaller asymptotic variances than single-equation methods. However, system methods require larger sample sizes, more complicated calculations, and more computing time.

The asymptotic advantage of system methods persists in small samples to some extent, although the improvement over single-equation methods is not so marked here. However, system methods, especially FIML, are more sensitive to mis-specification of the model (for example omitting a relevant explanatory variable from some of the equations). In particular, 2SLS is more robust than the other methods to this sort of error.

For more detailed comparisons, see, for example, Kmenta (1971, pp. 381–385) or Dhrymes (1970, pp. 372–380).

Exercises and Complements

7.2.1 (Inconsistency of OLS estimator) Consider the equation $\mathbf{y} = \mathbf{X\beta} + \mathbf{u}$. Suppose that the rows of $\mathbf{X}(n \times q)$ are a random sample from some distribution with mean $\mathbf{0}$ and covariance matrix $\mathbf{\Sigma} > 0$, and suppose that

$$V(\mathbf{u} \mid \mathbf{X}) = \sigma^2\mathbf{I}_n, \qquad E(\mathbf{u} \mid \mathbf{X}) = \mathbf{X}\boldsymbol{\theta},$$

where $\boldsymbol{\theta} = (\theta_1, \ldots, \theta_q)' \neq \mathbf{0}$ does not depend on n.

(a) Show that

$$E(\mathbf{u}) = \mathbf{0}, \qquad V(\mathbf{u}) = (\sigma^2 + \boldsymbol{\theta}'\boldsymbol{\Sigma}\boldsymbol{\theta})\mathbf{I}_n,$$

so that the random disturbances have mean $\mathbf{0}$, constant variance, and are uncorrelated with one another, but are correlated with $\mathbf{X}$.

(b) Show that the OLS estimator of $\boldsymbol{\beta}$ satisfies

$$\hat{\boldsymbol{\beta}} = \boldsymbol{\beta} + (\mathbf{X}'\mathbf{X})^{-1}\mathbf{X}'\mathbf{u}.$$

(c) Show that

$$E(\hat{\boldsymbol{\beta}} \mid \mathbf{X}) = \boldsymbol{\beta} + \boldsymbol{\theta}, \qquad V(\hat{\boldsymbol{\beta}} \mid \mathbf{X}) = \sigma^2(\mathbf{X}'\mathbf{X})^{-1}, \qquad \text{plim } n(\mathbf{X}'\mathbf{X})^{-1} = \boldsymbol{\Sigma}^{-1}.$$

Thus, using Exercise 6.6.5, deduce that plim $\hat{\boldsymbol{\beta}} = \boldsymbol{\beta} + \boldsymbol{\theta}$, and hence that $\hat{\boldsymbol{\beta}}$ is inconsistent.

7.2.2 (a) In Example 7.2.1, show that the OLS and IV estimators, respectively, of β and α are given by

$$\hat{\beta} = s_{xy}/s_{xx}, \qquad \hat{\alpha} = \bar{y} - \hat{\beta}\bar{x} \qquad \text{(OLS)},$$
$$\beta^* = s_{yz}/s_{xz}, \qquad \alpha^* = \bar{y} - \beta^*\bar{x} \qquad \text{(IV)}.$$

(b) Show that $C(y_i, x_i) = \beta\sigma_x^2$ and hence deduce that

$$\text{plim } \hat{\beta} = \beta\sigma_{\tilde{x}}^2/(\sigma_{\tilde{x}}^2 + \sigma_w^2),$$

so that $\hat{\beta}$ is inconsistent if $\beta \neq 0$.

(c) If $\beta \neq 0$ and $\mu \neq 0$, show that $\hat{\alpha}$ is also inconsistent.

(d) Supposing $C(x_i, z_i) = \sigma_{xz} \neq 0$ and $C(\varepsilon_i, z_i) = 0$, show that β^* and α^* are consistent estimates of β and α.

7.2.3 (a) If the "independent" variables $\mathbf{X}(n \times q)$ in (7.2.1) are regressed on new variables $\mathbf{Z}(n \times k)$ using the multivariate regression model $\mathbf{X} = \mathbf{Z}\boldsymbol{\Gamma} + \mathbf{V}$, where $\boldsymbol{\Gamma}(k \times q)$ is the parameter matrix, show that the OLS estimate of $\boldsymbol{\Gamma}$ is $\hat{\boldsymbol{\Gamma}} = (\mathbf{Z}'\mathbf{Z})^{-1}\mathbf{Z}'\mathbf{X}$ (see Section 6.6.3) and that the fitted value of $\mathbf{X}$ is given by (7.2.9).

(b) If $k = q$ and $\mathbf{Z}'\mathbf{X}$ is non-singular, show that the 2SLS estimator $\boldsymbol{\beta}^{**}$ in (7.2.10) reduces to the IV estimator $\boldsymbol{\beta}^*$ in (7.2.5).

(c) For $k \geq q$, show that $\boldsymbol{\beta}^{**}$ is consistent, provided that $\mathbf{Z}$ is uncorrelated with $\mathbf{u}$, plim $n^{-1}\mathbf{Z}'\mathbf{Z} = \boldsymbol{\Theta}(k \times k)$ non-singular, and plim $n^{-1}\mathbf{Z}'\mathbf{X} = \boldsymbol{\Psi}(k \times q)$ of rank q.

7.2.4 Show that the 2SLS estimator $\boldsymbol{\beta}^{**}$ in (7.2.10) is unaltered if $\mathbf{Z}(n \times k)$ is replaced by $\mathbf{ZA}$, where $\mathbf{A}$ is a $(k \times k)$ non-singular matrix.

7.3.1 Suppose the jth structural equation of an economic system includes only one endogenous variable. Show that this equation is identified. If it is also supposed that *all* the exogenous variables are included, show that this equation is exactly identified.

7.3.2 Partition the structural parameter matrices $\mathbf{B}$ and $\mathbf{\Gamma}$ as in (7.3.21) and define $\mathbf{D}=(\mathbf{B}_1', \mathbf{\Gamma}_1')'$ as in (7.3.22). Partition $\mathbf{B}^{-1}$ as

$$\mathbf{B}^{-1}=\begin{bmatrix}\boldsymbol{\beta}^{(0)\prime} & \boldsymbol{\beta}^{(1)\prime}\\ \mathbf{B}^{(0)} & \mathbf{B}^{(1)}\end{bmatrix},$$

where $\boldsymbol{\beta}^{(0)}$ is a p_1-vector. Expanding $\mathbf{B}\mathbf{B}^{-1}=\mathbf{I}$ and $\mathbf{\Gamma}\mathbf{B}^{-1}=\mathbf{\Pi}$, show that

$$\mathbf{B}_1\mathbf{B}^{(0)}=\mathbf{0}, \qquad \mathbf{B}_1\mathbf{B}^{(1)}=\mathbf{I}_{p-p_1}, \qquad \mathbf{\Gamma}_1\mathbf{B}^{(0)}=\mathbf{\Pi}_{10}, \qquad \mathbf{\Gamma}_1\mathbf{B}^{(1)}=\mathbf{\Pi}_{11}.$$

Hence

$$(\mathbf{0}, \mathbf{D})\mathbf{B}^{-1}=\begin{bmatrix}\mathbf{0} & \mathbf{I}_{p-p_1}\\ \mathbf{\Pi}_{10} & \mathbf{\Pi}_{11}\end{bmatrix}.$$

Since $\mathbf{B}^{-1}$ is non-singular, $\mathbf{D}$, $(\mathbf{0}, \mathbf{D})$ and the above matrix all have the same rank. Hence show that

$$\text{rank}\,(\mathbf{D})=p-p_1+\text{rank}\,(\mathbf{\Pi}_{10}).$$

7.3.3 Consider a model consisting of two simultaneous equations

$$\mathbf{y}_{(1)}=a_1\mathbf{x}_{(1)}+\mathbf{u}_{(1)},$$
$$\mathbf{y}_{(1)}=b_1\mathbf{y}_{(2)}+b_2\mathbf{x}_{(1)}+b_3\mathbf{x}_{(2)}+\mathbf{u}_{(2)},$$

where $\mathbf{y}_{(1)}, \mathbf{y}_{(2)}$ are endogenous and $\mathbf{x}_{(1)}, \mathbf{x}_{(2)}$ are exogenous.

(a) Describe the structural parameter matrices $\mathbf{B}$ and $\mathbf{\Gamma}$ for this model.

(b) Write these equations in reduced form.

(c) Show that the first equation is over-identified and the second is under-identified.

(d) What happens to the identifiability if a term $a_2\mathbf{x}_{(3)}$ is added to the first equation?

7.4.1 Suppose that the first structural equation of an economic system is exactly identified (so that $p_1-1=q-q_1$).

(a) Using Exercise 7.2.4 show that the instrumental variables $(\hat{\mathbf{Y}}_*, \mathbf{X}_0)$ in the 2SLS estimator (7.4.3) can be replaced by $\mathbf{X}=(\mathbf{X}_0, \mathbf{X}_1)$. Thus, we can write

$$-\hat{\boldsymbol{\beta}}_*=(\mathbf{I}, \mathbf{0})(\mathbf{X}'\mathbf{Y}_*, \mathbf{X}'\mathbf{X}_0)^{-1}\mathbf{X}'\mathbf{y}_{(1)},$$
$$-\hat{\boldsymbol{\gamma}}_0=(\mathbf{0}, \mathbf{I})(\mathbf{X}'\mathbf{Y}_*, \mathbf{X}'\mathbf{X}_0)^{-1}\mathbf{X}'\mathbf{y}_{(1)},$$

in the same way that (7.2.10) can be simplified to (7.2.5).

(b) Show that

$$\begin{pmatrix}\hat{\mathbf{\Pi}}_{00}\\ \hat{\mathbf{\Pi}}_{10}\end{pmatrix}\begin{pmatrix}1\\ \hat{\boldsymbol{\beta}}_*\end{pmatrix} = -\begin{pmatrix}\hat{\boldsymbol{\gamma}}_0\\ \mathbf{0}\end{pmatrix} = \begin{pmatrix}\mathbf{I}\\ \mathbf{0}\end{pmatrix}(\mathbf{0}, \mathbf{I})(\mathbf{X}'\mathbf{Y}_*, \mathbf{X}'\mathbf{X}_0)^{-1}\mathbf{X}'\mathbf{y}_{(1)},$$

where $(\hat{\mathbf{\Pi}}'_{00}, \hat{\mathbf{\Pi}}'_{10})' = (\mathbf{X}'\mathbf{X})^{-1}\mathbf{X}'(\mathbf{y}_{(1)}, \mathbf{Y}_*)$ is the OLS estimate of the reduced parameter matrix and $\hat{\boldsymbol{\beta}}_*, \hat{\boldsymbol{\gamma}}_0$ are given in (a). (Hint: pre-multiply this equation by $(\mathbf{X}'\mathbf{X})$ and, using $\mathbf{X}(\mathbf{I}, \mathbf{0})' = \mathbf{X}_0$, show that it reduces to an identity.)

(c) Hence deduce that the 2SLS estimates of $\boldsymbol{\beta}_0$ and $\boldsymbol{\gamma}_0$ coincide with ILS estimates (7.3.19) in this situation.

7.4.2 If the first structural equation is exactly identified, show that the restriction $\text{rank}(\mathbf{\Pi}_{10}) = p_1 - 1$ is irrelevant in maximizing the likelihood (7.4.4). Hence deduce that the LIML estimates of $\boldsymbol{\beta}$ and $\boldsymbol{\gamma}$ coincide with the ILS estimates in this case.

7.4.3 (a) In Example 7.4.1, show that if the covariance matrix of the structural form disturbances is

$$\mathbf{\Sigma} = \begin{bmatrix}3.125 & 3.725\\ 3.725 & 4.645\end{bmatrix},$$

then the covariance matrix of the reduced form disturbances is

$$\mathbf{\Psi} = \begin{bmatrix}4 & -2\\ -2 & 2\end{bmatrix}.$$

(b) Calculate the matrix of reduced form parameters using the equation $\mathbf{\Pi} = -\mathbf{\Gamma}\mathbf{B}^{-1}$.

7.5.1 Consider the system of seemingly unrelated regressions (7.5.1). If $\mathbf{\Sigma}$ is diagonal, then show that the OLS and GLS estimates of the parameters are the same.

7.5.2 (a) Consider the following two seemingly unrelated regressions

$$\mathbf{y}_{(1)} = \mathbf{X}_1\boldsymbol{\beta}_{(1)} + \mathbf{u}_{(1)},$$
$$\mathbf{y}_{(2)} = \mathbf{X}_1\boldsymbol{\beta}_{(2)} + \mathbf{X}_2\boldsymbol{\alpha} + \mathbf{u}_{(2)},$$

where $\mathbf{X}_1'\mathbf{X}_2 = \mathbf{0}$. Show that the OLS estimates of $\boldsymbol{\beta}_{(1)}$ and $\boldsymbol{\beta}_{(2)}$ equal the GLS estimates, but the OLS estimate of $\boldsymbol{\alpha}$ does not in general equal the GLS estimate. Also, show that the first equation is over-identified and the second is exactly identified Hint: let $\mathbf{\Sigma}(2\times 2)$ denote the known or estimated covariance matrix for the disturbances. Write

$$\mathbf{X} = \begin{bmatrix}\mathbf{X}_1 & \mathbf{0} & \mathbf{0}\\ \mathbf{0} & \mathbf{X}_1 & \mathbf{X}_2\end{bmatrix} \equiv \left[\mathbf{I}_2\otimes\mathbf{X}_1 \;\vdots\; \begin{matrix}\mathbf{0}\\ \mathbf{X}_2\end{matrix}\right] \quad \text{and} \quad \mathbf{\Omega} = \mathbf{\Sigma}\otimes\mathbf{I}_n.$$

Using (7.5.3), show that the GLS estimates of the parameters are

$$\begin{bmatrix} \boldsymbol{\beta}'_{(1)} \\ \boldsymbol{\beta}'_{(2)} \\ \boldsymbol{\alpha}' \end{bmatrix} = \begin{bmatrix} \mathbf{I}_2 \otimes (\mathbf{X}_1'\mathbf{X}_1)^{-1}\mathbf{X}_1' \\ \sigma^{21}(\sigma^{22})^{-1}(\mathbf{X}_2'\mathbf{X}_2)^{-1}\mathbf{X}_2', \ (\mathbf{X}_2'\mathbf{X}_2)^{-1}\mathbf{X}_2' \end{bmatrix} \begin{bmatrix} \mathbf{y}_{(1)} \\ \mathbf{y}_{(2)} \end{bmatrix}.$$

(b) Consider a general two-equation economic system where the first equation is over-identified and the second is exactly identified. Show that the 2SLS estimates equal the 3SLS estimates for the structural parameters of the *first* equation (but not the second). (Hint: see Exercise 7.4.1.)

(c) Generalize this result to a p-equation system with no identities, where the first equation is over-identified and all the others are exactly identified.

7.5.3 Consider a p-equation economic system with no identities, where the first equation is over-identified and all the others are exactly identified. Show that the LIML and FIML estimates of the structural parameters of the *first* equation are equal. (Hint: Write the likelihood as

$$L_0(\boldsymbol{\Pi}_{00}, \boldsymbol{\Pi}_{10}, \boldsymbol{\Psi}_{00})L_1(\boldsymbol{\Pi}_{01}, \boldsymbol{\Pi}_{11}, \boldsymbol{\Psi}),$$

where the first factor is the density of $\mathbf{V}_0 = (\mathbf{v}_{(1)}, \ldots, \mathbf{v}_{(p_1)})$, and the second factor is the density of $\mathbf{V}_1 \mid \mathbf{V}_0$. Show that in this situation the constraint (7.5.11) merely states that $\operatorname{rank}(\boldsymbol{\Pi}_{10}) = p_1 - 1$, and hence the second factor takes the same value when maximized over $\boldsymbol{\Pi}_{01}, \boldsymbol{\Pi}_{11}, \boldsymbol{\Psi}_{01}, \boldsymbol{\Psi}_{11}$, for any values of the parameters in the first factor. (See Exercise 6.3.3.) The first factor is simply the likelihood (7.4.4).)

7.5.4 (Kmenta, 1971, p. 530) A set of three "seemingly unrelated regression equations" is specified as

$$\mathbf{y}_{(1)} = \alpha_1 + \beta_1 \mathbf{x}_{(1)} + \mathbf{u}_{(1)},$$

$$\mathbf{y}_{(2)} = \alpha_2 + \beta_2 \mathbf{x}_{(2)} + \mathbf{u}_{(2)},$$

$$\mathbf{y}_{(3)} = \alpha_3 + \beta_3 \mathbf{x}_{(3)} + \mathbf{u}_{(3)}.$$

The covariance matrix of the disturbances is known. Consider the estimator of β_1 obtained by using Zellner's method on all three equations, and the estimator obtained by using this method on just the first two equations. Compare the variances of the two estimators.

8 Principal Component Analysis

8.1 Introduction

Chapter 1 has already introduced the open/closed book data, which involved the scores of 88 students on five examinations. This data was expressed in Table 1.2.1 as a matrix having 88 rows and five columns. One question which can be asked concerning this data is how the results on the five different examinations should be combined to produce an overall score. Various answers are possible. One obvious answer would be to use the overall mean, that is the linear combination $(x_1+x_2+x_3+x_4+x_5)/5$, or, equivalently, $\mathbf{l}'\mathbf{x}$, where $\mathbf{l}$ is the vector of weights $\frac{1}{5}\mathbf{1}=(\frac{1}{5}\frac{1}{5}\frac{1}{5}\frac{1}{5}\frac{1}{5})'$. But can one do better than this? That is one of the questions that principal component analysis seeks to answer. We call a linear combination $\mathbf{l}'\mathbf{x}$ a standardized linear combination (SLC) if $\sum l_i^2=1$. This technique was developed by Hotelling (1933) after its origin by Karl Pearson (1901).

As a first objective principal component analysis seeks the SLC of the original variables which has *maximal* variance. In the examination situation this might seem sensible—a large variance "separates out" the candidates, thereby easing consideration of differences between them. The students can then be ranked with respect to this SLC. Similar considerations apply in other situations, such as constructing an index of the cost of living.

More generally, principal component analysis looks for a *few* linear combinations which can be used to summarize the data, losing in the process as little information as possible. This attempt to reduce dimensionality can be described as "*parsimonious summarization*" of the data. For example, we might ask whether the important features of the open/closed book data can be summarized by, say, two linear combinations. If it were possible, we would be reducing the dimensionality from $p=5$ to $p=2$.

8.2 Definition and Properties of Principal Components

8.2.1 Population principal components

Section 1.5.3 introduced the so-called principal component transformation in the context of sample data. This was an orthogonal transformation which transforms any set of variables into a set of new variables which are uncorrelated with each other. Here we give its population counterpart.

Definition 8.2.1 *If* $\mathbf{x}$ *is a random vector with mean* $\boldsymbol{\mu}$ *and covariance matrix* $\boldsymbol{\Sigma}$, *then the* principal component transformation *is the transformation*

$$\mathbf{x} \rightarrow \mathbf{y} = \boldsymbol{\Gamma}'(\mathbf{x}-\boldsymbol{\mu}), \tag{8.2.1}$$

where $\boldsymbol{\Gamma}$ *is orthogonal,* $\boldsymbol{\Gamma}'\boldsymbol{\Sigma}\boldsymbol{\Gamma}=\boldsymbol{\Lambda}$ *is diagonal and* $\lambda_1 \geqslant \lambda_2 \geqslant \ldots \geqslant \lambda_p \geqslant 0$. *The strict positivity of the eigenvalues* λ_i *is guaranteed if* $\boldsymbol{\Sigma}$ *is positive definite. This representation of* $\boldsymbol{\Sigma}$ *follows from the spectral decomposition theorem (Theorem A.6.4). The* ith principal component *of* $\mathbf{x}$ *may be defined as the* ith *element of the vector* $\mathbf{y}$, *namely as*

$$y_i = \boldsymbol{\gamma}'_{(i)}(\mathbf{x}-\boldsymbol{\mu}). \tag{8.2.2}$$

Here $\boldsymbol{\gamma}_{(i)}$ *is the* ith *column of* $\boldsymbol{\Gamma}$, *and may be called the* ith *vector of* principal component loadings. *The function* y_p *may be called the last principal component of* $\mathbf{x}$.

Figure A.10.2 gives a pictorial representations in two dimensions of the principal components for $\boldsymbol{\mu}=\mathbf{0}$.

Example 8.2.1 Suppose that x_1 and x_2 are standardized to have mean 0 and variance 1, and have correlation ρ. The eigenvalues of this covariance matrix are $1\pm\rho$. If ρ is positive then the first eigenvalue is $\lambda_1=1+\rho$, and the first eigenvector is (1, 1). Hence the first principal component of $\mathbf{x}$ is

$$y_1 = 2^{-1/2}(x_1+x_2),$$

which is proportional to the mean of the elements of $\mathbf{x}$. Note that y_1 has variance $(1+\rho)$, which equals the first eigenvalue. The second principal component corresponding to the second eigenvalue $1-\rho$ is

$$y_2 = 2^{-1/2}(x_1-x_2),$$

which is proportional to the difference between the elements of $\mathbf{x}$, and has variance $(1-\rho)$, which equals the second eigenvalue. If ρ is negative, then the order of the principal components is reversed. For a p-variate extension of this example, see Example 8.2.2.

In the above example y_1 and y_2 are uncorrelated. This is a special case of a more general result, which is proved in part (c) of the following theorem.

Theorem 8.2.1 *If* $\mathbf{x} \sim (\boldsymbol{\mu}, \boldsymbol{\Sigma})$ *and* $\mathbf{y}$ *is as defined in* (8.2.1), *then*

(a) $E(y_i)=0$;

(b) $V(y_i)=\lambda_i$;

(c) $C(y_i, y_j)=0, i \neq j$;

(d) $V(y_1) \geq V(y_2) \geq \ldots \geq V(y_p) \geq 0$;

(e) $\sum_{i=1}^{p} V(y_i)=\operatorname{tr} \boldsymbol{\Sigma}$;

(f) $\prod_{i=1}^{p} V(y_i)=|\boldsymbol{\Sigma}|$.

Proof (a)–(d) follow from Definition 8.2.1 and the properties of the expectation operator. (e) follows from (b) and the fact that $\operatorname{tr} \boldsymbol{\Sigma}$ is the sum of the eigenvalues. (f) follows from (b) and the fact that $|\boldsymbol{\Sigma}|$ is the product of the eigenvalues. ■

It is instructive to confirm the results of the above theorem with respect to Example 8.2.1. Note particularly that $V(y_1)+V(y_2)=2$, and that $V(y_1) \times V(y_2)=(1-\rho^2)$, thus confirming parts (e) and (f) of the theorem.

Section (d) of Theorem 8.2.1 states that y_1 has a larger variance than any of the other principal components. (Here and elsewhere we shall use "larger" somewhat loosely to mean "not smaller".) However, we shall now prove a stronger result.

Theorem 8.2.2 *No SLC of* $\mathbf{x}$ *has a variance larger than* λ_1, *the variance of the first principal component.*

Proof Let the SLC be $\mathbf{a}'\mathbf{x}$, where $\mathbf{a}'\mathbf{a}=1$. We may write

$$\mathbf{a}=c_1\boldsymbol{\gamma}_{(1)}+\ldots+c_p\boldsymbol{\gamma}_{(p)}, \tag{8.2.4}$$

where $\boldsymbol{\gamma}_{(1)}, \ldots, \boldsymbol{\gamma}_{(p)}$ are the eigenvectors of $\boldsymbol{\Sigma}$. (Any vector can be written in this form since the eigenvectors constitute a basis for R^p.) Now, if $\alpha=\mathbf{a}'\mathbf{x}$, then

$$V(\alpha)=\mathbf{a}'\boldsymbol{\Sigma}\mathbf{a}=\mathbf{a}'\left(\sum_{i=1}^{p} \lambda_i \boldsymbol{\gamma}_{(i)}\boldsymbol{\gamma}'_{(i)}\right)\mathbf{a} \tag{8.2.5}$$

using the spectral decomposition of $\mathbf{\Sigma}$ (Theorem A.6.4). Now $\boldsymbol{\gamma}'_{(i)}\boldsymbol{\gamma}_{(j)}=\delta_{ij}$, the Kronecker delta. Inserting (8.2.4) into (8.2.5) we find that

$$V(\alpha)=\sum \lambda_i c_i^2. \tag{8.2.6}$$

But we know that

$$\sum c_i^2=1, \tag{8.2.7}$$

since $\mathbf{a}$ is given by (8.2.4) and satisfies $\mathbf{a}'\mathbf{a}=1$. Therefore, since λ_1 is the largest of the eigenvalues, the maximum of (8.2.6) subject to (8.2.7) is λ_1.

This maximum is obtained when $c_1=1$ and all the other c_i are zero. Therefore $V(\alpha)$ is maximized when $\mathbf{a}=\boldsymbol{\gamma}_{(1)}$. ■

For a multinormal random vector, the above theorem may be interpreted geometrically by saying that the first principal component represents the major axis of the ellipsoids of concentration. (See Figure 2.5.1).

A similar argument shows that the last principal component of $\mathbf{x}$ has a variance which is *smaller* than that of any other SLC. The intermediate components have a maximal variance property given by the following theorem.

Theorem 8.2.3 *If $\alpha=\mathbf{a}'\mathbf{x}$ is a SLC of $\mathbf{x}$ which is uncorrelated with the first k principal components of $\mathbf{x}$, then the variance of α is maximized when α is the* $(k+1)$th *principal component of* $\mathbf{x}$.

Proof We may write $\mathbf{a}$ in the form given by (8.2.4). Since α is uncorrelated with $\boldsymbol{\gamma}'_{(i)}\mathbf{x}$ for $i=1,\dots,k$, we know that $\mathbf{a}'\boldsymbol{\gamma}_{(i)}=0$ and therefore that $c_i=0$ for $i=1,\dots,k$. Therefore, using the same argument as in (8.2.6) and (8.2.7), the result follows. ■

Example 8.2.2 We now extend the bivariate example of Example 8.2.1. Suppose that $\mathbf{\Sigma}$ is the equicorrelation matrix, viz. $\mathbf{\Sigma}=(1-\rho)\mathbf{I}+\rho\mathbf{1}\mathbf{1}'$. If $\rho>0$ then $\lambda_1=1+(p-1)\rho$ and $\lambda_2=\lambda_3=\dots=\lambda_p=1-\rho$. The eigenvector corresponding to λ_1 is $\mathbf{1}$. Hence if $\rho>0$, the first principal component is proportional to $p^{-1}\mathbf{1}'\mathbf{x}$, the average of the x. This may be taken as a measure of overall "size", while "shape" may be defined in terms of the other eigenvectors which are all orthogonal to $\mathbf{1}$ (i.e. vectors containing some negative signs). (For a further discussion of size and shape see Section 8.6.) ■

It is clear that $\mathbf{a}'\mathbf{x}$ always has the same variance as $-\mathbf{a}'\mathbf{x}$, and therefore that in terms of variance we are uninterested in the difference between the two vectors $\mathbf{a}$ and $-\mathbf{a}$. As mentioned before, we consider only SLCs, so that $\sum a_i^2=1$. However, some authors scale $\mathbf{a}$ differently, e.g. in such a way that $\sum a_i^2$ is equal to the corresponding eigenvalue.

8.2.2 Sample principal components

We turn now to consider the sample-based counterpart of the previous section, thus building on Section 1.5.3. Let $\mathbf{X}=(\mathbf{x}_1,\ldots,\mathbf{x}_n)'$ be a sample data matrix, and $\mathbf{a}$ a standardized vector. Then $\mathbf{Xa}$ gives n observations on a new variable defined as a weighted sum of the columns of $\mathbf{X}$. The sample variance of this new variable is $\mathbf{a'Sa}$, where $\mathbf{S}$ is the sample covariance matrix of $\mathbf{X}$. We may then ask which such SLC has the largest variance. Not surprisingly the answer is that the SLC with largest variance is the first principal component defined by direct analogy with (8.2.2) as

$$\mathbf{y}_{(1)}=(\mathbf{X}-\mathbf{1}\bar{\mathbf{x}}')\mathbf{g}_{(1)}. \tag{8.2.8}$$

Here $\mathbf{g}_{(1)}$ is the standardized eigenvector corresponding to the largest eigenvalue of $\mathbf{S}$ (i.e. $\mathbf{S}=\mathbf{GLG}'$). This result may be proved in the same way as Theorem 8.2.2. Similarly, the ith sample principal component is defined as $\mathbf{y}_{(i)}=(\mathbf{X}-\mathbf{1}\bar{\mathbf{x}}')\mathbf{g}_{(i)}$, and these components satisfy the straightforward sample extensions of Theorems 8.2.1–8.2.3 (see Exercise 8.2.5).

Putting the principal components together we get

$$\mathbf{Y}=(\mathbf{X}-\mathbf{1}\bar{\mathbf{x}}')\mathbf{G}.$$

Also $\mathbf{X}-\mathbf{1}\bar{\mathbf{x}}'=\mathbf{YG}'$ since $\mathbf{G}$ is orthogonal; that is, $\mathbf{G}$ has transformed one $(n\times p)$ matrix $(\mathbf{X}-\mathbf{1}\bar{\mathbf{x}}')$ into another one of the same order $(\mathbf{Y})$. The covariance matrix of $\mathbf{Y}$ is given by

$$\begin{aligned}\mathbf{S}_Y=n^{-1}\mathbf{Y}'\mathbf{HY}&=n^{-1}\mathbf{G}'(\mathbf{X}-\mathbf{1}\bar{\mathbf{x}}')'\mathbf{H}(\mathbf{X}-\mathbf{1}\bar{\mathbf{x}}')\mathbf{G}\\&=n^{-1}\mathbf{G}'\mathbf{X}'\mathbf{HXG}=\mathbf{G}'\mathbf{SG}=\mathbf{L},\end{aligned}$$

where

$$\mathbf{H}=\mathbf{I}-(n^{-1}\mathbf{11}');$$

that is, the columns of $\mathbf{Y}$ are uncorrelated and the variance of $\mathbf{y}_{(j)}$ is l_j.

The rth element of $\mathbf{y}_{(i)}$, y_{ri}, represents the *score* of the ith principal component on the rth individual. In terms of individuals we can write the principal component transformation as

$$y_{ri}=\mathbf{g}'_{(i)}(\mathbf{x}_r-\bar{\mathbf{x}}).$$

Often we omit the subscript r and write $y_i=\mathbf{g}'_{(i)}(\mathbf{x}-\bar{\mathbf{x}})$ to emphasize the transformation rather than its effect on any particular individual.

Example 8.2.3 Consider the open/closed book examination data from Table 1.2.1. It has covariance matrix

$$\mathbf{S}=\begin{bmatrix} 302.3 & 125.8 & 100.4 & 105.1 & 116.1 \\ & 170.9 & 84.2 & 93.6 & 97.9 \\ & & 111.6 & 110.8 & 120.5 \\ & & & 217.9 & 153.8 \\ & & & & 294.4 \end{bmatrix}$$

and mean $\bar{\mathbf{x}}'=(39.0, 50.6, 50.6, 46.7, 42.3)$. A spectral decomposition of $\mathbf{S}$ yields principal components

$$y_1=0.51x_1+0.37x_2+0.35x_3+0.45x_4+0.53x_5-99.7,$$
$$y_2=0.75x_1+0.21x_2-0.08x_3-0.30x_4-0.55x_5+1.5,$$
$$y_3=-0.30x_1+0.42x_2+0.15x_3+0.60x_4-0.60x_5-19.8,$$
$$y_4=0.30x_1-0.78x_2-0.00x_3+0.52x_4-0.18x_5+11.1,$$
$$y_5=0.08x_1+0.19x_2-0.92x_3+0.29x_4+0.15x_5+13.9,$$

with variances 679.2, 199.8, 102.6, 83.7, and 31.8, respectively. Note that the first principal component gives positive weight to all the variables and thus represents an "average" grade. On the other hand, the second component represents a contrast between the open-book and closed-book examinations, with the first and last examinations weighted most heavily. Of course the difference in performance due to the two types of examination is also confounded with the differences between the individual subjects.

*8.2.3 Further properties of principal components

The most important properties of principal components have already been given in Theorems 8.2.1–8.2.3, along with the corresponding statements for sample principal components. We now turn to examine several other useful properties, each of which is briefly summarized as a heading at the beginning of the relevant section.

(*a*) *The sum of the first k eigenvalues divided by the sum of all the eigenvalues,* $(\lambda_1+\ldots+\lambda_k)/(\lambda_1+\ldots+\lambda_p)$, *represents the "proportion of total variation" explained by the first k principal components.*

The "total variation" in this statement is $\operatorname{tr}\mathbf{\Sigma}$, and the rationale for this statement is clear from part (e) of Theorem 8.2.1. A justification for the use of $\operatorname{tr}\mathbf{\Sigma}$ as a measure of multivariate scatter was given in Section 1.5.3, and the proportion of total variation defined here gives a quantitative measure of the amount of information retained in the reduction from p to k dimensions. For instance, we may say in Example 8.2.1 that the first component "explains" the proportion $\frac{1}{2}(1+\rho)$ of the total variation, and in Example 8.2.2 the proportion explained is $\{1+(p-1)\rho\}/p$.

In Example 8.2.3 the first principal component explains $679.2/1097.1=61.9\%$ of the total variation, and the first two components 80.1%. (Compare this with the fact that the largest of the original variables comprised only $302.3/1097.1=27.6\%$ of the total variation, and the largest two only 54.4%.)

(*b*) *The principal components of a random vector are not scale-invariant.* One disadvantage of principal component analysis is that it is *not* scale invariant. For example, given three variables, say weight in pounds, height in feet, and age in years, we may seek a principal component expressed say in ounces, inches, and decades. Two procedures seem feasible:

(a) multiply the data by 16, 12, and $\frac{1}{10}$, respectively, and then carry out a principal component analysis;
(b) carry out a principal component analysis and then multiply the elements of the relevant component by 16, 12, and $\frac{1}{10}$.

Unfortunately procedures (a) and (b) do *not* generally lead to the same result. This may be seen theoretically for $p=2$ by considering

$$\mathbf{\Sigma}=\begin{pmatrix} \sigma_1^2 & \rho\sigma_1\sigma_2 \\ \rho\sigma_1\sigma_2 & \sigma_2^2 \end{pmatrix},$$

where $\rho>0$. The larger eigenvalue is $\lambda_1=\frac{1}{2}(\sigma_1^2+\sigma_2^2)+\frac{1}{2}\Delta$, where $\Delta=\{(\sigma_1^2-\sigma_2^2)^2+4\sigma_1^2\sigma_2^2\rho^2\}^{1/2}$, with eigenvector proportional to

$$(a_1, a_2)=(\sigma_1^2-\sigma_2^2+\Delta, 2\rho\sigma_1\sigma_2) \tag{8.2.9}$$

(see Exercise 8.1.1). When $\sigma_1/\sigma_2=1$, the ratio a_2/a_1 given by (8.2.9) is unity. If $\sigma_1=\sigma_2$ and the first variable is multiplied by a factor k, then for scale-invariance we would like the new ratio a_2/a_1 to be k. However, changing σ_1 to $k\sigma_1$ in (8.2.9) shows that this is not the case.

Alternatively, the lack of scale-invariance can be examined empirically as shown in the following example. (See also Exercise 8.2.4.)

Example 8.2.4 From the correlation matrix of the open/closed data, the first principal component (after setting the sample mean to $\mathbf{0}$) is found to be

$$0.40u_1+0.43u_2+0.50u_3+0.46u_4+0.44u_5.$$

Here $u_1, \ldots, u_5$ are the standardized variables, so that $u_i = x_i/s_i$. Re-expressing the above components in terms of the original variables and standardizing, we get

$$y_1 = 0.023x_1 + 0.033x_2 + 0.048x_3 + 0.031x_4 + 0.026x_5,$$

or

$$y_1^* = 13.4y_1 = 0.31x_1 + 0.44x_2 + 0.64x_3 + 0.42x_4 + 0.35x_5.$$

The first component of the original variables obtained from the covariance matrix is given in Example 8.2.3 and has different coefficients. (The eigenvalues are also different—they account for 61.9, 18.2, 9.3, 7.6, and 2.9% of the variance when the covariance matrix is used and for 63.6, 14.8, 8.9, 7.8, and 4.9% when the correlation matrix is used.)

Algebraically, the lack of scale invariance can be explained as follows. Let $\mathbf{S}$ be the sample covariance matrix. Then if the ith variable is divided by d_i, the covariance matrix of the new variables is $\mathbf{DSD}$, where $\mathbf{D} = \text{diag}(d_i^{-1})$. However, if $\mathbf{x}$ is an eigenvector of $\mathbf{S}$, then $\mathbf{D}^{-1}\mathbf{x}$ is *not* an eigenvector of $\mathbf{DSD}$. In other words, the eigenvectors are not scale invariant.

The lack of scale-invariance illustrated above implies a certain sensitivity to the way scales are chosen. Two ways out of this dilemma are possible. First, one may seek so-called "natural" units, by ensuring that all variables measured are of the same type (for instance, all heights or all weights). Alternatively, one can standardize all variables so that they have unit variance, and find the principal components of the correlation matrix rather than the covariance matrix. The second option is the one most commonly employed, although this does complicate questions of hypothesis testing—see Section 8.4.4.

(*c*) *If the covariance matrix of* $\mathbf{x}$ *has rank* $r < p$, *then the total variation of* $\mathbf{x}$ *can be entirely explained by the first* r *principal components.*

This follows from the fact that if $\mathbf{\Sigma}$ has rank r, then the last $(p-r)$ eigenvalues of $\mathbf{\Sigma}$ are identically zero. Hence the result follows from (*a*) above.

(*d*) *The vector subspace spanned by the first* k *principal components* $(1 \leq k < p)$ *has smaller mean square deviation from the population (or sample) variables than any other* k*-dimensional subspace.*

If $\mathbf{x} \sim (\mathbf{0}, \mathbf{\Sigma})$ and a subspace $H \subset R^p$ is spanned by orthonormal vectors $\mathbf{h}_{(1)}, \ldots, \mathbf{h}_{(k)}$, then by projecting $\mathbf{x}$ onto this subspace we see that the squared distance d^2 from $\mathbf{x}$ to H has expectation

$$E(d^2) = E(\mathbf{x}'\mathbf{x}) - \sum_{j=1}^{k} E(\mathbf{h}'_{(j)}\mathbf{x}).$$

Let $\mathbf{f}_{(j)} = \mathbf{\Gamma}'\mathbf{h}_{(j)}$, $j = 1, \ldots, k$. We have

$$\begin{aligned} E(d^2) &= \operatorname{tr} \mathbf{\Sigma} - \sum_{j=1}^{k} E(\mathbf{f}'_{(j)}\mathbf{y}) \\ &= \operatorname{tr} \mathbf{\Sigma} - \sum_{i=1}^{p} \sum_{j=1}^{k} f_{ij}^2 \lambda_i \end{aligned} \tag{8.2.10}$$

since the $\mathbf{f}_{(j)}$ are also orthonormal. Exercise 8.2.7 then shows that (8.2.10) is minimized when $f_{ij} = 0$, $i = k+1, \ldots, p$, for each $j = 1, \ldots, k$; that is when the $\mathbf{f}_{(j)}$ span the subspace of the first k principal components. Note that the case $k = 1$ is just a reformulation of Theorem 8.2.2.

(e) *As a special case of* (d) *for* $k = p - 1$, *the plane perpendicular to the last principal component has a smaller mean square deviation from the population* (*or sample*) *variables than any other plane.*

(Of course, the plane perpendicular to the last principal component equals the subspace spanned by the first $(p-1)$ principal components.)

Consider the multiple regression $x_1 = \boldsymbol{\beta}'\mathbf{x}_2 + u$ of x_1 on $(x_2, \ldots, x_p)' = \mathbf{x}_2$, say, where u is uncorrelated with $\mathbf{x}_2$. The linear relationship $x_1 = \boldsymbol{\beta}'\mathbf{x}_2$ defines a plane in R^p and the parameter $\boldsymbol{\beta}$ may be found by minimizing $E(d^2)$ where $d = |x_1 - \boldsymbol{\beta}'\mathbf{x}_2|$ is the distance between the point $\mathbf{x}$ and the plane, with d measured *in the direction of the "dependent variable"*. On the other hand, the plane perpendicular to the last principal component can be found by minimizing $E(d^2)$, where d is measured *perpendicular to the plane.* Thus principal component analysis represents "orthogonal" regression in contrast to the "simple" regression of Chapter 6.

As an example where this point of view is helpful, consider the following bivariate regression-like problem, where both variables are subject to errors. (This problem has already been discussed in Example 7.2.1, using the method of instrumental variables.)

Example 8.2.5 Suppose an observed vector $\mathbf{x}'_r = (x_{r1}, x_{r2})$, $r = 1, \ldots, n$, is modelled by

$$x_{r1} = \xi_r + \varepsilon_{r1}, \qquad x_{r2} = \alpha + \beta\xi_r + \varepsilon_{r2}, \tag{8.2.11}$$

where ξ_r is the (unknown) "independent" variable, $\varepsilon_{r1} \sim N(0, \tau_1^2)$, $\varepsilon_{r2} \sim N(0, \tau_2^2)$, and the εs are independent of one another and of the ξs. The problem is to find the linear relationship (8.2.11) between x_1 and x_2, that is to estimate α and β. However, because x_1 is subject to error, a simple regression of x_2 on x_1 is *not* appropriate.

There are two ways to approach this problem. First the ξs can be viewed as n unknown additional parameters to be estimated, in which case (8.2.11) is called a *functional relationship.* Alternatively, the ξs can be regarded as a sample from $N(\mu, \sigma^2)$, where μ and σ^2 are two additional parameters to be estimated, in which case (8.2.11) is called a *structural relationship.* (See Kendall and Stuart, 1967, Chap. 29.)

If the ratio between the error variances $\lambda = \tau_2/\tau_1$ is known, then the parameters of both the functional and structural relationships can be estimated using maximum likelihood. In both cases it is appropriate to use the line through the sample mean $(\bar{x}_1, \bar{x}_2)$ whose slope is given by the first principal component of the covariance matrix of $(\lambda^{1/2}x_1, x_2)$. (With this scaling the error variance is spherically symmetric.) Explicit formulae for $\hat{\alpha}$ and $\hat{\beta}$ are given in Exercise 8.2.8.

Remarks (1) Unfortunately, if the error variances are completely unknown, it is *not possible* to estimate the parameters. In particular, maximum likelihood estimation breaks down in both cases. In the functional relationship case the likelihood becomes infinite for a wide choice of parameter values, whereas in the structural relationship case there is an inherent indeterminacy in the parameters; that is, different parameter values can produce the same distribution for the observations.

Thus, in this situation more information is needed before estimation can be carried out. For example, if it is possible to make replicated observations, then information can be obtained about the error variances.

Alternatively, it may be possible to make observations on new variables which also depend linearly on the unknown ξ. The use of one additional variable x_3 in a structural relationship was discussed in Example 7.2.1 in terms of instrumental variable estimation.

(2) More generally, the linear relationship in (8.2.11) can be extended to any number of dimensions $p>2$ to express variables $x_1, \ldots, x_p$ each as a linear function of an unknown ξ plus an error term, where all the error terms are independent of one another. If the ratios between the error variances are known, then the above comments on the use of the first principal component remain valid, but again, with unknown error variances, the estimation of the parameters of the functional relationship breaks down here also. (However, the structural relationship with unknown error variances can be considered as a special case of factor analysis, and estimation of the parameters is now possible. See Chapter 9, especially Exercise 9.2.7.)

8.2.4 Correlation structure

We now examine the correlations between $\mathbf{x}$ and its vector of principal components, $\mathbf{y}$, defined in Definition 8.2.1. For simplicity we assume that

x (and therefore **y**) has mean zero. The covariance between **x** and **y** is then

$$E(\mathbf{xy}') = E(\mathbf{xx}'\mathbf{\Gamma}) = \mathbf{\Sigma\Gamma} = \mathbf{\Gamma\Lambda\Gamma}'\mathbf{\Gamma} = \mathbf{\Gamma\Lambda}.$$

Therefore the covariance between x_i and y_j is $\gamma_{ij}\lambda_j$. Now x_i and y_i have variances σ_{ii} and λ_j, respectively, so if their correlation is ρ_{ij} then

$$\rho_{ij} = \gamma_{ij}\lambda_j/(\sigma_{ii}\lambda_j)^{1/2} = \gamma_{ij}(\lambda_j/\sigma_{ii})^{1/2}. \tag{8.2.12}$$

When $\mathbf{\Sigma}$ is a correlation matrix, $\sigma_{ii}=1$ so $\rho_{ij} = \gamma_{ij}\sqrt{\lambda_j}$.

We may say that the proportion of variation of x_i "explained" by y_j is ρ_{ij}^2. Then since the elements of **y** are uncorrelated, any set G of components explains a proportion

$$\rho_{iG}^2 = \sum_{j\in G} \rho_{ij}^2 = \frac{1}{\sigma_{ii}} \sum_{j\in G} \lambda_j\gamma_{ij}^2 \tag{8.2.13}$$

of the variation in x_i. The denominator of this expression represents the variation in x_i which is to be explained, and the numerator gives the amount explained by the set G. When G includes all the components the numerator is the (i, i)th element of $\mathbf{\Gamma\Lambda\Gamma}'$, which is of course just σ_{ii}, so that the ratio in (8.2.13) is one.

Note that the part of the *total variation* accounted for by the components in G can be expressed as the sum over all p variables of the proportion of *variation in each variable* explained by the components in G, where each variable is weighted by its variance; that is,

$$\sum_{j\in G} \lambda_j = \sum_{i=1}^{p} \sigma_{ii}\rho_{iG}^2. \tag{8.2.14}$$

See Exercise 8.2.9.

8.2.5 The effect of ignoring some components

In the open/closed book data of Example 8.2.3, the final two components explain barely 10% of the total variance. It therefore seems fair to ask what if anything would be lost by ignoring these components completely. This question can be broken down into several parts, by analogy with those considered in Section 8.2.4.

First we might ask what are the values of the correlation coefficients, which by analogy with (8.2.12) are given by

$$r_{ij}^2 = l_j g_{ij}^2/s_{ii}.$$

Calculations for the open/closed book data lead to the following values for r_{ij} and r_{ij}^2. (Note that the rows of the second table sum to 1 except for rounding error.)

	r_{ij}	Component j 1	2	3	4	5
	1	0.758	0.609	−0.175	0.156	0.026
	2	0.734	0.224	0.322	0.548	0.081
Variable i	3	0.853	−0.136	0.139	−0.003	−0.493
	4	0.796	−0.288	0.409	0.321	0.109
	5	0.812	−0.451	−0.354	−0.094	0.050

	r_{ij}^2	Component j 1	2	3	4	5
	1	0.574	0.371	0.030	0.024	0.001
	2	0.539	0.050	0.104	0.300	0.007
Variable i	3	0.727	0.018	0.019	0.000	0.243
	4	0.634	0.083	0.168	0.103	0.012
	5	0.660	0.204	0.125	0.009	0.002

In other words, the first component for this data explains $(0.758)^2 = 57.4\%$ of x_1, 53.9% of x_2, 72.7% of x_3, 63.4% of x_4, and 66.0% of x_5. The first two components taken together explain 94.5% of x_1, etc. Note also that whereas the last component explains only a small proportion of the variables 1, 2, 4, and 5, it accounts for $(0.493)^2 = 24.3\%$ of the variation in variable 3. Hence "throwing away" the last component is in fact rejecting much more information about variable 3 than it is about the others. Similarly, "throwing away" the last two components rejects most about variables 2 and 3, and least about variable 5.

Another example using these correlations is given in Exercise 8.2.2.

A common practical way of looking at the contributions of various principal components (due to Cattell, 1966) is to look at a "scree graph" such as Figure 8.2.1 (that is, one plots λ_j versus j). Such a diagram can often indicate clearly where "large" eigenvalues cease and "small" eigenvalues begin.

Various other "rules of thumb" for excluding principal components exist, including the following:

(a) include just enough components to explain say 90% of the total variation;
(b) (Kaiser) exclude those principal components whose eigenvalues are less than the average, i.e. less than one if a correlation matrix has been used.

It has been suggested that when $p \leq 20$, then Kaiser's criterion (b) tends to include too few components. Similarly, Cattell's scree test is said to include too many components. Thus a compromise is often used in practice.

The isotropy test (Section 8.4.3) can also give some indication of the number of components to include.

Example 8.2.6 Consider the Jeffers' (1967) pine pitprop data of Example 6.7.1. A principal component analysis of the correlation matrix yields the eigenvalues and eigenvectors given in Table 8.2.1 together with the associated percentages and cumulative percentages of variance explained. A look at Cattell's scree graph (Figure 8.2.1) suggests that three or six components should be retained whereas Kaiser's criterion suggests the number four. Note that seven components are needed to explain 90% of the variance. Jeffers decided to retain the first six components.

In this example the first six principal components turn out to have natural physical interpretations. For each component j consider those variables which have relatively high positive or negative weighting (say the variables i for which $g_{ij} \geq 0.70 \max_k g_{kj}$) as constituting an index of the combined action, or contrast, of the original variables. (However, it must be warned that this approach does not always lead to meaningful interpretations.) The first component weights highly variables

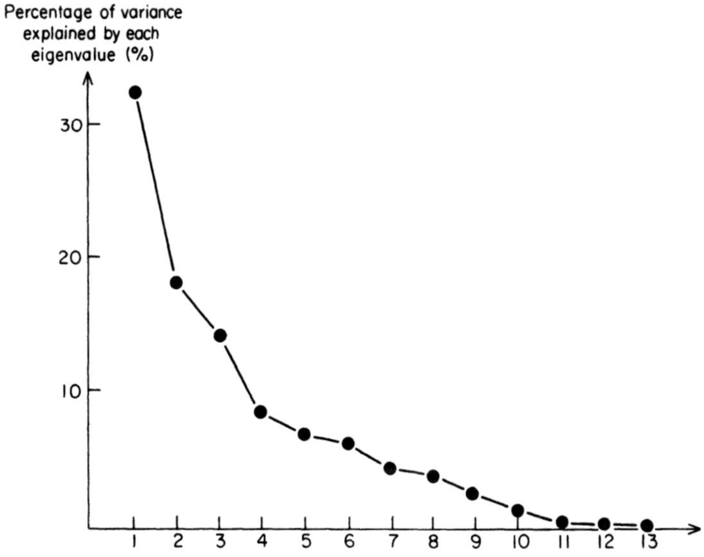

Figure 8.2.1 Scree graph for the Corsican pitprop data of Table 8.2.1.

Table 8.2.1 Eigenvectors and eigenvalues for the pitprop data

	Variable												
Eigenvector	$\mathbf{g}_{(1)}$	$\mathbf{g}_{(2)}$	$\mathbf{g}_{(3)}$	$\mathbf{g}_{(4)}$	$\mathbf{g}_{(5)}$	$\mathbf{g}_{(6)}$	$\mathbf{g}_{(7)}$	$\mathbf{g}_{(8)}$	$\mathbf{g}_{(9)}$	$\mathbf{g}_{(10)}$	$\mathbf{g}_{(11)}$	$\mathbf{g}_{(12)}$	$\mathbf{g}_{(13)}$
x_1	−0.40	0.22	−0.21	−0.09	−0.08	0.12	−0.11	0.14	0.33	−0.31	−0.00	0.39	−0.57
x_2	−0.41	0.19	−0.24	−0.10	−0.11	0.16	−0.08	0.02	0.32	−0.27	−0.05	−0.41	0.58
x_3	−0.12	0.54	0.14	0.08	0.35	−0.28	−0.02	0.00	−0.08	0.06	0.12	0.53	0.41
x_4	−0.17	0.46	0.35	0.05	0.36	−0.05	0.08	−0.02	−0.01	0.10	−0.02	−0.59	−0.38
x_5	−0.06	−0.17	0.48	0.05	0.18	0.63	0.42	−0.01	0.28	−0.00	0.01	0.20	0.12
x_6	−0.28	−0.01	0.48	−0.06	−0.32	0.05	−0.30	0.15	−0.41	−0.10	−0.54	0.08	0.06
x_7	−0.40	−0.19	0.25	−0.07	−0.22	0.00	−0.23	0.01	−0.13	0.19	0.76	−0.04	0.00
x_8	−0.29	−0.19	−0.24	0.29	0.19	−0.06	0.40	0.64	−0.35	−0.08	0.03	−0.05	0.02
x_9	−0.36	0.02	−0.21	0.10	−0.10	0.03	0.40	−0.70	−0.38	−0.06	−0.05	0.05	−0.06
x_{10}	−0.38	−0.25	−0.12	−0.21	0.16	−0.17	0.00	−0.01	0.27	0.71	−0.32	0.06	0.00
x_{11}	0.01	0.21	−0.07	0.80	−0.34	0.18	−0.14	0.01	0.15	0.34	−0.05	0.00	−0.01
x_{12}	0.12	0.34	0.09	−0.30	−0.60	−0.17	0.54	0.21	0.08	0.19	0.05	0.00	0.00
x_{13}	0.11	0.31	−0.33	−0.30	0.08	0.63	−0.16	0.11	−0.38	0.33	0.04	0.01	−0.01
Eigenvalue, λ_i	4.22	2.38	1.88	1.11	0.91	0.82	0.58	0.44	0.35	0.19	0.05	0.04	0.04
% of eigenvalues	32.5	18.3	14.5	8.5	7.0	6.3	4.5	3.4	2.7	1.5	0.4	0.3	0.3
Cumulative %	32.5	50.8	65.2	73.8	80.8	87.1	91.5	94.4	97.6	99.1	99.5	99.8	100

1, 2, 6, 7, 8, 9, and 10, and represents the overall size of the prop. The second component, weighting highly 3 and 4, measures the degree of seasoning. The third component, weighting 5 and 6, is a measure of the rate of growth of the timber. The fourth, fifth, and sixth components are most easily described as a measure of variable 11, a measure of variable 12, and an average of variables 5 and 13.

8.2.6 Graphical representation of principal components

The reduction in dimensions afforded by principal component analysis can be used graphically. Thus if the first two components explain "most" of the variance, then a scattergram showing the distribution of the objects on these two dimensions will often give a fair indication of the overall distribution of the data. Such plots can augment the graphical methods of Section 1.7 on the original variables.

Figure 8.2.2 shows the 88 individuals of open/closed book data plotted on the first two principal components of the covariance matrix (Example 8.2.3). Note that the variance along the y_1 axis is greater than the variance along the y_2 axis.

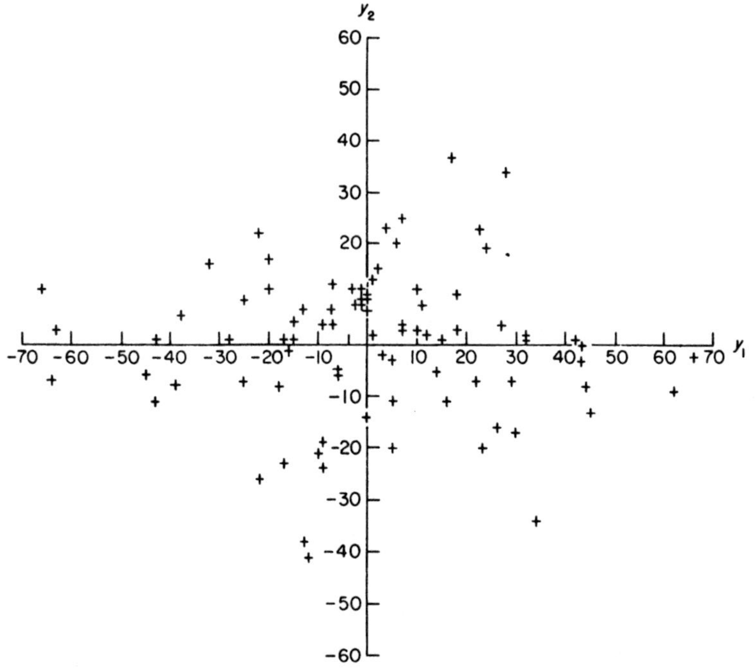

Figure 8.2.2 The 88 individuals of the open/closed book data plotted on the first two principal components.

As well as showing the distribution of *objects* on the first two principal components, the *variables* can also be represented on a similar diagram, using as coordinates the correlations derived in Section 8.2.5.

Example 8.2.7 The values of r_{ij} for the open/closed book data were calculated in Section 8.2.5. Hence variables 1–5 can be plotted on the first two principal axes, giving Figure 8.2.3. Note that all of the plotted points lie inside the unit circle and variables 1 and 5, which lie fairly close to the unit circle, are explained better by the first two principal components than variables 2, 3, 4.

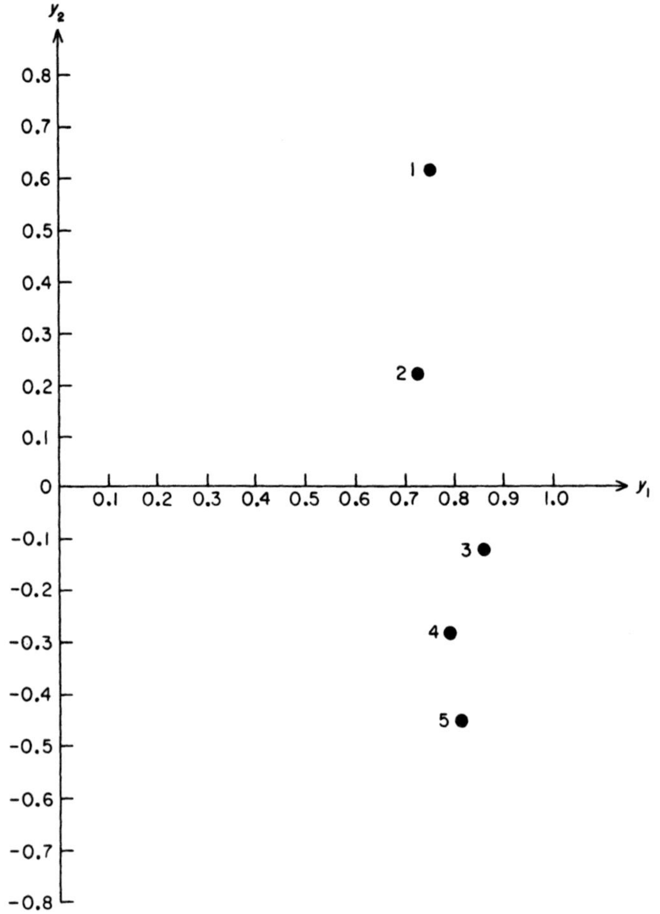

Figure 8.2.3 Correlations between the variables and principal components for the open/closed book data.

8.3 Sampling Properties of Principal Components

8.3.1 Maximum likelihood estimation for normal data

The small-sample distribution of eigenvalues and eigenvectors of a covariance matrix **S** is extremely complicated even when all parent correlations vanish. One reason is that the eigenvalues are *non-rational* functions of the elements of **S**. However, some large sample results are known, and many useful properties of the sample principal\components for normal data stem from the following maximum likelihood results.

Theorem 8.3.1 *For normal data when the eigenvalues of* $\boldsymbol{\Sigma}$ *are distinct, the sample principal components and eigenvalues are the maximum likelihood estimators of the corresponding population parameters.*

Proof The principal components and eigenvalues of $\boldsymbol{\Sigma}$ are related to $\boldsymbol{\Sigma}$ by means of a one-to-one function, except when $\boldsymbol{\Sigma}$ has coincident eigenvalues. Hence the theorem follows from the invariance of maximum likelihood estimators under one-to-one transformations. ■

When the eigenvalues of $\boldsymbol{\Sigma}$ are not distinct the above theorem does not hold. (This can easily be confirmed for the case $\boldsymbol{\Sigma}=\sigma^2\mathbf{I}$.) In such cases there is a certain arbitrariness in defining the eigenvectors of $\boldsymbol{\Sigma}$. Even if this is overcome, the eigenvalues of **S** are in general distinct, so the eigenvalues and eigenvectors of **S** would not be the same function of **S** as the eigenvalues and eigenvectors of $\boldsymbol{\Sigma}$ are of $\boldsymbol{\Sigma}$. Hence when the eigenvectors of $\boldsymbol{\Sigma}$ are not distinct, the sample principal components are *not* maximum likelihood estimators of their population counterparts. However, in such cases the following modified result may be used.

Theorem 8.3.2 *For normal data, when* $k>1$ *eigenvalues of* $\boldsymbol{\Sigma}$ *are* not *distinct, but take the common value* $\bar{\lambda}$, *then*

(a) *the m.l.e of* $\bar{\lambda}$ *is* $\bar{l}$, *the arithmetic mean of the corresponding sample eigenvalues, and*
(b) *the sample eigenvectors corresponding to the repeated eigenvalue are m.l.e.s, although they are not the unique m.l.e.s.*

Proof See Anderson (1963), and Kshirsagar (1972, p. 439). ■

Although we shall not prove the above theorem here, we may note that (a) appears sensible in the light of Theorem 8.3.1, since if λ_i is a distinct eigenvalue then the m.l.e. of λ_i is l_i, and since if the last k eigenvalues are equal to $\bar{\lambda}$, then

$$\operatorname{tr}\boldsymbol{\Sigma}=\lambda_1+\ldots+\lambda_{p-k}+k\bar{\lambda}$$

and

$$\operatorname{tr} \mathbf{S} = l_1 + \ldots + l_{p-k} + k\bar{l}.$$

Part (b) of the theorem also appears sensible in the light of Theorem 8.3.1.

Example 8.3.1 Let $\mathbf{\Sigma} = (1-\rho)\mathbf{I} + \rho\mathbf{1}\mathbf{1}'$, $\rho > 0$. We know from Example 8.2.2 that the largest eigenvalue of $\mathbf{\Sigma}$ is $\lambda_1 = 1 + (p-1)\rho$ and that the other $(p-1)$ eigenvalues all take the value $\bar{\lambda} = (1-\rho)$.

Suppose we assume nothing about $\mathbf{\Sigma}$ except that the smallest $p-1$ eigenvalues are equal. Then the m.l.e. of λ_1 is l_1, the first sample eigenvalue of $\mathbf{S}$, and the m.l.e. of $\bar{\lambda}$ is

$$\bar{l} = (\operatorname{tr} \mathbf{S} - l_1)/(p-1).$$

Note that the corresponding estimates of ρ, namely

$$(l_1 - 1)/(p-1) \quad \text{and} \quad 1 - \bar{l} \tag{8.3.2}$$

are not equal. (See Exercise 8.3.3.)

8.3.2 Asymptotic distributions for normal data

For large samples the following asymptotic result, based on the central limit theorem, provides useful distributions for the eigenvalues and eigenvectors of the sample covariance matrix.

Theorem 8.3.3 (Anderson, 1963) *Let $\mathbf{\Sigma}$ be a positive-definite matrix with distinct eigenvalues. Let $\mathbf{M} \sim W_p(\mathbf{\Sigma}, m)$ and set $\mathbf{U} = m^{-1}\mathbf{M}$. Consider spectral decompositions $\mathbf{\Sigma} = \mathbf{\Gamma}\mathbf{\Lambda}\mathbf{\Gamma}'$ and $\mathbf{U} = \mathbf{G}\mathbf{L}\mathbf{G}'$, and let $\boldsymbol{\lambda}$ and $\mathbf{l}$ be the vectors of diagonal elements in $\mathbf{\Lambda}$ and $\mathbf{L}$. Then the following asymptotic distributions hold as $m \to \infty$:*

(a) $\mathbf{l} \sim N_p(\boldsymbol{\lambda}, 2\mathbf{\Lambda}^2/m)$; *that is the eigenvalues of $\mathbf{U}$ are asymptotically normal, unbiased, and independent, with l_i having variance $2\lambda_i^2/m$.*

(b) $\mathbf{g}_{(i)} \sim N_p(\boldsymbol{\gamma}_i, \mathbf{V}_i/m)$, *where*

$$\mathbf{V}_i = \lambda_i \sum_{j \neq i} \frac{\lambda_j}{(\lambda_j - \lambda_i)^2} \boldsymbol{\gamma}_{(j)}\boldsymbol{\gamma}'_{(j)};$$

that is, the eigenvectors of $\mathbf{U}$ are asymptotically normal and unbiased, and have the stated asymptotic covariance matrix $\mathbf{V}_i/m$.

(c) *The covariance between the rth element of $\mathbf{g}_{(i)}$ and the sth element of $\mathbf{g}_{(j)}$ is $-\lambda_i\lambda_j\gamma_{rj}\gamma_{si}/m(\lambda_i - \lambda_j)^2$.*

(d) *The elements of $\mathbf{l}$ are asymptotically independent of the elements of $\mathbf{G}$.*

Proof We shall only give the proof of part (a) here, and we shall use the $O_p(\cdot)$ and $o_p(\cdot)$ notation given in (2.9.1) and (2.9.2).

If $\mathbf{M} \sim W_p(\mathbf{\Sigma}, m)$, then $\mathbf{M}^* = \mathbf{\Gamma}'\mathbf{M}\mathbf{\Gamma} \sim W_p(\mathbf{\Lambda}, m)$; that is, $\mathbf{M}^*$ has a Wishart distribution with diagonal scale matrix. Set $\mathbf{U}^* = m^{-1}\mathbf{M}^*$. Then using Exercise 3.4.17 it is easily seen that $E(u_{ii}^*) = \lambda_i$, $V(u_{ii}^*) = 2m^{-1}\lambda_i^2$, $E(u_{ij}^*) = 0$, $V(u_{ij}^*) = m^{-1}\lambda_i\lambda_j$, and $C(u_{ii}^*, u_{jj}^*) = 0$ *for* $i \neq j$. Since $\mathbf{M}^*$ can be represented as the sum of m independent matrices from $W_p(\mathbf{\Lambda}, 1)$, it follows from the central limit theorem that if we write $\mathbf{u}^* = (u_{11}^*, \ldots, u_{pp}^*)'$ and $\boldsymbol{\lambda} = (\lambda_1, \ldots, \lambda_p)'$, then

$$m^{1/2}(\mathbf{u}^* - \boldsymbol{\lambda}) \xrightarrow{D} N_p(\mathbf{0}, 2\mathbf{\Lambda}^2), \tag{8.3.3}$$

and

$$m^{1/2}u_{ij}^* \xrightarrow{D} N_1(0, \lambda_i\lambda_j), \; i \neq j.$$

Hence

$$u_{ii}^* = \lambda_i + O_p(m^{-1/2}), \qquad u_{ij}^* = O_p(m^{-1/2}), \qquad i \neq j. \tag{8.3.4}$$

Thus, the roots, $l_1 > \ldots > l_p$, of the polynomial $|\mathbf{U}^* - l\mathbf{I}|$ converge in probability to the roots, $\lambda_1 > \ldots > \lambda_p$, of the polynomial $|\mathbf{\Lambda} - l\mathbf{I}|$; that is

$$u_{ii}^* - l_i \xrightarrow{P} 0, \qquad i = 1, \ldots, p.$$

To get more precise bounds on $u_{ii}^* - l_i$, use (8.3.4) in the standard asymptotic expansion for determinants (A.2.3a) to write

$$|\mathbf{U}^* - v\mathbf{I}| = \prod_{j=1}^{p} (u_{jj}^* - v) + O_p(m^{-1}), \tag{8.3.5}$$

where v is any random variable bounded above in probability $v = O_p(1)$. In particular, because $l_i \leq \operatorname{tr} \mathbf{U}^* = \operatorname{tr} \mathbf{U}$ for all $i = 1, \ldots, p$, it is clear from (8.3.4) that $l_i = O_p(1)$. Since $l_1, \ldots, l_p$ are the eigenvalues of $\mathbf{U}^*$, setting $v = l_i$ in (8.3.5) yields

$$\prod_{j=1}^{p} (u_{jj}^* - l_i) = O_p(m^{-1}), \qquad i = 1, \ldots, p.$$

Now the true eigenvalues λ_i are distinct so for $i \neq j$, $u_{jj}^* - l_i$ is bounded in probability away from 0. Hence the nearness of the above product to 0 is due to only one factor, $u_{ii} - l_i$; that is

$$u_{ii}^* = l_i + O_p(m^{-1}), \qquad i = 1, \ldots, p.$$

Inserting this result in (8.3.3) gives

$$m^{1/2}(\mathbf{l}-\boldsymbol{\lambda}) = m^{1/2}(\mathbf{u}^* - \boldsymbol{\lambda}) + O_p(m^{-1/2}) \xrightarrow{D} N_p(\mathbf{0}, 2\boldsymbol{\Lambda}^2)$$

because the second term, $O_p(m^{-1/2})$, converges in probability to $\mathbf{0}$. Hence part (a) is proved. ■

For a sample of size n from a multinormal distribution, the unbiased covariance matrix $\mathbf{S}_u = (n-1)^{-1}n\mathbf{S}$ satisfies the assumptions on $\mathbf{U}$ in the above theorem with $m = n-1$ degrees of freedom. Thus, the eigenvalues $l_1 > \ldots > l_p$ of $\mathbf{S}_u$ with eigenvectors $\mathbf{g}_{(1)}, \ldots, \mathbf{g}_{(p)}$ have the asymptotic distribution stated in the theorem. Of course, asymptotically, it does not matter whether we work with $\mathbf{S}$ or $\mathbf{S}_u$.

Example 8.3.3 Let $l_1, \ldots, l_p$ denote the eigenvalues of $\mathbf{S}_u$ for a sample of size n. From part (a) of the theorem we deduce that, asymptotically, $l_i \sim N(\lambda_i, 2\lambda_i^2/(n-1))$. Therefore, using Theorem 2.9.2, asymptotically,

$$\log l_i \sim N(\log \lambda_i, 2/(n-1)).$$

This leads to the asymptotic confidence limits

$$\log \lambda_i = \log l_i \pm z(2/(n-1))^{1/2},$$

where z is the relevant critical value from the standard normal distribution.

For instance, consider the open/closed book data from Example 8.2.3 where $n = 88$. Putting $z = 1.96$ gives $(l_i e^{-0.297}, l_i e^{0.297})$ as an asymptotic 95% confidence interval for λ_i. Use of the five sample eigenvalues of $\mathbf{S}_u$, 687.0, 202.1, 103.7, 84.6, and 32.2, leads to the following intervals:

$$(510, 925), \quad (150, 272), \quad (77, 140), \quad (63, 114), \quad \text{and} \quad (24, 43).$$

Example 8.3.4 Theorem 8.3.3 can also be used for testing hypotheses on the eigenvectors of $\boldsymbol{\Sigma}$. Consider, for example, the null hypothesis that the ith eigenvector $\boldsymbol{\gamma}_{(i)}$ takes a given value $\mathbf{r}$, where $\mathbf{r}'\mathbf{r} = 1$ and we suppose that λ_i has multiplicity 1. If the null hypothesis $\boldsymbol{\gamma}_{(i)} = \mathbf{r}$ is true, then, by part (b) of Theorem 8.3.3, asymptotically,

$$(n-1)^{1/2}(\mathbf{g}_{(i)} - \mathbf{r}) \sim N(\mathbf{0}, \mathbf{V}_i).$$

Note that $\mathbf{V}_i$ has rank $(p-1)$; its eigenvectors are $\boldsymbol{\gamma}_{(i)}$ with eigenvalue 0 and $\boldsymbol{\gamma}_{(j)}$ with eigenvalue $(\lambda_i\lambda_j^{-1} + \lambda_i^{-1}\lambda_j - 2)^{-1}$ for $j \neq i$. Thus, (see Section A.8), a g-inverse $\mathbf{V}_i^-$ is defined by the same eigenvectors, and by eigenvalues zero and $\lambda_i\lambda_j^{-1} + \lambda_i^{-1}\lambda_j - 2$; that is,

$$\mathbf{V}_i^- = \sum_{j \neq i} (\lambda_i\lambda_j^{-1} + \lambda_i^{-1}\lambda_j - 2)\boldsymbol{\gamma}_{(j)}\boldsymbol{\gamma}_{(j)}' = \lambda_i\boldsymbol{\Sigma}^{-1} + \lambda_i^{-1}\boldsymbol{\Sigma} - 2\mathbf{I}.$$

Hence, by a modified version of Theorem 2.5.2 on Mahalanobis distances (see specifically Exercise 3.4.11), asymptotically,

$$(n-1)(\mathbf{g}_{(i)}-\mathbf{r})'(\lambda_i\boldsymbol{\Sigma}^{-1}+\lambda_i^{-1}\boldsymbol{\Sigma}-2\mathbf{I})(\mathbf{g}_{(i)}-\mathbf{r})\sim\chi^2_{p-1}. \qquad (8.3.6)$$

Let $\mathbf{W}_i^- = l_i\mathbf{S}^{-1}+l_i^{-1}\mathbf{S}-2\mathbf{I}$ denote the sample version of $\mathbf{V}_i^-$. Since $(n-1)^{1/2}(\mathbf{g}_{(i)}-\mathbf{r})=O_p(1)$ and $\mathbf{V}_i^-=\mathbf{W}_i^-+o_p(1)$, the limiting distribution in (8.3.6) remains the same if we replace $\mathbf{V}_i^-$ by $\mathbf{W}_i^-$. As $\mathbf{g}_{(i)}$ is an eigenvector of $\mathbf{W}_i^-$ with eigenvalue 0, (8.3.6) implies, asymptotically,

$$(n-1)(l_i\mathbf{r}'\mathbf{S}^{-1}\mathbf{r}+l_i^{-1}\mathbf{r}'\mathbf{S}\mathbf{r}-2)\sim\chi^2_{p-1}. \qquad (8.3.7)$$

8.4 Testing Hypotheses about Principal Components

8.4.1 Introduction

It is often useful to have a procedure for deciding whether the first k principal components include all the important variation in $\mathbf{x}$. Clearly one would be happy to ignore the remaining $(p-k)$ components if their corresponding population eigenvalues were all zero. However, this happens only if $\boldsymbol{\Sigma}$ has rank k, in which case $\mathbf{S}$ must also have rank k. Hence this situation is only trivially encountered in practice.

A second possible alternative would be to test the hypothesis that the proportion of variance explained by the last $(p-k)$ components is less than a certain value. This possibility is examined in Section 8.4.2.

A more convenient hypothesis is the question whether the last $(p-k)$ eigenvalues are equal. This would imply that the variation is equal in all directions of the $(p-k)$-dimensional space spanned by the last $(p-k)$ eigenvectors. This is the situation of *isotropic variation*, and would imply that if one component is discarded, then all the last k components should be discarded. The hypothesis of isotropic variation is examined in Section 8.4.3.

In all these sections we assume that we are given a random sample of normal data of size n.

8.4.2 The hypothesis that $(\lambda_1+\ldots+\lambda_k)/(\lambda_1+\ldots+\lambda_p)=\psi$

Let $l_1,\ldots,l_p$ be the eigenvalues of $\mathbf{S}_u$ and denote the sample counterpart of ψ by

$$\hat{\psi}=(l_1+\ldots+l_k)/(l_1+\ldots+l_p).$$

We seek the asymptotic distribution of $\hat{\psi}$. From Theorem 8.3.3 we know that the elements of $\mathbf{l}$ are asymptotically normal. Using Theorem 2.9.2

with $\mathbf{t}=\mathbf{l}$, $\boldsymbol{\mu}=\boldsymbol{\lambda}$, $f(\mathbf{t})=\Psi(\mathbf{1})$ and n replaced by $n-1$, we see that $\mathbf{V}=2\mathbf{\Lambda}^2$, and

$$d_i=\frac{\partial\hat{\psi}}{\partial l_i}=\begin{cases}(1-\psi)/\operatorname{tr}\mathbf{\Sigma} & \text{for } i=1,\ldots,k,\\ -\psi/\operatorname{tr}\mathbf{\Sigma} & \text{for } i=k+1,\ldots,p.\end{cases}$$

Therefore $\hat{\psi}$ is asymptotically normal with mean $f(\boldsymbol{\lambda})=\psi$, and with variance

$$\tau^2=\frac{2}{(n-1)(\operatorname{tr}\mathbf{\Sigma})^2}\{(1-\psi)^2(\lambda_1^2+\ldots+\lambda_k^2)+\psi^2(\lambda_{k+1}^2+\ldots+\lambda_p^2)\}$$

$$=\frac{2\operatorname{tr}\mathbf{\Sigma}^2}{(n-1)(\operatorname{tr}\mathbf{\Sigma})^2}(\psi^2-2\alpha\psi+\alpha), \tag{8.4.1}$$

where $\alpha=(\lambda_1^2+\ldots+\lambda_k^2)/(\lambda_1^2+\ldots+\lambda_p^2)$. Note that α may be interpreted as the proportion of variance explained by the first k principal components of a variable whose covariance matrix is $\mathbf{\Sigma}^2$. Results similar to the above have also been obtained by Sugiyama and Tong (1976), and Kshirsagar (1972, p. 454). Notice that since we are only interested in ratios of eigenvalues, it makes no difference whether we work with the eigenvalues of $\mathbf{S}$ or of $\mathbf{S}_u$.

Example 8.4.1 Equation (8.4.1) may be used to derive approximate confidence intervals for ψ. This may be illustrated using the open/closed book data. Let $k=1$ so that we seek a confidence interval for the proportion of variance explained by the first principal component. Here $n=88$ and we have,

$$\operatorname{tr}\mathbf{S}_u=l_1+\ldots+l_p=1109.6,\qquad \operatorname{tr}\mathbf{S}_u^2=l_1^2+\ldots+l_p^2=5.3176\times10^5,$$

$$\hat{\psi}=687.0/\operatorname{tr}\mathbf{S}_u=0.619.$$

If we estimate α and τ^2 by $\hat{\alpha}$ and $\hat{\tau}^2$, using the sample eigenvalues $l_1,\ldots,l_p$, then we find

$$\hat{\alpha}=0.888,\qquad \hat{\tau}^2=0.148/87=0.0017.$$

Thus, the standard error of $\hat{\psi}$ is $(0.0017)^{1/2}=0.041$, and an approximate 95% confidence interval is given by

$$0.619\pm1.96(0.041)=(0.539, 0.699).$$

In other words, although the point estimate seemed to indicate that the first principal component explains 61% of the variation, the 95% confidence interval suggests that the true value for the population lies between 54% and 70%.

8.4.3 The hypothesis that $(p-k)$ eigenvalues of $\mathbf{\Sigma}$ are equal

It should be noted that any non-zero eigenvalue, however small, is "significantly" different from zero, because if $\lambda_p = 0$ then the scatter of points lies in a lower-dimensional hyperplane and so $l_p = 0$ with probability one. Hence a test of the hypothesis $\lambda_p = 0$ has no meaning. However, a test of whether $\lambda_p = \lambda_{p-1}$ is meaningful. This would be testing the hypothesis that the scatter is partially isotropic in a two-dimensional subspace.

We consider the more general hypothesis that $\lambda_p = \lambda_{p-1} = \ldots = \lambda_{k+1}$. The isotropy test is useful in determining the number of principal components to use to describe the data. Suppose we have decided to include at least k components, and we wish to decide whether or not to include any more. The acceptance of the null hypothesis implies that if we are to include more than k principal components we should include all p of them because each of the remaining components contains the same amount of information.

Often one conducts a sequence of isotropy tests, starting with $k = 0$ and increasing k until the null hypothesis is accepted.

The likelihood ratio for this hypothesis can be found in the same manner as that used for the one-sample hypotheses on $\mathbf{\Sigma}$ which were discussed in Chapter 5. As Theorem 5.3.2 indicated, the LRT is given by

$$-2 \log \lambda = np(a - 1 - \log g),$$

where a and g are the arithmetic and geometric means of the eigenvalues of $\hat{\mathbf{\Sigma}}^{-1}\mathbf{S}$, where $\hat{\mathbf{\Sigma}}$ is the m.l.e. of $\mathbf{\Sigma}$ under the null hypothesis. Now in our case, by virtue of Theorem 8.3.2, $\hat{\mathbf{\Sigma}}$ and $\mathbf{S}$ have the same eigenvectors, and if the eigenvalues of $\mathbf{S}$ are $(l_1, \ldots, l_p)$, then the eigenvalues of $\hat{\mathbf{\Sigma}}$ are $(l_1, \ldots, l_k, a_0, \ldots, a_0)$, where $a_0 = (l_{k+1} + \ldots + l_p)/(p-k)$ denotes the arithmetic mean of the sample estimates of the repeated eigenvalue. (In our problem, we assume that the repeated eigenvalues come last, although there is no need for this to be so.) It follows that the eigenvalues of $\hat{\mathbf{\Sigma}}^{-1}\mathbf{S}$ are $(1, \ldots, 1, l_{k+1}/a_0, \ldots, l_p/a_0)$. Therefore $a = 1$ and $g = (g_0/a_0)^{(p-k)/p}$, where $g_0 = (l_{k+1} \times \ldots \times l_p)^{1/(p-k)}$ is the geometric mean of the sample estimates of the repeated eigenvalue. Inserting these values in the above equation we deduce that

$$-2 \log \lambda = n(p-k) \log (a_0/g_0). \tag{8.4.2}$$

Now the number of degrees of freedom in the asymptotic distribution of $-2 \log \lambda$ might appear to be $(p-k-1)$, the difference in the number of eigenvalues. However, this is not so because H_0 also affects the number of orthogonality conditions which $\mathbf{\Sigma}$ must satisfy. In fact (see Exercise

8.4.1) the relevant number of degrees of freedom is $\frac{1}{2}(p-k+2)(p-k-1)$. Also, a better chi-squared approximation is obtained if n in (8.4.2) is replaced by $n' = n-(2p+11)/6$ to give Bartlett's approximation,

$$\left(n-\frac{2p+11}{6}\right)(p-k)\log\left(\frac{a_0}{g_0}\right)\sim\chi^2_{(p-k+2)(p-k-1)/2}, \tag{8.4.3}$$

asymptotically. Note that the ratio a_0/g_0 is the same whether we work with the eigenvalues of $\mathbf{S}$ or $\mathbf{S}_u$.

Example 8.4.2 The three-dimensional data from Jolicoeur and Mosimann (1960) relating to $n = 24$ male turtles given in Example 5.3.4 may be used to test the hypothesis that $\lambda_2 = \lambda_3$. Using $\mathbf{S}_1$ given in Example 5.3.4, it is found that $l_1 = 187.14$, $l_2 = 3.54$, $l_3 = 1.05$. The values of a_0 and g_0 in (8.4.2) are

$$a_0 = \tfrac{1}{2}(3.54+1.05) = 2.29, \qquad g_0 = (3.54\times 1.05)^{1/2} = 1.93.$$

Also, $n' = 24-17/6 = 21.17$. Thus, the test statistic is given by

$$-2\log\lambda = (21.17)(2)\log(2.29/1.93) = 7.24,$$

which under H_0 has a chi-squared distribution with $\frac{1}{2}(p-k+2)\times(p-k-1) = 2$ degrees of freedom. The observed value lies between the 95th percentile ($=5.99$) and the 99th percentile ($=9.21$).

8.4.4 Hypotheses concerning correlation matrices

When the variables have been standarized, the sample correlation matrix $\mathbf{R}$ may be regarded as an estimate of the population correlation matrix $\mathbf{P}$. The hypothesis that all eigenvalues of $\mathbf{P}$ are equal is equivalent to the hypothesis that $\mathbf{P} = \mathbf{I}$. In (5.3.17) we described a test of this hypothesis due to Box (1949). This uses the statistic $-n'\log|\mathbf{R}|$, where $n' = n-(2p+11)/6$. This statistic has an asymptotic chi-squared distribution, with $\frac{1}{2}p(p-1)$ degrees of freedom.

Considerable difficulties arise if a test is required of the hypothesis that the $(p-k)$ smallest eigenvalues of $\mathbf{P}$ are equal, where $0<k<p-1$. Bartlett (1951) suggested a test statistic similar to (8.4.2), namely

$$(n-1)(p-k)\log(a_0/g_0), \tag{8.4.4}$$

where a_0 and g_0 are the arithmetic and geometric means of the $(p-k)$ smallest eigenvalues of $\mathbf{R}$. (See also Section 5.3.2a.) Unfortunately, this statistic does not have an asymptotic chi-squared distribution. Neverthe-less, if the first k components account for a fairly high proportion of the variance, or if we are prepared to accept a conservative test, we may treat the above expression as a chi-squared statistic with $\frac{1}{2}(p-k+2)(p-k-1)$ degrees of freedom (Dagnelie, 1975, p. 181; Anderson, 1963).

Example 8.4.3 The correlation matrix of the open/closed book data has eigenvalues 3.18, 0.74, 0.44, 0.39, and 0.25. Use of (8.4.4) leads to the following test statistics:

$$\text{for the hypothesis} \quad \lambda_2 = \lambda_3 = \lambda_4 = \lambda_5, \qquad \chi_9^2 = 26.11,$$

$$\text{for the hypothesis} \quad \lambda_3 = \lambda_4 = \lambda_5, \qquad \chi_5^2 = \ 7.30.$$

The former is significant at the 1% level, but the latter is not even significant at the 5% level. Thus we accept the hypothesis $\lambda_3 = \lambda_4 = \lambda_5$.

8.5 Correspondence Analysis

Correspondence analysis is a way of interpreting contingency tables which has several affinities with principal component analysis. We shall introduce this technique in the context of a botanical problem known as "gradient analysis". This concerns the quantification of the notion that certain species of flora prefer certain types of habitat, and that their presence in a particular location can be taken as an indicator of the local conditions. Thus one species of grass might prefer wet conditions, while another might prefer dry conditions. Other species may be indifferent. The classical approach to gradient analysis involves giving each species a "wet-preference score", according to its known preferences. Thus a wet-loving grass may score 10, and a dry-loving grass 1, with a fickle or ambivalent grass perhaps receiving a score of 5. The conditions in a given location may now be estimated by averaging the wet-preference scores of the species that are found there. To formalize this let $\mathbf{X}$ be the $(n \times p)$ one-zero matrix which represents the occurrences of n species in p locations; that is, $x_{ij} = 1$ if species i occurs in location j, and $x_{ij} = 0$ otherwise. If r_i is the wet-preference score allocated to the ith species, then the average wet-preference score of the species found in location j is

$$s_j \propto \sum_i x_{ij} r_i / x_{\cdot j}, \quad \text{where} \quad x_{\cdot j} = \sum_i x_{ij}.$$

This is the estimate of wetness in location j produced by the classical method of gradient analysis.

One drawback of the above method is that the r_i may be highly subjective. However, they themselves could be estimated by playing the same procedure in reverse—if s_j denotes the physical conditions in location j, then r_i could be estimated as the average score of the locations in which the ith species is found; that is

$$r_i \propto \sum_j x_{ij}s_j/x_{i\cdot}, \quad \text{where} \quad x_{i\cdot} = \sum_j x_{ij}.$$

The technique of correspondence analysis effectively takes both the above relationships simultaneously, and uses them to deduce scoring vectors **r** *and* **s** *which satisfy both the above equations. The vectors* **r** *and* **s** *are generated internally by the data, rather than being externally given.*

To see how correspondence analysis is related to principal component analysis, let us write the above equations in matrix form. Letting $\mathbf{A} = \text{diag}\,(x_{i\cdot})$, $\mathbf{B} = \text{diag}\,(x_{\cdot j})$, the above two equations are

$$\mathbf{s} \propto \mathbf{B}^{-1}\mathbf{X}'\mathbf{r}, \qquad \mathbf{r} \propto \mathbf{A}^{-1}\mathbf{X}\mathbf{s}.$$

If both these equations hold simultaneously then

$$\mathbf{r} \propto \mathbf{A}^{-1}\mathbf{X}\mathbf{B}^{-1}\mathbf{X}'\mathbf{r} \tag{8.5.1}$$

and

$$\mathbf{s} \propto \mathbf{B}^{-1}\mathbf{X}'\mathbf{A}^{-1}\mathbf{X}\mathbf{s}. \tag{8.5.2}$$

Therefore **r** is an eigenvector of $\mathbf{A}^{-1}\mathbf{X}\mathbf{B}^{-1}\mathbf{X}'$, a matrix which depends only on the elements of **X**. Similarly **s** is an eigenvector of $\mathbf{B}^{-1}\mathbf{X}'\mathbf{A}^{-1}\mathbf{X}$.

Of course one would like the coefficients of proportionality in each of the above equations to be unity, so that **r** and **s** can be interpreted as arithmetic means. This would lead one to choose **r** and **s** as the eigenvectors corresponding to an eigenvalue of unity. Unfortunately however, although unity is indeed an eigenvalue, the corresponding eigenvectors are each **1**. Hence this requirement would involve all scores being made identical, which is not a very sensible requirement.

Therefore we are led to consider solutions of (8.5.1) and (8.5.2) which involve an eigenvalue $\rho \neq 1$. The correspondence analysis technique is to choose the *largest* eigenvalue less than 1. One then uses the corresponding (right) eigenvectors of $\mathbf{A}^{-1}\mathbf{X}\mathbf{B}^{-1}\mathbf{X}'$ and $\mathbf{B}^{-1}\mathbf{X}'\mathbf{A}^{-1}\mathbf{X}$ to order the wet-preferences of the species and locations, respectively.

It is not difficult to show that if **r** is a standardized eigenvector of $\mathbf{A}^{-1}\mathbf{X}\mathbf{B}^{-1}\mathbf{X}'$ with eigenvalue ρ ($\rho > 0$), then $\mathbf{s} = \rho^{-1/2}\mathbf{B}^{-1}\mathbf{X}'\mathbf{r}$ is a standardized eigenvector of $\mathbf{B}^{-1}\mathbf{X}'\mathbf{A}^{-1}\mathbf{X}$ with the same eigenvalue. Further if $\rho < 1$ then $\mathbf{r}'\mathbf{A}\mathbf{1} = 0$ and $\mathbf{s}'\mathbf{B}\mathbf{1} = 0$. See Exercises 8.5.1–8.5.3.

For more details of the method of correspondence analysis see Hill (1974).

Example 8.5.1 In an archaeological study there are six graves containing five types of pottery between them. It is desired to order the graves and the pottery chronologically. Let $x_{ij} = 1$ if the ith grave contains the jth pottery and 0 otherwise. Suppose

$$\mathbf{X} = \begin{array}{c} \\ A \\ B \\ C \\ D \\ E \\ F \end{array} \begin{array}{c} \begin{array}{ccccc} 1 & 2 & 3 & 4 & 5 \end{array} \\ \begin{bmatrix} 0 & 0 & 1 & 1 & 0 \\ 1 & 1 & 0 & 0 & 1 \\ 0 & 1 & 1 & 1 & 1 \\ 0 & 0 & 1 & 1 & 0 \\ 1 & 0 & 0 & 0 & 1 \\ 1 & 0 & 1 & 1 & 1 \end{bmatrix} \end{array}.$$

The largest eigenvalue of $\mathbf{A}^{-1}\mathbf{X}\mathbf{B}^{-1}\mathbf{X}'$ less than one is 0.54, giving standardized eigenvectors $\mathbf{r}' = (0.515, \ -0.448, \ 0.068, \ 0.515, \ -0.513, \ 0.009)$ and $\mathbf{s}' = (-0.545, \ -0.326, \ 0.476, \ 0.476, \ -0.380)$, thus leading to chronological orders ((A, D), C, F, B, E) and ((3, 4), 2, 5, 1). Re-ordering the columns of $\mathbf{X}$, we get

$$\begin{array}{c} \\ A \\ D \\ C \\ F \\ B \\ E \end{array} \begin{array}{c} \begin{array}{ccccc} 3 & 4 & 2 & 5 & 1 \end{array} \\ \begin{bmatrix} 1 & 1 & 0 & 0 & 0 \\ 1 & 1 & 0 & 0 & 0 \\ 1 & 1 & 1 & 1 & 0 \\ 1 & 1 & 0 & 1 & 1 \\ 0 & 0 & 1 & 1 & 1 \\ 0 & 0 & 0 & 1 & 1 \end{bmatrix} \end{array}.$$

Columns 3 and 4 are indistinguishable, as are rows A and D. Also, the direction of time cannot be determined by this technique. Notice that the 1s are clustered about the line between the upper left-hand corner of the matrix and the lower right-hand corner, agreeing with one's intuition that later graves are associated with more recent types of pottery. Thus the correspondence analysis seems reasonable.

8.6 Allometry—the Measurement of Size and Shape

One problem of great concern to botanists and zoologists is the question of how bodily measurements are best summarized to give overall indications of size and shape. This is the problem of "allometry", which is described in considerable detail by Hopkins (1966), Sprent (1969, p. 30,

1972), and Mosimann (1970). We shall follow the approach of Rao (1971).

Define a *size factor* to be a linear function $s = \mathbf{a}'\mathbf{x}$ such that an increase in s results "on average" in an increase in each element of $\mathbf{x}$. Similarly, a *shape factor* is defined as a linear function $h = \mathbf{b}'\mathbf{x}$ for which an increase in h results "on average" in an increase in some of the elements of $\mathbf{x}$ and in a decrease in others.

The phrase "on average" is to be interpreted in the sense of a regression of $\mathbf{x}$ on s (and h). The covariance between s and x_i is $E(\mathbf{a}'\mathbf{x}x_i) = \mathbf{a}'\boldsymbol{\sigma}_i$, where $\boldsymbol{\sigma}_i$ is the ith row of the covariance matrix $\boldsymbol{\Sigma} = V(\mathbf{x})$. Therefore the regression coefficient of x_i on s is

$$\frac{C(x_i, s)}{V(s)} = \frac{\mathbf{a}'\boldsymbol{\sigma}_i}{\mathbf{a}'\boldsymbol{\Sigma}\mathbf{a}}.$$

If $\boldsymbol{\xi}$ is a pre-assigned vector of positive elements which specify the ratios of increases desired in the individual measurements, then to determine $\mathbf{a}$ such that $\mathbf{a}'\mathbf{x}$ represents size, we must solve the equation

$$\boldsymbol{\Sigma}\mathbf{a} \propto \boldsymbol{\xi}.$$

Then $\boldsymbol{\xi}$ is called a *size vector.* Note the difference between the size *vector* $\boldsymbol{\xi}$ and the size *factor* $\mathbf{a}'\mathbf{x}$. In practice we often use standarized vectors $\mathbf{x}$ so that $\boldsymbol{\Sigma}$ is replaced by the sample correlation matrix $\mathbf{R}$. In these circumstances it is plausible to take $\boldsymbol{\xi}$ to be a vector of ones, so that $\mathbf{a} \propto \mathbf{R}^{-1}\mathbf{1}$.

Using similar arguments Rao suggests that a shape factor should be $h = \mathbf{b}'\mathbf{x}$, where the elements of $\mathbf{b}$ are given by $\mathbf{R}^{-1}\boldsymbol{\eta}$, where $\boldsymbol{\eta}$ is a pre-assigned vector of plus and minus ones, with possibly some zeros. Of course, it is possible to consider several different shape factors. Note that the interpretation of a shape factor is given by looking at the components of the shape *vector* $\boldsymbol{\eta}$, not the coefficients $\mathbf{b}$ of the shape *factor* $\mathbf{b}'\mathbf{x}$. If the variables are uncorrelated so that $\mathbf{R} = \mathbf{I}$, then the two vectors coincide.

Since it is only the relative magnitude of the elements of $\boldsymbol{\xi}$ and $\boldsymbol{\eta}$ which are of interest, both s and h may be re-scaled without affecting their interpretive value. If $\mathbf{a}$ and $\mathbf{b}$ are scaled so that s and h have unit variance (and if each column of the data $\mathbf{X}$ is scaled to have mean 0 and unit variance), then the correlation between s and h is given by

$$\rho = \frac{1}{n}\sum_{i=1}^{n} \boldsymbol{\xi}'\mathbf{R}^{-1}\mathbf{x}_i\mathbf{x}_i'\mathbf{R}^{-1}\boldsymbol{\eta} = \boldsymbol{\xi}'\mathbf{R}^{-1}\boldsymbol{\eta}.$$

A shape factor corrected for size can be obtained by considering

$$h' = h - \rho s.$$

The values of s and h' are uncorrelated, and can be plotted on a two-dimensional chart.

An alternative approach to allometry based on principal components has been suggested by Jolicoeur and Mosimann (1960). If all the coefficients of the first principal component are positive, then it can be used to measure size. In this case the other principal components will have positive and negative coefficients, since they are orthogonal to the first principal component. Therefore second and subsequent principal components may be taken as measures of shape. (Note that if all the elements of the covariance matrix are positive, then the Perron–Frobenius theorem ensures that all the coefficients of the first principal component are positive—see Exercise 8.6.1.)

One advantage of principal components is that the principal component loadings have a dual interpretation in Rao's framework. First, if $\mathbf{g}_{(1)}$ is the first eigenvector of $\mathbf{R}$, that is $\mathbf{g}_{(1)}$ is the loading vector of the first principal component, then $\boldsymbol{\xi} = \mathbf{g}_{(1)}$ can play the role of a size *vector*. Moreover, since $\mathbf{a} = \mathbf{R}^{-1}\mathbf{g}_{(1)} \propto \mathbf{g}_{(1)}$, $\mathbf{g}_{(1)}$ can also be interpreted as the coefficient vector of a size *factor* $s = \mathbf{g}'_{(1)}\mathbf{x}$. The other eigenvectors can be used to measure shape, and the sizes of the eigenvalues of $\mathbf{R}$ give some indication of the relative magnitude of the various size and shape variations within the data.

Unfortunately, the interpretation of principal components as size and shape vectors is not always straightforward. One's notions of size and shape are usually given in terms of simply-defined $\boldsymbol{\xi}$ and $\boldsymbol{\eta}$ (usually vectors of ± 1s and 0s), but the principal component loadings can behave too arbitrarily to represent size and shape in a consistent way (Rao, 1964).

Example 8.6.1 In an analysis of 54 apple trees, Pearce (1965) considers the following three variables: total length of lateral shoots, circumference of the trunk, and height. Although the variables can all be measured in the same units, it is sensible in this case to use standardized variables because the variances are clearly of different orders of magnitude.

The correlation matrix is

$$\begin{bmatrix} 1 & 0.5792 & 0.2414 \\ & 1 & 0.5816 \\ & & 1 \end{bmatrix},$$

giving principal components

$$y_1 = 0.554x_1 + 0.651x_2 + 0.520x_3,$$

$$y_2 = -0.657x_1 - 0.042x_2 + 0.753x_3,$$

$$y_3 = -0.511x_1 + 0.758x_2 - 0.404x_3$$

with corresponding variances 1.94, 0.76, and 0.33 (64, 25, and 11% of the total, respectively). Note that the first principal component is effectively the sum of the three variables, the second principal component represents a contrast between x_1 and x_3, and the third is a contrast between x_2 on the one hand and x_1 and x_3 on the other.

If we use a size vector $\boldsymbol{\xi}=(1, 1, 1)'$, then we get a size factor

$$s=0.739x_1+0.142x_2+0.739x_3,$$

which is similar to the first principal component y_1, although the second coefficient is perhaps low. Similarly a shape vector $\boldsymbol{\eta}_1=(1, 0, -1)'$ contrasting x_1 with x_3 leads to a shape factor

$$h_1=1.32x_1+0.007x_2-1.32x_3,$$

which is similar to the second principal component. However, if we use a shape vector $\boldsymbol{\eta}_2=(1, -1, 0)'$ to study the contrast between x_1 and x_2, we obtain a shape factor

$$h_2=2.56x_1-3.21x_2+1.25x_3$$

which bears no resemblance to any of the principal components.

8.7 Discarding of Variables

In Section 6.7 we gave a method of discarding redundant variables based on regression analysis. We now give a method based on principal component analysis.

Suppose a principal component analysis is performed on the correlation matrix of all the p variables. Initially, let us assume that k variables are to be retained where k is known and $(p-k)$ variables are to be discarded. Consider the eigenvector corresponding to the smallest eigenvalue, and reject the variable with the largest coefficient (absolute value). Then the next smallest eigenvalue is considered. This process continues until the $(p-k)$ smallest eigenvalues have been considered.

In general, we discard the variable

(a) which has the largest coefficient (absolute value) in the component, and
(b) which has not been previously discarded.

This principle is consistent with the notion that we regard a component with small eigenvalue as of less importance and, consequently, the variable which dominates it should be of less importance or redundant.

However, the choice of k can be made more realistic as follows. Let λ_0 be a threshold value so that eigenvalues $\lambda \leq \lambda_0$ can be regarded as contributing too little to the data. The method in essence is due to Beale *et al.* (1967). Jolliffe (1972) recommends $\lambda_0 = 0.70$ from various numerical examples.

Example 8.7.1 Consider Jeffers' (1967) pine pitprop data of Example 6.7.1. From the principal component analysis on the correlation matrix (see Example 8.2.6), it can be seen that the first six components account for 87% of the variability. The other seven eigenvalues are below the threshold value of $\lambda_0 = 0.70$, so seven variables are to be rejected. The variable with the largest coefficient of $\mathbf{g}_{(13)}$, the eigenvector corresponding to the smallest eigenvalue, is the second and so x_2 is rejected. Similarly, the variables with largest coefficients in $\mathbf{g}_{(12)}, \ldots, \mathbf{g}_{(7)}$ are x_4, x_7, x_{10}, x_6, x_9, x_{12}, respectively, and as these are all distinct, these are successively rejected.

An alternative method of discarding variables based on the multiple correlation coefficient was given in Example 6.7.2, and the two methods give different sets of retained variables. For example, to retain six variables the above principal component method chooses variables 1, 3, 5, 8, 11, 13 whereas from Table 6.7.4, the multiple correlation method chooses variables 3, 4, 10, 11, 12, 13. Variables, 3, 11, and 13 are retained by both methods.

Remarks (1) Instead of using λ_0, k may be set equal to the number of components needed to account for more than some proportion, α_0, of total variation. Jolliffe found in his examples that λ_0 appeared to be better than α_0 as a criterion for deciding how many variables to reject. He finds $\alpha_0 = 0.80$ as the best value if it is to be used. Further, using either criterion, at least four variables should always be retained.

(2) A variant of this method of discarding variables is given by the following iterative procedure. At each stage of the procedure reject the variable associated with the smallest eigenvalue and re-perform the principal component analysis on the remaining variables. Carry on until all the eigenvalues are high.

(3) The relation between the "smallness" of $\Sigma l_i x_i$ and the rejection of one particular variable is rather open to criticism. However, present studies lead us to the belief that various different methods of rejecting variables in order to explain the variation within $\mathbf{X}$ produce in practice almost the same results (see Jolliffe, 1972, 1973). Of course, the method selected to deal with one's data depends upon the purpose of the study, as the next section makes clear.

8.8 Principal Component Analysis in Regression

In the multiple regression model of Chapter 6 it was noted that if the independent variables are highly dependent, then the estimates of the regression coefficients will be very imprecise. In this situation it is advantageous to disregard some of the variables in order to increase the stability of the regression coefficients.

An alternative way to reduce dimensionality is to use principal components (Massy, 1965). However, the choice of components in the regression context is somewhat different from the choice in Section 8.2.5. In that section the principal components with the *largest variances* are used in order to explain as much of the total variation of $\mathbf{X}$ as possible. On the other hand, in the context of multiple regression it is sensible to take those components having the *largest correlations* with the dependent variable because the purpose in a regression is to explain the dependent variable. In a *multivariate* regression the correlations with each of the dependent variables must be examined. Fortunately, there is often a tendency in data for the components with the largest variances to best explain the dependent variables.

If the principal components have a natural intuitive meaning, it is perhaps best to leave the regression equation expressed in terms of the components. In other cases, it is more convenient to transform back to the original variables.

More specifically, if n observations are made on a dependent variable and on p independent variables ($\mathbf{y}(n\times 1)$ and $\mathbf{X}(n\times p)$) and if the observations are centred so that $\bar{y}=\bar{x}_i=0$, $i=1,\ldots,p$, then the regression equation is

$$\mathbf{y}=\mathbf{X}\boldsymbol{\beta}+\boldsymbol{\varepsilon}, \quad \text{where} \quad \boldsymbol{\varepsilon}\sim N_n(\mathbf{0},\sigma^2\mathbf{H}) \quad \text{and} \quad \mathbf{H}=\mathbf{I}-n^{-1}\mathbf{1}\mathbf{1}'. \tag{8.8.1}$$

If $\mathbf{W}=\mathbf{XG}$ denotes the principal component transformation then the regression equation may be written

$$\mathbf{y}=\mathbf{W}\boldsymbol{\alpha}+\boldsymbol{\varepsilon}, \tag{8.8.2}$$

where $\boldsymbol{\alpha}=\mathbf{G}'\boldsymbol{\beta}$. Since the columns of $\mathbf{W}$ are orthogonal, the least squares estimators $\hat{\alpha}_i$ are unaltered if some of the columns of $\mathbf{W}$ are deleted from the regression. Further, their distribution is also unaltered. The full vector of least squares estimators is given by

$$\hat{\boldsymbol{\alpha}}=(\mathbf{W}'\mathbf{W})^{-1}\mathbf{W}'\mathbf{y} \quad \text{and} \quad \hat{\boldsymbol{\varepsilon}}=\mathbf{y}-\mathbf{W}\hat{\boldsymbol{\alpha}},$$

or

$$\hat{\alpha}_i=n^{-1}l_i^{-1}\mathbf{w}_{(i)}'\mathbf{y}, \qquad i=1,\ldots,p, \tag{8.8.3}$$

where l_i is the ith eigenvalue of the covariance matrix, $n^{-1}\mathbf{X}'\mathbf{X}$. Then $\hat{\alpha}_i$ has expectation α_i and variance $\sigma^2/(nl_i)$. The covariance between $\mathbf{w}_i$ and $\mathbf{y}$,

$$C(\mathbf{w}_i, \mathbf{y}) = n^{-1}\mathbf{w}'_{(i)}\mathbf{y} = n^{-1}\mathbf{g}'_{(i)}\mathbf{X}'\mathbf{y}, \tag{8.8.4}$$

can be used to test whether or not the contribution of $\mathbf{w}_i$ to the regression is significant. Under the null hypothesis, $\alpha_i = 0$,

$$\frac{\hat{\alpha}_i (nl_i)^{1/2}}{(\hat{\boldsymbol{\varepsilon}}'\hat{\boldsymbol{\varepsilon}}/(n-p-1))^{1/2}} = \frac{\mathbf{w}'_{(i)}\mathbf{y}}{(l_i n \hat{\boldsymbol{\varepsilon}}'\hat{\boldsymbol{\varepsilon}}/(n-p-1))^{1/2}} \sim t_{n-p-1}, \tag{8.8.5}$$

regardless of the true values of the other α_j.

Because the components are orthogonal to one another and orthogonal to the residual vector, we can write

$$\begin{aligned}\mathbf{y}'\mathbf{y} &= \sum_{i=1}^{p} \hat{\alpha}_i^2 \mathbf{w}'_{(i)}\mathbf{w}_{(i)} + \hat{\boldsymbol{\varepsilon}}'\hat{\boldsymbol{\varepsilon}} \\ &= \sum_{i=1}^{p} nl_i\hat{\alpha}_i^2 + \hat{\boldsymbol{\varepsilon}}'\hat{\boldsymbol{\varepsilon}}. \end{aligned} \tag{8.8.6}$$

Thus, $nl_i\hat{\alpha}_i^2/\mathbf{y}'\mathbf{y} = \text{corr}^2(\mathbf{y}, \mathbf{w}_{(i)})$ represents the proportion of variance of $\mathbf{y}$ accounted for by component i.

Selecting those components for which the statistic in (8.8.5) is significant is a straightforward way to choose which components to retain. Procedures which are more theoretically justifiable have also been developed (see Hill *et al.*, 1977).

Example 8.8.1 Consider the Jeffers' (1967) pitprop data of Examples 6.7.1, 6.7.2, and 8.2.6. The proportions of the variance of $\mathbf{y}$ accounted for by each of the components and by the residual vector (times 100%) are given by 7.56, 33.27, 3.05, 0.26, 13.70, 6.13, 3.24, 1.28, 0.03, 0.00, 0.07, 4.56, 0.00, and 26.84; and $\hat{\boldsymbol{\alpha}}$ is given by

$$\hat{\boldsymbol{\alpha}} = (0.134^{**}, -0.374^{**}, 0.128^{**}, -0.048, -0.388^{**}, 0.274^{**}, -0.237^{**}, \\ -0.170^{**}, 0.031, 0.000, -0.115, -1.054^{**}, 0.002).$$

Using the distributional result (8.8.5), we can check from Appendix C, Table C.2 whether each $\hat{\alpha}_i$ is significantly different from zero, and * (**) indicates significance at the 5% (1%) level.

By a procedure similar to Example 8.2.6, Jeffers decided to retain six components in his analysis. Of these components, 1, 2, 3, 5, and 6 had significant regression coefficients, so Jeffers carried out his regression on these five components.

However, from the above discussion a more valid analysis would be obtained by including components 7, 8, and 12 as well. Letting $J = \{1, 2, 3, 5, 6, 7, 8, 12\}$ denote the set of retained components, the restricted least squares estimator $\hat{\boldsymbol{\alpha}}$ for $\boldsymbol{\alpha}$ is given by $\tilde{\alpha}_j = \hat{\alpha}_j$ for $j \in J$ and $\tilde{\alpha}_j = 0$ for $j \notin J$. Then $\tilde{\boldsymbol{\beta}} = \mathbf{G}\tilde{\boldsymbol{\alpha}}$. If $\alpha_j = 0$ for $j \notin J$, then the variance of each $\tilde{\beta}_i$ is given by

$$V(\tilde{\beta}_i) = \sum_{j \in J} g_{ij}^2 \sigma^2 / n l_j.$$

Notice that $l_{12} = 0.04$ is very small compared to the other retained eigenvalues. Thus, for several variables (namely, 1, 2, 3, 4, and 5), $V(\tilde{\beta}_i)$ is drastically increased by including component 12. (However, these high variances are partly balanced by the resulting large correlations between these $\tilde{\beta}_i$.) Hence, for this data, the regression is more easily interpreted in terms of the principal components than in terms of the original variables.

The problem of redundant information in a multiple regression can also be approached by discarding variables, and Example 6.7.1 describes this technique applied to the Jeffers' data. Note that the regression based on the eight most significant principal components explains 72.8% of the variance of y, which is slightly better than the 72.5% obtained from Table 6.7.3, using the optimal choice for eight retained variables. However, for fewer than eight retained variables or components, the discarding variables approach explains a higher percentage of the variance of y than the principal components approach.

In general, it is perhaps preferable to use the discarding variables approach because it is conceptually simpler.

Another technique for dealing with redundant information in a regression analysis is the method of ridge regression (Hoerl and Kennard, 1970). See Exercise 8.8.1.

Exercises and Complements

8.1.1 For

$$\boldsymbol{\Sigma} = \begin{pmatrix} \sigma_1^2 & \rho\sigma_1\sigma_2 \\ \rho\sigma_1\sigma_2 & \sigma_2^2 \end{pmatrix}, \qquad \rho \neq 0,$$

show that the eigenvalues are

$$\tfrac{1}{2}(\sigma_1^2 + \sigma_2^2) \pm \tfrac{1}{2}\Delta, \quad \text{where} \quad \Delta = \{(\sigma_1^2 - \sigma_2^2)^2 + 4\sigma_1^2\sigma_2^2\rho^2\}^{1/2}$$

with eigenvectors proportional to $(2\rho\sigma_1\sigma_2, \sigma_2^2 - \sigma_1^2 + \Delta)$ and $(\sigma_2^2 - \sigma_1^2 + \Delta, -2\rho\sigma_1\sigma_2)$.

8.1.2 Find the eigenvalues and eigenvectors of

$$\boldsymbol{\Sigma}=\begin{bmatrix}\beta^2+\delta & \beta & \beta\\ \beta & 1+\delta & 1\\ \beta & 1 & 1+\delta\end{bmatrix}.$$

(Hint: $\boldsymbol{\Sigma}-\delta\mathbf{I}=(\beta,1,1)'(\beta,1,1)$ is of rank 1 with non-zero eigenvalue β^2+2 and eigenvector $(\beta,1,1)'$. Therefore $\boldsymbol{\Sigma}$ has eigenvalues $\beta^2+\delta+2$, δ, δ, and the same eigenvectors as $\boldsymbol{\Sigma}-\delta\mathbf{I}$.)

8.1.3 Suppose that $\mathbf{x}'=(x_1, x_2)$ has a bivariate multinomial distribution with $n=1$, so that $x_1=1$ with probability p, and $x_1=0$ with probability $q=1-p$, and x_2 always satisfies $x_2=1-x_1$.

(a) Show that the covariance matrix of $\mathbf{x}$ is

$$pq\begin{pmatrix}1 & -1\\ -1 & 1\end{pmatrix}.$$

(b) Show that the eigenvalues of this matrix are $2pq$ and zero, and that the eigenvectors are $(1,-1)'$ and $(1,1)'$, respectively.

8.2.1 (Dagnelie, 1975, p. 57) Berce and Wilbaux (1935) collected measurements on five meterological variables over an 11-year period. Hence $n=11$ and $p=5$. The variables were

x_1 = rainfall in November and December (in millimetres),
x_2 = average July temperature (in degrees Celsius)
x_3 = rainfall in July (in millimetres),
x_4 = radiation in July (in millilitres of alcohol),
x_5 = average harvest yield (in quintals per hectare).

The raw data was as follows:

Year	x_1	x_2	x_3	x_4	x_5
1920–21	87.9	19.6	1.0	1661	28.37
1921–22	89.9	15.2	90.1	968	23.77
1922–23	153.0	19.7	56.6	1353	26.04
1923–24	132.1	17.0	91.0	1293	25.74
1924–25	88.8	18.3	93.7	1153	26.68
1925–26	220.9	17.8	106.9	1286	24.29
1926–27	117.7	17.8	65.5	1104	28.00
1927–28	109.0	18.3	41.8	1574	28.37
1928–29	156.1	17.8	57.4	1222	24.96
1929–30	181.5	16.8	140.6	902	21.66
1930–31	181.4	17.0	74.3	1150	24.37

Show that this data leads to the mean vector, covariance matrix, and correlation matrix given below:

$$\bar{\mathbf{x}}' = (138.0, 17.75, 74.4, 1242, 25.66),$$

$$\mathbf{R}/\mathbf{S} = \begin{bmatrix} 1794 & -4.473 & 726.9 & -2218 & -52.01 \\ -0.087 & 1.488 & -26.62 & 197.5 & 1.577 \\ 0.491 & -0.624 & 1224 & -6203 & -56.44 \\ -0.239 & 0.738 & -0.808 & 48\,104 & 328.9 \\ -0.607 & 0.640 & -0.798 & 0.742 & 4.087 \end{bmatrix}.$$

(This matrix contains correlations below the main diagonal, variances on the main diagonal, and covariances above the main diagonal.)

8.2.2 (Dagnelie, 1975, p. 176) Using the correlation matrix for the meteorological variables x_1 to x_4 in Exercise 8.2.1, verify that the principal components y_1 to y_4 explain 65, 24, 7, and 4%, respectively, of the total variance, and that the eigenvectors (×1000) are, respectively,

$$(291, -506, 577, -571), \qquad (871, 425, 136, 205),$$
$$(-332, 742, 418, -405), \qquad (-214, -111, 688, 685).$$

Following Section 8.2.5, show that the correlations between x_i and y_j are given by the elements of the following matrix (÷1000):

$$\begin{bmatrix} 468 & 862 & -177 & -81 \\ -815 & 420 & 397 & -42 \\ 930 & 135 & 223 & 260 \\ -919 & 202 & -216 & 259 \end{bmatrix}.$$

Hence calculate the proportion of variance explained by y_3 and y_4 for each of x_1, x_2, x_3, and x_4.

8.2.3 (Dagnelie, 1975, p. 178) Using the covariance matrix for the meteorological variables x_1 to x_4 used in Exercise 8.2.2, show that eigenvalues are 49 023, 1817, 283, and 0.6, and that the eigenvectors (×1000) are, respectively,

$$(-49, 4, -129, 990), \qquad (954, 3, 288, 84),$$
$$(-296, -8, 949, 109), \qquad (-5, 1000, 8, -3).$$

Interpret the principal components, and compare them with those based on the correlation matrix found above. Which interpretation is more meaningful?

8.2.4 Exercise 8.2.2 gives the first principal component in terms of the standardized variables $u_i = x_i/s_i$. Express this linear combination in terms of the original variables x_i, and standardize the coefficients so their squares sum to 1. Compare this new linear combination with the first principal component using the covariance matrix (Exercise 8.2.3), to illustrate that principal component analysis is not scale invariant.

8.2.5 If **X** is a sample data matrix, show that no SLC of the observed variables has a variance which is larger than that of the first principal component. Also prove results corresponding to the other parts of Theorems 8.2.1–8.2.3.

8.2.6 (a) Extending Example 8.2.3, show that the first four students in Table 1.2.1 have the following scores on the first two principal components.

	Principal component	
Student	1	2
1	66.4	6.5
2	63.7	−6.8
3	63.0	3.1
4	44.6	−5.6

(b) Show that the score on the first principal component is zero if and only if

$$0.51x_1+0.37x_2+0.35x_3+0.45x_4+0.53x_5-99.7=0$$

and the score on the second principal component is zero if and only if

$$0.75x_1+0.21x_2-0.08x_3-0.30x_4-0.55x_5+1.5=0.$$

8.2.7 Suppose **F** is a $(p \times k)$ matrix $(k \leq p)$ with orthonormal columns and let $\lambda_1 \geq \lambda_2 \ldots \geq \lambda_p$. Set

$$h_i = \sum_{j=1}^{k} f_{ij}^2, \qquad i = 1, \ldots, p,$$

and set

$$\phi(\mathbf{h}) = \sum_{i=1}^{p} h_i\lambda_i = \sum_{i=1}^{p}\sum_{j=1}^{k} f_{ij}^2\lambda_i.$$

Show that

$$0 \leq h_i \leq 1 \quad \text{and} \quad \sum_{i=1}^{p} h_i = k.$$

Hence deduce that $\phi(\mathbf{h})$ is maximized when $h_1 = \ldots = h_k = 1$, and $h_{k+1} = \ldots = h_p = 0$; that is, when the columns of **F** span the subspace of the first k rows of **F**.

8.2.8 In Example 8.2.5, the regression with both variables subject to error, let **S** denote the (2×2) covariance matrix of the data **X**. Using Exercise 8.1.1, show that the first principal component of the covariance matrix of $(\lambda x_1, x_2)$ is given in terms of (x_1, x_2) by the line with slope

$$\hat{\beta} = [s_{22} - \lambda s_{11} + \{(s_{22} - \lambda s_{11})^2 + 4\lambda s_{12}^2\}^{1/2}]/2s_{12}.$$

Show that

$$|s_{12}|/s_{11} \leq |\hat{\beta}| \leq s_{22}/|s_{12}|;$$

that is, the slope lies between the slopes of the regression lines of x_2 on x_1 and of x_1 on x_2. Also show that the line goes through the sample mean $(\bar{x}_1, \bar{x}_2)$ if we take $\hat{\alpha} = \bar{x}_2 - \hat{\beta}\bar{x}_1$.

8.2.9 Verify formula (8.2.14) to show that the correlations between all the variables and a group G of components can be combined to give that part of the total variation accounted for by the components in G.

***8.3.1** (Girshick, 1939) The p.d.f. of the eigenvalues of a $W_p(\mathbf{I}, m)$ distribution is

$$f(l_1, \ldots, l_p) = c \prod_{i=1}^{p} l_i^{(m-p-1)/2} \prod_{i<j} (l_i - l_j) \exp\left(-\tfrac{1}{2} \sum_{i=1}^{p} l_i\right).$$

Show that this may also be written

$$f(l_1, \ldots, l_p) = c g^{p(m-p-1)/2} \exp(-\tfrac{1}{2}ap) \prod_{i<j} (l_i - l_j),$$

where a and g are the arithmetic and geometric means of the eigenvalues. Show that the eigenvectors of the Wishart matrix have the Haar invariant distribution on the space of orthogonal matrices.

8.3.2 Show that in the situation of Example 8.3.3, alternative confidence intervals could in principle be obtained from the fact that, asymptotically,

$$(\tfrac{1}{2}(n-1))^{1/2}(l_i - \lambda_i)/\lambda_i \sim N(0, 1).$$

Show that this leads to confidence intervals which can include negative values of λ_i, and hence that the variance-stabilizing logarithmic transform used in Example 8.3.3 is preferable.

8.3.3 If $\hat{\rho}_1$ and $\hat{\rho}_2$ are the two estimates defined in (8.3.2), show that

$$\hat{\rho}_2 = \hat{\rho}_1 + \{(p - \operatorname{tr} \mathbf{S})/(p-1)\}.$$

Hence the estimates are equal if and only if $\operatorname{tr} \mathbf{S} = p$.

Compare these estimators with the m.l.e. of ρ given in Exercise 4.2.12, where $\mathbf{\Sigma}$ is assumed to be of the form $\mathbf{\Sigma} = \sigma^2[(1-\rho)\mathbf{I} + \rho\mathbf{11}']$.

8.4.1 Show that $-2 \log \lambda$ given by (8.4.2) has an asymptotic chi-squared distribution with $\frac{1}{2}(p-k+2)(p-k-1)$ degrees of freedom. (Hint: Under H_0, $\mathbf{\Sigma}$ is determined by k distinct eigenvalues and one common eigenvalue, and the k eigenvectors each with p components which correspond to the distinct eigenvalues. This gives $k+1+kp$ parameters. However, the eigenvectors are constrained by $\frac{1}{2}k(k+1)$ orthogonality conditions, leaving $k+1+kp-\frac{1}{2}k(k+1)$ free parameters. Under H_1 the number of free parameters is obtained by putting $k=p-1$, viz. $p+(p-1)p-\frac{1}{2}(p-1)p = \frac{1}{2}p(p+1)$. By subtraction and using Theorem 5.2.1, we find that the number of degrees of freedom is $\frac{1}{2}(p-k+2)(p-k-1)$.)

8.4.2 (Dagnelie, 1975, p. 183) An (86×4) data matrix led to the covariance matrix

$$\mathbf{S} = \begin{bmatrix} 0.029\,004 & -0.008\,545 & 0.001\,143 & -0.006\,594 \\ & 0.003\,318 & 0.000\,533 & 0.003\,248 \\ & & 0.004\,898 & 0.005\,231 \\ & & & 0.008\,463 \end{bmatrix}.$$

Here the variables relate to the number of trees, height, surface area, and volume of 86 parcels of land. The eigenvalues of $\mathbf{S}$ are

$$l_1 = 0.033\,687, \quad l_2 = 0.011\,163, \quad l_3 = 0.000\,592, \quad \text{and} \quad l_4 = 0.000\,241.$$

Calculate the 'percentage of variance explained' by the various principal components and show that

$$\operatorname{tr} \mathbf{S} = 0.045\,683 \quad \text{and} \quad |\mathbf{S}| = 0.536\,41 \times 10^{-10}.$$

Show that if the hypothesis is $\lambda_1 = \lambda_2 = \lambda_3 = \lambda_4$, then using (8.4.3) gives as a test statistic

$$(86 - \tfrac{19}{6}) \log 317.2 = 477,$$

which is highly significant against χ^2_9.

Show similarly that the hypothesis $\lambda_2 = \lambda_3 = \lambda_4$ gives a χ^2_5 statistic of 306, and the hypothesis $\lambda_3 = \lambda_4$ gives a χ^2_2 statistic of 16.2. Since all these are highly significant we conclude that the eigenvalues of $\mathbf{\Sigma}$ are all distinct.

8.4.3 (See Kendall, 1975, p. 79). Let D_1 be the determinant of $\mathbf{S}$ so that $D_1 = l_1 l_2 \ldots l_p$. Let D_2 be the determinant of the m.l.e. of $\mathbf{\Sigma}$ that would have arisen had the last $(p-k)$ eigenvalues of $\mathbf{\Sigma}$ been assumed equal. Show that

$$D_2 = l_1 l_2 \ldots l_k \left(\frac{l_{k+1} + \ldots + l_p}{p-k}\right)^{p-k}.$$

Hence show that

$$D_2/D_1 = (a_0/g_0)^{p-k},$$

where a_0 and g_0 are the arithmetic and geometric means of the last $(p-k)$ eigenvalues of $\mathbf{S}$. Hence show that $n \log (D_2/D_1) = -2 \log \lambda$, where $-2 \log \lambda$ is given by (8.4.2).

8.4.4 Consider the special case of (8.4.1) where $k = 1$. Show that in this case $\alpha = \psi^2 \operatorname{tr}^2 \mathbf{\Sigma}/\operatorname{tr} \mathbf{\Sigma}^2$, so that

$$\tau^2 = \frac{2\psi^2}{n}\left(\frac{\operatorname{tr} \mathbf{\Sigma}^2}{\operatorname{tr}^2 \mathbf{\Sigma}} - 2\psi + 1\right).$$

Confirm that this formula leads to the same numerical results as obtained in Example 8.4.1.

8.5.1 Show that the matrices $\mathbf{A}^{-1}\mathbf{X}\mathbf{B}^{-1}\mathbf{X}'$ and $\mathbf{B}^{-1}\mathbf{X}'\mathbf{A}^{-1}\mathbf{X}$ in (8.5.1) and (8.5.2) have an eigenvalue of unity and that each of the corresponding eigenvectors is a vector of ones.

8.5.2 Show that each eigenvalue ρ of $\mathbf{A}^{-1}\mathbf{X}\mathbf{B}^{-1}\mathbf{X}'$ lies between 0 and 1, and if $\rho < 1$ then the corresponding eigenvector $\mathbf{r}$ satisfies $\mathbf{1}'\mathbf{A}\mathbf{r} = 0$. Prove a similar result for the matrix $\mathbf{B}^{-1}\mathbf{X}'\mathbf{A}^{-1}\mathbf{X}$. (Note that $\mathbf{A}^{1/2}\mathbf{1}$ is an eigenvector of the symmetric positive semi-definite matrix $(\mathbf{A}^{-1/2}\mathbf{X}\mathbf{B}^{-1/2})(\mathbf{A}^{-1/2}\mathbf{X}\mathbf{B}^{-1/2})'$ and all other eigenvectors must be orthogonal to this one.)

8.5.3 If $\mathbf{r}$ is a standardized eigenvector of $\mathbf{A}^{-1}\mathbf{X}\mathbf{B}^{-1}\mathbf{X}'$ with eigenvalue ρ $(\rho > 0)$, show that $\rho^{-1/2}\mathbf{B}^{-1}\mathbf{X}'\mathbf{r}$ is a standardized eigenvector of $\mathbf{B}^{-1}\mathbf{X}'\mathbf{A}^{-1}\mathbf{X}$ with the same eigenvalue.

8.6.1 (Perron–Frobenius theorem) If $\mathbf{A}$ is a symmetric matrix with positive elements, then all of the coefficients of the first eigenvector of $\mathbf{A}$ have the same sign. (Hint: Let $\mathbf{l}$ be the first eigenvector and define $\mathbf{m}$ by $m_i = |l_i|$. Then $\mathbf{m}'\mathbf{A}\mathbf{m} \geq \mathbf{l}'\mathbf{A}\mathbf{l}$.)

8.7.1 The following principal components were obtained from the correlation matrix of six bone measurements on 276 fowls (Wright, 1954). Show that by taking the threshold value $\lambda_0 = 0.70$ we reject variables x_4,

x_6, x_3, and x_1 in that order.

Measurements	Component numbers 1	2	3	4	5	6
x_1, skull length	0.35	0.53	0.76	−0.05	0.04	0.00
x_2, skull breadth	0.33	0.70	−0.64	0.00	0.00	−0.04
x_3, wing humerus	0.44	−0.19	−0.05	0.53	0.19	0.59
x_4, wing ulna	0.44	−0.25	0.02	0.48	−0.15	−0.63
x_5, leg	0.43	−0.28	−0.06	−0.51	0.67	−0.48
x_6, leg tibia	0.44	−0.22	−0.05	−0.48	−0.70	0.15
Eigenvalues	4.57	0.71	0.41	0.17	0.08	0.06

Show that the six components can be interpreted as overall size, head size versus body size, head shape, wing size versus leg size, leg shape, and wing shape (although these last two interpretations are perhaps dubious). Verify that the successive elimination of variables reduces the number of these features which can be discerned.

8.8.1 Hoerl and Kennard (1970) have proposed the method of *ridge regression* to improve the accuracy of the parameter estimates in the regression model

$$\mathbf{y}=\mathbf{X}\boldsymbol{\beta}+\mu\mathbf{1}+\mathbf{u}, \qquad \mathbf{u}\sim N_n(\mathbf{0}, \sigma^2\mathbf{I}).$$

Suppose the columns of $\mathbf{X}$ have been standardized to have mean 0 and variance 1. Then the *ridge estimate* of $\boldsymbol{\beta}$ is defined by

$$\boldsymbol{\beta}^*=(\mathbf{X}'\mathbf{X}+k\mathbf{I})^{-1}\mathbf{X}'\mathbf{y},$$

where for given $\mathbf{X}$, $k\geqslant 0$ is a (small) *fixed* number.

(a) Show that $\boldsymbol{\beta}^*$ reduces to the OLS estimate $\hat{\boldsymbol{\beta}}=(\mathbf{X}'\mathbf{X})^{-1}\mathbf{X}'\mathbf{y}$ when $k=0$.

(b) Let $\mathbf{X}'\mathbf{X}=\mathbf{GLG}'$ be a spectral decomposition of $\mathbf{X}'\mathbf{X}$ and let $\mathbf{W}=\mathbf{XG}$ be the principal component transformation given in (8.8.2). If $\boldsymbol{\alpha}=\mathbf{G}'\boldsymbol{\beta}$ represents the parameter vector for the principal components, show that the ridge estimate $\boldsymbol{\alpha}^*$ of $\boldsymbol{\alpha}$ can be simply related to the OLS estimate $\hat{\boldsymbol{\alpha}}$ by

$$\alpha_j^*=\frac{l_j}{l_j+k}\hat{\alpha}_j, \qquad j=1,\ldots,p,$$

and hence

$$\boldsymbol{\beta}^*=\mathbf{GDG}'\hat{\boldsymbol{\beta}}, \quad \text{where} \quad \mathbf{D}=\operatorname{diag}(l_j/(l_j+k)).$$

(c) One measure of the accuracy of $\boldsymbol{\beta}^*$ is given by the *trace mean square error*,

$$\phi(k) = \operatorname{tr} E[\boldsymbol{\beta}^* - \boldsymbol{\beta})(\boldsymbol{\beta}^* - \boldsymbol{\beta})'] = \sum_{i=1}^{p} E(\beta_i^* - \beta_i)^2.$$

Show that we can write $\phi(k) = \gamma_1(k) + \gamma_2(k)$, where

$$\gamma_1(k) = \sum_{i=1}^{p} V(\beta_i^*) = \sigma^2 \sum_{j=1}^{p} \frac{l_j}{(l_j + k)^2}$$

represents the sum of the variances of β_i^*, and

$$\gamma_2(k) = \sum_{i=1}^{p} [E(\beta_i^* - \beta_i)]^2 = k^2 \sum_{j=1}^{p} \frac{\alpha_j^2}{(l_j + k)^2}$$

represents the sum of the squared biases of β_i^*.

(d) Show that the first derivatives of $\gamma_1(k)$ and $\gamma_2(k)$ at $k = 0$ are

$$\gamma_1'(0) = -2\sigma^2 \sum \frac{1}{l_j^2}, \quad \gamma_2'(0) = 0.$$

Hence there exist values of $k > 0$ for which $\phi(k) < \phi(0)$, that is for which $\boldsymbol{\beta}^*$ has smaller trace mean square error than $\hat{\boldsymbol{\beta}}$. Note that the increase in accuracy is most pronounced when some of the eigenvalues l_j are near 0, that is, when the columns of $\mathbf{X}$ are nearly collinear. However, the optimal choice for k depends on the unknown value of $\boldsymbol{\beta} = \mathbf{G}\boldsymbol{\alpha}$.

9
Factor Analysis

9.1. Introduction

Factor analysis is a mathematical model which attempts to explain the correlation between a large set of variables in terms of a small number of underlying *factors*. A major assumption of factor analysis is that it is not possible to observe these factors directly; the variables depend upon the factors but are also subject to random errors. Such an assumption is particularly well-suited to subjects like psychology where it is not possible to measure exactly the concepts one is interested in (e.g. "intelligence"), and in fact it is often ambiguous just how to define these concepts.

Factor analysis was originally developed by psychologists. The subject was first put on a respectable statistical footing in the early 1940s by restricting attention to one particular form of factor analysis, that based on maximum likelihood estimation. In this chapter we shall concentrate on maximum likelihood factor analysis and also on a second method, principal factor analysis, which is closely related to the technique of principal component analysis.

In order to get a feel for the subject we first describe a simple example.

Example 9.1.1 (Spearman, 1904) An important early paper in factor analysis dealt with children's examination performance in Classics (x_1), French (x_2), and English (x_3). It is found that the correlation matrix is given by

$$\begin{pmatrix} 1 & 0.83 & 0.78 \\ & 1 & 0.67 \\ & & 1 \end{pmatrix}.$$

Although this matrix has full rank its dimensionality can be effectively reduced from $p=3$ to $p=1$ by expressing the three variables as follows:

$$x_1=\lambda_1 f+u_1, \qquad x_2=\lambda_2 f+u_2, \qquad x_3=\lambda_3 f+u_3. \tag{9.1.1}$$

In these equations f is an underlying "common factor" and λ_1, λ_2, and λ_3 are known as "factor loadings". The terms u_1, u_2, and u_3 represent random disturbance terms. The common factor may be interpreted as "general ability" and u_i will have small variance if x_i is closely related to general ability. The variation in u_i consists of two parts which we shall not try to disentangle in practice. First, this variance represents the extent to which an individual's ability at Classics, say, differs from his general ability, and second, it represents the fact that the examination is only an approximate measure of his ability in the subject.

The model defined in (9.1.1) will now be generalized to include $k>1$ common factors.

9.2 The Factor Model

9.2.1 Definition

Let $\mathbf{x}(p\times 1)$ be a random vector with mean $\boldsymbol{\mu}$ and covariance matrix $\boldsymbol{\Sigma}$. Then we say that the k-factor model holds for $\mathbf{x}$ if $\mathbf{x}$ can be written in the form

$$\mathbf{x}=\boldsymbol{\Lambda}\mathbf{f}+\mathbf{u}+\boldsymbol{\mu}, \tag{9.2.1}$$

where $\boldsymbol{\Lambda}(p\times k)$ is a matrix of constants and $\mathbf{f}(k\times 1)$ and $\mathbf{u}(p\times 1)$ are random vectors. The elements of $\mathbf{f}$ are called *common* factors and the elements of $\mathbf{u}$ *specific* or *unique* factors. We shall suppose

$$E(\mathbf{f})=\mathbf{0}, \qquad V(\mathbf{f})=\mathbf{I}, \tag{9.2.2}$$

$$E(\mathbf{u})=\mathbf{0}, \qquad C(u_i, u_j)=0, \qquad i\neq j, \tag{9.2.3}$$

and

$$C(\mathbf{f}, \mathbf{u})=\mathbf{0}. \tag{9.2.4}$$

Denote the covariance matrix of $\mathbf{u}$ by $V(\mathbf{u})=\boldsymbol{\Psi}=\operatorname{diag}(\psi_{11},\ldots,\psi_{pp})$. Thus, all of the factors are uncorrelated with one another and further the common factors are each standardized to have variance 1. It is sometimes convenient to suppose that $\mathbf{f}$ and $\mathbf{u}$ (and hence $\mathbf{x}$) are multinormally distributed.

Note that

$$x_i=\sum_{j=1}^{k}\lambda_{ij}f_j+u_i+\mu_i, \qquad i=1,\ldots,p,$$

so that

$$\sigma_{ii} = \sum_{j=1}^{k} \lambda_{ij}^2 + \psi_{ii}.$$

Thus, the variance of $\mathbf{x}$ can be split into two parts. First,

$$h_i^2 = \sum_{j=1}^{k} \lambda_{ij}^2$$

is called the *communality* and represents the variance of x_i which is shared with the other variables via the common factors. In particular $\lambda_{ij}^2 = C(x_i, f_j)$ represents the extent to which x_i depends on the jth common factor. On the other hand ψ_{ii} is called the *specific* or *unique variance* and is due to the unique factor u_i; it explains the variability in x_i not shared with the other variables.

The validity of the k-factor model can be expressed in terms of a simple condition on $\mathbf{\Sigma}$. Using (9.2.1)–(9.2.4) we get

$$\mathbf{\Sigma} = \mathbf{\Lambda}\mathbf{\Lambda}' + \mathbf{\Psi}. \tag{9.2.5}$$

The converse also holds. If $\mathbf{\Sigma}$ can be decomposed into the form (9.2.5) then the k-factor model holds for $\mathbf{x}$. (See Exercise 9.2.2) However, $\mathbf{f}$ and $\mathbf{u}$ are not uniquely determined by $\mathbf{x}$. We shall discuss this point further in Section 9.7 on factor scores.

9.2.2 Scale invariance

Re-scaling the variables of $\mathbf{x}$ is equivalent to letting $\mathbf{y} = \mathbf{C}\mathbf{x}$, where $\mathbf{C} = \text{diag}(c_i)$. If the k-factor model holds for $\mathbf{x}$ with $\mathbf{\Lambda} = \mathbf{\Lambda}_x$ and $\mathbf{\Psi} = \mathbf{\Psi}_x$, then

$$\mathbf{y} = \mathbf{C}\mathbf{\Lambda}_x\mathbf{f} + \mathbf{C}\mathbf{u} + \mathbf{C}\boldsymbol{\mu}$$

and

$$V(\mathbf{y}) = \mathbf{C}\mathbf{\Sigma}\mathbf{C} = \mathbf{C}\mathbf{\Lambda}_x\mathbf{\Lambda}_x'\mathbf{C} + \mathbf{C}\mathbf{\Psi}_x\mathbf{C}.$$

Thus the k-factor model also holds for $\mathbf{y}$ with factor loading matrix $\mathbf{\Lambda}_y = \mathbf{C}\mathbf{\Lambda}_x$ and specific variances $\mathbf{\Psi}_y = \mathbf{C}\mathbf{\Psi}_x\mathbf{C} = \text{diag}\,(c_i^2\psi_{ii})$.

Note that the factor loading matrix for the scaled variables $\mathbf{y}$ is obtained by scaling the factor loading matrix of the original variables (multiply the ith row of $\mathbf{\Lambda}_x$ by c_i). A similar comment holds for the specific variances. In other words, factor analysis (unlike principal component analysis) is unaffected by a re-scaling of the variables.

9.2.3. Non-uniqueness of factor loadings

If the k-factor model (9.2.1) holds, then it also holds if the factors are rotated; that is, if $\mathbf{G}$ is a $(k \times k)$ orthogonal matrix, then $\mathbf{x}$ can also be written as

$$\mathbf{x} = (\mathbf{\Lambda G})(\mathbf{G'f}) + \mathbf{u} + \boldsymbol{\mu}. \tag{9.2.6}$$

Since the random vector $\mathbf{G'f}$ also satisfies the conditions (9.2.2) and (9.2.4), we see that the k-factor model is valid with new factors $\mathbf{G'f}$ and new factor loadings $\mathbf{\Lambda G}$.

Thus, if (9.2.6) holds, we can also write $\mathbf{\Sigma}$ as $\mathbf{\Sigma} = (\mathbf{\Lambda G})(\mathbf{G'\Lambda'}) + \boldsymbol{\psi}$. In fact, for fixed $\mathbf{\Psi}$, this rotation is the only indeterminacy in the decomposition of $\mathbf{\Sigma}$ in terms of $\mathbf{\Lambda}$ and $\mathbf{\Psi}$; that is, if $\mathbf{\Sigma} = \mathbf{\Lambda\Lambda'} + \mathbf{\Psi} = \mathbf{\Lambda^*\Lambda^{*\prime}} + \mathbf{\Psi}$, then $\mathbf{\Lambda} = \mathbf{\Lambda^* G}$ for some orthogonal matrix $\mathbf{G}$ (Exercise 9.2.3).

This indeterminacy in the definition of factor loadings is usually resolved by rotating the factor loadings to satisfy an arbitrary constraint such as

$$\mathbf{\Lambda'\Psi^{-1}\Lambda} \text{ is diagonal}, \tag{9.2.7}$$

or

$$\mathbf{\Lambda'D^{-1}\Lambda} \text{ is diagonal}, \qquad \mathbf{D} = \operatorname{diag}(\sigma_{11}, \ldots, \sigma_{pp}), \tag{9.2.8}$$

where in either case, the diagonal elements are written in decreasing order, say. Both constraints are scale invariant and, except for possible changes of sign of the columns, $\mathbf{\Lambda}$ is then in general completely determined by either constraint. (See Exercise 9.2.4.) Note that when the number of factors $k = 1$, the constraint is irrelevant. Also, if some of the ψ_{ii} equal 0, then the constraint (9.2.7) cannot be used; an alternative is given in Exercise 9.2.8.

It is of interest to compare the number of parameters in $\mathbf{\Sigma}$ when $\mathbf{\Sigma}$ is unconstrained, with the number of free parameters in the factor model. Let s denote the difference. At first sight $\mathbf{\Lambda}$ and $\mathbf{\Psi}$ contain $pk + p$ free parameters. However (9.2.7) or (9.2.8) introduces $\frac{1}{2}k(k-1)$ constraints. Since the number of distinct elements of $\mathbf{\Sigma}$ is $\frac{1}{2}p(p+1)$ we see that

$$\begin{aligned} s &= \tfrac{1}{2}p(p+1) - \{pk + p - \tfrac{1}{2}k(k-1)\} \\ &= \tfrac{1}{2}(p-k)^2 - \tfrac{1}{2}(p+k). \end{aligned} \tag{9.2.9}$$

Usually it will be the case that $s > 0$. Then s will represent the extent to which the factor model offers a simpler interpretation for the behaviour of $\mathbf{x}$ than the alternative assumption that $V(\mathbf{x}) = \mathbf{\Sigma}$. If $s \geq 0$ and (9.2.6) holds and $\mathbf{\Lambda}$ and $\mathbf{\Psi}$ are known, then $\mathbf{\Sigma}$ can be written in terms of $\mathbf{\Lambda}$ and $\mathbf{\Psi}$, subject to the constraint (9.2.7) or (9.2.8) on $\mathbf{\Lambda}$.

9.2.4 Estimation of the parameters in factor analysis

In practice, we observe a data matrix $\mathbf{X}$ whose information is summarized by the sample mean $\bar{\mathbf{x}}$ and sample covariance matrix $\mathbf{S}$. The location parameter is not of interest here, and we shall estimate it by $\hat{\boldsymbol{\mu}} = \bar{\mathbf{x}}$. The interesting problem is how to estimate $\boldsymbol{\Lambda}$ and $\boldsymbol{\Psi}$ (and hence $\boldsymbol{\Sigma} = \boldsymbol{\Lambda}\boldsymbol{\Lambda}' + \boldsymbol{\Psi}$) from $\mathbf{S}$; that is, we wish to find estimates $\hat{\boldsymbol{\Lambda}}$ and $\hat{\boldsymbol{\Psi}}$ satisfying the constraint (9.2.7) or (9.2.8), for which the equation

$$\mathbf{S} \doteq \hat{\boldsymbol{\Lambda}}\hat{\boldsymbol{\Lambda}}' + \hat{\boldsymbol{\Psi}} \tag{9.2.10}$$

is satisfied, at least approximately. Given an estimate $\hat{\boldsymbol{\Lambda}}$, it is then natural to set

$$\hat{\psi}_{ii} = s_{ii} - \sum_{j=1}^{k} \hat{\lambda}_{ij}^2, \qquad i = 1, \ldots, p, \tag{9.2.11}$$

so that the diagonal equations in (9.2.10) always hold exactly. We shall only consider estimates for which (9.2.11) is satisfied and $\hat{\psi}_{ii} \geq 0$. Setting $\hat{\boldsymbol{\Sigma}} = \hat{\boldsymbol{\Lambda}}\hat{\boldsymbol{\Lambda}}' + \hat{\boldsymbol{\Psi}}$, we get

$$\hat{\sigma}_{ii} = \sum_{j=1}^{k} \hat{\lambda}_{ij}^2 + \hat{\psi}_{ii}$$

so that (9.2.11) is equivalent to the condition

$$\hat{\sigma}_{ii} = s_{ii}, \qquad i = 1, \ldots, p. \tag{9.2.12}$$

Three cases can occur in (9.2.10) depending on the value of s in (9.2.9). If $s < 0$ then (9.2.10) contains more parameters than equations. Then, in general, we expect to find an infinity of exact solutions for $\boldsymbol{\Lambda}$ and $\boldsymbol{\Psi}$, and hence, the factor model is *not* well-defined.

If $s = 0$ then (9.2.10) can generally be solved for $\boldsymbol{\Lambda}$ and $\boldsymbol{\Psi}$ exactly (subject to the constraint (9.2.7) or (9.2.8) on $\boldsymbol{\Lambda}$). The factor model contains as many parameters as $\boldsymbol{\Sigma}$ and hence offers no simplification of the original assumption that $V(\mathbf{x}) = \boldsymbol{\Sigma}$. However, the change in viewpoint can sometimes be very helpful. (See Exercise 9.2.7.)

If $s > 0$, as will usually be the case, then there will be more equations than parameters. Thus, it is not possible to solve (9.2.10) exactly in terms of $\hat{\boldsymbol{\Lambda}}$ and $\hat{\boldsymbol{\Psi}}$, and we must look for approximate solutions. In this case the factor model offers a simpler explanation for the behaviour of $\mathbf{x}$ than the full covariance matrix.

9.2.5 Use of the correlation matrix R in estimation

Because the factor model is scale invariant, we shall only consider estimates of $\boldsymbol{\Lambda} = \boldsymbol{\Lambda}_x$ and $\boldsymbol{\Psi} = \boldsymbol{\Psi}_x$ which are scale invariant. It is then convenient to consider the scaling separately from the relationships between the variables. Let $\mathbf{Y} = \mathbf{HXD}_S^{-1/2}$ where $\mathbf{D}_S = \operatorname{diag}(s_{11}, \ldots, s_{pp})$, denote the standardized variables so that

$$\sum_{r=1}^{n} y_{rj}=0 \quad \text{and} \quad \frac{1}{n}\sum_{r=1}^{n} y_{rj}^2=1, \qquad j=1,\ldots,p.$$

Then $\mathbf{Y}$ will have estimated factor loading matrix $\hat{\mathbf{\Lambda}}_y=\mathbf{D}_S^{-1/2}\hat{\mathbf{\Lambda}}_x$ and estimated specific variances $\hat{\mathbf{\Psi}}_y=\mathbf{D}_S^{-1}\mathbf{\Psi}_x$, and (9.2.10) can be written in terms of the correlation matrix of $\mathbf{x}$ as

$$\mathbf{R}\doteq\hat{\mathbf{\Lambda}}_y\hat{\mathbf{\Lambda}}_y'+\hat{\mathbf{\Psi}}_y. \tag{9.2.13}$$

Note that (9.2.11) becomes

$$\hat{\psi}_{yii}=1-\sum_{j=1}^{k}\hat{\lambda}_{yij}^2, \qquad i=1,\ldots,p \tag{9.2.14}$$

so that $\mathbf{\Psi}_y$ is not a parameter of the model any more, but a function of $\mathbf{\Lambda}_y$. However, $\mathbf{R}$ contains p fewer free parameters than $\mathbf{S}$ so that s, the difference between the number of equations and the number of free parameters in (9.2.13), is still given by (9.2.9). The p equations for the estimates of the scaling parameters are given by (9.2.12).

Since in practice it is the relationship between the variables which is of interest rather than their scaling, the data is often summarized by $\mathbf{R}$ rather than $\mathbf{S}$. The scaling estimates given in (9.2.12) are then not mentioned explicitly, and the estimated factor loadings and specific variances are presented in terms of the standardized variables.

Example 9.2.1 Let us return to the Spearman examination data with one factor. Since $s=\frac{1}{2}(3-1)^2-\frac{1}{2}(3+1)=0$, there is an exact solution to (9.2.13), which becomes

$$\mathbf{R}=\begin{bmatrix}\hat{\lambda}_1^2+\hat{\psi}_{11} & \hat{\lambda}_1\hat{\lambda}_2 & \hat{\lambda}_1\hat{\lambda}_3\\ & \hat{\lambda}_2^2+\hat{\psi}_{22} & \hat{\lambda}_2\hat{\lambda}_3\\ & & \hat{\lambda}_3^2+\hat{\psi}_{33}\end{bmatrix}. \tag{9.2.14}$$

The unique solution (except for the sign of $(\hat{\lambda}_1,\hat{\lambda}_2,\hat{\lambda}_3)$) is given by

$$\begin{gathered}\hat{\lambda}_1^2=r_{12}r_{13}/r_{23}, \qquad \hat{\lambda}_2^2=r_{12}r_{23}/r_{13}, \qquad \hat{\lambda}_3^2=r_{13}r_{23}/r_{12},\\ \hat{\psi}_{11}=1-\hat{\lambda}_1^2, \qquad \hat{\psi}_{22}=1-\hat{\lambda}_2^2, \qquad \hat{\psi}_{33}=1-\hat{\lambda}_3^2.\end{gathered} \tag{9.2.15}$$

With the value of $\mathbf{R}$ given in Example 9.1.1, we find that

$$\hat{\lambda}_1=0.983, \qquad \hat{\lambda}_2=0.844, \qquad \hat{\lambda}_3=0.794$$

$$\hat{\psi}_{11}=0.034, \qquad \hat{\psi}_{22}=0.287, \qquad \hat{\psi}_{33}=0.370.$$

Since $\hat{\psi}_{ii}=1-\hat{h}_i^2$, the model explains a higher proportion of the variance of x_1 than of x_2 and x_3.

For general (3×3) correlation matrices $\mathbf{R}$, it is possible for $\hat{\lambda}_1$ to exceed 1 (see Exercise 9.2.6). When this happens $\hat{\psi}_{11}$ is negative. However, this solution is unacceptable because ψ_{11} is a variance. Hence we adjust $\hat{\lambda}_1$ to be 1, and use an approximate solution for the other equations in (9.2.14). This situation is known as a "Heywood case".

We shall describe two methods of estimating the parameters of the factor model when $s>0$. The first is *principal factor analysis* and uses some ideas from principal component analysis. The second method, *maximum likelihood factor analysis,* is applicable when the data is assumed to be normally distributed, and enables a significance test to be made about the validity of the k-factor model. For further techniques see Lawley and Maxwell (1971).

9.3 Principal Factor Analysis

Principal factor analysis is one method of estimating the parameters of the k-factor model when s given by (9.2.9) is positive. Suppose the data is summarized by the correlation matrix $\mathbf{R}$ so that an estimate of $\mathbf{\Lambda}$ and $\mathbf{\Psi}$ is sought for the standardized variables. (We are implicitly assuming that the variances for the original variables are estimated by $\hat{\sigma}_{ii}=s_{ii}$ in (9.2.12).)

As first step, preliminary estimates $\tilde{h}_i^2$ of the communalities $h_i^2, i=1,\ldots,p$, are made. Two common estimates of the ith communality are

(a) the square of the multiple correlation coefficient of the ith variable with all the other variables, or
(b) the largest correlation coefficient between the ith variable and one of the other variables; that is, $\max_{j\neq i}|r_{ij}|$.

Note that the estimated communality $\tilde{h}_i^2$ is higher when x_i is highly correlated with the other variables, as we would expect.

The matrix $\mathbf{R}-\tilde{\mathbf{\Psi}}$ is called the *reduced correlation matrix* because the 1s on the diagonal have been replaced by the estimated communalities $\tilde{h}_i^2=1-\tilde{\psi}_{ii}$. By the spectral decomposition theorem (Theorem A.6.4), it can be written as

$$\mathbf{R}-\tilde{\mathbf{\Psi}}=\sum_{i=1}^{p} a_i\boldsymbol{\gamma}_{(i)}\boldsymbol{\gamma}'_{(i)}, \tag{9.3.1}$$

where $a_1\geq\ldots\geq a_p$ are eigenvalues of $\mathbf{R}-\tilde{\mathbf{\Psi}}$ with orthonormal eigenvectors $\boldsymbol{\gamma}_{(1)},\ldots,\boldsymbol{\gamma}_{(p)}$. Suppose the first k eigenvalues of $\mathbf{R}-\tilde{\mathbf{\Psi}}$ are positive.

Then the ith column of $\mathbf{\Lambda}$ is estimated by

$$\hat{\boldsymbol{\lambda}}_{(i)} = a_i^{1/2}\boldsymbol{\gamma}_{(i)}, \qquad i = 1, \ldots, k; \tag{9.3.2}$$

that is, $\hat{\boldsymbol{\lambda}}_{(i)}$ is proportional to the ith eigenvector of the reduced correlation matrix. In matrix form.

$$\hat{\mathbf{\Lambda}} = \mathbf{\Gamma}_1\mathbf{A}_1^{1/2},$$

where $\mathbf{\Gamma}_1 = (\boldsymbol{\gamma}_{(1)}, \ldots, \boldsymbol{\gamma}_{(k)})$ and $\mathbf{A}_1 = \text{diag}\,(a_1, \ldots, a_k)$. Since the eigenvectors are orthogonal, we see that $\hat{\mathbf{\Lambda}}'\hat{\mathbf{\Lambda}}$ is diagonal, so the constraint (9.2.8) is satisfied. (Recall that we are working with the standardized variables here, each of whose estimated true variance is 1.)

Finally, revised estimates of the specific variances are given in terms of $\hat{\mathbf{\Lambda}}$ by

$$\hat{\psi}_{ii} = 1 - \sum_{j=1}^{k} \hat{\lambda}_{ij}^2, \qquad i = 1, \ldots, p. \tag{9.3.3}$$

Then the principal factor solution is permissible if all the $\hat{\psi}_{ii}$ are non-negative.

As a motivation for the principal factor method, let us consider what happens when the communalities are known exactly and $\mathbf{R}$ equals the true correlation matrix. Then $\mathbf{R} - \mathbf{\Psi} = \mathbf{\Lambda\Lambda}'$ exactly. The constraint (9.2.8) implies that the columns of $\mathbf{\Lambda}$ are eigenvectors of $\mathbf{\Lambda\Lambda}'$, and hence $\mathbf{\Lambda}$ is given by (9.3.2). Further the last $(p-k)$ eigenvalues of $\mathbf{R} - \mathbf{\Psi}$ vanish, $a_{k+1} = \ldots = a_p = 0$. This property can be helpful in practice when $\mathbf{R}$ and $\mathbf{\Psi}$ are not exactly equal to their theoretical values and k is not known in advance. Hopefully, for some value of k, $a_1, \ldots, a_k$ will be "large" and positive, and the remaining eigenvalues, $a_{k+1}, \ldots, a_p$, will be near zero (some possibly negative). Of course, k must always be small enough that $s \geqslant 0$ in (9.2.9).

Example 9.3.1 Consider the open/closed book data of Table 1.2.1 with correlation matrix

$$\begin{bmatrix} 1 & 0.553 & 0.547 & 0.410 & 0.389 \\ & 1 & 0.610 & 0.485 & 0.437 \\ & & 1 & 0.711 & 0.665 \\ & & & 1 & 0.607 \\ & & & & 1 \end{bmatrix}.$$

If $k > 2$ then $s < 0$ and the factor model is not well defined. The principal factor solutions for $k = 1$ and $k = 2$, where we estimate the ith communality h_i^2 by $\max_j |r_{ij}|$, are given in Table 9.3.1. The eigenvalues of the reduced correlation matrix are 2.84, 0.38, 0.08, 0.02, and -0.05, suggesting that the two-factor solution fits the data well.

Table 9.3.1 Principal factor solutions for the open/closed book data with $k=1$ and $k=2$ factors

Variable	$k=1$		$k=2$		
	h_i^2	$\lambda_{(1)}$	h_i^2	$\lambda_{(1)}$	$\lambda_{(2)}$
1	0.417	0.646	0.543	0.646	0.354
2	0.506	0.711	0.597	0.711	0.303
3	0.746	0.864	0.749	0.864	−0.051
4	0.618	0.786	0.680	0.786	−0.249
5	0.551	0.742	0.627	0.742	−0.276

The first factor represents overall performance and for $k=2$, the second factor, which is much less important ($a_2=0.38 \ll 2.84=a_1$), represents a contrast across the range of examinations. Note that even for the two-factor solution $h_i^2 \ll 1$ for all i, and therefore a fair proportion of the variance of each variable is left unexplained by the common factors.

9.4 Maximum Likelihood Factor Analysis

When the data under analysis $\mathbf{X}$ is assumed to be normally distributed, then estimates of $\mathbf{\Lambda}$ and $\mathbf{\Psi}$ can be found by maximizing the likelihood. If $\boldsymbol{\mu}$ is replaced by its m.l.e. $\hat{\boldsymbol{\mu}}=\bar{\mathbf{x}}$, then, from (4.1.9), the log likelihood function becomes

$$l=-\tfrac{1}{2}n \log|2\pi\mathbf{\Sigma}|-\tfrac{1}{2}n \operatorname{tr} \mathbf{\Sigma}^{-1}\mathbf{S}. \tag{9.4.1}$$

Regarding $\mathbf{\Sigma}=\mathbf{\Lambda}\mathbf{\Lambda}'+\mathbf{\Psi}$ as a function of $\mathbf{\Lambda}$ and $\mathbf{\Psi}$, we can maximize (9.4.1) with respect to $\mathbf{\Lambda}$ and $\mathbf{\Psi}$. Let $\hat{\mathbf{\Lambda}}$ and $\hat{\mathbf{\Psi}}$ denote the resulting m.l.e.s subject to the constraint (9.2.7) on $\mathbf{\Lambda}$.

Note that in writing down the likelihood in (9.4.1) we have summarized the data using the covariance matrix $\mathbf{S}$ rather than the correlation matrix $\mathbf{R}$. However, it can be shown that the maximum likelihood estimates are scale invariant (Exercise 9.4.1) and that the m.l.e.s of the scale parameters are given by $\hat{\sigma}_{ii}=s_{ii}$ (see Theorem 9.4.2 below). Thus, (9.2.12) is satisfied, and the m.l.e.s for the parameters of the *standardized* variables can be found by replacing $\mathbf{S}$ by $\mathbf{R}$ in (9.4.1).

However, it is more convenient in our theoretical discussion to work with $\mathbf{S}$ rather than $\mathbf{R}$. Consider the function

$$F(\mathbf{\Lambda}, \mathbf{\Psi})=F(\mathbf{\Lambda}, \mathbf{\Psi}; \mathbf{S})=\operatorname{tr} \mathbf{\Sigma}^{-1}\mathbf{S}-\log|\mathbf{\Sigma}^{-1}\mathbf{S}|-p, \tag{9.4.2}$$

where

$$\mathbf{\Sigma}=\mathbf{\Lambda}\mathbf{\Lambda}'+\mathbf{\Psi}.$$

This is a linear function of the log likelihood l and a maximum in l corresponds to a minimum in F. Note that F can be expressed as

$$F = p(a - \log g - 1), \tag{9.4.3}$$

where a and g are the arithmetic and geometric means of the eigenvalues of $\mathbf{\Sigma}^{-1}\mathbf{S}$.

The minimization of this function $F(\mathbf{\Lambda}, \mathbf{\Psi})$ can be facilitated by proceeding in two stages. First, we minimize $F(\mathbf{\Lambda}, \mathbf{\Psi})$ over $\mathbf{\Lambda}$ for fixed $\mathbf{\Psi}$, and second, we minimize over $\mathbf{\Psi}$. This approach has the advantage that the first minimization can be carried out analytically although the minimization over $\mathbf{\Psi}$ must be done numerically. It was successfully developed by Joreskog (1967).

Theorem 9.4.1 *Let $\mathbf{\Psi} > 0$ be fixed and let $\mathbf{S}^* = \mathbf{\Psi}^{-1/2}\mathbf{S}\mathbf{\Psi}^{-1/2}$. Using the spectral decomposition theorem, write*

$$\mathbf{S}^* = \mathbf{\Gamma}\mathbf{\Theta}\mathbf{\Gamma}'.$$

Then the value of $\mathbf{\Lambda}$ satisfying the constraint (9.2.7), *which minimizes $F(\mathbf{\Lambda}, \mathbf{\Psi})$ occurs when the* ith *column of $\mathbf{\Lambda}^* = \mathbf{\Psi}^{-1/2}\mathbf{\Lambda}$ is given by $\boldsymbol{\lambda}^*_{(i)} = c_i\boldsymbol{\gamma}_{(i)}$, where $c_i = [\max(\theta_i - 1, 0)]^{1/2}$ for $i = 1, \ldots, k$.* ∎

Proof Since $\mathbf{\Lambda}^*\mathbf{\Lambda}^{*\prime}(p \times p)$ has rank at most k it may be written as

$$\mathbf{\Lambda}^*\mathbf{\Lambda}^{*\prime} = \mathbf{G}\mathbf{B}\mathbf{G}',$$

where $\mathbf{B} = \operatorname{diag}(b_1, \ldots, b_k)$ contains (non-negative) eigenvalues and $\mathbf{G} = (\mathbf{g}_{(1)}, \ldots, \mathbf{g}_{(k)})$ are standardized eigenvectors. Let

$$\mathbf{M} = \mathbf{G}'\mathbf{\Gamma},$$

so that $\mathbf{G}$ may be written in terms of $\mathbf{\Gamma}$ as $\mathbf{G} = \mathbf{\Gamma}\mathbf{M}'$. We shall minimize $F(\mathbf{\Lambda}, \mathbf{\Psi})$ over possible choices for $\mathbf{M}$ and $\mathbf{B}$, in that order.

Since $\mathbf{G}'\mathbf{G} = \mathbf{I}_k$, we see that $\mathbf{M}\mathbf{M}' = \mathbf{I}_k$. From (9.2.5),

$$\begin{aligned}\mathbf{\Sigma}^* = \mathbf{\Psi}^{-1/2}\mathbf{\Sigma}\mathbf{\Psi}^{-1/2} &= \mathbf{\Lambda}^*\mathbf{\Lambda}^{*\prime} + \mathbf{I}\\ &= \mathbf{\Gamma}\mathbf{M}'\mathbf{B}\mathbf{M}\mathbf{\Gamma}' + \mathbf{I} = \mathbf{\Gamma}(\mathbf{M}'\mathbf{B}\mathbf{M} + \mathbf{I})\mathbf{\Gamma}'.\end{aligned}$$

Using (A.2.4.f), we can write

$$\mathbf{\Sigma}^{*-1} = \mathbf{\Gamma}(\mathbf{I} - \mathbf{M}'\mathbf{B}(\mathbf{I} + \mathbf{B})^{-1}\mathbf{M})\mathbf{\Gamma}'.$$

Also

$$|\mathbf{\Sigma}^*| = \prod_{i=1}^{k} (b_i + 1).$$

Thus

$$F(\mathbf{\Lambda}, \mathbf{\Psi}) = C + \sum_i \log(b_i+1) - \operatorname{tr} \mathbf{M}'\mathbf{B}(\mathbf{I}+\mathbf{B})^{-1}\mathbf{M}\mathbf{\Theta}$$
$$= C + \sum_i \log(b_i+1) - \sum_{i,j} \frac{b_i\theta_j}{1+b_i} m_{ij}^2,$$

where C depends only on $\mathbf{S}$ and $\mathbf{\Psi}$. If the eigenvalues $\{b_i\}$ and $\{\theta_j\}$ are written in decreasing order, then, for fixed $\{b_i\}$ and $\{\theta_j\}$ this quantity is minimized over $\mathbf{M}$ when $m_{ii} = 1$ for $i = 1, \ldots, k$, and $m_{ij} = 0$ for $i \neq j$. Thus $\hat{\mathbf{M}} = (\mathbf{I}_k, \mathbf{0})$ and so $\mathbf{G} = \mathbf{\Gamma}\hat{\mathbf{M}}' = (\boldsymbol{\gamma}_{(1)}, \ldots, \boldsymbol{\gamma}_{(k)})$. Then

$$F(\mathbf{\Lambda}, \mathbf{\Psi}) = C + \sum \left[\log(b_i+1) - \frac{b_i\theta_i}{b_i+1}\right].$$

The minimum of this function over $b_i \geq 0$ occurs when

$$b_i = \max(\theta_i - 1, 0), \tag{9.4.4}$$

(although the value 0 in fact occurs rarely in practice), and so the minimizing value of $\mathbf{\Lambda}^*$ satisfies

$$\mathbf{\Lambda}^*\mathbf{\Lambda}^{*\prime} = \sum b_i \boldsymbol{\gamma}_{(i)}\boldsymbol{\gamma}_{(i)}'. \tag{9.4.5}$$

If the constraint (9.2.7) is satisfied then the columns of $\mathbf{\Lambda}^*$ are the eigenvectors of $\mathbf{\Lambda}^*\mathbf{\Lambda}^{*\prime}$, and the theorem follows. ∎

When $|\mathbf{\Psi}| = 0$, the constraint of Exercise 9.2.8 can be used and special care must be taken in the minimization of $F(\mathbf{\Lambda}, \mathbf{\Psi})$. As in Example 9.2.1, we shall term a situation in which $|\hat{\mathbf{\Psi}}| = 0$, a Heywood case. Usually, the estimate $\hat{\mathbf{\Psi}}$ will be positive definite, although Heywood cases are by no means uncommon (Joreskog, 1967).

An important property of the m.l.e.s $\hat{\mathbf{\Lambda}}$ and $\hat{\mathbf{\Psi}}$ is that the estimate of the variance of the ith variable,

$$\hat{\sigma}_{ii} = \sum_{j=1}^{k} \hat{\lambda}_{ij}^2 + \hat{\psi}_{ii},$$

equals the sample variance s_{ii} for $i = 1, \ldots, p$ (Joreskog, 1967). We shall prove this result in the special case when the solution is proper, that is when $\hat{\mathbf{\Psi}} > 0$, though it is true in general.

Theorem 9.4.2 *If* $\hat{\mathbf{\Psi}} > 0$, *then* $\operatorname{Diag}(\hat{\mathbf{\Lambda}}\hat{\mathbf{\Lambda}}' + \hat{\mathbf{\Psi}} - \mathbf{S}) = \mathbf{0}$. ∎

Proof Writing $\partial F/\partial \mathbf{\Psi}$ as a diagonal matrix and using Section A.9 and the chain rule, it is straightforward to show that

$$\frac{\partial F(\mathbf{\Lambda}, \mathbf{\Psi})}{\partial \mathbf{\Psi}} = \operatorname{Diag}(\mathbf{\Sigma}^{-1}(\mathbf{\Sigma} - \mathbf{S})\mathbf{\Sigma}^{-1}), \tag{9.4.6}$$

which equals $\mathbf{0}$ at $(\hat{\Lambda}, \hat{\Psi})$ when $\hat{\Psi}$ is proper. From (9.4.4) it is clear that at the m.l. solution, $b_i^{1/2}(1+b_i)=b_i^{1/2}\theta_i$ for all $i=1,\ldots,p$. Hence,

$$\mathbf{S}^*\hat{\boldsymbol{\lambda}}^*_{(i)}=b_i^{1/2}\mathbf{S}^*\boldsymbol{\gamma}^*_{(i)}=b_i^{1/2}\theta_i\boldsymbol{\gamma}^*_{(i)}=b_i^{1/2}(1+b_i)\boldsymbol{\gamma}^*_{(i)}$$
$$=(1+b_i)\hat{\boldsymbol{\lambda}}^*_{(i)}, \qquad i=1,\ldots,p.$$

Since $\mathbf{B}=\hat{\Lambda}^{*\prime}\hat{\Lambda}^*$ and $\hat{\Sigma}^*=\hat{\Lambda}^*\hat{\Lambda}^{*\prime}+\mathbf{I}$, we can write this in matrix form as

$$\mathbf{S}^*\hat{\Lambda}^*=\hat{\Lambda}^*(\mathbf{I}+\hat{\Lambda}^{*\prime}\hat{\Lambda}^*)=(\mathbf{I}+\hat{\Lambda}^*\hat{\Lambda}^{*\prime})\hat{\Lambda}^*=\hat{\Sigma}^*\hat{\Lambda}^*.$$

Thus

$$(\hat{\Sigma}^*-\mathbf{S}^*)\hat{\Lambda}^*=\mathbf{0}.$$

Using this result together with the expansion (A.2.4f) for $\hat{\Sigma}^{*-1}$, we get

$$(\hat{\Sigma}^*-\mathbf{S}^*)\hat{\Sigma}^{*-1}=(\hat{\Sigma}^*-\mathbf{S}^*)(\mathbf{I}-\hat{\Lambda}^*(\mathbf{I}+\hat{\Lambda}^{*\prime}\hat{\Lambda}^*)^{-1}\hat{\Lambda}^{*\prime})$$
$$=\hat{\Sigma}^*-\mathbf{S}^*.$$

Then pre-and post-multiplying by $\hat{\Psi}^{1/2}$ and $\hat{\Psi}^{-1/2}$, respectively, gives

$$(\hat{\Sigma}-\mathbf{S})\hat{\Sigma}^{-1}=(\hat{\Sigma}-\mathbf{S})\hat{\Psi}^{-1}.$$

Using this formula in (9.4.6), we get

$$\frac{\partial F(\hat{\Lambda}, \hat{\Psi})}{\partial \Psi}=\text{Diag}(\hat{\Sigma}^{-1}(\hat{\Sigma}-\mathbf{S})\hat{\Sigma}^{-1})$$
$$=\text{Diag}(\hat{\Psi}^{-1}(\hat{\Sigma}-\mathbf{S})\hat{\Psi}^{-1})$$
$$=\mathbf{0},$$

and hence Diag $(\hat{\Sigma}-\mathbf{S})=\mathbf{0}$ as required. ■

Example 9.4.1 Consider the open/closed book data of Example 9.3.1. A maximum likelihood factor analysis for $k=1$ and $k=2$ factors is presented in Table 9.4.1 The value and interpretations of the factor loadings

Table 9.4.1 Maximum likelihood factor solutions for the open/closed book data with $k=1$ and $k=2$ factors

Variable	$k=1$		$k=2$			$k=2$(rotated)	
	h_i^2	$\boldsymbol{\lambda}_{(1)}$	h_i^2	$\boldsymbol{\lambda}_{(1)}$	$\boldsymbol{\lambda}_{(2)}$	$\boldsymbol{\delta}_{(1)}$	$\boldsymbol{\delta}_{(2)}$
1	0.359	0.600	0.533	0.628	0.372	0.270	0.678
2	0.446	0.667	0.582	0.696	0.313	0.360	0.673
3	0.841	0.917	0.811	0.899	−0.050	0.743	0.510
4	0.596	0.772	0.648	0.779	−0.201	0.740	0.317
5	0.524	0.724	0.569	0.728	−0.200	0.698	0.286

are broadly similar to the results of the principal factor analysis. However, caution must be exercised when comparing the two sets of factor loadings when $k=2$ because they have been rotated to satisfy different constraints ((9.2.7) and (9.2.8), respectively). Note that, unlike principal factor analysis, the factor loadings when $k=1$ do not exactly equal the first column of factor loadings when $k=2$.

9.5 Goodness of Fit Test

One of the main advantages of the maximum likelihood technique is that it provides a test of the hypothesis H_k that k common factors are sufficient to describe the data against the alternative that $\mathbf{\Sigma}$ has no constraints. We showed in (5.3.8) that the likelihood ratio statistic λ is given by

$$-2\log\lambda = np(\hat{a}-\log\hat{g}-1), \tag{9.5.1}$$

where $\hat{a}$ and $\hat{g}$ are the arithmetic and geometric means of the eigenvalues of $\hat{\mathbf{\Sigma}}^{-1}\mathbf{S}$.

Comparing this with (9.4.3) we see that

$$-2\log\lambda = nF(\hat{\mathbf{\Lambda}}, \hat{\mathbf{\Psi}}). \tag{9.5.2}$$

The statistic $-2\log\lambda$ has an asymptotic χ_s^2 distribution under H_k, where s is given by (9.2.9).

In the trivial case where $k=0$, H_0 is testing the independence of the observed variables. The m.l.e. of $\mathbf{\Sigma}$ in this case is $\hat{\mathbf{\Sigma}}=\text{Diag}\,(\mathbf{S})$, and the eigenvalues of $\hat{\mathbf{\Sigma}}^{-1}\mathbf{S}$ are the same as those of $\hat{\mathbf{\Sigma}}^{-1/2}\mathbf{S}\hat{\mathbf{\Sigma}}^{-1/2}=\mathbf{R}$. In this case (9.5.1) becomes $-n\log|\mathbf{R}|$, a statistic which has already been discussed in (5.3.17) and Section 8.4.4.

Bartlett (1954) showed that the chi-squared approximation of (9.5.1) is improved if n is replaced by

$$n' = n-1-\tfrac{1}{6}(2p+5)-\tfrac{2}{3}k. \tag{9.5.3}$$

The chi-squared approximation to which this leads can probably be trusted if $n \geq p+50$. Hence for any specified k we shall test H_k with the statistic

$$U = n'F(\hat{\mathbf{\Lambda}}, \hat{\mathbf{\Psi}}), \tag{9.5.4}$$

where n' is given by (9.5.3), F is given by (9.4.2), and $\hat{\mathbf{\Lambda}}$ and $\hat{\mathbf{\Psi}}$ are maximum likelihood estimates. When H_k is true this statistic has an asymptotic chi-squared distribution with

$$s=\tfrac{1}{2}(p-k)^2-\tfrac{1}{2}(p+k) \tag{9.5.5}$$

degrees of freedom.

Of course, only rarely is k, the number of common factors, specified in advance. In practice the problem is usually to decide how many common factors it is worth fitting to the data. To cope with this, a sequential procedure for estimating k is needed. The usual method is to start with a small value of k, say $k=0$ or $k=1$, and increase the number of common factors one by one until H_k is not rejected. However, this procedure is open to criticism because the critical values of the test criterion have not been adjusted to allow for the fact that a set of hypotheses is being tested in sequence.

For some data it may happen that the k-factor model is rejected for *all* values of k for which $s \geq 0$ in (9.5.5). In such cases we conclude that there is *no* factor model which will fit the data. Of course, if there exists a k for which $s=0$, then it is usually possible to fit the data *exactly* using the factor model. But no reduction in parameters takes place in this case, so no hypothesis testing can be carried out, and it is up to the user to decide whether this reparameterization is helpful.

Example 9.5.1 Consider the open/closed book data for which we found the maximum likelihood factor solutions in Example 9.4.1. Here $n=88$. For $k=1$ factor, the goodness of fit statistic in (9.5.4) is given by

$$U=8.65\sim\chi_5^2.$$

Since $\chi^2_{5;0.05}=11.1$, we accept the one-factor solution as adequate for this data.

9.6 Rotation of Factors

9.6.1 Interpretation of factors

The constraint (9.2.7) or (9.2.8) on the factor loadings is a mathematical convenience to make the factor loadings unique; however, it can complicate the problem of interpretation. The interpretation of the factor loadings is the most straightforward if each variable is loaded highly on at most one factor, and if all the factor loadings are either large and positive or near zero, with few intermediate values. The variables are then split into disjoint sets, each of which is associated with one factor, and perhaps some variables are left over. A factor j can then be interpreted as an average quality over those variables i for which λ_{ij} is large.

The factors $\mathbf{f}$ in the factor model (9.2.1) are mathematical abstractions and do not necessarily have any intuitive meaning. In particular, the factors may be rotated using (9.2.6) without affecting the validity of the model and we are free to choose such a rotation to make the factors as intuitively meaningful as possible.

It is considered a disadvantage to choose a rotation subjectively because the factor analyst may try to force the factor loadings to fit his own preconceived pattern. A convenient analytical choice of rotation is given by the varimax method described below.

9.6.2 Varimax rotation

The varimax method of orthogonal rotation was proposed by Kaiser (1958). Its rationale is to provide axes with a few large loadings and as many near-zero loadings as possible. This is accomplished by an iterative maximization of a quadratic function of the loadings.

Let Λ be the $(p \times k)$ matrix of unrotated loadings, and let $\mathbf{G}$ be a $(k \times k)$ orthogonal matrix. The matrix of rotated loadings is

$$\mathbf{\Delta} = \Lambda \mathbf{G}; \tag{9.6.1}$$

that is, δ_{ij} represents the loadings of the ith variable on the jth factor.

The function ϕ that the varimax criterion maximizes is the sum of the variances of the *squared* loadings within each column of the loading matrix, where each row of loadings is normalized by its communality; that is,

$$\phi = \sum_{j=1}^{k} \sum_{i=1}^{p} (d_{ij}^2 - \bar{d}_j)^2 = \sum_{j=1}^{k} \sum_{i=1}^{p} d_{ij}^4 - p \sum_{j=1}^{k} \bar{d}_j^2, \tag{9.6.2}$$

where

$$d_{ij} = \frac{\delta_{ij}}{h_i} \quad \text{and} \quad \bar{d}_j = p^{-1} \sum_{i=1}^{p} d_{ij}^2.$$

The varimax criterion ϕ is a function of $\mathbf{G}$, and the iterative algorithm proposed by Kaiser finds the orthogonal matrix $\mathbf{G}$ which maximizes ϕ. See also Horst (1965, Chap. 18), and Lawley and Maxwell (1971, p. 72).

In the case where $k = 2$, the calculations simplify. For then $\mathbf{G}$ is given by

$$\mathbf{G} = \begin{bmatrix} \cos\theta & \sin\theta \\ -\sin\theta & \cos\theta \end{bmatrix}$$

and represents a rotation of the coordinate axes clockwise by an angle θ.

Thus,

$$d_{i1}=\{\lambda_{i1}\cos\theta-\lambda_{i2}\sin\theta\}/h_i,\qquad d_{i2}=\{\lambda_{i1}\sin\theta+\lambda_{i2}\cos\theta\}/h_i.$$

Let

$$G_{a,b}=\sum_{i=1}^{p}\frac{\lambda_{i1}^a\lambda_{i2}^b}{h_i^{a+b}}.$$

Substituting in (9.6.2) and using

$$4(\cos^4\theta+\sin^4\theta)=3+\cos 4\theta,\qquad \sin 2\theta=2\sin\theta\cos\theta,$$
$$\cos 2\theta=\cos^2\theta-\sin^2\theta,$$

it can be shown that

$$4\phi=(A^2+B^2)^{1/2}\cos(4\theta-\alpha)+C, \tag{9.6.3}$$

where

$$A=(G_{0,4}+G_{4,0}-6G_{2,2}-G_{0,2}^2-G_{2,0}^2+2G_{0,2}G_{2,0}+4G_{1,1}^2), \tag{9.6.4}$$
$$B=4(G_{1,3}-G_{3,1}-G_{1,1}G_{0,2}+G_{1,1}G_{2,0}), \tag{9.6.5}$$
$$C=p(3[G_{2,0}+G_{0,2}]^2-[3G_{0,2}^2+3G_{2,0}^2+2G_{0,2}G_{2,0}+4G_{1,1}^2]).$$

and

$$(A^2+B^2)^{1/2}\cos\alpha=A,\qquad (A^2+B^2)^{1/2}\sin\alpha=B. \tag{9.6.6}$$

In (9.6.3), the maximum value of ϕ is obtained when $4\theta=\alpha$. The value of α is obtained from (9.6.6) using

$$\tan\alpha=B/A \tag{9.6.7}$$

and α is uniquely determined from a consideration of the signs of A and B. Note that θ can take four possible values.

In the case of $k>2$ factors, an iterative solution for the rotation is used. The first and second factors are rotated by an angle determined by the above method. The new first factor is then rotated with the original third factor, and so on, until all the $\frac{1}{2}k(k-1)$ pairs of factors have been rotated. This sequence of rotations is called a cycle. These cycles are then repeated until one is completed in which all the angles have achieved some predetermined convergence criterion. Since the effect of each rotation is to increase the value of ϕ and the values of ϕ are bounded above, the iteration will converge (see Kaiser, 1958).

Example 9.6.1 Consider the two factor maximum likelihood solutions for the open/closed book data given in Example 9.4.1. Using (9.6.3)–9.6.7) it is found that $\theta=37.6°$ for the varimax rotation and the rotated factors are given in Table 9.4.1. The rotation can be represented graphically (Figure 9.4.1) by plotting the factor loadings as $p=5$ points in $k=2$ dimensions. The first factor now represents an "open-book" effect and the second factor a "closed-book" effect, although both factors influence the other exams as well.

In this simple example the overall interpretation is the same whether we use the rotated or the original factors. However, in more complicated situations, the benefits of rotation are much more noticeable.

Example 9.6.2 (Kendall, 1975). In a job interview, 48 applicants were each judged on 15 variables. The variables were

(1) Form of letter of application
(2) Appearance
(3) Academic ability
(4) Likeability
(5) Self-confidence
(6) Lucidity
(7) Honesty
(8) Salesmanship
(9) Experience
(10) Drive
(11) Ambition
(12) Grasp
(13) Potential
(14) Keenness to join
(15) Suitability

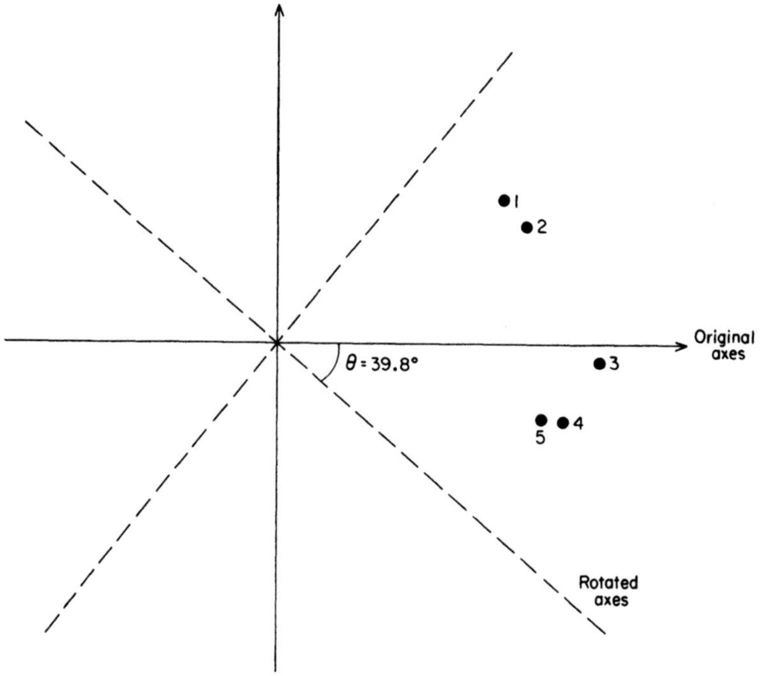

Figure 9.4.1 Plot of open/closed book factors before and after varimax rotation.

Table 9.6.1 Correlation matrix for the applicant data (Kendall, 1975)

	1	2	3	4	5	6	7	8	9	10	11	12	13	14	15
1	1.00	0.24	0.04	0.31	0.09	0.23	−0.11	0.27	0.55	0.35	0.28	0.34	0.37	0.47	0.59
2		1.00	0.12	0.38	0.43	0.37	0.35	0.48	0.14	0.34	0.55	0.51	0.51	0.28	0.38
3			1.00	0.00	0.00	0.08	−0.03	0.05	0.27	0.09	0.04	0.20	0.29	−0.32	0.14
4				1.00	0.30	0.48	0.65	0.35	0.14	0.39	0.35	0.50	0.61	0.69	0.33
5					1.00	0.81	0.41	0.82	0.02	0.70	0.84	0.72	0.67	0.48	0.25
6						1.00	0.36	0.83	0.15	0.70	0.76	0.88	0.78	0.53	0.42
7							1.00	0.23	−0.16	0.28	0.21	0.39	0.42	0.45	0.00
8								1.00	0.23	0.81	0.86	0.77	0.73	0.55	0.55
9									1.00	0.34	0.20	0.30	0.35	0.21	0.69
10										1.00	0.78	0.71	0.79	0.61	0.62
11											1.00	0.78	0.77	0.55	0.43
12												1.00	0.88	0.55	0.53
13													1.00	0.54	0.57
14														1.00	0.40
15															1.00

and the correlation matrix between them is given in Table 9.6.1. Because many of the correlations were quite high, it was felt that the judge might be confusing some of the variables in his own mind. Therefore a factor analysis was performed to discover the underlying types of variation in the data.

A maximum likelihood factor analysis was carried out and, using the

Table 9.6.2 Maximum likelihood factor solution of applicant data with $k = 7$ factors, unrotated

Variable	Factor loadings						
	1	2	3	4	5	6	7
1	0.090	−0.134	−0.338	0.400	0.411	−0.001	0.277
2	−0.466	0.171	0.037	−0.002	0.517	−0.194	0.167
3	−0.131	0.466	0.153	0.143	−0.031	0.330	0.316
4	0.004	−0.023	−0.318	−0.362	0.657	0.070	0.307
5	−0.093	0.017	0.434	−0.092	0.784	0.019	−0.213
6	0.281	0.212	0.330	−0.037	0.875	0.001	0.000
7	−0.133	0.234	−0.181	−0.807	0.494	0.001	−0.000
8	−0.018	0.055	0.258	0.207	0.853	0.019	−0.180
9	−0.043	0.173	−0.345	0.522	0.296	0.085	0.185
10	−0.079	−0.012	0.058	0.241	0.817	0.417	−0.221
11	−0.265	−0.131	0.411	0.201	0.839	−0.000	−0.001
12	0.037	0.202	0.188	0.025	0.875	0.077	0.200
13	−0.112	0.188	0.109	0.061	0.844	0.324	0.277
14	0.098	−0.462	−0.336	−0.116	0.807	−0.001	0.000
15	−0.056	0.293	−0.441	0.577	0.619	0.001	−0.000

Table 9.6.3 Maximum likelihood factor solution of applicant data with $k = 7$ factors, varimax rotation (Kendall, 1975)

Variable	Factor loadings						
	1	2	3	4	5	6	7
1	0.129	0.074	0.665	−0.096	0.017	−0.042	0.267
2	0.329	0.242	0.182	0.095	0.611	−0.013	−0.006
3	0.048	−0.017	0.097	0.688	0.043	0.007	0.008
4	0.249	0.759	0.252	−0.058	0.090	−0.096	0.204
5	0.882	0.184	−0.082	−0.074	0.190	0.059	−0.045
6	0.907	0.266	0.136	0.046	−0.042	−0.290	−0.016
7	0.199	0.911	−0.224	−0.013	0.174	−0.094	−0.204
8	0.875	0.082	0.264	−0.076	0.140	0.043	−0.058
9	0.073	−0.027	0.718	0.158	0.069	0.036	0.009
10	0.780	0.197	0.386	0.026	−0.051	0.398	−0.023
11	0.874	0.036	0.157	−0.052	0.382	0.142	0.205
12	0.775	0.346	0.286	0.172	0.143	−0.159	0.111
13	0.703	0.409	0.354	0.329	0.140	0.070	0.193
14	0.432	0.540	0.381	−0.540	−0.013	0.099	0.275
15	0.313	0.079	0.909	0.049	0.142	0.027	−0.214

hypothesis test described in Section 9.5, it was found that seven factors were needed to explain the data adequately. The original factor loadings are given in Table 9.6.2 and the loadings after the varimax rotation are given in Table 9.6.3.

It is very difficult to interpret the unrotated loadings. However, a pattern is clear from the rotated loadings. The first factor is loaded heavily on variables 5, 6, 8, 10, 11, 12, and 13 and represents perhaps an outward and salesmanlike personality. Factor 2, weighting variables 4 and 7, represents likeability and factor 3, weighting variables 1, 9, and 15, represents experience. Factors 4 and 5 each represent one variable, academic ability (3), and appearance (2), respectively. The last two factors have little importance and variable 14 (keenness) seemed to be associated with several of the factors.

Thus, the judge seems to have measured rather fewer qualities in his candidates than he thought.

9.7 Factor Scores

So far our study of the factor model has been concerned with the way in which the observed variables are functions of the (unknown) factors. For example, in Spearman's examination data we can describe the way in which a child's test scores will depend upon his overall intelligence.

However, it is also of interest to ask the converse question. Given a particular child's test scores, can we make any statement about his overall intelligence? For the general model we want to know how the factors depend upon the observed variables.

One way to approach the problem is to treat the unknown *common factor scores* as parameters to be estimated. Suppose $\mathbf{x}$ is a multinormal random vector from the factor model (9.2.1) and suppose that $\mathbf{\Lambda}$ and $\mathbf{\Psi}$ and $\boldsymbol{\mu}=\mathbf{0}$ are known. Given the vector $\mathbf{f}(p\times 1)$ of common factor scores, $\mathbf{x}$ is distributed as $N_p(\mathbf{\Lambda f}, \mathbf{\Psi})$. Thus, the log likelihood of $\mathbf{x}$ is given by

$$l(\mathbf{x};\mathbf{f}) = -\tfrac{1}{2}(\mathbf{x}-\mathbf{\Lambda f})'\mathbf{\Psi}^{-1}(\mathbf{x}-\mathbf{\Lambda f})-\tfrac{1}{2}\log|2\pi\mathbf{\Psi}|. \tag{9.7.1}$$

Setting the derivative with respect to $\mathbf{f}$ equal to $\mathbf{0}$ gives

$$\frac{\partial l}{\partial \mathbf{f}} = \mathbf{\Lambda}'\mathbf{\Psi}^{-1}(\mathbf{x}-\mathbf{\Lambda f}) = \mathbf{0},$$

so

$$\hat{\mathbf{f}} = (\mathbf{\Lambda}'\mathbf{\Psi}^{-1}\mathbf{\Lambda})^{-1}\mathbf{\Lambda}'\mathbf{\Psi}^{-1}\mathbf{x}. \tag{9.7.2}$$

The estimate in (9.7.2) is known as *Bartlett's factor score*. The specific factor scores can then be estimated by $\hat{\mathbf{u}} = \mathbf{x}-\mathbf{\Lambda}\hat{\mathbf{f}}$.

Note that (9.7.1) is the logarithm of the conditional density of $\mathbf{x}$ *given* $\mathbf{f}$. However, under the factor model, $\mathbf{f}$ can be considered as an $N_p(\mathbf{0}, \mathbf{I})$ random vector, thus giving $\mathbf{f}$ a prior distribution. Using this Bayesian approach, the posterior density of $\mathbf{f}$ is proportional to

$$\exp\left[-\tfrac{1}{2}(\mathbf{x}-\mathbf{\Lambda f})'\mathbf{\Psi}^{-1}(\mathbf{x}-\mathbf{\Lambda f})-\tfrac{1}{2}\mathbf{f}'\mathbf{f}\right], \tag{9.7.3}$$

which is a multinormal density whose mean

$$\mathbf{f}^* = (\mathbf{I}+\mathbf{\Lambda}'\mathbf{\Psi}^{-1}\mathbf{\Lambda})^{-1}\mathbf{\Lambda}'\mathbf{\Psi}^{-1}\mathbf{x} \tag{9.7.4}$$

is the Bayesian estimate of $\mathbf{f}$. The estimate in (9.7.4) is known as *Thompson's factor score*.

Each of these two factor scores has some favourable properties and there has been a long controversy as to which is better. For example

$$E(\hat{\mathbf{f}}\mid\mathbf{f}) = \mathbf{f}, \qquad E(\mathbf{f}^*\mid\mathbf{f}) = (\mathbf{I}+\mathbf{\Lambda}'\mathbf{\Psi}^{-1}\mathbf{\Lambda})^{-1}\mathbf{\Lambda}'\mathbf{\Psi}^{-1}\mathbf{\Lambda f} \tag{9.7.5}$$

so that Bartlett's factor score is an unbiased estimate of $\mathbf{f}$, whereas Thompson's score is biased. However, the average prediction errors are given by

$$E((\hat{\mathbf{f}}-\mathbf{f})(\hat{\mathbf{f}}-\mathbf{f})') = (\mathbf{\Lambda}'\mathbf{\Psi}^{-1}\mathbf{\Lambda})^{-1}, \tag{9.7.6}$$

$$E((\mathbf{f}^*-\mathbf{f})(\mathbf{f}^*-\mathbf{f})') = (\mathbf{I}+\mathbf{\Lambda}'\mathbf{\Psi}^{-1}\mathbf{\Lambda})^{-1}, \tag{9.7.7}$$

so that Thompson's score is more accurate. If the columns of $\mathbf{\Lambda}$ satisfy the constraint (9.2.7), then, for either factor score, the components of the factor score are uncorrelated with one another. Note that if the eigenvalues of $\boldsymbol{\Lambda}'\boldsymbol{\Psi}^{-1}\boldsymbol{\Lambda}$ are all large, then the prediction errors will be small, and also the two factor scores will be similar to one another.

Of course in practice $\boldsymbol{\Lambda}$, $\boldsymbol{\Psi}$, and $\boldsymbol{\mu}$ are not known in advance but are estimated from the same data for which we wish to know the factor scores. It would be theoretically attractive to estimate the factor scores, factor loadings, and specific variances all at the same time from the data, using maximum likelihood. However, there are too many parameters for this to be possible. There are many values of the parameters for which the likelihood becomes infinite. See Exercise 9.7.2.

9.8 Relationships Between Factor Analysis and Principal Component Analysis

Factor analysis, like principal component analysis, is an attempt to explain a set of data in a smaller number of dimensions than one starts with. Because the overall goals are quite similar, it is worth looking at the differences between these two approaches.

First, principal component analysis is merely a transformation of the data. No assumptions are made about the form of the covariance matrix from which the data comes. On the other hand, factor analysis supposes that the data comes from the well-defined model (9.2.1), where the underlying factors satisfy the assumptions (9.2.2)–(9.2.4). If these assumptions are not met, then factor analysis may give spurious results.

Second, in principal component analysis the emphasis is on a transformation from the observed variables to the principal components ($\mathbf{y}=\boldsymbol{\Gamma}'\mathbf{x}$), whereas in factor analysis the emphasis is on a transformation *from* the underlying factors *to* the observed variables. Of course, the principal component transformation is invertible ($\mathbf{x}=\boldsymbol{\Gamma}\mathbf{y}$), and if we have decided to retain the first k components, then $\mathbf{x}$ can be approximated by these components,

$$\mathbf{x}=\boldsymbol{\Gamma}\mathbf{y}=\boldsymbol{\Gamma}_1\mathbf{y}_1+\boldsymbol{\Gamma}_2\mathbf{y}_2 \doteq \boldsymbol{\Gamma}_1\mathbf{y}_1.$$

However, this point of view is less natural than in factor analysis where $\mathbf{x}$ can be approximated in terms of the common factors

$$\mathbf{x} \doteq \boldsymbol{\Lambda}\mathbf{f}$$

and the neglected specific factors are explicitly assumed to be "noise".

Note that in Section 9.3, when the specific variances are assumed to be 0, principal factor analysis is equivalent to principal component analysis.

Thus, if the factor model holds and if the specific variances are small, we expect principal component analysis and factor analysis to give similar results. However, if the specific variances are large, they will be absorbed into all the principal components, both retained and rejected, whereas factor analysis makes special provision for them.

9.9 Analysis of Covariance Structures

The factor model can be generalized in the following way (Joreskog, 1970). Let $\mathbf{X}(n \times p)$ be a random data matrix with mean

$$E(\mathbf{X}) = \mathbf{A}\boldsymbol{\Xi}\mathbf{P}, \tag{9.9.1}$$

where $\mathbf{A}(n \times g)$ and $\mathbf{P}(h \times p)$ are known matrices and $\boldsymbol{\Xi}(g \times h)$ is a matrix of parameters. Suppose that each row of $\mathbf{X}$ is normally distributed with the same covariance matrix

$$\boldsymbol{\Sigma} = \mathbf{B}(\boldsymbol{\Lambda}\boldsymbol{\Phi}\boldsymbol{\Lambda}' + \boldsymbol{\Psi})\mathbf{B} + \boldsymbol{\Theta}. \tag{9.9.2}$$

Here $\mathbf{B}(p \times q)$, $\Lambda(q \times k)$ the symmetric matrix $\boldsymbol{\Phi} \geq 0$, and the diagonal matrices $\boldsymbol{\Psi} \geq 0$ and $\boldsymbol{\Theta} \geq 0$ are all parameters. In applications some of the parameters may be fixed or constrained. For example in factor analysis all the rows of $\mathbf{X}$ have the same mean and we take $\mathbf{B} = \mathbf{I}$, $\boldsymbol{\Phi} = \mathbf{I}$, and $\boldsymbol{\Theta} = \mathbf{0}$.

The general model represented by (9.9.1) and (9.9.2) also includes other models where the means and covariances are structured in terms of parameters to be estimated, for instance multivariate regression, MANOVA, path analysis, and growth curve analysis. For details see Joreskog (1970). The practical importance of the general model lies in the fact that it can be successfully programmed. For a specific application and further details, see Joreskog and Sorbom (1977).

Exercises and Complements

9.2.1 In the factor model (9.2.1) prove that $\boldsymbol{\Sigma} = \boldsymbol{\Lambda}\boldsymbol{\Lambda}' + \boldsymbol{\Psi}$.

9.2.2 (a) If $\mathbf{x}$ is a random vector whose covariance matrix can be written in the form $\boldsymbol{\Sigma} = \boldsymbol{\Lambda}\boldsymbol{\Lambda}' + \boldsymbol{\Psi}$, show that there exist factors $\mathbf{f}, \mathbf{u}$ such that the k-factor model holds for $\mathbf{x}$. Hint: let $\mathbf{y} \sim N_k(\mathbf{0}, \mathbf{I} + \boldsymbol{\Lambda}'\boldsymbol{\Psi}^{-1}\boldsymbol{\Lambda})$ independently of $\mathbf{x}$ and define

$$\begin{pmatrix} \mathbf{u} \\ \mathbf{f} \end{pmatrix} = \begin{pmatrix} \mathbf{I}_p & \boldsymbol{\Lambda} \\ -\boldsymbol{\Lambda}'\boldsymbol{\Psi}^{-1} & \mathbf{I}_k \end{pmatrix}^{-1} \begin{pmatrix} \mathbf{x} - \boldsymbol{\mu} \\ \mathbf{y} \end{pmatrix}.$$

(*b*) If $\mathbf{x}$ is normally distributed, show that $(\mathbf{f}, \mathbf{u})$ may be assumed to be normally distributed.

(c) Show that $\mathbf{f}$ and $\mathbf{u}$ are not uniquely determined by $\mathbf{x}$.

9.2.3 If $\mathbf{\Lambda\Lambda}' = \mathbf{\Delta\Delta}'$ ($\mathbf{\Lambda}$ and $\mathbf{\Delta}$ are $(p \times k)$ matrices), show that $\mathbf{\Delta} = \mathbf{\Lambda G}$ for some $(k \times k)$ orthogonal matrix $\mathbf{G}$. (Hint: use Theorem A.6.5)

9.2.4 If $\mathbf{\Lambda\Lambda}' = \mathbf{\Delta\Delta}'$, and $\mathbf{\Lambda}'\mathbf{\Psi}^{-1}\mathbf{\Lambda}$ and $\mathbf{\Delta}'\mathbf{\Psi}^{-1}\mathbf{\Delta}$ are both diagonal matrices with distinct elements written in decreasing order, show that $\boldsymbol{\lambda}_{(i)} = \pm \boldsymbol{\delta}_{(i)}$, $i = 1, \ldots, k$. (Hint: Write

$$\mathbf{\Lambda\Lambda}' = \mathbf{\Psi}^{1/2}\left[\sum_{i=1}^{k} a_i \boldsymbol{\gamma}_{(i)}\boldsymbol{\gamma}'_{(i)}\right]\mathbf{\Psi}^{1/2},$$

where $\boldsymbol{\gamma}_{(i)}$ is the ith standardized eigenvector of $\mathbf{\Psi}^{-1/2}\mathbf{\Lambda\Lambda}'\mathbf{\Psi}^{-1/2}$ with eigenvalue a_i, and show that $\mathbf{\Psi}^{-1/2}\boldsymbol{\lambda}_{(i)} = \pm a_i^{1/2}\boldsymbol{\gamma}_{(i)}$. Now do the same thing for $\mathbf{\Delta}$.)

9.2.5 If $\mathbf{S}(3 \times 3)$ is a sample covariance matrix, solve $\mathbf{S} = \hat{\mathbf{\Lambda}}\hat{\mathbf{\Lambda}}' + \hat{\mathbf{\Psi}}$ for $\hat{\mathbf{\Lambda}}$ and $\hat{\mathbf{\Psi}}$. Compare the answer with (9.2.15) to show that the solution is scale invariant.

9.2.6 Let $\mathbf{R}$ be a (3×3) correlation matrix with $r_{12} = r_{13} = \frac{1}{3}$, $r_{23} = \frac{1}{10}$. Check that $\mathbf{R}$ is positive definite. Show that $\hat{\lambda}_1 > 1$ in (9.2.15) and hence that the exact factor solution is unacceptable.

9.2.7 (Structural relationships) Suppose a scientist has p instruments with which to measure a physical quantity ξ. He knows that each instrument measures a linear function of ξ but all the measurements are subject to (independent) errors. The scientist wishes to calibrate his instruments (that is, discover what the linear relationship between the different instruments would be if there were no error) and to discover the accuracy of each instrument.

If the scientist takes a sample of measurements for which the ξs can be regarded as realizations of a random variable, then show that his problem can be approached using the 1-factor model (9.2.1). In particular if $p = 2$, show that without more information about the error variances, he does not have enough information to solve his problem. (See Example 8.2.5.)

9.2.8 Suppose the factor model (9.2.1) is valid with some of the ψ_{ii} equal to 0. Then of course the constraint (9.2.7) is not appropriate and we consider an alternative constraint. For simplicity of notation suppose $\psi_{11} = \ldots = \psi_{gg} = 0$ and $\psi_{g+1,g+1}, \ldots, \psi_{pp} > 0$, and partition

$$\mathbf{\Sigma} = \begin{bmatrix} \mathbf{\Sigma}_{11} & \mathbf{\Sigma}_{12} \\ \mathbf{\Sigma}_{21} & \mathbf{\Sigma}_{22} \end{bmatrix} \quad \text{and} \quad \mathbf{\Lambda} = \begin{bmatrix} \mathbf{\Lambda}_{11} & \mathbf{\Lambda}_{12} \\ \mathbf{\Lambda}_{21} & \mathbf{\Lambda}_{22} \end{bmatrix}.$$

We shall explain the variation in the first g variables exactly in terms of the first g factors and explain the part of the variation of the last $(p-g)$ variables not explained by the first g variables $(\mathbf{\Sigma}_{22.1}=\mathbf{\Sigma}_{22}-\mathbf{\Sigma}_{21}\mathbf{\Sigma}_{11}^{-1}\mathbf{\Sigma}_{12})$ in terms of the remaining $k-g$ factors. Let $\mathbf{\Sigma}_{11}=\mathbf{\Gamma\Theta\Gamma}'$ be a spectral decomposition of $\mathbf{\Sigma}_{11}(g\times g)$, where $\mathbf{\Theta}$ is a diagonal matrix of eigenvalues and the columns of $\mathbf{\Gamma}$ are standardized eigenvectors. Suppose $\mathbf{\Lambda}^*$ satisfies

$$\mathbf{\Lambda}_{11}^*=\mathbf{\Gamma\Theta}^{1/2}, \qquad \mathbf{\Lambda}_{12}^*=\mathbf{0},$$

$$\mathbf{\Lambda}_{21}^*=\mathbf{\Sigma}_{21}\mathbf{\Gamma\Theta}^{-1/2}, \qquad \mathbf{\Lambda}_{22}^{*\prime}\mathbf{\Psi}_2^{-1}\mathbf{\Lambda}_{22}^* \text{ is diagonal,}$$

$$\mathbf{\Lambda}_{22}^*\mathbf{\Lambda}_{22}^{*\prime}+\mathbf{\Psi}_2=\mathbf{\Sigma}_{22.1},$$

where $\mathbf{\Psi}_2(p-g)\times(p-g)$ is a diagonal matrix with elements ψ_{ii}, $i>g$. Show that

(a) if $\mathbf{\Lambda}^*$ satisfies these conditions then

$$\mathbf{\Sigma}=\mathbf{\Lambda\Lambda}'+\mathbf{\Psi}=\mathbf{\Lambda}^*\mathbf{\Lambda}^{*\prime}+\mathbf{\Psi};$$

(b) there always exists an orthogonal matrix $\mathbf{G}(k\times k)$ such that

$$\mathbf{\Lambda}=\mathbf{\Lambda}^*\mathbf{G};$$

(c) except for the signs and order of the columns, $\mathbf{\Lambda}^*$ is uniquely determined by the above conditions.

9.3.1 Suppose in a principal factor analysis that all the preliminary estimates of the specific variances are given by the same number $\tilde{\psi}_{ii}=c$.

(a) If $c=0$, show that the columns of $\hat{\mathbf{\Lambda}}$ are the first k principal component loading vectors of the correlation matrix $\mathbf{R}$, scaled so that $\hat{\boldsymbol{\lambda}}_{(i)}'\hat{\boldsymbol{\lambda}}_{(i)}$ equals the variance of the ith principal component.

(b) If $c>0$ show that the columns of $\hat{\mathbf{\Lambda}}$ are still given by the principal component loading vectors of $\mathbf{R}$, but with a different scaling.

(c) If $k>1$ and $\mathbf{R}=(1-\rho)\mathbf{I}+\rho\mathbf{1}\mathbf{1}'$, $0<\rho<1$, is the equicorrelation matrix, show that the largest value of c which gives an acceptable principal factor solution is $c=1-\rho$.

9.4.1 Let $F(\mathbf{\Lambda},\mathbf{\Psi};\mathbf{S})$ be given by (9.4.2) and let $\hat{\mathbf{\Lambda}},\hat{\mathbf{\Psi}}$ denote the minimizing values of $\mathbf{\Lambda}$ and $\mathbf{\Psi}$. If $\mathbf{C}$ is a fixed diagonal matrix with positive elements, show that $\mathbf{\Lambda}=\mathbf{C}\hat{\mathbf{\Lambda}}$ and $\mathbf{\Psi}=\mathbf{C}\hat{\mathbf{\Psi}}\mathbf{C}$ minimize the function $F(\mathbf{\Lambda},\mathbf{\Psi};\mathbf{CSC})$. Hence deduce that the maximum likelihood estimators are scale invariant.

9.4.2 Explain why the hypothesis test described in Section 9.5 can be constructed with only a knowledge of $\mathbf{R}$ (the sample variances s_{ii} not being needed).

9.6.1 (Maxwell, 1961; Joreskog, 1967) A set of $p = 10$ psychological variables were measured on $n = 810$ normal children, with correlations given as follows:

Tests	2	3	4	5	6	7	8	9	10
1	0.345	0.594	0.404	0.579	−0.280	−0.449	−0.188	−0.303	−0.200
2		0.477	0.338	0.230	−0.159	−0.205	−0.120	−0.168	−0.145
3			0.498	0.505	−0.251	−0.377	−0.186	−0.273	−0.154
4				0.389	−0.168	−0.249	−0.173	−0.195	−0.055
5					−0.151	−0.285	−0.129	−0.159	−0.079
6						0.363	0.359	0.227	0.260
7							0.448	0.439	0.511
8								0.429	0.316
9									0.301

A maximum likelihood factor analysis was carried out with $k = 4$ factors yielding the following estimate $\hat{\mathbf{\Lambda}}$ of the factor loadings (rotated by varimax):

$$\begin{bmatrix} -0.03 & 0.59 & -0.31 & 0.41 \\ -0.04 & 0.09 & -0.12 & 0.59 \\ -0.06 & 0.42 & -0.20 & 0.69 \\ -0.11 & 0.33 & -0.08 & 0.48 \\ -0.06 & 0.76 & -0.07 & 0.24 \\ 0.23 & -0.11 & 0.36 & -0.17 \\ 0.15 & -0.24 & 0.78 & -0.16 \\ 0.93 & -0.04 & 0.37 & -0.06 \\ 0.27 & -0.11 & 0.44 & -0.18 \\ 0.09 & -0.00 & 0.63 & -0.04 \end{bmatrix}$$

(a) Give interpretations of the four factors using the factor loadings.

(b) The m.l. factor loadings give a value of $F(\hat{\mathbf{\Lambda}}, \hat{\mathbf{\Psi}}) = 0.0228$ in (9.4.2). Using (9.5.4), carry out a test of the hypothesis that four factors are adequate to describe the data.

9.7.1 Verify formulae (9.7.5)–(9.7.7) for the factor scores. (Hint: in (9.7.7) use (A.2.4f) to show that

$$\begin{aligned} E(\mathbf{f}^*\mathbf{f}^{*\prime}) = E(\mathbf{f}^*\mathbf{f}') &= (\mathbf{I}+\mathbf{\Lambda}'\mathbf{\Psi}^{-1}\mathbf{\Lambda})^{-1}\mathbf{\Lambda}'\mathbf{\Psi}^{-1}\mathbf{\Lambda} \\ &= \mathbf{I}-(\mathbf{I}+\mathbf{\Lambda}'\mathbf{\Psi}^{-1}\mathbf{\Lambda})^{-1}.) \end{aligned}$$

9.7.2 The following example shows that the specific variances and factor scores cannot be estimated together. For simplicity let $k=1$ and suppose that $\mathbf{\Lambda}=\boldsymbol{\lambda}(p\times 1)$ and $\boldsymbol{\mu}=\mathbf{0}$ are known. Let $\mathbf{F}=\mathbf{f}(n\times 1)$ denote the unknown factor scores. Then, extending (9.7.1), the likelihood of the data $\mathbf{X}$ can be written

$$L=|2\pi\mathbf{\Psi}|^{-1/2}\exp\left[-\tfrac{1}{2}\sum_{i=1}^{p}(\mathbf{x}_{(i)}-\lambda_i\mathbf{f})'(\mathbf{x}_{(i)}-\lambda_i\mathbf{f})/\psi_{ii}\right].$$

(Note that $\mathbf{f}(n\times 1)$ here represents the scores of n individuals on one factor whereas in (9.7.1) $\mathbf{f}(k\times 1)$ represented the scores on k factors of one individual.) Suppose $\lambda_1\neq 0$. Show that for any values of $\psi_{22},\ldots,\psi_{pp}$ if $\mathbf{f}=\lambda_1^{-1}\mathbf{x}_{(1)}$ and $\psi_{11}\to 0$, then the likelihood becomes infinite. Hence m.l.e.s do not exist for this problem

10
Canonical Correlation Analysis

10.1 Introduction

Canonical correlation analysis involves partitioning a collection of variables into two sets, an **x**-set and a **y**-set. The object is then to find linear combinations $\eta = \mathbf{a}'\mathbf{x}$ and $\phi = \mathbf{b}'\mathbf{y}$ such that η and ϕ have the *largest possible* correlation. Such linear combinations can give insight into the relationships between the two sets of variables.

Canonical correlation analysis has certain maximal properties similar to those of principal component analysis. However, whereas principal component analysis considers interrelationships *within* a set of variables, the focus of canonical correlation is on the relationship *between* two groups of variables.

One way to view canonical correlation analysis is as an extension of multiple regression. Recall that in multiple regression analysis the variables are partitioned into an **x**-set containing q variables and a y-set containing $p = 1$ variable. The regression solution involves finding the linear combination $\mathbf{a}'\mathbf{x}$ which is most highly correlated with y.

In canonical correlation analysis the **y**-set contains $p \geq 1$ variables and we look for vectors **a** and **b** for which the correlation between $\mathbf{a}'\mathbf{x}$ and $\mathbf{b}'\mathbf{y}$ is maximized. If **x** is interpreted as "causing" **y**, then $\mathbf{a}'\mathbf{x}$ may be called the "best predictor" and $\mathbf{b}'\mathbf{y}$ the "most predictable criterion". However, there is no assumption of causal asymmetry in the mathematics of canonical correlation analysis; **x** and **y** are treated symmetrically.

Example 10.1.1 Consider the open/closed book data of Table 1.2.1 and split the five variables into two sets—the closed-book exams (x_1, x_2) and the open-book exams (y_1, y_2, y_3). One possible quantity of interest here is how highly a student's ability on closed-book exams is correlated with his ability on open-book exams.

Alternatively, one might try to use the open-book exam results to predict the closed-book results (or vice versa). This data is explored further in Example 10.2.5.

10.2 Mathematical Development

10.2.1 Population canonical correlation analysis

Suppose that $\mathbf{x}$ is a q-dimensional random vector and that $\mathbf{y}$ is a p-dimensional random vector. Suppose further that $\mathbf{x}$ and $\mathbf{y}$ have means $\boldsymbol{\mu}$ and $\boldsymbol{\nu}$, and that

$$E\{(\mathbf{x}-\boldsymbol{\mu})(\mathbf{x}-\boldsymbol{\mu})'\}=\boldsymbol{\Sigma}_{11}, \qquad E\{(\mathbf{y}-\boldsymbol{\nu})(\mathbf{y}-\boldsymbol{\nu})'\}=\boldsymbol{\Sigma}_{22},$$
$$E\{(\mathbf{x}-\boldsymbol{\mu})(\mathbf{y}-\boldsymbol{\nu})'\}=\boldsymbol{\Sigma}_{12}=\boldsymbol{\Sigma}_{21}'.$$

Now consider the two linear combinations $\eta=\mathbf{a}'\mathbf{x}$ and $\phi=\mathbf{b}'\mathbf{y}$. The correlation between η and ϕ is

$$\rho(\mathbf{a},\mathbf{b})=\frac{\mathbf{a}'\boldsymbol{\Sigma}_{12}\mathbf{b}}{(\mathbf{a}'\boldsymbol{\Sigma}_{11}\mathbf{a}\mathbf{b}'\boldsymbol{\Sigma}_{22}\mathbf{b})^{1/2}}. \qquad (10.2.1)$$

We use the notation $\rho(\mathbf{a},\mathbf{b})$ to emphasize the fact that this correlation varies with different values of $\mathbf{a}$ and $\mathbf{b}$.

Now we may ask what values of $\mathbf{a}$ and $\mathbf{b}$ maximize $\rho(\mathbf{a},\mathbf{b})$. Equivalently we can solve the problem

$$\max_{\mathbf{a},\mathbf{b}} \mathbf{a}'\boldsymbol{\Sigma}_{12}\mathbf{b} \quad \text{subject to} \quad \mathbf{a}'\boldsymbol{\Sigma}_{11}\mathbf{a}=\mathbf{b}'\boldsymbol{\Sigma}_{22}\mathbf{b}=1, \qquad (10.2.2)$$

because (10.2.1) does not depend on the scaling of $\mathbf{a}$ and $\mathbf{b}$.

The solution to this problem is given below in Theorem 10.2.1. First, we need some notation.

Suppose that $\boldsymbol{\Sigma}_{11}$ and $\boldsymbol{\Sigma}_{22}$ are non-singular and let

$$\mathbf{K}=\boldsymbol{\Sigma}_{11}^{-1/2}\boldsymbol{\Sigma}_{12}\boldsymbol{\Sigma}_{22}^{-1/2}. \qquad (10.2.3)$$

Then set

$$\mathbf{N}_1=\mathbf{K}\mathbf{K}', \qquad \mathbf{N}_2=\mathbf{K}'\mathbf{K}, \qquad (10.2.4)$$

and

$$\begin{aligned}\mathbf{M}_1&=\boldsymbol{\Sigma}_{11}^{-1/2}\mathbf{N}_1\boldsymbol{\Sigma}_{11}^{1/2}=\boldsymbol{\Sigma}_{11}^{-1}\boldsymbol{\Sigma}_{12}\boldsymbol{\Sigma}_{22}^{-1}\boldsymbol{\Sigma}_{21},\\ \mathbf{M}_2&=\boldsymbol{\Sigma}_{22}^{-1/2}\mathbf{N}_2\boldsymbol{\Sigma}_{22}^{1/2}=\boldsymbol{\Sigma}_{22}^{-1}\boldsymbol{\Sigma}_{21}\boldsymbol{\Sigma}_{11}^{-1}\boldsymbol{\Sigma}_{12}.\end{aligned} \qquad (10.2.5)$$

Note that $\mathbf{N}_1$ and $\mathbf{M}_1$ are $(q \times q)$ matrices, and $\mathbf{N}_2$ and $\mathbf{M}_2$ are $(p \times p)$ matrices. From Theorem A.6.2, $\mathbf{M}_1, \mathbf{N}_1, \mathbf{M}_2$, and $\mathbf{N}_2$ all have the same non-zero eigenvalues. Further, since $\mathbf{N}_1 = \mathbf{K}\mathbf{K}'$ is p.s.d., all the non-zero eigenvalues are positive. Let k denote the number of non-zero eigenvalues. Then, by (A.4.2e) and (A.4.2f), $k = \text{rank}(\mathbf{K}) = \text{rank}(\mathbf{\Sigma}_{12})$. For simplicity suppose that the eigenvalues are all distinct, $\lambda_1 > \ldots > \lambda_k > 0$.

By the singular value decomposition theorem (Theorem A.6.5), $\mathbf{K}$ can be written in the form

$$\mathbf{K} = (\boldsymbol{\alpha}_1, \ldots, \boldsymbol{\alpha}_k)\mathbf{D}(\boldsymbol{\beta}_1, \ldots, \boldsymbol{\beta}_k)', \tag{10.2.6}$$

where $\boldsymbol{\alpha}_i$ and $\boldsymbol{\beta}_i$ are the standardized eigenvectors of $\mathbf{N}_1$ and $\mathbf{N}_2$, respectively, for λ_i, and $\mathbf{D} = \text{diag}(\lambda_1^{1/2}, \ldots, \lambda_k^{1/2})$. Since the λ_i are distinct, the eigenvectors are unique up to sign. (The signs of the eigenvectors are chosen so that the square roots in $\mathbf{D}$ are positive.) Also, since $\mathbf{N}_1$ and $\mathbf{N}_2$ are symmetric the eigenvectors are orthogonal. Thus

$$\boldsymbol{\alpha}_i'\boldsymbol{\alpha}_j = \delta_{ij}, \qquad \boldsymbol{\beta}_i'\boldsymbol{\beta}_j = \delta_{ij}, \tag{10.2.7}$$

where δ_{ij} is the Kronecker delta, equal to 1 for $i = j$ and 0 otherwise.

Definition *Using the above notation let*

$$\mathbf{a}_i = \mathbf{\Sigma}_{11}^{-1/2}\boldsymbol{\alpha}_i, \qquad \mathbf{b}_i = \mathbf{\Sigma}_{22}^{-1/2}\boldsymbol{\beta}_i, \qquad i = 1, \ldots, k. \tag{10.2.8}$$

Then

(a) *the vectors* $\mathbf{a}_i$ *and* $\mathbf{b}_i$ *are called the* ith canonical correlation vectors *for* $\mathbf{x}$ *and* $\mathbf{y}$, *respectively*;
(b) *the random variables* $\eta_i = \mathbf{a}_i'\mathbf{x}$ *and* $\phi_i = \mathbf{b}_i'\mathbf{y}$ *are called the* ith canonical correlation variables;
(c) $\rho_i = \lambda_i^{1/2}$ *is called the* ith canonical correlation coefficient.

Note that

$$\begin{aligned} C(\eta_i, \eta_j) &= \mathbf{a}_i'\mathbf{\Sigma}_{11}\mathbf{a}_j = \boldsymbol{\alpha}_i'\boldsymbol{\alpha}_j = \delta_{ij}, \\ C(\phi_i, \phi_j) &= \mathbf{b}_i'\mathbf{\Sigma}_{22}\mathbf{b}_j = \boldsymbol{\beta}_i'\boldsymbol{\beta}_j = \delta_{ij}. \end{aligned} \tag{10.2.9}$$

Thus the ith canonical correlation variables for $\mathbf{x}$ are uncorrelated and are standardized to have variance 1; similarly for the ith canonical correlation variables for $\mathbf{y}$. The main properties of canonical correlation analysis are given by the following theorem.

Theorem 10.2.1 *Using the above notation, fix* r, $1 \leq r \leq k$, *and let*

$$f_r = \max_{\mathbf{a}, \mathbf{b}} \mathbf{a}'\mathbf{\Sigma}_{12}\mathbf{b}$$

subject to

$$\mathbf{a}'\mathbf{\Sigma}_{11}\mathbf{a} = 1, \qquad \mathbf{b}'\mathbf{\Sigma}_{22}\mathbf{b} = 1, \qquad \mathbf{a}_i'\mathbf{\Sigma}_{11}\mathbf{a} = 0, \qquad i = 1, \ldots, r-1. \tag{10.2.10}$$

Then the maximum is given by $f_r = \rho_r$ *and is attained when* $\mathbf{a} = \mathbf{a}_r$, $\mathbf{b} = \mathbf{b}_r$.

Proof Before giving the proof, note that (10.2.10) is the largest correlation between linear combinations $\mathbf{a}'\mathbf{x}$ and $\mathbf{b}'\mathbf{y}$ subject to the restriction that $\mathbf{a}'\mathbf{x}$ is uncorrelated with the first $r-1$ canonical correlation variables for $\mathbf{x}$. Further, the case $r=1$ gives the required maximum in (10.2.2).

Note that the sign of $\mathbf{a}'\mathbf{\Sigma}_{12}\mathbf{b}$ is irrelevant because it can be changed by replacing $\mathbf{a}$ by $-\mathbf{a}$. Then it is convenient to solve (10.2.10) for f_r^2 instead of f_r. The proof splits naturally into three parts.

Step 1 Fix $\mathbf{a}$ and maximize over $\mathbf{b}$; that is, solve

$$\max_{\mathbf{b}} (\mathbf{a}'\mathbf{\Sigma}_{12}\mathbf{b})^2 = \max_{\mathbf{b}} \mathbf{b}'\mathbf{\Sigma}_{21}\mathbf{a}\mathbf{a}'\mathbf{\Sigma}_{12}\mathbf{b} \text{ subject to } \mathbf{b}'\mathbf{\Sigma}_{22}\mathbf{b} = 1.$$

By Theorem A.9.2 this maximum is given by the largest (and only non-zero) eigenvalue of the matrix $\mathbf{\Sigma}_{22}^{-1}\mathbf{\Sigma}_{21}\mathbf{a}\mathbf{a}'\mathbf{\Sigma}_{12}$. By Corollary A.6.2.1 this eigenvalue equals

$$\mathbf{a}'\mathbf{\Sigma}_{12}\mathbf{\Sigma}_{22}^{-1}\mathbf{\Sigma}_{21}\mathbf{a}. \tag{10.2.11}$$

Step 2 Now maximize (10.2.11) over $\mathbf{a}$ subject to the constraints in (10.2.10). Setting $\boldsymbol{\alpha} = \mathbf{\Sigma}_{11}^{1/2}\mathbf{a}$, the problem becomes

$$\max_{\boldsymbol{\alpha}} \boldsymbol{\alpha}'\mathbf{N}_1\boldsymbol{\alpha} \text{ subject to } \boldsymbol{\alpha}'\boldsymbol{\alpha} = 1,\ \boldsymbol{\alpha}_i'\boldsymbol{\alpha} = 0, \qquad i = 1, \ldots, r-1, \tag{10.2.12}$$

where $\mathbf{a}_i = \mathbf{\Sigma}^{-1/2}\boldsymbol{\alpha}_i$ is the ith canonical correlation vector. Note that the $\boldsymbol{\alpha}_i$ are the eigenvectors on $\mathbf{N}_1$ corresponding to the $(r-1)$ largest eigenvalues of $\mathbf{N}_1$. Thus, as in Theorem 8.2.3, the maximum in (10.2.12) is attained by setting $\boldsymbol{\alpha}$ equal to the eigenvector corresponding to the largest eigenvalue not forbidden to us; that is, take $\boldsymbol{\alpha} = \boldsymbol{\alpha}_r$, or equivalently $\mathbf{a} = \mathbf{a}_r$. Then

$$f_r^2 = \boldsymbol{\alpha}_r'\mathbf{N}_1\boldsymbol{\alpha}_r = \lambda_r\boldsymbol{\alpha}_r'\boldsymbol{\alpha}_r = \lambda_r.$$

Step 3 Lastly we show that this maximum is attained when $\mathbf{a} = \mathbf{a}_r$ and $\mathbf{b} = \mathbf{b}_r$. From the decomposition (10.2.6) note that $\mathbf{K}\boldsymbol{\beta}_r = \rho_r\boldsymbol{\alpha}_r$. Thus

$$\mathbf{a}_r'\mathbf{\Sigma}_{12}\mathbf{b}_r = \boldsymbol{\alpha}_r'\mathbf{K}\boldsymbol{\beta}_r = \rho_r\boldsymbol{\alpha}_r'\boldsymbol{\alpha}_r = \rho_r. \quad \blacksquare$$

As standardized eigenvectors of $\mathbf{N}_1$ and $\mathbf{N}_2$, respectively, $\boldsymbol{\alpha}_r$ and $\boldsymbol{\beta}_r$ are uniquely defined up to sign. Their signs are usually chosen to make the correlation $\mathbf{a}_r'\mathbf{\Sigma}_{12}\mathbf{b}_r > 0$, although in some situations it is more intuitive to think of a negative correlation.

The above theorem is sometimes stated in a weaker but more symmetric form as follows:

Theorem 10.2.2 *Let*

$$g_r = \max_{\mathbf{a},\mathbf{b}} \mathbf{a}'\mathbf{\Sigma}_{12}\mathbf{b}$$

subject to

$$\mathbf{a}'\mathbf{\Sigma}_{11}\mathbf{a} = \mathbf{b}'\mathbf{\Sigma}_{22}\mathbf{b} = 1, \qquad \mathbf{a}_i'\mathbf{\Sigma}_{11}\mathbf{a} = \mathbf{b}_i'\mathbf{\Sigma}_{22}\mathbf{b} = 0, \qquad i = 1, \ldots, r-1.$$

Then $g_r = \rho_r$ *and the maximum is attained when* $\mathbf{a} = \mathbf{a}_r$ *and* $\mathbf{b} = \mathbf{b}_r$.

Proof Since $\mathbf{a}_r$ and $\mathbf{b}_r$ satisfy the constraints and since these constraints are more restrictive than the constraints in Theorem 10.2.1, the result follows immediately from Theorem 10.2.1. ■

The correlations between the canonical correlation variables are summarized in the following theorem.

Theorem 10.2.3 *Let* η_i *and* ϕ_i *be the* ith *canonical correlation variables,* $i = 1, \ldots, k$, *and let* $\boldsymbol{\eta} = (\eta_1, \ldots, \eta_k)'$ *and* $\boldsymbol{\phi} = (\phi_1, \ldots, \phi_k)'$. *Then*

$$V\begin{pmatrix}\boldsymbol{\eta}\\ \boldsymbol{\phi}\end{pmatrix} = \begin{pmatrix}\mathbf{I} & \mathbf{\Lambda}^{1/2}\\ \mathbf{\Lambda}^{1/2} & \mathbf{I}\end{pmatrix}. \tag{10.2.13}$$

Proof From (10.2.9) we see that

$$V(\boldsymbol{\eta}) = V(\boldsymbol{\phi}) = \mathbf{I}.$$

From Theorem 10.2.1, $C(\eta_i, \phi_i) = \lambda_i^{1/2}$. Thus, the proof will be complete if we show that $C(\eta_i, \phi_j) = 0$ for $i \neq j$. This result follows from the decomposition (10.2.6) since

$$C(\eta_i, \phi_j) = \mathbf{a}_i'\mathbf{\Sigma}_{12}\mathbf{b}_j = \boldsymbol{\alpha}_i'\mathbf{K}\boldsymbol{\beta}_j = \lambda_j\boldsymbol{\alpha}_i'\boldsymbol{\alpha}_j = 0. \quad ■$$

Note that (10.2.13) is also the correlation matrix for $\boldsymbol{\eta}$ and $\boldsymbol{\phi}$.

The next theorem proves an important invariance property.

Theorem 10.2.4 *If* $\mathbf{x}^* = \mathbf{U}'\mathbf{x} + \mathbf{u}$ *and* $\mathbf{y}^* = \mathbf{V}'\mathbf{y} + \mathbf{v}$, *where* $U(q \times q)$ *and* $V(p \times p)$ *are non-singular matrices and* $\mathbf{u}(q \times 1)$, $\mathbf{v}(p \times 1)$ *are fixed vectors, then*

(a) *the canonical correlations between* $\mathbf{x}^*$ *and* $\mathbf{y}^*$ *are the same as those between* $\mathbf{x}$ *and* $\mathbf{y}$;

(b) *the canonical correlation vectors for* $\mathbf{x}^*$ *and* $\mathbf{y}^*$ *are given by* $\mathbf{a}_i^* = \mathbf{U}^{-1}\mathbf{a}_i$ *and* $\mathbf{b}_i^* = \mathbf{V}^{-1}\mathbf{b}_i$, $i = 1, \ldots, k$, *where* $\mathbf{a}_i$ *and* $\mathbf{b}_i$ *are the canonical correlation vectors for* $\mathbf{x}$ *and* $\mathbf{y}$.

Proof The matrix for $\mathbf{x}^*$ and $\mathbf{y}^*$ corresponding to $\mathbf{M}_1$ is

$$\mathbf{M}_1^* = (\mathbf{U}'\mathbf{\Sigma}_{11}\mathbf{U})^{-1}\mathbf{U}'\mathbf{\Sigma}_{12}\mathbf{V}(\mathbf{V}'\mathbf{\Sigma}_{22}\mathbf{V})^{-1}\mathbf{V}'\mathbf{\Sigma}_{21}\mathbf{U} = \mathbf{U}^{-1}\mathbf{M}_1\mathbf{U}.$$

Then $\mathbf{M}_1^*$ has the same eigenvalues as $\mathbf{M}_1$, and the relationship between the eigenvectors is given by Theorem A.6.2. Note that the standardization

$$\mathbf{a}_i^{*\prime}(\mathbf{U}'\mathbf{\Sigma}_{11}\mathbf{U})\mathbf{a}_i^* = \mathbf{a}_i'\mathbf{\Sigma}_{11}\mathbf{a}_i = 1,$$

remains unaltered. Working with $\mathbf{M}_2^*$ gives the eigenvectors $\mathbf{b}_i^*$. ■

If we put $\mathbf{U} = (\text{diag}\,\mathbf{\Sigma}_{11})^{-1/2}$ and $\mathbf{V} = (\text{diag}\,\mathbf{\Sigma}_{22})^{-1/2}$ in Theorem 10.2.2 then we obtain $\mathbf{M}^* = \mathbf{P}_{11}^{-1}\mathbf{P}_{12}\mathbf{P}_{22}^{-1}\mathbf{P}_{21}$, where $\mathbf{P}$ is the correlation matrix of $\mathbf{x}$ and $\mathbf{y}$. Hence we deduce that canonical correlation analysis using $\mathbf{P}$ leads to essentially the same results as canonical correlation analysis applied to $\mathbf{\Sigma}$. Recall that principal component analysis did not have this convenient invariance property.

Example 10.2.1 Working with the correlation matrix $\mathbf{P}$, consider the situation $q = p = 2$ with all the elements of $\mathbf{P}_{12}$ equal; that is, let

$$\mathbf{P}_{11} = \begin{bmatrix} 1 & \alpha \\ \alpha & 1 \end{bmatrix}, \qquad \mathbf{P}_{22} = \begin{bmatrix} 1 & \gamma \\ \gamma & 1 \end{bmatrix}, \qquad \mathbf{P}_{12} = \beta\mathbf{J},$$

where $\mathbf{J} = \mathbf{1}\mathbf{1}'$. Then

$$\mathbf{P}_{11}^{-1}\mathbf{P}_{12} = \frac{\beta}{1-\alpha^2}\begin{bmatrix} 1 & -\alpha \\ -\alpha & 1 \end{bmatrix}\begin{bmatrix} 1 & 1 \\ 1 & 1 \end{bmatrix} = \beta\mathbf{J}/(1+\alpha),$$

if $|\alpha| < 1$. Similarly, for $|\gamma| < 1$,

$$\mathbf{P}_{22}^{-1}\mathbf{P}_{21} = \beta\mathbf{J}/(1+\gamma).$$

Noting that $\mathbf{J}^2 = 2\mathbf{J}$, we have

$$\mathbf{P}_{11}^{-1}\mathbf{P}_{12}\mathbf{P}_{22}^{-1}\mathbf{P}_{21} = \{2\beta^2/(1+\alpha)(1+\gamma)\}\mathbf{J}.$$

Now the eigenvalues of $\mathbf{1}\mathbf{1}'$ are 2 and zero. Therefore the non-zero eigenvalue of the above matrix is

$$\lambda_1 = 4\beta^2/(1+\alpha)(1+\gamma).$$

Hence, the first canonical correlation coefficient is

$$\rho_1 = 2\beta/\{(1+\alpha)(1+\gamma)\}^{1/2}.$$

Note that $|\alpha|$, $|\gamma| < 1$ and therefore $\rho_1 > \beta$.

Example 10.2.2 In general, if $\mathbf{P}_{12}$ has rank one then we may write

$\mathbf{P}_{12}=\mathbf{ab}'$. Therefore

$$\mathbf{P}_{11}^{-1}\mathbf{P}_{12}\mathbf{P}_{22}^{-1}\mathbf{P}_{21}=\mathbf{P}_{11}^{-1}\mathbf{ab}'\mathbf{P}_{22}^{-1}\mathbf{ba}'.$$

The non-zero eigenvalue of this matrix is

$$\lambda_1=(\mathbf{a}'\mathbf{P}_{11}^{-1}\mathbf{a})(\mathbf{b}'\mathbf{P}_{22}^{-1}\mathbf{b}).$$

Note that this holds whatever the values of q and p. If $\mathbf{a}\propto\mathbf{1}$, $\mathbf{b}\propto\mathbf{1}$, then for $q=p=2$ we get the result already shown in Example 10.2.1. In general when $\mathbf{a}\propto\mathbf{1}$, $\mathbf{b}\propto\mathbf{1}$, λ_1 is proportional to the product of the sum of the elements of $\mathbf{P}_{11}^{-1}$ with the sum of the elements of $\mathbf{P}_{22}^{-1}$.

Note that in general $\mathbf{M}_1$ and $\mathbf{M}_2$ will have k non-zero eigenvalues, where k is the rank of $\mathbf{\Sigma}_{12}$. Usually $k=\min(p,q)$. Hence, in regression, where $p=1$, there is just one non-trivial canonical correlation vector, and this is indeed the (standardized) least squares regression vector. If the matrices $\mathbf{\Sigma}_{11}$ and $\mathbf{\Sigma}_{22}$ are not of full rank, then similar results can be developed using generalized inverses (Rao and Mitra, 1971).

10.2.2 Sample canonical correlation analysis

The above development for a population may be followed through for the analysis of sample data. All that is needed is that $\mathbf{S}_{ij}$ should be substituted for $\mathbf{\Sigma}_{ij}$ wherever it occurs $(i,j=1,2)$; and l_i for λ_i, r_i for ρ_i, etc. If the data is normal, then $\mathbf{S}$ is the maximum likelihood estimator of $\mathbf{\Sigma}$, so the sample canonical correlation values are maximum likelihood estimators of the corresponding population values, except when there are repeated population eigenvalues.

Example 10.2.3 For the head-length data of Table 5.1.1, we have the correlation matrix

$$\mathbf{R}_{11}=\begin{bmatrix}1 & 0.7346\\ 0.7346 & 1\end{bmatrix},\qquad \mathbf{R}_{22}=\begin{bmatrix}1 & 0.8392\\ 0.8392 & 1\end{bmatrix},$$

$$\mathbf{R}_{12}=\mathbf{R}_{21}'=\begin{bmatrix}0.7108 & 0.7040\\ 0.6932 & 0.7086\end{bmatrix}.$$

(See Example 5.1.1.) Note that all the elements of $\mathbf{R}_{12}$ are about 0.7, so that this matrix is nearly of rank one. This means that the second canonical correlation will be near zero, and the situation approximates that of Example 10.2.1. (See Exercise 10.2.12.) In fact the eigenvalues of $\mathbf{M}_1=\mathbf{R}_{11}^{-1}\mathbf{R}_{12}\mathbf{R}_{22}^{-1}\mathbf{R}_{21}$ are 0.6218 and 0.0029. Therefore the canonical correlation coefficients are $r_1=0.7886$ and $r_2=0.0539$. As expected, r_2 is close to zero, since $\mathbf{R}_{12}$ is almost of rank one. Note that r_1

exceeds any of the individual correlations between a variable of the first set and a variable of the second set (in particular $0.7886 > 0.7108$). The canonical correlation vectors for the standardized variables are obtained from the eigenvectors of $\mathbf{M}_1$ and $\mathbf{M}_2$. We have

$$\mathbf{a}_1 = \begin{bmatrix} 0.552 \\ 0.522 \end{bmatrix}, \qquad \mathbf{b}_1 = \begin{bmatrix} 0.505 \\ 0.538 \end{bmatrix}$$

and

$$\mathbf{a}_2 = \begin{bmatrix} 1.367 \\ -1.378 \end{bmatrix}, \qquad \mathbf{b}_2 = \begin{bmatrix} 1.767 \\ -1.757 \end{bmatrix}.$$

Hence the first canonical correlation variables are $\eta_1 = 0.552x_1 + 0.522x_2$ and $\phi_1 = 0.505y_1 + 0.538y_2$. These are approximately the sum of length and breadth of the head size of each brother, and may be interpreted as a measure of "girth". These variables are highly correlated between brothers. The second canonical correlation variables η_2 and ϕ_2 seem to be measuring the *difference* between length and breadth. This measure of "shape" would distinguish for instance between long thin heads and short fat heads. The head shape of first and second brothers appears therefore to have little correlation. (See also Section 8.6.)

10.2.3 Sampling properties and tests

The sampling distributions associated with canonical correlation analysis are very complicated, and we shall not go into them in detail here. The interested reader is referred to Kshirsagar (1972, pp. 261–277). We shall merely describe briefly an associated significance test.

First, consider the hypothesis $\mathbf{\Sigma}_{12} = \mathbf{0}$, which means the two sets of variables are uncorrelated with one another. From (5.3.11) (with $q = p_1$, $p = q_2$), we see that under normality, the likelihood ratio statistic for testing H_0 is given by

$$\lambda^{2/n} = |\mathbf{I} - \mathbf{S}_{22}^{-1}\mathbf{S}_{21}\mathbf{S}_{11}^{-1}\mathbf{S}_{12}| = \prod_{i=1}^{k} (1 - r_i^2),$$

which has a Wilks' $\Lambda(p, n-1-q, q)$ distribution. Here, $r_1, \ldots, r_k$ are the sample canonical correlation coefficients and $k = \min(p, q)$. Using Bartlett's approximation (3.7.11), we see that

$$-\{n - \tfrac{1}{2}(p+q+3)\} \log \prod_{i=1}^{k} (1 - r_i^2) \sim \chi^2_{pq},$$

asymptotically for large n.

Bartlett (1939) proposed a similar statistic to test the hypothesis that only s of the population canonical correlation coefficients are non-zero. This test is based on the statistic

$$-\{n-\tfrac{1}{2}(p+q+3)\}\log\prod_{i=s+1}^{k}(1-r_i^2)\sim\chi^2_{(p-s)(q-s)}, \qquad (10.2.14)$$

asymptotically. An alternative large sample test was proposed by Marriott (1952).

Example 10.2.4 Consider the head data of Example 10.2.3. There are $n=25$ observations and $q=p=2$. First, let us test whether or not the head measurements of one brother are independent of those of the other, that is, the hypothesis $\rho_1=\rho_2=0$. The LR test for this hypothesis was given in Example 5.3.3 and the null hypothesis was strongly rejected.

Second, to test whether $\rho_2=0$, we use (10.2.14) with $s=1$. This gives the statistic

$$-(25-\tfrac{7}{2})\log(1-0.0539^2)=0.063,$$

which, when tested against the χ_1^2 distribution, is clearly non-significant. Hence we accept the hypothesis $\rho_2=0$ for this data.

10.2.4 Scoring and prediction

Let $(\mathbf{X},\mathbf{Y})$ be a data matrix of n individuals on $(q+p)$ variables and let $\mathbf{a}_i$, $\mathbf{b}_i$ denote the ith canonical correlation vectors. Then the n-vectors $\mathbf{Xa}_i$ and $\mathbf{Yb}_i$ denote the *scores* of the n individuals on the ith canonical correlation variables for $\mathbf{x}$ and $\mathbf{y}$. In terms of the values of the variables for a particular individual, these scores take the form

$$\eta_i=\mathbf{a}_i'\mathbf{x}, \qquad \phi_i=\mathbf{b}_i'\mathbf{y}. \qquad (10.2.15)$$

Since correlations are unaltered by linear transformations, it is sometimes convenient to replace these scores by new scores

$$\eta_i^*=c_1\mathbf{a}_i'\mathbf{x}+d_1, \qquad \phi_i^*=c_2\mathbf{b}_i'\mathbf{y}+d_2, \qquad (10.2.16)$$

where $c_1, c_2>0$ and d_1, d_2 are real numbers. These scores are most important for the first canonical correlation vectors and they can be calculated for each of the n individuals.

Let r_i denote the ith canonical correlation coefficient. If the $\mathbf{x}$ and $\mathbf{y}$ variables are interpreted as the "predictor" and "predicted" variables, respectively, then the η_i score can be used to predict a value of the ϕ_i score using least squares regression. Since $\mathbf{Xa}_i$ and $\mathbf{Yb}_i$ each have sample variance 1, the predicted value of ϕ_i given η_i is

$$\hat{\phi}_i = r_i(\eta_i - \mathbf{a}_i'\bar{\mathbf{x}}) + \mathbf{b}_i'\bar{\mathbf{y}}, \tag{10.2.17}$$

or equivalently

$$\hat{\phi}_i^* = \frac{c_2 r_i}{c_1}(\eta_i^* - d_1 - c_1\mathbf{a}_i'\bar{\mathbf{x}}) + d_2 + c_2\mathbf{b}_i'\bar{\mathbf{y}}. \tag{10.2.18}$$

Note that r_i^2 represents the proportion of the variance of ϕ_i which is "explained" by the regression on $\mathbf{X}$. Of course, prediction is most important for the first canonical correlation variable.

Example 10.2.5 Consider again the open/closed book data discussed in Example 5.1.1. The covariance matrix for this data is given in Example 8.2.3.

The canonical correlations are $r_1 = 0.6630$ and $r_2 = 0.0412$ and the first canonical correlation vectors are

$$\eta_1 = 0.0260x_1 + 0.0518x_2,$$
$$\phi_1 = 0.0824y_1 + 0.0081y_2 + 0.0035y_3.$$

Thus the highest correlation between the open- and closed-book exams occurs between an average of x_1 and x_2 weighted on x_2, and an average of y_1, y_2, and y_3 heavily weighted on y_1.

The means of the exam results are

38.9545, 50.5909, 50.6023, 46.6818, 42.3068.

If we use these linear combinations to predict open-book results from the closed-book results we get the predictor

$$\hat{\phi}_1 = 0.0172x_1 + 0.0343x_2 + 2.2905.$$

Note that this function essentially predicts the value of y_1 from the values of x_1 and x_2.

10.3 Qualitative Data and Dummy Variables

Canonical correlation analysis can also be applied to *qualitative* data. Consider a two-way contingency table $\mathbf{N}(r \times c)$ in which individuals are classified according to each of two characteristics. The attributes on each characteristic are represented by the r row and the c column categories, respectively, and n_{ij} denotes the number of individuals with the ith row and jth column attributes. We wish to explore the relationship between the two characteristics. For example, Table 10.3.1 gives a (5×5)

Table 10.3.1 Social mobility contingency table (Glass, 1954; see also Goodman, 1972)

		Subject's status				
		1	2	3	4	5
Father's status	1	50	45	8	18	8
	2	28	174	84	154	55
	3	11	78	110	223	96
	4	14	150	185	714	447
	5	0	42	72	320	411

contingency table of $n = 3497$ individuals comparing their social status with the social status of their fathers.

Now a contingency table $\mathbf{N}$ is not a data matrix since it does not have the property that rows correspond to individuals while columns represent variables. However, it is possible to represent the data in an $(n \times (r+c))$ data matrix $\mathbf{Z} = (\mathbf{X}, \mathbf{Y})$, where the columns of $\mathbf{X}$ and $\mathbf{Y}$ are dummy zero–one variables for the row and column categories, respectively; that is, let

$$x_{ki} = \begin{cases} 1 & \text{if the } k\text{th individual belongs to the } i\text{th row category,} \\ 0 & \text{otherwise,} \end{cases}$$

and

$$y_{ki} = \begin{cases} 1 & \text{if the } k\text{th individual belongs to the } j\text{th column category,} \\ 0 & \text{otherwise,} \end{cases}$$

for $k = 1, \ldots, n$, and $i = 1, \ldots, r$; $j = 1, \ldots, c$. Note that $\mathbf{x}'_{(i)}\mathbf{y}_{(j)} = n_{ij}$. Also, the columns of $\mathbf{X}$ and the columns of $\mathbf{Y}$ each sum to $\mathbf{1}_n$.

The purpose of canonical correlation analysis is to find vectors $\mathbf{a}(r \times 1)$ and $\mathbf{b}(c \times 1)$ such that the variables $\eta = \mathbf{a}'\mathbf{x}$ and $\phi = \mathbf{b}'\mathbf{y}$ are maximally correlated. Since $\mathbf{x}$ has only one non-zero component, $\eta = a_1, a_2, \ldots,$ or a_r. Similarly $\phi = b_1, b_2, \ldots,$ or b_c. Thus, an individual in the ith row category and jth column category can be associated with his "score" (a_i, b_j). These scores may be plotted on a scattergram—there will be n_{ij} points at (a_i, b_j)—and the correlation represented by these points can be evaluated. Of course the correlation depends on the values of $\mathbf{a}$ and $\mathbf{b}$, and if these vectors maximize this correlation, then they give the vector of first canonical correlation loadings. Such scores are known as "canonical scores".

Of course, having calculated the first set of canonical scores, one could ask for a second set which maximizes the correlation among all sets which are uncorrelated with the first. This would correspond to the second canonical correlation.

Let $\mathbf{Z} = (\mathbf{X}, \mathbf{Y})$ be the data matrix corresponding to a contingency table $\mathbf{N}$. Let $f_i = \sum_j n_{ij}$ and $g_j = \sum_i n_{ij}$, so that $\mathbf{f}$ and $\mathbf{g}$ denote the marginal row and column sums, respectively, for the table. For simplicity, suppose $f_i > 0$, $g_j > 0$ for all i and j. Then a little calculation shows that

$$n\mathbf{S} = \mathbf{Z}'\mathbf{H}\mathbf{Z} = \mathbf{Z}'\mathbf{Z} - n\bar{\mathbf{z}}\bar{\mathbf{z}}'$$
$$= \begin{bmatrix} n\mathbf{S}_{11} & n\mathbf{S}_{12} \\ n\mathbf{S}_{21} & n\mathbf{S}_{22} \end{bmatrix} = \begin{bmatrix} \operatorname{diag}(\mathbf{f}) - n^{-1}\mathbf{f}\mathbf{f}' & \mathbf{N} - \hat{\mathbf{N}} \\ \mathbf{N}' - \hat{\mathbf{N}}' & \operatorname{diag}(\mathbf{g}) - n^{-1}\mathbf{g}\mathbf{g}' \end{bmatrix}, \quad (10.3.1)$$

where $\hat{\mathbf{N}} = \mathbf{f}\mathbf{g}'$ is the estimated value of $\mathbf{N}$ under the assumption that the row and column categories are independent.

Because the columns of $\mathbf{X}$ and $\mathbf{Y}$ each sum to $\mathbf{1}$, $\mathbf{S}_{11}^{-1}$ and $\mathbf{S}_{22}^{-1}$ do not exist. (See Exercise 10.3.1.) One way out of this difficulty is to drop a column of each of $\mathbf{X}$ and $\mathbf{Y}$, say the first column. Let $\mathbf{S}_{ij}^*$ denote the component of the covariance matrix obtained by deleting the first row and first column from $\mathbf{S}_{ij}$ for $i, j = 1, 2$. Similarly, let $\mathbf{f}^*$ and $\mathbf{g}^*$ denote the vectors obtained by deleting the first components of $\mathbf{f}$ and $\mathbf{g}$. Then it is easy to check that

$$(n\mathbf{S}_{11}^*)^{-1} = [\operatorname{diag}(\mathbf{f}^*)]^{-1} + f_1^{-1}\mathbf{1}\mathbf{1}', \qquad (n\mathbf{S}_{22}^*)^{-1} = [\operatorname{diag}(\mathbf{g}^*)]^{-1} + g_1^{-1}\mathbf{1}\mathbf{1}', \quad (10.3.2)$$

so that $\mathbf{M}_1^* = \mathbf{S}_{11}^{*-1}\mathbf{S}_{12}^*\mathbf{S}_{22}^{*-1}\mathbf{S}_{21}^*$ and $\mathbf{M}_2^*$ are straightforward to calculate for the reduced set of variables.

Note that the score associated with an individual who lies in the first row category of $\mathbf{N}$ is 0; and similarly for the first column category of $\mathbf{N}$.

Example 10.3.1 A canonical correlation analysis on Glass's social mobility data in Table 10.3.1 yields the following scores (coefficients of the first canonical correlation vectors) for the various social classes:

Father's status:	1	2	3	4	5
	0	3.15	4.12	4.55	4.96
Son's status:	1	2	3	4	5
	0	3.34	4.49	4.87	5.26

The first canonical correlation is $r_1 = 0.504$. Note that for both father's and son's social class, the scores appear in their natural order. Thus the social class of the father appears to be correlated with the social class of the son. Note that social classes 1 and 2 seem to be more distinct from one another than the other adjacent social classes, both for the son's and the father's status.

Hypothesis testing is difficult for canonical scores because the data matrix **Z** is clearly non-normal.

10.4 Qualitative and Quantitative Data

The ideas of the last section can be used when the data is a mixture of qualitative and quantitative characteristics. Each quantitative characteristic can of course be represented on one variable, and each qualitative characteristic, with say g attributes, can be represented by dummy zero–one values on $g-1$ variables.

We shall illustrate this procedure for an example on academic prediction of Barnett and Lewis (1963) and Lewis (1970). Lewis (1970) gives data collected from 382 university students on the number of GCE A-levels taken (a secondary school examination required for university entrance), and the student's average grade (scored 1 to 5, with 5 denoting the best score). These represented the "explanatory variables". The dependent variables concerned the student's final degree result at university which was classified as First, Upper Second, Lower Second, Third, and Pass. Alternatively, some students were classified as 3(4) if they took four years over a three-year course (the final degree results were not available for these students), or as $\rightarrow$ if they left without completing the course.

This information may be represented in terms of the following variables:

x_1 = average A-level grade;
$x_2 = 1$ if two A-levels taken, 0 otherwise;
$x_3 = 1$ if four A-levels taken, 0 otherwise;
$y_1 = 1$ if degree class II(i) obtained, 0 otherwise;
$y_2 = 1$ if degree class II(ii) obtained, 0 otherwise;
$y_3 = 1$ if degree class III obtained, 0 otherwise;
$y_4 = 1$ if Pass obtained, 0 otherwise;
$y_5 = 1$ if 3(4) obtained, 0 otherwise;
$y_6 = 1$ if $\rightarrow$ obtained, 0 otherwise.

Note that a student who obtains a First Class (I) degree, would score zero on *all* the **y**-variables. The data is summarized in Table 10.4.1. Note that most of these variables are "dummies" in the sense that they take either the value 0 or 1. Hence the assumptions of normality would be completely unwarranted in this example.

Since the **y**-variables are in some sense dependent on the **x**-variables, we may ask what scoring system maximizes the correlation between the ys and the xs. Lewis (1970) showed that the first canonical correlation

Table 10.4.1 Data from 382 Hull University students (Lewis, 1970)

class: I
y: (0,0,0,0,0,0)

x_1	x_2	x_3	Freq.
4.5	0	1	1
3.25	0	1	1
4.67	0	0	6
4.33	0	0	3
4.0	0	0	3
3.33	0	0	3
3.0	0	0	2
2.67	0	0	4
1.0	0	0	1
4.0	1	0	1

class: II(i)
y: (1,0,0,0,0,0)

x_1	x_2	x_3	Freq.
4.0	0	1	1
3.25	0	1	1
5.0	0	0	1
4.67	0	0	2
4.33	0	0	10
4.0	0	0	11
3.67	0	0	6
3.33	0	0	12
3.0	0	0	8
2.67	0	0	4
3.33	0	0	3
2.0	0	0	1
1.67	0	0	1
1.33	0	0	1
5.0	1	0	1
4.5	1	0	1
4.0	1	0	1
3.5	1	0	1
2.0	1	0	1

class: II(ii)
y: (0,1,0,0,0,0)

x_1	x_2	x_3	Freq.
4.25	0	1	1
3.25	0	1	3
3.75	0	1	1
4.67	0	0	1
4.33	0	0	6
4.0	0	0	10
3.67	0	0	7
3.33	0	0	18
3.0	0	0	14
2.67	0	0	15
2.33	0	0	11
2.0	0	0	5
1.67	0	0	1
1.33	0	0	2
5.0	1	0	2
4.5	1	0	1
4.0	1	0	6
3.5	1	0	3
3.0	1	0	6
2.5	1	0	1
2.0	1	0	1
1.5	1	0	2

class: III
y: (0,0,1,0,0,0)

x_1	x_2	x_3	Freq.
3.0	0	1	1
2.75	0	1	1
5.0	0	0	1
4.67	0	0	1
4.33	0	0	1
4.0	0	0	2
3.67	0	0	1
3.33	0	0	7
3.0	0	0	7
2.67	0	0	8
2.33	0	0	6
2.0	0	0	3
1.67	0	0	1
4.5	1	0	1
4.0	1	0	1
3.5	1	0	4
2.5	1	0	4
2.0	1	0	1

class: Pass
y: (0,0,0,1,0,0)

x_1	x_2	x_3	Freq.
3.75	0	1	1
3.25	0	1	1
2.75	0	1	2
2.0	0	1	2
4.67	0	0	2
4.33	0	0	1
4.0	0	0	1
3.67	0	0	3
3.33	0	0	7
3.0	0	0	12
2.67	0	0	7
2.33	0	0	7
2.0	0	0	8
1.67	0	0	3
1.33	0	0	1
1.0	0	0	1
4.5	1	0	1
4.0	1	0	1
3.5	1	0	3
3.0	1	0	4
2.5	1	0	4
2.0	1	0	2

class: 3(4)
y: (0,0,0,0,1,0)

x_1	x_2	x_3	Freq.
3.25	0	1	1
2.75	0	1	1
2.5	0	1	1
4.0	0	0	4
3.67	0	0	1
3.33	0	0	3
3.0	0	0	1
2.67	0	0	3
2.0	0	0	3
1.67	0	0	1
3.0	1	0	1
2.5	1	0	1
2.0	1	0	1

class: →
y: (0,0,0,0,0,1)

x_1	x_2	x_3	Freq.
2.25	0	1	1
3.67	0	0	3
3.0	0	0	3
2.67	0	0	5
2.33	0	0	2
2.0	0	0	1
1.33	0	0	1
1.0	0	0	1
4.0	1	0	1
3.5	1	0	3
3.0	1	0	4
2.5	1	0	1

vectors are given by $\mathbf{a}_1 = (-1.096, 1.313, 0.661)'$ and $\mathbf{b}_1 = (0.488, 1.877, 2.401, 2.971, 2.527, 3.310)$. Thus the scores corresponding to each of the degree results are

I	II(i)	II(ii)	III	Pass	3(4)	→
0	0.488	1.877	2.401	2.971	2.527	3.310

The first canonical correlation coefficient is $r_1 = 0.400$, which means that with the given scores only $r_1^2 = 0.160$ of the variation in the first canonical correlation variable for $\mathbf{y}$ is explained by variation in A-level scores. However, any other scoring system (such as a "natural" one, $1y_1 + 2y_2 + \ldots + 6y_6$) would explain less than 0.16 of the variance.

The above scores may be interpreted as follows. The scores for I, II(i), II(ii), III, and Pass come out in the "natural" order, but they are not equally spaced. Moreover the 3(4) group comes between III and Pass, while → scores higher than Pass. Note the large gap between II(i) and II(ii).

The canonical correlation vector on the $\mathbf{x}$-variables indicates that the higher one's A-level average, the better the degree one is likely to get. Also, those who take two or four A-levels are likely to get poorer degree results than those who take three A-levels. In fact, taking two A-levels instead of three is equivalent to a drop of about 0.5 in average grade. The fact that three A-levels is better than four A-levels is somewhat surprising. However, because the value of r_1^2 is not high, perhaps one should not read a great deal into these conclusions. (The test of Section 10.2.3 is not valid here because the population is clearly non-normal.)

The means for the nine variables are $(\bar{\mathbf{x}}', \bar{\mathbf{y}}') = (3.138, 0.173, 0.055, 0.175, 0.306, 0.134, 0.194, 0.058, 0.068)$. Thus we can use the A-level performance to predict degree results using

$$\hat{\phi} = -0.439x_1 + 0.525x_2 + 0.264x_3 + 3.199.$$

For a student taking three A-levels with an average grade of 3.75, $\hat{\phi} = 1.55$, so we would predict a degree result slightly better than II(ii).

Bartlett (1965) gives various other examples on the use of dummy variables.

Exercises and Complements

10.2.1 Show that for fixed $\mathbf{b}$

$$\max_{\mathbf{a}} \frac{(\mathbf{a}'\mathbf{\Sigma}_{12}\mathbf{b})^2}{(\mathbf{a}'\mathbf{\Sigma}_{11}\mathbf{a})(\mathbf{b}'\mathbf{\Sigma}_{22}\mathbf{b})} = \frac{\mathbf{b}'\mathbf{\Sigma}_{21}\mathbf{\Sigma}_{11}^{-1}\mathbf{\Sigma}_{12}\mathbf{b}}{\mathbf{b}'\mathbf{\Sigma}_{22}\mathbf{b}}$$

and that the maximum of this quantity over $\mathbf{b}$ is given by the largest eigenvalue of $\boldsymbol{\Sigma}_{11}^{-1}\boldsymbol{\Sigma}_{12}\boldsymbol{\Sigma}_{22}^{-1}\boldsymbol{\Sigma}_{21}$.

10.2.2 Show that $\mathbf{M}_1$, $\mathbf{M}_2$, $\mathbf{N}_1$, and $\mathbf{N}_2$ of Section 10.2.1 all have the same non-zero eigenvalues and that these eigenvalues are real and positive.

10.2.3 Show that

$$\mathbf{M}_1 = \mathbf{I}_p - \boldsymbol{\Sigma}_{11}^{-1}\boldsymbol{\Sigma}_{11.2}, \quad \text{where} \quad \boldsymbol{\Sigma}_{11.2} = \boldsymbol{\Sigma}_{11} - \boldsymbol{\Sigma}_{12}\boldsymbol{\Sigma}_{22}^{-1}\boldsymbol{\Sigma}_{21}.$$

Hence deduce that the first canonical correlation vector for $\mathbf{x}$ is given by the eigenvector corresponding to the smallest eigenvalue of $\boldsymbol{\Sigma}_{11.2}$.

10.2.4 Show that the squared canonical correlations are the roots of the equation

$$|\boldsymbol{\Sigma}_{12}\boldsymbol{\Sigma}_{22}^{-1}\boldsymbol{\Sigma}_{21} - \lambda\boldsymbol{\Sigma}_{11}| = 0$$

and that the canonical correlation vectors for $\mathbf{x}$ satisfy

$$\boldsymbol{\Sigma}_{12}\boldsymbol{\Sigma}_{22}^{-1}\boldsymbol{\Sigma}_{22}\mathbf{a}_i = \lambda_i\boldsymbol{\Sigma}_{11}\mathbf{a}_i.$$

10.2.5 (a) Show that the canonical correlation vectors $\mathbf{a}_i$ and $\mathbf{b}_i$ are eigenvectors of $\mathbf{M}_1$ and $\mathbf{M}_2$, respectively.

(b) Write $\mathbf{M}_1 = \mathbf{L}_1\mathbf{L}_2$, $\mathbf{M}_2 = \mathbf{L}_2\mathbf{L}_1$, where $\mathbf{L}_1 = \boldsymbol{\Sigma}_{11}^{-1}\boldsymbol{\Sigma}_{12}$ and $\mathbf{L}_2 = \boldsymbol{\Sigma}_{22}^{-1}\boldsymbol{\Sigma}_{21}$. Using Theorem A.6.2 show that the canonical correlation vectors can be expressed in terms of one another as

$$\mathbf{b}_i = \lambda_i^{-1/2}\mathbf{L}_2\mathbf{a}_i, \qquad \mathbf{a}_i = \lambda_i^{-1/2}\mathbf{L}_1\mathbf{b}_i.$$

10.2.6 Let $\mathbf{x}(q \times 1)$ and $\mathbf{y}(p \times 1)$ be random vectors such that for all $i, k = 1, \ldots, q$ and $j, l = 1, \ldots, p$, $i \neq k$ and $j \neq l$,

$$V(x_i) = 1, \qquad V(y_i) = 1, \qquad C(x_i, x_k) = \rho, \qquad C(y_j, y_l) = \rho', \qquad C(x_i, y_j) = \tau,$$

where ρ, ρ' and τ lie between 0 and 1. Show that the only canonical correlation variables are

$$\{q[(q-1)\rho + 1]\}^{-1/2} \sum_{i=1}^{q} x_i \quad \text{and} \quad \{p[(p-1)\rho' + 1]\}^{-1/2} \sum_{j=1}^{p} y_j.$$

10.2.7 When $p = 1$ show that the value of ρ_1^2 in Theorem 10.2.1 is the squared multiple correlation $\boldsymbol{\Sigma}_{21}\boldsymbol{\Sigma}_{11}^{-1}\boldsymbol{\Sigma}_{12}/\sigma_{22}$.

10.2.8 (a) The residual variance in predicting x_i by its linear regression on $\mathbf{b}'\mathbf{y}$ is given by

$$\delta_i^2 = V(x_i) - C(x_i, \mathbf{b}'\mathbf{y})^2/V(\mathbf{b}'\mathbf{y}).$$

Show that the vector **b** which minimizes $\sum_{i=1}^{q} \delta_i^2$ is the eigenvector corresponding to the largest eigenvalue of $\boldsymbol{\Sigma}_{22}^{-1}\boldsymbol{\Sigma}_{21}\boldsymbol{\Sigma}_{12}$.

(b) Show further that the best k linear functions of $y_1, \ldots, y_p$ for predicting $x_1, \ldots, x_q$ in the sense of minimizing the sum of residual variances correspond to the first k eigenvectors of $\boldsymbol{\Sigma}_{22}^{-1}\boldsymbol{\Sigma}_{21}\boldsymbol{\Sigma}_{12}$.

10.2.9 When $q = p = 2$ the squared canonical correlations can be computed explicitly. If

$$\boldsymbol{\Sigma}_{11} = \begin{pmatrix} 1 & \alpha \\ \alpha & 1 \end{pmatrix}, \quad \boldsymbol{\Sigma}_{22} = \begin{pmatrix} 1 & \beta \\ \beta & 1 \end{pmatrix}, \quad \boldsymbol{\Sigma}_{12} = \begin{pmatrix} a & b \\ c & d \end{pmatrix},$$

then show that the eigenvalues of $\mathbf{M}_1 = \boldsymbol{\Sigma}_{11}^{-1}\boldsymbol{\Sigma}_{12}\boldsymbol{\Sigma}_{22}^{-1}\boldsymbol{\Sigma}_{21}$ are given by

$$\lambda = \{B \pm (B^2 - 4C)^{1/2}\}/\{2(1-\alpha^2)(1-\beta^2)\},$$

where

$$B = a^2 + b^2 + c^2 + d^2 + 2(ad + bc)\alpha\beta - 2(ac + bd)\alpha - 2(ab + cd)\beta$$

and

$$C = (ad - bc)^2(1 + \alpha^2\beta^2 - \alpha^2 - \beta^2).$$

10.2.10 (Hotelling, 1936) Four examinations in reading speed, reading power, arithmetic speed, and arithmetic power were given to $n = 148$ children. The question of interest is whether reading ability is correlated with arithmetic ability. The correlations are given by

$$\mathbf{R}_{11} = \begin{pmatrix} 1.0 & 0.6328 \\ & 1.0 \end{pmatrix}, \quad \mathbf{R}_{22} = \begin{pmatrix} 1.0 & 0.4248 \\ & 1.0 \end{pmatrix}, \quad \mathbf{R}_{12} = \begin{pmatrix} 0.2412 & 0.0586 \\ -0.0553 & 0.0655 \end{pmatrix}.$$

Using Exercise 10.2.9, verify that the canonical correlations are given by

$$\rho_1 = 0.3945, \qquad \rho_2 = 0.0688.$$

10.2.11 Using (10.2.14) test whether $\rho_1 = \rho_2 = 0$ for Hotelling's examination data in Exercise 10.2.10. Show that one gets the test statistic

$$25.13 \sim \chi_4^2.$$

Since $\chi^2_{4;0.01} = 13.3$, we strongly reject the hypothesis that reading ability and arithmetic ability are independent for this data.

10.2.12 If in Example 10.2.1, $\beta = 0.7$, $\alpha = 0.74$, and $\gamma = 0.84$, show that $\rho_1 = 0.782$, thus approximating the situation in Example 10.2.3 where $r_1 = 0.7886$.

10.2.13 (a) Using the data matrix for the open/closed book data in Example 10 2.5 and Table 1.2.1, show that the scores of the first eight individuals on the first canonical correlation variables are as follows:

Subject	1	2	3	4	5	6	7	8
η_1	6.25	5.68	5.73	5.16	4.90	4.54	4.80	5.16
ϕ_1	6.35	7.44	6.67	6.00	6.14	6.71	6.12	6.30

(b) Plot the above eight points on a scattergram.

(c) Repeat the procedure for the second canonical correlation variable and analyze the difference in the correlations. (The second canonical correlation is $r_2 = 0.041$ and the corresponding loading vectors are given by

$$\mathbf{a}_2' = (-0.064, 0.076), \qquad \mathbf{b}_2' = (-0.091, 0.099, -0.014).)$$

10.2.14 The technique of ridge regression (Hoerl and Kennard, 1970) has been extended to canonical correlation analysis by Vinod (1976) giving what he calls the "canonical ridge" technique. This technique involves replacing the sample correlation matrix **R** by

$$\begin{bmatrix} \mathbf{R}_{11} + k_1\mathbf{I} & \mathbf{R}_{12} \\ \mathbf{R}_{21} & \mathbf{R}_{22} + k_2\mathbf{I} \end{bmatrix},$$

where k_1 and k_2 are small non-negative numbers, and carrying out a canonical correlation analysis on this new correlation matrix. For data which is nearly collinear (that is, $\mathbf{R}_{11}$ and/or $\mathbf{R}_{22}$ have eigenvalues near 0), show that small but non-zero values of k_1, k_2 lead to better estimates of the true canonical correlations and canonical correlation vectors than the usual analysis on **R** provides.

10.2.15 (Lawley, 1959) If **S** is based on a large number, n, of observations, then the following asymptotic results hold for the k non-zero canonical correlations, provided ρ_i^2 and $\rho_i^2 - \rho_j^2$ are not too close to zero, for all $i, j = 1, \ldots, k$, $i \neq j$:

$$2\rho_i E(r_i - \rho_i) = \frac{1}{(n-1)}(1-\rho_i^2)\left\{p+q-2-\rho_i^2+2(1-\rho_i^2)\sum_{\substack{s=1\\ s\neq i}}^{k}\frac{\rho_s^2}{\rho_i^2-\rho_s^2}\right\}+O(n^{-2}),$$

$$V(r_i) = \frac{1}{(n-1)}(1-\rho_i^2)^2 + O(n^{-2}),$$

$$\operatorname{corr}(r_i, r_j) = \frac{2\rho_i\rho_j(1-\rho_i^2)(1-\rho_j^2)}{(n-1)(\rho_i^2-\rho_j^2)^2} + O(n^{-2}).$$

10.3.1 Show that $\mathbf{S}_{11}\mathbf{1}=\mathbf{0}$ and $\mathbf{S}_{22}\mathbf{1}=\mathbf{0}$ in (10.3.1). Hence $\mathbf{S}_{11}^{-1}$ and $\mathbf{S}_{22}^{-1}$ do not exist.

10.3.2 The following example illustrates that it does not matter which row and column are deleted from a contingency table when constructing canonical scores. Let

$$\mathbf{N}=\begin{bmatrix}4 & 0\\ 1 & 2\\ 0 & 3\end{bmatrix}.$$

(a) Show that deleting the first row and first column from $\mathbf{N}$ leads in (10.3.2) to

$$\mathbf{M}_1^*=\frac{1}{15}\begin{bmatrix}2 & 6\\ 3 & 9\end{bmatrix}$$

and hence the canonical scores for the row categories are proportional to $(0, 2, 3)$.

(b) Similarly, show that deleting the second row and first column from $\mathbf{N}$ in (10.3.2) leads to

$$\mathbf{M}_1^*=\frac{1}{15}\begin{bmatrix}8 & -6\\ -4 & 3\end{bmatrix}$$

and hence to canonical scores proportional to $(-2, 0, 1)$. Since these scores are related to the scores in (a) by an additive constant, they are equivalent.

(c) Show that the canonical correlation between the row and column categories equals $(11/15)^{1/2}$.

(d) Note that because there are only two column categories, all scoring functions for column categories are equivalent (as long as they give distinct values to each of the two categories).

11
Discriminant Analysis

11.1 Introduction

Consider g populations or groups $\Pi_1, \ldots, \Pi_g$, $g \geq 2$. Suppose that associated with each population Π_j, there is a probability density $f_j(\mathbf{x})$ on R^p, so that if an individual comes from population Π_j, he has p.d.f. $f_j(\mathbf{x})$. Then the object of discriminant analysis is to allocate an individual to one of these g groups on the basis of his measurements $\mathbf{x}$. Of course, it is desirable to make as few "mistakes" as possible in a sense to be made precise later.

For example, the populations might consist of different diseases and $\mathbf{x}$ might measure the symptoms of a patient. Thus one is trying to diagnose a patient's disease on the basis of his symptoms. As another example, consider the samples from three species of iris given in Table 1.2.2. The object is then to allocate a new iris to one of these species on the basis of its measurements.

A *discriminant rule* d corresponds to a division of R^p into disjoint regions $R_1, \ldots, R_g$ $(\bigcup R_j = R^p)$. The rule d is defined by

$$\text{allocate } \mathbf{x} \text{ to } \Pi_j \quad \text{if} \quad \mathbf{x} \in R_j,$$

for $j = 1, \ldots, g$. Discrimination will be more accurate if Π_j has most of its probability concentrated in R_j for each j.

Usually, we have no prior information about which population an individual is likely to come from. However, if such information is available, it can be incorporated into a Bayesian approach.

The situation where the p.d.f.s $f_j(\mathbf{x})$ are known exactly is the simplest to analyse theoretically, although it is the least realistic in practice. We examine this case in Section 11.2.

A variant of this situation occurs when the form of the p.d.f. for each population is known, but there are parameters which must be estimated. The estimation is based on a sample data matrix $\mathbf{X}(n \times p)$ whose rows are partitioned into g groups,

$$\mathbf{X} = \begin{bmatrix} \mathbf{X}_1 \\ \cdot \\ \cdot \\ \cdot \\ \mathbf{X}_g \end{bmatrix}.$$

The $(n_j \times p)$ matrix $\mathbf{X}_j$ represents a sample of n_j individuals from the population Π_j. Note that in this chapter it is the *individuals* (rows) of $\mathbf{X}$ which are grouped into categories, whereas in the last chapter it was the variables (columns) which were grouped.

Finally, there is an empirical approach to discriminant analysis where we do not assume any particular form for the populations Π_j, but merely look for a "sensible" rule which will enable us to discriminate between them. One such rule is based on Fisher's linear discriminant function and is described in Section 11.5.

11.2 Discrimination when the Populations are Known

11.2.1 The maximum likelihood discriminant rule

Consider the situation where the exact distributions of the populations $\Pi_1, \ldots, \Pi_g$ are known. Of course this is extremely rare, although it may be possible to estimate the distributions fairly accurately provided the samples are large enough. In any case an examination of the situation where distributions are known serves as a useful framework against which other situations can be compared. The starting point for our analysis is the intuitively plausible maximum likelihood rule. We shall write the p.d.f. of the jth population as $f_j(\mathbf{x}) = L_j(\mathbf{x})$ to emphasize that we are thinking of the likelihood of the data point $\mathbf{x}$ as a function of the "parameter" j.

Definition 11.2.1 *The maximum likelihood discriminant rule for allocating an observation* $\mathbf{x}$ *to one of the populations* $\Pi_1, \ldots, \Pi_g$, *is to allocate* $\mathbf{x}$ *to the population which gives the largest likelihood to* $\mathbf{x}$.

That is, the maximum likelihood rule says one should allocate $\mathbf{x}$ to Π_j, where

$$L_j(\mathbf{x}) = \max_i L_i(\mathbf{x}). \tag{11.2.1}$$

If several likelihoods take the same maximum value, then any one of these may be chosen. This point will not always be repeated in what follows. Further, in the examples we consider, it will usually be the case that

$$P(L_i(\mathbf{x}) = L_k(\mathbf{x}) \text{ for some } i \neq k \mid \Pi_j) = 0,$$

for all $j = 1, \ldots, g$, so that the form of the allocation rule in the case of ties has no practical importance.

Example 11.2.1 If x is a 0–1 scalar random variable, and if Π_1 is the population with probabilities $(\frac{1}{2}, \frac{1}{2})$ and Π_2 is the population with probabilities $(\frac{1}{4}, \frac{3}{4})$, then the maximum likelihood discriminant rule allocates x to Π_1 when $x = 0$, and allocates x to Π_2 when $x = 1$. This is because $L_1(0) = \frac{1}{2} > L_2(0) = \frac{1}{4}$, and $L_2(1) = \frac{3}{4} > L_1(1) = \frac{1}{2}$.

Example 11.2.2 Suppose that $\mathbf{x}$ is a multinomial random vector, which comes *either* from Π_1, with multinomial probabilities $\alpha_1, \ldots, \alpha_k$ *or* from Π_2, with multinomial probabilities $\beta_1, \ldots, \beta_k$, where $\sum \alpha_i = \sum \beta_i = 1$ and $\sum x_i = n$, fixed. If $\mathbf{x}$ comes from Π_1, its likelihood is

$$\frac{n!}{x_1! \ldots x_k!} \alpha_1^{x_1} \ldots \alpha_k^{x_k}. \tag{11.2.2}$$

If $\mathbf{x}$ comes from Π_2 the likelihood is

$$\frac{n!}{x_1! \ldots x_k!} \beta_1^{x_1} \ldots \beta_k^{x_k}.$$

If λ is the ratio of these likelihoods, then the log likelihood ratio is

$$\log \lambda = \sum x_i \log \frac{\alpha_i}{\beta_i} = \sum x_i s_i, \tag{11.2.3}$$

where $s_i = \log(\alpha_i/\beta_i)$. The maximum likelihood discriminant rule allocates $\mathbf{x}$ to Π_1 if $\lambda > 1$, i.e. if $\log \lambda > 0$. We shall meet this rule again in the context of a linguistic seriation problem in Example 11.3.4.

Example 11.2.3 Suppose that Π_1 is the $N(\mu_1, \sigma_1^2)$ distribution, and Π_2 is the $N(\mu_2, \sigma_2^2)$ distribution. This situation is illustrated in Figure 11.2.1. for the case where $\mu_2 > \mu_1$ and $\sigma_1 > \sigma_2$. The likelihood L_i $(i = 1, 2)$ is

$$L_i(x) = (2\pi\sigma_i^2)^{-1/2} \exp\left\{-\tfrac{1}{2}\left(\frac{x-\mu_i}{\sigma_i}\right)^2\right\}.$$

Note $L_1(x)$ exceeds $L_2(x)$ if

$$\frac{\sigma_2}{\sigma_1} \exp\left\{-\tfrac{1}{2}\left[\left(\frac{x-\mu_1}{\sigma_1}\right)^2 - \left(\frac{x-\mu_2}{\sigma_2}\right)^2\right]\right\} > 1.$$

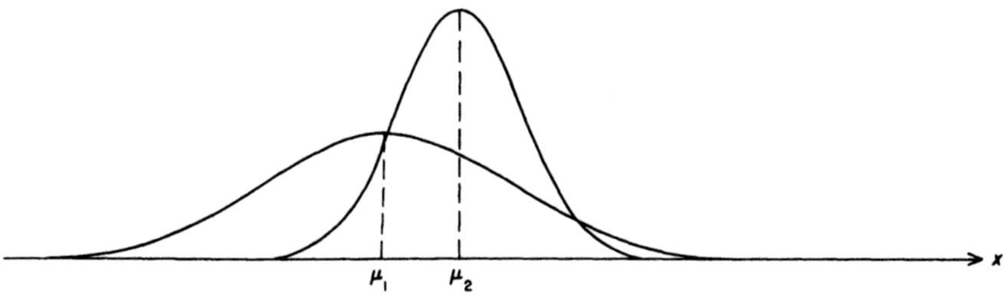

Figure 11.2.1 Normal likelihoods with unequal means and unequal variances (from Example 11.2.4).

On taking logarithms and rearranging, this inequality becomes

$$x^2\left(\frac{1}{\sigma_1^2}-\frac{1}{\sigma_2^2}\right)-2x\left(\frac{\mu_1}{\sigma_1^2}-\frac{\mu_2}{\sigma_2^2}\right)+\left(\frac{\mu_1^2}{\sigma_1^2}-\frac{\mu_2^2}{\sigma_2^2}\right)<2\log\frac{\sigma_2}{\sigma_1}.$$

If $\sigma_1>\sigma_2$ as in Figure 11.2.1, then the coefficient of x^2 is negative. Therefore the set of xs for which this inequality is satisfied falls in two distinct regions, one having low values of x and other having high values of x (see Exercise 11.2.8 and Anderson and Bahadur, 1962).

Example 11.2.4 An important special case of the previous example occurs when $\sigma_1=\sigma_2$, for then $L_1(x)$ exceeds $L_2(x)$ when

$$|x-\mu_2|>|x-\mu_1|.$$

In other words, if $\mu_2>\mu_1$ then the maximum likelihood discriminant rule allocates x to Π_2 if $x>\frac{1}{2}(\mu_1+\mu_2)$, and allocates x to Π_1 otherwise. If $\mu_1>\mu_2$ the position is reversed.

The multivariate generalization of this result is fundamental to this chapter and is given as a theorem.

Theorem 11.2.1 (a) *If Π_i is the $N_p(\boldsymbol{\mu}_i, \boldsymbol{\Sigma})$ population, $i=1,\ldots,g$, and $\boldsymbol{\Sigma}>0$, then the maximum likelihood discriminant rule allocates $\mathbf{x}$ to Π_j, where $j\in\{1,\ldots,g\}$ is that value of i which minimizes the square of the Mahalanobis distance*

$$(\mathbf{x}-\boldsymbol{\mu}_i)'\boldsymbol{\Sigma}^{-1}(\mathbf{x}-\boldsymbol{\mu}_i).$$

(b) *When $g=2$, the rule allocates $\mathbf{x}$ to Π_1 if*

$$\boldsymbol{\alpha}'(\mathbf{x}-\boldsymbol{\mu})>0, \tag{11.2.4}$$

where $\boldsymbol{\alpha}=\boldsymbol{\Sigma}^{-1}(\boldsymbol{\mu}_1-\boldsymbol{\mu}_2)$ and $\boldsymbol{\mu}=\frac{1}{2}(\boldsymbol{\mu}_1+\boldsymbol{\mu}_2)$, and to Π_2 otherwise.

Proof From (2.5.1), the ith likelihood is

$$L_i(\mathbf{x})=|2\pi\boldsymbol{\Sigma}|^{-1/2}\exp\{-\tfrac{1}{2}(\mathbf{x}-\boldsymbol{\mu}_i)'\boldsymbol{\Sigma}^{-1}(\mathbf{x}-\boldsymbol{\mu}_i)\}.$$

This is maximized when the exponent is minimized, which proves part (a) of the theorem.

For part (b) note that $L_1(\mathbf{x}) > L_2(\mathbf{x})$ if and only if

$$(\mathbf{x}-\boldsymbol{\mu}_1)'\boldsymbol{\Sigma}^{-1}(\mathbf{x}-\boldsymbol{\mu}_1) < (\mathbf{x}-\boldsymbol{\mu}_2)'\boldsymbol{\Sigma}^{-1}(\mathbf{x}-\boldsymbol{\mu}_2).$$

Cancelling and simplifying leads to the condition stated in the theorem. (See Exercise 11.2.1.) ■

Example 11.2.5 (Singular $\boldsymbol{\Sigma}$; Rao and Mitra, 1971, p. 204) Consider the situation of the above theorem when $\boldsymbol{\Sigma}$ is singular. Then the ML rule must be be modified. Note that Π_1 is concentrated on the hyperplane $\mathbf{N}'(\mathbf{x}-\boldsymbol{\mu}_1)=\mathbf{0}$, where the columns of $\mathbf{N}$ span the null space of $\boldsymbol{\Sigma}$, and that Π_2 is concentrated on the hyperplane $\mathbf{N}'(\mathbf{x}-\boldsymbol{\mu}_2)=\mathbf{0}$. (See Section 2.5.4.) If $\mathbf{N}'(\boldsymbol{\mu}_1-\boldsymbol{\mu}_2)\neq\mathbf{0}$, then these two hyperplanes are distinct, and discrimination can be carried out with perfect accuracy; namely if $\mathbf{N}'(\mathbf{x}-\boldsymbol{\mu}_1)=\mathbf{0}$ allocate to Π_1 and if $\mathbf{N}'(\mathbf{x}-\boldsymbol{\mu}_2)=\mathbf{0}$ allocate to Π_2.

The more interesting case occurs when $\boldsymbol{\mu}_1-\boldsymbol{\mu}_2$ is orthogonal to the null space of $\boldsymbol{\Sigma}$. If $\mathbf{N}'(\boldsymbol{\mu}_1-\boldsymbol{\mu}_2)=\mathbf{0}$, then the ML allocation rule is given by allocating $\mathbf{x}$ to Π_1 if

$$\boldsymbol{\alpha}'\mathbf{x} > \tfrac{1}{2}\boldsymbol{\alpha}'(\boldsymbol{\mu}_1+\boldsymbol{\mu}_2),$$

where $\boldsymbol{\alpha}=\boldsymbol{\Sigma}^{-}(\boldsymbol{\mu}_1-\boldsymbol{\mu}_2)$ and $\boldsymbol{\Sigma}^{-}$ is a g-inverse of $\boldsymbol{\Sigma}$. In the rest of this chapter we shall assume $\boldsymbol{\Sigma}>0$.

Note that when there are just $g=2$ groups, the ML discriminant rule is defined in terms of the *discriminant function*

$$h(\mathbf{x}) = \log L_1(\mathbf{x}) - \log L_2(\mathbf{x}). \qquad (11.2.5)$$

and the ML rule takes the form

$$\begin{aligned} &\text{allocate } \mathbf{x} \text{ to } \Pi_1 \text{ if } h(\mathbf{x})>0,\\ &\text{allocate } \mathbf{x} \text{ to } \Pi_2 \text{ if } h(\mathbf{x})<0. \end{aligned} \qquad (11.2.6)$$

In particular, note that the discriminant function given in (11.2.4) for two multinormal populations with the same covariance matrix is *linear*. Thus the boundary between the allocation regions in this case is a hyperplane passing through the midpoint of the line segment connecting the two group means, although the hyperplane is not necessarily perpendicular to this line segment. See Figure 11.3.1 for a picture in the sample case.

11.2.2 The Bayes discriminant rule

In certain situations it makes sense to suppose that the various populations have *prior probabilities*. For instance, in medical diagnosis we may regard flu as intrinsically more likely than polio. This information can be incorporated into the analysis using a Bayesian discriminant rule. For simplicity we shall suppose all prior probabilities π_j are strictly positive $j=1,\ldots,g$.

Definition 11.2.2 *If populations $\Pi_1,\ldots,\Pi_g$ have prior probabilities $(\pi_1,\ldots,\pi_g)=\boldsymbol{\pi}'$, then the* Bayes discriminant rule *(with respect to $\boldsymbol{\pi}$) allocates an observation* $\mathbf{x}$ *to the population for which*

$$\pi_j L_j(\mathbf{x}), \tag{11.2.7}$$

is maximized.

The function in (11.2.7) can be regarded as proportional to the *posterior* likelihood of Π_j given the data $\mathbf{x}$. Note that the ML rule is a special case of the Bayes rule when all the prior probabilities are equal.

In the case of discrimination between $g=2$ populations, the effect of introducing prior probabilities is simply to shift the critical value of the discriminant function by an amount $\log(\pi_2/\pi_1)$. The rule (11.2.6) becomes

$$\text{allocate } \mathbf{x} \text{ to } \Pi_1 \text{ if } h(\mathbf{x})>\log(\pi_2/\pi_1)$$

and to Π_2 otherwise. In particular for the case of two multinormal populations with the same covariance matrix, the boundary hyperplane is moved closer to the less likely population, but remains parallel to the boundary hyperplane of the ML allocation regions.

11.2.3 Optimal properties

The Bayes discriminant rules described above (including the ML rule) have certain optimal properties. First notice that the above rules are deterministic in the sense that if $\mathbf{x}_1=\mathbf{x}_2$ then $\mathbf{x}_1$ and $\mathbf{x}_2$ will always be allocated to the same population. However, for mathematical purposes it is convenient to define a wider class of discriminant rules. (In this wider class, it is possible to average two discriminant rules, and hence the set of all discriminant rules forms a convex set.)

Definition 11.2.3 A randomized discriminant rule d *involves allocating an observation* $\mathbf{x}$ *to a population j with probability $\phi_j(\mathbf{x})$, where $\phi_1,\ldots,\phi_g$ are non-negative functions each defined on R^p, which satisfy $\sum \phi_j(\mathbf{x})=1$ for all* $\mathbf{x}$.

It is clear that a deterministic allocation rule is a special case of a randomized allocation rule obtained by putting $\phi_j(\mathbf{x})=1$ for $\mathbf{x}\in R_j$ and $\phi_j(\mathbf{x})=0$ elsewhere. For example, the Bayes rule with respect to prior probabilities $\pi_1, \ldots, \pi_g$ is defined by

$$\phi_j(\mathbf{x}) = \begin{cases} 1 & \text{if} \quad \pi_j L_j(\mathbf{x}) = \max_i \pi_i L_i(\mathbf{x}), \\ 0 & \text{otherwise}, \end{cases} \tag{11.2.8}$$

except for those $\mathbf{x}$ where the maximum is attained by more than one population. (Since we are supposing that the set of such $\mathbf{x}$s has zero probability, whatever the true population, this ambiguity in $\phi_j(\mathbf{x})$ is irrelevant.)

The probability of allocating an individual to population Π_i, when in fact he comes from Π_j) is given by

$$p_{ij} = \int \phi_i(\mathbf{x}) L_j(\mathbf{x})\, d\mathbf{x}. \tag{11.2.9}$$

In particular, if an individual is in fact from Π_i, the probability of correctly allocating him is p_{ii} and the probability of misallocating him is $1 - p_{ii}$. The performance of the discriminant rule can be summarized by the numbers $p_{11}, \ldots, p_{gg}$. The following definition gives a partial order on the set of discrimination rules.

Definition 11.2.4 *Say that one discriminant rule d with probabilities of correct allocation* $\{p_{ii}\}$ *is* as good as *another rule d' with probabilities* $\{p'_{ii}\}$ *if*

$$p_{ii} \geq p'_{ii} \qquad \text{for all } i = 1, \ldots, g.$$

Say that d is better *than d' if at least one of the inequalities is strict. If d is a rule for which there is no better rule, say that d is* admissible.

Notice that it may not always be possible to compare two allocation rules using this criterion, for example if $p_{11} > p'_{11}$ but $p_{22} < p'_{22}$. However, we can prove the following optimal property of Bayes discriminant rules, which can be considered as a generalization of the Neyman–Pearson lemma (see Rao, 1973, p. 448).

Theorem 11.2.2 *All Bayes discriminant rules (including the ML rule) are admissible.*

Proof Let d^* denote the Bayes rule with respect to prior probabilities $\boldsymbol{\pi}$. Suppose there exists a rule d which is better than this rule. Let $\{p_{ii}\}$ and $\{p^*_{ii}\}$ denote the probabilities of correct classification for this rule and the Bayes rule, respectively. Then because d is better than d^* and since

$\pi_j > 0$ for all j, $\sum \pi_i p_{ii} > \sum \pi_i p_{ii}^*$. However, using (11.2.8) and (11.2.9),

$$\sum \pi_i p_{ii} = \sum \int \phi_i(\mathbf{x}) \pi_i L_i(\mathbf{x})\, d\mathbf{x}$$

$$\leq \sum_i \int \phi_i(\mathbf{x}) \max_j \pi_j L_j(\mathbf{x})\, d\mathbf{x} = \int \{\sum \phi_i(\mathbf{x})\} \max \pi_j L_j(\mathbf{x})\, d\mathbf{x}$$

$$= \int \max \pi_j L_j(\mathbf{x})\, d\mathbf{x}$$

$$= \int \sum \phi_i^*(\mathbf{x}) \pi_i L_i(\mathbf{x})\, d\mathbf{x}$$

$$= \sum \pi_i p_{ii}^*,$$

which contradicts the above statement. Hence the theorem is proved. ■

Note that in the above theorem, discriminant rules are judged on their g probabilities of correct allocation, $p_{11}, \ldots, p_{gg}$. However, if prior probabilities exist, then a discriminant rule can also be judged on the basis of a single number—the posterior probability of correct allocation, $\sum \pi_i p_{ii}$. Using this criterion, *any* two discriminant rules can be compared and we have the following result, first given in the case of $g = 2$ groups by Welch (1939).

Theorem 11.2.3 *If populations $\Pi_1, \ldots, \Pi_g$ have prior probabilities $\pi_1, \ldots, \pi_g$, then no discriminant rule has a larger posterior probability of correct allocation than the Bayes rule with respect to this prior.*

Proof Let d^* denote the Bayes rule with respect to the prior probabilities $\pi_1, \ldots, \pi_g$ with probabilities of correct allocation $\{p_{ii}^*\}$. Then, using the same argument as in Theorem 11.2.2, it is easily seen that for any other rule d with probabilities of correct classification $\{p_{ii}\}$, $\sum \pi_i p_{ii} \leq \sum \pi_i p_{ii}^*$; that is, the posterior probability of correct allocation is at least as large for the Bayes rule as for the other rule.

11.2.4 Decision theory and unequal costs

The discrimination problem described above can be phrased in the language of decision theory. Let

$$K(i, j) = \begin{cases} 0, & i = j, \\ c_{ij}, & i \neq j. \end{cases} \tag{11.2.10}$$

be a *loss function* representing the cost or loss incurred when an observation is allocated to Π_i when in fact it comes from Π_j. For this to be a sensible definition, suppose $c_{ij} > 0$ for all $i \neq j$. If d is a discriminant rule with allocation functions $\phi_i(\mathbf{x})$ given in (11.2.8), then the *risk function* is defined by

$$R(d, j) = E(K(d(\mathbf{x}), j) \mid \Pi_j)$$
$$= \sum_i K(i, j) \int \phi_i(\mathbf{x}) L_j(\mathbf{x})\, d\mathbf{x} \qquad (11.2.11)$$
$$= \sum_i c_{ij} p_{ij}$$

and represents the expected loss given that the observation comes from Π_j. In particular, if $c_{ij} = 1$ for $i \neq j$, then

$$R(d, j) = 1 - p_{jj}$$

represents the misallocation probabilities.

If prior probabilities exist then the *Bayes risk* can be defined by

$$r(d, \boldsymbol{\pi}) = \sum \pi_j R(d, j)$$

and represents the posterior expected loss.

As in Section 11.2.3, say that a discrimination rule d is *admissible* if there exists no other rule d' such that $R(d', j) \leq R(d, j)$ for all j, with at least one strict inequality.

Define the *Bayes rule* in this situation with respect to prior probabilities $\pi_1, \ldots, \pi_g$ as follows:

$$\text{allocate } \mathbf{x} \text{ to } \Pi_j \text{ if } \sum_{k \neq j} c_{jk} \pi_k L_k(\mathbf{x}) = \min_i \sum_{k \neq i} c_{ik} \pi_k L_k(\mathbf{x}). \qquad (11.2.12)$$

The following results can be proved in the same way as Theorem 11.2.2 and Theorem 11.2.3 (See also Exercise 11.2.4.)

Theorem 11.2.4 *All Bayes discrimination rules are admissible for the risk function R.* ■

Theorem 11.2.5 *If the populations $\Pi_1, \ldots, \Pi_g$ have prior probabilities $\pi_1, \ldots, \pi_g$, then no discriminant rule has smaller Bayes risk for the risk function R than the Bayes rule with respect to $\boldsymbol{\pi}$.* ■

The advantage of the decision theory approach is that it allows us to attach varying levels of importance to different sorts of error. For example, in medical diagnosis it might be regarded as more harmful to a patient's survival for polio to be misdiagnosed as flu than for flu to be misdiagnosed as polio.

In the remainder of this chapter we shall for simplicity place most of the emphasis on the ML discriminant rule (or equivalently, the Bayes rule with equal prior probabilities).

11.3 Discrimination under Estimation

11.3.1 The sample discriminant rule

The *sample ML discriminant rule* is useful when the forms of the distributions of $\Pi_1, \dots, \Pi_g$ are known, but their parameters must be estimated from a data matrix $\mathbf{X}(n \times p)$. We suppose that the rows of $\mathbf{X}$ are partitioned into g groups, $\mathbf{X}' = (\mathbf{X}_1', \dots, \mathbf{X}_g')$ and that $\mathbf{X}_i$ contains n_i observations from Π_i.

For example, suppose the groups are assumed to be samples from the multinormal distribution with different means and the same covariance matrix. Let $\bar{\mathbf{x}}_i$ and $\mathbf{S}_i$ denote the sample mean and covariance matrix of the ith group. Then unbiased estimates of $\boldsymbol{\mu}_1, \dots, \boldsymbol{\mu}_g$ and $\boldsymbol{\Sigma}$ are $\bar{\mathbf{x}}_1, \dots, \bar{\mathbf{x}}_g$ and $\mathbf{S}_u = \sum n_i \mathbf{S}_i/(n-g)$. The sample ML discriminant rule is then obtained by inserting these estimates in Theorem 11.2.1. In particular when $g = 2$ the sample ML discriminant rule allocates $\mathbf{x}$ to Π_1 if and only if

$$\mathbf{a}'\{\mathbf{x} - \tfrac{1}{2}(\bar{\mathbf{x}}_1 + \bar{\mathbf{x}}_2)\} > 0, \qquad (11.3.1)$$

where $\mathbf{a} = \mathbf{S}_u^{-1}(\bar{\mathbf{x}}_1 - \bar{\mathbf{x}}_2)$.

Example 11.3.1 Consider the $n_1 = n_2 = 50$ observations on two species of iris, *I. setosa* and *I. versicolour*, given in Table 1.2.2. For simplicity of exposition, we shall discriminate between them only on the basis of the first two variables, sepal length and sepal width. Then the sample means and variances for each group are given by

$$\bar{\mathbf{x}}_1 = (5.006, 3.428)', \qquad \bar{\mathbf{x}}_2 = (5.936, 2.770)',$$

$$\mathbf{S}_1 = \begin{pmatrix} 0.1218 & 0.0972 \\ & 0.1408 \end{pmatrix}, \qquad \mathbf{S}_2 = \begin{pmatrix} 0.2611 & 0.0835 \\ & 0.0965 \end{pmatrix}.$$

Thus,

$$\mathbf{a} = [(50\mathbf{S}_1 + 50\mathbf{S}_2)/98]^{-1}(\bar{\mathbf{x}}_1 - \bar{\mathbf{x}}_2)$$

$$= \begin{pmatrix} 0.1953 & 0.0922 \\ 0.0922 & 0.1211 \end{pmatrix}^{-1} \begin{pmatrix} -0.930 \\ 0.658 \end{pmatrix} = \begin{pmatrix} -11.436 \\ 14.143 \end{pmatrix},$$

and the discriminant rule is given by allocating to Π_1 if

$$h(\mathbf{x}) = (-11.436, 14.143)\begin{pmatrix} x_1 - \frac{1}{2}(5.006 + 5.936) \\ x_2 - \frac{1}{2}(3.428 + 2.770) \end{pmatrix}$$

$$= -11.436x_1 + 14.143x_2 + 18.739 > 0$$

and to Π_2 otherwise. A picture of the allocation regions is given in Figure

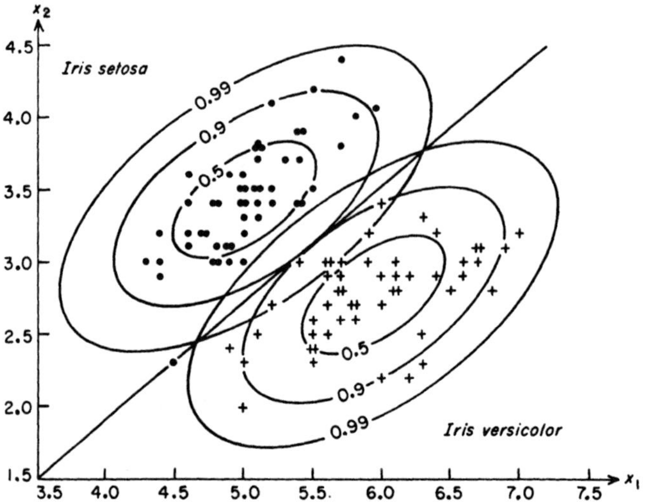

Figure 11.3.1 Discrimination between two species of iris: a plot of the data points in two dimensions (x_1 = sepal length, x_2 = sepal width) together with ellipses of concentration for probabilities 0.5, 0.9, and 0.99, and the boundary line defined by the discriminant function (Dagnelie, 1975, p. 313).

11.3.1. The estimated ellipses containing 50, 90, and 99% of the probability mass within each group have been drawn to give a visual impression of the accuracy of the discrimination. See also Example 11.6.1.

Note that the sample means themselves have scores $h(\bar{\mathbf{x}}_1) = 9.97$ and $h(\bar{\mathbf{x}}_2) = -9.97$, so that the boundary line passes through the midpoint of the line segment connecting the two means.

Consider a new iris with measurements $\mathbf{x} = (5.296, 3.213)'$. Since $h(\mathbf{x}) = 3.615 > 0$ we allocate $\mathbf{x}$ to Π_1.

Example 11.3.2 Extend the last example now to incude the $n_3 = 50$ observations on the third species of iris, *I. virginica*, in Table 1.2.2, which has sample mean and variance

$$\bar{\mathbf{x}}_3 = \begin{pmatrix} 6.588 \\ 2.974 \end{pmatrix}, \qquad \mathbf{S}_3 = \begin{pmatrix} 0.3963 & 0.0919 \\ & 0.1019 \end{pmatrix}.$$

In this case $\mathbf{\Sigma}$ is estimated by

$$\mathbf{S}_u = (50\mathbf{S}_1 + 50\mathbf{S}_2 + 50\mathbf{S}_3)/147 = \begin{pmatrix} 0.2650 & 0.0927 \\ 0.0927 & 0.1154 \end{pmatrix}$$

and discrimination is based on the three functions

$$h_{12}(\mathbf{x}) = (\bar{\mathbf{x}}_1 - \bar{\mathbf{x}}_2)'\mathbf{S}_u^{-1}\{\mathbf{x} - \tfrac{1}{2}(\bar{\mathbf{x}}_1 + \bar{\mathbf{x}}_2)\}$$
$$= -7.657x_1 + 11.856x_2 + 5.153,$$
$$h_{13}(\mathbf{x}) = -10.220x_1 + 12.147x_2 + 20.362,$$
$$h_{23}(\mathbf{x}) = -2.562x_1 + 0.291x_2 + 15.208.$$

Notice that $h_{12}(\mathbf{x})$ is not identical to the discriminant function $h(\mathbf{x})$ of Example 11.3.1, because we are using a slightly different estimate of $\mathbf{\Sigma}$.

Then the allocation regions are defined by

$$\text{allocate } \mathbf{x} \text{ to} \begin{cases} \Pi_1 \text{ if } h_{12}(\mathbf{x}) > 0 \text{ and } h_{13}(\mathbf{x}) > 0, \\ \Pi_2 \text{ if } h_{12}(\mathbf{x}) < 0 \text{ and } h_{23}(\mathbf{x}) > 0, \\ \Pi_3 \text{ if } h_{13}(\mathbf{x}) < 0 \text{ and } h_{23}(\mathbf{x}) < 0. \end{cases}$$

Writing

$$h_{ij}(\mathbf{x}) = (\bar{\mathbf{x}}_i - \bar{\mathbf{x}}_j)'\mathbf{S}_u^{-1}\mathbf{x} - \tfrac{1}{2}\bar{\mathbf{x}}_i'\mathbf{S}_u^{-1}\bar{\mathbf{x}}_i + \tfrac{1}{2}\bar{\mathbf{x}}_j\mathbf{S}_u^{-1}\bar{\mathbf{x}}_j$$

for $i \neq j$, it is easy to see that the discriminant functions are linearly related by

$$h_{12}(\mathbf{x}) + h_{23}(\mathbf{x}) = h_{13}(\mathbf{x}).$$

(See Exercise 11.3.2.) Thus the boundary consists of three lines meeting at the point where $h_{12}(\mathbf{x}) = h_{23}(\mathbf{x}) = h_{13}(\mathbf{x}) = 0$.

A picture of the allocation regions for the iris data is given in Figure 11.3.2. Notice that it is more difficult to discriminate accurately between

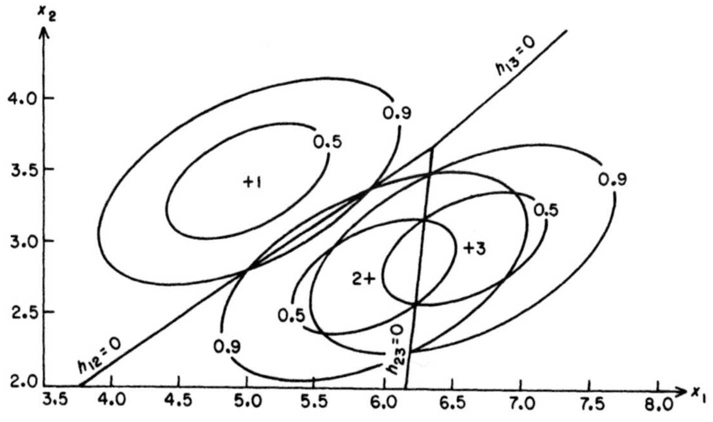

Figure 11.3.2 Discrimination between three species of iris using two variables (x_1 = sepal length, x_2 = sepal width): a plot of the sample means together with the ellipses of concentration for probabilities 0.5 and 0.9, and the boundaries between the allocation regions. Here **1** *= I. setosa,* **2** *= I. versicolour,* **3** *= I. virginica (Dagnelie, 1975, p. 322).*

I. versicolour and *I. virginica* than it is to discriminate between either of these species and *I. setosa*.

Example 11.3.3 Smith (1947) collected information on normal individuals and psychotic individuals. The data on two of the variables is illustrated in Figure 11.3.3. The diagonal line in this figure represents the boundary given by the sample discriminant rule defined in (11.3.1), which assumes equal covariance matrices for the populations. However, from the figure it is clear that the variances of the variables for the normal individuals are smaller than for the psychotics. Also, the correlation between the variables within each population appears low. For this reason Smith tried a sample ML discriminant rule assuming bivariate normal populations with *unequal* covariance matrices *but* with zero correlation between the variables in each population. These assumptions lead to a quadratic allocation rule with an elliptical boundary (Exercise 11.2.8). (In Figure 11.3.3 the axes have been scaled to make the boundary circular.)

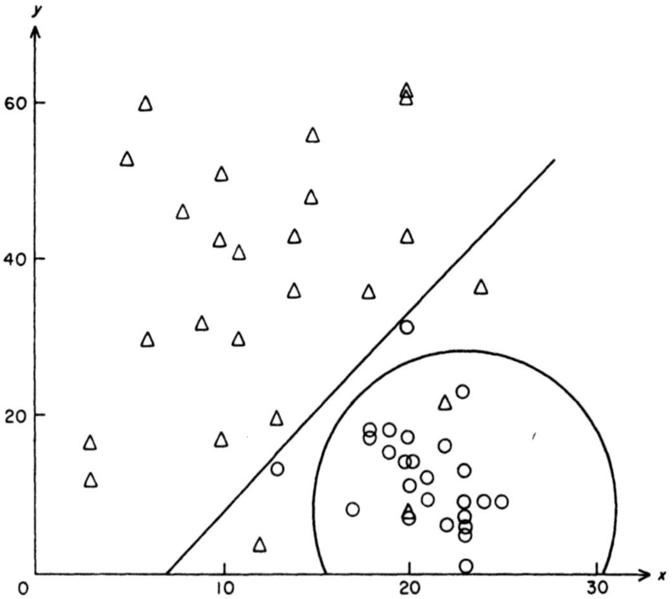

Figure 11.3.3 Discrimination between normal individuals (O) *and psychotics*(Δ) *on the basis of two variables x and y: a plot of the data points plus a linear boundary and a circular boundary between the groups* (*Smith*, 1947; *Bartlett, 1965, p. 208*).

For the given sample this rule appears to give a slight improvement over the use of a linear boundary.

Discriminant analysis can also be used as an aid to seriation (putting objects into chronological order), as seen by the following example.

Example 11.3.4 Cox and Brandwood (1959) used discriminant analysis to place in chronological order seven works of Plato—*Republic, Laws, Critias, Philebus, Politicus, Sophist,* and *Timaeus,* where it is known that *Republic* was written before *Laws* and that the other five works were written in between. However, the order of these five works is not known. The stylistic property on which the statistical analysis is based is the type of sentence ending. The last five syllables of a sentence are each either long or short, leading to $2^5 = 32$ types of sentence ending. For each work the percentage of sentences having each type of ending is given in Table 11.3.1.

It is assumed that sentence style varied systematically over time, and a function measuring this change of style is sought in order to order the works chronologically. (Note that this problem is not really discriminant analysis as defined in Section 11.1, because we do not want to allocate these intermediate works to *Republic* or *Laws.*)

Suppose each of the works has a multinomial distribution and in particular suppose the parameters for *Republic* and *Laws* are $\alpha_1, \ldots, \alpha_{32}$, and $\beta_1, \ldots, \beta_{32}$, respectively. Then, from Example 11.2.2, the ML discriminant function between *Republic* and *Laws* (standardized by the number of sentences) is given by

$$h(\mathbf{x}) = \sum x_i s_i \Big/ \sum x_i,$$

where $s_i = \log(\alpha_i/\beta_i)$ and $x_1, \ldots, x_{32}$ are the number of sentences with each type of ending in a particular work. We do not know the parameters α_i and β_i, $i = 1, \ldots, 32$. However, since the number of sentences in *Republic* and *Laws* is much larger than in the other works, we shall replace α_i and β_i by their sample estimates from Table 11.3.1.

The scores of each work on this discriminant function are given in Table 11.3.2. The table also gives their standard errors; the formulae for these are given in Exercise 11.3.3. From these scores it appears that the most likely order is *Republic, Timaeus, Sophist, Critias, Politicus, Philebus, Laws.* This order is not in accord with the view held by the majority of classical scholars, although there is a minority group who reached a similar ordering by apparently independent arguments. For a discussion of questions related to the statistical significance of this ordering, see Exercise 11.3.3, and Cox and Brandwood (1959).

Table 11.3.1 Percentage distribution of sentence endings in seven works of Plato (Cox and Brandwood, 1959)

Type of ending	Π_0 *Rep.*	Π_1 *Laws*	*Crit.*	*Phil.*	*Pol.*	*Soph.*	*Tim.*
∪ ∪ ∪ ∪ ∪	1.1	2.4	3.3	2.5	1.7	2.8	2.4
– ∪ ∪ ∪ ∪	1.6	3.8	2.0	2.8	2.5	3.6	3.9
∪ – ∪ ∪ ∪	1.7	1.9	2.0	2.1	3.1	3.4	6.0
∪ ∪ – ∪ ∪	1.9	2.6	1.3	2.6	2.6	2.6	1.8
∪ ∪ ∪ – ∪	2.1	3.0	6.7	4.0	3.3	2.4	3.4
∪ ∪ ∪ ∪ –	2.0	3.8	4.0	4.8	2.9	2.5	3.5
– – ∪ ∪ ∪	2.1	2.7	3.3	4.3	3.3	3.3	3.4
– ∪ – ∪ ∪	2.2	1.8	2.0	1.5	2.3	4.0	3.4
– ∪ ∪ – ∪	2.8	0.6	1.3	0.7	0.4	2.1	1.7
– ∪ ∪ ∪ –	4.6	8.8	6.0	6.5	4.0	2.3	3.3
∪ – – ∪ ∪	3.3	3.4	2.7	6.7	5.3	3.3	3.4
∪ – ∪ – ∪	2.6	1.0	2.7	0.6	0.9	1.6	2.2
∪ – ∪ ∪ –	4.6	1.1	2.0	0.7	1.0	3.0	2.7
∪ ∪ – – ∪	2.6	1.5	2.7	3.1	3.1	3.0	3.0
∪ ∪ – ∪ –	4.4	3.0	3.3	1.9	3.0	3.0	2.2
∪ ∪ ∪ – –	2.5	5.7	6.7	5.4	4.4	5.1	3.9
– – – ∪ ∪	2.9	4.2	2.7	5.5	6.9	5.2	3.0
– – ∪ – ∪	3.0	1.4	2.0	0.7	2.7	2.6	3.3
– – ∪ ∪ –	3.4	1.0	0.7	0.4	0.7	2.3	3.3
– ∪ – – ∪	2.0	2.3	2.0	1.2	3.4	3.7	3.3
– ∪ – ∪ –	6.4	2.4	1.3	2.8	1.8	2.1	3.0
– ∪ ∪ – –	4.2	0.6	4.7	0.7	0.8	3.0	2.8
∪ ∪ – – –	2.8	2.9	1.3	2.6	4.6	3.4	3.0
∪ – ∪ – –	4.2	1.2	2.7	1.3	1.0	1.3	3.3
∪ – – ∪ –	4.8	8.2	5.3	5.3	4.5	4.6	3.0
∪ – – – ∪	2.4	1.9	3.3	3.3	2.5	2.5	2.2
∪ – – – –	3.5	4.1	2.0	3.3	3.8	2.9	2.4
– ∪ – – –	4.0	3.7	4.7	3.3	4.9	3.5	3.0
– – ∪ – –	4.1	2.1	6.0	2.3	2.1	4.1	6.4
– – – ∪ –	4.1	8.8	2.0	9.0	6.8	4.7	3.8
– – – – ∪	2.0	3.0	3.3	2.9	2.9	2.6	2.2
– – – – –	4.2	5.2	4.0	4.9	7.3	3.4	1.8
Number of sentences	3778	3783	150	958	770	919	762

Table 11.3.2 Mean scores and their standard errors from seven works of Plato (Cox and Brandwood, 1959)

	Crit.	*Phil.*	*Pol.*	*Soph.*	*Tim.*	*Rep.*	*Laws.*
Mean score	−0.0346	0.1996	0.1303	−0.0407	−0.1170	−0.2652	0.2176
Estimated variance	0.003 799	0.000 334 2	0.000 397 3	0.000 571 9	0.000 721 8		
Estimated standard error	0.0616	0.0183	0.01993	0.0239	0.0269		

11.3.2 The likelihood ratio discriminant rule

An alternative rule to the sample ML allocation rule uses the *likelihood ratio criterion*, due to Anderson (1958, p. 141). This rule involves calculating the likelihoods of the hypotheses

H_i : $\mathbf{x}$ and the rows of $\mathbf{X}_i$ are from Π_i, and the rows of $\mathbf{X}_j$ are from Π_j, $j \neq i$,

for $i = 1,\ldots, g$. Then $\mathbf{x}$ is allocated to the group for which the hypothesis H_i has the largest likelihood.

For example, consider the discrimination problem between $g = 2$ multinormal populations with the same covariance matrix.

The m.l.e.s of $\boldsymbol{\mu}_1$, $\boldsymbol{\mu}_2$, and $\boldsymbol{\Sigma}$ under H_1 are $(n_1\bar{\mathbf{x}}_1+\mathbf{x})/(n_1+1)$, $\bar{\mathbf{x}}_2$, and

$$\tilde{\boldsymbol{\Sigma}}_1 = \frac{1}{n_1+n_2+1}\left\{\mathbf{W}+\frac{n_1}{1+n_1}(\mathbf{x}-\bar{\mathbf{x}}_1)(\mathbf{x}-\bar{\mathbf{x}}_1)'\right\},$$

where $\mathbf{W} = n_1\mathbf{S}_1 + n_2\mathbf{S}_2$. Similarly, under H_2 the m.l.e.s are $\bar{\mathbf{x}}_1$, $(n_2\bar{\mathbf{x}}_2+\mathbf{x})/(n_2+1)$, and

$$\tilde{\boldsymbol{\Sigma}}_2 = \frac{1}{n_1+n_2+1}\left\{\mathbf{W}+\frac{n_2}{1+n_2}(\mathbf{x}-\bar{\mathbf{x}}_2)(\mathbf{x}-\bar{\mathbf{x}}_2)'\right\}.$$

The likelihood ratio statistic, λ; is the $\frac{1}{2}(n_1+n_2+1)$th power of

$$\frac{|\tilde{\boldsymbol{\Sigma}}_2|}{|\tilde{\boldsymbol{\Sigma}}_1|} = \frac{1+(n_2/(1+n_2))(\mathbf{x}-\bar{\mathbf{x}}_2)'\mathbf{W}^{-1}(\mathbf{x}-\bar{\mathbf{x}}_2)}{1+(n_1/(1+n_1))(\mathbf{x}-\bar{\mathbf{x}}_1)'\mathbf{W}^{-1}(\mathbf{x}-\bar{\mathbf{x}}_1)}.$$

We accept H_1, and allocate $\mathbf{x}$ to Π_1 if and only if this expression exceeds one. This occurs if and only if

$$\frac{n_2}{1+n_2}(\mathbf{x}-\bar{\mathbf{x}}_2)'\,\mathbf{W}^{-1}(\mathbf{x}-\bar{\mathbf{x}}_2) > \frac{n_1}{1+n_1}(\mathbf{x}-\bar{\mathbf{x}}_1)'\mathbf{W}^{-1}(\mathbf{x}-\bar{\mathbf{x}}_1).$$

In the special case where $n_1 = n_2$ this likelihood ratio criterion is equivalent to the sample ML discriminant rule defined above. The procedures are also asymptotically equivalent if n_1 and n_2 are both large. However, when $n_1 \neq n_2$ this likelihood ratio criterion has a slight tendency to allocate $\mathbf{x}$ to the population which has the larger sample size.

11.3.3 A Bayesian approach

A problem with the sample ML rule is that it does not allow for sampling variation in the estimates of the parameters. This problem may not be serious for large n but can give misleading results if n is not much bigger than p.

A Bayesian method of dealing with this problem is to put a prior density $\pi(\boldsymbol{\theta})$ on the parameters $\boldsymbol{\theta}$ (see, for example, Aitchison *et al.*, 1977). Then the likelihood of an observation $\mathbf{x}$ given the data $\mathbf{X}$, on the assumption that $\mathbf{x}$ comes from the jth population, is given by averaging the p.d.f. of $\mathbf{x}$ given $\boldsymbol{\theta}$ with respect to the posterior density of $\boldsymbol{\theta}$ given $\mathbf{X}$; that is

$$L_j(\mathbf{x} \mid \mathbf{X}) = \int f_j(\mathbf{x} \mid \boldsymbol{\theta}) f(\boldsymbol{\theta} \mid \mathbf{X})\, d\boldsymbol{\theta},$$

where $f(\boldsymbol{\theta} \mid \mathbf{X}) \propto f(\mathbf{X} \mid \boldsymbol{\theta})\pi(\boldsymbol{\theta})$.

For example, if we are comparing two multinormal populations with the same covariance matrix, let $\boldsymbol{\theta} = (\boldsymbol{\mu}_1, \boldsymbol{\mu}_2, \boldsymbol{\Sigma})$ and let $\pi(\boldsymbol{\theta})$ be the vague prior of Example 4.3.1. Then

$$L_1(\mathbf{x} \mid \mathbf{X}) \propto \int |\boldsymbol{\Sigma}|^{-(p+1+n+1)/2} \exp\{-\tfrac{1}{2} \operatorname{tr} \boldsymbol{\Sigma}^{-1}[n_1\mathbf{S}_1 + n_2\mathbf{S}_2 + n_1(\bar{\mathbf{x}}_1 - \boldsymbol{\mu}_1)(\bar{\mathbf{x}}_1 - \boldsymbol{\mu})'$$
$$+ n_2(\bar{\mathbf{x}}_2 - \boldsymbol{\mu}_2)(\bar{\mathbf{x}}_2 - \boldsymbol{\mu}_2)' + (\mathbf{x} - \boldsymbol{\mu}_1)(\mathbf{x} - \boldsymbol{\mu}_1)']\}\, d\boldsymbol{\mu}_1\, d\boldsymbol{\mu}_2\, d\boldsymbol{\Sigma}. \quad (11.3.2)$$

Using the identity,

$$n_1(\bar{\mathbf{x}}_1 - \boldsymbol{\mu}_1)(\bar{\mathbf{x}}_1 - \boldsymbol{\mu}_1)' + (\mathbf{x} - \boldsymbol{\mu}_1)(\mathbf{x} - \boldsymbol{\mu}_1)'$$
$$= (n_1 + 1)\left(\boldsymbol{\mu}_1 - \frac{n_1\bar{\mathbf{x}}_1 + \mathbf{x}}{n_1 + 1}\right)\left(\boldsymbol{\mu}_1 - \frac{n_1\bar{\mathbf{x}}_1 + \mathbf{x}}{n_1 + 1}\right)' + \frac{n_1}{n_1 + 1}(\bar{\mathbf{x}}_1 - \mathbf{x})(\bar{\mathbf{x}}_1 - \mathbf{x})' \quad (11.3.3)$$

and integrating (11.3.2) over $d\boldsymbol{\mu}_1$ and $d\boldsymbol{\mu}_2$ first, we get

$$L_1(\mathbf{x} \mid \mathbf{X}) \propto \int |\boldsymbol{\Sigma}|^{-(p+1+n-1)/2} \exp\left\{-\tfrac{1}{2}\operatorname{tr}\boldsymbol{\Sigma}^{-1}[n\mathbf{S} + \frac{n_1}{n_1+1}(\bar{\mathbf{x}}_1 - \mathbf{x})(\bar{\mathbf{x}}_1 - \mathbf{x})']\right\} d\boldsymbol{\Sigma}$$
$$\propto \left|n\mathbf{S} + \frac{n_1}{n_1+1}(\mathbf{x} - \bar{\mathbf{x}}_1)(\mathbf{x} - \bar{\mathbf{x}}_1)'\right|^{-(n-1)/2}$$
$$\propto \{1 + (n-1-p)^{-1}(\mathbf{x} - \bar{\mathbf{x}}_1)'[n(1+1/n_1)/(n-1-p)\mathbf{S}]^{-1}$$
$$\times (\mathbf{x} - \bar{\mathbf{x}}_1)\}^{-(n-1-p+p)/2}.$$

Thus, using Exercise 2.6.5, $\mathbf{x}$ has a multivariate Student's t distribution with parameters $n-p-1$, $\bar{\mathbf{x}}_1$, and $\{n(1+1/n_1)/(n-1-p)\}\mathbf{S}$. Here $n = n_1 + n_2$ and $n\mathbf{S} = n_1\mathbf{S}_1 + n_2\mathbf{S}_2$. Similarly, $L_2(\mathbf{x} \mid \mathbf{X})$ is a multivariate Student's t distribution with parameters $n-p-1$, $\bar{\mathbf{x}}_2$, and $\{n(1+1/n_2)/(n-1-p)\}\mathbf{S}$.

As with the sample ML rule, we allocate $\mathbf{x}$ to the population with the larger likelihood. Note that if $n_1 = n_2$, the two rules are the same.

This Bayesian approach can also be used with unequal covariance matrices. See Exercise 11.3.5.

11.4 Is Discrimination Worthwhile?

Consider the situation where we are discriminating between g multinormal populations with the same covariance matrix, and the parameters are estimated from the data $\mathbf{X}=(\mathbf{X}_1',\ldots,\mathbf{X}_g')'$. If all the true means are equal, $\boldsymbol{\mu}_1=\ldots=\boldsymbol{\mu}_g$, then of course it is pointless to try to discriminate between the groups on the basis of these variables.

However, even if the true means are equal, the sample means $\bar{\mathbf{x}}_1,\ldots,\bar{\mathbf{x}}_g$ will be different from one another, so that an apparently plausible discriminant analysis can be carried out. Thus to check whether or not the discriminant analysis is worthwhile, it is interesting to test the hypothesis $\boldsymbol{\mu}_1=\ldots=\boldsymbol{\mu}_g$ given $\boldsymbol{\Sigma}_1=\ldots=\boldsymbol{\Sigma}_g$. This hypothesis is exactly the one-way multivariate analysis of variance described in Section 5.3.3a.

Recall that two possible tests of this hypothesis are obtained by partitioning the "total" sum of squares and products (SSP) matrix $\mathbf{T}=\mathbf{X}'\mathbf{H}\mathbf{X}$ as $\mathbf{T}=\mathbf{W}+\mathbf{B}$, where $\mathbf{W}$ and $\mathbf{B}$ are the "within-groups" and "between-groups" SSP matrices, repectively.

Then the Wilks' Λ test and the greatest root test are given by functions of the eigenvalues of $\mathbf{W}^{-1}\mathbf{B}$. In particular, if $g=2$, then $\mathbf{W}^{-1}\mathbf{B}$ has only one non-zero eigenvalue and the two tests are the same and are equivalent to the two-sample Hotelling's T^2 test. Then under the null hypothesis,

$$\left\{\frac{n_1n_2(n-2)}{n}\right\}\mathbf{d}'\mathbf{W}^{-1}\mathbf{d}\sim T^2(p,n-2)=\{(n-2)p/(n-p-1)\}F_{p,n-p-1}$$

and the null hypothesis is rejected for large values of this statistic.

11.5 Fisher's Linear Discriminant Function

Another approach to the discrimination problem based on a data matrix $\mathbf{X}$ can be made by not assuming any particular parametric form for the distribution of the populations $\Pi_1,\ldots,\Pi_g$, but by merely looking for a "sensible" rule to discriminate between them. Fisher's suggestion was to look for the *linear function* $\mathbf{a}'\mathbf{x}$ which *maximized* the ratio of the between-groups sum of squares to the within-groups sum of squares; that is, let

$$\mathbf{y}=\mathbf{X}\mathbf{a}=\begin{bmatrix}\mathbf{X}_1\mathbf{a}\\ \cdot\\ \cdot\\ \cdot\\ \mathbf{X}_g\mathbf{a}\end{bmatrix}=\begin{bmatrix}\mathbf{y}_1\\ \cdot\\ \cdot\\ \cdot\\ \mathbf{y}_g\end{bmatrix}$$

be a linear combination of the columns of $\mathbf{X}$. Then $\mathbf{y}$ has total sum of squares

$$\mathbf{y}'\mathbf{H}\mathbf{y} = \mathbf{a}'\mathbf{X}'\mathbf{H}\mathbf{X}\mathbf{a} = \mathbf{a}'\mathbf{T}\mathbf{a}, \tag{11.5.1}$$

which can be partitioned as a sum of the within-groups sum of squares,

$$\sum \mathbf{y}_i'\mathbf{H}_i\mathbf{y}_i = \sum \mathbf{a}'\mathbf{X}_i'\mathbf{H}_i\mathbf{X}_i\mathbf{a} = \mathbf{a}'\mathbf{W}\mathbf{a}, \tag{11.5.2}$$

plus the between-groups sum of squares,

$$\sum n_i(\bar{y}_i - \bar{y})^2 = \sum n_i\{\mathbf{a}'(\bar{\mathbf{x}}_i - \bar{\mathbf{x}})\}^2 = \mathbf{a}'\mathbf{B}\mathbf{a}, \tag{11.5.3}$$

where $\bar{\mathbf{y}}_i$ is the mean of the ith sub-vector $\mathbf{y}_i$ of $\mathbf{y}$, and $\mathbf{H}_i$ is the $(n_i \times n_i)$ centring matrix.

Fisher's criterion is intuitively attractive because it is easier to tell the groups apart if the between-groups sum of squares for $\mathbf{y}$ is large relative to the within-groups sum of squares. The ratio is given by

$$\mathbf{a}'\mathbf{B}\mathbf{a}/\mathbf{a}'\mathbf{W}\mathbf{a}. \tag{11.5.4}$$

If $\mathbf{a}$ is the vector which maximizes (11.5.4) we shall call the linear function $\mathbf{a}'\mathbf{x}$, *Fisher's linear discriminant function* or the *first canonical variate*. Notice that the vector $\mathbf{a}$ can be re-scaled without affecting the ratio (11.5.4).

Theorem 11.5.1 *The vector* $\mathbf{a}$ *in Fisher's linear discriminant function is the eigenvector of* $\mathbf{W}^{-1}\mathbf{B}$ *corresponding to the largest eigenvalue.*

Proof This result follows by an application of Theorem A.9.2. ■

Once the linear discriminant function has been calculated, an observation $\mathbf{x}$ can be allocated to one of the g populations on the basis of its "discriminant score" $\mathbf{a}'\mathbf{x}$. The sample means $\bar{\mathbf{x}}_i$ have scores $\mathbf{a}'\bar{\mathbf{x}}_i = \bar{y}_i$. Then $\mathbf{x}$ is allocated to that population whose mean score is closest to $\mathbf{a}'\mathbf{x}$; that is, allocate $\mathbf{x}$ to Π_j if

$$|\mathbf{a}'\mathbf{x} - \mathbf{a}'\bar{\mathbf{x}}_j| < |\mathbf{a}'\mathbf{x} - \mathbf{a}'\bar{\mathbf{x}}_i| \quad \text{for all } i \neq j. \tag{11.5.5}$$

Fisher's discriminant function is most important in the special case of $g = 2$ groups. Then $\mathbf{B}$ has rank one and can be written as

$$\mathbf{B} = \left(\frac{n_1 n_2}{n}\right)\mathbf{d}\mathbf{d}',$$

where $\mathbf{d} = \bar{\mathbf{x}}_1 - \bar{\mathbf{x}}_2$. Thus, $\mathbf{W}^{-1}\mathbf{B}$ has only one non-zero eigenvalue which can be found explicitly. This eigenvalue equals

$$\operatorname{tr} \mathbf{W}^{-1}\mathbf{B} = \left(\frac{n_1 n_2}{n}\right)\mathbf{d}'\mathbf{W}^{-1}\mathbf{d}$$

and the corresponding eigenvector is

$$\mathbf{a} = \mathbf{W}^{-1}\mathbf{d}. \tag{11.5.6}$$

(See Exercise 11.5.1.) Then the discriminant rule becomes

$$\text{allocate } \mathbf{x} \text{ to } \Pi_1 \text{ if } \mathbf{d}'\mathbf{W}^{-1}\{\mathbf{x}-\tfrac{1}{2}(\bar{\mathbf{x}}_1+\bar{\mathbf{x}}_2)\}>0 \tag{11.5.7}$$

and to Π_2 otherwise.

Note that the allocation rule given by (11.5.7) is exactly the same as the sample ML rule for two groups from the multinormal distribution with the same covariance matrix, given in (11.3.1). However, the justifications for this rule are quite different in the two cases. In (11.3.1) there is an explicit assumption of normality, whereas in (11.5.7) we have merely sought a "sensible" rule based on a linear function of $\mathbf{x}$. Thus, we might hope that this rule will be appropriate for populations where the hypothesis of normality is not exactly satisfied.

For $g \geq 3$ groups the allocation rule based on Fisher's linear discriminant function and the sample ML rule for multinormal populations with the same covariance matrix will not be the same unless the sample means are collinear (although the two rules will be similar if the means are nearly collinear). See Exercise 11.5.3.

In general, $\mathbf{W}^{-1}\mathbf{B}$ has $\min(p, g-1)$ non-zero eigenvalues. The corresponding eigenvectors define the second, third, and subsequent "canonical variates". (Canonical variates have an important connection with the canonical correlations of the last chapter, see Exercise 11.5.4.). The first k canonical variates, $k \leq \min(p, g-1)$, are useful when it is hoped to summarize the difference between the groups in k dimensions. See Section 12.5.

11.6 Probabilities of Misclassification

11.6.1 Probabilities when the parameters are estimated

Formally, the probabilities of misclassification p_{ij} are given in (11.2.9). If the parameters of the underlying distributions are estimated from the data, then we get estimated probabilities $\hat{p}_{ij}$.

For example consider the case of two normal populations $N_p(\boldsymbol{\mu}_1, \boldsymbol{\Sigma})$ and $N_p(\boldsymbol{\mu}_2, \boldsymbol{\Sigma})$. If $\boldsymbol{\mu} = \frac{1}{2}(\boldsymbol{\mu}_1+\boldsymbol{\mu}_2)$, then when $\mathbf{x}$ comes from Π_1, $\boldsymbol{\alpha}'(\mathbf{x}-\boldsymbol{\mu}) \sim N(\frac{1}{2}\boldsymbol{\alpha}'(\boldsymbol{\mu}_1-\boldsymbol{\mu}_2), \boldsymbol{\alpha}'\boldsymbol{\Sigma}\boldsymbol{\alpha})$. Since the discriminant function is given by $h(\mathbf{x}) = \boldsymbol{\alpha}'(\mathbf{x}-\boldsymbol{\mu})$ with $\boldsymbol{\alpha} = \boldsymbol{\Sigma}^{-1}(\boldsymbol{\mu}_1-\boldsymbol{\mu}_2)$, we see that if $\mathbf{x}$ comes from Π_1, $h(\mathbf{x}) \sim N(\frac{1}{2}\Delta^2, \Delta^2)$, where

$$\Delta^2 = (\boldsymbol{\mu}_1-\boldsymbol{\mu}_2)'\boldsymbol{\Sigma}^{-1}(\boldsymbol{\mu}_1-\boldsymbol{\mu}_2)$$

is the squared Mahalanobis distance between the populations. Similarly, if $\mathbf{x}$ comes from Π_2, $h(\mathbf{x}) \sim N(-\frac{1}{2}\Delta^2, \Delta^2)$.

Thus, the misclassification probabilities are given by

$$p_{12} = P(h(\mathbf{x}) > 0 \mid \Pi_2)$$
$$= \Phi(-E(h)/\sqrt{V(h)}) = \Phi(-\tfrac{1}{2}\Delta), \qquad (11.6.1)$$

where Φ is the standard normal distribution function. Similarly, $p_{21} = \Phi(-\frac{1}{2}\Delta)$ also.

If the parameters are estimated from the data then a natural estimate of Δ^2 is

$$D^2 = (\bar{\mathbf{x}}_1 - \bar{\mathbf{x}}_2)'\mathbf{S}_u^{-1}(\bar{\mathbf{x}}_1 - \bar{\mathbf{x}}_2),$$

and the estimated probabilities of misclassification are $\hat{p}_{12} = \hat{p}_{21} = \Phi(-\frac{1}{2}D)$. Unfortunately this approach tends to be optimistic; that is, it tends to underestimate the true misclassification probabilities when n is small. Aitchison *et al.* (1977) have argued that probabilities based on the Bayes rule of Section 11.3.3 are more realistic.

Example 11.6.1 In discriminating between *I. setosa* and *I. versicolour* in Example 11.3.1 we find $D^2 = 19.94$. Therefore the estimated probability of misclassification is

$$\Phi(-\tfrac{1}{2}D) = 0.013 \quad \text{or} \quad 1.3\%.$$

This value is confirmed visually by looking at the confidence ellipsoids in Figure 11.3.1.

11.6.2 Resubstitution method

Suppose that the discrimination is based on a data matrix $\mathbf{X}$ of which n_j individuals come from population j. If the discriminant rule is defined by allocation regions R_i, let n_{ij} be the number of individuals from Π_j which lie in R_i (so $\sum_i n_{ij} = n_j$). Then $\hat{p}_{ij} = n_{ij}/n_j$ is an estimate of p_{ij}. Unfortunately, this method also tends to be optimistic about misallocation probabilities.

Example 11.6.2 Consider the *I. setosa* (Π_1) and *I. versicolour* (Π_2) data again. From Figure 11.3.1 it is clear that one observation from Π_1 will be allocated to Π_2 and no observations from Π_2 will be allocated to Π_1. Thus

$$\hat{p}_{12} = 0, \qquad \hat{p}_{21} = 1/50.$$

If we assume that $p_{12} = p_{21}$, thus we get the single estimate $\hat{\hat{p}}_{12} = \hat{\hat{p}}_{21} = 1/100$, which is in general agreement with Example 11.6.1. However, note that for such a small probability of misclassification, these sample sizes are too small to get accurate estimates.

11.6.3 The U-method of jack-knifing

The problem with the resubstitution method is that the same observations are used to define the discriminant rule and to judge its accuracy. One way to avoid this problem is the U-method of jack-knifing, defined as follows.

Let $\mathbf{x}_r$, $r = 1,\ldots, n_1$, be the first n_1 rows of the $\mathbf{X}$ matrix representing the individuals from Π_1. For each r, let $R_1^{(r)},\ldots, R_g^{(r)}$ denote the allocation regions for some discriminant rule based on the $(n-1)\times p$ matrix obtained by deleting the rth row from $\mathbf{X}$. Then $\mathbf{x}_r$ can be judged on the basis of this rule, which is not derived using $\mathbf{x}_r$.

If we let n_{i1}^* denote the number of the first n_1 individuals for which $\mathbf{x}_r \in R_i^{(r)}$, then $p_{i1}^* = n_{i1}/n_1$ is an estimate of the misallocation rates. By repeating this procedure for each of the other populations $j = 2,\ldots, g$, we get estimates p_{ij}^*.

For two multinormal populations with the same covariance matrix, this approach leads to more reliable estimates of the misallocation probabilities than either of the above two methods. For further details and more methods of estimating the misallocation probabilities, see Kshirsagar (1972, pp. 218–225).

11.7 Discarding of Variables

Consider the discrimination problem between two multinormal populations with means $\boldsymbol{\mu}_1, \boldsymbol{\mu}_2$ and common covariance matrix $\boldsymbol{\Sigma}$. The coefficients of the theoretical ML discriminant function $\boldsymbol{\alpha}'\mathbf{x}$ are given by

$$\boldsymbol{\alpha} = \boldsymbol{\Sigma}^{-1}\boldsymbol{\delta} \quad \text{where} \quad \boldsymbol{\delta} = \boldsymbol{\mu}_1 - \boldsymbol{\mu}_2. \tag{11.7.1}$$

In practice of course the parameters are estimated by $\bar{\mathbf{x}}_1, \bar{\mathbf{x}}_2$, and $\mathbf{S}_u = m^{-1}(n_1\mathbf{S}_1 + n_2\mathbf{S}_2) = m^{-1}\mathbf{W}$, where $m = n_1 + n_2 - 2$. Letting $\mathbf{d} = \bar{\mathbf{x}}_1 - \bar{\mathbf{x}}_2$, the coefficients of the sample ML rule are given by

$$\mathbf{a} = m\mathbf{W}^{-1}\mathbf{d}.$$

Partition $\boldsymbol{\alpha}' = (\boldsymbol{\alpha}_1', \boldsymbol{\alpha}_2')$ and $\boldsymbol{\delta}' = (\boldsymbol{\delta}_1', \boldsymbol{\delta}_2')$, where $\boldsymbol{\alpha}_1$ and $\boldsymbol{\delta}_1$ have k components, and suppose $\boldsymbol{\alpha}_2 = \mathbf{0}$; that is suppose the variables $x_{k+1},\ldots, x_p$ have no discriminating power once the other variables have been taken into account, and hence may be safely discarded. Note that the hypothesis $\boldsymbol{\alpha}_2 = \mathbf{0}$ is equivalent to $\boldsymbol{\delta}_{2.1} = \mathbf{0}$, where $\boldsymbol{\delta}_{2.1} = \boldsymbol{\delta}_2 - \boldsymbol{\Sigma}_{21}\boldsymbol{\Sigma}_{11}^{-1}\boldsymbol{\delta}_1$. (See Exercise

11.7.1) It is also equivalent to $\Delta_p^2 = \Delta_k^2$, where

$$\Delta_p^2 = \boldsymbol{\delta}'\boldsymbol{\Sigma}^{-1}\boldsymbol{\delta}, \qquad \Delta_k^2 = \boldsymbol{\delta}_1\boldsymbol{\Sigma}_{11}^{-1}\boldsymbol{\delta}_1;$$

that is, the Mahalanobis distance between the populations is the same whether based on the first k components or on all p components.

A test of the hypothesis $H_0: \boldsymbol{\alpha}_2 = \mathbf{0}$ using the sample Mahalanobis distances

$$D_p^2 = m\mathbf{d}'\mathbf{W}^{-1}\mathbf{d} \quad \text{and} \quad D_k^2 = m\mathbf{d}_1'\mathbf{W}_{11}^{-1}\mathbf{d}_1$$

has been proposed by Rao (1973, p. 568). This test uses the statistic

$$\{(m-p+1)/(p-k)\}c^2(D_p^2 - D_k^2)/(m + c^2D_k^2), \qquad (11.7.2)$$

where $c^2 = n_1n_2/n$. Under the null hypothesis, (11.7.2) has the $F_{p-k,m-p+1}$ distribution and we reject H_0 for large values of this statistic. See Theorem 3.6.2.

The most important application of this test occurs with $k = p-1$, when we are testing the importance of one particular variable once all the other variables have been taken into account. In this case the statistic in (11.7.2) can be simplified using the inverse of the total SSP matrix $\mathbf{T}^{-1} = (t^{ij})$. Consider the hypothesis $H_0: \alpha_i = 0$, where i is fixed. Then (11.7.2), with D_k now representing the Mahalanobis distance based on all the variables except the ith, equals

$$(m-p+1)c^2a_i^2/\{mt^{ii}(m + c^2D_p^2)\}, \qquad (11.7.3)$$

and has the $F_{1,m-p+1}$ distribution when $\alpha_i = 0$. (See Exercise 11.7.2.)

Of course the statistic (11.7.3) is strictly valid only if the variable i is selected ahead of time. However, it is often convenient to look at the value of this statistic on all the variables to see which ones are important.

Example 11.7.1 Consider the two species of iris of Example 11.3.1, using the first two variables, sepal length and sepal width. Here $n_1 = n_2 = 50$, $m = 100 - 2 = 98$, and the discriminant function has coefficients $\mathbf{a} = (-11.436, 14.143)'$. It is easily shown that

$$\mathbf{T}^{-1} = \begin{bmatrix} 0.025\,62 & 0.007\,07 \\ & 0.046\,02 \end{bmatrix}$$

and that $D_p^2 = 19.94$. Thus the two F statistics testing the importance of one variable given the other from (11.7.3) are given by

$$211.8 \quad \text{and} \quad 180.3,$$

respectively. Since $F_{1,97;0.01} = 7.0$, we conclude that both variables are highly useful in the discrimination.

Table 11.7.1 Discriminant coefficients for iris data with F statistics

a_i	F statistic
3.053	0.716
18.023	25.7
−21.766	24.6
−30.844	10.7

However, a different conclusion follows if we discriminate on the basis of all four variables, sepal length and width and petal length and width. Then the discriminant coefficients and corresponding F statistics are given by Table 11.7.1. Since $F_{1,95;0.01} = 7.0$, the F statistic for sepal length is not significant, so we conclude that sepal length is redundant if all the other variables are present. Notice that the F statistics in Table 11.7.1 are all smaller than the F statistics based on two variables alone. This feature reflects the empirical fact that as the number of variables increases, the information carried by any one variable, not carried by the other variables, tends to decrease.

11.8 When Does Correlation Improve Discrimination?

It might be thought that a linear combination of two variables would provide a better discriminator if they were correlated than if they were uncorrelated. However, this is not necessarily so, as is shown in the following example.

Example 11.8.1 (Cochran, 1962) Let Π_1 and Π_2 be two bivariate normal populations. Suppose that Π_1 is $N_2(\mathbf{0}, \mathbf{\Sigma})$ and Π_2 is $N_2(\boldsymbol{\mu}, \mathbf{\Sigma})$, where $\boldsymbol{\mu} = (\mu_1, \mu_2)'$ and

$$\mathbf{\Sigma} = \begin{bmatrix} 1 & \rho \\ \rho & 1 \end{bmatrix}.$$

Now the Mahalanobis distance between Π_1 and Π_2 is

$$\Delta^2 = \boldsymbol{\mu}'\mathbf{\Sigma}^{-1}\boldsymbol{\mu} = (\mu_1^2 + \mu_2^2 - 2\rho\mu_1\mu_2)/(1-\rho^2).$$

If the variables are uncorrelated then

$$\Delta^2 = \mu_1^2 + \mu_2^2 = \Delta_0^2, \quad \text{say.}$$

Now the correlation will improve discrimination (i.e. reduce the probability of misclassification given by (11.6.1)) if and only if $\Delta^2 > \Delta_0^2$. This happens if and only if

$$\rho\{(1+f^2)\rho - 2f\} > 0, \quad \text{where} \quad f = \mu_2/\mu_1.$$

In other words, discrimination is improved unless ρ lies between zero and $2f/(1+f^2)$, but a small value of ρ can actually harm discrimination. Note that if $\mu_1 = \mu_2$ then *any* positive correlation reduces the power of discrimination

Exercises and Complements

11.2.1 If Π_i is the $N_p(\boldsymbol{\mu}_i, \boldsymbol{\Sigma})$ population for $i = 1, 2, \boldsymbol{\Sigma} > 0$, show that the ML discriminant rule is given by

$$\text{allocate } \mathbf{x} \text{ to } \Pi_1 \text{ if } \boldsymbol{\alpha}'\{\mathbf{x} - \tfrac{1}{2}(\boldsymbol{\mu}_1 + \boldsymbol{\mu}_2)\} > 0,$$

where $\boldsymbol{\alpha} = \boldsymbol{\Sigma}^{-1}(\boldsymbol{\mu}_1 - \boldsymbol{\mu}_2)$.

11.2.2 Consider three bivariate normal populations with the same covariance matrix, given by

$$\boldsymbol{\mu}_1 = \begin{bmatrix} 0 \\ 0 \end{bmatrix}, \quad \boldsymbol{\mu}_2 = \begin{bmatrix} 0 \\ 1 \end{bmatrix}, \quad \boldsymbol{\mu}_3 = \begin{bmatrix} 1 \\ 0 \end{bmatrix}, \quad \boldsymbol{\Sigma} = \begin{bmatrix} 5 & 2 \\ 2 & 1 \end{bmatrix}.$$

(a) Draw the ML allocation regions and show that the three boundary lines meet at the point $(\frac{15}{2}, \frac{7}{2})$.

(b) Suppose the three populations have prior probabilities $\frac{1}{2}$, $\frac{1}{3}$, and $\frac{1}{6}$. Draw the Bayes allocation regions and find the point where the three boundary lines meet.

11.2.3 Show that the boundary hyperplane for the ML allocation rule between $N_p(\boldsymbol{\mu}_1, \boldsymbol{\Sigma})$ and $N_p(\boldsymbol{\mu}_2 \boldsymbol{\Sigma})$, $\boldsymbol{\Sigma} > 0$, is orthogonal to $\boldsymbol{\delta} = \boldsymbol{\mu}_1 - \boldsymbol{\mu}_2$ if and only if $\boldsymbol{\delta}$ is an eigenvector of $\boldsymbol{\Sigma}$.

11.2.4 Show that the Bayes allocation rule with unequal costs given in (11.2.12) reduces to the rule given in (11.2.8) when the costs are all equal. Prove Theorems 11.2.4 and 11.2.5.

11.2.5 (See Bartlett, 1965) (a) Consider populations $N_p(\boldsymbol{\mu}_1, \boldsymbol{\Sigma})$, $N_p(\boldsymbol{\mu}_2, \boldsymbol{\Sigma})$ and let $\boldsymbol{\delta} = \boldsymbol{\mu}_1 - \boldsymbol{\mu}_2$. In the case of biological data it is sometimes found that the correlation between each pair of variables is approximately the same, so that scaling each variable to have unit variance, we can write $\boldsymbol{\Sigma} = \mathbf{E}$, where $\mathbf{E} = (1-\rho)\mathbf{I} + \rho\mathbf{1}\mathbf{1}'$ is the equicorrelation matrix. Using

(A.3.2b) show that the ML discriminant function is proportional to

$$h(\mathbf{x}) = \sum_{i=1}^{p} \delta_i x_i - p\rho\{1+(p-1)\rho\}^{-1}\bar{\delta} \sum_{i=1}^{p} x_i + \text{const.},$$

where $\bar{\delta} = p^{-1} \sum \delta_i$. Calculate the corresponding allocation regions for the ML rule. Discuss the cases ρ small, ρ near 1. (This discriminant function is very useful in practice, even when the correlations are not exactly equal, because it can be easily calculated without the use of a computer.)

(b) Write $h(\mathbf{x}) = h_1(\mathbf{x}) + h_2(\mathbf{x}) + \text{const.}$, where

$$h_1(\mathbf{x}) = p\bar{\delta}\{p^{-1} - \rho[1+(p-1)\rho]^{-1}\} \sum x_i$$

is proportional to the first principal component, and

$$h_2(\mathbf{x}) = \sum (\delta_i - \bar{\delta})x_i$$

lies in the isotropic $(p-1)$-dimensional subspace of principal components corresponding to the smaller eigenvalue of $\mathbf{\Sigma}$. Interpret $h_1(\mathbf{x})$ and $h_2(\mathbf{x})$ as *size* and *shape* factors, in the sense of Section 8.6.

11.2.6 (See Bartlett, 1965) The following problem involves two multinormal populations with the *same* means but *different* covariance matrices. In discriminating between monozygotic and dizygotic twins of like sex on the basis of simple physical measurements such as weight, height, etc., the observations recorded are the differences $x_1, \ldots, x_p$ between corresponding measurements on each set of twins. As either twin might have been measured first, the expected mean differences are automatically zero. Let the covariance matrices for the two types of twins be denoted $\mathbf{\Sigma}_1$ and $\mathbf{\Sigma}_2$, and assume for simplicity that

$$\mathbf{\Sigma}_1 = \sigma_1^2\{(1-\rho)\mathbf{I} + \rho\mathbf{1}\mathbf{1}'\},$$
$$\mathbf{\Sigma}_2 = \sigma_2^2\{(1-\rho)\mathbf{I} + \rho\mathbf{1}\mathbf{1}'\}.$$

Under the assumption of multivariate normality, show that the ML discriminant function is proportional to

$$z_1 - \rho\{1+(p-1)\rho\}^{-1} z_2 + \text{const.},$$

where $z_1 = x_1^2 + \cdots + x_p^2$ and $z_2 = (x_1 + \cdots + x_p)^2$. How would the boundary between the allocation regions be determined so that the two types of misclassification have equal probability?

11.2.7 There is a prior probability, π_j, that an observation $\mathbf{x} = (x_1, x_2)'$ comes from population Π_j, where $j = 1$ or 2 and $\pi_1 + \pi_2 = 1$. Within Π_j, $\mathbf{x}$

has a bivariate Cauchy distribution, with density function

$$f_j(\mathbf{x}) = (2\pi)^{-1}k_j(k_j^2 + x_1^2 + x_2^2)^{-3/2}, \qquad -\infty < x_1, x_2 < \infty, \qquad j = 1, 2.$$

The cost of wrongly deciding that $\mathbf{x}$ comes from Π_j is c_j $(j = 1, 2)$ and there are no other costs. Find the decision rule which minimizes the expected cost of misallocating $\mathbf{x}$.

11.2.8 (a) In Example 11.2.3 with $\sigma_1 > \sigma_2$ calculate the set of xs for which we would allocate to Π_1 and compare with Figure 11.2.1.

(b) Consider two bivariate normal populations $N_2(\boldsymbol{\mu}, \boldsymbol{\Sigma}_1)$, $N_2(\boldsymbol{\mu}, \boldsymbol{\Sigma}_2)$, where we suppose the correlation between the two variables is 0 in each population; that is

$$\boldsymbol{\Sigma}_1 = \text{diag}\,(\sigma_1^2, \sigma_2^2), \qquad \boldsymbol{\Sigma}_2 = \text{diag}\,(\tau_1^2, \tau_2^2).$$

If $\sigma_1^2 > \sigma_2^2$ and $\tau_1^2 > \tau_2^2$, show that the boundary of the ML allocation rule is an ellipse and find its equation.

11.3.1 Why is the discriminant function $h(\mathbf{x})$ used in Example 11.3.1 to discriminate between two species of iris different from the discriminant function $h_{12}(\mathbf{x})$ which is used in Example 11.3.2 to help discriminate between three species of iris.

11.3.2 Let $h_{12}(\mathbf{x})$, $h_{13}(\mathbf{x})$, and $h_{23}(\mathbf{x})$ be the three sample discriminant functions used to discriminate between three multinormal populations with the same covariance matrix, as in Example 11.3.2. Show that unless $\bar{\mathbf{x}}_1$, $\bar{\mathbf{x}}_2$, and $\bar{\mathbf{x}}_3$ are collinear, the solutions to the three equations,

$$h_{12}(\mathbf{x}) = 0, \qquad h_{13}(\mathbf{x}) = 0, \qquad h_{23}(\mathbf{x}) = 0,$$

have a point in common. Draw a picture describing what happens if $\bar{\mathbf{x}}_1$, $\bar{\mathbf{x}}_2$, and $\bar{\mathbf{x}}_3$ are collinear.

11.3.3 In Example 11.3.4 suppose that the multinomial parameters $\alpha_1, \ldots, \alpha_{32}$ and $\beta_1, \ldots, \beta_{32}$ for Plato's works, *Republic* (Π_0) and *Laws* (Π_1), are known exactly. Suppose the distribution of sentence ending for each of the other works also follows a multinomial distribution, with parameters

$$\gamma_{\lambda i} = \alpha_i^{1-\lambda}\beta_i^{\lambda} \Big/ \sum_{j=1}^{32} \alpha_j^{1-\lambda}\beta_j^{\lambda},$$

where λ is a parameter which is different for each of the five intermediate works.

(a) Show that this family of multinomial distributions varies from $\Pi_0(\lambda=0)$ to $\Pi_1(\lambda=1)$ and hence can be used to represent a gradual change in populations from Π_0 to Π_1.

(b) Show that $\phi(\lambda)=\sum \gamma_{\lambda i}s_i$, where $s_i=\log(\alpha_i/\beta_i)$ is a monotone function of $\lambda, 0\leqslant\lambda\leqslant 1$. Thus $\phi(\lambda)$ can be used instead of λ to parametrize this family of distributions.

(c) Suppose that for a particular work containing N sentences, there are x_i endings of type i. Show that the m.l.e. of $\phi(\lambda)$ is given by

$$\bar{s}=N^{-1}\sum x_i s_i,$$

which is the same as the discriminant score of Example 11.3.4. Thus the discriminant score estimates a function which gives a measure of the location of the work between Π_0 and Π_1.

(d) Show that the variance of $\bar{s}$ equals

$$V(\bar{s})=N^{-1}\{\sum \gamma_{\lambda i}s_i^2-\phi(\lambda)^2\}$$

and that an unbiased estimate of $V(\bar{s})$ is

$$V_e(\bar{s})=[N(N-1)]^{-1}\{\sum x_i s_i^2-N\bar{s}^2\}.$$

(e) For two works of sizes N' and N'', let the corresponding mean scores and estimated variances be $\bar{s}'$, $\bar{s}''$ and $V_e(\bar{s}')$, $V_e(\bar{s}'')$. Using the fact that for large N' and N'', $\bar{s}'-\bar{s}''$ will be approximately normally distributed with mean $\phi(\lambda')-\phi(\lambda'')$ and variance $V_e(\bar{s}')+V_e(\bar{s}'')$, show that an approximate significance test of chronological ordering of these two works is given by the statistic

$$\psi=(\bar{s}'-\bar{s}'')/\{V_e(\bar{s}')+V_e(\bar{s}'')\}^{1/2}.$$

If $|\psi|$ is significantly large for an $N(0,1)$ variate then the observed ordering of the works is significant.

(f) For the two works *Critias* and *Timaeus* we have $\bar{s}'=-0.0346$ $\bar{s}''=-0.1170$, $V_e(\bar{s}')=.003\,799$, $V_e(\bar{s}'')=0.000\,721\,8$. Test the hypothesis $\lambda'=\lambda''$ and hence assess the significance of the ordering given by the discriminant scores.

11.3.4 The genus *Chaetocnema* (*a genus of* flea-beetles) contains two allied species, *Ch. concinna* and *Ch. heikertingeri*, that were long confused with one another. Lubischew (1962) gave the following measurements of characters for samples of males of the two species:

Ch. concinna x_1	*Ch. concinna* x_2	*Ch. heikertingeri* x_1	*Ch. heikertingeri* x_2
191	131	186	107
185	134	211	122
200	137	201	114
173	127	242	131
171	118	184	108
160	118	211	118
188	134	217	122
186	129	223	127
174	131	208	125
163	115	199	124
190	143	211	129
174	131	218	126
201	130	203	122
190	133	192	116
182	130	195	123
184	131	211	122
177	127	187	123
178	126	192	109

Here, x_1 is the sum of the widths (in micrometres) of the first joints of the first two tarsi ("feet"); x_2 is the corresponding measurement for the second joints.

Find the sample ML discriminant function for the two species.

Suppose that we have the pair of observations $(x_1, x_2)'$ for a new specimen, but do not know to which of the two species the specimen belongs. Calculate the equation of the line (in the (x_1, x_2) plane) such that, if $(x_1, x_2)'$ lies on one side, the specimen seems more likely to belong to *Ch. concinna*, whereas if $(x_1, x_2)'$ lies on the other side, the specimen seems more likely to belong to *Ch. heikertingeri*. Allocate a new observation (190, 125) to one of these species. Plot these data on graph paper together with the line and comment.

11.3.5 Using the Bayesian approach of Section 11.3.3, for two multinormal populations with unequal covariance matrices, show that $f_1(\mathbf{x} \mid \mathbf{X})$ is a multivariate Student's t density with parameters $n_1 - p$, $\bar{\mathbf{x}}_1$, and $\{(n_1+1)/(n_1-p)\}\mathbf{S}_1$. Find $f_2(\mathbf{x} \mid \mathbf{X})$.

11.5.1 In Section 11.5 when $g=2$, show that the matrix $\mathbf{W}^{-1}\mathbf{B}$ has only one non-zero eigenvalue, which equals $\{n_1 n_2/n\}\mathbf{d}'\mathbf{W}\mathbf{d}$, and find the corresponding eigenvector.

11.5.2 Show that the following eigenvectors are equivalent (assuming $\mathbf{W}$ has rank p):

(a) the eigenvector corresponding to the largest eigenvalue of $\mathbf{W}^{-1}\mathbf{B}$;
(b) the eigenvector corresponding to the largest eigenvalue of $\mathbf{W}^{-1}\mathbf{T}$;
(c) the eigenvector corresponding to the smallest eigenvalue of $\mathbf{T}^{-1}\mathbf{W}$.

11.5.3 (a) When the number of groups $g=3$, show that the allocation rule based on Fisher's linear discriminant function is different from the sample ML allocation rule, unless the sample means are collinear. (However, if the means are nearly collinear the two rules will be very similar.)

(b) Calculate Fisher's discriminant function for the first two variables of the three species of iris, and compare the allocation regions with the three given in Example 11.3.2. (The largest eigenvalue of $\mathbf{W}^{-1}\mathbf{B}$ is 4.17, with eigenvector $(1, -1.29)'$.

11.5.4 Let $\mathbf{X}(n \times p)$ be a data matrix partitioned into g groups. Define a new dummy zero–one data matrix $\mathbf{Y}(n \times (g-1))$ by

$$y_{ij} = \begin{cases} 1 & \text{if } \mathbf{x}_i \text{ is in the } j\text{th group,} \\ 0 & \text{otherwise,} \end{cases}$$

for $j=1,\dots, g-1$; $i=1,\dots, n$. Let $\mathbf{S}$ denote the covariance matrix of $(\mathbf{X}, \mathbf{Y})$. Show that

$$n\mathbf{S}_{11} = \mathbf{T} = \text{“total” SSP matrix}$$

and

$$n\mathbf{S}_{12}\mathbf{S}_{22}^{-1}\mathbf{S}_{21} = \mathbf{B} = \text{“between-groups” SSP matrix.}$$

Hence, carry out a canonical correlation analysis between $\mathbf{X}$ and $\mathbf{Y}$, and deduce that the canonical correlation variables for $\mathbf{X}$ equal the canonical variates of Section 11.5. (Hint: the canonical correlation variables $\mathbf{a}_i'\mathbf{x}$ of $\mathbf{X}$ are given by $\mathbf{T}^{-1}\mathbf{B}\mathbf{a}_i = \lambda_i \mathbf{a}_i$, or equivalently by $\mathbf{W}^{-1}\mathbf{B}\mathbf{a}_i = \{\lambda_i/(1-\lambda_i)\}\mathbf{a}_i$.)

11.6.1 If $\mathbf{x} \sim N_p(\boldsymbol{\mu}_1, \boldsymbol{\Sigma})$, show that $\boldsymbol{\alpha}'(\mathbf{x}-\boldsymbol{\mu}) \sim N_p(-\frac{1}{2}\Delta^2, \Delta^2)$, where

$$\boldsymbol{\mu} = \tfrac{1}{2}(\boldsymbol{\mu}_1 + \boldsymbol{\mu}_2), \qquad \boldsymbol{\alpha} = \boldsymbol{\Sigma}^{-1}(\boldsymbol{\mu}_1 - \boldsymbol{\mu}_2), \qquad \Delta^2 = (\boldsymbol{\mu}_1 - \boldsymbol{\mu}_2)'\boldsymbol{\Sigma}^{-1}(\boldsymbol{\mu}_1 - \boldsymbol{\mu}_2).$$

11.7.1 let $\boldsymbol{\delta}_{2.1} = \boldsymbol{\delta}_2 - \boldsymbol{\Sigma}_{21}\boldsymbol{\Sigma}_{11}^{-1}\boldsymbol{\delta}_1$ and let $\boldsymbol{\alpha} = \boldsymbol{\Sigma}^{-1}\boldsymbol{\delta}$. Show that $\boldsymbol{\delta}_{2.1} = \mathbf{0}$ if and only if $\boldsymbol{\alpha} = \mathbf{0}$ if and only if $\boldsymbol{\delta}'\boldsymbol{\Sigma}^{-1}\boldsymbol{\delta} = \boldsymbol{\delta}_1'\boldsymbol{\Sigma}_{11}^{-1}\boldsymbol{\delta}_1$. (Hint: Using (A.2.4g) and (A.2.4f), show that $\boldsymbol{\delta}'\boldsymbol{\Sigma}^{-1}\boldsymbol{\delta} = \boldsymbol{\delta}_1\boldsymbol{\Sigma}_{11}^{-1}\boldsymbol{\delta}_1 + \boldsymbol{\delta}_{2.1}'\boldsymbol{\Sigma}^{22}\boldsymbol{\delta}_{2.1}$.)

11.7.2 Show that when $k = p-1$, (11.7.2) can be expressed in the form (11.7.3) (with $i = p$). (Hint: Partition $\mathbf{W}^{-1}$ so that

$$\mathbf{W}^{22} = (\mathbf{W}_{22} - \mathbf{W}_{21}\mathbf{W}_{11}^{-1}\mathbf{W}_{12})^{-1} = \mathbf{w}^{pp}.$$

Using (A.2.4b) and (A.2.3m) show that

$$t^{pp} = \frac{|\mathbf{T}_{11}|}{|\mathbf{T}|} = \frac{|\mathbf{W}_{11}|(1+c^2D_k^2/m)}{|\mathbf{W}|(1+c^2D_p^2/m)} = w^{pp}\frac{1+c^2D_k^2/m}{1+c^2D_p^2/m}.$$

Using (A.2.4g) and (A.2.4f), show that

$$\mathbf{d}'\mathbf{W}^{-1}\mathbf{d} - \mathbf{d}_1'\mathbf{W}_{11}^{-1}\mathbf{d}_1 = \mathbf{d}'\mathbf{V}\mathbf{d}.$$

where

$$\mathbf{V} = w^{pp}\begin{pmatrix} -\boldsymbol{\beta} \\ 1 \end{pmatrix}(-\boldsymbol{\beta}', 1) \quad \text{and} \quad \boldsymbol{\beta} = \mathbf{W}_{11}^{-1}\mathbf{W}_{12}.$$

Finally note that

$$a_p = m(\mathbf{W}^{21}\mathbf{d}_1 + \mathbf{W}^{22}\mathbf{d}_2) = mw^{pp}(\mathbf{d}_2 - \mathbf{W}_{21}\mathbf{W}_{11}^{-1}\mathbf{d}_1),$$

where $\mathbf{d}_2 = d_p$, and hence

$$a_p^2 = m^2 w^{pp}\mathbf{d}_{2.1}\mathbf{W}^{22}\mathbf{d}_{2.1} = mw^{pp}(D_p^2 - D_k^2).)$$

11.7.3 Verify that formulae (11.7.2) and (11.7.3) give the same values for the F statistics based on the two variables sepal length and sepal width in Example 11.7.1.

11.7.4 A random sample of 49 old men participating in a study of aging were classified by psychiatric examination into one of two categories: senile or non-senile. Morrison (1976, pp. 138–139).

An independently administered adult intelligence test revealed large differences between the two groups in certain standard subsets of the test. The results for these sections of the test are given below.

The group means are as follows:

Subtest	Senile ($N_1 = 37$)	Non-senile ($N_2 = 12$)
x_1 Information	12.57	8.75
x_2 Similarities	9.57	5.35
x_3 Arithmetic	11.49	8.50
x_4 Picture completion	7.97	4.75

The "within-group" covariance matrix $\mathbf{S}_u$ and its inverse are given by

$$\mathbf{S}_u = \begin{bmatrix} 11.2553 & 9.4042 & 7.1489 & 3.3830 \\ & 13.5318 & 7.3830 & 2.5532 \\ & & 11.5744 & 2.6170 \\ & & & 5.8085 \end{bmatrix}$$

$$\mathbf{S}_u^{-1} = \begin{bmatrix} 0.2591 & -0.1358 & -0.0588 & -0.0647 \\ & 0.1865 & -0.0383 & -0.0144 \\ & & 0.1510 & -0.0170 \\ & & & 0.2112 \end{bmatrix}.$$

Calculate the linear discriminant function between the two groups based on the data and investigate the errors of misclassification. Do the subtests "information" and "arithmetic" provide additional discrimination once the other two subtests are taken into account?

12
Multivariate Analysis of Variance

12.1 Introduction

When there is more than one variable measured per plot in the design of an experiment, the design is analysed by multivariate analysis of variance techniques (MANOVA techniques in short). Thus we have the direct multivariate extension of every univariate design, but we will confine our main attention to the multivariate one-way classification which is an extension of the univariate one-way classification.

12.2 Formulation of Multivariate One-way Classification

First consider a formulation resulting from agricultural experiments. Let there be k treatments which are assigned in a completely random order to some agricultural land. Suppose there are n_j plots receiving the jth treatment, $j=1,\ldots,k$. Let us assume that $\mathbf{x}_{ij}$ is the $(p\times 1)$ yield-vector of the ith plot receiving the jth treatment.

In general terminology, the plots are *experimental designs*, the treatments are *conditions*, and the yield is a *response* or *outcome*.

We assume that the $\mathbf{x}_{ij}$ are generated from the model

$$\mathbf{x}_{ij}=\boldsymbol{\mu}+\boldsymbol{\tau}_j+\boldsymbol{\varepsilon}_{ij}, \qquad i=1,\ldots,n_j, \quad j=1,\ldots,k, \tag{12.2.1}$$

where $\boldsymbol{\varepsilon}_{ij}=$ independent $N_p(\mathbf{0},\boldsymbol{\Sigma})$, $\boldsymbol{\mu}=$ overall effect on the yield-vector, and $\boldsymbol{\tau}_j=$ effect due to the jth treatment.

This design can be viewed as a multi-sample problem, i.e. we can regard $\mathbf{x}_{ij}$, $i=1,\ldots,n_j$, as a random sample from $N_p(\boldsymbol{\mu}_j,\boldsymbol{\Sigma})$, $j=1,\ldots,k$, where

$$\boldsymbol{\mu}_j=\boldsymbol{\mu}+\boldsymbol{\tau}_j, \qquad j=1,\ldots,k. \tag{12.2.2}$$

We usually wish to test the hypothesis

$$H_0 : \boldsymbol{\mu}_1 = \ldots = \boldsymbol{\mu}_k, \tag{12.2.3}$$

which is equivalent to testing that there is no difference between the treatments $\boldsymbol{\tau}_1, \ldots, \boldsymbol{\tau}_k$. This problem has already been treated in Section 5.3.3a and Example 6.4.1. We now give some further details.

12.3 The Likelihood Ratio Principle

The test To test H_0 against $H_1 : \boldsymbol{\mu}_i \neq \boldsymbol{\mu}_j$, for some $i \neq j$, we have from Section 5.3.3a that the likelihood ratio criterion is

$$\Lambda = |\mathbf{W}|/|\mathbf{T}|, \tag{12.3.1}$$

where

$$\mathbf{W} = \sum_{j=1}^{k} \sum_{i=1}^{n_j} (\mathbf{x}_{ij} - \bar{\mathbf{x}}_j)(\mathbf{x}_{ij} - \bar{\mathbf{x}}_j)' \tag{12.3.2}$$

and

$$\mathbf{T} = \sum_{j=1}^{k} \sum_{i=1}^{n_j} (\mathbf{x}_{ij} - \bar{\mathbf{x}})(\mathbf{x}_{ij} - \bar{\mathbf{x}})', \tag{12.3.3}$$

with

$$\bar{\mathbf{x}}_j = \frac{\sum_{i=1}^{n_j} \mathbf{x}_{ij}}{n_j}, \qquad \bar{\mathbf{x}} = \frac{\sum_{j=1}^{k} \sum_{i=1}^{n_j} \mathbf{x}_{ij}}{n}, \qquad n = \sum_{j=1}^{k} n_j.$$

Recall that **W** and **T** are respectively the "within-samples" and "total" sum of squares and products (SSP) matrices, respectively.

We can further show that

$$\mathbf{T} = \mathbf{W} + \mathbf{B}, \tag{12.3.4}$$

where

$$\mathbf{B} = \sum_{j=1}^{k} n_j (\bar{\mathbf{x}}_j - \bar{\mathbf{x}})(\bar{\mathbf{x}}_j - \bar{\mathbf{x}})'$$

is the "between-samples" SSP matrix. The identity (12.3.4) is the MANOVA identity.

Under H_0, it was shown in Section 5.3.3a that

$$\mathbf{W} \sim W_p(\boldsymbol{\Sigma}, n-k), \qquad \mathbf{B} \sim W_p(\boldsymbol{\Sigma}, k-1), \tag{12.3.5}$$

where $\mathbf{W}$ and $\mathbf{B}$ are independent. Further, if $n \geq p+k$,

$$\Lambda = |\mathbf{W}|/|\mathbf{W}+\mathbf{B}| \sim \Lambda(p, n-k, k-1),$$

where Λ is a Wilks' lambda variable. We reject H_0 for small values of Λ. H_0 can be tested by forming the MANOVA table as set out in Table 12.3.1.

For calculations of Λ, the following result is helpful:

$$\Lambda = \prod_{j=1}^{p} (1+\lambda_j)^{-1}, \tag{12.3.6}$$

where $\lambda_1, \ldots, \lambda_p$ are the eigenvalues of $\mathbf{W}^{-1}\mathbf{B}$. This result follows on noting that if $\lambda_1, \ldots, \lambda_p$ are the eigenvalues of $\mathbf{W}^{-1}\mathbf{B}$, then (λ_i+1), $i=1, \ldots, p$, are the eigenvalues of $\mathbf{W}^{-1}(\mathbf{B}+\mathbf{W})$. More details are given in Section 3.7. In particular, we shall need (3.7.9), namely,

$$\frac{(m-p+1)\{1-\sqrt{\Lambda(p, m, 2)}\}}{p\sqrt{\Lambda(p, m, 2)}} \sim F_{2p, 2(m-p+1)}. \tag{12.3.7}$$

Example 12.3.1 Consider measurements published by Reeve (1941) on the skulls of $n=13$ ant-eaters belonging to the subspecies *chapadensis*, deposited in the British Museum from $k=3$ different localities. On each skull, $p=3$ measurements were taken:

x_1, the basal length excluding the premaxilla,
x_2, the occipito-nasal length,
x_3, the maximum nasal length.

Table 12.3.2 shows the common logarithms of the original measurements

Table 12.3.1 Multivariate one-way classification

Source	d.f.	SSP matrix	Wilks' criterion
Between samples	$k-1$	$\mathbf{B} = \sum_{j=1}^{k} n_j(\bar{\mathbf{x}}_j - \bar{\mathbf{x}})(\bar{\mathbf{x}}_j - \bar{\mathbf{x}})'$	$|\mathbf{W}|/|\mathbf{W}+\mathbf{B}| \sim \Lambda(p, n-k, k-1)$
Within samples	$n-k$	$\mathbf{W} = \mathbf{T} - \mathbf{B}$†	
Total	$n-1$	$\mathbf{T} = \sum_{j=1}^{k} \sum_{i=1}^{n_j} (\mathbf{x}_{ij} - \bar{\mathbf{x}})(\mathbf{x}_{ij} - \bar{\mathbf{x}})'$	

† Difference

Table 12.3.2 Logarithms of multiple measurements on ant-eater skulls at three localities (Reeve, 1941)

	Minas Graes, Brazil			Matto Grosso, Brazil			Santa Cruz, Bolivia		
	x_1	x_2	x_3	x_1	x_2	x_3	x_1	x_2	x_3
	2.068	2.070	1.580	2.045	2.054	1.580	2.093	2.098	1.653
	2.068	2.074	1.602	2.076	2.088	1.602	2.100	2.106	1.623
	2.090	2.090	1.613	2.090	2.093	1.643	2.104	2.101	1.653
	2.097	2.093	1.613	2.111	2.114	1.643	—	—	—
	2.117	2.125	1.663	—	—	—	—	—	—
	2.140	2.146	1.681	—	—	—	—	—	—
Means	2.097	2.100	1.625	2.080	2.087	1.617	2.099	2.102	1.643

in millimetres and their means. We form the MANOVA table as given in Table 12.3.3.

We find that

$$\Lambda(3, 10, 2) = |\mathbf{W}|/|\mathbf{T}| = 0.6014.$$

We have from (12.3.7) that

$$\tfrac{8}{3}(1-\Lambda^{1/2})/\Lambda^{1/2} = 0.772 \sim F_{6,16}$$

which, from Appendix C, Table C.3, is not significant. Therefore, we conclude there are no significant differences between localities.

Mardia (1971) gives a modified test which can be used for moderately non-normal data.

Table 12.3.3 Matrices in MANOVA table for Reeve's data $\times 10^{-7}$

Source	d.f.	a_{11}	a_{12}	a_{13}	a_{22}	a_{23}	a_{33}
Between	2	8060	6233	7498	4820	5859	11 844
Within	1	63 423	62 418	76 157	63 528	76 127	109 673
Total	12	71 483	68 651	83 655	68 348	81 986	121 517

12.4 Testing Fixed Contrasts

There are problems where the interest is not so much in testing the equality of the means but testing the significance of a fixed *contrast*, i.e.

$$H_0': a_1\boldsymbol{\mu}_1 + \ldots + a_k\boldsymbol{\mu}_k = \mathbf{0}, \tag{12.4.1}$$

where $a_1, a_2, \ldots, a_k$ are given constants such that $\sum a_i = 0$. Thus we wish to test

$$H_0': a_1\boldsymbol{\mu}_1 + \ldots + a_k\boldsymbol{\mu}_k = \mathbf{0} \quad \text{and no other contrast null}$$

against

$$H_1: \boldsymbol{\mu}_i \neq \boldsymbol{\mu}_j, \qquad i \neq j.$$

We show that the likelihood ratio leads to the criterion

$$|\mathbf{W}|/|\mathbf{W}+\mathbf{C}| \sim \Lambda(p, n-k, 1), \tag{12.4.2}$$

where

$$\mathbf{C} = \left(\sum_{j=1}^{k} a_j\bar{\mathbf{x}}_j\right)\left(\sum_{j=1}^{k} a_j\bar{\mathbf{x}}_j'\right)\Big/\left(\sum_{j=1}^{k} \frac{a_j^2}{n_j}\right). \tag{12.4.3}$$

The value of $\max \log L$ under H_1 is the same as in Section 5.3.3a. Under H_0, let $\boldsymbol{\lambda}$ be a vector of Lagrange multipliers. Then

$$\log L = -\frac{n}{2}\log|\boldsymbol{\Sigma}| - \frac{1}{2}\sum_{j=1}^{k}\sum_{i=1}^{n_j}(\mathbf{x}_{ij}-\boldsymbol{\mu}_j)'\boldsymbol{\Sigma}^{-1}(\mathbf{x}_{ij}-\boldsymbol{\mu}_j) + \boldsymbol{\lambda}'(a_1\boldsymbol{\mu}_1 + \ldots + a_k\boldsymbol{\mu}_k). \tag{12.4.4}$$

On differentiating with respect to $\boldsymbol{\mu}_j$, it is found that

$$\boldsymbol{\mu}_j = \bar{\mathbf{x}}_j - n_j^{-1}a_j\boldsymbol{\Sigma}\boldsymbol{\lambda}, \tag{12.4.5}$$

which on using the constraint

$$\sum_{j=1}^{k} a_j\boldsymbol{\mu}_j = \mathbf{0}$$

leads us to

$$\boldsymbol{\lambda} = \boldsymbol{\Sigma}^{-1}\sum_{1j=1}^{k} a_j\bar{\mathbf{x}}_j\Big/\sum_{j=1}^{k}\frac{a_j^2}{n_j}.$$

On substituting this result in the right-hand side of (12.4.5), we get $\hat{\boldsymbol{\mu}}_j$. Inserting the $\hat{\boldsymbol{\mu}}_j$ in (12.4.4) and maximizing over $\boldsymbol{\Sigma}$ it is found using Theorem 4.2.1 that

$$n\hat{\boldsymbol{\Sigma}} = \mathbf{W} + \mathbf{C}. \tag{12.4.6}$$

Hence, we obtain (12.4.2). Note that H_0' implies $\sum a_j \boldsymbol{\tau}_j = \mathbf{0}$ because $\sum a_j = 0$.

We can describe $\mathbf{C}$ as the SSP matrix due to the contrast $\sum a_j \boldsymbol{\mu}_j$. If there is a set of mutually orthogonal contrasts, such matrices will be additive (see Exercise 12.4.1).

12.5 Canonical Variables and a Test of Dimensionality

12.5.1 The problem

We can regard $\boldsymbol{\mu}_i$ as the coordinates of a point in p-dimensional space for $i = 1, \ldots, k$. When the null hypothesis H_0 of (12.2.3) is true, the vectors $\boldsymbol{\mu}_1, \ldots, \boldsymbol{\mu}_k$ are identical. Thus if r is the dimension of the hyperplane spanned by $\boldsymbol{\mu}_1, \ldots, \boldsymbol{\mu}_k$, then H_0 is equivalent to $r = 0$. It can be seen that in any case

$$r \leq \min(p, k-1) = t, \quad \text{say.} \tag{12.5.1}$$

If H_0 is rejected, it is of importance to determine the actual dimensionality r, where $r = 0, 1, \ldots, t$. If $r = t$ there is no restriction on the $\boldsymbol{\mu}$s, and $r < t$ occurs if and only if there are exactly $s = t - r$ linearly independent relationships between the k mean vectors.

Note that such a problem does not arise in the univariate case because for $p = 1$ we have $t = 1$, so either $r = 0$ (i.e. H_0) or $r = 1$ (i.e. the alternative).

Thus we wish to test a new hypothesis

$$H_0: \text{the } \boldsymbol{\mu}_i \text{ lie in an } r\text{-dimensional hyperplane} \tag{12.5.2}$$

against

$$H_1: \text{the } \boldsymbol{\mu}_i \text{ are unrestricted.} \tag{12.5.3}$$

where $i = 1, \ldots, k$.

12.5.2 The LR test ($\boldsymbol{\Sigma}$ known)

Let us now assume that $\bar{\mathbf{x}}_i$, $i = 1, \ldots, k$, are k sample means for samples of sizes $n_1, \ldots, n_k$ from $N_p(\boldsymbol{\mu}_i, \boldsymbol{\Sigma})$, respectively. We assume that $\boldsymbol{\Sigma}$ is known. Since $\bar{\mathbf{x}}_i \sim N_p(\boldsymbol{\mu}_i, \boldsymbol{\Sigma}/n_i)$, the log likelihood function is

$$l(\boldsymbol{\mu}_1, \ldots, \boldsymbol{\mu}_k) = c - \tfrac{1}{2} \sum_{i=1}^{k} n_i (\bar{\mathbf{x}}_i - \boldsymbol{\mu}_i)' \boldsymbol{\Sigma}^{-1} (\bar{\mathbf{x}}_i - \boldsymbol{\mu}_i), \tag{12.5.4}$$

where c is a constant. Under H_1, $\hat{\boldsymbol{\mu}}_i = \bar{\mathbf{x}}_i$ and we have

$$\max_{H_1} l(\boldsymbol{\mu}_1, \ldots, \boldsymbol{\mu}_k) = c. \tag{12.5.5}$$

Under H_0 the maximum of l is given in the following theorem:

Theorem 12.5.1 *We have*

$$\max_{H_0} l(\boldsymbol{\mu}_1, \ldots, \boldsymbol{\mu}_k) = c - \tfrac{1}{2}(\gamma_{r+1} + \ldots + \gamma_p), \tag{12.5.6}$$

where $\gamma_1 \geq \gamma_2 \geq \ldots \geq \gamma_p$ *are the eigenvalues of*

$$|\mathbf{B} - \gamma \boldsymbol{\Sigma}| = 0. \tag{12.5.7}$$

Proof (Rao, 1973, p. 559) We need to show

$$\min_{H_0} \sum_{i=1}^{k} n_i(\bar{\mathbf{x}}_i - \boldsymbol{\mu}_i)'\boldsymbol{\Sigma}^{-1}(\bar{\mathbf{x}}_i - \boldsymbol{\mu}_i) = \gamma_{r+1} + \ldots + \gamma_p. \tag{12.5.8}$$

Let

$$\mathbf{y}_i = \boldsymbol{\Sigma}^{-1/2}\bar{\mathbf{x}}_i, \qquad \boldsymbol{\nu}_i = \boldsymbol{\Sigma}^{-1/2}\boldsymbol{\mu}_i. \tag{12.5.9}$$

Then H_0 implies that the $\boldsymbol{\nu}_i$ are confined to an r-dimensional hyperplane. We can determine an r-dimensional hyperplane by a point $\mathbf{z}_0$, and r orthonormal direction vectors $\mathbf{z}_1, \ldots, \mathbf{z}_r$, so that the points $\boldsymbol{\nu}_i$ on this plane can be represented as

$$\boldsymbol{\nu}_i = \mathbf{z}_0 + d_{i1}\mathbf{z}_1 + \ldots + d_{ir}\mathbf{z}_r, \qquad i = 1, \ldots, k, \tag{12.5.10}$$

where the d_{ij} are constants. Hence (12.5.8) is equivalent to minimizing

$$\sum_{i=1}^{k} n_i\left(\mathbf{y}_i - \mathbf{z}_0 - \sum_{j=1}^{r} d_{ij}\mathbf{z}_j\right)'\left(\mathbf{y}_i - \mathbf{z}_0 - \sum_{j=1}^{r} d_{ij}\mathbf{z}_j\right) \tag{12.5.11}$$

subject to $\mathbf{z}_j'\mathbf{z}_j = 1$, $\mathbf{z}_j'\mathbf{z}_l = 0$, $j \neq l$, for $j, l = 1, 2, \ldots, r$. On differentiating with respect to d_{ij} for fixed $\mathbf{z}_0, \mathbf{z}_1, \ldots, \mathbf{z}_r$, the minimum of (12.5.11) is found to occur when

$$d_{ij} = \mathbf{y}_i'\mathbf{z}_j - \mathbf{z}_0'\mathbf{z}_j. \tag{12.5.12}$$

Then (12.5.11) reduces to

$$\sum_{i=1}^{k} n_i(\mathbf{y}_i - \mathbf{z}_0)'(\mathbf{y}_i - \mathbf{z}_0) - \sum_{i=1}^{k}\sum_{j=1}^{r} n_i[(\mathbf{y}_i - \mathbf{z}_0)'\mathbf{z}_j]^2, \tag{12.5.13}$$

which is to be minimized with respect to $\mathbf{z}_0, \mathbf{z}_1, \ldots, \mathbf{z}_r$. Let $\bar{\mathbf{y}} = \sum n_i\mathbf{y}_i/n$. Then (12.5.13) can be written as

$$\sum_{i=1}^{k} n_i(\mathbf{y}_i - \bar{\mathbf{y}})'(\mathbf{y}_i - \bar{\mathbf{y}}) - \sum_{i=1}^{k}\sum_{j=1}^{r} n_i[(\mathbf{y}_i - \bar{\mathbf{y}})'\mathbf{z}_j]^2 + n(\bar{\mathbf{y}} - \mathbf{z}_0)'\mathbf{F}(\bar{\mathbf{y}} - \mathbf{z}_0), \tag{12.5.14}$$

where

$$\mathbf{F}=\mathbf{I}-\sum_{j=1}^{r}\mathbf{z}_j\mathbf{z}_j'.$$

Since $\mathbf{F}$ is positive semi-definite, we see that (12.5.14) is minimized for fixed $\mathbf{z}_1, \ldots, \mathbf{z}_r$ by

$$\mathbf{z}_0=\bar{\mathbf{y}}. \tag{12.5.15}$$

Also, minimizing (12.5.14) with respect to $\mathbf{z}_1, \ldots, \mathbf{z}_r$ is equivalent to maximizing

$$\sum_{i=1}^{k}\sum_{j=1}^{r} n_i[(\mathbf{y}_i-\bar{\mathbf{y}})'\mathbf{z}_j]^2=\sum_{j=1}^{r}\mathbf{z}_j'\mathbf{A}\mathbf{z}_j, \tag{12.5.16}$$

where

$$\mathbf{A}=\sum n_i(\mathbf{y}_i-\bar{\mathbf{y}})(\mathbf{y}_i-\bar{\mathbf{y}})'.$$

Using (12.5.9), it is seen that

$$\mathbf{A}=\mathbf{\Sigma}^{-1/2}\mathbf{B}\mathbf{\Sigma}^{-1/2}.$$

Hence $|\mathbf{A}-\gamma\mathbf{I}|=0$ if and only if $|\mathbf{B}-\gamma\mathbf{\Sigma}|=0$, so that $\mathbf{A}$ and $\mathbf{\Sigma}^{-1}\mathbf{B}$ have the same eigenvalues, namely $\gamma_1 \geqslant \ldots \geqslant \gamma_p$, given by (12.5.7).

Let $\mathbf{g}_1, \ldots, \mathbf{g}_p$ be the corresponding standardized eigenvectors of $\mathbf{A}$. Then, following the same argument as in Section 8.2.3(d), it is easy to see that

$$\max\sum_{j=1}^{r}\mathbf{z}_j'\mathbf{A}\mathbf{z}_j=\gamma_1+\ldots+\gamma_r, \tag{12.5.17}$$

and that this maximum is attained for $\mathbf{z}_j=\mathbf{g}_j$, $j=1, \ldots, r$. Further,

$$\sum n_i(\mathbf{y}_i-\bar{\mathbf{y}})'(\mathbf{y}_i-\bar{\mathbf{y}})=\operatorname{tr}\mathbf{A}=\gamma_1+\ldots+\gamma_p. \tag{12.5.18}$$

Hence, from (12.5.13)–(12.5.18) we obtain

$$\text{left-hand side of (12.5.8)}=\gamma_{r+1}+\ldots+\gamma_p, \tag{12.5.19}$$

and the proof of (12.5.8) is complete. ■

Now the log likelihood ratio criterion λ is (12.5.6) minus (12.5.5) and thus

$$-2\log\lambda=\gamma_{r+1}+\ldots+\gamma_p \tag{12.5.20}$$

is the test criterion. The hypothesis H_0 is rejected for large values of (12.5.20).

12.5.3 Asymptotic distribution of the likelihood ratio criterion

For large values of $n_1, \ldots, n_k$, (12.5.20) is distributed as a chi-squared variable with f d.f. where f is the number of restrictions on the pk parameters $\boldsymbol{\mu}_i$, $i = 1, \ldots, k$, under H_0. It can be seen that an r-dimensional hyperplane is specified by $r+1$ points, say $\boldsymbol{\mu}_1, \ldots, \boldsymbol{\mu}_{r+1}$. The rest of the points can be expressed as $a_1\boldsymbol{\mu}_1 + \ldots + a_{r+1}\boldsymbol{\mu}_{r+1}$, where $a_1 + \ldots + a_{r+1} = 1$ say. Thus the number of unrestricted parameters is $p(r+1) + (k-r-1)r$. Consequently,

$$\gamma_{r+1} + \ldots + \gamma_p \sim \chi_f^2, \tag{12.5.21}$$

asymptotically, where

$$f = pk - p(r+1) - (k-r-1)r = (p-r)(k-r-1).$$

12.5.4 The estimated plane

Write

$$\mathbf{Z}' = (\mathbf{z}_1, \ldots, \mathbf{z}_r), \qquad \boldsymbol{\nu}' = (\boldsymbol{\nu}_1, \ldots, \boldsymbol{\nu}_k), \qquad \mathbf{Y}' = (\mathbf{y}_1, \ldots, \mathbf{y}_k).$$

(Note that here a lower case letter $\boldsymbol{\nu}$ is used to represent a $(k \times p)$ matrix.) From (12.5.15), $\mathbf{z}_0 = \bar{\mathbf{y}} = \boldsymbol{\Sigma}^{-1/2}\bar{\mathbf{x}}$, so, using (12.5.10) and (12.5.12), we can write

$$\hat{\boldsymbol{\nu}}' = \mathbf{Z}'\mathbf{Z}(\mathbf{Y}' - \bar{\mathbf{y}}\mathbf{1}') + \bar{\mathbf{y}}\mathbf{1}', \tag{12.5.22}$$

where $\hat{\boldsymbol{\nu}}$ is the m.l.e. of $\boldsymbol{\nu}$ under H_0.

Let $\mathbf{q}_j = \boldsymbol{\Sigma}^{-1/2}\mathbf{z}_j$ denote the jth eigenvector of $\boldsymbol{\Sigma}^{-1}\mathbf{B} = \boldsymbol{\Sigma}^{-1/2}\mathbf{A}\boldsymbol{\Sigma}^{1/2}$, so that

$$\mathbf{B}\mathbf{q}_j = \gamma_j \boldsymbol{\Sigma}\mathbf{q}_j, \qquad \mathbf{q}_j'\boldsymbol{\Sigma}\mathbf{q}_j = 1, \qquad \mathbf{q}_j'\boldsymbol{\Sigma}\mathbf{q}_l = 0, \qquad j \neq l, \tag{12.5.23}$$

and write $\mathbf{Q}' = (\mathbf{q}_1, \ldots, \mathbf{q}_r)$. Then since

$$\mathbf{Y}' = \boldsymbol{\Sigma}^{-1/2}\bar{\mathbf{X}}, \qquad \mathbf{Q}' = \boldsymbol{\Sigma}^{-1/2}\mathbf{Z}', \qquad \boldsymbol{\mu}' = \boldsymbol{\Sigma}^{1/2}\boldsymbol{\nu}',$$

we find from (12.5.22) that

$$\hat{\boldsymbol{\mu}}' = \boldsymbol{\Sigma}\mathbf{Q}'\mathbf{Q}(\bar{\mathbf{X}}' - \bar{\mathbf{x}}\mathbf{1}') + \bar{\mathbf{x}}\mathbf{1}', \tag{12.5.24}$$

where $\hat{\boldsymbol{\mu}}' = (\hat{\boldsymbol{\mu}}_1, \ldots, \hat{\boldsymbol{\mu}}_k)$ is the m.l.e. of $\boldsymbol{\mu}$ under H_0 and $\bar{\mathbf{X}}' = (\bar{\mathbf{x}}_1, \ldots, \bar{\mathbf{x}}_k)$.

Note that (12.5.24) is of the form $\hat{\boldsymbol{\mu}}_i = \mathbf{D}\bar{\mathbf{x}}_i + (\mathbf{I} - \mathbf{D})\bar{\mathbf{x}}$ for $i = 1, \ldots, k$, where $\mathbf{D}$ is of rank r; that is, $\hat{\boldsymbol{\mu}}_i$ is an estimator of $\boldsymbol{\mu}_i$ in an r-dimensional plane passing through $\bar{\mathbf{x}}$. It is convenient to represent points $\mathbf{x}$ in this plane using the coordinates

$$\mathbf{x}^* = (\mathbf{q}_1'\mathbf{x}, \ldots, \mathbf{q}_r'\mathbf{x})',$$

which are called *canonical coordinates*. The linear function $y_j = \mathbf{q}_j'\mathbf{x}$ is called the jth *canonical variable* (or variate) and the vector $\mathbf{q}_j$ is called the jth *canonical vector.*

Using $\mathbf{Q\Sigma Q}'=\mathbf{I}_r$ from (12.5.23), we observe from (12.5.24) that

$$\mathbf{Q}\hat{\boldsymbol{\mu}}'=\mathbf{Q}\bar{\mathbf{X}}'. \tag{12.5.25}$$

Thus, the canonical coordinates of $\hat{\boldsymbol{\mu}}_i$ are given by

$$(\mathbf{q}_1'\bar{\mathbf{x}}_i,\ldots,\mathbf{q}_r'\bar{\mathbf{x}}_i)'=\hat{\boldsymbol{\mu}}_i^*,\quad \text{say,}$$

for $i=1,\ldots,k$, and are called *canonical means.* Note that the $\hat{\boldsymbol{\mu}}_i^*$ are r-dimensional vectors whereas the $\boldsymbol{\mu}_i$ are p-dimensional vectors.

Suppose for the moment that $\mathbf{Q}$ does not depend on $\bar{\mathbf{x}}_i-\boldsymbol{\mu}_i$. Since $\bar{\mathbf{x}}_i\sim N_p(\boldsymbol{\mu}_i, n_i^{-1}\boldsymbol{\Sigma})$ and $\mathbf{q}_j'\boldsymbol{\Sigma}\mathbf{q}_j=1$, $\mathbf{q}_j'\boldsymbol{\Sigma}\mathbf{q}_l=0$ for $j\neq l$, it can be seen that

$$\hat{\boldsymbol{\mu}}_i^*\sim N_r(\boldsymbol{\mu}_i^*, n_i^{-1}\mathbf{I}),$$

where $\boldsymbol{\mu}_i^*=(\mathbf{q}_1'\boldsymbol{\mu}_i,\ldots,\mathbf{q}_r'\boldsymbol{\mu}_i)'$ is the canonical coordinate representation of the true mean $\boldsymbol{\mu}_i$ (under H_0 or H_1). Hence

$$n_i(\hat{\boldsymbol{\mu}}_i^*-\boldsymbol{\mu}_i^*)'(\hat{\boldsymbol{\mu}}_i^*-\boldsymbol{\mu}_i^*)\sim\chi_r^2. \tag{12.5.26}$$

This formula can be used to construct confidence ellipsoids for the $\boldsymbol{\mu}_i^*$. Of course, since $\mathbf{Q}$ depends to some extent on $\bar{\mathbf{x}}_i-\boldsymbol{\mu}_i$, these confidence regions are only approximate.

12.5.5 The LR test (unknown $\boldsymbol{\Sigma}$)

Consider the hypotheses H_0 versus H_1 in (12.5.2) and (12.5.3) where $\boldsymbol{\Sigma}$ is now unknown. Unfortunately, the LRT in this situation is quite complicated. However, an alternative test can be constructed if we replace $\boldsymbol{\Sigma}$ by the unbiased estimate $\mathbf{W}/(n-k)$, and use the test of Section 12.5.2, treating $\boldsymbol{\Sigma}$ as known. It can be shown that for large n this test is asymptotically equivalent to the LRT. (See Cox and Hinkley, 1974, p. 361.) Then the test given by (12.5.21) becomes in this context, asymptotically,

$$(n-k)(\lambda_{r+1}+\ldots+\lambda_p)\sim\chi_f^2,\qquad f=(p-r)(k-r-1), \tag{12.5.27}$$

where $\lambda_1,\ldots,\lambda_p$ are now the roots of

$$|\mathbf{B}-\lambda\mathbf{W}|=0. \tag{12.5.28}$$

Note that $(n-k)\lambda_j\doteq\gamma_j$ of the last section because $\mathbf{W}/(n-k)\doteq\boldsymbol{\Sigma}$. Estimates of the k group means and their r-dimensional plane can be constructed as in Section 12.5.4, and will be explored further in the next section.

Bartlett (1947) suggested the alternative statistic

$$D_r^2 = \{n-1-\tfrac{1}{2}(p+k)\} \sum_{j=r+1}^{p} \log(1+\lambda_j) \sim \chi_f^2, \tag{12.5.29}$$

which improves the chi-squared approximation in (12.5.27) and therefore (12.5.29) will be preferred. Note that for large n, (12.5.29) reduces to (12.5.27).

We can now perform the tests of dimensionality $r = 0, 1, 2, \ldots, t$ sequentially. First test $H_0: r = 0$. If D_0^2 is significant then test $H_0: r = 1$, and so on. In general, if $D_0^2, D_1^2, \ldots, D_{r-1}^2$ are significant but D_r^2 is not significant then we may infer the dimensionality to be r.

Although we have assumed p roots of (12.5.28), there will be at most t non-zero roots.

12.5.6 The estimated plane (unknown $\boldsymbol{\Sigma}$)

Assume that r is the dimension of the plane spanned by the true group means. As in Section 12.5.4 we can look more deeply into the separation of the groups in this plane (now assuming $\boldsymbol{\Sigma}$ is unknown). Let $\mathbf{l}_j$ be the eigenvector of $\mathbf{W}^{-1}\mathbf{B}$ corresponding to λ_j, normalized by

$$\mathbf{l}_j'[\mathbf{W}/(n-k)]\mathbf{l}_j = 1. \tag{12.5.30}$$

Then the $\mathbf{l}_1, \ldots, \mathbf{l}_r$ are canonical vectors analogous to $\mathbf{q}_1, \ldots, \mathbf{q}_r$ of Section 12.5.4 (where we assumed $\boldsymbol{\Sigma}$ was known). As in that section, these can be used to estimate the plane of the true group means and to represent points within this estimated plane.

The projection of a point $\mathbf{x}$ onto the estimated plane can be represented in terms of the r-dimensional *canonical coordinates* $(\mathbf{l}_1'\mathbf{x}, \ldots, \mathbf{l}_r'\mathbf{x})$. In particular the *canonical means* of the k groups, $\mathbf{m}_i = (\mathbf{l}_1'\bar{\mathbf{x}}_i, \ldots, \mathbf{l}_r'\bar{\mathbf{x}}_i)'$, $i = 1, \ldots, k$, represent the projection of the group means onto this plane and can be used to study the differences between the groups.

The vector $\mathbf{l}_j$ is the *canonical vector* for the jth *canonical variable* $y_j = \mathbf{l}_j'\mathbf{x}$. Note that from Section 11.5, the canonical variables are optimal discriminant functions; that is, for the data matrix $\mathbf{X}$, the jth canonical variable is that linear function which maximizes the between group variance relative to within group variance, subject to the constraint that it is uncorrelated with the preceding canonical variables. Consequently, for any value $r \leq t$, the canonical variables $y_1, \ldots, y_r$ are those linear functions which separate the k sample means as much as possible.

A graph of the canonical means in $r = 1$ or $r = 2$ dimensions can give a useful picture of the data. Note that on such a graph, the estimated variance matrix for the canonical coordinates of an observation coming from any of the k groups is the identity. (The canonical variables are uncorrelated with one another and their estimated variance is normalized by (12.5.30) to equal 1.) Thus the units of each axis on the graph represent one standard deviation for each canonical coordinate of an observation, and the canonical coordinates are uncorrelated with one another. Hence, we can get some idea of the strength of separation between the groups.

Also, we can estimate the accuracy of each of the canonical means on such a graph. From (12.5.26) a rough $100(1-\alpha)\%$ confidence region for the ith true canonical mean $\boldsymbol{\mu}_i^* = (\mathbf{l}_1'\boldsymbol{\mu}_i, \ldots, \mathbf{l}_r'\boldsymbol{\mu}_i)'$ is given by the disk of radius $n_i^{-1/2}\chi_{r;\alpha}$ about the sample canonical mean $\mathbf{m}_i = (\mathbf{l}_1'\bar{\mathbf{x}}_i, \ldots, \mathbf{l}_r'\bar{\mathbf{x}}_i)'$, where $\chi_{r;\alpha}^2$ is the upper α critical point of a χ_r^2 variable. Note that $(\lambda_1+\ldots+\lambda_r)/(\operatorname{tr} \mathbf{W}^{-1}\mathbf{B}) = (\lambda_1+\ldots+\lambda_r)/(\lambda_1+\ldots+\lambda_p)$ represents the proportion of the between groups variation which is explained by the first r canonical variates.

Canonical analysis is the analogue for *grouped* data of principal component analysis for *ungrouped* data. Further, since an estimate of the inherent variability of the data is given by $\mathbf{W}$, canonical coordinates are invariant under changes of scale of the original variables. (This property is not shared by principal component analysis.)

Example 12.5.1 Consider the iris data of Table 1.2.2 with $p=4$. Let x_1 = sepal length, x_2 = sepal width, x_3 = petal length, and x_4 = petal width. Here, the three samples are the three species *I. setosa*, *I. versicolour*, and *I. virginica*. Further, $n_1 = n_2 = n_3 = 50$. It is found that the mean vectors are

$$\bar{\mathbf{x}}_1 = \begin{bmatrix} 5.006 \\ 3.428 \\ 1.462 \\ 0.246 \end{bmatrix}, \quad \bar{\mathbf{x}}_2 = \begin{bmatrix} 5.936 \\ 2.770 \\ 4.260 \\ 1.326 \end{bmatrix}, \quad \bar{\mathbf{x}}_3 = \begin{bmatrix} 6.588 \\ 2.974 \\ 5.552 \\ 2.026 \end{bmatrix}.$$

The SSP matrix between species is

$$\mathbf{B} = \begin{bmatrix} 63.21 & -19.95 & 165.25 & 71.28 \\ & 11.35 & -57.24 & -22.93 \\ & & 437.11 & 186.78 \\ & & & 80.41 \end{bmatrix}.$$

The SSP matrix within species is

$$\mathbf{W}=\begin{bmatrix} 38.96 & 13.63 & 24.62 & 5.64 \\ & 16.96 & 8.12 & 4.81 \\ & & 27.22 & 6.27 \\ & & & 6.16 \end{bmatrix}.$$

It is found that the eigenvalues of $\mathbf{W}^{-1}\mathbf{B}$ are

$$\lambda_1 = 32.1877, \qquad \lambda_2 = 0.2853$$

and the corresponding eigenvectors normalized by (12.5.30) are

$$\mathbf{l}_1' = (0.83, 1.54, -2.20, -2.81),$$
$$\mathbf{l}_2' = (-0.02, -2.17, 0.92, -2.83).$$

In this case, (12.5.29) becomes

$$D_0^2 = 546.01 \sim \chi_8^2$$

which by Appendix C, Table C.1 is highly significant. Thus the mean differences are significant. Now

$$D_1^2 = 36.52 \sim \chi_3^2,$$

which is again highly significant. Hence, since $t=2$, both canonical variables are necessary, and the dimension cannot be reduced. In fact, the canonical means for *I. setosa* are

$$\mathbf{m}_1' = (\mathbf{l}_1'\bar{\mathbf{x}}_1, \mathbf{l}_2'\bar{\mathbf{x}}_1) = (5.50, -6.89).$$

Similarly, for the *I. versicolour* and *I. virginica* varieties, we have, respectively,

$$\mathbf{m}_2' = (-3.93, -5.94), \qquad \mathbf{m}_3' = (-7.89, -7.18).$$

Figure 12.5.1 shows a plot together with the approximate 99% confidence circles of radius $(\chi^2_{2;0.01}/50)^{1/2} = (9.21/50)^{1/2} = 0.429$, for the three canonical means. It is quite clear from the figure that the three species are widely different but *I. versicolour* and *I. virginica* are nearer than *I. setosa*.

Example 12.5.2 (Delany and Healy, 1966) White-toothed shrews of the genus *Crocidura* occur in the Channel and Scilly Islands of the British Isles and the French mainland. From $p=10$ measurements on each of $n=399$ skulls obtained from the $k=10$ localities, Tresco, Bryher, St Agnes, St Martin's, St Mary's, Sark, Jersey, Alderney, Guernsey, and Cap Gris Nez, the between and within SSP matrices are as given in Exercise

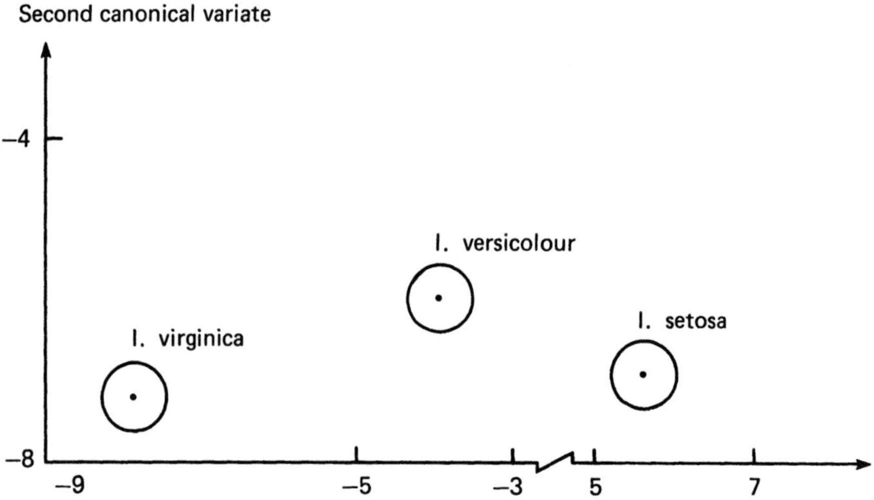

Figure 12.5.1 First canonical variate

12.5.1. The sample sizes for the data from the localities are, respectively, 144, 16, 12, 7, 90, 25, 6, 26, 53, 20.

The first two canonical vectors are found to be

$$\mathbf{l}_1' = (-1.48, 0.37, 2.00, 8.59, -3.93, 0.90, 8.10, 8.78, 0.00, 0.44),$$

$$\mathbf{l}_2' = (0.44, 1.81, -0.58, -9.65, -7.01, 1.97, 3.97, 7.10, -7.61, 0.34).$$

The first two canonical variables account for 93.7% of the between-samples variation (see Exercise 12.5.1). The sample canonical means, $\mathbf{m}_i' = (\mathbf{l}_1'\bar{\mathbf{x}}_i, \mathbf{l}_2'\bar{\mathbf{x}}_i)$, $i = 1, \ldots, n$, (centred to have overall weighted mean $\mathbf{0}$) and the 99% confidence circles for the true canonical means are shown in Figure 12.5.2. This suggests that the three populations of *C. russula* (Alderney, Guernsey, and Cap Gris Nez) show more differentiation than the five populations of *C. suaveolens* (Tresco, Bryher, St Agnes, St Mary's, St Martin's) in the Scilly Isles. The shrews from Sark are distinct from those in the five Scilly Isles and Jersey. Further discussion of this data is given in Exercise 12.5.1 and in Chapter 13 on Cluster Analysis.

It should be noted that the present analysis differs slightly from the one given by Delany and Healy (1966). In their analysis the SSP matrix was calculated without using the n_i as weights, i.e.

$$\mathbf{B} = \sum_{i=1}^{k} (\mathbf{x}_i - \bar{\mathbf{x}})(\mathbf{x}_i - \bar{\mathbf{x}})'.$$

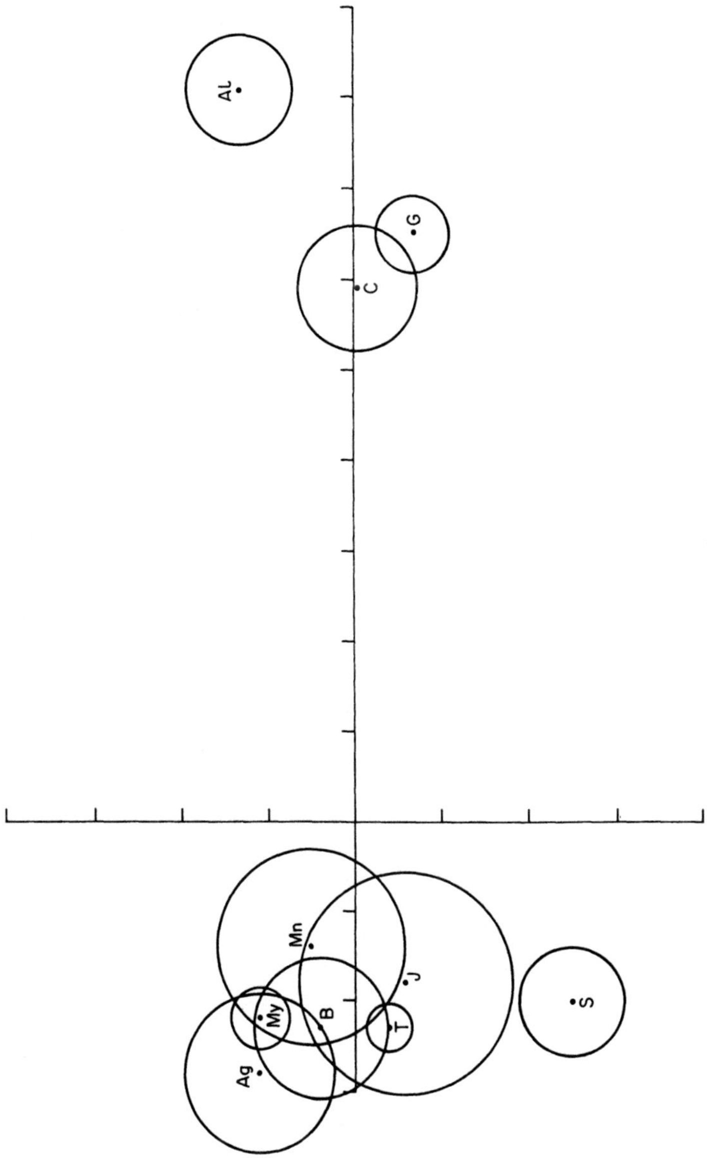

Figure 12.5.2 *Plot of the 10 groups of shrews with respect to the first two canonical variates. Ag, St Agnes; Al, Alderney; B, Bryher; C, Cap Griz Nez; G. Guernsey; J, Jersey; Mn, St Martin's; My, St Mary's; S, Sark; T, Tresco.*

Consequently the canonical variables obtained differ from those given above, and their diagram corresponding to Figure 12.5.2 differs mainly in that the canonical means for the Jersey sample become very close to those of Sark. However, the conclusions are broadly the same.

12.5.7 Profile analysis

A related hypothesis to the above dimensionality hypothesis is the *profile hypothesis*, which assumes that the group means lie on a line and that the direction of this line is given by the vector $\mathbf{1}$; that is, the difference between each pair of group means is a vector whose components are all equal.

This hypothesis is of interest when all of the variables measure similar quantities and it is expected that the difference between the groups will be reflected equally in all the variables.

If the model is parametrized in terms of the k group means $\boldsymbol{\mu}_1, \ldots, \boldsymbol{\mu}_k$ then the profile hypothesis can be written in the form of Section 6.3 as

$$\mathbf{C}_1\boldsymbol{\mu}\mathbf{M}_1 = \mathbf{0},$$

where $\boldsymbol{\mu}' = (\boldsymbol{\mu}_1, \ldots, \boldsymbol{\mu}_k)$ and $\mathbf{C}_1((k-1)\times k)$ and $\mathbf{M}_1(p\times(p-1))$ are given by

$$\mathbf{C}_1 = \begin{bmatrix} 1 & -1 & 0 & \cdots & 0 \\ 0 & 1 & -1 & \cdots & 0 \\ \cdot & \cdot & \cdot & & \cdot \\ \cdot & \cdot & \cdot & & \cdot \\ \cdot & \cdot & \cdot & & \cdot \\ 0 & 0 & 0 & \cdots & -1 \end{bmatrix} \quad \text{and} \quad \mathbf{M}_1 = \begin{bmatrix} 1 & 0 & \cdots & 0 \\ -1 & 1 & \cdots & 0 \\ 0 & -1 & \cdots & 0 \\ \cdot & \cdot & & \cdot \\ \cdot & \cdot & & \cdot \\ \cdot & \cdot & & \cdot \\ 0 & 0 & \cdots & -1 \end{bmatrix}.$$

This hypothesis is of interest because $\mathbf{M}_1$ is *not* equal to the identity; that is, a relationship between the variables is assumed as well as a relationship between the groups. For further details see Morrison (1976, pp. 205–216).

12.6 The union intersection approach

For the union intersection approach we concentrate on the one-way classification given in (12.2.1).

We wish to test the hypothesis $\boldsymbol{\tau}_1 = \ldots = \boldsymbol{\tau}_k$. This multivariate hypothesis is true if and only if all of the univariate hypotheses

$$H_{0\mathbf{a},\mathbf{c}}: \sum_{i=1}^{k} c_i \mathbf{a}'\boldsymbol{\tau}_i = 0$$

are true for all p-vectors $\mathbf{a}$ and all k-vector contrasts $\mathbf{c}$. Following Section 5.3.3a, it is found that a suitable test criterion is λ_1, the largest eigenvalue of $\mathbf{W}^{-1}\mathbf{B}$. We reject H_0 for large values of λ_1.

This approach has the advantage that it allows us to construct simultaneous confidence regions for contrasts between the $\boldsymbol{\tau}_i$. Percentage points are given in Appendix C, Table C.4 of $\theta = \lambda_1/(1+\lambda_1)$ for $p=2$ and selected critical values. We may use these to form $100(1-\alpha)\%$ simultaneous confidence intervals which include intervals of the form

$$\mathbf{a}'(\boldsymbol{\tau}_j - \boldsymbol{\tau}_k) \in \mathbf{a}'(\bar{\mathbf{x}}_j - \bar{\mathbf{x}}_k) \pm \left\{\frac{\theta_\alpha}{1-\theta_\alpha} \mathbf{a}'\mathbf{W}\mathbf{a}\left(\frac{1}{n_j} + \frac{1}{n_k}\right)\right\}^{1/2},$$

where θ_α denotes the upper α critical value of θ. In general we denote the random variable corresponding to the distribution of the largest eigenvalue of $|\mathbf{B} - \theta(\mathbf{B}+\mathbf{W})| = 0$ by $\theta(p, \nu_1, \nu_2)$, where $\mathbf{B} \sim \mathbf{W}_p(\boldsymbol{\Sigma}, \nu_2)$ independently of $\mathbf{W} \sim W_p(\boldsymbol{\Sigma}, \nu_1)$. Here ν_1 is the error degrees of freedom and ν_2 is the hypothesis degrees of freedom. More details are given in Section 3.7. This test is related to the LRT in as much as both are functions of the eigenvalues of $\mathbf{W}^{-1}\mathbf{B}$.

Example 12.6.1 In Reeve's data given in Example 12.3.1 we find that the non-zero eigenvalues of $\mathbf{W}^{-1}\mathbf{B}$ are

$$\lambda_1 = 0.3553, \qquad \lambda_2 = 0.2268.$$

(Note rank $(\mathbf{W}^{-1}\mathbf{B}) = 2$.) From Appendix C, Table C.4 the 1% critical value of $\theta(3, 10, 2) = \theta(2, 9, 3)$ is $\theta_{0.01} = 0.8074$. Since $\lambda_1/(1+\lambda_1) = 0.2622 < \theta_{0.01}$ we accept the hypothesis that there are no differences, just as in Example 12.3.1.

Even if we look at differences between x_3 for groups 2 and 3 which look plausibly different, by forming a 99% CI with $\mathbf{a} = (0, 0, 1)'$, $j = 2$, $k = 3$, we get an interval

$$\tau_{23} - \tau_{33} \in 1.617 - 1.643 \pm \left\{\frac{0.01097\,\theta_{0.01}}{1-\theta_{0.01}}\left(\frac{1}{4}+\frac{1}{3}\right)\right\}^{1/2}$$

$$= -0.026 \pm 0.164$$

which contains 0. Because the null hypothesis is accepted, it follows from the construction of the test that *this and all other* contrasts are non-significant.

Example 12.6.2 We now examine Example 12.5.1 by this method. The greatest eigenvalue of $\mathbf{W}^{-1}\mathbf{B}$ was $\lambda_1 = 32.188$ and so $\theta = \lambda_1/(1+\lambda_1) = 0.9699 \sim \theta(4, 147, 2) = \theta(2, 145, 4)$. From Appendix C, Table C.4, $\theta_{0.01} = 0.1098$, so we strongly reject the hypothesis of equal mean vectors.

Suppose we are interested in sepal width (x_2) differences. Choosing $\mathbf{a}' = (0, 1, 0, 0)$, the $\pm$ term is always

$$\left\{\frac{16.96\theta_{0.01}}{1-\theta_{0.01}}\left(\frac{1}{50}+\frac{1}{50}\right)\right\}^{1/2} = 0.29 \quad \text{as} \quad n_1 = n_2 = n_3 = 50.$$

This leads to the following 99% SCIs for sepal width differences:

I. virginica and *I. versicolour*	$\tau_{12} - \tau_{22} \in 2.97 - 2.77 \pm 0.29 = 0.20 \pm 0.29;$
I. setosa and *I. versicolour*	$\tau_{32} - \tau_{22} \in 3.43 - 2.77 \pm 0.29 = 0.66 \pm 0.29;$
I. setosa and *I. virginica*	$\tau_{32} - \tau_{12} \in 3.43 - 2.97 \pm 0.29 = 0.46 \pm 0.29.$

Thus, we accept the hypothesis of equal sepal width means at the 1% level for *I. virginica* and *I. versicolour.* It is easily verified that all other confidence intervals on all other variable differences do not contain zero, so apart from this single sepal width difference, the three species are different on all four variables.

Note that simultaneous confidence intevals are helpful because they provide detailed information about each variable and about the differences in means pair-wise. However, there is an advantage in canonical analysis because it gives us a global picture of the group differences.

12.7 Two-way Classification

The ANOVA has a direct generalization for vector variables leading to an analysis of a matrix of sums of squares and products, as in the single classification problem. Now consider a two-way layout. Let us suppose we have nrc independent observations generated by the model

$$\mathbf{x}_{ijk} = \boldsymbol{\mu} + \boldsymbol{\alpha}_i + \boldsymbol{\tau}_j + \boldsymbol{\eta}_{ij} + \boldsymbol{\varepsilon}_{ijk}, \qquad i = 1, \ldots, r, \quad j = 1, \ldots, c, \quad k = 1, \ldots, n,$$

where $\boldsymbol{\alpha}_i$ is the ith row effect, $\boldsymbol{\tau}_j$ is the jth column effect, $\boldsymbol{\eta}_{ij}$ is the interaction effect between the ith row and the jth column, and $\boldsymbol{\varepsilon}_{ijk}$ is the error term which is assumed to be independent $N_p(\mathbf{0}, \boldsymbol{\Sigma})$ for all i, j, k. We require that the number of observations in each (i, j)-cell should be the same, so that the total sum of squares and products matrix can be suitably decomposed. We are interested in testing the null hypotheses of equality of the $\boldsymbol{\alpha}_i$, equality of the $\boldsymbol{\tau}_j$, and equality of the $\boldsymbol{\eta}_{ij}$. We partition the total sum of squares and products matrix in an exactly analogous manner to univariate analysis.

Let $\mathbf{T}, \mathbf{R}, \mathbf{C}$, and $\mathbf{E}$ be the total, rows, columns, and errors SSP matrices, respectively. As in the univariate case, we can show that the following MANOVA identity holds, i.e.

$$\mathbf{T} = \mathbf{R} + \mathbf{C} + \mathbf{E}, \tag{12.7.1}$$

where

$$\mathbf{T} = \sum_{i=1}^{r} \sum_{j=1}^{c} \sum_{k=1}^{n} (\mathbf{x}_{ijk} - \bar{\mathbf{x}}_{\cdots})(\mathbf{x}_{ijk} - \bar{\mathbf{x}}_{\cdots})',$$

$$\mathbf{R} = cn \sum_{i=1}^{r} (\bar{\mathbf{x}}_{i\cdot\cdot} - \bar{\mathbf{x}}_{\cdots})(\bar{\mathbf{x}}_{i\cdot\cdot} - \bar{\mathbf{x}}_{\cdots})',$$

$$\mathbf{C} = rn \sum_{j=1}^{c} (\bar{\mathbf{x}}_{\cdot j\cdot} - \bar{\mathbf{x}}_{\cdots})(\bar{\mathbf{x}}_{\cdot j\cdot} - \bar{\mathbf{x}}_{\cdots})',$$

$$\mathbf{E} = \sum_{i=1}^{r} \sum_{j=1}^{c} \sum_{k=1}^{n} (\mathbf{x}_{ijk} - \bar{\mathbf{x}}_{i\cdot\cdot} - \bar{\mathbf{x}}_{\cdot j\cdot} + \bar{\mathbf{x}}_{\cdots})(\mathbf{x}_{ijk} - \bar{\mathbf{x}}_{i\cdot\cdot} - \bar{\mathbf{x}}_{\cdot j\cdot} + \bar{\mathbf{x}}_{\cdots})',$$

with

$$\bar{\mathbf{x}}_{i\cdot\cdot} = \frac{1}{cn} \sum_{j=1}^{c} \sum_{k=1}^{n} \mathbf{x}_{ijk}, \qquad \bar{\mathbf{x}}_{\cdot j\cdot} = \frac{1}{rn} \sum_{i=1}^{r} \sum_{k=1}^{n} \mathbf{x}_{ijk}$$

and

$$\bar{\mathbf{x}}_{\cdots} = \frac{1}{rcn} \sum_{i=1}^{r} \sum_{j=1}^{c} \sum_{k=1}^{n} \mathbf{x}_{ijk}.$$

We may further decompose $\mathbf{E}$ into

$$\mathbf{E} = \mathbf{I} + \mathbf{W}, \tag{12.7.2}$$

where

$$\mathbf{I} = n \sum_{i=1}^{r} \sum_{j=1}^{c} (\bar{\mathbf{x}}_{ij\cdot} - \bar{\mathbf{x}}_{i\cdot\cdot} - \bar{\mathbf{x}}_{\cdot j\cdot} + \bar{\mathbf{x}}_{\cdots})(\bar{\mathbf{x}}_{ij\cdot} - \bar{\mathbf{x}}_{i\cdot\cdot} - \bar{\mathbf{x}}_{\cdot j\cdot} + \bar{\mathbf{x}}_{\cdots})'$$

and

$$\mathbf{W} = \sum_{i=1}^{r} \sum_{j=1}^{c} \sum_{k=1}^{n} (\mathbf{x}_{ijk} - \bar{\mathbf{x}}_{ij\cdot})(\mathbf{x}_{ijk} - \bar{\mathbf{x}}_{ij\cdot})',$$

with

$$\bar{\mathbf{x}}_{ij\cdot} = \frac{1}{n} \sum_{k=1}^{n} \mathbf{x}_{ijk}.$$

Here $\mathbf{I}$ is the SSP matrix due to interaction and $\mathbf{W}$ is the residual SSP matrix.

Tests for interactions
Thus from (12.7.1) and (12.7.2) we may partition $\mathbf{T}$ into

$$\mathbf{T}=\mathbf{R}+\mathbf{C}+\mathbf{I}+\mathbf{W}. \tag{12.7.3}$$

Clearly under the hypothesis H_0 of all $\boldsymbol{\alpha}_i$, $\boldsymbol{\tau}_j$, and $\boldsymbol{\eta}_{ij}$ being zero, $\mathbf{T}$ must have the $W_p(\boldsymbol{\Sigma}, rcn-1)$ distribution.

Also, we can write

$$\mathbf{W}=\sum_{i=1}^{r}\sum_{j=1}^{c}\mathbf{A}_{ij}$$

say, and, whether or not H_0 holds, the $\mathbf{A}_{ij}$ are i.i.d. $W_p(\boldsymbol{\Sigma}, n-1)$.

Thus, as in Section 12.3,

$$\mathbf{W}\sim W_p(\boldsymbol{\Sigma}, rc(n-1)).$$

In the same spirit as in univariate analysis, whether or not the $\boldsymbol{\alpha}$s and $\boldsymbol{\tau}$s vanish,

$$\mathbf{I}\sim W_p(\boldsymbol{\Sigma}, (r-1)(c-1))$$

if the $\boldsymbol{\eta}_{ij}$ are equal.

It can be shown that the matrices $\mathbf{R}$, $\mathbf{C}$, $\mathbf{I}$, and $\mathbf{W}$ are distributed independently of one another, and further we can show that the LR statistic for testing the equality of the interaction terms is

$$|\mathbf{W}|/|\mathbf{W}+\mathbf{I}|=|\mathbf{W}|/|\mathbf{E}|\sim \Lambda(p, \nu_1, \nu_2),$$

where

$$\nu_1=rc(n-1), \qquad \nu_2=(r-1)(c-1). \tag{12.7.4}$$

We reject the hypothesis of no interaction for low values of Λ. We may alternatively look at the largest eigenvalue θ of $\mathbf{I}(\mathbf{W}+\mathbf{I})^{-1}$ as in Section 12.6, which is distributed as $\theta(p, \nu_1, \nu_2)$.

Note that if $n=1$, i.e. there is one observation per cell, then $\mathbf{W}$ has zero d.f., so we cannot make any test for the presence of interaction.

Tests for main effects
If the column effects vanish, then $\mathbf{C}$ can be shown to have the W_p $(\boldsymbol{\Sigma}, c-1)$ distribution. The LR statistic to test the equality of the $\boldsymbol{\tau}_j$ irrespective of the $\boldsymbol{\alpha}_i$ and $\boldsymbol{\eta}_{ij}$ can be shown to be

$$|\mathbf{W}|/|\mathbf{W}+\mathbf{C}|\sim \Lambda(p, \nu_1, \nu_2),$$

where

$$\nu_1 = rc(n-1), \qquad \nu_2 = c-1. \tag{12.7.5}$$

Equality is again rejected for low values of Λ.

Alternatively we can look at the largest eigenvalue θ of $\mathbf{C}(\mathbf{W}+\mathbf{C})^{-1}$ which is distributed as $\theta(p, \nu_1, \nu_2)$, where ν_1 and ν_2 are given by (12.7.5).

Similarly, to test for the equality for the r row effects $\boldsymbol{\alpha}_i$ irrespective of the column effects $\boldsymbol{\tau}_j$, we replace $\mathbf{C}$ by $\mathbf{R}$, and interchange r and c in the above.

Note that if significant interactions are present then it does not make much sense to test for row and column effects. One possibility in this situation is to make tests separately on each of the rc row and column categories.

We might alternatively decide to ignore interaction effects completely, either because we have tested for them and shown them to be non-significant, or because $n = 1$, or because for various reasons we may have no $\boldsymbol{\eta}_{ij}$ term in our model. In such cases we work on the error matrix $\mathbf{E}$ instead of $\mathbf{W}$ and our test statistic for column effects is

$$|\mathbf{E}|/|\mathbf{E}+\mathbf{C}| \sim \Lambda(p, \nu_1, \nu_2),$$

where

$$\nu_1 = rcn - r - c + 1, \qquad \nu_2 = c-1. \tag{12.7.6}$$

We may alternatively look at the largest eigenvalue θ of $\mathbf{C}(\mathbf{E}+\mathbf{C})^{-1}$ which is distributed as $\theta(p, \nu_1, \nu_2)$ with the values of ν_1 and ν_2 given by (12.7.6).

Example 12.7.1 (Morrison, 1976, p. 190) We wish to compare the weight losses of male and female rats ($r = 2$ sexes) under $c = 3$ drugs where $n = 4$ rats of each sex are assigned at random to each drug. Weight losses are observed for the first and second weeks ($p = 2$) and the data is given in Table 12.7.1. We wish to compare the effects of the drugs, the effect of sex, and whether there is any interaction.

We first test for interaction. We construct the MANOVA table (Table 12.7.2), using the totals in Table 12.7.1. From Table 12.7.2, we find that

$$|\mathbf{W}| = 4920.75, \qquad |\mathbf{W}+\mathbf{I}| = 6281.42.$$

Hence,

$$\Lambda = |\mathbf{W}|/|\mathbf{W}+\mathbf{I}| = 0.7834 \sim \Lambda(2, 18, 2).$$

From (12.3.7),

$$(17/2)(1-\sqrt{0.7834})/\sqrt{0.7834} = 1.10 \sim F_{4,34}.$$

Table 12.7.1 Weight losses (in grams) for the first and second weeks for rats of each sex under drugs A, B, and C (Morrison, 1976, p. 190)

Sex	Drug A	Drug B	Drug C	Row sums
Male	(5, 6)	(7, 6)	(21, 15)	(33, 27)
	(5, 4)	(7, 7)	(14, 11)	(26, 22)
	(9, 9)	(9, 12)	(17, 12)	(35, 33)
	(7, 6)	(6, 8)	(12, 10)	(25, 24)
Column sums	(26, 25)	(29, 33)	(64, 48)	(119, 106)
Female	(7, 10)	(10, 13)	(16, 12)	(33, 35)
	(6, 6)	(8, 7)	(14, 9)	(28, 22)
	(9, 7)	(7, 6)	(14, 8)	(30, 21)
	(8, 10)	(6, 9)	(10, 5)	(24, 24)
Column sums	(30, 33)	(31, 35)	(54, 34)	(115, 102)
Treatment sums	(56, 58)	(60, 68)	(118, 82)	
Grand total		(234, 208)		

This is clearly not significant so we conclude there are no interactions and proceed to test for main effects.

First, for drugs, $|\mathbf{W}+\mathbf{C}| = 29\,180.83$. Therefore

$$\Lambda = 4920.75/(29\,180.83) = 0.1686 \sim \Lambda(2, 18, 2).$$

From (12.3.7),

$$(17/2)(1-\sqrt{0.1686})/\sqrt{0.1686} = 12.20 \sim F_{4,34}.$$

Table 12.7.2 MANOVA table for the data in Table 12.7.1

Source	d.f.	SSP matrix **A** a_{11}	a_{12}	a_{22}
Sex (**R**)	1	0.667	0.667	0.667
Drugs (**C**)	2	301.0	97.5	36.333
Interaction (**I**)	2	14.333	21.333	32.333
Residual (**W**)	18	94.5	76.5	114.0
Total (**T**)	23	410.5	196.0	183.333

This is significant at 0.1%, so we conclude that there are very highly significant differences between drugs.

Finally, for sex, $|\mathbf{W}+\mathbf{R}|=4957.75$. Thus

$$\Lambda = 4920.75/4957.75 = 0.9925 \sim \Lambda(2, 18, 1).$$

Again from (3.7.10), the observed value of $F_{2,34}$ is 0.06. This is not significant so we conclude there are no differences in weight loss between the sexes.

Simultaneous confidence regions

We begin by requiring the interaction parameters $\boldsymbol{\eta}_{ij}$ in the two-way model to vanish, otherwise comparisons among row and column treatments are meaningless. If so, then, in a similar manner to Section 12.6, the $100(1-\alpha)\%$ simultaneous confidence intervals for linear compounds of the differences of the i_1th and i_2th row effects are given by

$$\mathbf{a}'(\boldsymbol{\alpha}_{i_1}-\boldsymbol{\alpha}_{i_2}) \in \mathbf{a}'(\bar{\mathbf{x}}_{i_1\cdot\cdot}-\bar{\mathbf{x}}_{i_2\cdot\cdot}) \pm \left\{\frac{2\theta_\alpha}{cn(1-\theta_\alpha)}\mathbf{a}'\mathbf{W}\mathbf{a}\right\}^{1/2}, \tag{12.7.7}$$

where θ_α is obtained from Appendix C, Table C.4, with $\nu_1 = rc(n-1)$, $\nu_2 = r-1$, and $\mathbf{a}$ is any vector.

Similarly, for the columns

$$\mathbf{a}'(\boldsymbol{\tau}_{j_1}-\boldsymbol{\tau}_{j_2}) \in \mathbf{a}'(\bar{\mathbf{x}}_{\cdot j_1\cdot}-\bar{\mathbf{x}}_{\cdot j_2\cdot}) \pm \left\{\frac{2\theta_\alpha}{rn(1-\theta_\alpha)}\mathbf{a}'\mathbf{W}\mathbf{a}\right\}^{1/2}, \tag{12.7.8}$$

where θ_α is again obtained from tables with r and c interchanged in ν_1 and ν_2 above.

Example 12.7.2 (Morrison, 1976, p. 190) For the drug data from Table 12.7.2,

$$\mathbf{W} = \begin{bmatrix} 94.5, & 76.5 \\ 76.5, & 114.0 \end{bmatrix}.$$

From Appendix C, Table C.4, it is found for $p=2$, $v_1=18$, $v_2=2$, and $\alpha=0.01$ that $\theta_\alpha=0.502$. Taking $\mathbf{a}'=(1,0)$, for the first week, from (12.7.8), the 99% simultaneous confidence intervals for the B–A, C–B, and C–A differences are

$$-4.36 \le \tau_{B1}-\tau_{A1} \le 5.36, \qquad 2.39 \le \tau_{C1}-\tau_{B1} \le 12.11$$
$$2.89 \le \tau_{C1}-\tau_{A1} \le 12.61,$$

where we have used the above value of $\mathbf{W}$ and drug means (totals) from Table 12.7.1. Hence at the 1% significance level, the drugs A and B are not different with respect to their effects on weight during the first week of the trial. However, the effect of drug C is different from drugs A and B.

Extension to higher designs

By following the same principle of partitioning the total SSP matrix, we are able by analogy with univariate work to analyse many more complex and higher order designs. We refer to Bock (1975) for further work.

Exercises and Complements

12.3.1 (a) Prove the multivariate analysis of variance identity (12.3.4) after writing

$$\mathbf{x}_{ij}-\bar{\mathbf{x}}=(\mathbf{x}_{ij}-\bar{\mathbf{x}}_j)+(\bar{\mathbf{x}}_j-\bar{\mathbf{x}}).$$

(b) Show from Section 5.3.3a, that under H_0,

$$\mathbf{W}\sim W_p(\mathbf{\Sigma}, n-k), \qquad \mathbf{B}\sim W_p(\mathbf{\Sigma}, k-1).$$

Further $\mathbf{W}$ and $\mathbf{B}$ are independent.

12.3.2 (See, for example, Kshirsagar, 1972, p. 345) Under $H_1:\boldsymbol{\mu}_i\neq\boldsymbol{\mu}_j$ for some $i\neq j$, show that

(a) $E(\mathbf{W})=(n-k)\mathbf{\Sigma}$,

(b) $E(\mathbf{B})=(k-1)\mathbf{\Sigma}+\mathbf{\Delta}$

where

$$\mathbf{\Delta}=\sum_{j=1}^{k} n_j(\boldsymbol{\mu}_j-\bar{\boldsymbol{\mu}})(\boldsymbol{\mu}_j-\bar{\boldsymbol{\mu}})', \qquad \bar{\boldsymbol{\mu}}=\sum_{j=1}^{k}\frac{n_j\boldsymbol{\mu}_j}{n}.$$

12.3.3 The data considered in Example 12.3.1 is a subset of Reeve's data (1941). A fuller data leads to the following information on skulls at six localities:

Locality	Subspecies	Sample size	Mean vector, $\bar{\mathbf{x}}_i'$
Sta. Mata, Colombia	*instabilis*	21	(2.054, 2.066, 1.621)
Minas Geraes, Brazil	*chapadensis*	6	(2.097, 2.100, 1.625)
Matto Grosso, Brazil	*chapadensis*	9	(2.091, 2.095, 1.624)
Sta. Cruz, Bolivia	*chapadensis*	3	(2.099, 2.102, 1.643)
Panama	*chiriquensis*	4	(2.092, 2.110, 1.703)
Mexico	*mexicana*	5	(2.099, 2.107, 1.671)
Total		48	(2.077, 2.086, 1.636)

Show that the "between-groups" SSP matrix is

$$\mathbf{B}=\begin{bmatrix} 0.020\,021\,1, & 0.017\,444\,8, & 0.013\,081\,1 \\ & 0.015\,851\,7, & 0.015\,066\,5 \\ & & 0.030\,681\,8 \end{bmatrix}.$$

Assuming the "within-groups" SSP matrix to be

$$\mathbf{W}=\begin{bmatrix} 0.013\,630\,9, & 0.012\,769\,1, & 0.016\,437\,9 \\ & 0.012\,922\,7, & 0.017\,135\,5 \\ & & 0.036\,151\,9 \end{bmatrix}$$

show that

$$\lambda_1=2.4001, \qquad \lambda_2=0.9050, \qquad \lambda_3=0.0515.$$

Hence show that there are differences between the mean vectors of the six groups.

12.3.4 Measurements are taken on the head lengths u_1, head breadths u_2, and weights u_3, of 140 schoolboys of almost the same age belonging to six different schools in an Indian city. The between-schools and total SSP matrices were

$$\mathbf{B}=\begin{bmatrix} 752.0 & 214.2 & 521.3 \\ & 151.3 & 401.2 \\ & & 1612.7 \end{bmatrix}$$

and

$$\mathbf{T}=\begin{bmatrix} 13\,561.3 & 1217.9 & 3192.5 \\ & 1650.9 & 4524.8 \\ & & 22\,622.3 \end{bmatrix}.$$

Show that $|\mathbf{T}|=213\,629\,309\,844$. Obtain $\mathbf{W}=\mathbf{T}-\mathbf{B}$. Hence or otherwise, show that $|\mathbf{W}|=176\,005\,396\,253$. Consequently $|\mathbf{W}|/|\mathbf{T}|=0.823\,88$, which is distributed as $\Lambda(3, 134, 5)$. Show that Bartlett's approximation (3.7.11) gives $26.06\sim\chi^2_{15}$. Hence conclude at the 5% level of significance that there are differences between schools.

12.4.1 Let $\mathbf{A}=(a_{ij})$ be a known $(k\times r)$ matrix such that $\mathbf{A}'\mathbf{D}^{-1}\mathbf{A}=\mathbf{I}_r$, where $\mathbf{D}=\text{diag}\,(n_1,\ldots,n_k)$. Suppose we wish to test $H_0:\mathbf{A}'\boldsymbol{\mu}=\mathbf{0}$, $\boldsymbol{\mu}'=(\boldsymbol{\mu}_1,\ldots,\boldsymbol{\mu}_k)$ against $H_1:\boldsymbol{\mu}_i\neq\boldsymbol{\mu}_j$, $i\neq j$. Following the method of Section 12.4 show, under H_0, that

$$\hat{\boldsymbol{\mu}}_j=\bar{\mathbf{x}}_j-n_j^{-1}\sum_{i=1}^{r} a_{ji}\boldsymbol{\Sigma}\boldsymbol{\lambda}_i,$$

where

$$\mathbf{\Sigma\lambda}_i = \sum_{j=1}^{k} a_{ji}\bar{\mathbf{x}}_j.$$

Hence prove that

$$n\hat{\mathbf{\Sigma}} = \mathbf{W} + \sum_{i=1}^{r} \left(\sum_{j=1}^{k} a_{ji}\bar{\mathbf{x}}_j\right)\left(\sum_{j=1}^{k} a_{ji}\bar{\mathbf{x}}_j'\right)$$

and derive the likelihood ratio test.

12.5.1 For the shrew data of Delany and Healy (1966) referred to in Example 12.5.2 we have, from a total sample of 399,

$$\mathbf{B} = \begin{bmatrix} 208.02 & 88.56 & 101.39 & 41.16 & 10.06 & 98.35 & 57.46 & 33.01 & 55.12 & 63.99 \\ & 39.25 & 42.14 & 17.20 & 4.16 & 41.90 & 24.35 & 14.16 & 22.66 & 27.32 \\ & & 50.26 & 20.23 & 4.98 & 47.75 & 28.04 & 16.04 & 27.54 & 31.14 \\ & & & 8.76 & 2.14 & 19.33 & 11.66 & 6.41 & 11.03 & 12.88 \\ & & & & 0.78 & 4.73 & 2.73 & 1.45 & 3.05 & 3.13 \\ & & & & & 47.47 & 27.97 & 15.76 & 25.92 & 30.10 \\ & & & & & & 17.48 & 9.47 & 15.11 & 17.68 \\ & & & & & & & 5.44 & 8.58 & 10.06 \\ & & & & & & & & 15.59 & 16.88 \\ & & & & & & & & & 19.87 \end{bmatrix},$$

$$\mathbf{W} = \begin{bmatrix} 57.06 & 21.68 & 21.29 & 4.29 & 2.68 & 22.31 & 2.87 & 3.14 & 11.78 & 14.54 \\ & 20.76 & 6.08 & 2.65 & 1.78 & 8.63 & 0.73 & 1.06 & 3.66 & 6.91 \\ & & 13.75 & 0.98 & 0.41 & 7.16 & 1.14 & 1.25 & 6.20 & 5.35 \\ & & & 1.59 & 0.68 & 2.06 & 0.28 & 0.21 & 0.62 & 1.49 \\ & & & & 1.50 & 1.53 & 0.36 & 0.13 & 0.22 & 0.94 \\ & & & & & 14.34 & 1.38 & 1.21 & 4.39 & 5.92 \\ & & & & & & 2.13 & 0.27 & 0.72 & 0.77 \\ & & & & & & & 0.93 & 0.73 & 0.74 \\ & & & & & & & & 4.97 & 3.16 \\ & & & & & & & & & 5.79 \end{bmatrix}.$$

The eigenvalues of $\mathbf{W}^{-1}\mathbf{B}$ are

$$\lambda_1 = 15.98, \quad \lambda_2 = 0.99, \quad \lambda_3 = 0.48, \quad \lambda_4 = 0.36, \quad \lambda_5 = 0.15,$$
$$\lambda_6 = 0.10, \quad \lambda_7 = 0.05, \quad \lambda_8 = 0.01, \quad \lambda_9 = 0.0006, \quad \lambda_{10} = 0$$

and the first two canonical variables are as given in Example 12.5.2.

Show that $D_0^2 = 1751.2$, $D_1^2 = 652.4$, $D_2^2 = 385.6$, $D_3^2 = 233.5$, $D_4^2 = 114.2$, $D_5^2 = 60.0$, $D_6^2 = 23.0$, $D_7^2 = 4.1$, $D_8^2 = 0.2$. Hence show that the dimension of the data should be six if we assume the approximation (12.5.29) for the distribution of D_r^2, and use a test size $\alpha = 0.01$. Comment why this approximation could be inadequate.

12.7.1 For the drug data of Example 12.7.1, obtain the 99% simultaneous confidence intervals for the drug differences B–A, C–B, and C–A for the second week. Show that the pair-wise differences between the drugs for the second week are not significant at the 1% level. Comment on whether the significance of the original MANOVA could be due to the effect of drug C in the first week.

13
Cluster Analysis

13.1 Introduction

13.1.1 Single-sample case

If instead of the categorized iris data of Table 1.2.2, we were presented with the 150 observations in an unclassified manner, then the aim might have been to group the data into homogeneous classes or *clusters*. Such would have been the goal before the species of iris were established. For illustrative purposes, Table 13.2.1 gives a small mixed sub-sample from the iris data and the aim is to divide it into groups.

This kind of problem appears in any new investigation where one wishes to establish not only the identity but the affinity of new specimens. Before Darwin, the evolutionary tree was such an unsolved problem. In general we can summarize the problem of cluster analysis as follows.

Single-sample problem Let $\mathbf{x}_1, \ldots, \mathbf{x}_n$ be measurements of p variables on each of n objects which are believed to be heterogeneous. Then the aim of cluster analysis is to group these objects into g homogeneous classes where g is also unknown (but usually assumed to be much smaller than n).

Some points should be noted.

We call a group "homogeneous" if its members are close to each other but the members of that group differ considerably from those of another. This leads to the idea of setting up a metric between the points to quantify the notion of "nearness". For various possible choices of distance, see Section 13.4.

The techniques are usually applied in two rather different situations. In one case, the purpose of the analysis is purely descriptive. There are no assumptions about the form of the underlying population and the clustering is simply a useful condensation of the data. In other cases, there is an underlying model where each observation in the sample may arise from any one of a small number of different distributions (see Section 13.2.1). We will pursue both aspects.

The term "clustering" is taken to be synonymous to "numerical taxonomy", "classification". Other terms used in this context are Q-analysis, typology, pattern recognition, and clumping.

13.1.2 Multi-sample case

Another context where the problem of cluster analysis arises is when we are given a collection of *samples*, and the aim is to group the samples into homogeneous groups. The problem can be better understood through a specific situation in zoology already described in Example 12.5.2. It is known that white-toothed shrews of the known genus *Crocidura* occur in the Channel and Scilly Islands of the British Isles and the French mainland. A large number of observations from 10 localities in the Channel and Scilly Islands were obtained to examine the belief that there may be two species of *Crocidura*. The localities were geographically close, but it is assumed that only one sub-species was present in any one place. Thus the problem here is to group "samples" rather than "objects," as in the single-sample case. In general, the problem can be summarized as follows.

Multi-sample problem Let $\mathbf{x}_{ij}$, $i = 1, \ldots, n_j$, be the observations in the jth (random) sample, $j = 1, 2, \ldots, m$. The aim of cluster analysis is to group the m samples into g homogeneous classes where g is unknown, $g \leq m$.

13.2 A Probabilistic Formulation

13.2.1 Single-sample case

General case

Let us assume that $\mathbf{x}_1, \ldots, \mathbf{x}_n$ are independent. Further, each may arise from any one of g possible sub-populations with p.d.f. $f(\mathbf{x}; \boldsymbol{\theta}_k)$, $k = 1, \ldots, g$, where we assume g is known. These assumptions would be the same as for the standard discrimination problem (see Chapter 11) if it were known from which sub-population each $\mathbf{x}_i$ came.

Let $\boldsymbol{\gamma} = (\gamma_1, \ldots, \gamma_n)'$ be a set of identifying labels so that $\gamma_i = k \Rightarrow \mathbf{x}_j$ comes from kth sub-population, $i = 1, \ldots, n$; $k = 1, \ldots, g$. Suppose that

C_k = the set of $\mathbf{x}_i$ assigned to the kth group by $\boldsymbol{\gamma}$, $k = 1, \ldots, g$. The likelihood function is

$$L(\boldsymbol{\gamma}; \boldsymbol{\theta}_1, \ldots, \boldsymbol{\theta}_g) = \prod_{\mathbf{x} \in C_1} f(\mathbf{x}; \boldsymbol{\theta}_1) \cdots \prod_{\mathbf{x} \in C_g} f(\mathbf{x}; \boldsymbol{\theta}_g). \tag{13.2.1}$$

We can show that the maximum likelihood method possesses an important allocation propertv. Let $\hat{\boldsymbol{\gamma}}, \hat{\boldsymbol{\theta}}_1, \ldots, \hat{\boldsymbol{\theta}}_g$ be the m.l.e.s of $\boldsymbol{\gamma}$ and the $\boldsymbol{\theta}$s respectively. Let $\hat{C}_1, \ldots, \hat{C}_k$ be the partition under $\hat{\boldsymbol{\gamma}}$. Since moving a sample point from $\hat{C}_k$ to $\hat{C}_l$ will reduce the likelihood, we have

$$L(\hat{\boldsymbol{\gamma}}; \hat{\boldsymbol{\theta}}_1, \ldots, \hat{\boldsymbol{\theta}}_g) f(\mathbf{x}; \hat{\boldsymbol{\theta}}_l)/f(\mathbf{x}; \hat{\boldsymbol{\theta}}_k) \leq L(\hat{\boldsymbol{\gamma}}; \hat{\boldsymbol{\theta}}_1, \ldots, \hat{\boldsymbol{\theta}}_g).$$

Thus

$$f(\mathbf{x}; \hat{\boldsymbol{\theta}}_l) \leq f(\mathbf{x}; \hat{\boldsymbol{\theta}}_k) \quad \text{for} \quad \mathbf{x} \in \hat{C}_k, \qquad l \neq k, \qquad l = 1, \ldots, g. \tag{13.2.2}$$

This is quite a familiar allocation rule *per se* (see Chapter 11).

The normal case

Let us assume that $f(\mathbf{x}; \boldsymbol{\theta}_k)$ denotes the p.d.f. of $N_p(\boldsymbol{\mu}_k, \boldsymbol{\Sigma}_k)$, $k = 1, \ldots, g$. Then the log likelihood function is

$$l(\boldsymbol{\gamma}; \boldsymbol{\theta}) = \text{const} - \frac{1}{2} \sum_{k=1}^{g} \sum_{\mathbf{x}_i \in C_k} (\mathbf{x}_i - \boldsymbol{\mu}_k)' \boldsymbol{\Sigma}_k^{-1} (\mathbf{x}_i - \boldsymbol{\mu}_k) - \frac{1}{2} \sum_{k=1}^{g} n_k \log |\boldsymbol{\Sigma}_k|, \tag{13.2.3}$$

where there are n_k observations in C_k. Hence, for a given $\boldsymbol{\gamma}$, the likelihood is maximized by the ordinary m.l.e. of $\boldsymbol{\mu}$ and $\boldsymbol{\Sigma}$, i.e.

$$\hat{\boldsymbol{\mu}}_k(\boldsymbol{\gamma}) = \bar{\mathbf{x}}_k, \qquad \hat{\boldsymbol{\Sigma}}_k(\boldsymbol{\gamma}) = \mathbf{S}_k, \tag{13.2.4}$$

where $\bar{\mathbf{x}}_k$ is the mean and $\mathbf{S}_k$ is the covariance matrix of the n_k observations in C_k. On substituting (13.2.4) in (13.2.3), we obtain

$$l(\boldsymbol{\gamma}; \hat{\boldsymbol{\theta}}(\boldsymbol{\gamma})) = \text{const} - \tfrac{1}{2} \sum n_k \log |\mathbf{S}_k|.$$

Hence the m.l.e. of $\boldsymbol{\gamma}$ is the grouping that minimizes

$$\prod_{k=1}^{g} |\mathbf{S}_k|^{n_k}. \tag{13.2.5}$$

To avoid the degenerate case of infinite likelihood, we assume that there are at least $p+1$ observations assigned to each group so that $n_k \geq p+1$ and $n \geq g(p+1)$.

If, however, we assume that the groups have identical covariance matrices, $\boldsymbol{\Sigma}_1 = \ldots = \boldsymbol{\Sigma}_g = \boldsymbol{\Sigma}$ (unknown), the same method leads to the grouping that minimizes

$$|\mathbf{W}|, \tag{13.2.6}$$

where

$$\mathbf{W} = \sum_{k=1}^{g} \sum_{C_k} (\mathbf{x}_i - \bar{\mathbf{x}}_k)(\mathbf{x}_i - \bar{\mathbf{x}}_k)',$$

is the pooled within-groups sums of squares and products (SSP) matrix, and the second summation is taken over $\mathbf{x}_i \in C_k$.

Example 13.2.1 In the iris data, suppose that we were given the table unclassified according to varieties. For our illustrative purpose, let us assume from Table 1.2.2 that a sub-sample was given as in Table 13.2.1.

In Table 13.2.1, the first three observations are the first three readings from *I. versicolour*, whereas the last three observations are the first three readings from *I. virginica*, both on the first two variables only. To satisfy $n \geqslant 3g$, we shall take $g = 2$, with three observations in each group. For all possible partitions, Table 13.2.2 gives the values of $|\mathbf{S}_1|$, $|\mathbf{S}_2|$, and $|\mathbf{W}| = |\mathbf{S}_1 + \mathbf{S}_2|$. Both ML methods lead to the clusters (1, 3, 6) and (2, 4, 5). Note that under the correct grouping (1, 2, 3) and (4, 5, 6), the values of $|\mathbf{S}_1| = 0.0012$, $|\mathbf{S}_2| = 0.1323$ are disparate. From a scatter diagram of these points given in Figure 13.2.1 we can understand the ML grouping better.

Table 13.2.1 A sub-sample of iris data

	1	2	3	4	5	6
Sepal length	7.0	6.4	6.9	6.3	5.8	7.1
Sepal width	3.2	3.2	3.1	3.3	2.7	3.0

Table 13.2.2 Clusters for the data in Table 13.2.1

Group 1	Group 2	$\lvert\mathbf{S}_1\rvert$	$\lvert\mathbf{S}_2\rvert$	$10^3\lvert\mathbf{S}_1\rvert\,\lvert\mathbf{S}_2\rvert$	$\lvert\mathbf{S}_1+\mathbf{S}_2\rvert$
1 2 3	4 5 6	0.001 200	0.132 300	0.1588	0.180
1 2 4	2 5 6	0.001 200	0.073 633	0.0884	0.127
1 2 5	3 4 6	0.030 000	0.000 133	0.0040	0.198
1 2 6	3 4 5	0.004 800	0.070 533	0.3386	0.167
1 3 4	2 5 6	0.002 133	0.073 633	0.1571	0.151
1 3 5	2 4 6	0.001 633	0.000 833	0.0014	0.188
1 3 6	2 4 5	0.000 300	0.004 033	0.0012	0.017
1 4 5	2 3 6	0.073 633	0.000 300	0.0221	0.181
1 4 6	2 3 5	0.005 633	0.032 033	0.1805	0.170
1 5 6	2 3 4	0.028 033	0.000 533	0.0150	0.114

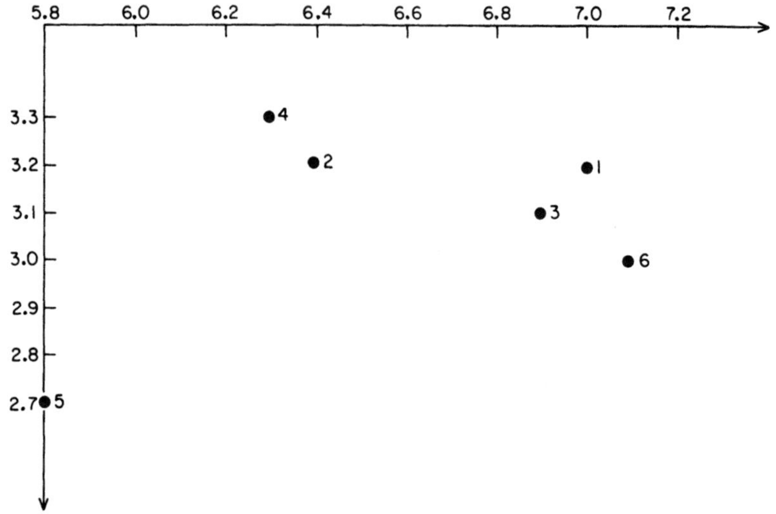

Figure 13.2.1 Scatter diagram for the iris data in Table 13.2.1.

It is known for this data (see Example 11.3.2) that there is considerable overlap between the varieties *I. versicolour* and *I. virginica*, but these are clearly separated from *I. setosa*.

A test for a single cluster

Before starting clustering, we may wish to look into whether the data is unstructured, that is, whether there is just one cluster. Hence, the hypothesis of interest is

$$H_0 : \gamma_1 = \cdots = \gamma_n$$

against the alternative that not all the γ_i are equal. Assuming normality with equal covariance matrices, we get

$$-2 \log \lambda = n \log \{\max_{\boldsymbol{\gamma}} (|\mathbf{T}|/|\mathbf{W}|)\}, \qquad (13.2.7)$$

where

$$\mathbf{T} = \sum_{i=1}^{n} (\mathbf{x}_i - \bar{\mathbf{x}})(\mathbf{x}_i - \bar{\mathbf{x}})'$$

is the total sums of squares and products matrix. Thus if the number of groups g is unknown, we need to minimize $|\mathbf{W}|$ over all permissible partitions of $\boldsymbol{\gamma}$ (those partitions for which $g \leq n - p$, so that rank $(\mathbf{W}) = p$). Unfortunately, even for large n, the distribution of (13.2.7) is not known.

(However, some progress has been made in the case $p=1$; see Hartigan, 1978.) A more tractable test is given below in (13.2.9).

It may also be worthwhile to check that the data is not spherically symmetric. One method would be to apply the test of uniformity on the directions $\mathbf{y}_i/\|\mathbf{y}_i\|$ given in Chapter 15, where $\mathbf{y}_i=\mathbf{S}^{-1/2}(\mathbf{x}_i-\bar{\mathbf{x}})$ is the Mahalanobis transformation of Section 1.5.2.

The criterion of minimizing $|\mathbf{W}|$ was first put forward on an intuitive basis by Friedman and Rubin (1967), whereas the maximum likelihood justification was given by Scott and Symons (1971). An alternative procedure based on minimizing $\operatorname{tr}(\mathbf{W})$ was developed intuitively by Edwards and Cavalli-Sforza (1965), and this idea can also be given a maximum likelihood interpretation (see Exercise 13.2.1).

Remarks (1) Let $\mathbf{y}_j=\mathbf{A}\mathbf{x}_j+\mathbf{b}$, where $\mathbf{A}$ is a non-singular matrix. Then (13.2.5) and (13.2.6) can be written as

$$|\mathbf{A}|^{-2n}\prod_{k=1}^{g}|\mathbf{S}_{k,y}|^{n_k}, \qquad |\mathbf{A}|^{-2}\,|\mathbf{W}_y|,$$

respectively, where $\mathbf{S}_{k,y}$ and $\mathbf{W}_y$ denote $\mathbf{S}_k$ and $\mathbf{W}$ for the transformed variables $\mathbf{y}_i$. Hence, in particular, as far as minimization of these criteria is concerned, the scale of the variables is immaterial.

(2) The task of minimizing (13.2.5) or (13.2.6) is formidable even for a computer. For example, if $g=2$, we need to look at 2^{n-1} combinations in general. In practice, to circumvent this problem, a relative minimum is used which has the property that any reassignment of one or two observations results in a larger value, although it may not be an absolute minimum. It is worthwhile to examine individually the partitions in the neighbourhood of the (estimated) true split.

(3) *Estimation of g* We have assumed g to be preassigned. In practice, if g is allowed to vary, the maximum likelihood methods will always partition the data into the maximum number of partitions allowed. Thus, g must be chosen by some other method. For equal covariance matrices, Marriott (1971) has suggested taking the correct number of groups to be the value of g for which

$$g^2\,|\mathbf{W}| \tag{13.2.8}$$

is minimum. Everitt (1974, pp. 59–60, 92) from his study of various criteria finds (13.2.8) to be the most useful. The most suitable algorithms for implementing the optimization of the criterion (13.2.8) are those of Friedman and Rubin (1967) and McRae (1971). (See also Hartigan, 1975.) A rule of thumb is to use $g\sim(n/2)^{1/2}$.

(4) *Mixtures* The problem of cluster analysis can also be studied using

a mixture of p.d.f.s. If we assume that each $\mathbf{x}_i$ has probability p_k of coming from the kth population $k = 1, \ldots, g$, then $\mathbf{x}_1, \ldots, \mathbf{x}_n$ is a sample from

$$f(\mathbf{x}) = \sum_{k=1}^{g} p_k f(\mathbf{x}; \boldsymbol{\mu}_k, \boldsymbol{\Sigma}_k),$$

where $f(\cdot\,; \boldsymbol{\mu}, \boldsymbol{\Sigma})$ is the p.d.f. of $N_p(\boldsymbol{\mu}, \boldsymbol{\Sigma})$, $\sum p_k = 1$.

For the maximum likelihood estimates of p_k, $\boldsymbol{\mu}_k$, *and* $\boldsymbol{\Sigma}_k$ when $k = 2$, see Exercise 13.2.3. Once these estimates are known, we can regard each distribution as indicating a separate group, and individuals are then assigned by the Bayes allocation rule of Section 11.2.2, i.e. assign $\mathbf{x}_i$ to the kth distribution when

$$\hat{p}_l f(\mathbf{x}_i; \hat{\boldsymbol{\mu}}_l, \hat{\boldsymbol{\Sigma}}_l) \leq \hat{p}_k f(\mathbf{x}_i; \hat{\boldsymbol{\mu}}_k, \hat{\boldsymbol{\Sigma}}_k) \quad \text{for all} \quad l \neq k.$$

For this model it is possible to carry out asymptotic hypothesis tests. Let $\lambda = L_g/L_{g'}$ where λ is the ratio of the likelihood of g groups against that of g' groups $(g < g')$. From a Monte Carlo investigation, Wolfe (1971) recommends the approximation

$$-\frac{2}{n}(n - 1 - p - \tfrac{1}{2}g') \log \lambda \sim \chi_f^2, \qquad f = 2p(g' - g), \tag{13.2.9}$$

for the distribution of λ under the null hypothesis that there are g groups.

Note that this mixture model is equivalent to the first model of this section with the additional assumption that $\boldsymbol{\gamma}$ is an (unobservable) random variable whose components are the outcomes of n independent multinomial trials. Scott and Symons (1971) give a Bayesian approach in which all the parameters are random variables.

(5) *Criticisms* These techniques usually require large amounts of computer time, and cannot be recommended for use with large data sets. In many applications the number of parameters to be estimated increases indefinitely with the sample size and therefore the estimates are not "consistent" (Marriott, 1971). Further, even if g is fixed and the data consists of a sample from a mixture of unimodal distributions, the groups will be the truncated centres of these distributions, mixed with tails of other distributions. Thus even for $\boldsymbol{\Sigma}_1 = \ldots = \boldsymbol{\Sigma}_g = \boldsymbol{\Sigma}$, $\mathbf{W}$ will not be a consistent estimate of $\boldsymbol{\Sigma}$. Further, in this case the criterion (13.2.6) has a tendency to partition the sample into groups of about the same size even when the true clusters are of unequal sizes (Scott and Symons, 1971). When the modes are near together and the distributions overlap considerably, separation may be impossible even for very large sample sizes.

For our discussion we have assumed $\mathbf{x}_1, \ldots, \mathbf{x}_n$ to be independent random variables, but this may not be true in some cases, e.g. when

$\mathbf{x}_1, \ldots, \mathbf{x}_n$ are measurements of the various stages of evolution of a particular creature.

13.2.2 Multi-sample case

For the multi-sample problem described in Section 13.1.2, we shall limit our discussion to normal samples with equal covariance matrices. Each sample can be summarized by its mean $\bar{\mathbf{x}}_i$ and SSP matrix $\mathbf{W}_i$. Then $\bar{\mathbf{x}}_i \sim N_p(\boldsymbol{\mu}_i, n_i^{-1}\boldsymbol{\Sigma})$, independently for $i = 1, \ldots, m$, and a consistent estimate of $\boldsymbol{\Sigma}$ is given by

$$\hat{\boldsymbol{\Sigma}} = \mathbf{W}/(n-m), \tag{13.2.10}$$

where, in the notation of Chapter 12, $\mathbf{W} = \mathbf{W}_1 + \cdots + \mathbf{W}_m$ is the within-samples SSP matrix and $n = n_1 + \cdots + n_m$. Assuming n to be large, we shall now suppose $\boldsymbol{\Sigma}$ is known.

Suppose these m samples have come from only g populations ($g \leq m$), i.e.

$$\boldsymbol{\mu}_i \in (\boldsymbol{\nu}_1, \ldots, \boldsymbol{\nu}_g), \qquad g \leq m.$$

Thus, there are only g distinct means out of $\boldsymbol{\mu}_1, \ldots, \boldsymbol{\mu}_m$. Let $\boldsymbol{\beta}$ be the set of identifying labels, and let C_k be the set of $\bar{\mathbf{x}}_j$ assigned to the kth group by $\boldsymbol{\beta}$, $k = 1, \ldots, g$. Now, maximizing the likelihood implies minimizing

$$\sum_{k=1}^{g} \sum_{\mathbf{x}_j \in C_k} n_j(\bar{\mathbf{x}}_j - \boldsymbol{\nu}_k)'\boldsymbol{\Sigma}^{-1}(\bar{\mathbf{x}}_j - \boldsymbol{\nu}_k). \tag{13.2.11}$$

Let

$$N_k = \sum_{\mathbf{x}_j \in C_k} n_j, \qquad \bar{\bar{\mathbf{x}}}_k = \sum_{\bar{\mathbf{x}}_j \in C_k} \frac{n_j \bar{\mathbf{x}}_j}{N_k}$$

so that $\bar{\bar{\mathbf{x}}}_k$ is the weighted mean vector of the means in the kth group. From (13.2.11), $\hat{\boldsymbol{\nu}}_k = \bar{\bar{\mathbf{x}}}_k$, so that the m.l.e. of $\boldsymbol{\beta}$ is the grouping that minimizes

$$w_g^2 = \sum_{k=1}^{g} \sum_{\bar{\mathbf{x}}_j \in C_k} n_j(\bar{\mathbf{x}}_j - \bar{\bar{\mathbf{x}}}_k)'\boldsymbol{\Sigma}^{-1}(\bar{\mathbf{x}}_j - \bar{\bar{\mathbf{x}}}_k). \tag{13.2.12}$$

It is equivalent to maximizing

$$b_g^2 = \sum_{k=1}^{g} N_k(\bar{\bar{\mathbf{x}}}_k - \bar{\mathbf{x}})'\boldsymbol{\Sigma}^{-1}(\bar{\bar{\mathbf{x}}}_k - \bar{\mathbf{x}}) \tag{13.2.13}$$

since the sum of (13.2.12) and (13.2.13) is fixed. Note that $\bar{\mathbf{x}}$ is the mean vector of the pooled sample. Thus we arrive at (13.2.13) as our criterion.

For computational purposes, the following forms of (13.2.12) and (13.2.13) are useful. Let

$$D_{ij}^2=(\bar{\mathbf{x}}_i-\bar{\mathbf{x}}_j)'\mathbf{\Sigma}^{-1}(\bar{\mathbf{x}}_i-\bar{\mathbf{x}}_j), \qquad B_{ij}=(\bar{\mathbf{x}}_i-\bar{\mathbf{x}})'\mathbf{\Sigma}^{-1}(\bar{\mathbf{x}}_j-\bar{\mathbf{x}}) \quad (13.2.14)$$

so that D_{ij} is the Mahalanobis distance between $\bar{\mathbf{x}}_i$ and $\bar{\mathbf{x}}_j$. It can be seen that (13.2.12) and (13.2.13) can be written, respectively, as (see Exercise 13.2.4)

$$w_g^2=\frac{1}{2}\sum_{k=1}^{g} N_k^{-1}\sum_{C_k} n_i n_j D_{ij}^2 \quad (13.2.15)$$

and

$$b_g^2=\sum_{k=1}^{g} N_k^{-1}\sum_{C_k} n_i n_j B_{ij} \quad (13.2.16)$$

where the second summation is extended over $\bar{\mathbf{x}}_i, \bar{\mathbf{x}}_j \in C_k$. If the matrix of Mahalanobis distances D_{ij} is available, then (13.2.15) is computationally convenient. When $\mathbf{\Sigma}$ is unknown, replace it by (13.2.10) in these criteria.

Example 13.2.2 For the shrew data described in Section 13.1.2, the Mahalanobis distances D_{ij} are given in Table 13.2.3, where $\mathbf{\Sigma}$ is estimated from (13.2.10). Here $k=10$, and

$$n_1=144,\quad n_2=16,\quad n_3=12,\quad n_4=7,\quad n_5=90,$$
$$n_6=25,\quad n_7=6,\quad n_8=26,\quad n_9=53,\quad n_{10}=20.$$

We assume *a priori* (say from Figure 12.5.2) that we are looking for two clusters C_1 and C_2 with 3 and 7 elements, respectively. We wish to

Table 13.2.3 Mahalanobis distances D_{ij} between 10 island races of white-toothed shrews (from Delany and Healy, 1966; Gower and Ross, 1969)

		Scilly Islands					Channel Islands				France
		1	2	3	4	5	6	7	8	9	10
Scilly Islands	1. Tresco	0									
	2. Bryher	1.61	0								
	3. St Agnes	1.97	2.02	0							
	4. St Martin's	1.97	2.51	2.88	0						
	5. St Mary's	1.40	1.70	1.35	2.21	0					
Channel Islands	6. Sark	2.45	3.49	3.34	3.83	3.19	0				
	7. Jersey	2.83	3.94	3.64	2.89	3.01	3.00	0			
	8. Alderney	9.58	9.59	10.05	8.78	9.30	9.74	9.23	0		
	9. Guernsey	7.79	7.82	8.43	7.08	7.76	7.86	7.76	2.64	0	
French mainland	10. Cap Gris Nez	7.86	7.92	8.36	7.44	7.79	7.90	8.26	3.38	2.56	0

Table 13.2.4 Value of clustering criterion w_g^2 in ascending order for the shrew data and C_1; C_2 is the complement of C_1

Partition C_1	w_g^2
(8, 9, 10)	643
(7, 8, 9)	1997
(4, 8, 9)	2014
(3, 8, 9)	2457
(1, 5, 6)	2499
.	.
.	.
.	.
(1, 3, 9)	5404
(1, 8, 10)	5446

minimize w_g^2 over 120 partitions. Using values of D_{ij} in Table 13.2.3 with the n_j in (13.2.15) on the computer (we give a brief summary in Table 13.2.4), it is seen that the clusters are (8)–(10) and (1)–(7). Note that the difference between the two smallest values of w_g^2 is large. This fact is seen in Figure 12.5.2, which is a projection onto two dimensions, and the analysis confirms this finding. We defer further discussion until Example 13.3.1.

13.3 Hierarchical Methods

The clustering methods in Section 13.2 can be described as *optimization partitioning techniques* since the clusters are formed by optimizing a clustering criterion. We now consider *hierarchical methods* in which the clusterings into g and $g+1$ groups have the property that (i) they have $g-1$ identical groups and (ii) the remaining single group in the g groups is divided into two in the $g+1$ groups.

By the very nature of these techniques, once an object is allocated to a group, it cannot be reallocated as g decreases, unlike the optimization techniques of Section 13.2. The end product of these techniques is a tree diagram ("dendrogram") (see Section 13.3.1).

These techniques operate on a matrix $\mathbf{D}=(d_{ij})$ of distances between the points $\mathbf{x}_1, \ldots, \mathbf{x}_n$ rather than the points themselves. Possible choices for the distance matrix will be discussed in Section 13.4.

We shall only consider two important hierarchical methods to give the flavour of the area. It should be emphasized that the methods are basically descriptive.

13.3.1 Nearest neighbour single linkage cluster analysis

This method groups the points which are nearest to one another in the following way. First, order the $\frac{1}{2}n(n-1)$ interpoint distances into ascending order.

(a) Let $C_1, \ldots, C_n$ be the starting clusters each containing one point, namely $\mathbf{x}_i \in C_i$.

(b) Without any loss of generality, let $d_{r_1 s_1} = \min_{rs}$ so that $\mathbf{x}_{r_1}$ and $\mathbf{x}_{s_1}$ are nearest. Then these two points are grouped into a cluster, so we have $n-1$ clusters, where $C_{r_1} + C_{s_1}$ is a new cluster.

(c) Let $d_{r_2 s_2}$ be the next smallest distance. If neither r_1 nor s_1 equals r_2 or s_2, the new $n-2$ clusters are $C_{r_1} + C_{s_1}$, $C_{r_2} + C_{s_2}$ plus the remaining old clusters. If $r_2 = r_1$ and $s_1 \neq s_2$ the new $n-2$ clusters are $C_{r_1} + C_{s_1} + C_{s_2}$, plus the remaining old clusters.

(d) The process continues as in (c) through all $\frac{1}{2}n(n-1)$ distances. At the ith stage let $d_{r_i s_i}$ denote the ith smallest distance. Then the cluster containing r_i is joined with the cluster containing s_i. (Note that if r_i and s_i are already in the same cluster, then no new groups are formed in this stage.)

(e) The clustering process can be halted before all the clusters have been joined into one group by stopping when the inter-cluster distances are all greater than d_0, where d_0 is an arbitrary *threshold level.* Let $C_1^*, \ldots, C_g^*$ be the resulting clusters. These clusters have the property that if d_0' $(>d_0)$ is a higher threshold, then two clusters C_j, C_k will be joined at the threshold d_0' if at least one distance d_{rs} (or a single link) exists between r and s with $\mathbf{x}_r \in C_j$, $\mathbf{x}_s \in C_k$, and $d_0 < d_{rs} \leq d_0'$.

This property has led the method to be called *single linkage* cluster analysis. Note that once links are established between objects, they cannot be broken.

Example 13.3.1 In the shrew data, let $d_0 = 3.0$. We can order the distances as shown in column 2 of Table 13.3.1. In the first step, (3, 5) becomes a cluster. Note that at stages 4, 6, 7, and 8 no new clusters result. By the 16th step, we have clusters at the level $d_{6,7} = 3.00$,

$$(1, 2, 3, 4, 5, 6, 7), \qquad (8, 9, 10).$$

Of course, if we carry on further, we get a single cluster.

Table 13.3.1 Single linkage procedure

Order	Distances (ordered)	Clusters
1	$d_{35}=1.35$	(1), (2), (3, 5), (4), (6), (7), (8), (9), (10)
2	$d_{15}=1.40$	(1, 3, 5), (2), (4), (6), (7), (8), (9), (10)
3	$d_{12}=1.61$	(1, 2, 3, 5), (4), (6), (7), (8), (9), (10)
4	$d_{25}=1.70$	(1, 2, 3, 5), (4), (6), (7), (8), (9), (10)†
5‡	$d_{14}=1.969$	(1, 2, 3, 4, 5), (6), (7), (8), (9), (10)
6‡	$d_{13}=1.972$	(1, 2, 3, 4, 5), (6), (7), (8), (9), (10)†
7	$d_{23}=2.02$	(1, 2, 3, 4, 5), (6), (7), (8), (9), (10)†
8	$d_{45}=2.21$	(1, 2, 3, 4, 5), (6), (7), (8), (9), (10)†
9	$d_{16}=2.45$	(1, 2, 3, 4, 5, 6), (7), (8), (9), (10)
10	$d_{24}=2.51$	(1, 2, 3, 4, 5, 6), (7), (8), (9), (10)†
11	$d_{9,10}=2.56$	(1, 2, 3, 4, 5, 6), (7), (8), (9, 10)
12	$d_{89}=2.64$	(1, 2, 3, 4, 5, 6), (7), (8, 9, 10)
13	$d_{17}=2.84$	(1, 2, 3, 4, 5, 6, 7), (8, 9, 10)
14	$d_{34}=2.88$	(1, 2, 3, 4, 5, 6, 7), (8, 9, 10)†
15	$d_{47}=2.89$	(1, 2, 3, 4, 5, 6, 7), (8, 9, 10)†
16	$d_{67}=3.00$	(1, 2, 3, 4, 5, 6, 7), (8, 9, 10)†
·	·	·
·	·	·
·	·	·
45	$d_{49}=7.08$	(1, 2, 3, 4, 5, 6, 7, 8, 9, 10)

† No new clusters.
‡ More accurate values of distances to break the tie.

At $d_0=3.00$, we infer that there are two species, one containing (1)–(7) and the other (8)–(10). In fact, the species containing shrews from Alderney (8), Guernsey (9), and Cap Gris Nez (10) is named *Crocidura russcula* and the other is named *C. snaveolens*. Some of these facts were only appreciated in 1958 (see Delany and Healy, 1966).

Algorithm An alternative but equivalent description of the computation, involving only $n-1$ steps, can be given as follows:

(a) Assuming $d_{12}=\min_{i,j}(d_{ij})\leq d_0$, let $C_2^*, \ldots, C_n^*$ be the groups after joining the pair 1, 2 to form C_2^*.
(b) Define the new $((n-1)\times(n-1))$ distance matrix $\mathbf{D}^*=(d_{ij}^*)$ with $d_{2j}^*=\min(d_{1j}, d_{2j})$ for $j=3, \ldots, n$; $d_{ij}^*=d_{ij}$ for $i, j=3, \ldots, n$. Find $\min(d_{ij}^*)$, $i, j=2, \ldots, n$ and then proceed as in (a).

Note that this algorithm uses the recurrence relation

$$d_{k(ij)}=\min(d_{ik}, d_{jk}) \tag{13.3.1}$$

to define the distance between group k and the group (ij) formed by the fusion of groups i and j.

In hierarchical methods, clustering is generally obtained through two types of algorithm:

(a) *agglomerative*—a successive pooling of subsets of the set of objects;
(b) *divisive*—successive partitions of the set of objects.

In the above discussion, we have used an agglomerative procedure, *but* it should be noted that the single-linkage method can also be implemented by a divisive algorithm, and by an algorithm which belongs to neither category (Jardine, 1970). In general, a method should not be confused with its algorithm. Indeed, Lance and Williams (1967) have given a general agglomerative algorithm with which many of the common hierarchical methods can be described (see Exercise 13.3.1). The importance of a method lies in the fact that the generated clusters have some desired property.

Properties (1) *Chaining* The m.l.e. methods tend to lead to spherical or elliptical clusters each one around a nucleus. However, the above method leads to "rod" type elongated clusters partly because the links once made cannot be broken. Also, the clusters have no nuclei, thus leading to a chaining effect. (See Example 13.3.2.) Thus, it is expected that single linkage will not give satisfactory results if intermediates (as a result of random noise) are present between clusters.

(2) *Monotonicity* It can be seen that the single linkage method gives clustering of identical topology for any monotonic transformation of d_{ij}.

(3) *Ties* It is easily checked that it does not matter which choice is made if, at some stage in the algorithm, there is a tie for the smallest distance between two clusters. However, other hierarchical methods are not always well defined in this situation.

(4) *Other properties* Jardine and Sibson (1971) show that this is the only method consistent with a particular set of axioms for hierarchical clustering. In particular, it optimizes "connected sets of points". Hartigan (1973) uses this method in an attempt to minimize the number of mutations in a reconstruction model of evolutionary history. Gower and Ross (1969) have shown the relation of this method to the minimum spanning tree, and a computer algorithm is given by Ross (1969).

Dendrogram A dendrogram is a tree diagram in which the x axis represents the "objects" while the lower y axis represents distances. The branching of the tree gives the order of the $n-1$ links; the first fork represents the first link, the second fork the second link, and so on until all join together at the trunk. We illustrate this definition by an example.

Example 13.3.2 For the data in Example 13.3.1, a dendrogram from

Table 13.3.1 is given in Figure 13.3.1. The first link joins (3, 5), etc. The level of the horizontal lines shows the order in which the links were formed. We have arranged the order of the objects along the x axis as arising sequentially in clustering in Table 13.3.1 so that links do not "overlap". The dendrogram indicates that the distances between animals from the Scilly Isles are very small, suggesting origin from a common stock. The distances between shrews from Sark (6) and Jersey (7) is greater than between any pair of Scilly Island populations. The three populations of *Crocidura russcula* (8, 9, 10) do not form as close a cluster as those of the Scilly Islands.

The value of the *threshold* (d_0) is arbitrary, and no probabilistic work has been done here. For the complete dendrogram take $d_0 = \infty$. Obviously for a given d_0, the clusters can be read off from a dendrogram, e.g. for $d_0 = 3.50$ we obtain from Figure 13.3.1 (1, 2, 3, 4, 5, 6, 7) and (8, 9, 10). If we know g in advance, again the dendrogram gives the clusters. Broadly speaking, d_0 and g are functionally related.

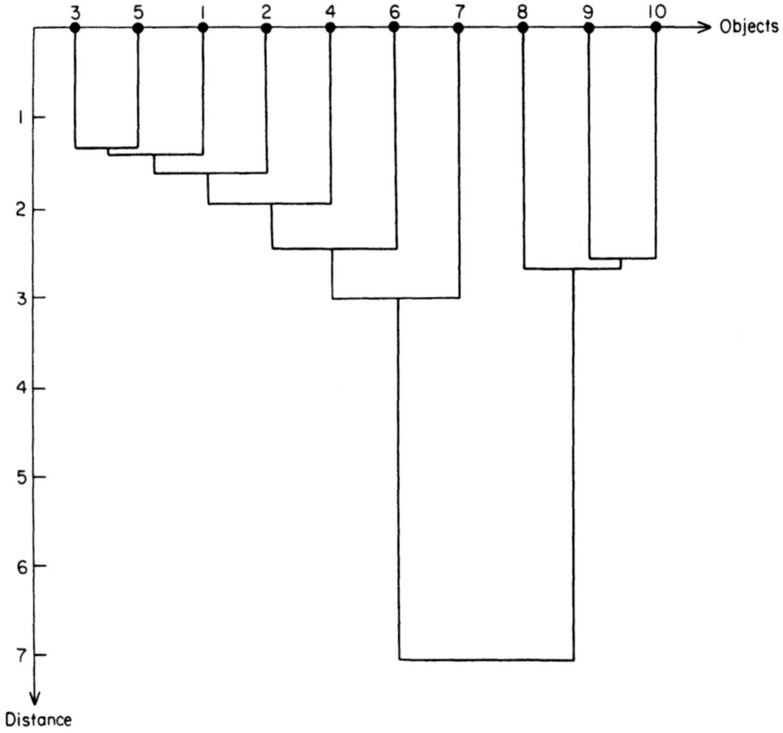

Figure 13.3.1 Dendrogram for shrew data (single linkage).

The dendrogram can be used to construct a new distance matrix between the objects. For two objects i and j, let u_{ij} be the *smallest* threshold d_0 for which i and j lie in the same cluster. The horizontal lines in the dendrogram are called *nodes*. Then u_{ij} can be found from the dendrogram as the smallest node which is linked to both i and j. For the single linkage dendrogram (although not in general for other dendrograms), the u_{ij} satisfy the *ultrametric inequality*

$$u_{ij} \leq \max(u_{ik}, u_{kj}) \qquad \text{for all } i, j, k \tag{13.3.2}$$

(see Exercises 13.3.4 and 13.3.5).

13.3.2 Complete linkage (furthest neighbour) method

The complete linkage method is similar to the single linkage method except that the distance between two clusters is now defined as the *largest* distance between pairs of elements in each cluster, rather than the smallest. The method can be described by the following algorithm.

(a) Start with clusters $C_1, \ldots, C_n$ containing $\mathbf{x}_1, \ldots, \mathbf{x}_n$, respectively.
(b) Assuming $d_{12} = \min(d_{ij})$ over all i and j, let $C_2^*, \ldots, C_n^*$ be the groups after joining the pair 1, 2 to form C_2^*.
(c) Define a new $((n-1)\times(n-1))$ distance matrix $D^* = (d_{ij}^*)$ with $d_{2j}^* = \max(d_{1j}, d_{2j})$ for $j = 3, \ldots, n$; $d_{ij}^* = d_{ij}$ for $i, j = 3, \ldots, n$. Find $\min d_{ij}^*$, $i, j = 2, \ldots, n$; then proceed as in (b).

Continue until all the distances between clusters are greater than d_0, where d_0 is an arbitrary threshold value. Note that when the clustering is completed we will have $\max_{i,j \in C} d_{ij} \leq d_0$ for each cluster C. Thus, this method tends to produce compact clusters with no chaining effect. As in single-linkage, it can be seen that computation can be carried out using a recurrence relation for the distance between group k and a group (ij) formed by the fusion of groups i and j; namely,

$$d_{k(ij)} = \max(d_{ik}, d_{jk}). \tag{13.3.3}$$

Most of the comments made for the single linkage method apply to this method except that its optimization properties are not known. Of course, the final groups in complete linkage do possess the property that all within-group distances are less than the threshold value, but the converse is not true, i.e. the method is not guaranteed to find all such groups. Like

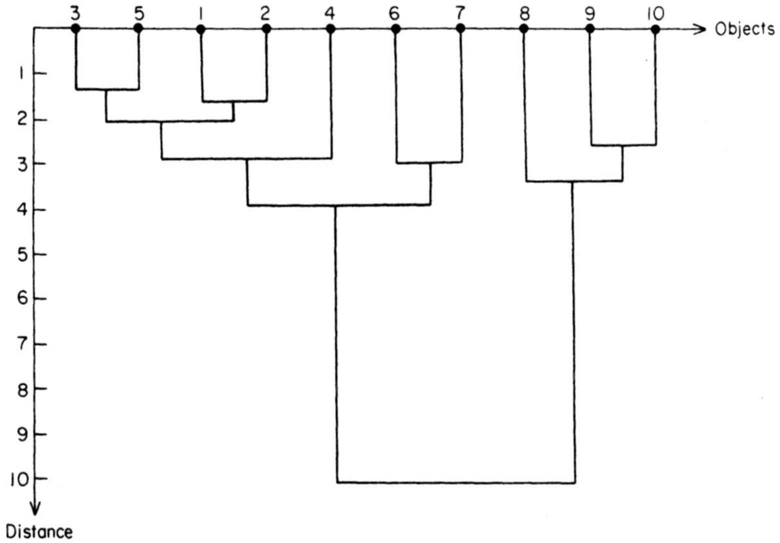

Figure 13.3.2 Dendrogram for the shrew data (complete linkage).

single linkage, the method has the invariance property under monotone transformations of the d_{ij}. However, the complete linkage method may not be well defined if at some stage in the algorithm there is a tie for the smallest distance (unlike single linkage).

Example 13.3.3 For the shrew data, the smallest distance is $d_{3,5} = 1.35$, so that these are clustered. Although $d_{1,5} = 1.40$ we cannot cluster these next because $d_{1,3} = 1.98$. The next cluster is $(1, 2)$ since $d_{1,2} = 1.61$. Following this procedure, we find the dendrogram as given in Figure 13.3.2. Note that there is no chaining as in the single linkage case (Figure 13.3.1). At $d_0 = 3.38$, we have groups $(1, 2, 3, 4, 5)$, $(6, 7)$, $(8, 9, 10)$ but at each stage the groups are *not* chained as seen in Figure 13.3.2.

13.4 Distances and Similarities

13.4.1 Distances

Definition *Let P and Q be two points where these may represent measurements* $\mathbf{x}$ *and* $\mathbf{y}$ *on two objects. A real-valued function* $d(P, Q)$ *is a* distance function *if it has the following properties:*

(I) symmetry, $d(P, Q) = d(Q, P)$;
(II) non-negativity, $d(P, Q) \geq 0$;
(III) identification mark, $d(P, P) = 0$.

For many distance functions the following properties also hold:

(IV) definiteness, $d(P, Q) = 0$ if and only if $P = Q$;
(V) triangle inequality, $d(P, Q) \leq d(P, R) + d(R, Q)$.

If (I)–(V) *hold* d *is called a* metric.

For some purposes, it is sufficient to consider distance functions satisfying only (I)–(III), but we will only consider distances for which (I)–(V) are satisfied unless otherwise mentioned. Note that (I) need not always be true: in sociology perhaps, subject P's opinion of subject Q measured by $d(P, Q)$ may differ from subject Q's opinion of subject P, $d(Q, P)$. Further, (V) is not satisfied for some distances (see Exercise 13.4.6). On the other hand, the distance may satisfy some stronger condition than (V), such as the ultrametric inequality (13.3.2).

One would expect $d(P, Q)$ to increase as "dissimilarity" or "divergence" between P and Q increases. Thus $d(P, Q)$ is also described as a coefficient of dissimilarity even when it does not satisfy the metric properties (IV) and (V).

Quantitative data

Euclidean distance Let $\mathbf{X}$ be an $(n \times p)$ data matrix with rows $\mathbf{x}_1', \ldots, \mathbf{x}_n'$. Then the Euclidean distance between the points $\mathbf{x}_i$ and $\mathbf{x}_j$ is d_{ij}, where

$$d_{ij}^2 = \sum_{k=1}^{p} (x_{ik} - x_{jk})^2 = \|\mathbf{x}_i - \mathbf{x}_j\|^2. \tag{13.4.1}$$

This distance function satisfies properties (I)–(V). It also satisfies the following properties:

(a) *Positive semi-definite property* Let $\mathbf{A} = (-\frac{1}{2}d_{ij}^2)$. Then $\mathbf{HAH}$ is p.s.d., where $\mathbf{H} = \mathbf{I} - n^{-1}\mathbf{11}'$ is the centring matrix. (For a proof of this result, see Section 14.2. In fact, Theorem 14.2.1 also gives the converse result.) This property will be of importance when we examine similarity coefficients.
(b) d_{ij} is invariant under orthogonal transformations of the $\mathbf{x}$s.
(c) We have the cosine law

$$d_{ij}^2 = b_{ii} + b_{jj} - 2b_{ij}, \tag{13.4.2}$$

where $b_{ij} = (\mathbf{x}_i - \bar{\mathbf{x}})'(\mathbf{x}_j - \bar{\mathbf{x}})$ is the centred inner product between $\mathbf{x}_i$ and $\mathbf{x}_j$. Another useful identity for calculation purposes is given by

$$\sum_{i=1}^{n} \sum_{j=1}^{n} d_{ij}^2 = 2n \sum_{i=1}^{n} b_{ii}. \tag{13.4.3}$$

Karl Pearson distance When the variables are not commensurable, it is desirable to standardize (13.4.1); that is, we can use

$$d_{ij}^2 = \sum_{k=1}^{p} \frac{(x_{ik} - x_{jk})^2}{s_k^2}, \tag{13.4.4}$$

where s_k^2 is the variance of the kth variable. We shall call such a standardized distance the "Karl Pearson distance" and denote it by K^2. This distance is then invariant under changes of scale. Another way to scale is to replace s_k by the range

$$R_k = \max_{i,j} |x_{ik} - x_{jk}|. \tag{13.4.5}$$

For cluster analysis, one usually uses (13.4.4) except when the difference in scale between two variables is intrinsic, when one uses Euclidean distance.

Mahalanobis distance We define squared Mahalanobis distance between points $\mathbf{x}_i$ and $\mathbf{x}_j$ as

$$D_{ij}^2 = (\mathbf{x}_i - \mathbf{x}_j)'\hat{\mathbf{\Sigma}}^{-1}(\mathbf{x}_i - \mathbf{x}_j). \tag{13.4.6}$$

Its variants and properties have already been discussed in Sections 13.2.2 and 1.6.

Qualitative data (multinomial populations)

Consider a classification of individuals into p categories. For each $r = 1, \ldots, g$, let $(x_{r1}, \ldots, x_{rp}) = \mathbf{x}_r'$ denote the observed proportions from a population of size n_r lying in each of these categories. For example, $\mathbf{x}_r$ might denote the proportions of people having blood group types A_1, A_2, B, and O in each of g countries. Of course,

$$\sum_{i=1}^{p} x_{ri} = 1, \qquad r = 1, \ldots, g.$$

We shall now consider several choices of a distance measure between these populations.

Euclidean distance The points $\mathbf{x}_r$ lie on the hyperplane $x_1 + \ldots + x_p = 1$ in the positive orthant of R^p. An obvious measure of distance between these points is Euclidean distance, given by

$$d_{rs}^2 = \sum_{i=1}^{p} (x_{ri} - x_{si})^2.$$

This distance may be suitable if the proportions are merely measured quantities for which no model of stochastic variation is being considered. However, if the proportions are thought of as random vectors, then a Mahalanobis-like distance is more appropriate.

A Mahalanobis-like distance Suppose that for each $r = 1, \ldots, g$, $\mathbf{x}_r$ represents the proportions based on a sample n_r from a multinomial distribution with parameter $\mathbf{a} = (a_1, \ldots, a_p)'$ (the same parameter for each r). Then $\mathbf{x}_r$ has mean $\mathbf{a}$ and covariance matrix $\mathbf{\Sigma}_r = n_r^{-1}\mathbf{\Sigma}$, where $\mathbf{\Sigma} = (\sigma_{ij})$ is given by

$$\sigma_{ij} = \begin{cases} a_i(1-a_i), & i = j, \\ -a_i a_j, & i \neq j; \end{cases} \tag{13.4.7}$$

that is, $\mathbf{\Sigma} = \text{diag}(\mathbf{a}) - \mathbf{a}\mathbf{a}'$. Since $\mathbf{x}_r$ lies on a hyperplane, $\mathbf{\Sigma}$ is singular. However, it is easily checked that a g-inverse of $\mathbf{\Sigma}$ is given by

$$\mathbf{\Sigma}^- = \text{diag}(a_1^{-1}, \ldots, a_p^{-1}), \tag{13.4.8}$$

Thus (see Example 11.2.5), we can define a (generalized) Mahalanobis distance between $\mathbf{x}_r$ and $\mathbf{x}_s$ as $n_r n_s/(n_r + n_s)$ times

$$\sum_{i=1}^{p} \frac{(x_{ri} - x_{si})^2}{a_i}. \tag{13.4.9}$$

Remarks (1) Unfortunately, there are two problems with this approach. First, in practice we wish to compare multinomial populations with *different* parameters (and hence different covariance matrices). Thus (13.4.9) can only be viewed as an "approximate" Mahalanobis distance and $\mathbf{a}$ must be thought of as an "average" parameter for the populations. To reduce the effect of the differences between populations on sample size, we shall drop the factor $n_r n_s/(n_r + n_s)$ given before (13.4.9).

(2) The second problem involves the estimate of the "average" parameter $\mathbf{a}$. A common procedure is to estimate $\mathbf{a}$ pairwise using

$$\hat{a}_i(r, s) = \tfrac{1}{2}(x_{ri} + x_{si})$$

(so $\hat{\mathbf{a}}(r, s)$ depends on r and s). Then (13.4.9) becomes the Mahalanobis-like distance

$$D_{1;rs}^2 = 2 \sum_{i=1}^{p} \frac{(x_{ri} - x_{si})^2}{(x_{ri} + x_{si})}. \tag{13.4.10}$$

This form was suggested by Mahalanobis (see Bhattacharyya, 1946) and rediscovered by Sanghvi (1953).

(3) Other possibilities for estimating $\mathbf{a}$ include a global average of all the proportions ($\hat{a}_i = g^{-1}(x_{1i} + \ldots + x_{gi})$) and/or weighting each proportion x_{ri} by its sample size n_r (see Balakrishnan and Sanghvi, 1968).

(4) The distance in (13.4.10) is obtained by using a pooled estimate of the parameter $\mathbf{a}$. An alternative procedure is to use a pooled estimate of $\mathbf{\Sigma}$, based on the sample covariance matrices thus giving yet another Mahalanobis-like distance. See Exercise 13.4.1 and Balakrishnan and Sanghvi (1968).

Bhattacharyya distance Let $\mathbf{v}_r = (x_{r1}^{1/2}, \ldots, x_{rp}^{1/2})'$, $r = 1, \ldots, g$, so that the vectors $\mathbf{v}_r$ are points on the unit sphere in R^p with centre at the origin. The cosine of the angle between $\mathbf{v}_r$ and $\mathbf{v}_s$ is

$$\cos B_{rs} = \sum_{i=1}^{p} v_{ri}v_{si} = \sum_{i=1}^{p} (x_{ri}x_{si})^{1/2}, \qquad (13.4.11)$$

so that the angle B_{rs} is the great circle distance between $\mathbf{v}_r$ and $\mathbf{v}_s$. The Euclidean distance of the chord between $\mathbf{v}_r$ and $\mathbf{v}_s$ is given by

$$D_{2;rs}^2 = \sum_{i=1}^{p} (x_{ri}^{1/2} - x_{si}^{1/2})^2. \qquad (13.4.12)$$

We shall call $D_{2;rs}$ the *Bhattacharyya distance* (although sometimes the angle B_{rs} is given this name). The two measures are connected by

$$D_{2;rs}^2 = 4 \sin^2 \tfrac{1}{2} B_{rs}.$$

Remarks (1) If $\mathbf{x}_r$ and $\mathbf{x}_s$ come from multinomial populations with the *same* parameter $\mathbf{a}$, then D_1^2 and $4D_2^2$ are asymptotically the same for large n_r, n_s. See Exercise 13.4.2.

(2) Bhattacharyya distance can be interpreted as an asymptotic Mahalanobis distance. From Example 2.9.1, we see that $\mathbf{v}_r \sim N_p(\mathbf{b}, (4n_r)^{-1}\mathbf{\Sigma})$ for large n_r, where $b_i = a_i^{1/2}$, $i = 1, \ldots, p$, and $\mathbf{\Sigma} = \mathbf{I} - \mathbf{bb}'$. Although $\mathbf{\Sigma}$ is singular, it is easy to see that a g-inverse of $\mathbf{\Sigma}$ is given by $\mathbf{\Sigma}^- = \mathbf{I}$. Thus if v_r and v_s are proportions from multinomial distributions with the same parameter, the asymptotic squared Mahalanobis distance between them is given by

$$4n_r n_s(n_r + n_s)D_{2;rs}^2.$$

(3) Note that with both D_1^2 and D_2^2, differences between very small proportions are given more weight than differences between intermediate or large proportions. However, with Euclidean distance all such differences are weighted equally.

(4) Practical studies (Sanghvi and Balakrishnan, 1972) have shown that D_1^2 and D_2^2 are very similar to one another and in practice it hardly matters which one is chosen. Bhattacharyya distance is perhaps preferable because there is no need to estimate unknown parameters and because it has a simple geometric interpretation.

Example 13.4.1 Table 13.4.1 gives the relative gene frequencies for the blood-group systems with types A_1, A_2, B, and O for large samples from four human populations: (1) Eskimo, (2) Bantu, (3) English, and (4) Korean. The object is to assess the affinities between the populations.

The Bhattacharyya distance matrix is found to be

	Eskimo	Bantu	English	Korean
Eskimo	0	23.26	16.34	16.87
Bantu		0	9.85	20.43
English			0	19.60
Korean				0

Use of the complete linkage clustering method suggests the two clusters Bantu–English and Eskimo–Korean. Cavalli-Sforza and Edwards (1967) came to this conclusion by a maximum likelihood method. However, the single linkage method, which one might think would be appropriate in this situation, does not support this conclusion.

Multi-classification case

If there are t classifications instead of a single one, e.g. if there are gene frequencies for each of the five blood-group systems of classification, then we can sum the distances using

$$D^2 = \sum_{k=1}^{t} D_{f,k}^2$$

Table 13.4.1 Relative frequencies of blood groups A_1, A_2, B, O for four populations (Cavalli-Sforza and Edwards, 1967)

Blood groups	Populations			
	Eskimo	Bantu	English	Korean
A_1	0.2914	0.1034	0.2090	0.2208
A_2	0.0000	0.0866	0.0696	0.0000
B	0.0316	0.1200	0.0612	0.2069
O	0.6770	0.6900	0.6602	0.5723

Table 13.4.2 Distances

1. Euclidean distance: $\left\{\sum_{k=1}^{p} w_k(x_{rk}-x_{sk})^2\right\}^{1/2}$.
   (a) Unstandardized, $w_k = 1$.
   (b) Standardized by s.d., $w_k = 1/s_k^2$ (Karl Pearson distance).
   (c) Standardized by range $w_k = 1/R_k^2$.
2. Mahalanobis distance: $\{(\mathbf{x}_r-\mathbf{x}_s)'\mathbf{\Sigma}^{-1}(\mathbf{x}_r-\mathbf{x}_s)\}^{1/2}$
   $\mathbf{\Sigma}$ = any transforming positive definite matrix.
3. City-block metric (Manhattan metric): $\sum_{k=1}^{p} w_k\,|x_{rk}-x_{sk}|$.
   Mean character difference $w_k = 1/p$
4. Minkowski metric: $\left\{\sum_{k=1}^{p} w_k\,|x_{rk}-x_{sk}|^{\lambda}\right\}^{1/\lambda}$, $\lambda \geq 1$.
5. Canberra metric: $\sum_{k=1}^{p} \frac{|x_{rk}-x_{sk}|}{(x_{rk}+x_{sk})}$.
   (Scaling does not depend on whole range of the variable.)
6. Bhattacharyya distance (proportions): $\left\{\sum_{i=1}^{p} (x_i^{1/2}-y_i^{1/2})^2\right\}^{1/2}$
7. Distances between groups:
   (a) Karl Pearson dissimilarity coefficient:

$$\left\{\frac{1}{p}\sum_{k=1}^{p}\frac{(\bar{x}_{rk}-\bar{x}_{sk})^2}{(s_{rk}^2/n_r)+(s_{sk}^2/n_s)}\right\}^{1/2},$$

   where n_j = size of jth sample $j = r, s$; $\bar{x}_{jk}$, s_{jk}^2 = mean and variance of kth variable for the jth sample,
   (b) Mahalanobis distance: $\{(\bar{\mathbf{x}}_r-\bar{\mathbf{x}}_s)'\hat{\mathbf{\Sigma}}^{-1}(\bar{\mathbf{x}}_r-\bar{\mathbf{x}}_s)\}^{1/2}$.

where $D_{f,k}^2$, $f = 1, 2$, is either the Mahalanobis distance or the Bhattacharyya distance between two populations on the kth classification.

The process of summing up distances in this way is meaningful when the t types of classification are independent.

A list of various distances is given in Table 13.4.2.

13.4.2 Similarity coefficients

So far, we have concentrated on measures of distance or dissimilarity, but there are situations as in taxonomy where it is often common to use measures of similarity between points A and B.

Definition *A reasonable measure of* similarity, $s(A, B)$, *should have the following properties:*

(i) $s(A, B) = s(B, A)$,
(ii) $s(A, B) > 0$,
(iii) $s(A, B)$ *increases as the similarity between A and B increases.*

Because greater similarity means less dissimilarity, similarity coefficients can be used in any of the hierarchical techniques of Section 13.3 simply by changing the signs of all the inequalities.

We now consider some examples.

Qualitative variables

Let the presence or absence of p attributes on two objects P and Q be denoted $(x_1, \ldots, x_p)$ and $(y_1, \ldots, y_p)$, where $x_i = 1$ or 0 depending on whether the ith attribute is present or absent for object P.

Set

$$\begin{aligned} a &= \sum x_i y_i, \\ b &= \sum (1 - x_i) y_i, \\ c &= \sum x_i (1 - y_i), \\ d &= \sum (1 - x_i)(1 - y_i); \end{aligned} \tag{13.4.13}$$

that is, a, b, c, d are the frequencies of $(x_i, y_i) = (1, 1)$, $(0, 1)$, $(1, 0)$, and $(0, 0)$, respectively.

The simplest measure of similarity between P and Q is

$$s_1(P, Q) = \frac{a}{p}. \tag{13.4.14}$$

An alternative is the *simple matching coefficient* (Sokal and Michener, 1958) defined as

$$s_2(P, Q) = \frac{(a + d)}{p}, \tag{13.4.15}$$

which satisfies $s_2(P, P) = 1$.

It is not clear in practice whether to use s_1, s_2, or some other association coefficient. In s_2 all matched pairs of variables are equally weighted, whereas in s_1 negative matches are excluded. In both s_1 and s_2 every attribute is given equal weight, but in some applications it might be preferable to use a differential weighting of attributes. For a discussion of these problems, see Jardine and Sibson (1971, Chapter 4) and Everitt (1974, pp. 49–50).

Let $\mathbf{X}$ be the data matrix containing the presence/absence information on p attributes for n objects. The matrix $\mathbf{X}$ is called an *incidence matrix* since $x_{ij} = 1$ or 0. In view of (13.4.13), the matrix of similarities based on (13.4.14) is simply

$$\mathbf{S}_1 = \frac{(\mathbf{XX}')}{p}, \qquad \mathbf{S}_1 = (s_{ij}^{(1)}) \tag{13.4.16}$$

whereas, the matrix of similarities based on (13.4.15) is

$$\mathbf{S}_2 = \frac{\{\mathbf{XX}' + (\mathbf{J}-\mathbf{X})(\mathbf{J}-\mathbf{X})'\}}{p}, \tag{13.4.17}$$

where $\mathbf{J} = \mathbf{11}'$, $\mathbf{S}_2 = (s_{ij}^{(2)})$. Note that the diagonal elements of $\mathbf{S}_2$ are 1. Clearly, (13.4.16) and (13.4.17) are p.s.d.

Example 13.4.2 Consider the (6×5) data matrix of Example 8.5.1 in which $x_{ij} = 1$ if the ith grave contains the jth variety of pottery and 0 otherwise. The aim here is to see which graves have similar varieties of pottery to one another.

It is found that $5\mathbf{S}_2$ for $A, \ldots, F$ is

$$\begin{bmatrix} 5 & 0 & 3 & 5 & 1 & 3 \\ 0 & 5 & 2 & 0 & 4 & 2 \\ 3 & 2 & 5 & 3 & 1 & 3 \\ 5 & 0 & 3 & 5 & 1 & 3 \\ 1 & 4 & 1 & 1 & 5 & 3 \\ 3 & 2 & 3 & 3 & 3 & 5 \end{bmatrix}.$$

We can now use any clustering method of Section 13.3 to see if there are any clusters. For example, using single linkage, A and D are grouped together first (which is not surprising since they are identical), and then B and E. The next link joins all the graves into a single group.

Mixed variables

If there are qualitative as well as quantitative variables, Gower (1971*a*) has proposed the following similarity coefficient between ith and jth points:

$$s_{ij}^{(3)} = 1 - \frac{1}{p} \sum_{k=1}^{p} w_k \, |x_{ik} - x_{jk}|, \tag{13.4.18}$$

where $w_k = 1$ if k is qualitative, $w_k = 1/R_k$ if k is quantitative where R_k is the range of the kth variable.

It can be shown that the matrix $(s_{ij}^{(3)})$ is positive semi-definite, but if R_k is replaced by the sample standard deviation s_k, this may not be so (see Exercise 13.4.3).

In some applications the data consists of an $(n \times n)$ matrix $\mathbf{D}$ consisting of distances (or similarities) between the points, rather than an $(n \times p)$ data matrix $\mathbf{X}$. In this situation, the choice of distance has already been made and one can immediately apply any of the hierarchical techniques of Section 13.3. Examples of such data are quite common in the context of multidimensional scaling (see Chapter 14).

13.5 Other Methods and Comparative Approach

We have only considered a few approaches to cluster analysis in the previous sections. However, there is a wide range of different approaches none of which falls neatly into a single framework; for example, clumping techniques in which the classes or clumps can overlap and mode analysis, in which one searches for natural groupings of data, by assuming disjoint density surfaces in the sample distribution. For a good discussion of these methods, we refer the reader to Everitt (1974) and Jardine and Sibson (1971). Other useful references are Cormack (1971) and Sneath and Sokal (1973). For clustering algorithms, see Hartigan (1975).

13.5.1 Comparison of methods

Optimization techniques usually require large amounts of computing time and consequently their use is limited to small data sets. For the various likelihood criteria mentioned, the underlying assumptions about the shape of the distributions are important (e.g. whether or not $\mathbf{\Sigma}$ is the same for all groups, and whether or not $\mathbf{\Sigma}$ is known). Also it is important to try several different starting configurations when carrying out the optimization. Possible starting configurations include a partially optimized configuration arising from some simple cluster technique or a configuration based on *a priori* knowledge.

Hierarchical techniques are more suitable for use in the analysis of biological or zoological data because for such data a hierarchical structure can safely be assumed to exist; e.g. in the shrew data where shrews can be grouped into *species* and these species themselves can be grouped into *genera*, etc. These methods have the considerable advantage in requiring far less computing time and therefore they can be used for larger data sets. The major difficulty with these techniques is making a choice of a distance measure. Of various hierarchical techniques, single linkage is the only one to satisfy various analytical properties (see Jardine and Sibson, 1971; Fisher and van Ness, 1971, 1973).

As has been already mentioned, in practice single linkage may not provide useful solutions because of its sensitivity to the noise present between relatively distinct clusters and the subsequent chaining effect. On the other hand, the complete linkage method gives compact clusters, but it does not necessarily guarantee to find all groups where within-group distances are less than some value.

The use of any hierarchical method entails a loss of information. For example the single linkage method effectively replaces the original distance matrix **D** by a new set of distances which satisfy the ultra-metric inequality. In general, hierarchical techniques should be regarded as a useful descriptive method for an initial investigation of the data. For a fuller discussion, see Everitt (1974).

Spurious solutions All the cluster methods make implicit assumptions about the type of structure present, and when these assumptions are not satisfied spurious solutions are likely to be obtained; that is, each clustering method imposes a certain amount of structure on the data and it should be ensured that the conclusions we reach are not just an artefact of the method used. For example, most of the optimization methods are biased towards finding elliptical and spherical clusters. If the data contains clusters of other shapes (say snake-like), these may not be found by such methods and consequently important information will have been lost, and in some cases a misleading solution will result. On the other hand, hierarchical methods impose hierarchy when it may not be present. Everitt (1974) considers examples to elucidate these points.

13.5.2 Comparative and Graphical Approach

In general, several different techniques should be applied for clustering as the variety helps to prevent misleading solutions being accepted. These methods should be supported by graphical techniques; we have already used canonical variates for the shrew data for the multi-sample case and, similarly, principal coordinates could be used for the single sample case. In some cases, when the only data is an inter-point distance matrix, multidimensional scaling techniques could be used, see Chapter 14. The harmonic curves given in Section 1.7.3 also give some idea of clustering.

Example 13.5.1 Consider again the shrew data of Section 13.1.2. The differences between the 10 races of shrew are conveniently summarized

Table 13.5.1 Canonical means for shrew data

Sample	Canonical variate 1	2	3	4	5	6	7	8	9
1	−2.33	−0.37	−0.39	−0.07	0.26	0.00	0.06	0.05	−0.01
2	−2.29	0.39	−1.12	0.26	−0.49	−1.05	−0.53	−0.03	0.02
3	−2.77	1.06	0.34	0.46	−1.01	0.79	−0.54	0.28	−0.04
4	−1.37	0.53	−0.85	−1.37	0.59	0.85	−0.75	−0.52	−0.03
5	−2.23	1.14	0.36	0.07	−0.19	0.02	0.17	−0.07	0.01
6	−2.01	−2.54	1.52	0.39	−0.31	−0.16	−0.06	−0.11	−0.02
7	−1.76	−0.59	2.15	−1.92	0.76	0.23	−0.50	0.24	0.13
8	8.10	1.27	0.98	−0.33	0.43	−0.39	−0.10	0.05	−0.04
9	6.54	−0.69	−0.53	−0.49	−0.45	0.12	0.11	−0.00	0.01
10	5.95	−0.05	−0.19	2.01	0.52	0.30	−0.14	−0.03	0.03

in Table 13.5,1 using the canonical means **M** of Example 12.5.2, where m_{ij} denotes the jth canonical coordinate of the ith race of shrew, $j=1,\ldots,9$; $i=1,\ldots,10$. For convenience each canonical coordinate is centred to have weighted sample mean 0.

A plot of the 10 harmonic curves is given in Figure 13.5.1. Note that these curves are dominated by their lower frequencies because the canonical means vary most in their first coordinates. Further, note that the L_2 distance between the harmonic curves equals the Mahalanobis distance between the corresponding races of shrew; hence the distances between the curves agree with Table 13.2.3. In particular, the curves seem to form two main clusters, 1–7 and 8–10, in agreement with the single linkage cluster analysis of Example 13.3.1.

Exercises and Complements

13.2.1 (Scott and Symons, 1971) Let us suppose $\mathbf{x}_1,\ldots,\mathbf{x}_n$ to be a mixed sample from the $N_p(\boldsymbol{\mu}_k, \boldsymbol{\Sigma}_k)$, $k=1,\ldots,g$, populations where the $\boldsymbol{\mu}_k$ are unknown. If $\boldsymbol{\Sigma}=\ldots=\boldsymbol{\Sigma}_g=\boldsymbol{\Sigma}$ where $\boldsymbol{\Sigma}$ is known, show that the ML partition $\hat{\boldsymbol{\gamma}}$ minimizes $\text{tr}(\mathbf{W}\boldsymbol{\Sigma}^{-1})$. Show that it is equivalent to maximizing the weighted between groups sum of squares

$$\sum_{k=1}^{g} n_k(\bar{\mathbf{x}}_k-\bar{\mathbf{x}})'\boldsymbol{\Sigma}^{-1}(\bar{\mathbf{x}}_k-\bar{\mathbf{x}}),$$

where $\bar{\mathbf{x}}_k$ is the mean for a cluster C_k.

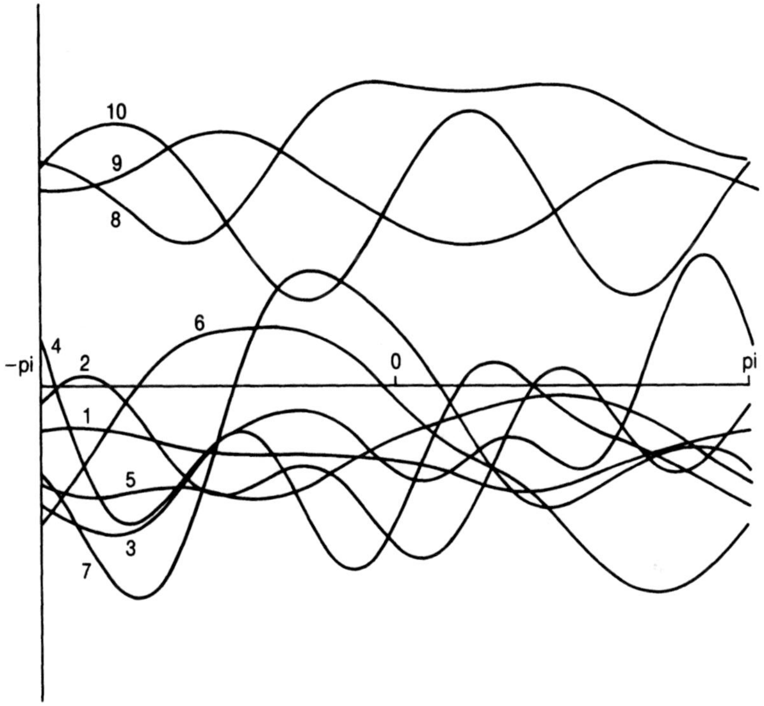

Figure 13.5.1 Harmonic curves for the shrew data.

13.2.2 Let us suppose $\mathbf{x}_1, \ldots, \mathbf{x}_n$ to be a mixed sample from the $N_p(\boldsymbol{\mu}_k, \boldsymbol{\Sigma}_k)$, $k = 1, \ldots, g$, populations where $\boldsymbol{\mu}_k$ and $\boldsymbol{\Sigma}_k$ are unknown. Further, let $\mathbf{y}_{k1}, \ldots, \mathbf{y}_{km_k} = 1, \ldots, g$, be previous samples of independent observations known to come from $N_p(\boldsymbol{\mu}_k, \boldsymbol{\Sigma}_k)$, $k = 1, \ldots, g$, respectively. Show that the m.l.e.s of $\boldsymbol{\mu}_k, \boldsymbol{\Sigma}_k$ and the partition $\boldsymbol{\gamma}$ of the $\mathbf{x}$s are given by

$$\hat{\boldsymbol{\mu}}_k = \frac{(n_k \bar{\mathbf{x}}_k + m_k \bar{\mathbf{y}}_k)}{(m_k + n_k)}, \qquad \hat{\boldsymbol{\Sigma}}_k = \frac{(\mathbf{W}_{kx} + \mathbf{W}_{ky} + \mathbf{W}_{kxy})}{(m_k + n_k)},$$

and $\hat{\gamma}$ minimizes

$$\prod_{k=1}^{g} |\hat{\boldsymbol{\Sigma}}_k|^{m_k+n_k},$$

where

$$\mathbf{W}_{kx} = \sum_{C_k} (\mathbf{x}_i - \bar{\mathbf{x}}_k)(\mathbf{x}_i - \bar{\mathbf{x}}_k)', \qquad \mathbf{W}_{ky} = \sum_{C_k} (\mathbf{y}_{ki} - \bar{\mathbf{y}}_k)(\mathbf{y}_{ki} - \bar{\mathbf{y}}_k)',$$

and

$$\mathbf{W}_{kxy} = \frac{m_k n_k}{m_k + n_k} (\bar{\mathbf{x}}_k - \bar{\mathbf{y}}_k)(\bar{\mathbf{x}}_k - \bar{\mathbf{y}}_k)'.$$

13.2.3 (Day, 1969) Let $\mathbf{x}_1, \ldots, \mathbf{x}_n$ be drawn from the mixture

$$f(\mathbf{x}) = p_1 f_1(\mathbf{x}) + p_2 f_2(\mathbf{x}),$$

where f_i is the p.d.f. of $N_p(\boldsymbol{\mu}_i, \boldsymbol{\Sigma})$, $i = 1, 2$, and $p_1 + p_2 = 1$.

(a) Show that the mean $\mathbf{m}$ and the covariance matrix $\mathbf{V}$ of the mixture are given by

$$\mathbf{m} = p_1\boldsymbol{\mu}_1 + p_2\boldsymbol{\mu}_2, \qquad \mathbf{V} = \boldsymbol{\Sigma} + p_1 p_2(\boldsymbol{\mu}_1 - \boldsymbol{\mu}_2)(\boldsymbol{\mu}_1 - \boldsymbol{\mu}_2)'.$$

Show that the m.l.e.s of $\mathbf{m}$ and $\mathbf{V}$ are given by

$$\hat{\mathbf{m}} = \bar{\mathbf{x}}, \qquad \hat{\mathbf{V}} = \mathbf{S}.$$

Let

$$\mathbf{a} = \boldsymbol{\Sigma}^{-1}(\boldsymbol{\mu}_2 - \boldsymbol{\mu}_1), \qquad b = \tfrac{1}{2}(\boldsymbol{\mu}_1'\boldsymbol{\Sigma}^{-1}\boldsymbol{\mu}_1 - \boldsymbol{\mu}_2'\boldsymbol{\Sigma}^{-1}\boldsymbol{\mu}_2) + \log(p_2/p_1).$$

Show that

$$\hat{\mathbf{a}} = \hat{\mathbf{V}}^{-1}(\hat{\boldsymbol{\mu}}_1 - \hat{\boldsymbol{\mu}}_2)/\{1 - \hat{p}_1\hat{p}_2(\hat{\boldsymbol{\mu}}_1 - \hat{\boldsymbol{\mu}}_2)'\hat{\mathbf{V}}^{-1}(\hat{\boldsymbol{\mu}}_1 - \hat{\boldsymbol{\mu}}_2)\}, \qquad (*)$$

$$\hat{b} = -\tfrac{1}{2}\hat{\mathbf{a}}'(\hat{\boldsymbol{\mu}}_1 + \hat{\boldsymbol{\mu}}_2) + \log(\hat{p}_2/\hat{p}_1). \qquad (**)$$

(b) If $P(k \mid \mathbf{x}_j)$ is the probability that observation $\mathbf{x}_j$ arises from the component k then the m.l.e.s of $P(k \mid x_j)$ are

$$\hat{P}(1 \mid \mathbf{x}_j) = \hat{p}_1 e_{1j}/\{\hat{p}_1 e_{1j} + \hat{p}_2 e_{2j}\},$$

$$\hat{P}(2 \mid \mathbf{x}_j) = 1 - \hat{P}(1 \mid \mathbf{x}_j),$$

where

$$e_{ij} = \exp\{-\tfrac{1}{2}(\mathbf{x}_j - \hat{\boldsymbol{\mu}}_i)'\hat{\boldsymbol{\Sigma}}^{-1}(\mathbf{x}_j - \hat{\mu}_i)\}.$$

Hence show that the m.l. equations can be written as

$$\frac{\hat{p}_2}{\hat{p}_1} = \sum_{j=1}^{n} \hat{P}(2 \mid \mathbf{x}_j) \Big/ \sum_{j=1}^{n} \hat{P}(1 \mid \mathbf{x}_j),$$

$$\hat{\boldsymbol{\mu}}_1 = \left\{\sum_j \mathbf{x}_j \hat{P}(1 \mid \mathbf{x}_j)\right\} \Big/ \sum_j \hat{P}(1 \mid \mathbf{x}_j),$$

$$\hat{\boldsymbol{\mu}}_2 = \left\{\sum_j \mathbf{x}_j \hat{P}(2 \mid \mathbf{x}_j)\right\} \Big/ \sum_j \hat{P}(2 \mid \mathbf{x}_j),$$

$$\hat{\boldsymbol{\Sigma}} = \frac{1}{n} \sum_j \{(\mathbf{x}_j - \hat{\boldsymbol{\mu}}_1)(\mathbf{x}_j - \hat{\boldsymbol{\mu}}_1)' \hat{P}(1 \mid \mathbf{x}_j) + (\mathbf{x}_j - \hat{\boldsymbol{\mu}}_2)(\mathbf{x}_j - \hat{\boldsymbol{\mu}}_2)' P(2 \mid \mathbf{x}_j)\}.$$

Hence, deduce that $\hat{p}_1$, $\hat{\boldsymbol{\mu}}_1$, and $\hat{\boldsymbol{\mu}}_2$ are some functions of the $\mathbf{x}$s and of $\hat{\mathbf{a}}$ and $\hat{b}$, whereas $\hat{\mathbf{V}}$ is a function of the $\mathbf{x}$s only. Hence conclude that (*) and (**) form a set of equations of the form

$$\hat{\mathbf{a}} = \phi_1(\hat{\mathbf{a}}, \hat{b}; \mathbf{x}_1, \ldots, \mathbf{x}_n), \qquad \hat{b} = \phi_2(\hat{\mathbf{a}}, \hat{b}; \mathbf{x}_1, \ldots, \mathbf{x}_n)$$

which can be solved numerically by iteration.

13.2.4 (a) Substituting

$$\bar{\mathbf{x}}_j - \bar{\bar{\mathbf{x}}}_k = N_k^{-1} \sum_{\bar{\mathbf{x}}_r \in C_k} n_r(\bar{\mathbf{x}}_j - \bar{\mathbf{x}}_r)$$

in w_g^2 given by (13.2.12), and then, using

$$(\bar{\mathbf{x}}_j - \bar{\mathbf{x}}_r)' \boldsymbol{\Sigma}^{-1} (\bar{\mathbf{x}}_j - \bar{\mathbf{x}}_s) = \tfrac{1}{2}(D_{jr}^2 + D_{js}^2 - D_{rs}^2), \qquad j, r, s \in C_i,$$

prove (13.2.15).

(b) Using

$$\bar{\bar{\mathbf{x}}}_k - \bar{\bar{\mathbf{x}}} = N_k^{-1} \sum_{\bar{\mathbf{x}}_j \in C_k} n_j(\bar{\mathbf{x}}_j - \bar{\mathbf{x}})$$

in (13.2.13), prove (13.2.16). For $g = 2$, use $N_2(\bar{\bar{\mathbf{x}}}_2 - \bar{\mathbf{x}}) = -N_1(\bar{\bar{\mathbf{x}}}_1 - \bar{\mathbf{x}})$ to show that

$$b_g^2 = \left\{\frac{(N_1 + N_2)}{N_1 N_2}\right\} \sum_{j, j' \in C_1} n_j n_{j'} B_{jj'}.$$

13.3.1 (Lance and Williams, 1967) Let $\mathbf{D} = (d_{ij})$ be a distance matrix. Show that the distance between group k and a group (ij) formed by the merger of groups i and j can be written

$$d_{k(ij)} = \alpha d_{ki} + \alpha d_{kj} + \gamma\, |d_{ki} - d_{kj}|$$

with

(i) $\alpha = \frac{1}{2}$, $\gamma = -\frac{1}{2}$ for single linkage,
(ii) $\alpha = \frac{1}{2}$, $\gamma = \frac{1}{2}$ for complete linkage,
(iii) $\alpha = \frac{1}{2}$, $\gamma = 0$ for a "weighted average".

13.3.2 Show that the squared Euclidean distance matrix (d_{ij}^2) for the iris data of Table 1.2.2 is

$$\begin{bmatrix} 0 & 0.36 & 0.02 & 0.50 & 1.69 & 0.05 \\ & 0 & 0.26 & 0.02 & 0.61 & 0.53 \\ & & 0 & 0.40 & 1.37 & 0.05 \\ & & & 0 & 0.61 & 0.73 \\ & & & & 0 & 1.78 \\ & & & & & 0 \end{bmatrix}.$$

Draw the dendrograms obtained on applying single linkage and complete linkage clustering methods. Show that (i) at threshold 0.30, the single linkage method leads to clusters (1, 3, 6), (2, 4), (5) and (ii) at threshold 0.75, the complete linkage method leads to clusters (1, 2, 3, 4, 6), (5). From the scatter diagram of the points in Example 13.2.1, compare and contrast these results with the clusters (1, 3, 6) and (2, 4, 5) obtained in Example 13.2.1.

13.3.3 Suppose two bivariate normal populations A_1, A_2 have means

$$\begin{pmatrix} 0 \\ 0 \end{pmatrix} \quad \text{and} \quad \begin{pmatrix} 3 \\ 2 \end{pmatrix}$$

and each has dispersion matrix

$$\begin{bmatrix} 1 & 0 \\ 0 & 1 \end{bmatrix}.$$

The following observations were drawn at random from A_1 and A_2, the first six from A_1 and the last three from A_2.

$$\begin{pmatrix} 1.14 \\ 1.01 \end{pmatrix}, \begin{pmatrix} 0.98 \\ -0.94 \end{pmatrix}, \begin{pmatrix} 0.49 \\ 1.02 \end{pmatrix}, \begin{pmatrix} -0.31 \\ 0.76 \end{pmatrix}, \begin{pmatrix} 1.78 \\ 0.69 \end{pmatrix}, \begin{pmatrix} 0.40 \\ -0.43 \end{pmatrix},$$

$$\begin{pmatrix} 2.42 \\ 1.37 \end{pmatrix}, \begin{pmatrix} 2.55 \\ 1.06 \end{pmatrix}, \begin{pmatrix} 2.97 \\ 2.37 \end{pmatrix}.$$

Using single linkage cluster analysis, and the Euclidean distance matrix (unsquared distances) for these nine observations given below, divide the observations into two groups and comment on the result.

	2	3	4	5	6	7	8	9
1	1.96	0.65	1.47	0.72	1.62	1.33	1.41	2.28
2		2.02	2.13	1.82	0.77	2.72	2.54	3.86
3			0.84	1.33	1.45	1.96	2.06	2.82
4				2.09	1.39	2.80	2.88	3.65
5					1.78	0.93	0.85	2.06
6						2.71	2.62	3.80
7							0.34	1.14
8								1.38

13.3.4 (Ultrametric distances) If d is a distance which satisfies the ultrametric inequality

$$d(P, Q) \leqslant \max \{d(P, R), d(Q, R)\},$$

for all points P, Q, and R, then d is called an *ultrametric* distance.

(a) For three points P, Q, and R, let $a = d(P, Q)$, $b = d(P, R)$, and $c = d(Q, R)$ and suppose $a \leqslant b \leqslant c$. Show that $b = c$, i.e. the two larger distances are equal.

(b) Using (a), show that d satisfies the triangle inequality.

13.3.5 Let $\mathbf{D} = (d_{ij})$ be a distance matrix and let $\mathbf{U} = (u_{ij})$ denote the distance matrix obtained from the corresponding single linkage dendrogram; that is, u_{ij} is the smallest threshold d_0 for which objects i and j lie in the same cluster.

(a) Show that $\mathbf{U}$ is an ultrametric distance matrix; that is,

$$u_{ij} \leqslant \max (u_{ik}, u_{kj}) \qquad \text{for all } i, j, k.$$

(b) Show that $\mathbf{D} = \mathbf{U}$ if and only if $\mathbf{D}$ is ultrametric.

(c) Consider the matrix of distances between five objects:

$$\begin{bmatrix} 0 & 4 & 1 & 4 & 3 \\ & 0 & 4 & 2 & 4 \\ & & 0 & 4 & 3 \\ & & & 0 & 4 \\ & & & & 0 \end{bmatrix}$$

Draw the dendrogram for the single linkage method. Verify that $\mathbf{D} = \mathbf{U}$ and hence that $\mathbf{D}$ is ultrametric.

13.4.1 (Balakrishnan and Sanghvi, 1968) Let $x_{r1}, \ldots, x_{rp}$ be proportions based on n_r observations for $r = 1, \ldots, g$. Let

$$\hat{\sigma}_{ij}^{(r)} = \begin{cases} x_{ri}(1-x_{ri}), & i=j, \\ -x_{ri}x_{rj}, & i \neq j. \end{cases}$$

Define

$$\hat{\mathbf{\Sigma}} = \sum_{r=1}^{g} n_r \mathbf{\Sigma}^{(r)} \Big/ \sum_{r=1}^{g} n_r.$$

Why might $(\mathbf{x}_r - \mathbf{x}_s)'\hat{\mathbf{\Sigma}}^{-1}(\mathbf{x}_r - \mathbf{x}_s)$ be a suitable measure of distance when it is suspected that the $\mathbf{\Sigma}^{(r)}$ are not all equal. Is $\hat{\mathbf{\Sigma}}$ singular?

13.4.2 Suppose $\mathbf{x}$ and $\mathbf{y}$ are proportions based on samples of sizes n_1 and n_2 from multinomial distributions with the same parameter $\mathbf{a}$. Let $m = \min(n_1, n_2)$. Using (3.4.10) and (3.4.12), show that the ratio between the Mahalanobis distance and the Bhattacharyya distance is given by

$$\frac{D_1^2}{D_2^2} = 2 \sum_{i=1}^{p} \frac{(x_i^{1/2} + y_i^{1/2})^2}{(x_i + y_i)}.$$

Let $\mathbf{b} = (a_1^{1/2}, \ldots, a_p^{1/2})'$. Using the fact that $x_i^{1/2} - b_i$ and $y_i^{1/2} - b_i$ have order $O_p(m^{-1/2})$, show that $D_1^2/D_2^2 - 4$ has order $O_p(m^{-1})$ and hence D_1^2 is asymptotically equivalent to $4D_2^2$ as $m \to \infty$. (The notation, $O_p(\cdot)$, order in probability, is defined in (2.9.1).)

13.4.3 (Gower, 1971*a*) (a) For quantitative observations with values $x_1, \ldots, x_n$, consider the similarity coefficient $s_{ij} = 1 - \{|x_i - x_j|/R\}$, where R is the range. Show that we can assume $R = 1$ by rescaling, and $1 = x_1 \geq x_2 \geq \ldots \geq x_{n-1} \geq x_n = 0$ by permuting the rows of the data matrix and shifting the origin so that $x_n = 0$. Thus the similarity matrix $\mathbf{S}_n = (s_{ij})$ is

$$\begin{bmatrix} 1 & 1-(x_1-x_2) & 1-(x_1-x_3) & \cdots & 1-(x_1-x_n) \\ & 1 & 1-(x_2-x_3) & \cdots & 1-(x_2-x_n) \\ & & 1 & \cdots & 1-(x_3-x_n) \\ & & & & \vdots \\ & & & & 1 \end{bmatrix}.$$

By a series of elementary transformations, show that for $1 \leq p \leq n$, the principal $p \times p$ leading minor is

$$\Delta_{pp} = 2^{p-1}[1 - \tfrac{1}{2}(x_1 - x_p)] \prod_{i=1}^{p-1} (x_i - x_{i+1}).$$

Since $x_1 - x_p \leqslant x_1 - x_n = 1$ and $x_i - x_{i+1} \geqslant 0$ show that $\Delta_{pp} \geqslant 0$ for all p, and therefore $\mathbf{S}_n$ is a p.s.d. matrix. Extend this result to t quantitative variates, noting that the average of t p.s.d. matrices is again p.s.d. Hence deduce that $(s_{ij}^{(3)})$ is p.s.d.

(b) Suppose $\mathbf{S}_n^*$ is defined by $s_{ij}^* = 1 - \{|x_i - x_j|/T\}$, where $T > R$. Show that again $\Delta_{pp} \geqslant 0$ and so $\mathbf{S}_n^*$ is p.s.d.

(c) If R is replaced by the standard deviation s, then we may have $s < R$, so $1 - \frac{1}{2}(x_1 - x_p)$ need not be positive and therefore Δ_{pp} need not be positive. Hence $\mathbf{S}_n$ obtained from $s_{ij} = 1 - \{|x_i - x_j|/s\}$ need not be p.s.d.

13.4.4 Obtain $\mathbf{S}_1$ for the data in Example 13.4.2 and apply the single linkage method to obtain clusters. Why is it an unsatisfactory approach?

13.4.5 (D. G. Kendall, 1971.) Let $\mathbf{X}$ be a $(n \times p)$ matrix of actual numbers of occurrences (or proportions) called an *abundance* matrix; that is, x_{ik} is a non-negative integer representing the number of artefacts of type k found on individual i; or x_{ik} is a fraction between 0 and 1 representing the fraction of artefacts of type k which are found on individual i (so $\sum_{k=1}^{p} x_{ik} = 1$). Define a similarity measure

$$s_{ij} = \sum_{k=1}^{p} \min(x_{ik}, x_{jk}).$$

If $\sum_{k=1}^{p} x_{ik} = 1$, show that

$$s_{ij} = 1 - \frac{1}{2} \sum_{k=1}^{p} |x_{ik} - x_{jk}|$$

and that (s_{ij}) need not be p.s.d.

13.4.6 Show that the measure of divergence between two p.d.f.s f_1 and f_2 given by

$$\int (f_1 - f_2) \log\left(\frac{f_2}{f_1}\right) dx$$

does not satisfy the triangular inequality.

13.4.7 Let $\mathbf{x}_i \sim N_p(\boldsymbol{\mu}_i, \boldsymbol{\Sigma})$, $i = 1, 2$. Show that the squared Bhattacharyya distance

$$\int (f_1^{1/2} - f_2^{1/2})^2 \, dx = 2 - 2\int (f_1 f_2)^{1/2} \, dx$$

between the points is a function of the squared Mahalanobis distance,

$$D^2 = (\boldsymbol{\mu}_1 - \boldsymbol{\mu}_2)' \boldsymbol{\Sigma}^{-1} (\boldsymbol{\mu}_1 - \boldsymbol{\mu}_2).$$

14
Multidimensional Scaling

14.1 Introduction

Multidimensional scaling (MDS) is concerned with the problem of constructing a configuration of n points in Euclidean space using information about the distances between the n objects. The interpoint distances themselves may be subject to error. Note that this technique differs from those described in earlier chapters in an important way. Previously, the data points were directly observable as n points in p-space, but here we can only observe a function of the data points, namely the $\frac{1}{2}n(n-1)$ distances.

A simple test example is given in Table 14.1.1, where we are given the road distances (not the "shortest distances") between towns, and the aim is to construct a geographical map of England based on this information. Since these road distances equal the true distances subject to small perturbations, we expect that any sensible MDS method will produce a configuration which is "close" to the true map of these towns.

However, the distances need not be based on Euclidean distances, and can represent many types of dissimilarities between objects. Also in some cases, we start not with dissimilarities but with a set of *similarities* between objects. (We have already given a general account of different types of distances and similarities in Section 13.4.) An example of similarity between two Morse code signals could be the percentage of people who think the Morse code sequences corresponding to the pair of characters are identical after hearing them in rapid succession. Table 14.1.2 gives such data for characters consisting of the 10 numbers in Morse code. These "similarities" can then be used to plot the signals in two-dimensional space. The purpose of this plot is to observe which signals were "like", i.e. near, and which were "unlike", i.e. far from each other, and also to observe the general interrelationship between signals.

Table 14.1.1 Road distances in miles between 12 British towns †

	1	2	3	4	5	6	7	8	9	10	11	12
1												
2	244											
3	218	350										
4	284	77	369									
5	197	167	347	242								
6	312	444	94	463	441							
7	215	221	150	236	279	245						
8	469	583	251	598	598	169	380					
9	166	242	116	257	269	210	55	349				
10	212	53	298	72	170	392	168	531	190			
11	253	325	57	340	359	143	117	264	91	273		
12	270	168	284	164	277	378	143	514	173	111	256	

† 1 = Aberystwyth, 2 = Brighton, 3 = Carlisle, 4 = Dover, 5 = Exeter, 6 = Glasgow, 7 = Hull, 8 = Inverness, 9 = Leeds, 10 = London, 11 = Newcastle, 12 = Norwich.

Definition *An* $(n \times n)$ *matrix* $\mathbf{D}$ *is called a distance matrix if it is symmetric and*

$$d_{rr} = 0, \qquad d_{rs} \geqslant 0, \qquad r \neq s.$$

Starting with a distance matrix $\mathbf{D}$, the object of MDS is to find points $P_1, \ldots, P_n$ in k dimensions such that if $\hat{d}_{rs}$ denotes the Euclidean distance between P_r and P_s, then $\hat{\mathbf{D}}$ is "similar" in some sense to $\mathbf{D}$. The points P_r are unknown and usually the dimension k is also unknown. In practice

Table 14.1.2 Percentage of times that the pairs of Morse code signals for two numbers were declared to be the same by 598 subjects (Rothkopf, 1957; the reference contains entries for 26 letters as well)

	1	2	3	4	5	6	7	8	9	0
1	84									
2	62	89								
3	16	59	86							
4	6	23	38	89						
5	12	8	27	56	90					
6	12	14	33	34	30	86				
7	20	25	17	24	18	65	85			
8	37	25	16	13	10	22	65	88		
9	57	28	9	7	5	8	31	58	91	
0	52	18	9	7	5	18	15	39	79	94

one usually limits the dimension to $k = 1, 2$, or 3 in order to facilitate the interpretation of the solution.

Nature of the solution It is important to realize that the configuration produced by any MDS method is indeterminate with respect to translation, rotation, and reflection. In the two-dimensional case of road distances (Table 14.1.1), the whole configuration of points can be "shifted" from one place in the plane to another and the whole configuration can be "rotated" or "reflected".

In general, if $P_1, \ldots, P_n$ with coordinates $\mathbf{x}_i' = (x_{i1}, \ldots, x_{ip})$, $i = 1, \ldots, n$, represents an MDS solution in p dimensions, then

$$\mathbf{y}_i = \mathbf{A}\mathbf{x}_i + \mathbf{b}, \qquad i = 1, \ldots, n,$$

is also a solution, where $\mathbf{A}$ is an orthogonal matrix and $\mathbf{b}$ is any vector.

Types of solution Methods of solution using only the rank order of the distances

$$d_{r_1 s_1} < d_{r_2 s_2} < \cdots < d_{r_m s_m}, \qquad m = \tfrac{1}{2}n(n-1), \tag{14.1.1}$$

where $(r_1, s_1), \ldots, (r_m, s_m)$ denotes all pairs of subscripts of r and s, $r < s$, are termed *non-metric methods of multidimensional scaling.*

The rank orders are invariant under monotone increasing transformations f of the d_{rs}, i.e.

$$d_{r_1 s_1} < d_{r_2 s_2} < \cdots \Leftrightarrow f(d_{r_1 s_1}) < f(d_{r_2 s_2}) < \cdots.$$

Therefore the configurations which arise from non-metric scaling are indeterminate not only with respect to translation, rotation, and reflection, but also with respect to uniform expansion or contraction.

Solutions which try to obtain P_i directly from the given distances are called *metric methods*. These methods derive P_r such that, in some sense, the new distances $\hat{d}_{rs}$ between points P_r and P_s are as close to the original d_{rs} as possible.

In general the purpose of MDS is to provide a "picture" which can be used to give a meaningful interpretation of the data. Hopefully, the picture will convey useful information about the relationships between the objects. Note that this chapter differs from most of the earlier chapters in that no probabilistic framework is set up; the technique is purely data-analytic.

One important use of MDS is seriation. The aim here is to order a set of objects chronologically on the basis of dissimilarities or similarities between them. Suppose the points in the MDS configuration in $k = 2$ dimensions lie nearly on a smooth curve. This property then suggests that the differences in the data are in fact one-dimensional and the ordering of the points along this curve can be used to seriate the data. (See D. G. Kendall, 1971.)

14.2 Classical Solution

14.2.1 Some theoretical results

Definition *A distance matrix* $\mathbf{D}$ *is called* Euclidean *if there exists a configuration of points in some Euclidean space whose interpoint distances are given by* $\mathbf{D}$; *that is, if for some p, there exists points* $\mathbf{x}_1, \ldots, \mathbf{x}_n \in R^p$ *such that*

$$d_{rs}^2 = (\mathbf{x}_r - \mathbf{x}_s)'(\mathbf{x}_r - \mathbf{x}_s). \tag{14.2.1}$$

The following theorem enables us to tell whether $\mathbf{D}$ is Euclidean, and, if so, how to find a corresponding configuration of points. First we need some notation. For any distance matrix $\mathbf{D}$, let

$$\mathbf{A} = (a_{rs}), \qquad a_{rs} = -\tfrac{1}{2}d_{rs}^2 \tag{14.2.2}$$

and set

$$\mathbf{B} = \mathbf{HAH}, \tag{14.2.3}$$

where $\mathbf{H} = \mathbf{I} - n^{-1}\mathbf{1}\mathbf{1}'$ is the $(n \times n)$ centring matrix.

Theorem 14.2.1 *Let* $\mathbf{D}$ *be a distance matrix and define* $\mathbf{B}$ *by* (14.2.3). *Then* $\mathbf{D}$ *is Euclidean if and only if* $\mathbf{B}$ *is p.s.d. In particular, the following results hold*:

(a) *If* $\mathbf{D}$ *is the matrix of Euclidean interpoint distances for a configuration* $\mathbf{Z} = (\mathbf{z}_1, \ldots, \mathbf{z}_n)'$, *then*

$$b_{rs} = (\mathbf{z}_r - \bar{\mathbf{z}})'(\mathbf{z}_s - \bar{\mathbf{z}}), \qquad r, s = 1, \ldots, n. \tag{14.2.4}$$

In matrix form (14.2.4) *becomes* $\mathbf{B} = (\mathbf{HZ})(\mathbf{HZ})'$ *so* $\mathbf{B} \geq 0$. *Note that* $\mathbf{B}$ *can be interpreted as the "centred inner product matrix" for the configuration* $\mathbf{Z}$.

(b) *Conversely, if* $\mathbf{B}$ *is p.s.d. of rank p then a configuration corresponding to* $\mathbf{B}$ *can be constructed as follows. Let* $\lambda_1 > \cdots > \lambda_p$ *denote the positive eigenvalues of* $\mathbf{B}$ *with corresponding eigenvectors* $\mathbf{X} = (\mathbf{x}_{(1)}, \ldots, \mathbf{x}_{(p)})$ *normalized by*

$$\mathbf{x}_{(i)}'\mathbf{x}_{(i)} = \lambda_i, \qquad i = 1, \ldots, p. \tag{14.2.5}$$

Then the points P_r *in* R^p *with coordinates* $\mathbf{x}_r = (x_{r1}, \ldots, x_{rp})'$ *(so* $\mathbf{x}_r$ *is the rth* row *of* $\mathbf{X}$*) have interpoint distances given by* $\mathbf{D}$. *Further, this* *configuration has centre of gravity* $\bar{\mathbf{x}} = \mathbf{0}$, *and* $\mathbf{B}$ *represents the inner product matrix for this configuration.*

Proof We first prove (a). Suppose

$$d_{rs}^2 = -2a_{rs} = (\mathbf{z}_r - \mathbf{z}_s)'(\mathbf{z}_r - \mathbf{z}_s). \tag{14.2.6}$$

We can write

$$\mathbf{B} = \mathbf{HAH} = \mathbf{A} - n^{-1}\mathbf{AJ} - n^{-1}\mathbf{JA} + n^{-2}\mathbf{JAJ}, \tag{14.2.7}$$

where $\mathbf{J} = \mathbf{11}'$. Now

$$\frac{1}{n}\mathbf{AJ} = \begin{bmatrix} \bar{a}_{1\cdot} & \dots & \bar{a}_{1\cdot} \\ \cdot & & \cdot \\ \cdot & & \cdot \\ \cdot & & \cdot \\ \bar{a}_{n\cdot} & \dots & \bar{a}_{n\cdot} \end{bmatrix}, \quad \frac{1}{n}\mathbf{JA} = \begin{bmatrix} \bar{a}_{\cdot 1} & \dots & \bar{a}_{\cdot n} \\ \cdot & & \cdot \\ \cdot & & \cdot \\ \cdot & & \cdot \\ \bar{a}_{\cdot 1} & \dots & \bar{a}_{\cdot n} \end{bmatrix}, \quad \frac{1}{n^2}\mathbf{JAJ} = \begin{bmatrix} \bar{a}_{\cdot\cdot} & \dots & \bar{a}_{\cdot\cdot} \\ \cdot & & \cdot \\ \cdot & & \cdot \\ \cdot & & \cdot \\ \bar{a}_{\cdot\cdot} & \dots & \bar{a}_{\cdot\cdot} \end{bmatrix},$$

where

$$\bar{a}_{r\cdot} = \frac{1}{n}\sum_{s=1}^{n} a_{rs}, \qquad \bar{a}_{\cdot s} = \frac{1}{n}\sum_{r=1}^{n} a_{rs}, \qquad \bar{a}_{\cdot\cdot} = \frac{1}{n^2}\sum_{r,s=1}^{n} a_{rs}. \tag{14.2.8}$$

Thus

$$b_{rs} = a_{rs} - \bar{a}_{r\cdot} - \bar{a}_{\cdot s} + \bar{a}_{\cdot\cdot}. \tag{14.2.9}$$

After substituting for a_{rs} from (14.2.6) and using (14.2.8), this formula simplifies to

$$b_{rs} = (\mathbf{z}_r - \bar{\mathbf{z}})'(\mathbf{z}_s - \bar{\mathbf{z}}). \tag{14.2.10}$$

(See Exercise 14.2.1 for further details.) Thus (a) is proved.

Conversely, to prove (b) suppose $\mathbf{B} \geq 0$ and consider the configuration given in the thorem. Let $\mathbf{\Lambda} = \operatorname{diag}(\lambda_1, \dots, \lambda_p)$ and let $\mathbf{\Gamma} = \mathbf{X}\mathbf{\Lambda}^{-1/2}$, so that the columns of $\mathbf{\Gamma}$, $\boldsymbol{\gamma}_{(i)} = \lambda_i^{-1/2}\mathbf{x}_{(i)}$ are *standardized* eigenvectors of $\mathbf{B}$. Then by the spectral decomposition theorem (Remark 4 after Theorem A.6.4),

$$\mathbf{B} = \mathbf{\Gamma\Lambda\Gamma}' = \mathbf{XX}';$$

that is, $b_{rs} = \mathbf{x}_r'\mathbf{x}_s$, so $\mathbf{B}$ represents the inner product matrix for this configuration.

We must now show that $\mathbf{D}$ represents the matrix of interpoint distances for this configuration. Using (14.2.9) to write $\mathbf{B}$ in terms of $\mathbf{A}$, we get

$$\begin{aligned}(\mathbf{x}_r - \mathbf{x}_s)'(\mathbf{x}_r - \mathbf{x}_s) &= \mathbf{x}_r'\mathbf{x}_r - 2\mathbf{x}_r'\mathbf{x}_s + \mathbf{x}_s'\mathbf{x}_s \\ &= b_{rr} - 2b_{rs} + b_{ss} \\ &= a_{rr} - 2a_{rs} + a_{ss} \\ &= -2a_{rs} = d_{rs}^2 \qquad (14.2.11)\end{aligned}$$

because $a_{rr} = -\frac{1}{2}d_{rr}^2 = 0$ and $-2a_{rs} = d_{rs}^2$.

Finally, note that $\mathbf{B1} = \mathbf{HAH1} = \mathbf{0}$, so that $\mathbf{1}$ is an eigenvector of $\mathbf{B}$ corresponding to the eigenvalue 0. Thus $\mathbf{1}$ is orthogonal to the columns of $\mathbf{X}$, $\mathbf{x}_{(i)}'\mathbf{1} = 0$, $i = 1, \ldots, p$. Hence

$$n\bar{\mathbf{x}} = \sum_{r=1}^{n} \mathbf{x}_i = \mathbf{X}'\mathbf{1} = (\mathbf{x}_{(1)}'\mathbf{1}, \ldots, \mathbf{x}_{(p)}'\mathbf{1})' = \mathbf{0}$$

so that the centre of gravity of this configuration lies at the origin. ■

Remarks (1) The matrix $\mathbf{X}$ can be visualized in the following way in terms of the eigenvectors of $\mathbf{B}$ and the corresponding points:

$$\begin{array}{ll|cccc|c} & & \multicolumn{4}{c|}{\text{Eigenvalues}} & \text{Vector} \\ & & \lambda_1 & \lambda_2 & & \lambda_p & \text{notation} \\ \text{Points} & P_1 & x_{11} & x_{12} & \cdots & x_{1p} & \mathbf{x}_1' \\ & P_2 & x_{21} & x_{22} & \cdots & x_{2p} & \mathbf{x}_2' \\ & \vdots & \vdots & \vdots & & \vdots & \vdots \\ & P_n & x_{n1} & x_{n2} & \cdots & x_{np} & \mathbf{x}_n' \\ \text{Vector notation.} & & \mathbf{x}_{(1)} & \mathbf{x}_{(2)} & \cdots & \mathbf{x}_{(p)} & \end{array}$$

Centre of gravity:

$$\bar{x}_1 = 0, \bar{x}_2 = 0, \ldots, \bar{x}_p = 0, \qquad \bar{\mathbf{x}} = \frac{1}{n}\sum \mathbf{x}_r = \mathbf{0}.$$

In short, the rth *row* of $\mathbf{X}$ contains the coordinates of the rth point, whereas the ith *column* of $\mathbf{X}$ contains the eigenvector corresponding to

(2) Geometrically, if $\mathbf{B}$ is the centred inner product matrix for a configuration $\mathbf{Z}$, then $b_{rr}^{1/2}$ equals the distance between $\mathbf{z}_r$ and $\bar{\mathbf{z}}$, and $b_{rs}/(b_{rr}b_{ss})^{1/2}$ equals the cosine of the angle subtended at $\bar{\mathbf{z}}$ between $\mathbf{z}_r$ and $\mathbf{z}_s$.

(3) Note that $\mathbf{1}$ is an eigenvector of $\mathbf{B}$ whether $\mathbf{D}$ is Euclidean or not.

(4) The theorem does not hold if $\mathbf{B}$ has negative eigenvalues. The reason can be found in (14.2.5) because it is impossible to normalize a vector to have a negative squared norm.

(5) *History* This result was first proved by Schoenberg (1935) and Young and Householder (1938). Its use as a basis for multidimensional scaling was put forward by Torgerson (1958) and the ideas were substantially amplified by Gower (1966).

14.2.2 A practical algorithm

Suppose we are given a distance matrix $\mathbf{D}$ which we hope can approximately represent the interpoint distances of a configuration in a Euclidean space of low dimension k (usually $k=1$, 2, or 3). The matrix $\mathbf{D}$ may or may not be Euclidean; however, even if $\mathbf{D}$ is Euclidean, the dimension of the space in which it can be represented will usually be too large to be of practical interest.

One possible choice of configuration in k dimensions is suggested by Theorem 14.2.1. *Choose the configuration in R^k whose coordinates are determined by the first k eigenvectors of* $\mathbf{B}$. If the first k eigenvalues of $\mathbf{B}$ are "large" and positive and the other eigenvalues are near 0 (positive or negative), then hopefully, the interpoint distances of this configuration will closely approximate $\mathbf{D}$.

This configuration is called the *classical solution to the MDS problem in k dimensions.* It is a metric solution and its optimal properties are discussed in Section 14.4. For computational purposes we shall summarize the calculations involved:

(a) From $\mathbf{D}$ construct the matrix $\mathbf{A}=(-\frac{1}{2}d_{rs}^2)$.

(b) Obtain the matrix $\mathbf{B}$ with elements $b_{rs}=a_{rs}-\bar{a}_{r\cdot}-\bar{a}_{\cdot s}+\bar{a}_{\cdot\cdot}$.

(c) Find the k largest eigenvalues $\lambda_1>\cdots>\lambda_k$ of $\mathbf{B}$ (k chosen ahead of time), with corresponding eigenvectors $\mathbf{X}=(\mathbf{x}_{(1)},\ldots,\mathbf{x}_{(k)})$ which are normalized by $\mathbf{x}_{(i)}'\mathbf{x}_{(i)}=\lambda_i$, $i=1,\ldots,k$. (We are supposing here that the first k eigenvalues are all positive.)

(d) The required coordinates of the points P_r are $\mathbf{x}_r=(x_{r1},\ldots,x_{rp})'$, $r=1,\ldots,k$, the *rows* of $\mathbf{X}$.

Example 14.2.1 To illustrate the algorithm, consider a (7×7) distance matrix

$$\mathbf{D}=\begin{bmatrix} 0 & 1 & \sqrt{3} & 2 & \sqrt{3} & 1 & 1 \\ & 0 & 1 & \sqrt{3} & 2 & \sqrt{3} & 1 \\ & & 0 & 1 & \sqrt{3} & 2 & 1 \\ & & & 0 & 1 & \sqrt{3} & 1 \\ & & & & 0 & 1 & 1 \\ & & & & & 0 & 1 \\ & & & & & & 0 \end{bmatrix}.$$

Constructing the matrix **A** from (14.2.2), it is found that

$$\bar{a}_{r\cdot}=-\tfrac{13}{14}, \qquad r=1,\ldots,6, \quad \bar{a}_{7\cdot}=-\tfrac{3}{7};$$
$$\bar{a}_{r\cdot}=\bar{a}_{\cdot r}, \qquad \bar{a}_{\cdot\cdot}=-\tfrac{6}{7}.$$

Hence from (14.2.9) the matrix **B** is given by

$$\mathbf{B}=\frac{1}{2}\begin{bmatrix} 2 & 1 & -1 & -2 & -1 & 1 & 0 \\ & 2 & 1 & -1 & -2 & -1 & 0 \\ & & 2 & 1 & -1 & -2 & 0 \\ & & & 2 & 1 & -1 & 0 \\ & & & & 2 & 1 & 0 \\ & & & & & 2 & 0 \\ & & & & & & 0 \end{bmatrix}.$$

The columns of **B** are linearly dependent. It can be seen that

$$\mathbf{b}_{(3)}=\mathbf{b}_{(2)}-\mathbf{b}_{(1)}, \qquad \mathbf{b}_{(4)}=-\mathbf{b}_{(1)}, \qquad \mathbf{b}_{(5)}=-\mathbf{b}_{(2)},$$
$$\mathbf{b}_{(6)}=\mathbf{b}_{(1)}-\mathbf{b}_{(2)}, \qquad \mathbf{b}_{(7)}=\mathbf{0}. \qquad (14.2.12)$$

Hence the rank of matrix **B** is at the most 2. From the leading (2×2) matrix it is clear that the rank is 2. Thus, a configuration exactly fitting the distance matrix can be constructed in $k=2$ dimensions.

The eigenvalues of **B** are found to be

$$\lambda_1=3, \qquad \lambda_2=3, \qquad \lambda_3=\cdots=\lambda_7=0.$$

The configuration can be constructed using any two orthogonal vectors for the eigenspace corresponding to $\lambda = 3$, such as

$$\mathbf{x}'_{(1)} = (a, a, 0, -a, -a, 0, 0), \qquad a = \tfrac{1}{2}\sqrt{3},$$
$$\mathbf{x}'_{(2)} = (b, -b, -2b, -b, b, 2b, 0), \qquad b = \tfrac{1}{2}.$$

Then the coordinates of the seven points are

A	B	C	D	E	F	G
$(\frac{1}{2}\sqrt{3}, \frac{1}{2})$	$(\frac{1}{2}\sqrt{3}, -\frac{1}{2})$	$(0, -1)$	$(-\frac{1}{2}\sqrt{3}, -\frac{1}{2})$	$(-\frac{1}{2}\sqrt{3}, \frac{1}{2})$	$(0, 1)$	$(0, 0)$.

The centre of gravity of these points is of course $(0, 0)$, and it can be verified that the distance matrix for these points is $\mathbf{D}$. In fact, A to F are vertices of a hexagon with each side of length 1, and the line FC is the y axis. Its centre is G. (Indeed, $\mathbf{D}$ was constructed with the help of these points.) A similar configuration based on a non-Euclidean distance is described in Exercise 14.2.7.

14.2.3 Similarities

In some situations we start not with distances between n objects, but with similarities. Recall that an $(n \times n)$ matrix $\mathbf{C}$ is called a similarity matrix if $c_{rs} = c_{sr}$ and if

$$c_{rs} \leq c_{rr} \quad \text{for all } r, s. \tag{14.2.13}$$

Examples of possible similarity matrices were given in Section 13.4.

To use the techniques of the preceding sections, it is necessary to transform the similarities to distances. A useful transformation is the following.

Definition *The* standard transformation *from a similarity matrix* $\mathbf{C}$ *to a distance matrix* $\mathbf{D}$ *is defined by*

$$d_{rs} = (c_{rr} - 2c_{rs} + c_{ss})^{1/2}. \tag{14.2.14}$$

Note that if (14.2.13) holds, then the quantity under the square root in (14.2.14) must be non-negative, and that $d_{rr} = 0$. Hence $\mathbf{D}$ is a distance matrix.

Many of the similarity matrices discussed in Section 13.5 were p.s.d. This property is attractive because the resulting distance matrix, using the standard transformation, is Euclidean.

Theorem 14.2.2 *If* $\mathbf{C} \geq 0$, *then the distance matrix* $\mathbf{D}$ *defined by the standard transformation* (14.2.14) *is Euclidean, with centred inner product matrix* $\mathbf{B} = \mathbf{HCH}$.

Proof First note that since $\mathbf{C} \geq 0$,

$$d_{rs}^2 = c_{rr} - 2c_{rs} + c_{ss} = \mathbf{x}'\mathbf{C}\mathbf{x} \geq 0,$$

where $\mathbf{x}$ is a vector with $+1$ in the rth place and -1 in the sth place, for $r \neq s$. Thus, the standard transformation is well defined and $\mathbf{D}$ is a distance matrix.

Let $\mathbf{A}$ and $\mathbf{B}$ be defined by (14.2.2) and (14.2.3). Since $\mathbf{HCH}$ is also p.s.d., it is sufficient to prove that $\mathbf{B}=\mathbf{HCH}$ in order to conclude that $\mathbf{D}$ is Euclidean with centred inner product matrix $\mathbf{HCH}$.

Now $\mathbf{B}=\mathbf{HAH}$ can be written elementwise using (14.2.9). Substituting for $a_{rs} = -\frac{1}{2}d_{rs}^2$ using (14.2.14) gives

$$\begin{aligned}-2b_{rs} &= d_{rs}^2 - \frac{1}{n}\sum_{i=1}^{n} d_{ri}^2 - \frac{1}{n}\sum_{j=1}^{n} d_{js}^2 + \frac{1}{n^2}\sum_{i,j=1}^{n} d_{ij}^2 \\ &= c_{rr} - 2c_{rs} + c_{ss} - \frac{1}{n}\sum_{i=1}^{n}(c_{rr} - 2c_{ri} + c_{ii}) \\ &\quad - \frac{1}{n}\sum_{j=1}^{n}(c_{jj} - 2c_{js} + c_{ss}) + \frac{1}{n^2}\sum_{i,j=1}^{n}(c_{ii} - 2c_{ij} + c_{jj}) \\ &= -2c_{rs} + 2\bar{c}_{r\cdot} + 2\bar{c}_{\cdot s} - 2c_{\cdot\cdot}.\end{aligned}$$

Hence

$$b_{rs} = c_{rs} - \bar{c}_{r\cdot} - \bar{c}_{\cdot s} + c_{\cdot\cdot}$$

or, in matrix form, $\mathbf{B}=\mathbf{HCH}$. Thus the theorem is proved. ■

Example 14.2.2 We now consider the Morse code data given in Table 14.1.2 and described in Section 14.1. The data is presented as a similarity matrix $\mathbf{C}=(c_{rs})$. Using the standard transformation from similarities to distances, take

$$d_{rs} = (c_{rr} + c_{ss} - 2c_{rs})^{1/2}.$$

We obtain the eigenvectors and eigenvalues of $\mathbf{HCH}$ in accordance with Theorem 14.2.2. It is found that

$$\lambda_1 = 187.4, \quad \lambda_2 = 121.0, \quad \lambda_3 = 95.4, \quad \lambda_4 = 55.4, \quad \lambda_5 = 46.6,$$
$$\lambda_6 = 31.5, \quad \lambda_7 = 9.6, \quad \lambda_8 = 4.5, \quad \lambda_9 = 0.0, \quad \lambda_{10} = -4.1.$$

The first two eigenvectors appropriately normalized are

$$(-4.2, -0.3, 3.7, 5.6, 5.4, 3.8, 0.9, -3.0, -6.2, -5.7),$$
$$(-3.2, -5.8, -4.3, -0.6, 0.0, 4.0, 5.5, 3.6, 0.6, 0.2).$$

However, the first two principal coordinates account for only $100(\lambda_1 + \lambda_2)/\Sigma|\lambda_i|$ percent $= 56\%$ of the total configuration. The points P_r are plotted in Figure 14.2.1. It can be seen that the x_1 axis measures the increasing number of dots whereas the x_2 axis measures the heterogeneity

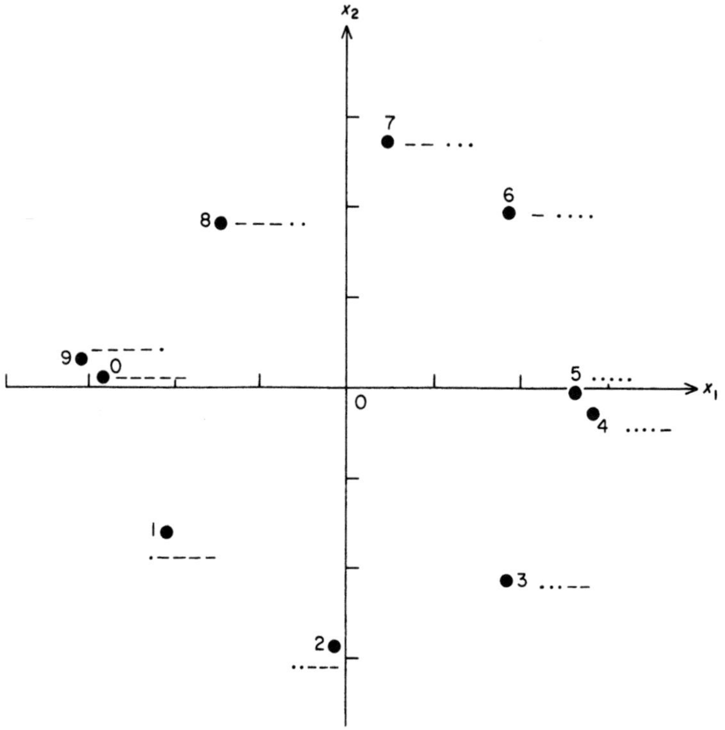

Figure 14.2.1 Classical solution for Morse code data in Table 14.1.1.

of the signal. If we regard the OP_rs as vectors, then angles between consecutive vectors are about 45° except between the vectors from 0 and 9, and 4 and 5. The small separation between these latter points might be expected because the change in just the last character may not make much impact on the untrained ear. Thus the configuration brings out the main features of the general interrelationship between these signals. The points are roughly on a circle in the order $0, 1, \ldots, 9$.

14.3 Duality Between Principal Coordinate Analysis and Principal Component Analysis

So far in this chapter we have treated the $(n \times n)$ matrix $\mathbf{D}$ of distances between n objects as the starting point for our analysis. However, in many situations, we start with a data matrix $\mathbf{X}(n \times p)$ and must make a choice of distance function.

Several possibilities for a distance function were discussed in Section 13.4. The simplest choice is, of course, Euclidean distance. In this case there is a close connection between the work in Section 14.2 and principal component analysis.

Let $\mathbf{X}(n \times p)$ be a data matrix and let $\lambda_1 \geq \cdots \geq \lambda_p$ be the eigenvalues of $n\mathbf{S} = \mathbf{X}'\mathbf{H}\mathbf{X}$, where $\mathbf{S}$ is the sample covariance matrix. For simplicity we shall suppose the eigenvalues are all non-zero and distinct. Then $\lambda_1, \ldots, \lambda_p$ are also the non-zero eigenvalues of $\mathbf{B} = \mathbf{H}\mathbf{X}\mathbf{X}'\mathbf{H}$. Note that the rows of $\mathbf{HX}$ are just the centred rows of $\mathbf{X}$, so that $\mathbf{B}$ represents the centred inner product matrix,

$$b_{rs} = (\mathbf{x}_r - \bar{\mathbf{x}})'(\mathbf{x}_s - \bar{\mathbf{x}}).$$

Definition *Let $\mathbf{v}_{(i)}$ be the ith eigenvector of $\mathbf{B}$,*

$$\mathbf{B}\mathbf{v}_{(i)} = \lambda_i \mathbf{v}_{(i)},$$

normalized by $\mathbf{v}'_{(i)}\mathbf{v}_{(i)} = \lambda_i$, $i = 1, \ldots, p$. For fixed k $(1 \leq k \leq p)$, the rows of $\mathbf{V}_k = (\mathbf{v}_{(1)}, \ldots, \mathbf{v}_{(k)})$ are called the principal coordinates of $\mathbf{X}$ in k dimensions.

Thus, from Theorem 14.2.1, if $\mathbf{D}$ is the Euclidean distance matrix between the rows of $\mathbf{X}$, then the k-dimensional classical solution to the MDS problem is given by the principal coordinates of $\mathbf{X}$ in k dimensions. Principal coordinates are closely linked to principal components, as the following result shows.

Theorem 14.3.1 *The principal coordinates of $\mathbf{X}$ in k dimensions are given by the centred* scores *of the n objects on the first k principal components.*

Proof Let $\boldsymbol{\gamma}_{(i)}$ denote the ith principal component loading vector, standardized by $\boldsymbol{\gamma}'_{(i)}\boldsymbol{\gamma}_{(i)} = 1$, so that by the spectral decomposition theorem (Theorem A.6.4),

$$\mathbf{X}'\mathbf{H}\mathbf{X} = \boldsymbol{\Gamma}\boldsymbol{\Lambda}\boldsymbol{\Gamma}',$$

where $\boldsymbol{\Gamma} = (\boldsymbol{\gamma}_{(1)}, \ldots, \boldsymbol{\gamma}_{(p)})$ and $\boldsymbol{\Lambda} = \text{diag}(\lambda_1, \ldots, \lambda_p)$. By the singular value decomposition theorem (Theorem A.6.5), we can choose the signs of $\boldsymbol{\gamma}_{(i)}$ and $\mathbf{v}_{(i)}$, so that $\mathbf{HX}$ can be written in terms of these eigenvectors as

$$\mathbf{H}\mathbf{X} = \mathbf{V}\boldsymbol{\Gamma}',$$

where $\mathbf{V} = \mathbf{V}_p = (\mathbf{v}_{(1)}, \ldots, \mathbf{v}_{(p)})$.

The scores of the n rows of $\mathbf{HX}$ on the ith principal component are given by the n elements of $\mathbf{H}\mathbf{X}\boldsymbol{\gamma}_{(i)}$. Thus, writing $\boldsymbol{\Gamma}_k = (\boldsymbol{\gamma}_{(1)}, \ldots, \boldsymbol{\gamma}_{(k)})$, the scores on the first k principal components are given by

$$\mathbf{H}\mathbf{X}\boldsymbol{\Gamma}_k = \mathbf{V}\boldsymbol{\Gamma}'\boldsymbol{\Gamma}_k = \mathbf{V}(\mathbf{I}_k, \mathbf{0})' = \mathbf{V}_k,$$

since the columns of $\mathbf{\Gamma}$ are orthogonal to one another. Hence the theorem is proved. ■

Since the columns of $\mathbf{\Gamma}_k$ are orthogonal to one another, $\mathbf{\Gamma}_k'\mathbf{\Gamma}_k = \mathbf{I}_k$, we see that $\mathbf{V}_k = \mathbf{X}\mathbf{\Gamma}_k$ represents a projection of $\mathbf{X}$ onto a k-dimensional subspace of R^p. The projection onto principal coordinates is optimal out of all k-dimensional projections because it is closest to the original p-dimensional configuration. (See Theorem 14.4.1.)

This result is dual to the result in principal component analysis that the sum of the variances of the first k principal components is larger than the sum of the variances of any other k uncorrelated linear combination of the columns of $\mathbf{X}$. (See Exercise 8.2.5.)

14.4 Optimal Properties of the Classical Solution and Goodness of Fit

Given a distance matrix $\mathbf{D}$, the object of MDS is to find a configuration $\hat{\mathbf{X}}$ in a low-dimensional Euclidean space R^k whose interpoint distances, $\hat{d}_{rs}^2 = (\hat{\mathbf{x}}_r - \hat{\mathbf{x}}_s)'(\hat{\mathbf{x}}_r - \hat{\mathbf{x}}_s)$ say, closely match $\mathbf{D}$. The circumflex or "hat" will be used in this section to indicate that the interpoint distances $\hat{\mathbf{D}}$ for the configuration $\hat{\mathbf{X}}$ are "fitted" to the original distances $\mathbf{D}$. Similarly, let $\hat{\mathbf{B}}$ denote the fitted centred inner product matrix.

Now let $\mathbf{X}$ be a configuration in R^p and let $\mathbf{L} = (\mathbf{L}_1, \mathbf{L}_2)$ be a $(p \times p)$ orthogonal matrix where $\mathbf{L}_1$ is $(p \times k)$. Then $\mathbf{XL}$ represents a projection of the configuration $\mathbf{X}$ onto the subspace of R^p spanned by the columns of $\mathbf{L}_1$. We can think of $\hat{\mathbf{X}} = \mathbf{XL}$ as a "fitted" configuration in k dimensions.

Since $\mathbf{L}$ is orthogonal, the distances between the rows of $\mathbf{X}$ are the same as the distances between the rows of $\mathbf{XL}$,

$$d_{rs}^2 = \sum_{i=1}^{p} (x_{ri} - x_{si})^2 = \sum_{i=1}^{p} (\mathbf{x}_r'\mathbf{l}_{(i)} - \mathbf{x}_s'\mathbf{l}_{(i)})^2. \tag{14.4.1}$$

If we denote the distances between the rows of $\mathbf{XL}_1$ by $\hat{D}$, then

$$\hat{d}_{rs}^2 = \sum_{i=1}^{k} (\mathbf{x}_r'\mathbf{l}_{(i)} - \mathbf{x}_s'\mathbf{l}_{(i)})^2. \tag{14.4.2}$$

Thus, $\hat{d}_{rs} \leq d_{rs}$; that is, projecting a configuration reduces the interpoint distances. Hence, a measure of the discrepancy between the original configuration $\mathbf{X}$ and the projected configuration $\hat{\mathbf{X}}$ is given by

$$\phi = \sum_{r,s=1}^{n} (d_{rs}^2 - \hat{d}_{rs}^2). \tag{14.4.3}$$

Then the classical solution to the MDS problem in k dimensions has the following optimal property:

Theorem 14.4.1 *Let* $\mathbf{D}$ *be a Euclidean distance matrix corresponding to a configuration* $\mathbf{X}$ *in* R^p, *and fix* k $(1 \leq k < p)$. *Then amongst all projections* $\mathbf{XL}_1$ *of* $\mathbf{X}$ *onto* k*-dimensional subspaces of* R^p, *the quantity* (14.4.3) *is minimized when* $\mathbf{X}$ *is projected onto its principal coordinates in* k *dimensions.*

Proof Using (14.4.1) and (14.4.2) we see that

$$\phi = \sum_{r,s=1}^{n} \sum_{i=k+1}^{p} (\mathbf{x}_r'\mathbf{l}_{(i)} - \mathbf{x}_s'\mathbf{l}_{(i)})^2$$
$$= \operatorname{tr} \mathbf{L}_2' \left\{ \sum_{r,s=1}^{n} (\mathbf{x}_r - \mathbf{x}_s)(\mathbf{x}_r - \mathbf{x}_s)' \right\} \mathbf{L}_2$$
$$= 2n^2 \operatorname{tr} \mathbf{L}_2'\mathbf{S}\mathbf{L}_2$$

since

$$\sum_{r,s=1}^{n} (\mathbf{x}_r - \mathbf{x}_s)(\mathbf{x}_r - \mathbf{x}_s)' = 2n \sum_{r=1}^{n} (\mathbf{x}_r - \bar{\mathbf{x}})(\mathbf{x}_r - \bar{\mathbf{x}})' - 2 \sum_{r=1}^{n} (\mathbf{x}_r - \bar{\mathbf{x}}) \sum_{s=1}^{n} (\mathbf{x}_s - \bar{\mathbf{x}})'$$
$$= 2n^2\mathbf{S}.$$

Letting $\lambda_1 \geq \cdots \geq \lambda_p$ denote the eigenvalues of $n\mathbf{S}$ with standardized eigenvectors $\mathbf{\Gamma} = (\boldsymbol{\gamma}_{(1)}, \ldots, \boldsymbol{\gamma}_{(p)})$, we can write

$$\phi = 2n \operatorname{tr} \mathbf{F}_2' \mathbf{\Lambda} \mathbf{F}_2,$$

where $\mathbf{F}_2 = \mathbf{\Gamma}'\mathbf{L}_2$ is a column orthonormal matrix ($\mathbf{F}_2'\mathbf{F}_2 = \mathbf{I}_{p-k}$). Using Exercise 8.2.7, we see that ϕ is minimized when $\mathbf{F}_2 = (\mathbf{0}, \mathbf{I}_{p-k})'$; that is, when $\mathbf{L}_2 = (\boldsymbol{\gamma}_{(k+1)}, \ldots, \boldsymbol{\gamma}_{(p)})$. Thus the columns of $\mathbf{L}_1$ span the space of the first k eigenvectors of $n\mathbf{S}$ and so $\mathbf{XL}_1$ represents the principal coordinates of $\mathbf{X}$ in k dimensions. Note that for this principal coordinate projection,

$$\phi = 2n(\lambda_{k+1} + \cdots + \lambda_p). \quad \blacksquare \tag{14.4.4}$$

When $\mathbf{D}$ is not necessarily Euclidean, it is more convenient to work with the matrix $\mathbf{B} = \mathbf{HAH}$. If $\hat{\mathbf{X}}$ is a fitted configuration with centred inner product matrix $\hat{\mathbf{B}}$, then a measure of the discrepancy between $\mathbf{B}$ and $\hat{\mathbf{B}}$ is given (Mardia, 1978) by

$$\psi = \sum_{r,s=1}^{n} (b_{rs} - \hat{b}_{rs})^2 = \operatorname{tr} (\mathbf{B} - \hat{\mathbf{B}})^2. \tag{14.4.5}$$

For this measure also, we can prove that the classical solution to the MDS problem is optimal.

Theorem 14.4.2 *If* $\mathbf{D}$ *is a distance matrix (not necessarily Euclidean), then for fixed* k, (14.4.5) *is minimized over all configurations* $\hat{\mathbf{X}}$ *in* k *dimensions when* $\hat{\mathbf{X}}$ *is the classical solution to the MDS problem.*

Proof Let $\lambda_1 \geq \cdots \geq \lambda_n$ denote the eigenvalues of $\mathbf{B}$, some of which might be negative, and let $\mathbf{\Gamma}$ denote the corresponding standardized eigenvectors. For simplicity, suppose $\lambda_k > 0$ (the situation $\lambda_k < 0$ is discussed in Exercise 14.4.2). Let $\hat{\lambda}_1 \geq \cdots \geq \hat{\lambda}_n \geq 0$ denote the eigenvalues of $\hat{\mathbf{B}}$. By the spectral decomposition theorem (Theorem A.6.4) we can write the symmetric matrix $\mathbf{\Gamma}'\hat{\mathbf{B}}\mathbf{\Gamma}$ as

$$\mathbf{\Gamma}'\hat{\mathbf{B}}\mathbf{\Gamma} = \mathbf{G}\hat{\mathbf{\Lambda}}\mathbf{G}',$$

where $\mathbf{G}$ is orthogonal. Then

$$\psi = \operatorname{tr}(\mathbf{B}-\hat{\mathbf{B}})^2 = \operatorname{tr}\mathbf{\Gamma}'(\mathbf{B}-\hat{\mathbf{B}})\mathbf{\Gamma}\mathbf{\Gamma}'(\mathbf{B}-\hat{\mathbf{B}})\mathbf{\Gamma} = \operatorname{tr}(\mathbf{\Lambda}-\mathbf{G}\hat{\mathbf{\Lambda}}\mathbf{G}')(\mathbf{\Lambda}-\mathbf{G}\hat{\mathbf{\Lambda}}\mathbf{G}').$$

We see that for fixed $\hat{\mathbf{\Lambda}}$ (see Exercise 14.4.2), ψ is minimized when $\mathbf{G} = \mathbf{I}$, so that

$$\psi = \sum_{i=1}^{n} (\lambda_i - \hat{\lambda}_i)^2.$$

Since $\hat{\mathbf{X}}$ lies in R^k, $\mathbf{B} = \mathbf{H}\hat{\mathbf{X}}\hat{\mathbf{X}}'\mathbf{H}$ will have at most k non-zero eigenvalues, which must be non-negative. Thus, it is easy to see that ψ is minimized when

$$\hat{\lambda}_i = \begin{cases} \lambda_i, & i = 1, \ldots, k, \\ 0, & i = k+1, \ldots, n. \end{cases}$$

Hence $\hat{\mathbf{B}} = \mathbf{\Gamma}_1\mathbf{\Lambda}_1\mathbf{\Gamma}_1'$, where $\mathbf{\Gamma}_1 = (\boldsymbol{\gamma}_{(1)}, \ldots, \boldsymbol{\gamma}_{(k)})$ and $\mathbf{\Lambda}_1 = \operatorname{diag}(\lambda_1, \ldots, \lambda_k)$ so that $\hat{\mathbf{X}}$ can be taken to equal $\mathbf{\Gamma}_1\mathbf{\Lambda}_1^{1/2}$, the classical solution to the MDS problem in k dimensions. Note that the minimum value of ψ is given by

$$\psi = \lambda_{k+1}^2 + \cdots + \lambda_p^2. \quad \blacksquare \tag{14.4.6}$$

The above two theorems suggest possible *agreement measures* for the "proportion of a distance matrix $\mathbf{D}$ explained" by the k-dimensional classical MDS solution. Supposing $\lambda_k > 0$, these measures are (Mardia, 1978)

$$\alpha_{1,k} = \left(\sum_{i=1}^{k} \lambda_i \Big/ \sum_{i=1}^{n} |\lambda_i|\right) \times 100\%, \tag{14.4.7}$$

and

$$\alpha_{2,k} = \left(\sum_{i=1}^{k} \lambda_k^2 \Big/ \sum_{i=1}^{n} \lambda_i^2\right) \times 100\%. \tag{14.4.8}$$

We need to use absolute values in (14.4.7) because some of the smaller eigenvalues might be negative.

Example 14.4.1 We now consider the example of constructing a map of Britain from the road distances between 12 towns (Table 14.1.1). From this data, it is found that

$$\lambda_1 = 394\,473,\quad \lambda_2 = 63\,634,\quad \lambda_3 = 13\,544,\quad \lambda_4 = 10\,245,$$
$$\lambda_5 = 2465,\quad \lambda_6 = 1450,\quad \lambda_7 = 501,\quad \lambda_8 = 0,$$
$$\lambda_9 = -17,\quad \lambda_{10} = -214,\quad \lambda_{11} = -1141,\quad \lambda_{12} = -7063.$$

We note that the last four eigenvalues are negative, but they are small in relation to $\lambda_1, \ldots, \lambda_4$. We know from Theorem 14.2.1 that some negative values are expected because the distance matrix is not Euclidean.

The percentage variation explained by the first two eigenvectors is

$$\alpha_{1,2} = 92.6\% \quad \text{or} \quad \alpha_{2,2} = 99.8\%.$$

The first two eigenvectors, standardized so that $\mathbf{x}_{(i)}'\mathbf{x}_{(i)} = \lambda_i$, are

$$(45, 203, -138, 212, 189, -234, -8, -382, -32, 153, -120, 112)$$
$$(140, -18, 31, -76, 140, 31, -50, -26, -5, -27, -34, -106).$$

Since the MDS solution is invariant under rotations and translations, the coordinates have been superimposed on the true map in Figure 14.4.1 by Procrustes rotation with scaling (see Example 14.7.1). We find that the two eigenvectors closely reproduce the true map.

14.5 Seriation

14.5.1 Description

Multidimensional scaling can be used to pick out one-dimensional structure in a data set; that is, we expect the data to be parametrized by a single axis. The most common example is seriation, where we want to ascertain the chronological ordering of the data. Note that although MDS can be used to order the data in time, the *direction* of time must be determined independently.

Example 14.5.1 Consider the archeological problem of Example 8.5.1 where the similarity between graves is measured by the number of types of pottery they have in common. Using the similarity matrix $\mathbf{S}_2$ of Example 13.4.2 and the standard transformation of Theorem 14.2.2 (see Exercise 14.2.6), it is found that

$$\lambda_1 = 1.75,\quad \lambda_2 = 0.59,\quad \lambda_3 = 0.35,\quad \lambda_4 = 0.05,\quad \lambda_5 = \lambda_6 = 0,$$

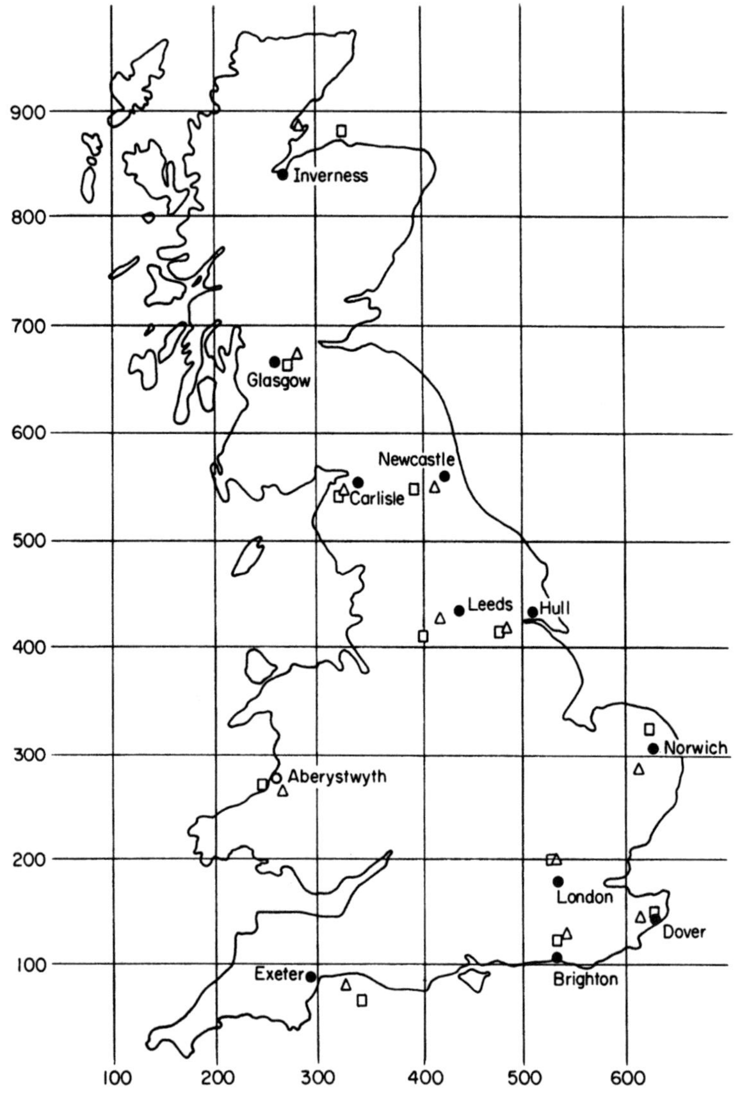

Figure 14.4.1 MDS solutions for the road data in Table 1.2.2. ●, *original points;* △, *classical solution;* □, *Shepard–Kruskal solution.*

with coordinates in two dimensions

$$(-0.60, 0.77, -0.19, -0.60, 0.64, -0.01),$$

and

$$(-0.15, 0.20, 0.60, -0.15, -0.35, -0.14).$$

See Figure 14.5.1. The coordinates in one dimension suggest the order (A, D), C, F, E, B, which is similar but not identical with the ordering given by correspondence analysis (Example 8.5.1).

It is often a good idea to plot the data in more than one dimension to see if the data is in fact one-dimensional. For example, the artificial data in the above example does not particularly seem to lie in one dimension. However, even when the data is truly one-dimensional, it need not lie on the axis of the first dimension but can sometimes lie on a curve as the following section shows.

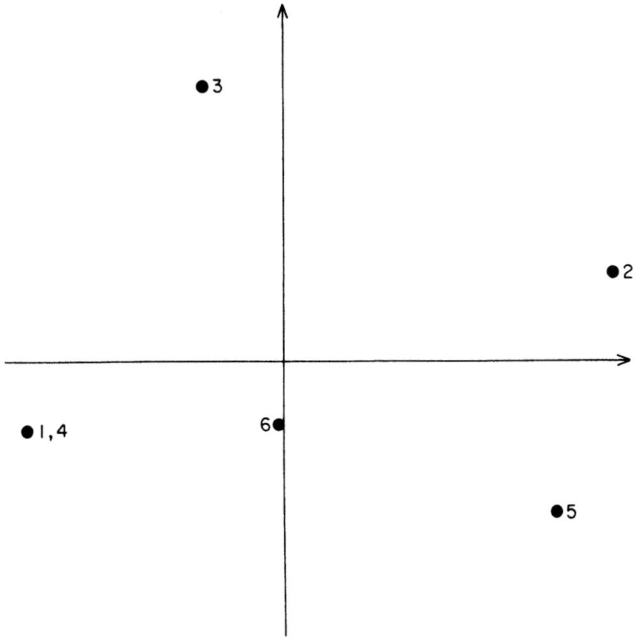

Figure 14.5.1 Classical MDS solution in two dimensions using similarity matrix $\mathbf{S}_2$ *for grave data in Example 14.5.1.*

14.5.2 Horseshoe effect

In some situations we can measure accurately the distance between two objects when they are close together but not when they are far apart. Distances which are "moderate" and those which are "large" appear to be the same. For example, consider the archeology example described above. Graves which are close together in time will have some varieties of pottery in common, but those which are separated by more than a certain gap of time will have *no* varieties in common.

This merging of all "large" distances tends to pull the farthest objects closer together and has been labelled by D. G. Kendall (1971) the "horseshoe effect". This effect can be observed clearly in the following artificial example.

Example 14.5.2 (D. G. Kendall, 1971) Consider the (51×51) similarity matrix **C** defined by

$$c_{rr}=9, \qquad c_{rs}=\begin{cases} 8 & \text{if } 1\leqslant |r-s|\leqslant 3,\\ \cdots, & \\ 1 & \text{if } 22\leqslant |r-s|\leqslant 24,\\ 0 & \text{if } |r-s|\geqslant 25.\end{cases}$$

Using the standard transformation from similarities to distances leads to eight negative eigenvalues (varying from −0.09 to −2.07) and 43 non-negative eigenvalues,

$$126.09, 65.94, 18.17, 7.82, 7.61, 7.38, 7.02, 5.28, 3.44, \ldots, 0.10, 0.$$

A plot of the configuration in two dimensions is given in Figure 14.5.2.

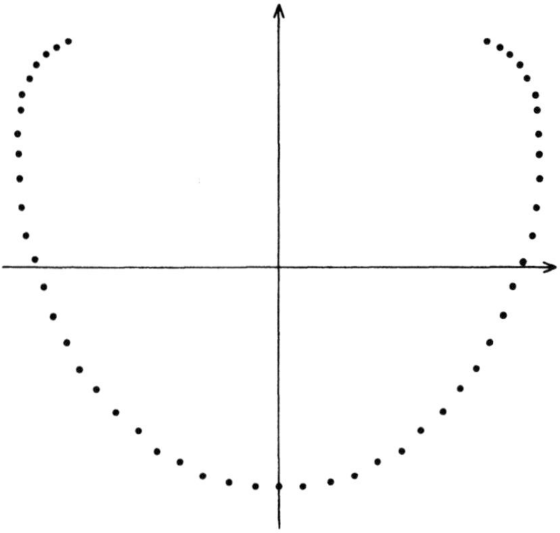

Figure 14.5.2 Two-dimensional representation of Kendall's matrix.

The furthest points are pulled together so that the configuration looks roughly like part of the circumference of a circle. Note that while the ordering is clear from this figure, it is not ascertainable from the one-dimensional classical solution.

14.6 Non-metric Methods

Implicit in the preceding sections is the assumption that there is a "true" configuration in k dimensions with interpoint distances δ_{rs}. We wish to reconstruct this configuration using an observed distance matrix $\mathbf{D}$ whose elements are of the form

$$d_{rs} = \delta_{rs} + e_{rs}. \tag{14.6.1}$$

Here the e_{rs} represent errors of measurement plus distortion errors arising because the distances do not exactly correspond to a configuration in R^k.

However, in some situations it is more realistic to hypothesize a less rigid relationship between d_{rs} and δ_{rs}; namely, suppose

$$d_{rs} = f(\delta_{rs} + e_{rs}), \tag{14.6.2}$$

where f is an unknown monotone increasing function. For this "model", the only information we can use to reconstruct the δ_{rs} is the *rank order* of the d_{rs}. For example, for the road map data, we could try to reconstruct the map of England using the information

the quickest journey is that from Brighton to London;

the next quickest journey is that from Hull to Leeds;

...

the longest journey is that from Dover to Inverness.

In this non-metric approach $\mathbf{D}$ is not thought of as a "distance" matrix but as a "dissimilarity" matrix. In fact the non-metric approach is often most appropriate when the data is presented as a similarity matrix. For in this situation the transformation from similarities to distances is somewhat arbitrary and perhaps the strongest statement one should make is that greater similarity implies less dissimilarity.

An algorithm to construct a configuration based on the rank order information has been developed by Shepard (1962*a*,*b*) and Kruskal (1964).

Shepard–Kruskal algorithm

(a) Given a dissimilarity matrix $\mathbf{D}$, order the off-diagonal elements so that

$$d_{r_1s_1} \leqslant \cdots \leqslant d_{r_ms_m}, \qquad m = \frac{1}{2}n(n-1), \tag{14.6.3}$$

where $(r_1, s_1), \ldots, (r_m, s_m)$ denote all pairs of unequal subscripts, $r_i < s_i$. Say that numbers d^*_{rs} are monotonically related to the d_{rs} (and write $d^*_{rs} \overset{\text{mon}}{\sim} d_{rs}$) if

$$d_{rs} < d_{uv} \Rightarrow d^*_{rs} \leqslant d^*_{uv} \qquad \text{for all } r<s,\ u<v. \tag{14.6.4}$$

(b) Let $\hat{\mathbf{X}}(n \times k)$ be a configuration in R^k with interpoint distances $\hat{d}_{rs}$. Define the (squared) *stress* of $\hat{\mathbf{X}}$ by

$$S^2(\hat{\mathbf{X}}) = \min \sum_{r<s} (d^*_{rs} - \hat{d}_{rs})^2 \Big/ \sum_{r<s} \hat{d}^2_{rs}, \tag{14.6.5}$$

where the minimum is taken over d^*_{rs} such that $d^*_{rs} \overset{\text{mon}}{\sim} d_{rs}$. The d^*_{rs} which minimizes (14.6.5) represent the *least squares monotone regression* of $\hat{d}_{rs}$ on d_{rs}. Thus (14.6.5) represents the extent to which the rank order of the $\hat{d}_{rs}$ disagrees with the rank order of the d_{rs}. If the rank orders match exactly (which is very rare in practice), then $S(\hat{\mathbf{X}}) = 0$. The presence of the denominator in (14.6.5) standardizes the stress and makes it invariant under transformations of the sort $\mathbf{y}_r = c\mathbf{x}_r$, $r = 1, \ldots, n$, $c \neq 0$. The stress is also invariant under transformations of the form $\mathbf{y}_r = \mathbf{A}\mathbf{x}_r + \mathbf{b}$ when $\mathbf{A}$ is orthogonal.

(c) For each dimension k, the configuration which has the smallest stress is called the *best fitting configuration in k dimensions.* Let

$$S_k = \min_{\hat{\mathbf{X}}(n \times k)} S(\hat{\mathbf{X}})$$

denote this minimal stress.

(d) To choose the correct dimension, calculate $S_1, S_2, \ldots,$ until the value becomes low. Say, for example, S_k is low for $k = k_0$. Since S_k is a decreasing function of k, $k = k_0$ is the "right dimension". A rule of thumb is provided by Kruskal to judge the tolerability of S_k; $S_k \geqslant 20\%$, poor; $S_k = 10\%$, fair; $S_k \leqslant 5\%$, good; $S_k = 0$, perfect.

Remarks (1) The "best configuration" starting from an arbitrary initial configuration can be obtained by using a computer routine developed by

Kruskal (1964) which utilizes the method of steepest descent to find the local minimum. The initial configuration can be taken as the classical solution. Unfortunately, there is no way of distinguishing in practice between a local minimum and the global minimum.

(2) The Shepard–Kruskal solution is invariant under rotation, translation, and uniform expansion or contraction of the best-fitting configuration.

(3) The Shepard–Kruskal solution is non-metric since it utilizes only the rank orders (14.6.3). However, we still need a sufficiently objective numerical measure of distance to determine the rank order of the d_{rs}.

(4) *Similarities* The non-metric method works just as well with similarities as with dissimilarities. One simply changes the direction of the inequalities.

(5) *Missing values* The Shepard–Kruskal method is easily adapted to the situation where there are missing values. One simply omits the missing dissimilarities in the ordering (14.6.3) and deletes the corresponding terms from the numerator and denominator of (14.6.5). As long as not too many values are missing, the method still seems to work well.

(6) *Treatment of ties* The constraint given by (14.6.4) is called the *primary treatment of ties* (PTT). If $d_{rs} = d_{uv}$ then no constraint is made on d^*_{rs} and d^*_{uv}. An alternative constraint, called the *secondary treatment of ties* (STT) is given by

$$d_{rs} \leqslant d_{uv} \Rightarrow d^*_{rs} \leqslant d^*_{uv},$$

which has the property that $d_{rs} = d_{uv} \Rightarrow d^*_{rs} = d^*_{uv}$. However, one must be cautious when using STT on data with a large number of ties. The use of STT on data such as Example 14.5.2 leads to the horseshoe effect (see, D. G. Kendall, 1971).

(7) *Comparison of methods* The computation is simpler for the classical method than it is for the non-metric method. It is not known how robust the classical method is to monotone transformations of the distance function; however, both methods seem to give similar answers when applied to well-known examples in the field. Figure 14.4.1 gives the two solutions for the road data. For the Shepard–Kruskal solution for the Morse code data, see Shepard (1963).

(8) We have not commented on the non-metric method of Guttman (1968). For a mathematical and empirical analysis of multidimensional scaling algorithms of Kruskal's M-D-SCA1 and Guttman–Lingoes' SAA-I, we refer to Lingoes and Roskam (1973) which also contains certain recommendations for improvement of these algorithms.

14.7 Goodness of Fit Measure: Procrustes Rotation

We now describe a *goodness of fit* measure (Green, 1952; Gower, 1971*b*), used to compare two configurations. Let $\mathbf{X}$ be the $(n \times p)$ matrix of the coordinates of n points obtained from $\mathbf{D}$ by one technique. Suppose that $\mathbf{Y}$ is the $(n \times q)$ matrix of coordinates of another set of points obtained by another technique, or using another measure of distance. Let $q \leq p$. By adding columns of zeros to $\mathbf{Y}$, we may also assume $\mathbf{Y}$ to be $(n \times p)$.

The measure of goodness of fit adopted is obtained by moving the points $\mathbf{y}_r$ relative to the points $\mathbf{x}_r$ until the "residual" sum of squares

$$\sum_{r=1}^{n} (\mathbf{x}_r - \mathbf{y}_r)'(\mathbf{x}_r - \mathbf{y}_r) \tag{14.7.1}$$

is minimal. We can move $\mathbf{y}_r$ relative to $\mathbf{x}_r$ through rotation, reflection, and translation, i.e. by

$$\mathbf{A}'\mathbf{y}_r + \mathbf{b}, \qquad r = 1, \ldots, n, \tag{14.7.2}$$

where $\mathbf{A}'$ is a $(p \times p)$ orthogonal matrix. Hence, we wish to solve

$$R^2 = \min_{\mathbf{A},\mathbf{b}} \sum_{r=1}^{n} (\mathbf{x}_r - \mathbf{A}'\mathbf{y}_r - \mathbf{b})'(\mathbf{x}_r - \mathbf{A}'\mathbf{y}_r - \mathbf{b}) \tag{14.7.3}$$

for $\mathbf{A}$ and $\mathbf{b}$. Note that $\mathbf{A}$ and $\mathbf{b}$ are found by least squares. Their values are given in the following theórem.

Theorem 14.7.1 *Let $\mathbf{X}(n \times p)$ and $\mathbf{Y}(n \times p)$ be two configurations of n points, for convenience centred at the origin, so $\bar{\mathbf{x}} = \bar{\mathbf{y}} = \mathbf{0}$. Let $\mathbf{Z} = \mathbf{Y}'\mathbf{X}$ and using the singular value decomposition theorem (Theorem A.6.5), write*

$$\mathbf{Z} = \mathbf{V}\mathbf{\Gamma}\mathbf{U}', \tag{14.7.4}$$

where $\mathbf{V}$ and $\mathbf{U}$ are orthogonal $(p \times p)$ matrices and $\mathbf{\Gamma}$ is a diagonal matrix of non-negative elements. Then the minimizing values of $\mathbf{A}$ and $\mathbf{b}$ in (14.7.3) are given by

$$\hat{\mathbf{b}} = \mathbf{0}, \qquad \hat{\mathbf{A}} = \mathbf{V}\mathbf{U}', \tag{14.7.5}$$

and further

$$R^2 = \operatorname{tr} \mathbf{X}\mathbf{X}' + \operatorname{tr} \mathbf{Y}\mathbf{Y}' - 2 \operatorname{tr} \mathbf{\Gamma}. \tag{14.7.6}$$

Proof On differentiating with respect to $\mathbf{b}$, we have

$$\hat{\mathbf{b}} = \bar{\mathbf{x}} - \mathbf{A}'\bar{\mathbf{y}} \tag{14.7.7}$$

where $\bar{\mathbf{y}} = \sum \mathbf{y}_r/n$, $\bar{\mathbf{x}} = \sum \mathbf{x}_r/n$. Since both configurations are centred, $\hat{\mathbf{b}} = \mathbf{0}$.

Then we can rewrite (14.7.3) as

$$R^2 = \min_{\mathbf{A}} \operatorname{tr}(\mathbf{X}-\mathbf{YA})(\mathbf{X}-\mathbf{YA})' = \operatorname{tr}\mathbf{XX}' + \operatorname{tr}\mathbf{YY}' - 2\max_{\mathbf{A}} \operatorname{tr}\mathbf{X}'\mathbf{YA}. \tag{14.7.8}$$

The constraints on $\mathbf{A}$ are $\mathbf{AA}' = \mathbf{I}$, i.e. $\mathbf{a}_i'\mathbf{a}_i = 1$, $\mathbf{a}_i'\mathbf{a}_j = 0$, $i \neq j$, where $\mathbf{a}_i'$ is the ith row of $\mathbf{A}$. Hence there are $p(p+1)/2$ constraints.

Let $\frac{1}{2}\mathbf{\Lambda}$ be a $(p \times p)$ symmetric matrix of Lagrange multipliers for these constraints. The aim is to maximize

$$\operatorname{tr}\{\mathbf{Z}'\mathbf{A} - \tfrac{1}{2}\mathbf{\Lambda}(\mathbf{AA}'-\mathbf{I})\}, \tag{14.7.9}$$

where $\mathbf{Z}' = \mathbf{X}'\mathbf{Y}$. By direct differentiation it can be shown that

$$\frac{\partial}{\partial \mathbf{A}} \operatorname{tr}(\mathbf{Z}'\mathbf{A}) = \mathbf{Z}, \qquad \frac{\partial}{\partial \mathbf{A}} \operatorname{tr}(\mathbf{\Lambda AA}') = 2\mathbf{\Lambda A}. \tag{14.7.10}$$

Hence on differentiating (14.7.9) and equating the derivatives to zero, we find that $\mathbf{A}$ must satisfy

$$\mathbf{Z} = \mathbf{\Lambda A}. \tag{14.7.11}$$

Write $\mathbf{Z}$ using (14.7.4). Noting that $\mathbf{\Lambda}$ is symmetric and that $\mathbf{A}$ is to be orthogonal, we get, from (14.7.11),

$$\mathbf{\Lambda}^2 = \mathbf{ZA}'\mathbf{AZ} = \mathbf{ZZ}' = (\mathbf{V\Gamma U}')(\mathbf{U\Gamma V}').$$

Thus we can take $\mathbf{\Lambda} = \mathbf{V\Gamma V}'$. Substituting this value of $\mathbf{\Lambda}$ in (14.7.11) we see that

$$\hat{\mathbf{A}} = \mathbf{VU}' \tag{14.7.12}$$

is a solution of (14.7.11). Note that $\hat{\mathbf{A}}$ is orthogonal. Using this value of $\hat{\mathbf{A}}$ in (14.7.8) gives (14.7.6).

Finally, to verify that $\hat{\mathbf{A}}$ maximizes (14.7.9) (and is not just a stationary point) we must differentiate (14.7.9) with respect to $\mathbf{A}$ a second time. For this purpose it is convenient to write $\mathbf{A}$ as a vector $\mathbf{a} = (\mathbf{a}_{(1)}', \ldots, \mathbf{a}_{(p)}')'$. Then (14.7.9) is a quadratic function of the elements of $\mathbf{a}$ and the second derivative of (14.7.9) with respect to $\mathbf{a}$ can be expressed as the matrix $-\mathbf{I}_p \otimes \mathbf{\Lambda}$. Since $\mathbf{\Lambda} = \mathbf{V\Gamma V}'$, and the diagonal elements of $\mathbf{\Gamma}$ are non-negative, we see that the second derivative matrix is negative semi-definite. Hence $\hat{\mathbf{A}}$ maximizes (14.7.9). ■

We have assumed that the column means of $\mathbf{X}$ and $\mathbf{Y}$ are zero. Then the "best" rotation of $\mathbf{Y}$ relative to $\mathbf{X}$ is $\mathbf{Y}\hat{\mathbf{A}}$, where $\hat{\mathbf{A}}$ is given by (14.7.12), and $\hat{\mathbf{A}}$ is called the *Procrustes rotation* of $\mathbf{Y}$ relative to $\mathbf{X}$. Noting from (14.7.4) that $\mathbf{X}'\mathbf{YY}'\mathbf{X} = \mathbf{U\Gamma}^2\mathbf{U}'$, we can rewrite (14.7.8) as

$$R^2 = \operatorname{tr}\mathbf{XX}' + \operatorname{tr}\mathbf{YY}' - 2\operatorname{tr}(\mathbf{X}'\mathbf{YY}'\mathbf{X})^{1/2}. \tag{14.7.12}$$

It can be seen that (14.7.8) is zero if and only if the $\mathbf{y}_r$ can be rotated to the $\mathbf{x}_r$ exactly.

Scale factor If the scales of two configurations are different, then the transformation (14.7.2) should be of the form

$$c\mathbf{A}'\mathbf{y}_r+\mathbf{b},$$

where $c>0$. Following the above procedure, it can be seen that

$$\hat{c}=(\text{tr}\,\mathbf{\Gamma})/(\text{tr}\,\mathbf{YY}') \tag{14.7.14}$$

and the other estimates remain as before. This transformation is called the *Procrustes rotation with scaling* of $\mathbf{Y}$ relative to $\mathbf{X}$. Then the new minimum residual sum of squares is given by

$$R^2=\text{tr}\,(\mathbf{XX}')+\hat{c}^2\,\text{tr}\,(\mathbf{YY}')-2\hat{c}\,\text{tr}\,(\mathbf{X}'\mathbf{YY}'\mathbf{X})^{1/2}, \tag{14.7.15}$$

where $\hat{c}$ is given by (14.7.14). Note that this procedure is not symmetrical with respect to $\mathbf{X}$ and $\mathbf{Y}$. Symmetry can be obtained by selecting scaling so that

$$\text{tr}\,(\mathbf{XX}')=\text{tr}\,(\mathbf{YY}').$$

For an excellent review of this topic, see Sibson (1978).

Example 14.7.1 The actual Ordnance Survey coordinates of the 12 towns in Table 14.1.1 are

	1	2	3	4	5	6	7	8	9	10	11	12
E	257	529	339	629	292	259	508	265	433	533	420	627
N	279	104	554	142	90	665	433	842	438	183	563	308

Treating these quantities as planar coordinates $\mathbf{X}(12\times 2)$ (the curvature of the Earth has little effect), and the first two eigenvectors from the classical MDS solution for the data given in Example 14.4.1 as $\mathbf{Y}$, it is found that

$$\mathbf{X}'\mathbf{Y}=\begin{pmatrix}182\,119.068 & -91\,647.926\\ -495\,108.159 & -25\,629.185\end{pmatrix},\quad \mathbf{U}=\begin{pmatrix}-0.347\,729 & -0.937\,595\\ 0.937\,595 & -0.347\,729\end{pmatrix},$$

$$\mathbf{\Gamma}=\begin{pmatrix}527\,597.29 & 0\\ 0 & 94\,851.13\end{pmatrix},\qquad \mathbf{V}=\begin{pmatrix}-0.999\,890 & 0.014\,858\\ 0.014\,858 & 0.999\,890\end{pmatrix}.$$

This leads to the transforming of the $\mathbf{y}$s to match the $\mathbf{x}$s by

$$\mathbf{y}_r^*=c\mathbf{A}'\mathbf{y}_r+\mathbf{b},$$

where

$$c = 1.358\,740, \quad \mathbf{A} = \mathbf{VU}' = \begin{pmatrix} 0.333\,760, & -0.942\,658 \\ -0.942\,658, & -0.333\,760 \end{pmatrix}, \quad \mathbf{b} = \begin{pmatrix} 424.250 \\ 383.417 \end{pmatrix}.$$

The transformation has been used on $\mathbf{y}_r$ to obtain $\mathbf{y}_r^*$, and these $\mathbf{y}_r^*$ are plotted in Figure 14.4.1 together with the $\mathbf{x}_r$. We have

$$\operatorname{tr} \mathbf{XX}' = 853\,917, \qquad \operatorname{tr} \mathbf{YY}' = 458\,107.$$

Hence from (14.7.15), the residual is $R^2 = 8172$.

The Shepard–Kruskal solution has a stress of 0.0404. Using this solution as $\mathbf{Y}$ and the Ordnance Survey coordinates again as $\mathbf{X}$ leads to a residual of $R^2 = 13\,749$. Hence the classical solution fits a bit better for this data. Of course Figure 14.4.1 shows little difference between the two solutions.

If we are given two distance matrices $\mathbf{D}_1$ and $\mathbf{D}_2$ *but* not the corresponding points, (14.7.6) cannot be computed without using some method to compute the "points". The first two terms are expressible in terms of $\mathbf{D}_1$ and $\mathbf{D}_2$ but not $\operatorname{tr} \boldsymbol{\Gamma}$.

*14.8 Multi-sample Problem and Canonical Variates

Consider the case of g p-variate populations with means $\boldsymbol{\mu}_r$, $r = 1, \ldots, g$, and common covariance matrix $\boldsymbol{\Sigma}$. If we are given a data matrix $\mathbf{Y}$ representing samples of size n_r from the rth group, $r = 1, \ldots, g$, let $\bar{\mathbf{y}}_r$ denote the rth sample mean, and estimate the common covariance matrix $\boldsymbol{\Sigma}$ by $\mathbf{W}/\nu$, where $\mathbf{W}$ is the within-groups sum of squares and products (SSP) matrix with ν degrees of freedom.

Assume that the overall (unweighted) mean $\bar{\mathbf{y}}$ is

$$\bar{\mathbf{y}} = \sum_{r=1}^{n} \bar{\mathbf{y}}_r = \mathbf{0}. \tag{14.8.1}$$

We shall work with the Mahalanobis distances

$$d_{rs}^2 = \nu(\bar{\mathbf{y}}_r - \bar{\mathbf{y}}_s)'\mathbf{W}^{-1}(\bar{\mathbf{y}}_r - \bar{\mathbf{y}}_s). \tag{14.8.2}$$

It is easily checked that if $\mathbf{B}$ is defined by (14.2.3), then

$$\mathbf{B} = \nu\bar{\mathbf{Y}}\mathbf{W}^{-1}\bar{\mathbf{Y}}',$$

where $\bar{\mathbf{Y}}' = (\bar{\mathbf{y}}_1, \ldots, \bar{\mathbf{y}}_g)$.

Thus $\mathbf{B} \geq 0$ and so $\mathbf{D}$ is Euclidean. Let $\mathbf{X}$ be the configuration for $\mathbf{B}$ defined in Theorem 14.2.1, and fix k, $1 \leq k \leq p$. Then, the first k columns of $\mathbf{X}$ can be regarded as the coordinates of points representing the g means in k dimensions ($k \leq p$). This configuration has the optimal property that it is the "best" representation in k dimensions.

Note that $\bar{\mathbf{Y}}'\bar{\mathbf{Y}}$ is the (unweighted) between-groups SSP matrix. Let $\mathbf{l}_i$ denote the ith canonical vector of Section 12.5 using this unweighted between-groups SSP matrix; that is, define $\mathbf{l}_i$ by

$$\nu \mathbf{W}^{-1}\bar{\mathbf{Y}}'\bar{\mathbf{Y}}\mathbf{l}_i = \lambda_i \mathbf{l}_i, \qquad \nu^{-1}\mathbf{l}_i'\mathbf{W}\mathbf{l}_i = 1, \qquad i = 1, \ldots, \min(p, g)$$

where λ_i is the ith eigenvalue of $\nu\mathbf{W}^{-1}\bar{\mathbf{Y}}'\bar{\mathbf{Y}}$, which is the same as the ith eigenvalue of $\mathbf{B}$. Then the scores of the g groups on the ith canonical coordinate are given by

$$\bar{\mathbf{Y}}\mathbf{l}_i.$$

Since $\mathbf{B}\bar{\mathbf{Y}}\mathbf{l}_i = \lambda_i \bar{\mathbf{Y}}\mathbf{l}_i$ and $\mathbf{l}_i'\bar{\mathbf{Y}}'\bar{\mathbf{Y}}\mathbf{l}_i = \lambda_i$, we see from (14.2.5) that

$$\bar{\mathbf{Y}}\mathbf{l}_i = \mathbf{x}_{(i)},$$

so that $\bar{\mathbf{Y}}\mathbf{l}_i$ is also the ith eigenvector of $\mathbf{B}$. Thus, the canonical means in k dimensions, that is, the scores of the first k canonical variates on the g groups, are the same as the coordinates given by Theorem 14.2.1.

Exercises and Complements

14.2.1 Using (14.2.6), show that (14.2.9) can be written in the form (14.2.10).

14.2.2 In the notation of Theorem 14.2.1 show that

(a) $b_{rr} = \bar{a}_{..} - 2\bar{a}_{r.}, \; b_{rs} = a_{rs} - \bar{a}_{r.} - \bar{a}_{.s} + \bar{a}_{..}, \; r \neq s;$

(b) $\mathbf{B} = \sum_{i=1}^{p} \mathbf{x}_{(i)}\mathbf{x}_{(i)}';$

(c) $\sum_{r=1}^{n} \lambda_r = \sum_{r=1}^{n} b_{rr} = \frac{1}{2n} \sum_{r,s=1}^{n} d_{rs}^2.$

14.2.3 (Gower, 1968) Let $\mathbf{D} = (d_{rs})$ be an $(n \times n)$ Euclidean distance matrix with configuration $\mathbf{X} = (\mathbf{x}_1, \ldots, \mathbf{x}_n)'$ in p-dimensional principal coordinates, given by Theorem 14.2.1. Suppose we wish to add an additional point to the configuration using distances $d_{r,n+1}$, $r = 1, \ldots, n$ (which we know to be Euclidean), allowing for a $(p+1)$th dimension. If the first n points are represented by $(x_{r1}, \ldots, x_{rp}, 0)'$, $r = 1, \ldots, n$, then show that the $(n+1)$th point is given by

$$\mathbf{x}_{n+1}=(x_{n+1,1},\ldots,x_{n+1,p},x_{n+1,p+1})'=(\mathbf{x}',y)', \text{ say}$$

where

$$\mathbf{x}=\tfrac{1}{2}\mathbf{\Lambda}^{-1}\mathbf{X}'\mathbf{f},\qquad \mathbf{f}=(f_1,\ldots,f_n)',\qquad f_r=b_{rr}-d^2_{r,n+1},$$

and

$$y^2=\frac{1}{n}\sum_{r=1}^{n}d^2_{r,n+1}-\frac{1}{n}\sum_{r=1}^{n}b_{rr}-\mathbf{x}'\mathbf{x}.$$

Hence $\mathbf{x}$ is uniquely determined but y is determined only in value, not in sign. Give the reason. (Hint: substitute $f_r=2\mathbf{x}_r'\mathbf{x}_{n+1}-\mathbf{x}'_{n+1}\mathbf{x}_{n+1}$ for f_r in terms of $\mathbf{x}_1,\ldots,\mathbf{x}_{n+1}$ to verify the formula for $\mathbf{x}$.)

14.2.4 If $\mathbf{C}$ is p.s.d. then show that $c_{ii}+c_{jj}-2c_{ij}>0$. Show that the distance d_{ij} defined by $d_{ij}^2=c_{ii}+c_{jj}-2c_{ij}$ satisfies the triangle inequality.

14.2.5 For the Bhattacharyya distance matrix $\mathbf{D}$ given in Example 13.4.1, the eigenvalues of $\mathbf{B}$ are

$$\lambda_1=318.97,\qquad \lambda_2=167.72,\qquad \lambda_3=11.11,\qquad \lambda_4=0.$$

Hence, $\mathbf{D}$ is Euclidean and the two-dimensional representation accounts for $\alpha_{1,2}=98\%$ of the variation.

Show that the principal coordinates in two dimensions for Eskimo, Bantu, English, and Korean are, respectively,

$$(9.69, 7.29),\qquad (-11.39,-2.51),\qquad (-6.00, 4.57),\qquad (7.70,-9.34).$$

Plot these points and comment on the conclusions drawn in Example 13.4.1.

14.2.6 For the grave data (Example 13.4.2) using the similarity matrix $\mathbf{S}_2$ show that the distance matrix given by the standard transformation is

$$\mathbf{D}=\begin{bmatrix}0 & \sqrt{10} & 2 & 0 & \sqrt{8} & 2\\ & 0 & \sqrt{6} & \sqrt{10} & \sqrt{2} & \sqrt{6}\\ & & 0 & 2 & \sqrt{8} & 2\\ & & & 0 & \sqrt{8} & 2\\ & & & & 0 & 2\\ & & & & & 0\end{bmatrix}.$$

14.2.7 Suppose that $1, 2,\ldots, 7$ are regions (enclosed by unbroken lines) in a country arranged as in Exercise Figure 1. Let the distance matrix be

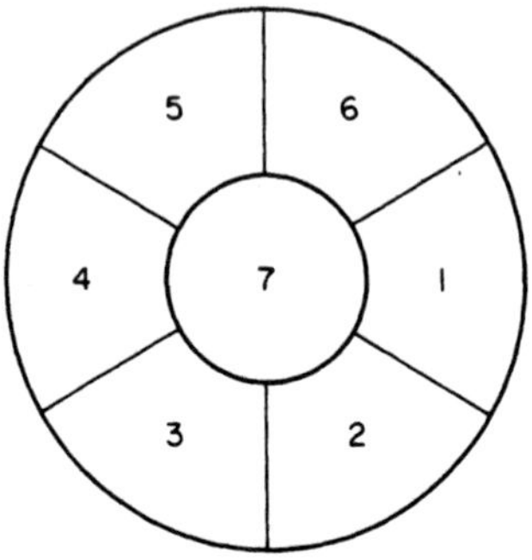

Exercise Figure 1 Seven regions in a country.

constructed by counting the minimum number of boundaries crossed to pass from region i to region j.

Show that the distance matrix is given by

$$\begin{bmatrix} 0 & 1 & 2 & 2 & 2 & 1 & 1 \\ & 0 & 1 & 2 & 2 & 2 & 1 \\ & & 0 & 1 & 2 & 2 & 1 \\ & & & 0 & 1 & 2 & 1 \\ & & & & 0 & 1 & 1 \\ & & & & & 0 & 1 \\ & & & & & & 0 \end{bmatrix}$$

Show that the distances constructed in this way obey the triangle inequality $d_{ik} \leq d_{ij} + d_{jk}$, but by showing that the eigenvalues of the matrix $\mathbf{B}$ are

$$\lambda_1 = \lambda_2 = \tfrac{7}{2}, \quad \lambda_3 = \lambda_4 = \tfrac{1}{2}, \quad \lambda_5 = 0, \quad \lambda_6 = -\tfrac{1}{7}, \quad \lambda_7 = -1,$$

deduce that this metric is non-Euclidean.

Since $\lambda_1 = \lambda_2$, select any two orthogonal eigenvectors corresponding to λ_1 and λ_2 and, by plotting the seven points so obtained, show that the original map is reconstructed. As in Example 14.2.1, the points are vertices of a hexagon with centre at the origin.

14.2.8 (Lingoes, 1971; Mardia, 1978) Let $\mathbf{D}$ be a distance matrix. Show that for some real number a, there exists a Euclidean configuration in $p \leq n-2$ dimensions with interpoint distances d^*_{rs} satisfying

$$d^{*2}_{rs} = d^2_{rs} - 2a, \quad r \neq s; \qquad d^*_{rr} = 0.$$

Thus d^{*2}_{rs} is a linear function of d^2_{rs}, so the configuration preserves the rank order of the distances. (Hint: show that the matrix $\mathbf{D}^*$ leads to $\mathbf{A}^*$ and $\mathbf{B}^*$ given by

$$\mathbf{A}^* = (-\tfrac{1}{2}d_{rs}^{*2}) = \mathbf{A} - a(\mathbf{I}-\mathbf{J}), \qquad \mathbf{B}^* = \mathbf{HA}^*\mathbf{H} = \mathbf{B} - a\mathbf{H}.$$

If $\mathbf{B}$ has eigenvalues $\lambda_1 \geq \cdots \geq \lambda_u > 0 \geq \lambda_1' \geq \cdots \geq \lambda_v'$, then $\mathbf{B}^*$ has eigenvalues $\lambda_r - a$, $r = 1, \ldots, u$; 0; and $\lambda_r' - a$, $r = 1, \ldots, v$. Then the choice $a = \lambda_v'$ makes $\mathbf{B}^*$ p.s.d. of rank at most $n-2$.)

14.4.1 Let $\mathbf{l}_1, \ldots, \mathbf{l}_p$ be orthonormal vectors in $R^q (p \leq q)$ and let $\mathbf{z}_r$, $r = 1, \ldots, n$, be points in R^q. Let H_r denote the foot of the perpendicular of $\mathbf{z}_r$ on the subspace spanned by $\mathbf{l}_1, \ldots, \mathbf{l}_p$. Show that with respect to the new coordinate system with axes $\mathbf{l}_1, \ldots, \mathbf{l}_p$, the coordinates of H_r are $(\mathbf{l}_1'\mathbf{z}_r, \ldots, \mathbf{l}_p'\mathbf{z}_r)'$. What modification must be made if the $\mathbf{l}_i$ are orthogonal but not orthonormal?

14.4.2 Let $\mathbf{\Lambda} = \operatorname{diag}(\lambda_1, \ldots, \lambda_p)$, where $\lambda_1 \geq \cdots \geq \lambda_p$ are real numbers, and let $\hat{\mathbf{\Lambda}} = \operatorname{diag}(\hat{\lambda}_1, \ldots, \hat{\lambda}_p)$, where $\hat{\lambda}_1 \geq \cdots \geq \hat{\lambda}_p \geq 0$ are non-negative numbers. Show that minimizing

$$\operatorname{tr}(\mathbf{\Lambda} - \mathbf{G}\hat{\mathbf{\Lambda}}\mathbf{G}')^2$$

over orthogonal matrices $\mathbf{G}$ is equivalent to maximizing

$$\operatorname{tr}(\mathbf{\Lambda}\mathbf{G}\hat{\mathbf{\Lambda}}\mathbf{G}') = \sum_{i,j=1}^{p} \lambda_i \hat{\lambda}_j g_{ij}^2 = \sum_{i=1}^{p} \lambda_i h_i = \phi(\mathbf{h}) \quad \text{say,}$$

where

$$h_i = \sum_{j=1}^{p} \hat{\lambda}_j g_{ij}^2 \geq 0 \quad \text{and} \quad \sum_{i=1}^{p} h_i = \sum_{j=1}^{p} \hat{\lambda}_j.$$

Show that $\phi(\mathbf{h})$ is maximized over such vectors $\mathbf{h}$ when $h_i = \hat{\lambda}_i$ for $i = 1, \ldots, p$; that is, when $\mathbf{G} = \mathbf{I}$.

15
Directional Data

15.1 Introduction

We have so far considered data where we are interested in both the direction and the magnitude of the random vector $\mathbf{x} = (\mathbf{x}_1, \ldots, \mathbf{x}_p)'$. There are various statistical problems which arise when the observations themselves are directions, or we are only interested in the direction of $\mathbf{x}$. We can regard directions as points on the circumference of a circle in two dimensions or on the surface of a sphere in three dimensions. In general, directions may be visualized as points on the surface of a hypersphere.

We will denote a random direction in p dimensions by $\mathbf{l}$, where $\mathbf{l}'\mathbf{l} = 1$. The unit vector $\mathbf{l}$ takes values on the surface of a p-dimensional hypersphere S_p of unit radius and having its centre at the origin. (Note that this notation differs from the notation in topology where S_p is usually written S_{p-1}.)

For $p = 2$, we can equivalently consider the circular variable θ defined by (see Figure 15.1.1a)

$$l_1 = \cos\theta, \qquad l_2 = \sin\theta, \qquad 0 \leqslant \theta < 2\pi, \tag{15.1.1}$$

whereas for $p = 3$ we have the spherical variable (θ, ϕ) defined by (see Figure 15.1.1b))

$$l_1 = \cos\theta, \quad l_2 = \sin\theta\cos\phi, \quad l_3 = \sin\theta\sin\phi, \quad 0 \leqslant \theta \leqslant \pi, \quad 0 \leqslant \phi < 2\pi \tag{15.1.2}$$

In general, it is convenient to consider the density of $\mathbf{l}$ in terms of the

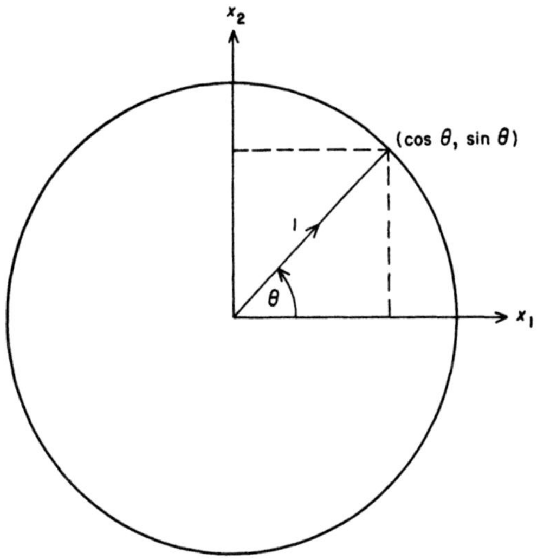

Figure 15.1.1a Representation for circular variable.

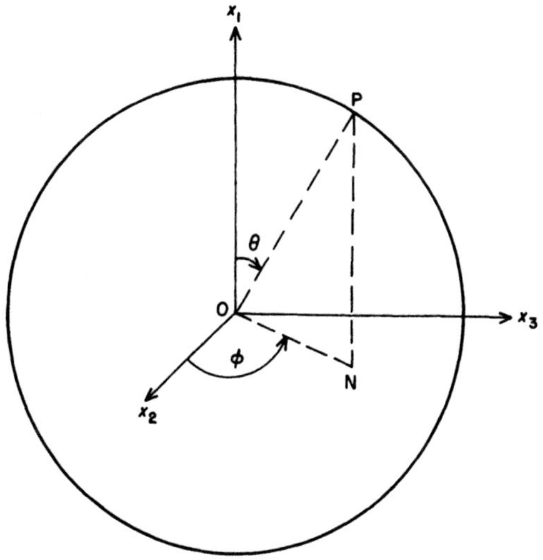

Figure 15.1.1b Representation for spherical variable.

spherical polar coordinates $\boldsymbol{\theta}' = (\theta_1, \ldots, \theta_{p-1})$ with the help of the transformation

$$\mathbf{l} = \mathbf{u}(\boldsymbol{\theta}), \qquad 0 \leq \theta_i \leq \pi, \quad i = 1, \ldots, p-2, \quad 0 \leq \theta_{p-1} < 2\pi, \tag{15.1.3}$$

where

$$u_i(\boldsymbol{\theta}) = \cos\theta_i \prod_{j=0}^{i-1} \sin\theta_j, \qquad i = 1, \ldots, p, \quad \sin\theta_0 = \cos\theta_p = 1.$$

15.2 Descriptive Measures

Let $\mathbf{l}_i$, $i = 1, \ldots, n$, be a random sample on S_p. For $p = 2$, these points can be plotted on a circle, whereas for $p = 3$, they can be plotted on a globe. Let

$$\bar{\mathbf{l}} = \frac{1}{n} \sum_{i=1}^{n} \mathbf{l}_i$$

denote the sample mean vector. The direction of the vector $\bar{\mathbf{l}}$ is

$$\mathbf{l}_0 = \bar{\mathbf{l}}/\bar{R}, \qquad \bar{R} = (\bar{\mathbf{l}}'\bar{\mathbf{l}})^{1/2}, \tag{15.2.1}$$

which can be regarded as the *mean direction* of the sample. The points $\mathbf{l}_i$ regarded as of unit mass, have centre of gravity $\bar{\mathbf{l}}$ which has direction $\mathbf{l}_0$ and distance from the origin $\bar{R}$ (see Figure 15.2.1).

The mean direction has the following *locational* property similar to $\bar{\mathbf{x}}$. Under an orthogonal transformation $\mathbf{l}^* = \mathbf{A}\mathbf{l}$, i.e. under an arbitrary rotation or reflection, the new mean direction is seen to be

$$\mathbf{l}_0^* = \mathbf{A}\mathbf{l}_0;$$

that is, the transformation amounts to simply rotating $\mathbf{l}_0$ by $\mathbf{A}$. Further, it has the following *minimization* property. Consider $S(\mathbf{a})$, the arithmetic mean of the squared chordal distances between $\mathbf{l}_i$ and $\mathbf{a}$, where $\mathbf{a}'\mathbf{a} = 1$. We have

$$S(\mathbf{a}) = \frac{1}{n} \sum_{i=1}^{n} |\mathbf{l}_i - \mathbf{a}|^2 = 2(1 - \bar{\mathbf{l}}'\mathbf{a}). \tag{15.2.2}$$

Using a Lagrange multiplier, it can be seen that (15.2.2) is minimized when $\mathbf{a} = \mathbf{l}_0$. Further,

$$\min S(\mathbf{a}) = 2(1 - \bar{R}). \tag{15.2.3}$$

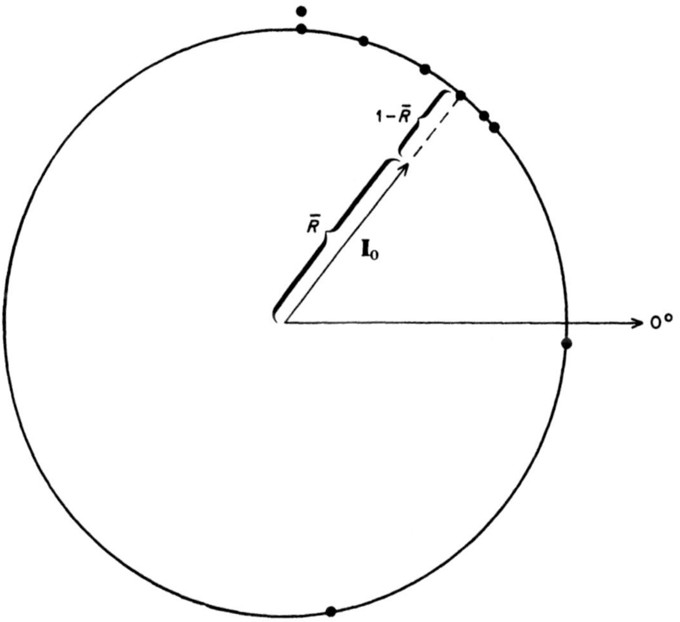

Figure 15.2.1 Basic circular statistics $\bar{R}$ (distance from origin), $\mathbf{l}_0$ (mean direction). (Data in Exercise 15.2.3.)

Since $S(\mathbf{a})$ is non-negative and $\bar{R}$ is positive, we have

$$0 \leqslant \bar{R} \leqslant 1. \tag{15.2.4}$$

In view of (15.2.3) the quantity $2(1-\bar{R})$ may be called the *spherical variance*. However, it is usual to work with $\bar{R}$ in place of $2(1-\bar{R})$ in directional data, and it measures clustering around the mean direction. Note that $\bar{R} \doteq 0$ when the points are uniformly distributed (i.e. widely dispersed) and $\bar{R} \doteq 1$ when the points are heavily concentrated near a point. Following the terminology of mechanics, the quantity $R = n\bar{R}$ is called the "resultant length", and the vector

$$\mathbf{r} = n\bar{\mathbf{l}} = \sum_{i=1}^{n} \mathbf{l}_i \tag{15.2.5}$$

is called the "resultant vector".

Example 15.2.1 The declinations and dips of remnant magnetization in nine specimens of Icelandic lava flows of 1947–48 given in Fisher (1953)

Table 15.2.1 Measurements from the magnetized lava flows in 1947–48 in Iceland (Hosper data cited in Fisher, 1953)

l_1	l_2	l_3
0.2909	−0.0290	−0.9563
0.1225	−0.0624	−0.9905
0.2722	−0.2044	−0.9403
0.4488	−0.4334	−0.7815
0.2253	−0.2724	−0.9354
0.1705	−0.3207	−0.9317
0.3878	0.1171	−0.9143
0.5176	0.0217	−0.8554
0.1822	0.0032	−0.9833

lead to the values given in Table 15.2.1.We find that

$$\sum_{i=1}^{n} l_{i1} = 2.6178, \qquad \sum_{i=1}^{n} l_{i2} = -1.1803, \qquad \sum_{i=1}^{n} l_{i3} = -8.2887.$$

Hence

$$R^2 = \left(\sum_{i=1}^{n} l_{i1}\right)^2 + \left(\sum_{i=1}^{n} l_{i2}\right)^2 + \left(\sum_{i=1}^{n} l_{13}\right)^2 = 76.9485.$$

Consequently, $\bar{R} = 0.9747$, which indicates heavy concentration around the mean direction. Further, the mean direction from (15.2.1) is given by

$$\mathbf{l}_0' = (0.2984, -0.1346, -0.9449).$$

15.3 Basic Distributions

Following Section 15.2, we define the population *resultant length* ρ by

$$\rho = \left[\sum_{i=1}^{p} \{E(l_i)\}^2\right]^{1/2} = \{E(\mathbf{l}')E(\mathbf{l})\}^{1/2}, \tag{15.3.1}$$

and when $\rho > 0$, the population mean direction is defined by

$$\boldsymbol{\mu} = E(\mathbf{l})/\rho. \tag{15.3.2}$$

We will denote the probability element (p.e.) of $\mathbf{l}$ on S_p by dS_p. The Jacobian of the transformation from $(r, \boldsymbol{\theta})$ to $\mathbf{x}$ is given by (2.4.7). On separating the parts for r and $\boldsymbol{\theta}$, in the Jacobian, it is seen that

$$dS_p = a_p(\boldsymbol{\theta})\, d\boldsymbol{\theta}, \tag{15.3.5}$$

where

$$a_p(\boldsymbol{\theta}) = \prod_{j=2}^{p-1} \sin^{p-j} \theta_{j-1}, \qquad a_2(\theta) = 1.$$

15.3.1 The uniform distribution

If a direction $\mathbf{l}$ is *uniformly distributed* on S_p, then its probability element is $c_p\, dS_p$. Thus, on using (15.3.3), the p.d.f. of $\boldsymbol{\theta}$ is

$$c_p a_p(\boldsymbol{\theta}). \tag{15.3.4}$$

Integrating over $\boldsymbol{\theta}$, it is found that

$$c_p = \Gamma(\tfrac{1}{2}p)/(2\pi^{p/2}). \tag{15.3.5}$$

From (15.1.3) and (15.3.4), it can be seen that $E(\mathbf{l}) = \mathbf{0}$ for this distribution. Hence we have $\rho = 0$. Consequently, $\boldsymbol{\mu}$ is not defined for this distribution. Similarly,

$$E(l_i l_j) = 0 \quad \text{for} \quad i \neq j, \qquad E(l_i^2) = 1/p.$$

15.3.2 The von Mises distribution

A unit random vector $\mathbf{l}$ is said to have a p-variate *von Mises–Fisher* distribution if its p.e. is

$$c_p(\kappa) e^{\kappa \boldsymbol{\mu}'\mathbf{l}}\, dS_p, \qquad \mathbf{l} \in S_p \tag{15.3.6}$$

where $\kappa \geq 0$ and $\boldsymbol{\mu}'\boldsymbol{\mu} = 1$. Here κ is called the concentration parameter (for reasons given below) and $c_p(\kappa)$ is the normalizing constant. If a random vector $\mathbf{l}$ has its p.e. of the form (15.3.6), we will say that it is distributed as $M_p(\boldsymbol{\mu}, \kappa)$. For the cases of $p = 2$ and $p = 3$, the distributions are called von Mises and Fisher distributions after the names of the originators (see von Mises, 1918; Fisher, 1953), and these particular cases explain the common nomenclature for $p > 3$. However, the distribution (15.3.6) was introduced by Watson and Williams (1956).

The normalizing constant

We show that

$$c_p(\kappa) = \kappa^{(p-1)/2} / \{(2\pi)^{p/2} I_{(p-1)/2}(\kappa)\}, \tag{15.3.7}$$

where $I_r(\kappa)$ denotes the modified Bessel function of the first kind and order r, which can be defined by the integral formula

$$I_r(\kappa)=\frac{(\frac{1}{2}\kappa)^r}{\Gamma(r+\frac{1}{2})\Gamma(\frac{1}{2})}\int_0^{\pi} e^{\pm\kappa\cos\theta}\sin^{2r}\theta\, d\theta. \tag{15.3.8}$$

The proofs of the results assumed about Bessel functions in this chapter can be found in G. N. Watson (1944). We have

$$\{c_p(\kappa)\}^{-1}=\int_{\mathbf{l}\in S_p} e^{\kappa\boldsymbol{\mu}'\mathbf{l}}\, dS_p=\int_{\mathbf{l}^*\in S_p} e^{\kappa l_1^*}\, dS_p, \tag{15.3.9}$$

where we have transformed $\mathbf{l}$ to $\mathbf{l}^*$ by an orthogonal transformation so that $l_1^*=\boldsymbol{\mu}'\mathbf{l}$. Under this transformation dS_p remains invariant. On using the polar transformation $\mathbf{l}^*=\mathbf{u}(\boldsymbol{\theta})$, we find with the help of (15.3.3) that

$$\{c_p(\kappa)\}^{-1}=\int_0^{\pi}\cdots\int_0^{\pi}\int_0^{2\pi} e^{\kappa\cos\theta_1}\prod_{j=2}^{p-1}\sin^{p-j}\theta_{j-1}\, d\theta_1\ldots d\theta_{p-1}.$$

Using (15.3.8) in carrying out the integration on the right-hand side, we obtain (15.3.7).

Mean direction

We show that the mean direction of $\mathbf{l}$ is $\boldsymbol{\mu}$ and

$$\rho=A(\kappa)=\frac{I_{p/2}(\kappa)}{I_{p/2-1}(\kappa)}. \tag{15.3.10}$$

Let $\mathbf{a}=\kappa\boldsymbol{\mu}$. From (15.3.6), we have

$$\int_{\mathbf{l}\in S_p} e^{\mathbf{a}'\mathbf{l}}\, dS_p=1/c_p(|\mathbf{a}'\mathbf{a}|^{1/2}).$$

Differentiating both sides with respect to $\mathbf{a}$ and using the recurrence relation

$$\kappa I_{s+1}(\kappa)=\kappa I_s'(\kappa)-sI_s(\kappa), \tag{15.3.11}$$

we find that

$$E(\mathbf{l})=A(\kappa)\boldsymbol{\mu}. \tag{15.3.12}$$

Hence (15.3.10) follows.

On differentiating the logarithm of (15.3.6) twice with respect to $\boldsymbol{\mu}$, we note that the mode of the distribution is also $\boldsymbol{\mu}$ if $\kappa > 0$. It has an antimode at $\mathbf{l} = -\boldsymbol{\mu}$.

Closure property

Under the orthogonal transformation $\mathbf{l}^* = \mathbf{A}\mathbf{l}$, $\mathbf{l}^*$ is again distributed as $M_p(\mathbf{A}\boldsymbol{\mu}, \kappa)$.

Particular cases

Let $\boldsymbol{\alpha}' = (\alpha_1, \ldots, \alpha_{p-1})$ be the spherical polar coordinates of $\boldsymbol{\mu}$. Using the polar transformation (15.1.3), we find that the p.d.f. of $\boldsymbol{\theta}$ is

$$g(\boldsymbol{\theta}; \boldsymbol{\alpha}, \kappa) = c_p(\kappa)[\exp\{\kappa \mathbf{u}'(\boldsymbol{\alpha})\mathbf{u}(\boldsymbol{\theta})\}]a_p(\boldsymbol{\theta}),$$
$$0 \leq \theta_i \leq \pi, \quad i = 1, \ldots, p-2, \quad 0 \leq \theta_{p-1} < 2\pi. \tag{15.3.13}$$

Thus the density of the von Mises distribution ($p = 2$) is

$$g(\theta; \alpha, \kappa) = \{2\pi I_0(\kappa)\}^{-1} e^{\kappa \cos(\theta - \alpha)}, \qquad 0 \leq \theta < 2\pi, \tag{15.3.14}$$

whereas the density for the Fisher distribution ($p = 3$) is

$$g(\theta, \phi; \alpha, \beta, \kappa) = \{\kappa/(4\pi \sinh \kappa)\} \exp[\kappa\{\cos \alpha \cos \theta$$
$$+ \sin \alpha \sin \theta \cos(\phi - \beta)\}] \sin \theta, \qquad 0 \leq \theta \leq \pi, \quad 0 \leq \phi < 2\pi, \tag{15.3.15}$$

where $\kappa > 0$ and (see Exercise 15.3.2 for details)

$$I_{1/2}(\kappa) = (2 \sinh \kappa)/(2\pi\kappa)^{1/2}. \tag{15.3.16}$$

If $\boldsymbol{\alpha} = \mathbf{0}$, we have, from (15.3.13),

$$g(\boldsymbol{\theta}; \boldsymbol{\alpha}, \kappa) = c_p(\kappa)\{\exp(\kappa \cos \theta_1)\}a_p(\boldsymbol{\theta}),$$
$$0 \leq \theta_i \leq \pi, \quad i = 1, \ldots, p-2, \quad 0 \leq \theta_{p-1} < 2\pi, \tag{15.3.17}$$

where θ_1 is the angle between $\mathbf{l}$ and $(1, 0, \ldots, 0)'$. Hence θ_1 and $(\theta_2, \ldots, \theta_{p-1})$ are independently distributed. Further, from (15.3.4) $(\theta_2, \ldots, \theta_{p-1})$ has a uniform distribution on S_{p-1} and the p.d.f. of θ_1 is given by

$$g(\theta_1; \kappa) = \{c_p(\kappa)/c_{p-1}\} e^{\kappa \cos \theta_1} \sin^{p-2} \theta_1, \qquad 0 \leq \theta_1 \leq \pi. \tag{15.3.18}$$

Values of κ

For large κ we have, approximately,

$$2\kappa(1 - \mathbf{l}'\boldsymbol{\mu}) \sim \chi^2_{p-1}. \tag{15.3.19}$$

To prove this result, suppose $\boldsymbol{\mu} = (1, 0, \ldots, 0)'$ so that $1 - \mathbf{l}'\boldsymbol{\mu} = 1 - \cos \theta_1$, and note that, with large probability, θ_1 will be near 0. Then substitute

$$\cos \theta_1 \doteq 1 - \tfrac{1}{2}\theta_1^2, \qquad \sin \theta_1 \doteq \theta_1, \qquad I_p(\kappa) \doteq (2\pi\kappa)^{-1/2} e^{\kappa}$$

in (15.3.18) to conclude that, approximately,

$$\kappa\theta_1^2 \sim \chi^2_{p-1},$$

which is equivalent to (15.3.19) since $\cos\theta_1 \doteq 1-\frac{1}{2}\theta_1^2$.

For $\kappa = 0$, the distribution reduces to the uniform distribution given by (15.3.4), where (see Exercise 15.3.1)

$$\lim_{\kappa\to 0} c_p(\kappa) = c_p.$$

Shape of the distribution

Using the foregoing results, we can now look into the behaviour of the distribution. For $\kappa > 0$, the distribution has a mode at the mean direction, whereas when $\kappa = 0$ the distribution is uniform. The larger the value of κ, the greater is the clustering around the mean direction. This behaviour explains the reason for describing κ as the concentration parameter.

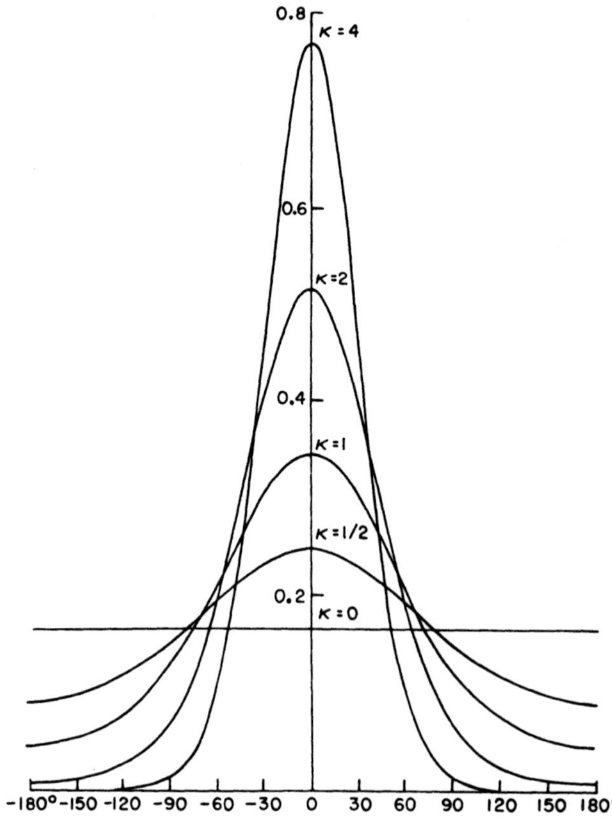

Figure 15.3.1a Linear representation of the density of $M_2((1, 0), \kappa)$, $\kappa = 0, \frac{1}{2}, 1, 2, 4$.

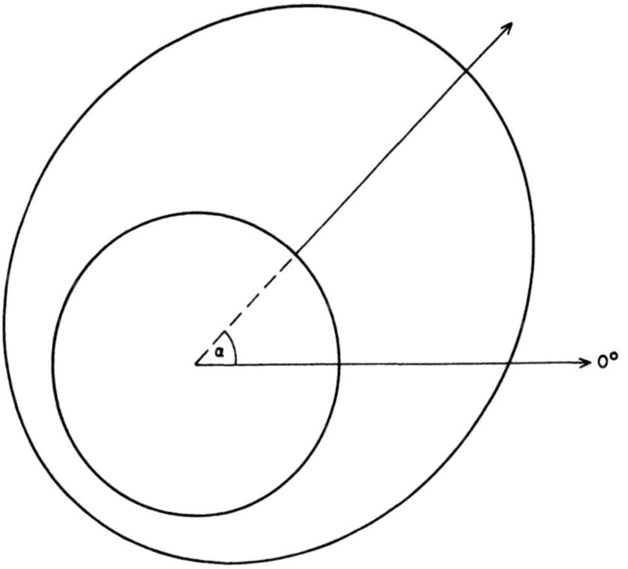

Figure 15.3.1b Circular representation of the density of $M_2((\cos\alpha, \sin\alpha)', 1)$.

Figures 15.3.1a and 15.3.1b give the shape of the distribution for varying values of κ and μ. For $p=3$ and $\boldsymbol{\mu}=(1,0,0)'$ the distribution is rotationally symmetric about the x_1 axis, having a maximum at the north pole $\theta=0$ and a minimum at the south pole $\theta=\pi$.

15.3.3 The Bingham distribution

Sometimes the observations are not *directions* but *axes*; that is, the unit vectors $\mathbf{l}$ and $-\mathbf{l}$ define the same axis. In this context it is appropriate to consider p.d.f.s for $\mathbf{l}$ which satisfy the antipodal symmetric property

$$f(\mathbf{l})=f(-\mathbf{l}).$$

The observations in such cases can be regarded as being on a hyperhemisphere, or on a sphere with opposite points not distinguishable.

An important distribution for dealing with directional data with antipodal symmetry and for dealing with axial data is the Bingham distribution. The unit vector $\mathbf{l}$ is said to have a Bingham distribution, $B_p(\mathbf{A}, \boldsymbol{\kappa})$, if its p.e. is of the form

$$d(\boldsymbol{\kappa}) \exp\{\operatorname{tr}(\mathbf{KA'll'A})\}\, dS_p, \qquad \mathbf{l}\in S_p, \tag{15.3.20}$$

where $\mathbf{A}$ now denotes an orthogonal matrix, $\mathbf{K}$ is a diagonal matrix of constants $\boldsymbol{\kappa} = (\kappa_1, \ldots, \kappa_p)'$ and $d(\boldsymbol{\kappa})$ is the normalizing constant depending only on $\boldsymbol{\kappa}$ and is given (see Exercise 15.3.6) by

$$\frac{1}{d(\boldsymbol{\kappa})} = 2 \sum_{r_1=0}^{\infty} \cdots \sum_{r_p=0}^{\infty} \frac{\prod_{j=1}^{p} \Gamma(r_j + \frac{1}{2})}{\Gamma\left(\sum_{j=1}^{p} r_j + \frac{p}{2}\right)} \prod_{j=1}^{p} \frac{\kappa_j^{r_j}}{r_j!}. \tag{15.3.21}$$

Since $\operatorname{tr}(\mathbf{A}'\mathbf{l}\mathbf{l}'\mathbf{A}) = 1$, the sum of the parameters κ_i in (15.3.20) is arbitrary, and it is usual to take $\kappa_p = 0$. For $\kappa_1 = \ldots = \kappa_p$, the distribution reduces to the uniform distribution. We have

$$\operatorname{tr}(\mathbf{K}\mathbf{A}'\mathbf{l}\mathbf{l}'\mathbf{A}) = \sum_{i=1}^{p} \kappa_i (\mathbf{l}'\mathbf{a}_{(i)})^2, \qquad \mathbf{A} = (\mathbf{a}_{(1)}, \ldots, \mathbf{a}_{(p)}).$$

Using the polar transformation (15.1.3) as in Section 15.3.2 when $\kappa_2 = \ldots = \kappa_p = 0$, $\kappa_1 = \kappa$, the joint p.d.f. of $\boldsymbol{\theta}$ becomes

$$\text{const.}\, e^{-\kappa \cos^2 \theta_1} a_p(\boldsymbol{\theta}), \qquad 0 \le \theta_i \le \pi, \quad i = 1, \ldots, p-2, \quad 0 \le \theta_{p-1} < 2\pi. \tag{15.3.22}$$

For $\kappa > 0$, this distribution is symmetric with high concentration around the equator (for $p = 3$) in the form of a "girdle", while for $\kappa < 0$ the distribution is bimodal with poles at $\theta_1 = 0, \pi$ and is rotationally symmetric about the x_1 axis. For $\kappa = 0$, the random variables $\theta_1, \ldots, \theta_{p-1}$ are uniformly distributed. In general, the larger the elements of $\boldsymbol{\kappa}$, the greater the clustering around the axes with direction cosines $\mathbf{a}_{(i)}$, $i = 1, \ldots, p$. Further, different values of $\boldsymbol{\kappa}$ in (15.3.20) give the uniform distribution, symmetric and asymmetric girdle distributions (i.e. the probability mass is distributed in the form of an elliptic pattern like a girdle around a great circle) and bimodal distributions. Figure 15.3.2 shows common configurations of sample points on a sphere. The sample points in Figure 15.3.2(a) could arise from a Fisher distribution while the sample points in Figure 15.3.2(b) and (c) could have arisen from the Bingham distribution.

For $p = 2$, the p.d.f. (15.3.20) reduces in polar coordinates to

$$f(\theta; \alpha, \kappa) = \{2\pi I_0(\kappa)\}^{-1} e^{\kappa \cos 2(\theta - \alpha)}, \qquad \kappa > 0, \quad 0 \le \theta < 2\pi.$$

The distribution is of course bimodal with the two modes situated 180° apart, and is called the *bimodal distribution of von Mises type*. The transformation $\theta^* = \theta \bmod \pi$ represents an axis by an angle in $(0, \pi)$, and the distribution of θ^* is called the *von Mises-type axial* distribution. Note that the distribution of $\theta' = 2\theta^* = 2\theta$ is $M((\cos 2\alpha, \sin 2\alpha)', \kappa)$. Thus the process of "doubling the angles" of θ (or θ^*) leads to the von Mises

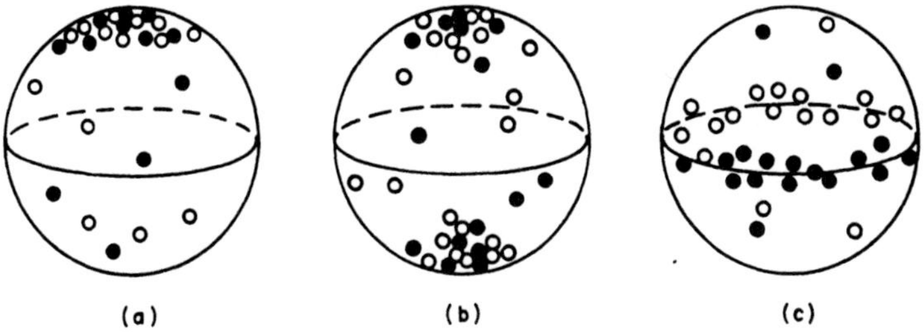

Figure 15.3.2 Configuration of sample points from (a) unimodal, (b) bimodal, and (c) girdle distributions. An open circle denotes a point on the other side.

distribution. Unfortunately there is no simple analogue to this process for $p \leq 3$ related to the distribution $M_p(\boldsymbol{\mu}, \kappa)$.

The distribution for $p=3$ was put forward by Bingham (1964, 1974) who investigated its statistical properties extensively. For an elaborate discussion of the distribution for $p=2$ and for $p=3$, we refer to Mardia (1972a).

15.4 Distribution Theory

Let $\mathbf{l}_1, \ldots, \mathbf{l}_n$ be a random sample from $M_p(\boldsymbol{\mu}, \kappa)$. For inference problems, the sample resultant length R and the sample mean direction vector $\bar{\boldsymbol{\theta}}$ (in polar coordinates) are of importance; that is,

$$\sum_{i=1}^{n} \mathbf{l}_i = (r_1, \ldots, r_p)' = \mathbf{r} = R\mathbf{u}(\bar{\boldsymbol{\theta}}). \tag{15.4.1}$$

The statistics $r_1, \ldots, r_p$ are the components of the resultant vector $\mathbf{r} = \sum \mathbf{l}_i$. In most cases, the sampling distribution of R and $\bar{\boldsymbol{\theta}}$ from the von Mises–Fisher population can be derived with the help of the corresponding distribution for the uniform case, which we consider first.

15.4.1 The characteristic function method and the isotropic case

Let $\phi(\mathbf{t}) = E\{\exp(i\mathbf{t}'\mathbf{l})\}$ be the characteristic function of $\mathbf{l}$. Then from the inversion theorem, the p.d.f. of $\mathbf{r}$ for sample size n is given by

$$(2\pi)^{-p} \int_{-\infty}^{\infty} e^{-i\mathbf{t}'\mathbf{r}} \phi^n(\mathbf{t})\, d\mathbf{t}, \tag{15.4.2}$$

provided $\phi^n(\mathbf{t})$ is an absolutely integrable function. On using the polar transformations $\mathbf{t}=\rho\mathbf{u}(\boldsymbol{\alpha})$ and $\mathbf{r}=R\mathbf{u}(\bar{\boldsymbol{\theta}})$ in (15.4.2), we find that the p.d.f. of $(R,\bar{\boldsymbol{\theta}})$ is

$$g_0(R,\bar{\boldsymbol{\theta}})=(2\pi)^{-p}R^{p-1}a_p(\bar{\boldsymbol{\theta}})\int_0^\infty\int_{S_p}\psi^n(\rho,\boldsymbol{\alpha})e^{-i\rho R\mathbf{u}'(\boldsymbol{\alpha})\mathbf{u}(\boldsymbol{\theta})}\rho^{p-1}a_p(\boldsymbol{\alpha})\,d\rho\,d\boldsymbol{\alpha}, \tag{15.4.3}$$

where

$$\psi(\rho,\boldsymbol{\alpha})=[\phi(\mathbf{t})]_{\mathbf{t}=\rho\mathbf{u}(\boldsymbol{\alpha})},$$

and the second integral is taken over $0\leqslant\phi_i\leqslant\pi$, $i=1,\ldots,p-2$; $0\leqslant\phi_{p-1}<2\pi$.

We now deduce the following results for the *isotropic* random walk with n equal steps, starting from the origin, in p dimensions; that is, $\mathbf{l}_1,\ldots,\mathbf{l}_n$ is a random sample from the uniform population with p.d.f. given by (15.3.4). In this case it can be shown that (15.4.2) is valid for all $n\geqslant2$ at all continuity points of the p.d.f. of $\mathbf{R}$, provided that when $\phi^n(\mathbf{t})$ is not absolutely integrable, the integral is interpreted as a limit over increasing discs centred at $\mathbf{0}$. (See Watson, 1944, pp. 419–421.)

Since (15.3.9) holds for complex variables (see Exercise 15.3.3), the c.f. of $\mathbf{l}$ is

$$\psi(\rho,\boldsymbol{\alpha})=c_p\int_{S_p}e^{i\mathbf{t}'\mathbf{l}}\,d\mathbf{l}=c_p/c_p(i\rho)=\Gamma(\tfrac{1}{2}p)(\tfrac{1}{2}\rho)^{1-p/2}J_{(p-1)/2}(\rho), \tag{15.4.4}$$

where $J_r(\rho)$ is the ordinary Bessel function of rth order related to the modified Bessel function by

$$J_r(\rho)=i^{-r}I_r(i\rho). \tag{15.4.5}$$

On substituting (15.4.4) in (15.4.3), we deduce from (15.4.3), after integrating over $\boldsymbol{\alpha}$ with the help of (15.3.9), that

(i) R and $\bar{\boldsymbol{\theta}}$ are independently distributed;
(ii) $\bar{\boldsymbol{\theta}}$ is uniformly distributed on S_p;
(iii) the p.d.f. of R is given by

$$\begin{aligned}h_n(R)&=(2\pi)^{-p}c_p^{n-1}R^{p-1}\int_0^\infty\rho^{p-1}\{c_p(i\rho R)c_p^n(i\rho)\}^{-1}\,d\rho\\&=2^{(1-p/2)(n-1)}\Gamma(\tfrac{1}{2}p)^{n-1}R\int_0^\infty\rho^{n-p(n-1)/2}\\&\qquad\times J_{p/2-1}(\rho R)J^n_{p/2-1}(\rho)\,d\rho,\qquad 0<R<n\end{aligned} \tag{15.4.6}$$

Of course, R denotes the distance covered from the origin after n steps in the random walk.

15.4.2 The von Mises–Fisher case

Let us now assume that $\mathbf{l}_1, \ldots, \mathbf{l}_n$ is a random sample from $M_p(\boldsymbol{\mu}, \kappa)$. On integrating the joint density over constant values of $(R, \bar{\boldsymbol{\theta}})$, we find that the p.d.f. of $(R, \bar{\boldsymbol{\theta}})$ is

$$g(R, \bar{\boldsymbol{\theta}}; \boldsymbol{\mu}, \kappa) = \{c_p^n(\kappa)/c_p^n\} e^{\kappa R \mathbf{u}'(\boldsymbol{\mu})\mathbf{u}(\bar{\boldsymbol{\theta}})} g_0(R, \bar{\boldsymbol{\theta}}),$$

where g_0 is the p.d.f. of $(R, \bar{\boldsymbol{\theta}})$ under uniformity. From Section 15.4.1, we get

$$g(R, \bar{\boldsymbol{\theta}}; \boldsymbol{\mu}, \kappa) = \{c_p^n(\kappa)/c_p^n\} e^{\kappa R \mathbf{u}'(\boldsymbol{\mu})\mathbf{u}(\bar{\boldsymbol{\theta}})} h_n(R) c_p a_p(\bar{\boldsymbol{\theta}}), \qquad (15.4.7)$$

where $h_n(R)$ is the p.d.f. of R for the uniform case given by (15.4.5). On integrating (15.4.7) over $\bar{\boldsymbol{\theta}}$ with the help of (15.3.9), we see that the marginal density of R is

$$h_n^{(\kappa)}(R) = c_p^{1-n} c_p^n(\kappa) h_n(R)/c_p(\kappa R). \qquad (15.4.8)$$

This result is due to Watson and Williams (1956). From (15.4.7) and (15.4.8) it follows (Mardia, 1975*b*) that

$$\bar{\boldsymbol{\theta}} \mid R \sim M_p(\boldsymbol{\mu}, \kappa R). \qquad (15.4.9)$$

Notice that the distribution of $\bar{\boldsymbol{\theta}}$ depends on R unless $\kappa = 0$, which is in contrast to the normal case where the mean vector and the covariance matrix are independently distributed. This fact influences inference on $\boldsymbol{\mu}$.

For other distributional problems, see Exercises 15.4.1–15.4.3.

15.5 Maximum Likelihood Estimators for the von Mises–Fisher Distribution

Let $\mathbf{l}_i$, $i = 1, \ldots, n$, be a random sample from $M_p(\boldsymbol{\mu}, \kappa)$. We assume $\kappa > 0$ so that the population is non-uniform. The logarithm of the likelihood function is

$$\text{const.} + n \log c_p(\kappa) + \kappa \boldsymbol{\mu}'\mathbf{r} + \lambda(1 - \boldsymbol{\mu}'\boldsymbol{\mu}), \qquad (15.5.1)$$

where λ is a Lagrange multiplier. On differentiating with respect to $\boldsymbol{\mu}$, we find that

$$\kappa \mathbf{r} = 2\lambda \boldsymbol{\mu}.$$

Using $\boldsymbol{\mu}'\boldsymbol{\mu}=1$, we find that

$$\hat{\boldsymbol{\mu}}=\mathbf{r}/R=\mathbf{l}_0. \qquad (15.5.2)$$

Differentiating (15.5.1) with respect to κ and substituting (15.5.2) gives the m.l.e. of κ as the solution of

$$-c_p'(\kappa)/c_p(\kappa)=\bar{R}, \qquad (15.5.3)$$

where $\bar{R}=n^{-1}R$. On using (15.3.7) and (15.3.11), we find that

$$c_p'(\kappa)/c_p(\kappa)=-A(\kappa), \qquad (15.5.4)$$

where $A(\kappa)$ is defined by (15.3.10). Hence (15.5.3) becomes

$$A(\hat{\kappa})=\bar{R} \quad \text{or} \quad \hat{\kappa}=A^{-1}(\bar{R}). \qquad (15.5.5)$$

The functions $A(\kappa)$ and $A^{-1}(\kappa)$ are tabulated for $p=2$ and $p=3$ (see, for example, Mardia, 1972). Using for large κ the asymptotic formula

$$I_p(\kappa)\doteq(2\pi\kappa)^{-1/2}e^{\kappa}[1-\{(4p^2-1)/8\kappa\}], \qquad (15.5.6)$$

we have

$$A(\kappa)\doteq 1-\{(p-1)/2\kappa\}. \qquad (15.5.7)$$

Hence, for large κ,

$$\hat{\kappa}\doteq\tfrac{1}{2}(p-1)/(1-\bar{R}). \qquad (15.5.8)$$

This approximation can be trusted for $\bar{R}\geqslant 0.8$ when $p=2$ and $\bar{R}\geqslant 0.9$ when $p=3$.

Since $I_p(\kappa)$ can be written in a series as

$$I_p(\kappa)=\sum_{r=0}^{\infty}\frac{1}{\Gamma(p+r+1)\Gamma(r+1)}\left(\frac{\kappa}{2}\right)^{2r+p}, \qquad (15.5.9)$$

it follows that, for small κ,

$$A(\kappa)\doteq\kappa/p \qquad (15.5.10)$$

so that

$$\hat{\kappa}\doteq p\bar{R}. \qquad (15.5.11)$$

This approximation can be used for $\bar{R}\leqslant 0.1$ when $p=2$ and for $\bar{R}<0.05$ when $p=3$.

It can be seen that the m.l.e. of $\boldsymbol{\alpha}$ is $\bar{\boldsymbol{\theta}}$. Further, differentiating the log likelihood, log L, twice and using the following result from (15.3.12),

$$E(\mathbf{r})=nA(\kappa)\boldsymbol{\mu}=nA(\kappa)\mathbf{u}(\boldsymbol{\alpha}),$$

we obtain (Mardia, 1975*a*)

$$E\left\{-\frac{\partial^2 \log L}{\partial \kappa^2}\right\}=nA'(\kappa), \qquad E\left\{-\frac{\partial^2 \log L}{\partial \kappa \partial \alpha_i}\right\}=0, \qquad i=1,\ldots,p-1, \tag{15.5.12}$$

$$E\left\{-\frac{\partial^2 \log L}{\partial \alpha_i \partial \alpha_j}\right\}=0, \qquad i\neq j,$$

$$E\left\{-\frac{\partial^2 \log L}{\partial \alpha_i^2}\right\}=n\kappa A(\kappa)\prod_{j=0}^{i-1}\sin^2\alpha_j, \qquad i=1,\ldots,p-1. \tag{15.5.13}$$

Hence, for large n, we conclude from Section 4.2.1 that $\hat{\kappa}$, $\bar{\theta}_1, \ldots, \bar{\theta}_{p-1}$, are asymptotically independently normally distributed with means κ, $\alpha_1, \ldots, \alpha_{p-1}$, and

$$\operatorname{var}(\hat{\kappa})=[nA'(\kappa)]^{-1}, \tag{15.5.14}$$

$$\operatorname{var}(\bar{\theta}_i)=\frac{1}{n}\left\{\kappa A(\kappa)\prod_{j=0}^{i-1}\sin^2\alpha_j\right\}^{-1}, \qquad =1,\ldots,p-1. \tag{15.5.15}$$

15.6 Test of Uniformity: The Rayleigh Test

Let $\mathbf{l}_i$, $i=1,\ldots,n$, be a random sample of $M_p(\boldsymbol{\mu}, \kappa)$. Consider

$$H_0:\kappa=0 \quad \text{against} \quad H_1:\kappa\neq 0,$$

where $\boldsymbol{\mu}$ is unknown. Let λ be the LR statistic. Under H_0, the likelihood function is

$$L_0=c_p^n, \tag{15.6.1}$$

whereas on using the m.l.e.s of κ and $\boldsymbol{\mu}$ given (15.5.2) and (15.5.5), we find that

$$L_1=\{c_p(\hat{\kappa})\}^n \exp(n\hat{\kappa}\bar{R}), \tag{15.6.2}$$

where

$$\hat{\kappa}=A^{-1}(\bar{R}). \tag{15.6.3}$$

Let λ be the likelihood ratio statistic for the problem. From (15.6.1)–(15.6.3), we find that

$$\log\lambda=-n\{\log(c_p(\hat{\kappa})/c_p)+\hat{\kappa}A(\hat{\kappa})\}.$$

Note that λ is a function of $\bar{R}$ alone.

We now show that λ is a monotonically decreasing function of $\bar{R}$. First think of $\log \lambda$ as a function of $\hat{\kappa}$. Using the following result from (15.5.4),

$$c_p'(\kappa) = -c_p(\kappa)A(\kappa), \tag{15.6.4}$$

we obtain

$$\frac{d(\log \lambda)}{d\hat{\kappa}} = -n\hat{\kappa}A'(\hat{\kappa}). \tag{15.6.5}$$

Since $\hat{\kappa} \geq 0$, to show that (15.6.5) is negative it remains to establish

$$A'(\kappa) \geq 0. \tag{15.6.6}$$

If $\boldsymbol{u}(\boldsymbol{\theta})$ is distributed as $M_p((1, 0, \ldots, 0)', \kappa)$, we have, from (15.3.12),

$$E(\cos \theta_1) = A(\kappa); \tag{15.6.7}$$

that is, from (15.3.18),

$$A(\kappa) = \frac{c_p(\kappa)}{c_{p-1}} \int_0^{\pi} \cos \theta_1 e^{\kappa \cos \theta_1} \sin^{p-2} \theta_1 \, d\theta_1.$$

On differentiating both sides with respect to κ and using (15.6.4) and (15.6.7), we find that

$$A'(\kappa) = V(\cos \theta_1) \geq 0. \tag{15.6.8}$$

Hence λ is a monotonically decreasing function of $\hat{\kappa}$. Finally, on differentiating (15.6.3) with respect to $\hat{\kappa}$ and using (15.6.6), it is found that $\hat{\kappa}$ is a monotonically increasing function of $\bar{R}$.

Therefore the critical region of the Rayleigh test reduces to

$$\bar{R} > K. \tag{15.6.9}$$

This test is what we should expect intuitively since, under the hypothesis of uniformity, the value of $\bar{R}$ will be small.

The p.d.f.s of R under H_0 and H_1 are given by (15.4.6) and (15.4.8), respectively. For large n, we show that asymptotically under H_0,

$$pn\bar{R}^2 \sim \chi_p^2. \tag{15.6.10}$$

From Section 15.3.1, $E(\mathbf{l}) = \mathbf{0}$, $V(\mathbf{l}) = p^{-1}\mathbf{I}$. Hence $\mathbf{r}$ is distributed asymptotically as $N_p(\mathbf{0}, (n/p)\mathbf{I})$. Consequently, the asymptotic distribution of $R^2 = \mathbf{r}'\mathbf{r}$ can be obtained which leads to (15.6.10).

The result for $p = 3$ was first proved by Rayleigh (1919). For appropriate tables and further discussion, see Mardia (1972a). This test is the uniformly most powerful invariant test (see Exercise 15.6.2) for testing uniformity against the alternative of a von Mises–Fisher distribution.

Example 15.6.1 From the vanishing angles of 209 homing pigeons in a clock resetting experiment (Schmidt–Koenig, 1965; Mardia, 1972a, p. 123), it is found that

$$r_1 = -7.2098, \qquad r_2 = 67.3374, \qquad \bar{R} = 0.3240.$$

It is expected that the pigeons have a preferred direction. We have $2n\bar{R}^2 = 43.88$. The upper 1% value of χ_2^2 is 13.82. Hence we reject the hypothesis of uniformity.

15.7 Some Other Problems

There are various inference problems which arise in directional data.

15.7.1 Test for the mean direction

Consider the problem of testing

$$H_0: \boldsymbol{\mu} = (1, 0, 0, \ldots, 0)' = \mathbf{e}_1 \quad \text{against} \quad H_1: \boldsymbol{\mu} \neq \mathbf{e}_1, \tag{15.7.1}$$

where κ is known and is non-zero. Let λ be the likelihood ratio test for the problem. Using (15.5.2), we find that the critical region is

$$-2 \log \lambda = 2\kappa(R - r_1) > K. \tag{15.7.2}$$

When H_0 is true, we obtain from Wilks' theorem that, asymptotically as $n \to \infty$,

$$2\kappa(R - r_1) \sim \chi_{p-1}^2. \tag{15.7.3}$$

For another proof of (15.7.3) for large κ, see Exercise 15.7.1. This approximation is valid for $n \geq 20$ and $\kappa \geq 5$ if $p = 2$, $\kappa \geq 3$ if $p = 3$. Better approximations for $p = 2, 3$ are given in Mardia (1972, p. 115 and p. 247).

When κ is unknown, we can again obtain the likelihood ratio test. For a discussion of this and other exact tests see Mardia (1972). However, for large κ, the following procedure of Watson and Williams (1956) is adequate. Since

$$r_1 = \sum_{i=1}^{n} \cos \theta_{i1}$$

in polar coordinates, we have from (15.3.19) that approximately

$$2\kappa(n - r_1) = 2\kappa \sum_{i=1}^{n} (1 - \cos \theta_{i1}) \sim \chi_{n(p-1)}^2. \tag{15.7.4}$$

Now

$$2\kappa(n-r_1)=2\kappa(R-r_1)+2\kappa(n-R). \tag{15.7.5}$$

From an application of Cochran's theorem to (15.7.3) and (15.7.4) it can be shown that

$$2\kappa(n-R)\sim\chi^2_{(n-1)(p-1)} \tag{15.7.6}$$

approximately for large κ. Further the two terms on the right-hand side of (15.7.5) are approximately independently distributed. Hence for testing (15.7.1) with large but unknown κ we can use the statistic

$$F_{p-1,(n-1)(p-1)}=(n-1)(R-r_1)/(n-R), \tag{15.7.7}$$

where H_0 is rejected for large values of F.

For a general mean direction $\boldsymbol{\mu}_0$, r_1 should be replaced by $\mathbf{r}'\boldsymbol{\mu}_0$ in (15.7.7). Then this test for testing $H_0:\boldsymbol{\mu}=\boldsymbol{\mu}_0$ against $H_1:\boldsymbol{\mu}\neq\boldsymbol{\mu}_0$ becomes

$$F_{p-1,(n-1)(p-1)}=(n-1)(R-\mathbf{r}'\boldsymbol{\mu}_0)/(n-R)$$

and H_0 is rejected for large values of F. This test can be used when $\mathbf{r}'\boldsymbol{\mu}_0\geqslant\frac{5}{6}n$ if $p=2$ and when $\mathbf{r}'\boldsymbol{\mu}_0\geqslant\frac{3}{5}n$ if $p=3$, for all sample sizes $n\geqslant 2$.

Thus, under H_0,

$$P\{(n-1)(R-\mathbf{r}'\boldsymbol{\mu}_0)/(n-R)\leqslant F_{p-1,(n-1)(p-1);\alpha}\}=1-\alpha,$$

where $F_{p-1,(n-1)(p-1);\alpha}$ is the upper α percentage point of the $F_{p-1,(n-1)(p-1)}$ distribution. Writing $\mathbf{r}'\boldsymbol{\mu}_0=R\cos\delta$, the above inequality can be rearranged to give

$$\cos\delta\geqslant 1-(n-R)F_{p-1,(n-1)(p-1);\alpha}/[(n-1)R],$$

which gives a $100(1-\alpha)\%$ confidence interval for δ and hence for $\boldsymbol{\mu}_0$. In three dimensions the vectors $\boldsymbol{\mu}_0$ allowed by this value of δ lie within a cone of semiangle δ and axis through the sample mean $\bar{\mathbf{l}}$ (see Figure 15.7.1).

Example 15.7.1 In an early Quaternary zone of lava flows from Western Iceland, it was found for 45 readings (Hospers' data cited in Fisher, 1953) that the sums of the observed directions were

$$r_1=-37.2193,\qquad r_2=-11.6127,\qquad r_3=+0.6710. \tag{15.7.8}$$

The sample dipole field appropriate to the geographical latitude is found to have the direction

$$0.9724,\qquad 0.2334,\qquad 0. \tag{15.7.9}$$

It is enquired whether the direction of magnetization differs from the reversed dipole field.

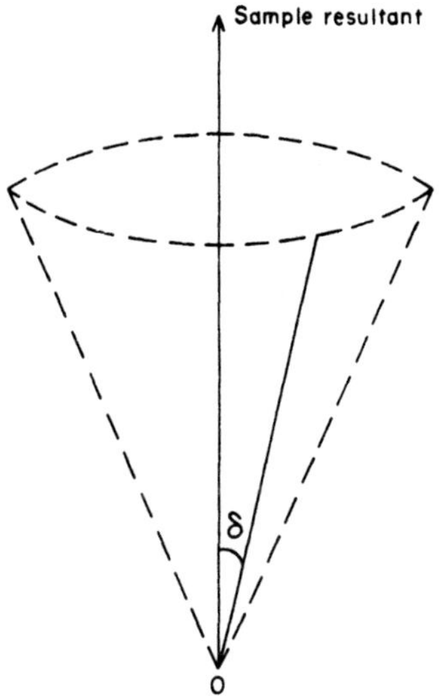

Figure 15.7.1 Confidence cone for the true mean direction.

Under H_0, we may take the true mean direction as the reversed dipole field, i.e. from (15.7.9) the components of $\boldsymbol{\mu}_0$ are

$$\mu_1 = -0.9724, \qquad \mu_2 = -0.2334, \qquad \mu_3 = 0. \qquad (15.7.10)$$

We have $R = 38.9946$. Since $\bar{R}$ is large, we use the approximation (15.5.8) for estimating $\hat{\kappa}$. It is found that $\hat{\kappa} = 7.3$. Also,

$$\mu_1 r_1 + \mu_2 r_2 + \mu_3 r_3 = 38.9025.$$

Hence the observed value of (15.7.7) is 0.67. The 5% value of $F_{2,88}$ is 3.10. Hence the null hypothesis is accepted at the 5% level of significance.

For this data the confidence interval for δ becomes at the 95% level

$$(n-1)R(1-\cos\delta)/(n-R) \leqslant 3.10.$$

This gives $\cos\delta \geqslant 0.9891$, or $\delta \leqslant 8.4°$. Thus the semiangle of the 95% confidence cone is 8.4°.

15.7.2 Principal component analysis

Let $\mathbf{l}_i$, $i = 1, \ldots, n$, be a random sample of unit vectors. We can sensibly define the *first principal direction* as the direction $\mathbf{a}$ ($\mathbf{a}'\mathbf{a} = 1$) which minimizes the angles θ_i between $\mathbf{a}$ and $\mathbf{l}_i$, $i=1,\ldots,n$; that is, if

$$Q=\sum_{i=1}^{n}\cos^2\theta_i=\sum_{i=1}^{n}(\mathbf{a}'\mathbf{l}_i)^2 \tag{15.7.11}$$

is maximum. The square of the angle ensures that we are talking of angles between two lines $\mathbf{a}$ and $\mathbf{l}_i$ passing through the origin, so that for this purpose each $\mathbf{l}_i$ is considered as an axis rather than a direction. Let

$$\mathbf{T}=\sum_{i=1}^{n}\mathbf{l}_i\mathbf{l}_i' \tag{15.7.12}$$

be the matrix of sums of squares and products. (Notice that the mean has not been subtracted off.) Suppose that $\lambda_1,\ldots,\lambda_p$ are the eigenvalues of $\mathbf{T}$ $(\lambda_1>\ldots>\lambda_p)$ and $\boldsymbol{\gamma}_{(1)},\ldots,\boldsymbol{\gamma}_{(p)}$ are the corresponding eigenvectors. In view of $\mathbf{l}_i'\mathbf{l}_i=1$,

$$\operatorname{tr}\mathbf{T}=n=\lambda_1+\ldots+\lambda_p. \tag{15.7.13}$$

Further maximizing (15.7.11) is equivalent to maximizing

$$Q=\mathbf{a}'\mathbf{T}\mathbf{a}$$

subject to $\mathbf{a}'\mathbf{a}=1$. Using the argument of principal component analysis (Chapter 8), we conclude that the maximizing value of $\mathbf{a}$ is $\mathbf{a}=\boldsymbol{\gamma}_{(1)}$, which represents the first principal *direction*, and max $Q=\lambda_1$, which measures the *concentration* around this axis.

We can similarly talk of the second *principal direction* $\mathbf{b}$, where $\mathbf{b}$ is perpendicular to $\mathbf{a}$. Of course $\mathbf{b}=\boldsymbol{\gamma}_{(2)}$ and so on. However, note that λ_p is specified from (15.7.13) as soon as $\lambda_1,\ldots,\lambda_{p-1}$ are known.

If $\lambda_1\doteq\ldots\doteq\lambda_p$, the angular distances will be almost the same for all $\mathbf{a}$. Hence, in this case, the sample distribution is approximately uniform.

We have not dealt with the distributional problems, but the maximum likelihood estimators of the parameters in the Bingham distribution neatly tie up with these eigenvalues and eigenvectors. (See Exercise 15.7.5) This leads to the solution of various other inference problems. For example, it is natural to use the following criterion for testing uniformity for axial data,.

$$\tfrac{1}{2}p(p+2)\sum_{i=1}^{n}\left(\lambda_i-\frac{n}{p}\right)^2\sim\chi^2_{p(p+1)/2-1}.$$

This has been studied by Bingham (1964) and Mardia (1975*d*). There is a wealth of recent statistical thinking in this area and we refer to Mardia (1975*d*) and Downs (1972) for a review of recent developments.

Example 15.7.2 Mardia (1975*d*). Tyror (1957) analysed the distribution of the directions of perihelia of 448 long-period comets (with periods greater than 200 years up to and including comet 1952f). From the accretion theory of Lyttleton (1953, 1961), one expects the following behaviour of the distribution of these directions: the distribution should be non-uniform, and, in particular, the perihelion points should exhibit a preference for lying near the galactic plane.

The various statistics are as follows.

$$R = 63.18, \qquad \mathbf{l}_0' = (-0.0541, -0.3316, 0.9419),$$

$$\mathbf{T}/n = \begin{bmatrix} 0.289\,988 & -0.012\,393 & 0.030\,863 \\ -0.012\,393 & 0.389\,926 & -0.011\,153 \\ 0.030\,863 & -0.011\,153 & 0.320\,086 \end{bmatrix},$$

where $\mathbf{T}$ denotes the matrix of sum of squares and products. It may be noted that if we apply the Rayleigh test which assumes the alternative of a Fisher distribution, we have $3\,n\bar{R}^2 \sim \chi_3^2$. The observed value $3\,n\bar{R}^2 = 26.3$, which is significant at the 0.1% level.

The values of the eigenvalues and the eigenvectors of $\mathbf{T}/n$ are as follows where $\bar{\lambda}_i = \lambda_i/n$:

$$\bar{\lambda}_1 = 0.3947, \qquad \boldsymbol{\gamma}_{(1)} = (-0.1774, 0.9600, -0.2168)',$$
$$\bar{\lambda}_2 = 0.3347, \qquad \boldsymbol{\gamma}_{(2)} = (0.4922, 0.2773, 0.8251)',$$
$$\bar{\lambda}_3 = 0.2705, \qquad \boldsymbol{\gamma}_{(3)} = (0.8522, 0.0397, -0.5217)'.$$

Disparate values of $\bar{\lambda}_i$ indicate a lop-sided girdle distribution. Further, the normal to the preferred great circle (containing the girdle) has the direction cosines

$$l = 0.8522, \qquad m = 0.0397, \qquad n = -0.5217. \qquad (15.7.14)$$

Now the direction of the galactic pole is

$$(0.8772, -0.0536, -0.4772)$$

and the angle between this direction and the direction of the eigenvector (15.7.14) corresponding to λ_1 is only 6.08°. Hence there is strong evidence that the Lyttleton theory is true.

The large value of $\bar{R}$ and the disparate values of λ_i suggest that this data comes from some distribution which combines the features of both the von Mises–Fisher and Bingham models. For further statistical analysis of this data, see Mardia (1975*d*).

Exercises and Complements

15.2.1 Show that $\bar{R}$ is invariant under orthogonal transformations $\mathbf{l}_i^* = \mathbf{A}\mathbf{l}_i$, $i = 1, \ldots, n$.

15.2.2 Show that $\bar{R} = 0$ for the points $\{\cos(2\pi i/n), \sin(2\pi i/n)\}$, $i = 1, \ldots, n$. (These points are called "uniform scores" on the circle.)

15.2.3 The following stopping positions were observed for nine spins of a roulette wheel marked $0°, 1°, \ldots, 359°$.

$$43°,\quad 45°,\quad 52°,\quad 61°,\quad 75°,\quad 88°,\quad 88°,\quad 279°,\quad 357°.$$

Show that

$$\frac{1}{n}\sum \cos\theta_i = 0.4470, \qquad \frac{1}{n}\sum \sin\theta_i = 0.5529.$$

Hence, show that $\bar{R} = 0.71$, $\bar{\theta} = 51°$. Does the data indicate a preferred direction?

15.3.1 Using (15.3.7) and (15.5.9), show that $\lim_{\kappa\to 0} c_p(\kappa) = c_p$.

15.3.2 For $p = 3$, show that (15.3.18) gives (with $\theta_1 = \theta$)

$$g(\theta; \kappa) = \{c_3(\kappa)/c_2\}e^{\kappa\cos\theta}\sin\theta, \qquad 0 \leq \theta \leq \pi.$$

On integrating over $(0, \pi)$, prove that the normalizing constant is

$$c_3(\kappa)/c_2 = \kappa/(2\sinh\kappa).$$

Hence deduce (15.3.16).

15.3.3 (Mardia, 1975*b*) Define an analytic function

$$I_\nu^*(z) = \sum_{k=0}^{\infty} \frac{(z/4)^k}{[k!\,(\nu+1)\ldots(\nu+k)]}.$$

(a) Using (15.5.9) show that $I_\nu^*(z)$ is related to the modified Bessel function by

$$I_\nu(z) = \Gamma(\nu+1)^{-1}(z/2)^\nu I_\nu^*(z^2).$$

(b) Show that the normalization constant for the von Mises–Fisher distribution can also be written

$$c_p(\kappa)^{-1} = 2\pi^{p/2} I_{p/2-1}^*(\kappa^2)/\Gamma(\tfrac{1}{2}p).$$

(c) Using analytic continuation on the normalization constant, show that the characteristic function of the von Mises–Fisher distribution is

$$\psi(\mathbf{t}; \boldsymbol{\mu}, \kappa) = \int_{\mathbf{l} \in S_p} c_p(\kappa) \exp(i\mathbf{t}'\mathbf{l} + \kappa\boldsymbol{\mu}'\mathbf{l})\, dS_p$$

$$= I^*_{p/2-1}(\kappa^2 - \mathbf{t}'\mathbf{t} + 2i\kappa\mathbf{t}'\boldsymbol{\mu})/I^*_{p/2-1}(\kappa^2).$$

(d) Deduce that

$$E(\mathbf{l}) = A(\kappa)\boldsymbol{\mu}, \qquad V(\mathbf{l}) = \boldsymbol{\mu}\boldsymbol{\mu}'\{1 - p^{-1}A(\kappa) - A^2(\kappa)\} + \kappa^{-1}A(\kappa)\mathbf{I}.$$

15.3.4 Let $\mathbf{x}$ be distributed as $N_p(\boldsymbol{\mu}, \kappa^{-1}\mathbf{I})$ with $\boldsymbol{\mu}'\boldsymbol{\mu} = 1$. Show that the conditional distribution of $\mathbf{x}$ given $\mathbf{x}'\mathbf{x} = 1$ is $M_p(\boldsymbol{\mu}, \kappa)$.

15.3.5 Let $\mathbf{x}$ be $N_p(\mathbf{0}, \boldsymbol{\Sigma})$. Show that the conditional distribution of $\mathbf{x}$ given $\mathbf{x}'\mathbf{x} = 1$ leads to a density of Bingham form.

15.3.6 If $\mathbf{l}$ is distributed as $B_p(\mathbf{A}, \boldsymbol{\kappa})$, and $\mathbf{G}$ is an orthogonal matrix, show that $\mathbf{l}^* = \mathbf{G}\mathbf{l}$ is distributed as $B_p(\mathbf{GA}, \boldsymbol{\kappa})$. Hence the normalizing constant depends only on $\boldsymbol{\kappa}$. Further, using the polar transformation, show that

$$[d(\boldsymbol{\kappa})]^{-1} = \int_0^{\pi} \cdots \int_0^{\pi} \int_0^{2\pi} \prod_{i=1}^{p} \exp\{\kappa_i u_i^2(\boldsymbol{\theta})\} a_p(\boldsymbol{\theta})\, d\boldsymbol{\theta}.$$

On expanding the exponentials and interchanging the summation and integration, show that $d(\boldsymbol{\kappa})$ reduces to (15.3.21).

15.3.7 If $\mathbf{l}$ is uniformly distributed on S_p, show that $\mathbf{l}^* = \mathbf{G}\mathbf{l}$ is also uniformly distributed where $\mathbf{G}$ is an orthogonal matrix.

15.3.8 (See, for example, Mardia, 1972*a*) (a) Let θ be uniformly distributed on $(0, 2\pi)$. Letting $E(e^{ip\theta}) = \phi_p$ denote the Fourier coefficients of θ for integer p, show that $\phi_p = 1$ if $p = 0$, and $\phi_p = 0$ otherwise.

(b) Let two independent variables θ_1 and θ_2 be distributed on a circle with θ_1 uniform. Using the Fourier coefficients or otherwise, show that $(\theta_1 + \theta_2) \bmod 2\pi$ is again uniformly distributed.

15.3.9 (Mardia, 1975*c, d*) *Maximum entropy characterizations* Let $f(\mathbf{l})$ be the p.d.f. of $\mathbf{l}$. The entropy of the distribution is defined by

$$E\{-\log f(\mathbf{l})\}.$$

Show that the maximum entropy distribution

(i) with no constraints on the moments of $\mathbf{l}$, is the uniform distribution;
(ii) with $E(\mathbf{l})$ fixed, non-zero, is $M_p(\boldsymbol{\mu}, \kappa)$;
(iii) with $E(\mathbf{l}\mathbf{l}')$ fixed is $B_p(\mathbf{A}, \kappa)$.

15.3.10 (von Mises, 1918; M. S. Bingham and Mardia, 1975) *Maximum likelihood characterizations* Let $f(\mathbf{l}; \boldsymbol{\mu})$ be a p.d.f. on S_p with mean direction $\boldsymbol{\mu}$ and $\rho > 0$. If

(i) for all random samples with $R > 0$, $\mathbf{l}_0$ is a maximum likelihood estimator of $\boldsymbol{\mu}$, and
(ii) $f(\mathbf{l}; \boldsymbol{\mu}) = g(\mathbf{l}'\boldsymbol{\mu})$ for all $\mathbf{l} \in S_p$, where g is a function of one variable and is lower semi-continuous from the left at 1,

then $\mathbf{l}$ has a von Mises–Fisher distribution.

15.3.11 (Mardia, 1972*a*, p. 114; Kent, 1978) Show that as $\kappa \to \infty$, $M_p(\boldsymbol{\mu}, \kappa)$ converges to a multinormal distribution with mean $\boldsymbol{\mu}$ and covariance matrix κ^{-1} times the identity on the hyperplane tangent to the sphere at $\boldsymbol{\mu}$. (Hint: using Exercise 15.3.3 and the asymptotic formula

$$I_\nu(z) = (2\pi z)^{-1/2} e^z [1 + O(|z|^{-1})], \qquad z \to \infty, \quad |\arg z| < \tfrac{1}{2}\pi - \delta,$$

where δ is an arbitrary positive number, show that

$$\exp(-i\kappa^{1/2}\mathbf{t}'\boldsymbol{\mu})\psi(\kappa^{1/2}\mathbf{t}; \boldsymbol{\mu}, \kappa) \to \exp(-\tfrac{1}{2}\mathbf{t}'\boldsymbol{\Sigma}\mathbf{t})$$

as $\kappa \to \infty$ for each $\mathbf{t}$, where $\boldsymbol{\Sigma} = \mathbf{I} - \boldsymbol{\mu}\boldsymbol{\mu}'$.)

15.4.1 Let $\mathbf{l}$ be uniformly distributed on S_p. Show that the p.d.f. of $\mathbf{r}$ is

$$c_p h_n(R)/R^{p-1},$$

where $h_n(R)$ is the p.d.f. of $R = (\mathbf{r}'\mathbf{r})^{1/2}$ given by (15.4.6).

15.4.2 (Mardia, 1975*b*) If in the isotropic random walk, the successive steps are of different lengths $\beta_1, \ldots, \beta_n$ then show that the c.f. of $\mathbf{l}_j$ is $c_p/c_p(i\rho\beta_j)$ rather than (15.4.4). As in Section 15.4.1, prove that

(i) $\bar{\boldsymbol{\theta}}$ and R are independent;
(ii) $\bar{\boldsymbol{\theta}}$ is uniformly distributed and the p.d.f. of R is

$$h_n(R, \boldsymbol{\beta}) = (2\pi)^{-p} c_p^{n-1} R^{p-1} \int_0^\infty \rho^{p-1} \{c_p(i\rho R) \prod_{j=1}^n c_p(i\rho\beta_j)\}^{-1} \, d\rho.$$

Using (15.4.4)–(15.4.5), express the p.d.f. of R in terms of ordinary Bessel functions.

15.4.3 For $\boldsymbol{\mu} = (1, 0, \ldots, 0)'$ show that r_1 is a sufficient statistic for κ. Hence deduce that the distribution of $R \mid r_1$ does not depend on κ. (For the p.d.f. of $R \mid r_1$, see Stephens (1962), Mardia (1975*b*).

15.5.1 For $\boldsymbol{\mu}=(1, 0, \ldots, 0)'$,

(i) show from the Rao–Blackwell theorem that r_1/n is the best unbiased estimator of $A(\kappa)$, and
(ii) show that the m.l.e. of κ is $A^{-1}(r_1)$.

15.5.2 Prove that

$$A'(\kappa)=[\kappa-(p-1)\rho-\kappa\rho^2]/\kappa$$

where $\rho=A(\kappa)$ and express (15.5.14) and (15.5.15) in terms of ρ and κ. (Hence, for tabulated values of $\rho=A(\kappa)$, the asymptotic variances of $\hat{\kappa}$ and $\bar{\boldsymbol{\theta}}$ can be obtained.) Using (15.5.7) and (15.5.10), examine the effect as $\kappa \to 0$ (uniformity) and $\kappa \to \infty$ (normality) on these variances.

15.5.3 Writing the log likelihood (15.5.1) as

$$\log L = \textit{constant} + n \log c(\kappa) + n\kappa\{\bar{l}_1\mu_1+\ldots+\bar{l}_{p-1}\mu_{p-1} \pm \bar{l}_p(1-\mu_1^2-\ldots-\mu_{p-1}^2)^{1/2}\}$$

show that

$$E\left\{\frac{\partial^2 \log L}{\partial \kappa^2}\right\}=-nA'(\kappa), \qquad E\left\{\frac{\partial^2 \log L}{\partial\kappa\,\partial\mu_1}\right\}=0,$$

$$E\left\{\frac{\partial^2 \log L}{\partial \mu_1^2}\right\}=-\kappa A(\kappa)(\mu_1^2+\mu_p^2)/\mu_p^2, \qquad E\left\{\frac{\partial^2 \log L}{\partial\mu_1\,\partial\mu_2}\right\}=-\kappa A(\kappa)\mu_1\mu_2/\mu_p^2.$$

Hence, show that the distributions of $\hat{\kappa}$ and $(\hat{\mu}_1, \ldots, \hat{\mu}_{p-1})$ are asymptotically independent normal but $\hat{\mu}_1, \ldots, \hat{\mu}_{p-1}$ are no longer independent.

15.6.1 Let $\mathbf{l}_1, \ldots, \mathbf{l}_n$ be a random sample from $M_p(\mathbf{e}_1, \kappa)$ where $\mathbf{e}_1=(1, 0, \ldots, 0)'$. For testing

$$H_0: \kappa=0 \quad \text{against} \quad H_1: \kappa \neq 0,$$

show that the Neyman–Pearson lemma leads to the best critical region of the form $r_1 > \kappa$. Why is this critical region expected intuitively? What is the asymptotic distribution of r_1 under H_0?

15.6.2 (Beran, 1968; Mardia, 1975*b*) From the Neyman–Pearson lemma show that the uniformly most powerful test for uniformity invariant under rotation with $M_p(\boldsymbol{\mu}, \kappa)$ as the alternative has a critical region of the form

$$[c_p(\kappa)]^n \int_{S_p} \prod_{i=1}^{n} \{\exp(\kappa \mathbf{l}'\mathbf{l}_i)\}\, d\mathbf{l} > K;$$

that is, the test is defined by

$$[c_p(\kappa)]^n/c_p(\kappa R) > K.$$

Using $A(\kappa) \geq 0$, $c_p(\kappa) \geq 0$, and $c_p'(\kappa) = -A(\kappa)c_p(\kappa)$, show that $c_p(\kappa)$ is a decreasing function of κ. Hence show that this test is the same as the Rayleigh test.

15.7.1 Let $\mathbf{l}_1, \ldots, \mathbf{l}_n$ be a random sample from $M_p(\mathbf{e}_1, \kappa)$, where $\mathbf{e}_1 = (1, 0, \ldots, 0)'$.

(a) Show that

$$2\kappa(R - r_1) = 2\kappa R(1 - \cos \bar{\theta}_1).$$

From (15.3.19), deduce that, for large κ, $2\kappa(R - r_1)$ is approximately a χ^2_{p-1} variable.

(b) For large n, show that, approximately $\bar{R} \sim N(\rho, A'(\kappa)/n)$.

15.7.2 (Mardia, 1975*b*) For a sample of size n from $M_p(\boldsymbol{\mu}, \kappa)$ where κ is known, show that a $(1-\alpha)100\%$ confidence cone for $\boldsymbol{\mu}$ can be constructed from (15.4.9) as follows. Since $\cos^{-1}(\bar{\mathbf{l}}'\boldsymbol{\mu}) \mid R$ is distributed as (15.3.17) with κ replaced by κR, let $P(\cos^{-1}(\bar{\mathbf{l}}'\boldsymbol{\mu}) > \delta \mid R) = \alpha$. Then show that for given κ and R, the probability that the true mean direction lies within a cone with vertex at the origin, axis through $\bar{\mathbf{l}}$ and semi-vertical angle δ is $1-\alpha$.

If κ is unknown, it can be replaced by $\hat{\kappa}$ and the method provides an approximate confidence cone for $\boldsymbol{\mu}$.

15.7.3 (Watson, 1956) Let F_α be the upper α critical value of the F-distribution with degrees of freedom $p-1$ and $(n-1)(p-1)$. For large κ, from (15.7.7) show that a $100(1-\alpha)\%$ confidence cone for $\boldsymbol{\mu}$ is a cone with vertex at $\mathbf{0}$, axis through the sample mean direction and semi-vertical angle δ, where

$$\delta = \cos^{-1}[1 - \{(n-R)F_\alpha/(n-1)R\}].$$

15.7.4 (Watson and Williams, 1956) Let $R_1, \ldots, R_q$ denote the resultant lengths for q independent random samples of size $n_1, \ldots, n_q$ drawn from $M_p(\boldsymbol{\mu}, \kappa)$. Let $n = \sum n_i$. Suppose that R is the resultant of the combined sample. For large κ, using the Watson–Williams approximations (15.7.4) and (15.7.6), show that, approximately,

$$2\kappa\left(n - \sum R_i\right) \sim \chi^2_{(p-1)(n-q)}, \qquad 2\kappa\left(\sum R_i - R\right) \sim \chi^2_{(p-1)(q-1)},$$

where these two random variables are independently distributed. Hence, for large κ, construct a test of equality of the mean directions.

15.7.5 (For $p=3$ see Bingham, 1972; also Mardia, 1972*a*) Let $\mathbf{l}_1, \ldots, \mathbf{l}_p$ be distributed as $B_p(\mathbf{A}, \boldsymbol{\kappa})$. Suppose that $\kappa_1 \leq \kappa_2 \leq \ldots \leq \kappa_{p-1} \leq \kappa_p$ and $\kappa_p = 0$. Show that the m.l.e. of $\mathbf{a}_{(i)}$ equals $\boldsymbol{\gamma}_{(i)}$. Further, the m.l.e.s of $\kappa_1, \ldots, \kappa_{p-1}$ are the solutions of

$$-\frac{\partial \log d(\hat{\kappa})}{\partial \kappa_i} = \frac{1}{n} \lambda_i, \qquad i = 1, \ldots, p-1,$$

where $\lambda_1 \geq \ldots \geq \lambda_p$ are the eigenvalues of $\mathbf{T}$ and $\boldsymbol{\gamma}_{(1)}, \ldots, \boldsymbol{\gamma}_{(p)}$ are the corresponding eigenvectors.

Appendix A
Matrix Algebra

A.1 Introduction

This appendix gives (i) a summary of basic definitions and results in matrix algebra with comments and (ii) details of those results and proofs which are used in this book but normally not treated in undergraduate Mathematics courses. It is designed as a convenient source of reference to be used in the rest of the book. A geometrical interpretation of some of the results is also given. If the reader is unfamiliar with any of the results not proved here he should consult a text such as Graybill (1969, especially pp. 4–52, 163–196, and 222–235) or Rao (1973, pp. 1–78). For the computational aspects of matrix operations see for example Wilkinson (1965).

Definition *A matrix* $\mathbf{A}$ *is a rectangular array of numbers. If* $\mathbf{A}$ *has* n *rows and* p *columns we say it is of* order $n \times p$. *For example,* n *observations on* p *random variables are arranged in this way.*

Notation 1 We write matrix $\mathbf{A}$ of order $n \times p$ as

$$\mathbf{A} = \begin{bmatrix} a_{11} & a_{12} & \cdots & a_{1p} \\ a_{21} & a_{22} & \cdots & a_{2p} \\ \cdot & \cdot & & \cdot \\ \cdot & \cdot & & \cdot \\ \cdot & \cdot & & \cdot \\ a_{n1} & a_{n2} & \cdots & a_{np} \end{bmatrix} = (a_{ij}), \qquad \text{(A.1.1)}$$

where a_{ij} is the element in row i and column j of the matrix $\mathbf{A}$, $i = 1, \ldots, n$; $j = 1, \ldots, p$. Sometimes, we write $(\mathbf{A})_{ij}$ for a_{ij}.

We may write the matrix $\mathbf{A}$ as $\mathbf{A}(n \times p)$ to emphasize the row and column order. In general, matrices are represented by boldface upper case letters throughout this book, e.g. $\mathbf{A}$, $\mathbf{B}$, $\mathbf{X}$, $\mathbf{Y}$, $\mathbf{Z}$. Their elements are represented by small letters with subscripts.

Definition *The* transpose *of a matrix* $\mathbf{A}$ *is formed by interchanging the rows and columns:*

$$\mathbf{A}' = \begin{pmatrix} a_{11} & a_{21} & \cdots & a_{n1} \\ a_{12} & a_{22} & \cdots & a_{n2} \\ \cdot & \cdot & & \cdot \\ \cdot & \cdot & & \cdot \\ \cdot & \cdot & & \cdot \\ a_{1p} & a_{2p} & \cdots & a_{np} \end{pmatrix}.$$

Definition *A matrix with column-order one is called a* column vector. *Thus*

$$\mathbf{a} = \begin{pmatrix} a_1 \\ a_2 \\ \cdot \\ \cdot \\ \cdot \\ a_n \end{pmatrix}$$

is a column vector with n components.

In general, boldface lower case letters represent column vectors. Row vectors are written as column vectors transposed, i.e.

$$\mathbf{a}' = (a_1, \ldots, a_n).$$

Notation 2 We write the columns of the matrix $\mathbf{A}$ as $\mathbf{a}_{(1)}, \mathbf{a}_{(2)}, \ldots, \mathbf{a}_{(p)}$ and the rows (if written as column vectors) as $\mathbf{a}_1, \mathbf{a}_2, \ldots, \mathbf{a}_n$ so that

$$\mathbf{A} = (\mathbf{a}_{(1)}, \mathbf{a}_{(2)}, \ldots, \mathbf{a}_{(p)}) = \begin{bmatrix} \mathbf{a}'_1 \\ \mathbf{a}'_2 \\ \cdot \\ \cdot \\ \cdot \\ \mathbf{a}'_n \end{bmatrix}, \qquad \text{(A.1.2)}$$

where

$$\mathbf{a}_{(j)} = \begin{bmatrix} a_{1j} \\ \cdot \\ \cdot \\ \cdot \\ a_{nj} \end{bmatrix}, \qquad \mathbf{a}_i = \begin{bmatrix} a_{i1} \\ \cdot \\ \cdot \\ \cdot \\ a_{ip} \end{bmatrix}.$$

Definition *A matrix written in terms of its sub-matrices is called a* **partitioned** matrix.

Notation 3 Let $\mathbf{A}_{11}$, $\mathbf{A}_{12}$, $\mathbf{A}_{21}$, and $\mathbf{A}_{22}$ be submatrices such that $\mathbf{A}_{11}(r\times s)$ has elements a_{ij}, $i=1,\ldots,r$; $j=1,\ldots,s$ and so on. Then we write

$$\mathbf{A}(n\times p)=\begin{bmatrix}\mathbf{A}_{11}(r\times s) & \mathbf{A}_{12}(r\times(p-s))\\ \mathbf{A}_{21}((n-r)\times s) & \mathbf{A}_{22}((n-r)\times(p-s))\end{bmatrix}.$$

Obviously, this notation can be extended to contain further partitions of $\mathbf{A}_{11}$, $\mathbf{A}_{12}$, etc.

A list of some important types of particular matrices is given in Table A.1.1. Another list which depends on the next section appears in Table A.3.1.

Table A.1.1 Particular matrices and types of matrix (List 1). For List 2 see Table A.3.1.

Name	Definiton	Notation	Trivial Examples
1 Scalar	$p=n=1$	a, b	(1)
2*a* Column vector	$p=1$	$\mathbf{a}, \mathbf{b}, \ldots$	$\begin{pmatrix}1\\2\end{pmatrix}$
2b Unit vector	$(1,\ldots,1)'$	$\mathbf{1}$ or $\mathbf{1}_p$	$\begin{pmatrix}1\\1\end{pmatrix}$
3 Rectangular	$p\times n$	$\mathbf{A}(n\times p)$	
4 Square	$p=n$	$\mathbf{A}(p\times p)$	$\begin{pmatrix}1&3\\4&5\end{pmatrix}$
4a Diagonal	$p=n$, $a_{ij}=0$, $i\neq j$	$\operatorname{diag}(a_{ii})$	$\begin{pmatrix}2&0\\0&1\end{pmatrix}$
4b Identity	$\operatorname{diag}(\mathbf{1})$	$\mathbf{I}$ or $\mathbf{I}_p$	$\begin{pmatrix}1&0\\0&1\end{pmatrix}$
4c Symmetric	$a_{ij}=a_{ji}$		$\begin{pmatrix}3&2\\2&5\end{pmatrix}$
4d Unit matrix	$p=n$, $a_{ij}=1$	$\mathbf{J}_p=\mathbf{11}'$	$\begin{pmatrix}1&1\\1&1\end{pmatrix}$
4e Triangular matrix (upper)	$a_{ij}=0$ below the diagonal	$\mathbf{\Delta}'$	$\begin{pmatrix}1&0&0\\2&2&0\\3&2&5\end{pmatrix}$
Triangular matrix (lower)	$a_{ij}=0$ above the diagonal	$\mathbf{\Delta}$	
5 Asymmetric	$a_{ij}\neq a_{ji}$		$\begin{pmatrix}1&1\\2&3\end{pmatrix}$
6 Null	$a_{ij}=0$	$\mathbf{0}$	$\begin{pmatrix}0&0&0\\0&0&0\end{pmatrix}$

As shown in Table A.1.1 a square matrix $\mathbf{A}(p\times p)$ is *diagonal* if $a_{ij}=0$ for all $i\neq j$. There are two convenient ways to construct diagonal matrices. If $\mathbf{a}=(a_1,\ldots,a_p)'$ is any vector and $\mathbf{B}(p\times p)$ is any square matrix then

$$\operatorname{diag}(\mathbf{a})=\operatorname{diag}(a_i)=\operatorname{diag}(a_1,\ldots,a_p)=\begin{pmatrix} a_1 & \cdots & 0\\ \cdot & & \cdot\\ \cdot & & \cdot\\ \cdot & & \cdot\\ 0 & \cdots & a_p\end{pmatrix}$$

and

$$\operatorname{Diag}(\mathbf{B})=\begin{pmatrix} b_{11} & \cdots & 0\\ \cdot & & \cdot\\ \cdot & & \cdot\\ \cdot & & \cdot\\ 0 & \cdots & b_{pp}\end{pmatrix}$$

each defines a diagonal matrix.

A.2 Matrix Operations

Table A.2.1. gives a summary of various important matrix operations. We deal with some of these in detail, assuming the definitions in the table.

Table A.2.1 Basic matrix operations

Operation	Restrictions	Definitions	Remarks
1 Addition	$\mathbf{A}$, $\mathbf{B}$ of the same order	$\mathbf{A}+\mathbf{B}=(a_{ij}+b_{ij})$	
2 Subtraction	$\mathbf{A}$, $\mathbf{B}$ of the same order	$\mathbf{A}-\mathbf{B}=(a_{ij}-b_{ij})$	
3a Scalar multiplication		$c\mathbf{A}=(ca_{ij})$	
3b Inner product	$\mathbf{a}$, $\mathbf{b}$ of the same order	$\mathbf{a}'\mathbf{b}=\sum a_ib_i$	
3c Multiplication	Number of columns of $\mathbf{A}$ equals number of rows of $\mathbf{B}$	$\mathbf{AB}=(\mathbf{a}_i'\mathbf{b}_{(j)})$	$\mathbf{AB}\neq\mathbf{BA}$
4 Transpose		$\mathbf{A}'=(\mathbf{a}_1,\mathbf{a}_2,\ldots,\mathbf{a}_n)$	Section A.2.1.
5 Trace	$\mathbf{A}$ square	$\operatorname{tr}\mathbf{A}=\sum a_{ii}$	Section A.2.2.
6 Determinant	$\mathbf{A}$ square	$\lvert\mathbf{A}\rvert$	Section A.2.3.
7 Inverse	$\mathbf{A}$ square and $\lvert\mathbf{A}\rvert\neq 0$	$\mathbf{AA}^{-1}=\mathbf{A}^{-1}\mathbf{A}=\mathbf{I}$	$(\mathbf{A}+\mathbf{B})^{-1}\neq\mathbf{A}^{-1}+\mathbf{B}^{-1}$, Section A.2.4
8 g-inverse ($\mathbf{A}^-$)	$\mathbf{A}(n\times p)$	$\mathbf{AA}^-\mathbf{A}=\mathbf{A}$	Section A.8

A.2.1 Transpose

The transpose satisfies the simple properties

$$(\mathbf{A}')'=\mathbf{A},\qquad (\mathbf{A}+\mathbf{B})'=\mathbf{A}'+\mathbf{B}',\qquad (\mathbf{AB})'=\mathbf{B}'\mathbf{A}'. \tag{A.2.1}$$

For partitioned $\mathbf{A}$,

$$\mathbf{A}' = \begin{bmatrix} \mathbf{A}'_{11} & \mathbf{A}'_{21} \\ \mathbf{A}'_{12} & \mathbf{A}'_{22} \end{bmatrix}.$$

If $\mathbf{A}$ is a symmetric matrix, $a_{ij} = a_{ji}$, so that

$$\mathbf{A}' = \mathbf{A}.$$

A.2.2 Trace

The trace function, $\text{tr}\,\mathbf{A} = \Sigma a_{ii}$, satisfies the following properties for $\mathbf{A}(p \times p)$, $\mathbf{B}(p \times p)$, $\mathbf{C}(p \times n)$, $\mathbf{D}(n \times p)$, and scalar α:

$$\text{tr}\,\alpha = \alpha, \qquad \text{tr}\,\mathbf{A} \pm \mathbf{B} = \text{tr}\,\mathbf{A} \pm \text{tr}\,\mathbf{B}, \qquad \text{tr}\,\alpha\mathbf{A} = \alpha\,\text{tr}\,\mathbf{A} \tag{A.2.2a}$$

$$\text{tr}\,\mathbf{CD} = \text{tr}\,\mathbf{DC} = \sum_{i,j} c_{ij} d_{ji}, \tag{A.2.2b}$$

$$\sum \mathbf{x}_i'\mathbf{A}\mathbf{x}_i = \text{tr}\,(\mathbf{AT}), \quad \text{where} \quad \mathbf{T} = \sum \mathbf{x}_i\mathbf{x}_i'. \tag{A.2.2c}$$

To prove this last property, note that since $\sum \mathbf{x}_i'\mathbf{A}\mathbf{x}_i$ is a scalar, the left-hand side of (A.2.2c) is

$$\begin{aligned} \text{tr} \sum \mathbf{x}_i'\mathbf{A}\mathbf{x}_i &= \sum \text{tr}\,\mathbf{x}_i'\mathbf{A}\mathbf{x}_i \quad \text{by (A.2.2a)} \\ &= \sum \text{tr}\,\mathbf{A}\mathbf{x}_i\mathbf{x}_i' \quad \text{by (A.2.2b)} \\ &= \text{tr}\,\mathbf{A} \sum \mathbf{x}_i\mathbf{x}_i' \quad \text{by (A.2.2a).} \end{aligned}$$

As a special case of (A.2.2b) note that

$$\text{tr}\,\mathbf{CC}' = \text{tr}\,\mathbf{C}'\mathbf{C} = \sum c_{ij}^2. \tag{A.2.2d}$$

A.2.3 Determinants and cofactors

Definition *The determinant of a square matrix* $\mathbf{A}$ *is defined as*

$$|\mathbf{A}| = \sum (-1)^{|\tau|} a_{1\tau(1)} \dots a_{p\tau(p)}, \tag{A.2.3a}$$

where the summation is taken over all permutations τ *of* $(1, 2, \dots, p)$, *and* $|\tau|$ *equals* $+1$ *or* -1, *depending on whether* τ *can be written as the product of an even or odd number of transpositions.*

For $p = 2$,

$$|\mathbf{A}| = a_{11}a_{22} - a_{12}a_{21}. \tag{A.2.3b}$$

Definition *The* cofactor *of* a_{ij} *is defined by* $(-1)^{i+j}$ *times the* minor *of* a_{ij}, *where the* minor *of* a_{ij} *is the value of the determinant obtained after deleting the ith row and the jth column of* $\mathbf{A}$.

We denote the cofactor of a_{ij} by A_{ij}. Thus for $p = 3$,

$$A_{11} = \begin{vmatrix} a_{22} & a_{23} \\ a_{32} & a_{33} \end{vmatrix}, \qquad A_{12} = -\begin{vmatrix} a_{21} & a_{23} \\ a_{31} & a_{33} \end{vmatrix}, \qquad A_{13} = \begin{vmatrix} a_{21} & a_{22} \\ a_{31} & a_{32} \end{vmatrix}. \tag{A.2.3c}$$

Definition *A square matrix is* non-singular *if* $|\mathbf{A}| \neq 0$; *otherwise it is singular.*

We have the following results:

(I) $$|\mathbf{A}| = \sum_{j=1}^{p} a_{ij}A_{ij} = \sum_{i=1}^{p} a_{ij}A_{ij}, \qquad \text{any } i, j, \tag{A.2.3d}$$

but

$$\sum_{k=1}^{p} a_{ik}A_{jk} = 0, \qquad i \neq j. \tag{A.2.3e}$$

(II) If $\mathbf{A}$ is triangular or diagonal,

$$|\mathbf{A}| = \prod a_{ii}. \tag{A.2.3f}$$

(III) $|c\mathbf{A}| = c^p\,|\mathbf{A}|$. (A.2.3g)

(IV) $|\mathbf{AB}| = |\mathbf{A}|\,|\mathbf{B}|$. (A.2.3h)

(V) For square submatrices $\mathbf{A}(p \times p)$ and $\mathbf{B}(q \times q)$,

$$\begin{vmatrix} \mathbf{A} & \mathbf{C} \\ \mathbf{0} & \mathbf{B} \end{vmatrix} = |\mathbf{A}|\,|\mathbf{B}|. \tag{A.2.3i}$$

(VI) $$\begin{vmatrix} \mathbf{A}_{11} & \mathbf{A}_{12} \\ \mathbf{A}_{21} & \mathbf{A}_{22} \end{vmatrix} = |\mathbf{A}_{11}|\,|\mathbf{A}_{22} - \mathbf{A}_{21}\mathbf{A}_{11}^{-1}\mathbf{A}_{12}| = |\mathbf{A}_{22}|\,|\mathbf{A}_{11} - \mathbf{A}_{12}\mathbf{A}_{22}^{-1}\mathbf{A}_{21}|, \tag{A.2.3j}$$

$$\begin{vmatrix} \mathbf{A} & \mathbf{a} \\ \mathbf{a}' & b \end{vmatrix} = |\mathbf{A}|\,(b - \mathbf{a}'\,\mathbf{A}^{-1}\,\mathbf{a}).$$

(VII) For $\mathbf{B}(p \times n)$ and $\mathbf{C}(n \times p)$, and non-singular $\mathbf{A}(p \times p)$,

$$|\mathbf{A} + \mathbf{BC}| = |\mathbf{A}|\,|\mathbf{I}_p + \mathbf{A}^{-1}\mathbf{BC}| = |\mathbf{A}|\,|\mathbf{I}_n + \mathbf{CA}^{-1}\mathbf{B}|, \tag{A.2.3k}$$

$$|\mathbf{A} + \mathbf{b}'\mathbf{a}| = |\mathbf{A}|\,(1 + \mathbf{b}'\mathbf{A}^{-1}\,\mathbf{a}).$$

Remarks (1) Properties (I)–(III) follow easily from the definition (A.2.3a). As an application of (I), from (A.2.3b), (A.2.3c), and (A.2.3d), we have, for $p = 3$,

$$|\mathbf{A}| = a_{11}(a_{22}a_{33} - a_{23}a_{32}) - a_{12}(a_{21}a_{33} - a_{23}a_{31}) + a_{13}(a_{21}a_{32} - a_{31}a_{22}).$$

(2) To prove (V), note that the only permutations giving non-zero terms in the summation (A.2.3a) are those taking $\{1, \ldots, p\}$ to $\{1, \ldots, p\}$ and $\{p+1, \ldots, p+q\}$ to $\{p+1, \ldots, p+q\}$.

(3) To prove (VI), simplify $\mathbf{BAB}'$ and then take its determinant where

$$\mathbf{B} = \begin{bmatrix} \mathbf{I} & -\mathbf{A}_{12}\mathbf{A}_{22}^{-1} \\ \mathbf{0} & \mathbf{I} \end{bmatrix}.$$

From (VI), we deduce, after putting $\mathbf{A}^{11} = \mathbf{A}$, $\mathbf{A}^{12} = \mathbf{x}'$, etc.,

$$\begin{vmatrix} \mathbf{A} & \mathbf{x} \\ \mathbf{x} & c \end{vmatrix} = |\mathbf{A}|\,\{c - \mathbf{x}'\mathbf{A}^{-1}\mathbf{x}\}. \tag{A.2.3l}$$

(4) To prove the second part of (VII), simplify

$$\begin{vmatrix} \mathbf{I}_p & -\mathbf{A}^{-1}\mathbf{B} \\ \mathbf{C} & \mathbf{I}_n \end{vmatrix}$$

using (VI). As special cases of (VII) we see that, for non-singular $\mathbf{A}$,

$$|\mathbf{A}+\mathbf{bb}'| = |\mathbf{A}|\,(1+\mathbf{b}'\mathbf{A}^{-1}\mathbf{b}), \tag{A.2.3m}$$

and that, for $\mathbf{B}(p\times n)$ and $\mathbf{C}(n\times p)$,

$$|\mathbf{I}_p+\mathbf{BC}| = |\mathbf{I}_n+\mathbf{CB}|. \tag{A.2.3n}$$

In practice, we can simplify determinants using the property that the value of a determinant is unaltered if a linear combination of some of the columns (rows) is added to another column (row).

(5) Determinants are usually evaluated on computers as follows. $\mathbf{A}$ is decomposed into upper and lower triangular matrices $\mathbf{A}=\mathbf{LU}$. If $\mathbf{A}>0$, then the Cholesky decomposition is used (i.e. $\mathbf{U}=\mathbf{L}'$ so $\mathbf{A}=\mathbf{LL}'$). Otherwise the Crout decomposition is used where the diagonal elements of $\mathbf{U}$ are ones.

A.2.4 Inverse

Definition *As already defined in Table* A.1.1, *the* inverse *of* $\mathbf{A}$ *is the unique matrix* $\mathbf{A}^{-1}$ *satisfying*

$$\mathbf{AA}^{-1} = \mathbf{A}^{-1}\mathbf{A} = \mathbf{I}. \tag{A.2.4a}$$

The inverse exists if and only if $\mathbf{A}$ *is non-singular, that is, if and only if* $|\mathbf{A}|\neq 0$.

We write the (i, j)th element of $\mathbf{A}^{-1}$ by a^{ij}. For partitioned $\mathbf{A}$, we write

$$\mathbf{A}^{-1} = \begin{bmatrix} \mathbf{A}^{11} & \mathbf{A}^{12} \\ \mathbf{A}^{21} & \mathbf{A}^{22} \end{bmatrix}.$$

The following properties hold:

(I) $\mathbf{A}^{-1} = \dfrac{1}{|\mathbf{A}|}(A_{ij})'$. (A.2.4b)

(II) $(c\mathbf{A})^{-1} = c^{-1}\mathbf{A}^{-1}$. (A.2.4c)

(III) $(\mathbf{AB})^{-1} = \mathbf{B}^{-1}\mathbf{A}^{-1}$. (A.2.4d)

(IV) The unique solution of $\mathbf{Ax}=\mathbf{b}$ is $\mathbf{x}=\mathbf{A}^{-1}\mathbf{b}$. (A.2.4e)

(V) If all the necessary inverses exist, then for $\mathbf{A}(p\times p)$, $\mathbf{B}(p\times n)$, $\mathbf{C}(n\times n)$, and $\mathbf{D}(n\times p)$,

$$(\mathbf{A}+\mathbf{BCD})^{-1} = \mathbf{A}^{-1} - \mathbf{A}^{-1}\mathbf{B}(\mathbf{C}^{-1}+\mathbf{DA}^{-1}\mathbf{B})^{-1}\mathbf{DA}^{-1}, \tag{A.2.4f}$$

$$(\mathbf{A}+\mathbf{a}\,\mathbf{b}')^{-1} = \mathbf{A}^{-1} - \{(\mathbf{A}^{-1}\,\mathbf{a})(\mathbf{b}'\mathbf{A}^{-1})(1+\mathbf{b}'\mathbf{A}^{-1}\,\mathbf{a})^{-1}\}.$$

(VI) If all the necessary inverses exist, then for partitioned $\mathbf{A}$, the elements of $\mathbf{A}^{-1}$ are

$$\left.\begin{aligned} &\mathbf{A}^{11}=(\mathbf{A}_{11}-\mathbf{A}_{12}\mathbf{A}_{22}^{-1}\mathbf{A}_{21})^{-1}, &&\mathbf{A}^{22}=(\mathbf{A}_{22}-\mathbf{A}_{21}\mathbf{A}_{11}^{-1}\mathbf{A}_{12})^{-1},\\ &\mathbf{A}^{12}=-\mathbf{A}^{11}\mathbf{A}_{12}\mathbf{A}_{22}^{-1}, &&\mathbf{A}^{21}=-\mathbf{A}_{22}^{-1}\mathbf{A}_{21}\mathbf{A}^{11}. \end{aligned}\right\} \tag{A.2.4g}$$

Alternatively, $\mathbf{A}^{12}$ and $\mathbf{A}^{21}$ can be defined by

$$\mathbf{A}^{12}=-\mathbf{A}_{11}^{-1}\mathbf{A}_{12}\mathbf{A}^{22}, \qquad \mathbf{A}^{21}=-\mathbf{A}^{22}\mathbf{A}_{21}\mathbf{A}_{11}^{-1}.$$

(VII) For symmetrical matrices $\mathbf{A}$ and $\mathbf{D}$, we have, if all necessary inverses exist

$$\begin{pmatrix} \mathbf{A} & \mathbf{B} \\ \mathbf{B}' & \mathbf{D} \end{pmatrix}^{-1}=\begin{pmatrix} \mathbf{A}^{-1} & 0 \\ 0 & 0 \end{pmatrix}+\begin{pmatrix} -\mathbf{E} \\ \mathbf{I} \end{pmatrix}(\mathbf{D}-\mathbf{B}'\mathbf{A}^{-1}\mathbf{B})^{-1}(-\mathbf{E}',\mathbf{I})$$

where $\mathbf{E}=\mathbf{A}^{-1}\mathbf{B}$.

Remarks (1) The result (I) follows on using (A.2.3d), (A.2.3e). As a simple application, note that, for $p=2$, we have

$$\mathbf{A}^{-1}=\frac{1}{a_{11}a_{22}-a_{12}a_{21}}\begin{pmatrix} a_{22} & -a_{12} \\ -a_{21} & a_{11} \end{pmatrix}.$$

(2) Formulae (II)–(VI) can be verified by checking that the product of the matrix and its inverse reduces to the identity matrix, e.g. to verify (III), we proceed

$$(\mathbf{AB})^{-1}(\mathbf{AB})=\mathbf{B}^{-1}\mathbf{A}^{-1}(\mathbf{AB})=\mathbf{B}^{-1}\mathbf{IB}=\mathbf{I}.$$

(3) We have assumed $\mathbf{A}$ to be a square matrix with $|\mathbf{A}|\neq 0$ in defining $\mathbf{A}^{-1}$. For $\mathbf{A}(n\times p)$, a generalized inverse is defined in Section A.8.

(4) In computer algorithms for evaluating $\mathbf{A}^{-1}$, the following methods are commonly used. If $\mathbf{A}$ is symmetric, the Cholesky method is used, namely, decomposing $\mathbf{A}$ into the form $\mathbf{LL}'$ where $\mathbf{L}$ is lower triangular and then using $\mathbf{A}^{-1}=(\mathbf{L}^{-1})'\mathbf{L}^{-1}$. For non-symmetric matrices, Crout's method is used, which is a modification of Gaussian elimination.

A.2.5 Kronecker products

Definition *Let* $\mathbf{A}=(a_{ij})$ *be an* $(m\times n)$ *matrix and* $\mathbf{B}=(b_{kl})$ *be a* $(p\times q)$ *matrix. Then the* Kronecker product *of* $\mathbf{A}$ *and* $\mathbf{B}$ *is defined as*

$$\begin{bmatrix} a_{11}\mathbf{B} & a_{12}\mathbf{B} & \cdots & a_{1n}\mathbf{B} \\ a_{21}\mathbf{B} & a_{22}\mathbf{B} & \cdots & a_{2n}\mathbf{B} \\ \cdot & \cdot & & \cdot \\ \cdot & \cdot & & \cdot \\ \cdot & \cdot & & \cdot \\ a_{m1}\mathbf{B} & a_{m2}\mathbf{B} & \cdots & a_{mn}\mathbf{B} \end{bmatrix},$$

which is an $(mp\times nq)$ *matrix. It is denoted by* $\mathbf{A}\otimes\mathbf{B}$.

Definition *If* $\mathbf{X}$ *is an* $(n \times p)$ *matrix let* $\mathbf{X}^V$ *denote the np-vector obtained by "vectorizing"* $\mathbf{X}$; *that is, by stacking the columns of* $\mathbf{X}$ *on top of one another so that*

$$\mathbf{X}^V = \begin{bmatrix} \mathbf{x}_{(1)} \\ \mathbf{x}_{(2)} \\ \cdot \\ \cdot \\ \cdot \\ \mathbf{x}_{(p)} \end{bmatrix}.$$

From these definitions the elementary properties given below easily follow:

(I) $\alpha(\mathbf{A} \otimes \mathbf{B}) = (\alpha\mathbf{A}) \otimes \mathbf{B} = \mathbf{A} \otimes (\alpha\mathbf{B})$ for all scalar α, and hence can be written without ambiguity as $\alpha\mathbf{A} \otimes \mathbf{B}$.
(II) $\mathbf{A} \otimes (\mathbf{B} \otimes \mathbf{C}) = (\mathbf{A} \otimes \mathbf{B}) \otimes \mathbf{C}$. Hence this can be written as $\mathbf{A} \otimes \mathbf{B} \otimes \mathbf{C}$.
(III) $(\mathbf{A} \otimes \mathbf{B})' = \mathbf{A}' \otimes \mathbf{B}'$.
(IV) $(\mathbf{A} \otimes \mathbf{B})(\mathbf{F} \otimes \mathbf{G}) = (\mathbf{AF}) \otimes (\mathbf{BG})$. Here parentheses are necessary.
(V) $(\mathbf{A} \otimes \mathbf{B})^{-1} = \mathbf{A}^{-1} \otimes \mathbf{B}^{-1}$ for non-singular $\mathbf{A}$ and $\mathbf{B}$.
(VI) $(\mathbf{A} + \mathbf{B}) \otimes \mathbf{C} = \mathbf{A} \otimes \mathbf{C} + \mathbf{B} \otimes \mathbf{C}$.
(VII) $\mathbf{A} \otimes (\mathbf{B} + \mathbf{C}) = \mathbf{A} \otimes \mathbf{B} + \mathbf{A} \otimes \mathbf{C}$.
(VIII) $(\mathbf{AXB})^V = (\mathbf{B}' \otimes \mathbf{A})\mathbf{X}^V$.
(IX) $\operatorname{tr}(\mathbf{A} \otimes \mathbf{B}) = (\operatorname{tr} \mathbf{A})(\operatorname{tr} \mathbf{B})$.

A.3 Further Particular Matrices and Types of Matrix

Table A.3.1 gives another list of some important types of matrices. We consider a few in more detail.

A.3.1 Orthogonal matrices

A square matrix $\mathbf{A}(n \times n)$ is *orthogonal* if $\mathbf{AA}' = \mathbf{I}$. The following properties hold:

(I) $\mathbf{A}^{-1} = \mathbf{A}'$.
(II) $\mathbf{A}'\mathbf{A} = \mathbf{I}$.
(III) $|\mathbf{A}| = \pm 1$.
(IV) $\mathbf{a}_i'\mathbf{a}_j = 0,\ i \neq j$; $\mathbf{a}_i'\mathbf{a}_i = 1$. $\mathbf{a}_{(i)}'\mathbf{a}_{(j)} = 0,\ i \neq j$; $\mathbf{a}_{(i)}'\mathbf{a}_{(i)} = 1$.
(V) $\mathbf{C} = \mathbf{AB}$ is orthogonal if $\mathbf{A}$ and $\mathbf{B}$ are orthogonal.

Remarks (1) All of these properties follow easily from the definition $\mathbf{AA}' = \mathbf{I}$. Result (IV) states that the sum of squares of the elements in each

Table A.3.1 Particular types of matrices (List 2)

Name	Definition	Examples	Details in
Non-singular	$\|\mathbf{A}\| \neq 0$	$\begin{bmatrix} 1 & 2 \\ 0 & 1 \end{bmatrix}$	Section A.2.3.
Singular	$\|\mathbf{A}\| = 0$	$\begin{bmatrix} 1 & 2 \\ 1 & 2 \end{bmatrix}$	Section A.2.3.
Orthogonal	$\mathbf{AA}' = \mathbf{A}'\mathbf{A} = \mathbf{I}$	$\begin{bmatrix} \cos\theta & -\sin\theta \\ \sin\theta & \cos\theta \end{bmatrix}$	Section A.3.1.
Equicorrelation	$\mathbf{E} = (1-\rho)\mathbf{I} + \rho\mathbf{J}$	$\begin{bmatrix} 1 & \rho \\ \rho & 1 \end{bmatrix}$	Section A.3.2.
Idempotent	$\mathbf{A}^2 = \mathbf{A}$	$\frac{1}{2}\begin{bmatrix} 1 & -1 \\ -1 & 1 \end{bmatrix}$	
Centring matrix, H_n	$\mathbf{H}_n = \mathbf{I}_n - n^{-1}\mathbf{J}_n$		Section A.3.3.
Positive definite (p.d.)	$\mathbf{x}'\mathbf{A}\mathbf{x} > 0$ for all $\mathbf{x} \neq \mathbf{0}$	$x_1^2 + x_2^2$	Section A.7.
Positive semi-definite (p.s.d.)	$\mathbf{x}'\mathbf{A}\mathbf{x} \geq 0$ for all $\mathbf{x} \neq \mathbf{0}$	$(x_1 - x_2)^2$	Section A.7.

row (column) is unity whereas the sum of the cross-products of the elements in any two rows (columns) is zero.

(2) The *Helmert matrix* is a particular orthogonal matrix whose columns are defined by

$$\mathbf{a}'_{(1)} = (n^{-1/2}, \ldots, n^{-1/2}),$$

$$\mathbf{a}'_{(j)} = (d_j, \ldots, d_j, -(j-1)d_j, 0, \ldots, 0), \qquad j = 2, \ldots, n,$$

where $d_j = \{j(j-1)\}^{-1/2}$, is repeated $j-1$ times.

(3) Orthogonal matrices can be used to represent a change of basis, or rotation. See Section A.5.

A.3.2 Equicorrelation matrix

Consider the $(p \times p)$ matrix defined by

$$\mathbf{E} = (1-\rho)\mathbf{I} + \rho\mathbf{J}, \tag{A.3.2a}$$

where ρ is any real number. Then $e_{ii} = 1$, $e_{ij} = \rho$, for $i \neq j$. For statistical purposes this matrix is most useful for $-(p-1)^{-1} < \rho < 1$, when it is called the *equicorrelation* matrix.

Direct verification shows that, provided $\rho \neq 1, -(p-1)^{-1}$, then $\mathbf{E}^{-1}$ exists and is given by

$$\mathbf{E}^{-1} = (1-\rho)^{-1}[\mathbf{I} - \rho\{1 + (p-1)\rho\}^{-1}\mathbf{J}]. \tag{A.3.2b}$$

Its determinant is given by

$$|\mathbf{E}| = (1-\rho)^{p-1}\{1+\rho(p-1)\}. \tag{A.3.2c}$$

This formula is most easily verified using the eigenvalues given in Remark 6 of Section A.6.

A.3.3 Centring matrix

The $(n \times n)$ centring matrix is defined by $\mathbf{H} = \mathbf{H}_n = \mathbf{I} - n^{-1}\mathbf{J}$, where $\mathbf{J} = \mathbf{1}\mathbf{1}'$. Then

(I) $\mathbf{H}' = \mathbf{H}$, $\mathbf{H}^2 = \mathbf{H}$.
(II) $\mathbf{H1} = \mathbf{0}$, $\mathbf{HJ} = \mathbf{JH} = \mathbf{0}$.
(III) $\mathbf{Hx} = \mathbf{x} - \bar{x}\mathbf{1}$, where $\bar{x} = n^{-1}\sum x_i$.
(IV) $\mathbf{x}'\mathbf{Hx} = n^{-1}\sum (x_i - \bar{x})^2$.

Remark (1) Property (I) states that $\mathbf{H}$ is symmetric and idempotent.

(2) Property (III) is most important in data analysis. The ith element of $\mathbf{Hx}$ is $x_i - \bar{x}$. Therefore, premultiplying a column vector by $\mathbf{H}$ has the effect of re-expressing the elements of the vector as *deviations from the mean*. Similarly, premultiplying a matrix by $\mathbf{H}$ re-expresses each element of the matrix as a *deviation from its column mean*, i.e. $\mathbf{HX}$ has its (i, j)th element $x_{ij} - \bar{x}_j$, where $\bar{x}_j$ is the mean of the jth column of $\mathbf{X}$. This "centring" property explains the nomenclature for $\mathbf{H}$.

A.4 Vector Spaces, Rank, and Linear Equations

A.4.1 Vector spaces

The set of vectors in R^n satisfies the following properties. For all $\mathbf{x}, \mathbf{y} \in R^n$ and all $\lambda, \mu \in R$,

(1) $\lambda(\mathbf{x}+\mathbf{y}) = \lambda\mathbf{x} + \lambda\mathbf{y}$,
(2) $(\lambda+\mu)\mathbf{x} = \lambda\mathbf{x} + \mu\mathbf{x}$,
(3) $(\lambda\mu)\mathbf{x} = \lambda(\mu\mathbf{x})$,
(4) $1\mathbf{x} = \mathbf{x}$.

Thus R^n can be considered as a *vector space* over the real numbers R.

Definition *If W is a subset of R^n such that for all $\mathbf{x}, \mathbf{y} \in W$ and $\lambda \in R$*

$$\lambda(\mathbf{x}+\mathbf{y}) \in W,$$

then W is called a vector subspace *of R^n.*

Two simple examples of subspaces of R^n are $\{\mathbf{0}\}$ and R^n itself.

Definition *Vectors* $\mathbf{x}_1, \ldots, \mathbf{x}_k$ *are called* linearly dependent *if there exist numbers* $\lambda_1, \ldots, \lambda_k$, *not all zero, such that*

$$\lambda_1\mathbf{x}_1 + \ldots + \lambda_k\mathbf{x}_k = \mathbf{0}.$$

Otherwise the k *vectors are* linearly independent.

Definition *Let* W *be a subspace of* R^n. *Then a* basis *of* W *is a maximal linearly independent set of vectors.*

The following properties hold for a basis of W:

(I) Every basis of W contains the same (finite) number of elements. This number is called the *dimension* of W and denoted dim W. In particular dim $R^n = n$.

(II) If $\mathbf{x}_1, \ldots, \mathbf{x}_k$ is a basis for W then every element $\mathbf{x}$ in W can be expressed as a linearly combination of $\mathbf{x}_1, \ldots, \mathbf{x}_k$; that is, $\mathbf{x} = \lambda_1\mathbf{x}_1 + \ldots + \lambda_k\mathbf{x}_k$ for some numbers $\lambda_1, \ldots, \lambda_k$.

Definition *The* inner (*or* scalar *or* dot) product *between two vectors* $\mathbf{x}, \mathbf{y} \in R^n$ *is defined by*

$$\mathbf{x} \cdot \mathbf{y} = \mathbf{x}'\mathbf{y} = \sum_{i=1}^{n} x_i y_i.$$

The vectors $\mathbf{x}$ *and* $\mathbf{y}$ *are called* orthogonal *if* $\mathbf{x} \cdot \mathbf{y} = 0$.

Definition *The* norm *of a vector* $\mathbf{x} \in R^n$ *is given by*

$$\|\mathbf{x}\| = (\mathbf{x} \cdot \mathbf{x})^{1/2} = \left(\sum x_i^2\right)^{1/2}.$$

Then the distance *between two vectors* $\mathbf{x}$ *and* $\mathbf{y}$ *is given by*

$$\|\mathbf{x} - \mathbf{y}\|.$$

Definition *A basis* $\mathbf{x}_1, \ldots, \mathbf{x}_k$ *of a subspace* W *of* R^n *is called* orthonormal *if all the elements have norm* 1 *and are orthogonal to one another; that is, if*

$$\mathbf{x}_i'\mathbf{x}_j = \begin{cases} 1, & i = j, \\ 0, & i \neq j. \end{cases}$$

In particular, if $\mathbf{A}(n \times n)$ is an orthogonal matrix then the columns of $\mathbf{A}$ form an orthonormal basis of R^n.

A.4.2 Rank

Definition *The* rank *of a matrix* $\mathbf{A}(n \times p)$ *is defined as the maximum number of linearly independent rows (columns) in* $\mathbf{A}$.

We denote it by $r(\mathbf{A})$ or rank $(\mathbf{A})$.

The following properties hold:

(I) $0 \leq r(\mathbf{A}) \leq \min(n, p)$. (A.4.2a)

(II) $r(\mathbf{A}) = r(\mathbf{A}')$. (A.4.2b)

(III) $r(\mathbf{A}+\mathbf{B}) \leq r(\mathbf{A}) + r(\mathbf{B})$. (A.4.2c)

(IV) $r(\mathbf{AB}) \leq \min\{r(\mathbf{A}), r(\mathbf{B})\}$. (A.4.2d)

(V) $r(\mathbf{A}'\mathbf{A}) = r(\mathbf{AA}') = r(\mathbf{A})$. (A.4.2e)

(VI) If $\mathbf{B}(n \times n)$ and $\mathbf{C}(p \times p)$ are non-singular then $r(\mathbf{BAC}) = r(\mathbf{A})$. (A.4.2f)

(VII) If $n = p$ then $r(\mathbf{A}) = p$ if and only if $\mathbf{A}$ is non-singular. (A.4.2g)

Table A.4.1 gives the ranks of some particular matrices.

Remarks (1) Another definition of $r(\mathbf{A})$ is $r(\mathbf{A})$ = the largest order of those (square) submatrices which have non-vanishing determinants.

(2) If we define $M(\mathbf{A})$ as the vector subspace in R^n spanned by the columns of $\mathbf{A}$, then $r(\mathbf{A}) = \dim M(\mathbf{A})$ and we may choose linearly independent columns of $\mathbf{A}$ as a basis for $M(\mathbf{A})$. Note that for any p-vector $\mathbf{x}$, $\mathbf{Ax} = x_1\mathbf{a}_{(1)} + \ldots + x_p\mathbf{a}_{(p)}$ is a linear combination of the columns of $\mathbf{A}$ and hence $\mathbf{Ax}$ lies in $M(\mathbf{A})$.

(3) Define the *null space* of $\mathbf{A}(n \times p)$ by

$$N(\mathbf{A}) = \{\mathbf{x} \in R^p : \mathbf{Ax} = \mathbf{0}\}.$$

Then $N(\mathbf{A})$ is a vector subspace of R^p of dimension k, say. Let $\mathbf{e}_1, \ldots, \mathbf{e}_p$ be a basis of R^p for which $\mathbf{e}_1, \ldots, \mathbf{e}_k$ are a basis of $N(\mathbf{A})$. Then $\mathbf{Ae}_{k+1}, \ldots, \mathbf{Ae}_p$ form a maximally linearly independent set of vectors in $M(\mathbf{A})$, and hence are a basis for $M(\mathbf{A})$. Thus, we get the important result

$$\dim N(\mathbf{A}) + \dim M(\mathbf{A}) = p. \qquad \text{(A.4.2h)}$$

(4) To prove (V) note that if $\mathbf{Ax} = \mathbf{0}$, then $\mathbf{A}'\mathbf{Ax} = \mathbf{0}$; conversely if $\mathbf{A}'\mathbf{Ax} = \mathbf{0}$ then $\mathbf{x}'\mathbf{A}'\mathbf{Ax} = \|\mathbf{Ax}\|^2 = 0$ and so $\mathbf{Ax} = \mathbf{0}$. Thus $N(\mathbf{A}) = N(\mathbf{A}'\mathbf{A})$. Since $\mathbf{A}$ and $\mathbf{A}'\mathbf{A}$ each have p columns, we see from (A.4.2h) that $\dim M(\mathbf{A}) = \dim M(\mathbf{A}'\mathbf{A})$ so that $r(\mathbf{A}) = r(\mathbf{A}'\mathbf{A})$.

Table A.4.1 Rank of some matrices

Matrix	Rank
Non-singular $\mathbf{A}(p \times p)$	p
diag (a_i)	Number of non-zero a_i
$\mathbf{H}_n$	$n-1$
Idempotent $\mathbf{A}$	tr $\mathbf{A}$
$\mathbf{CAB}$, non-singular $\mathbf{B}$, $\mathbf{C}$	$r(\mathbf{A})$

(5) If $\mathbf{A}$ is symmetric, its rank equals the number of non-zero eigenvalues of $\mathbf{A}$. For general $\mathbf{A}(n \times p)$, the rank is given by the number of non-zero eigenvalues of $\mathbf{A}'\mathbf{A}$. See Section A.6.

A.4.3 Linear equations

For the n linear equations

$$x_1\mathbf{a}_{(1)} + \ldots + x_p\mathbf{a}_{(p)} = \mathbf{b} \tag{A.4.3a}$$

or

$$\mathbf{A}\mathbf{x} = \mathbf{b} \tag{A.4.3b}$$

with the coefficient matrix $\mathbf{A}(n \times p)$, we note the following results:

(I) If $n = p$ and $\mathbf{A}$ is non-singular, the unique solution is

$$\mathbf{x} = \mathbf{A}^{-1}\mathbf{b} = \frac{1}{|\mathbf{A}|}[A_{ij}]'\mathbf{b}. \tag{A.4.3c}$$

(II) The equation is *consistent* (i.e. admits at least one solution) if and only if

$$r(\mathbf{A}) = r[(\mathbf{A}, \mathbf{b})]. \tag{A.4.3d}$$

(III) For $\mathbf{b} = \mathbf{0}$, there exists a non-trivial solution (i.e. $\mathbf{x} \neq \mathbf{0}$) if and only if $r(\mathbf{A}) < p$.

(IV) The equation $\mathbf{A}'\mathbf{A} = \mathbf{A}'\mathbf{b}$ is always consistent. (A.4.3e)

Remarks (1) To prove (II) note that the vector $\mathbf{A}\mathbf{x}$ is a linear combination of the columns of $\mathbf{A}$. Thus the equation $\mathbf{A}\mathbf{x} = \mathbf{b}$ has a solution if and only if $\mathbf{b}$ can be expressed as a linear combination of the columns of $\mathbf{A}$.

(2) The proof of (III) is immediate from the definition of rank.

(3) To prove (IV) note that $M(\mathbf{A}'\mathbf{A}) \subseteq M(\mathbf{A}')$ because $\mathbf{A}'\mathbf{A}$ is a matrix whose columns are linear combinations of the columns of $\mathbf{A}'$. From Remark 4 of Section A.4.2 we see that $\dim M(\mathbf{A}'\mathbf{A}) = \dim M(\mathbf{A}) = \dim M(\mathbf{A}')$ and hence $M(\mathbf{A}'\mathbf{A}) = M(\mathbf{A}')$. Thus, $\mathbf{A}'\mathbf{b} \in M(\mathbf{A}'\mathbf{A})$, and so $r(\mathbf{A}'\mathbf{A}) = r(\mathbf{A}'\mathbf{A}, \mathbf{A}'\mathbf{b})$.

A.5 Linear Transformations

Definitions *The transformation from* $\mathbf{x}(p \times 1)$ *to* $\mathbf{y}(n \times 1)$ *given by*

$$\mathbf{y} = \mathbf{A}\mathbf{x} + \mathbf{b}, \tag{A.5.1}$$

where $\mathbf{A}$ *is an* $(n \times p)$ *matrix is called a* linear transformation. *For* $n = p$, *the transformation is called* non-singular *if* **A** *is non-singular and in this case the inverse transformation is*

$$\mathbf{x}=\mathbf{A}^{-1}(\mathbf{y}-\mathbf{b}).$$

An orthogonal *transformation is defined by*

$$\mathbf{y}=\mathbf{A}\mathbf{x}, \tag{A.5.2}$$

where **A** *is an orthogonal matrix. Geometrically, an orthogonal matrix represents a rotation of the coordinate axes. See Section* A.10.

A.6 Eigenvalues and Eigenvectors

A.6.1 General results

If $\mathbf{A}(p\times p)$ is any square matrix then

$$q(\lambda)=|\mathbf{A}-\lambda\mathbf{I}| \tag{A.6.1}$$

is a pth order polynomial in λ. The p roots of $q(\lambda)$, $\lambda_1,\dots,\lambda_p$, possibly complex numbers, are called *eigenvalues* of **A**. Some of the λ_i will be equal if $q(\lambda)$ has multiple roots.

For each $i=1,\dots,p$, $|\mathbf{A}-\lambda_i\mathbf{I}|=0$, so $\mathbf{A}-\lambda_i\mathbf{I}$ is singular. Hence, there exists a non-zero vector $\boldsymbol{\gamma}$ satisfying

$$\mathbf{A}\boldsymbol{\gamma}=\lambda_i\boldsymbol{\gamma}. \tag{A.6.2}$$

Any vector satisfying (A.6.2) is called a (*right*) *eigenvector* of **A** for the eigenvalue λ_i. If λ_i is complex, then $\boldsymbol{\gamma}$ may have complex entries. An eigenvector $\boldsymbol{\gamma}$ with real entries is called *standardized* if

$$\boldsymbol{\gamma}'\boldsymbol{\gamma}=1. \tag{A.6.3}$$

If $\mathbf{x}$ and $\mathbf{y}$ are eigenvectors for λ_i and $\alpha\in R$, then $\mathbf{x}+\mathbf{y}$ and $\alpha\mathbf{x}$ are also eigenvectors for λ_i. Thus, the set of all eigenvectors for λ_i forms a subspace which is called the *eigenspace* of **A** for λ_i.

Since the coefficient of λ^p in $q(\lambda)$ is $(-1)^p$, we can write $q(\lambda)$ in terms of its roots as

$$q(\lambda)=\prod_{i=1}^{p}(\lambda_i-\lambda). \tag{A.6.4}$$

Setting $\lambda=0$ in (A.6.1) and (A.6.4) gives

$$|\mathbf{A}|=\prod\lambda_i; \tag{A.6.5}$$

that is, $|\mathbf{A}|$ is the product of the eigenvalues of **A**. Similarly, matching the coefficient of λ in (A.6.1) and (A.6.4) gives

$$\sum a_{ii} = \operatorname{tr} \mathbf{A} = \sum \lambda_i; \tag{A.6.6}$$

that is, tr **A** is the sum of the eigenvalues of **A**.

Let $\mathbf{C}(p \times p)$ be a non-singular matrix. Then

$$|\mathbf{A} - \lambda \mathbf{I}| = |\mathbf{C}|\,|\mathbf{A} - \lambda \mathbf{C}^{-1}\mathbf{C}|\,|\mathbf{C}^{-1}| = |\mathbf{CAC}^{-1} - \lambda \mathbf{I}|. \tag{A.6.7}$$

Thus **A** and $\mathbf{CAC}^{-1}$ have the same eigenvalues. Further, if $\boldsymbol{\gamma}$ is an eigenvector of **A** for λ_i, then $\mathbf{CAC}^{-1}(\mathbf{C}\boldsymbol{\gamma}) = \lambda_i \mathbf{C}\boldsymbol{\gamma}$, so that

$$\boldsymbol{\nu} = \mathbf{C}\boldsymbol{\gamma}$$

is an eigenvector of $\mathbf{CAC}^{-1}$ for λ_i.

Let $\alpha \in R$. Then $|\mathbf{A} + \alpha\mathbf{I} - \lambda \mathbf{I}| = |\mathbf{A} - (\lambda - \alpha)\mathbf{I}|$, so that $\mathbf{A} + \alpha \mathbf{I}$ has eigenvalues $\lambda_i + \alpha$. Further, if $\mathbf{A}\boldsymbol{\gamma} = \lambda_i \boldsymbol{\gamma}$, then $(\mathbf{A} + \alpha\mathbf{I})\boldsymbol{\gamma} = (\lambda_i + \alpha)\boldsymbol{\gamma}$, so that **A** and $\mathbf{A} + \alpha \mathbf{I}$ have the same eigenvectors.

Bounds on the dimension of the eigenspace of **A** for λ_i are given by the following theorem.

Theorem A.6.1 *Let λ_1 denote any particular eigenvalue of $\mathbf{A}(p \times p)$, with eigenspace* H *of dimension r. If k denotes the multiplicity of λ_1 in $q(\lambda)$, then $1 \leq r \leq k$.*

Proof Since λ_1 is an eigenvalue, there is at least one non-trivial eigenvector so $r \geq 1$.

Let $\mathbf{e}_1, \ldots, \mathbf{e}_r$ be an orthonormal basis of H and extend it so that $\mathbf{e}_1, \ldots, \mathbf{e}_r, \mathbf{f}_1, \ldots, \mathbf{f}_{p-r}$ is an orthonormal basis of R^p. Write $\mathbf{E} = (\mathbf{e}_1, \ldots, \mathbf{e}_r)$, $\mathbf{F} = (\mathbf{f}_1, \ldots, \mathbf{f}_{p-r})$. Then $(\mathbf{E}, \mathbf{F})$ is an orthogonal matrix so that $\mathbf{I}_p = (\mathbf{E}, \mathbf{F})(\mathbf{E}, \mathbf{F})' = \mathbf{EE}' + \mathbf{FF}'$ and $|(\mathbf{E}, \mathbf{F})| = 1$. Also $\mathbf{E}'\mathbf{AE} = \lambda_1 \mathbf{E}'\mathbf{E} = \lambda_1 \mathbf{I}_r$, $\mathbf{F}'\mathbf{F} = \mathbf{I}_{p-r}$, and $\mathbf{F}'\mathbf{AE} = \lambda_1 \mathbf{F}'\mathbf{E} = \mathbf{0}$. Thus

$$\begin{aligned}
q(\lambda) &= |\mathbf{A} - \lambda \mathbf{I}| = |(\mathbf{E}, \mathbf{F})'|\,|\mathbf{A} - \lambda \mathbf{I}|\,|(\mathbf{E}, \mathbf{F})| \\
&= |(\mathbf{E}, \mathbf{F})'[\mathbf{AEE}' + \mathbf{AFF}' - \lambda \mathbf{EE}' - \lambda \mathbf{FF}'](\mathbf{E}, \mathbf{F})| \\
&= \begin{vmatrix} (\lambda_1 - \lambda)\mathbf{I}_r & \mathbf{E}'\mathbf{AF} \\ \mathbf{0} & \mathbf{F}'\mathbf{AF} - \lambda \mathbf{I}_{p-r} \end{vmatrix} \\
&= (\lambda_1 - \lambda)^r q_1(\lambda), \text{ say,}
\end{aligned}$$

using (A.2.3i). Thus the multiplicity of λ_1 as a root of $q(\lambda)$ is at least r.

Remarks (1) If **A** is symmetric then $r = k$; see Section A.6.2. However, if **A** is not symmetric, it is possible that $r < k$. For example,

$$\mathbf{A} = \begin{pmatrix} 0 & 1 \\ 0 & 0 \end{pmatrix}$$

has eigenvalue 0 with multiplicity 2; however, the corresponding eigenspace which is generated by $(1, 0)'$ only has dimension 1.

(2) If $r = 1$, then the eigenspace for λ_1 has dimension 1 and the standardized eigenvector for λ_1 is unique (up to sign).

Now let $\mathbf{A}(n \times p)$ and $\mathbf{B}(p \times n)$ be any two matrices and suppose $n \geq p$. Then from (A.2.3j)

$$\begin{vmatrix} -\lambda \mathbf{I}_n & -\mathbf{A} \\ \mathbf{B} & \mathbf{I}_p \end{vmatrix} = (-\lambda)^{n-p}\,|\mathbf{BA} - \lambda \mathbf{I}_p| = |\mathbf{AB} - \lambda \mathbf{I}_n|. \tag{A.6.8}$$

Hence the n eigenvalues of $\mathbf{AB}$ equal the p eigenvalues of $\mathbf{BA}$, plus the eigenvalue 0, $n - p$ times. The following theorem describes the relationship between the eigenvectors.

Theorem A.6.2 *For $\mathbf{A}(n \times p)$ and $\mathbf{B}(p \times n)$, the non-zero eigenvalues of $\mathbf{AB}$ and $\mathbf{BA}$ are the same and have the same multiplicity. If $\mathbf{x}$ is a non-trivial eigenvector of $\mathbf{AB}$ for an eigenvalue $\lambda \neq 0$, then $\mathbf{y} = \mathbf{Bx}$ is a non-trivial eigenvector of $\mathbf{BA}$.*

Proof The first part follows from (A.6.8). For the second part substituting $\mathbf{y} = \mathbf{Bx}$ in the equation $\mathbf{B}(\mathbf{ABx}) = \lambda \mathbf{Bx}$ gives $\mathbf{BAy} = \lambda \mathbf{y}$. The vector $\mathbf{x}$ is non-trivial if $\mathbf{x} \neq \mathbf{0}$. Since $\mathbf{Ay} = \mathbf{ABx} = \lambda \mathbf{x} \neq \mathbf{0}$, it follows that $\mathbf{y} \neq \mathbf{0}$ also. ■

Corollary A.6.2.1 *For $\mathbf{A}(n \times p)$, $\mathbf{B}(q \times n)$, $\mathbf{a}(p \times 1)$, and $\mathbf{b}(q \times 1)$, the matrix $\mathbf{Aab'B}$ has rank at most 1. The non-zero eigenvalue, if present, equals $\mathbf{b'BAa}$, with eigenvector $\mathbf{Aa}$.*

Proof The non-zero eigenvalue of $\mathbf{Aab'B}$ equals that of $\mathbf{b'BAa}$, which is a scalar, and hence is its own eigenvalue. The fact that $\mathbf{Aa}$ is a corresponding eigenvector is easily checked. ■

A.6.2 Symmetric matrices

If $\mathbf{A}$ is symmetric, it is possible to give more detailed information about its eigenvalues and eigenvectors.

Theorem A.6.3 *All the eigenvalues of a symmetric matrix $\mathbf{A}(p \times p)$ are real.*

Proof If possible, let

$$\boldsymbol{\gamma} = \mathbf{x} + i\mathbf{y}, \qquad \lambda = a + ib, \qquad \boldsymbol{\gamma} \neq \mathbf{0}. \tag{A.6.9}$$

From (A.6.2), after equating real and imaginary parts, we have

$$\mathbf{Ax} = a\mathbf{x} - b\mathbf{y}, \qquad \mathbf{Ay} = b\mathbf{x} + a\mathbf{y}.$$

On premultiplying by $\mathbf{y}'$ and $\mathbf{x}'$, respectively, and subtracting, we obtain $b=0$. Hence from (A.6.9), λ is real. ■

In the above discussion, we can choose $\mathbf{y}=\mathbf{0}$ so we can assume $\boldsymbol{\gamma}$ to be real.

Theorem A.6.4 (Spectral decomposition theorem, or Jordan decomposition theorem) *Any symmetric matrix* $\mathbf{A}(p\times p)$ *can be written as*

$$\mathbf{A}=\boldsymbol{\Gamma}\boldsymbol{\Lambda}\boldsymbol{\Gamma}'=\sum \lambda_i \boldsymbol{\gamma}_{(i)}\boldsymbol{\gamma}'_{(i)}, \tag{A.6.10}$$

where $\boldsymbol{\Lambda}$ *is a diagonal matrix of eigenvalues of* $\mathbf{A}$, *and* $\boldsymbol{\Gamma}$ *is an orthogonal matrix whose columns are standardized eigenvectors.*

Proof Suppose we can find orthonormal vectors $\boldsymbol{\gamma}_{(1)},\ldots,\boldsymbol{\gamma}_{(p)}$ such that $\mathbf{A}\boldsymbol{\gamma}_{(i)}=\lambda_i\boldsymbol{\gamma}_{(i)}$ for some numbers λ_i. Then

$$\boldsymbol{\gamma}'_{(i)}\mathbf{A}\boldsymbol{\gamma}_{(j)}=\lambda_j\boldsymbol{\gamma}'_{(i)}\boldsymbol{\gamma}_{(j)}=\begin{cases}\lambda_i, & i=j,\\ 0, & i\neq j,\end{cases}$$

or in matrix form

$$\boldsymbol{\Gamma}'\mathbf{A}\boldsymbol{\Gamma}=\boldsymbol{\Lambda}. \tag{A.6.11}$$

Pre- and post-multiplying by $\boldsymbol{\Gamma}$ and $\boldsymbol{\Gamma}'$ gives (A.6.10). From (A.6.7), $\mathbf{A}$ and $\boldsymbol{\Lambda}$ have the same eigenvalues, so the elements of $\boldsymbol{\Lambda}$ are exactly the eigenvalues of $\mathbf{A}$ with the same multiplicities.

Thus we must find an orthonormal basis of eigenvectors. Note that if $\lambda_i\neq\lambda_j$ are distinct eigenvalues with eigenvectors $\mathbf{x}+\mathbf{y}$, respectively, then $\lambda_i\mathbf{x}'\mathbf{y}=\mathbf{x}'\mathbf{A}\mathbf{y}=\mathbf{y}'\mathbf{A}\mathbf{x}=\lambda_j\mathbf{y}'\mathbf{x}$, so that $\mathbf{y}'\mathbf{x}=0$. *Hence for a symmetric matrix, eigenvectors corresponding to distinct eigenvalues are orthogonal to one another.*

Suppose there are k distinct eigenvalues of $\mathbf{A}$ with corresponding eigenspaces $H_1,\ldots,H_k$ of dimensions $r_1,\ldots,r_k$. Let

$$r=\sum_{j=1}^{k} r_j.$$

Since distinct eigenspaces are orthogonal, there exists an orthonormal set of vectors $\mathbf{e}_1,\ldots,\mathbf{e}_r$ such that the vectors labelled

$$\sum_{i=1}^{j-1} r_i+1,\ldots,\sum_{i=1}^{j} r_i$$

form a basis for H_j. From Theorem A.6.1, r_j is less than or equal to the multiplicity of the corresponding eigenvalue. Hence by re-ordering the eigenvalues λ_i if necessary, we may suppose

$$\mathbf{A}\mathbf{e}_i=\lambda_i\mathbf{e}_i, \qquad i=1,\ldots,r,$$

and $r \leq p$. (If all p eigenvalues are distinct, then we know from Theorem A.6.1 that $r = p$).

If $r = p$, set $\boldsymbol{\gamma}_{(i)} = \mathbf{e}_i$ and the proof follows. We shall show that the situation $r < p$ leads to a contradiction, and therefore cannot arise.

Without loss of generality we may suppose that all of the eigenvalues of $\mathbf{A}$ are strictly positive. (If not, we can replace $\mathbf{A}$ by $\mathbf{A} + \alpha\mathbf{I}$ for a suitable α, because $\mathbf{A}$ and $\mathbf{A} + \alpha\mathbf{I}$ have the same eigenvectors). Set

$$\mathbf{B} = \mathbf{A} - \sum_{i=1}^{r} \lambda_i \mathbf{e}_i \mathbf{e}_i'.$$

Then

$$\operatorname{tr} \mathbf{B} = \operatorname{tr} \mathbf{A} - \sum_{i=1}^{r} \lambda_i (\mathbf{e}_i' \mathbf{e}_i) = \sum_{i=r+1}^{p} \lambda_i > 0,$$

since $r < p$. Thus $\mathbf{B}$ has at least one non-zero eigenvalue, say θ. Let $\mathbf{x} \neq \mathbf{0}$ be a corresponding eigenvector. Then for $1 \leq j \leq r$,

$$\theta \mathbf{e}_j' \mathbf{x} = \mathbf{e}_j' \mathbf{B} \mathbf{x} = \left\{ \lambda_j \mathbf{e}_j' - \sum_{i=1}^{r} \lambda_i (\mathbf{e}_j' \mathbf{e}_i) \mathbf{e}_i' \right\} \mathbf{x} = 0,$$

so that $\mathbf{x}$ is orthogonal to $\mathbf{e}_j$, $j = 1, \ldots, r$. Therefore,

$$\theta \mathbf{x} = \mathbf{B} \mathbf{x} = \left(\mathbf{A} - \sum \lambda_i \mathbf{e}_i \mathbf{e}_i' \right) \mathbf{x} = \mathbf{A} \mathbf{x} - \sum \lambda_i (\mathbf{e}_i' \mathbf{x}) \mathbf{e}_i = \mathbf{A} \mathbf{x}$$

so that $\mathbf{x}$ is an eigenvector of $\mathbf{A}$ also. Thus $\theta = \lambda_i$ for some i and $\mathbf{x}$ is a linear combination of some of the $\mathbf{e}_i$, which contradicts the orthogonality between $\mathbf{x}$ and the $\mathbf{e}_i$. ■

Corollary A.6.4.1 *If $\mathbf{A}$ is a non-singular symmetric matrix, then for any integer n,*

$$\boldsymbol{\Lambda}^n = \operatorname{diag}(\lambda_i^n) \quad \text{and} \quad \mathbf{A}^n = \boldsymbol{\Gamma} \boldsymbol{\Lambda}^n \boldsymbol{\Gamma}'. \tag{A.6.12}$$

If all the eigenvalues of $\mathbf{A}$ are positive then we can define the rational powers

$$\mathbf{A}^{r/s} = \boldsymbol{\Gamma} \boldsymbol{\Lambda}^{r/s} \boldsymbol{\Gamma}', \quad \text{where} \quad \boldsymbol{\Lambda}^{r/s} = \operatorname{diag}(\lambda_i^{r/s}), \tag{A.6.13}$$

for integers $s > 0$ and r. If some of the eigenvalues of $\mathbf{A}$ are zero, then (A.6.12) and (A.6.13) hold if the exponents are restricted to be non-negative.

Proof Since

$$\mathbf{A}^2 = (\boldsymbol{\Gamma} \boldsymbol{\Lambda} \boldsymbol{\Gamma}')^2 = \boldsymbol{\Gamma} \boldsymbol{\Lambda} \boldsymbol{\Gamma}' \boldsymbol{\Gamma} \boldsymbol{\Lambda} \boldsymbol{\Gamma}' = \boldsymbol{\Gamma} \boldsymbol{\Lambda}^2 \boldsymbol{\Gamma}'$$

and

$$\mathbf{A}^{-1} = \boldsymbol{\Gamma} \boldsymbol{\Lambda}^{-1} \boldsymbol{\Gamma}', \qquad \boldsymbol{\Lambda}^{-1} = \operatorname{diag}(\lambda_i^{-1}),$$

we see that (A.6.12) can be easily proved by induction. To check that rational powers make sense note that

$$(\mathbf{A}^{r/s})^s = \mathbf{\Gamma}\mathbf{\Lambda}^{r/s}\mathbf{\Gamma}' \ldots \mathbf{\Gamma}\mathbf{\Lambda}^{r/s}\mathbf{\Gamma}' = \mathbf{\Gamma}\mathbf{\Lambda}^r\mathbf{\Gamma}' = \mathbf{A}^r. \quad \blacksquare$$

Motivated by (A.6.13), we can define powers of $\mathbf{A}$ for real-valued exponents.

Important special cases of (A.6.13) are

$$\mathbf{A}^{1/2} = \mathbf{\Gamma}\mathbf{\Lambda}^{1/2}\mathbf{\Gamma}', \qquad \mathbf{\Lambda}^{1/2} = \operatorname{diag}(\lambda_i^{1/2}) \tag{A.6.14}$$

when $\lambda_i \geq 0$ for all i and

$$\mathbf{A}^{-1/2} = \mathbf{\Gamma}\mathbf{\Lambda}^{-1/2}\mathbf{\Gamma}', \qquad \mathbf{\Lambda}^{-1/2} = \operatorname{diag}(\lambda_i^{-1/2}) \tag{A.6.15}$$

when $\lambda_i > 0$ for all i. The decomposition (A.6.14) is called the *symmetric square root decomposition* of $\mathbf{A}$.

Corollary A.6.4.2 *The rank of* $\mathbf{A}$ *equals the number of non-zero eigenvalues.*

Proof By (A.4.2f), $r(\mathbf{A}) = r(\mathbf{\Lambda})$, whose rank is easily seen to equal the number of non-zero diagonal elements. $\blacksquare$

Remarks (1) Theorem A.6.4 shows that a symmetric matrix $\mathbf{A}$ is uniquely determined by its eigenvalues and eigenvectors, or more specifically by its distinct eigenvalues and corresponding eigenspaces.

(2) Since $\mathbf{A}^{1/2}$ has the same eigenvectors as $\mathbf{A}$ and has eigenvalues which are given functions of the eigenvalues of $\mathbf{A}$, we see that the symmetric square root is uniquely defined.

(3) If the λ_i are all distinct and written in decreasing order say, then $\mathbf{\Gamma}$ is uniquely determined, up to the signs of its columns.

(4) If $\lambda_{k+1} = \ldots = \lambda_p = 0$ then (A.6.10) can be written more compactly as

$$\mathbf{A} = \mathbf{\Gamma}_1\mathbf{\Lambda}_1\mathbf{\Gamma}_1' = \sum_{i=1}^{k} \lambda_i \boldsymbol{\gamma}_{(i)}\boldsymbol{\gamma}_{(i)}',$$

where $\mathbf{\Lambda}_1 = \operatorname{diag}(\lambda_1, \ldots, \lambda_k)$ and $\mathbf{\Gamma}_1 = (\boldsymbol{\gamma}_{(1)}, \ldots, \boldsymbol{\gamma}_{(k)})$.

(5) A symmetric matrix $\mathbf{A}$ has rank 1 if and only if

$$\mathbf{A} = \mathbf{x}\mathbf{x}'$$

for some $\mathbf{x}$. Then the only non-zero eigenvalue of $\mathbf{A}$ is given by

$$\operatorname{tr} \mathbf{A} = \operatorname{tr} \mathbf{x}\mathbf{x}' = \mathbf{x}'\mathbf{x}$$

and the corresponding eigenspace is generated by $\mathbf{x}$.

(6) Since $\mathbf{J} = \mathbf{1}\mathbf{1}'$ has rank 1 with eigenvalue p and corresponding eigenvector $\mathbf{1}$, we see that the equicorrelation matrix $\mathbf{E} = (1-\rho)\mathbf{I} + \rho\mathbf{J}$ has

eigenvalues $\lambda_1 = 1+(p-1)\rho$ and $\lambda_2 = \ldots = \lambda_p = 1-\rho$, and the same eigenvectors as $\mathbf{J}$. For the eigenvectors $\boldsymbol{\gamma}_{(2)}, \ldots, \boldsymbol{\gamma}_{(p)}$, we can select any standardized set of vectors orthogonal to $\mathbf{1}$ and each other. A possible choice for $\boldsymbol{\Gamma}$ is the Helmert matrix of Section A.3.1. Multiplying the eigenvalues together yields the formula for $|\mathbf{E}|$ given in (A.3.2c).

(7) If $\mathbf{A}$ is symmetric and idempotent (that is, $\mathbf{A} = \mathbf{A}'$ and $\mathbf{A}^2 = \mathbf{A}$), then $\lambda_i = 0$ or 1 for all i, because $\mathbf{A} = \mathbf{A}^2$ implies $\boldsymbol{\Lambda} = \boldsymbol{\Lambda}^2$.

(8) If $\mathbf{A}$ is symmetric and idempotent then $r(\mathbf{A}) = \operatorname{tr} \mathbf{A}$. This result follows easily from (A.6.6) and Corollary A.6.4.2.

(9) As an example, consider

$$\mathbf{A} = \begin{pmatrix} 1 & \rho \\ \rho & 1 \end{pmatrix}. \tag{A.6.16}$$

The eigenvalues of $\mathbf{A}$ from (A.6.1) are the solutions of

$$\begin{vmatrix} 1-\lambda & \rho \\ \rho & 1-\lambda \end{vmatrix} = 0,$$

namely, $\lambda_1 = 1+\rho$ and $\lambda_2 = 1-\rho$. Thus,

$$\boldsymbol{\Lambda} = \operatorname{diag}(1+\rho, 1-\rho). \tag{A.6.17}$$

For $\rho \neq 0$, the eigenvector corresponding to $\lambda_1 = 1+\rho$ from (A.6.2) is

$$\begin{pmatrix} 1 & \rho \\ \rho & 1 \end{pmatrix}\begin{pmatrix} x_1 \\ x_2 \end{pmatrix} = (1+\rho)\begin{pmatrix} x_1 \\ x_2 \end{pmatrix},$$

which leads to $x_1 = x_2$, therefore the first standardized eigenvector is

$$\boldsymbol{\gamma}_{(1)} = \begin{pmatrix} 1/\sqrt{2} \\ 1/\sqrt{2} \end{pmatrix}.$$

Similarly, the eigenvector corresponding to $\lambda_2 = 1-\rho$ is

$$\boldsymbol{\gamma}_{(2)} = \begin{pmatrix} 1/\sqrt{2} \\ -1/\sqrt{2} \end{pmatrix}.$$

Hence,

$$\boldsymbol{\Gamma} = \begin{pmatrix} 1/\sqrt{2} & 1/\sqrt{2} \\ 1/\sqrt{2} & -1/\sqrt{2} \end{pmatrix}. \tag{A.6.18}$$

If $\rho = 0$ then $\mathbf{A} = \mathbf{I}$ and any orthonormal basis will do.

(10) Formula (A.6.14) suggests a method for calculating the symmetric *square root* of a matrix. For example, for the matrix in (A.6.16) with

$\rho^2 < 1$, we find on using $\mathbf{\Lambda}$ and $\mathbf{\Gamma}$ from (A.6.11) and (A.6.14) that

$$\mathbf{A}^{1/2} = \mathbf{\Gamma}\mathbf{\Lambda}^{1/2}\mathbf{\Gamma} = \begin{pmatrix} a & b \\ b & a \end{pmatrix},$$

where

$$2a = (1+\rho)^{1/2} + (1-\rho)^{1/2}, \qquad 2b = (1+\rho)^{1/2} - (1-\rho)^{1/2}.$$

*(11) The following methods are commonly used to calculate eigenvalues and eigenvectors on computers. For symmetric matrices, the Householder reduction to tri-diagonal form (i.e. $a_{ij} = 0$, for $i \geq j+2$ and $i \leq j-2$) is used followed by the QL algorithm. For non-symmetric matrices, reduction to upper Hessenberg form (i.e. $a_{ij} = 0$ for $i \geq j+2$) is used followed by the QR algorithm.

(12) For general matrices $\mathbf{A}(n \times p)$, we can use the spectral decomposition theorem to derive the following result.

Theorem A.6.5 (Singular value decomposition theorem) *If* $\mathbf{A}$ *is an* $(n \times p)$ *matrix of rank* r, *then* $\mathbf{A}$ *can be written as*

$$\mathbf{A} = \mathbf{ULV}' \tag{A.6.19}$$

where $\mathbf{U}(n \times r)$ *and* $\mathbf{V}(p \times r)$ *are column orthonormal matrices* ($\mathbf{U}'\mathbf{U} = \mathbf{V}'\mathbf{V} = \mathbf{I}_r$) *and* $\mathbf{L}$ *is a diagonal matrix with positive elements.*

Proof Since $\mathbf{A}'\mathbf{A}$ is a symmetric matrix which also has rank r, we can use the spectral decomposition theorem to write

$$\mathbf{A}'\mathbf{A} = \mathbf{V}\mathbf{\Lambda}\mathbf{V}', \tag{A.6.20}$$

where $\mathbf{V}(p \times r)$ is a column orthonormal matrix of eigenvectors of $\mathbf{A}'\mathbf{A}$ and $\mathbf{\Lambda} = \text{diag}(\lambda_1, \ldots, \lambda_r)$ contains the non-zero eigenvalues. Note that all the λ_i are positive because $\lambda_i = \mathbf{v}'_{(i)}\mathbf{A}'\mathbf{A}\mathbf{v}_{(i)} = \|\mathbf{A}\mathbf{v}_{(i)}\|^2 > 0$. Let

$$l_i = \lambda_i^{1/2}, \qquad i = 1, \ldots, r, \tag{A.6.21}$$

and set $\mathbf{L} = \text{diag}(l_1, \ldots, l_r)$. Define $\mathbf{U}(n \times r)$ by

$$\mathbf{u}_{(i)} = l_i^{-1}\mathbf{A}\mathbf{v}_{(i)}, \qquad i = 1, \ldots, r. \tag{A.6.22}$$

Then

$$\mathbf{u}'_{(j)}\mathbf{u}_{(i)} = l_i^{-1}l_j^{-1}\mathbf{v}'_{(j)}\mathbf{A}'\mathbf{A}\mathbf{v}_{(i)} = \lambda_i l_i^{-1}l_j^{-1}\mathbf{v}'_{(j)}\mathbf{v}_{(i)} = \begin{cases} 1, & i = j, \\ 0, & i \neq j. \end{cases}$$

Thus $\mathbf{U}$ is also a column orthonormal matrix.

Any p-vector $\mathbf{x}$ can be written as $\mathbf{x} = \sum \alpha_i \mathbf{v}_{(i)} + \mathbf{y}$ where $\mathbf{y} \in N(\mathbf{A})$, the null space of $\mathbf{A}$. Note that $N(\mathbf{A}) = N(\mathbf{A}'\mathbf{A})$ is the eigenspace of $\mathbf{A}'\mathbf{A}$ for the eigenvalue 0, so that $\mathbf{y}$ is orthogonal to the eigenvectors $\mathbf{v}_{(i)}$. Let $\mathbf{e}_i$

denote the r-vector with 1 in the ith place and 0 elsewhere. Then

$$\begin{aligned}\mathbf{ULV'x} &= \sum \alpha_i \mathbf{ULe}_i + \mathbf{0} \\ &= \sum \alpha_i l_i \mathbf{u}_{(i)} + \mathbf{0} \\ &= \sum \alpha_i \mathbf{Av}_{(i)} + \mathbf{Ay} = \mathbf{Ax}.\end{aligned}$$

Since this formula holds for all $\mathbf{x}$ it follows that $\mathbf{ULV'} = \mathbf{A}$. ■

Note that the columns of $\mathbf{U}$ are eigenvectors of $\mathbf{AA'}$ and the columns of $\mathbf{V}$ are eigenvectors of $\mathbf{A'A}$. Also, from Theorem A.6.2, the eigenvalues of $\mathbf{AA'}$ and $\mathbf{A'A}$ are the same.

A.7 Quadratic Forms and Definiteness

Definition *A* quadratic form *in the vector* $\mathbf{x}$ *is a function of the form*

$$Q(\mathbf{x}) \equiv \mathbf{x'Ax} = \sum_{i=1}^{p} \sum_{j=1}^{p} a_{ij} x_i x_j, \tag{A.7.1}$$

where $\mathbf{A}$ *is a symmetric matrix; that is,*

$$Q(\mathbf{x}) = a_{11}x_1^2 + \ldots + a_{pp}x_p^2 + 2a_{12}x_1x_2 + \ldots + 2a_{p-1,p}x_{p-1}x_p.$$

Clearly, $Q(\mathbf{0}) = 0$.

Definition (1) $Q(\mathbf{x})$ *is called a* positive definite (p.d.) quadratic form *if* $Q(\mathbf{x}) > 0$ *for all* $\mathbf{x} \neq \mathbf{0}$.

(2) $Q(\mathbf{x})$ *is called a* positive semi-definite (p.s.d) quadratic form *if* $Q(\mathbf{x}) \geqslant 0$ *for all* $\mathbf{x} \neq \mathbf{0}$.

(3) *A symmetric matrix* $\mathbf{A}$ *is called p.d. (p.s.d) if* $Q(\mathbf{x})$ *is p.d. (p.s.d.) and we write* $\mathbf{A} > 0$ *or* $\mathbf{A} \geqslant 0$ *for* $\mathbf{A}$ *positive definite or positive semi-definite, respectively.*

Negative definite and negative semi-definite quadratic forms are similarly defined.

For $p = 2$, $Q(\mathbf{x}) = x_1^2 + x_2^2$ *is p.d. while* $Q(\mathbf{x}) = (x_1 - x_2)^2$ *is p.s.d.*

Canonical form Any quadratic form can be converted into a weighted sum of squares without cross-product terms with the help of the following theorem.

Theorem A.7.1 *For any symmetric matrix* $\mathbf{A}$, *there exists an orthogonal transformation*

$$\mathbf{y} = \mathbf{\Gamma' x} \tag{A.7.2}$$

such that

$$\mathbf{x}'\mathbf{A}\mathbf{x}=\sum \lambda_i y_i^2. \tag{A.7.3}$$

Proof *Consider the spectral decomposition given in Theorem* A.6.4:

$$\mathbf{A}=\mathbf{\Gamma}\mathbf{\Lambda}\mathbf{\Gamma}'. \tag{A.7.4}$$

From (A.7.2),

$$\mathbf{x}'\mathbf{A}\mathbf{x}=\mathbf{y}'\mathbf{\Gamma}'\mathbf{A}\mathbf{\Gamma}\mathbf{y}=\mathbf{y}'\mathbf{\Gamma}'\mathbf{\Gamma}\mathbf{\Lambda}\mathbf{\Gamma}'\mathbf{\Gamma}\mathbf{y}=\mathbf{y}'\mathbf{\Lambda}\mathbf{y}.$$

Hence (A.7.3) follows. ■

It is important to recall that $\mathbf{\Gamma}$ has as its columns the eigenvectors of $\mathbf{A}$ and that $\lambda_1, \ldots, \lambda_p$ are the eigenvalues of $\mathbf{A}$. Using this theorem, we can deduce the following results for a matrix $\mathbf{A}>0$.

Theorem A.7.2 *If* $\mathbf{A}>0$ *then* $\lambda_i>0$ *for* $i=1,\ldots,p$. *If* $\mathbf{A}\geqslant 0$, *then* $\lambda_i \geqslant 0$.

Proof If $\mathbf{A}>0$, we have, for all $\mathbf{x}\neq \mathbf{0}$,

$$0<\mathbf{x}'\mathbf{A}\mathbf{x}=\lambda_1 y_1^2+\ldots+\lambda_p y_p^2.$$

From (A.7.2), $\mathbf{x}\neq\mathbf{0}$ implies $\mathbf{y}\neq\mathbf{0}$. Choosing $y_1=1$, $y_2=\ldots=y_p=0$, we deduce that $\lambda_1>0$. Similarly $\lambda_i>0$ for all i. If $\mathbf{A}\geqslant 0$ the above inequalities are weak. ■

Corollary A.7.2.1 *If* $\mathbf{A}>0$, *then* $\mathbf{A}$ *is non-singular and* $|\mathbf{A}|>0$.

Proof Use the determinant of (A.7.4) with $\lambda_i>0$. ■

Corollary A.7.2.2 *If* $\mathbf{A}>0$, *then* $\mathbf{A}^{-1}>0$.

Proof From (A.7.3), we have

$$\mathbf{x}'\mathbf{A}^{-1}\mathbf{x}=\sum y_i^2/\lambda_i. \quad ■ \tag{A.7.5}$$

Corollary A.7.2.3 (Symmetric decomposition) *Any matrix* $\mathbf{A}\geqslant 0$ *can be written as*

$$\mathbf{A}=\mathbf{B}^2, \tag{A.7.6}$$

where $\mathbf{B}$ *is a symmetric matrix.*

Proof Take $\mathbf{B}=\mathbf{\Gamma}\mathbf{\Lambda}^{1/2}\mathbf{\Gamma}'$ in (A.7.4). ■

Theorem A.7.3 *If* $\mathbf{A}\geqslant 0$ *is a* $(p\times p)$ *matrix, then for any* $(p\times n)$ *matrix* $\mathbf{C}$, $\mathbf{C}'\mathbf{A}\mathbf{C}\geqslant 0$. *If* $\mathbf{A}>0$ *and* $\mathbf{C}$ *is non-singular (so* $p=n$*), then* $\mathbf{C}'\mathbf{A}\mathbf{C}>0$.

Proof If $\mathbf{A}\geqslant 0$ then for any n-vector $\mathbf{x}\neq\mathbf{0}$,

$$\mathbf{x}'\mathbf{C}'\mathbf{A}\mathbf{C}\mathbf{x}=(\mathbf{C}\mathbf{x})'\mathbf{A}(\mathbf{C}\mathbf{x})\geqslant 0, \quad \text{so} \quad \mathbf{C}'\mathbf{A}\mathbf{C}\geqslant 0.$$

If $\mathbf{A}>0$ and $\mathbf{C}$ is non-singular, the $\mathbf{Cx}\neq\mathbf{0}$, so $(\mathbf{Cx})'\mathbf{A}(\mathbf{Cx})>0$, and hence $\mathbf{C}'\mathbf{AC}>0$. ■

Corollary A.7.3.1 *If $\mathbf{A}\geq 0$ and $\mathbf{B}>0$ are $(p\times p)$ matrices, then all of the non-zero eigenvalues of $\mathbf{B}^{-1}\mathbf{A}$ are positive.*

Proof Since $\mathbf{B}>0$, $\mathbf{B}^{-1/2}$ exists and, by Theorem A.6.2, $\mathbf{B}^{-1/2}\mathbf{AB}^{-1/2}$, $\mathbf{B}^{-1}\mathbf{A}$, and $\mathbf{AB}^{-1}$ have the same eigenvalues. By Theorem A.7.3, $\mathbf{B}^{-1/2}\mathbf{AB}^{-1/2}\geq 0$, so all of the non-zero eigenvalues are positive. ■

Remarks (1) There are other forms of interest:

(a) *Linear form.* $\mathbf{a}'\mathbf{x}=a_1x_1+\ldots+a_px_p$. Generally called a linear combination.

(b) *Bilinear form.* $\mathbf{x}'\mathbf{Ay}=\sum\sum a_{ij}x_iy_j$.

(2) We have noted in Corollary A.7.2.1 that $|\mathbf{A}|>0$ for $\mathbf{A}>0$. In fact, $|\mathbf{A}_{11}|>0$ for all partitions of $\mathbf{A}$. The proof follows on considering $\mathbf{x}'\mathbf{Ax}>0$ for all $\mathbf{x}$ with $x_{i+1}=\ldots=x_p=0$. The converse is also true.

(3) For

$$\mathbf{\Sigma}=\begin{pmatrix}1 & \rho\\ \rho & 1\end{pmatrix},\qquad \rho^2<1$$

the transformation (A.7.2) is given by (A.6.18),

$$y_1=(x_1+x_2)/\sqrt{2},\qquad y_2=(x_1-x_2)/\sqrt{2}.$$

Thus, from (A.7.3) and (A.7.5),

$$\mathbf{x}'\mathbf{\Sigma x}=x_1^2+2\rho x_1x_2+x_2^2=(1+\rho)y_1^2+(1-\rho)y_2^2,$$

$$\mathbf{x}'\mathbf{\Sigma}^{-1}\mathbf{x}=\frac{1}{(1-\rho^2)}(x_1^2-2\rho x_1x_2+x_2^2)=\frac{y_1^2}{1+\rho}+\frac{y_2^2}{1-\rho}.$$

A geometrical interpretation of these results will be found in Section A.10.4.

(4) Note that the centring matrix $\mathbf{H}\geq 0$ because $\mathbf{x}'\mathbf{Hx}=\sum(x_i-\bar{x})^2\geq 0$.

(5) For any matrix $\mathbf{A}$, $\mathbf{AA}'\geq 0$ and $\mathbf{A}'\mathbf{A}\geq 0$. Further, $r(\mathbf{AA}')=r(\mathbf{A}'\mathbf{A})=r(\mathbf{A})$.

*A.8 Generalized Inverse

We now consider a method of defining an inverse for any matrix.

Definition *For a matrix $\mathbf{A}(n\times p)$, $\mathbf{A}^-$ is called a* g-inverse (generalized inverse) *of* $\mathbf{A}$ *if*

$$\mathbf{A}\mathbf{A}^{-}\mathbf{A}=\mathbf{A}. \qquad (A.8.1)$$

A generalized inverse always exists although in general it is not unique. ∎

Methods of construction

(1) Using the singular value decomposition theorem, (Theorem A.6.5) for $\mathbf{A}(n\times p)$, write $\mathbf{A}=\mathbf{ULV}'$. Then it is easily checked that

$$\mathbf{A}^{-}=\mathbf{VL}^{-1}\mathbf{U}' \qquad (A.8.2)$$

defines a g-inverse.

(2) If $r(\mathbf{A})=r$, re-arrange the rows and columns of $\mathbf{A}(n\times p)$ and partition $\mathbf{A}$ so that $\mathbf{A}_{11}$ is an $(r\times r)$ non-singular matrix. Then it can be verified that

$$\mathbf{A}^{-}=\begin{pmatrix}\mathbf{A}_{11}^{-1} & \mathbf{0}\\ \mathbf{0} & \mathbf{0}\end{pmatrix} \qquad (A.8.3)$$

is a g-inverse.

The result follows on noting that there exist $\mathbf{B}$ and $\mathbf{C}$ such that

$$\mathbf{A}_{12}=\mathbf{A}_{11}\mathbf{B},\ \mathbf{A}_{21}=\mathbf{C}\,\mathbf{A}_{11}\text{ and }\mathbf{A}_{22}=\mathbf{C}\,\mathbf{A}_{11}\mathbf{B}.$$

(3) If $\mathbf{A}(p\times p)$ is non-singular then $\mathbf{A}^{-}=\mathbf{A}^{-1}$ is uniquely defined.

(4) If $\mathbf{A}(p\times p)$ is symmetric of rank r, then, using Remark 4 after Theorem A.6.4, $\mathbf{A}$ can be written as $\mathbf{A}=\boldsymbol{\Gamma}_1\boldsymbol{\Lambda}_1\boldsymbol{\Gamma}_1'$, where $\boldsymbol{\Gamma}_1$ is a column orthonormal matrix of eigenvectors corresponding to the non-zero eigenvalues $\boldsymbol{\Lambda}_1=\operatorname{diag}(\lambda_1,\ldots,\lambda_r)$ of $\mathbf{A}$. Then it is easily checked that

$$\mathbf{A}^{-}=\boldsymbol{\Gamma}_1\boldsymbol{\Lambda}_1^{-1}\boldsymbol{\Gamma}_1' \qquad (A.8.4)$$

is a g-inverse.

Applications

(1) *Linear equations.* A particular solution of the consistent equations

$$\mathbf{Ax}=\mathbf{b}, \qquad (A.8.5)$$

is

$$\mathbf{x}=\mathbf{A}^{-}\mathbf{b}. \qquad (A.8.6)$$

Proof From (A.8.1),

$$\mathbf{A}\mathbf{A}^{-}\mathbf{Ax}=\mathbf{Ax}\Rightarrow\mathbf{A}(\mathbf{A}^{-}\mathbf{b})=\mathbf{b}$$

which when compared with (A.8.5) leads to (A.8.6). ∎

It can be shown that a general solution of a consistent equation is

$$\mathbf{x}=\mathbf{A}^{-}\mathbf{b}+(\mathbf{I}-\mathbf{G})\mathbf{z},$$

where $\mathbf{z}$ is arbitrary and $\mathbf{G}=\mathbf{A}^{-}\mathbf{A}$. For $\mathbf{b}=\mathbf{0}$, a general solution is $(\mathbf{I}-\mathbf{G})\mathbf{z}$.

(2) *Quadratic forms.* Let $\mathbf{A}(p \times p)$ be a symmetric matrix of rank $r \leq p$. Then there exists an orthogonal transformation such that for $\mathbf{x}$ restricted to $M(\mathbf{A})$ the subspace spanned by the columns of $\mathbf{A}$, $\mathbf{x}'\mathbf{A}^-\mathbf{x}$ can be written as

$$\mathbf{x}'\mathbf{A}^-\mathbf{x} = \sum u_i^2/\lambda_i, \tag{A.8.7}$$

where $\lambda_1, \ldots, \lambda_r$ are the non-zero eigenvalues of $\mathbf{A}$.

Proof First note that if $\mathbf{x}$ lies in $M(\mathbf{A})$ we can write $\mathbf{x} = \mathbf{Ay}$ for some $\mathbf{y}$, so that

$$\mathbf{x}'\mathbf{A}^-\mathbf{x} = \mathbf{y}'\mathbf{AA}^-\mathbf{Ay} = \mathbf{y}'\mathbf{Ay}$$

does not depend upon the particular g-inverse chosen. From the spectral decomposition of $\mathbf{A}$ we see that $M(\mathbf{A})$ is spanned by the eigenvectors of $\mathbf{A}$ corresponding to non-zero eigenvalues, say by $(\boldsymbol{\gamma}_{(1)}, \ldots, \boldsymbol{\gamma}_{(r)}) = \boldsymbol{\Gamma}_1$. Then if $\mathbf{x} \in M(\mathbf{A})$, it can be written as $\mathbf{x} = \boldsymbol{\Gamma}_1\mathbf{u}$ for some r-vector $\mathbf{u}$. Defining $\mathbf{A}^-$ by (A.8.4), we see that (A.8.7) follows.

Remarks (1) For the equicorrelation matrix $\mathbf{E}$, if $1+(p-1)\rho = 0$, then $(1-\rho)^{-1}\mathbf{I}$ is a g-inverse of $\mathbf{E}$.

(2) Under the following conditions $\mathbf{A}^-$ is defined uniquely:

$$\mathbf{AA}^-\mathbf{A} = \mathbf{A}, \qquad \mathbf{AA}^- \text{ and } \mathbf{A}^-\mathbf{A} \text{ symmetric}, \qquad \mathbf{A}^-\mathbf{AA}^- = \mathbf{A}^-.$$

*(3) For $\mathbf{A} \geq 0$, $\mathbf{A}^-$ is normally computed by using Cholesky decomposition (see Remark 4, Section A.2.4.).

A.9 Matrix Differentiation and Maximization Problems

Let us define the derivative of $f(\mathbf{X})$ with respect to $\mathbf{X}(n \times p)$ as the matrix

$$\frac{\partial f(\mathbf{X})}{\partial \mathbf{X}} = \left(\frac{\partial f(\mathbf{X})}{\partial x_{ij}}\right).$$

We have the following results:

$$\text{(I)} \quad \frac{\partial \mathbf{a}'\mathbf{x}}{\partial \mathbf{x}} = \mathbf{a}. \tag{A.9.1}$$

$$\text{(II)} \quad \frac{\partial \mathbf{x}'\mathbf{x}}{\partial \mathbf{x}} = 2\mathbf{x}, \quad \frac{\partial \mathbf{x}'\mathbf{Ax}}{\partial \mathbf{x}} = (\mathbf{A}+\mathbf{A}')\mathbf{x}, \quad \frac{\partial \mathbf{x}'\mathbf{Ay}}{\partial \mathbf{x}} = \mathbf{Ay}. \tag{A.9.2}$$

(III) $\dfrac{\partial |\mathbf{X}|}{\partial x_{ij}} = X_{ij}$ if all elements of $\mathbf{X}(n \times n)$ are distinct

$$= \begin{cases} X_{ii}, & i=j \\ 2X_{ij}, & i \neq j \end{cases} \text{ if } \mathbf{X} \text{ is symmetric,} \tag{A.9.3}$$

where X_{ij} is the (i, j)th cofactor of $\mathbf{X}$.

(IV) $\dfrac{\partial \operatorname{tr} \mathbf{XY}}{\partial \mathbf{X}} = \mathbf{Y}'$ if all elements of $\mathbf{X}(n \times p)$ are distinct,

$$= \mathbf{Y} + \mathbf{Y}' - \operatorname{Diag}(\mathbf{Y}) \text{ if } \mathbf{X}(n \times n) \text{ is symmetric.} \tag{A.9.4}$$

(V) $\dfrac{\partial \mathbf{X}^{-1}}{\partial x_{ij}} = -\mathbf{X}^{-1}\mathbf{J}_{ij}\mathbf{X}^{-1}$ if all elements of $\mathbf{X}(n \times n)$ are distinct

$$= \begin{cases} -\mathbf{X}^{-1}\mathbf{J}_{ii}\mathbf{X}^{-1}, & i=j, \\ -\mathbf{X}^{-1}(\mathbf{J}_{ij} + \mathbf{J}_{ji})\mathbf{X}^{-1}, & i \neq j \end{cases} \text{ if } \mathbf{X} \text{ is symmetric,} \tag{A.9.5}$$

where $\mathbf{J}_{ij}$ denotes a matrix with a 1 in the (i, j)th place and zeros elsewhere.

We now consider some applications of these results to some stationary value problems.

Theorem A.9.1 *The vector* $\mathbf{x}$ *which minimizes*

$$f(\mathbf{x}) = (\mathbf{y} - \mathbf{A}\mathbf{x})'(\mathbf{y} - \mathbf{A}\mathbf{x})$$

is given by

$$\mathbf{A}'\mathbf{A}\mathbf{x} = \mathbf{A}'\mathbf{y}. \tag{A.9.6}$$

Proof Differentiate $f(\mathbf{x})$ and set the derivative equal to $\mathbf{0}$. Note that the second derivative matrix $2\mathbf{A}'\mathbf{A} \geq 0$ so that the solution to (A.9.6) will give a minimum. Also note that from (A.4.3e), (A.9.6) is a consistent set of equations. ■

Theorem A.9.2 *Let* $\mathbf{A}$ *and* $\mathbf{B}$ *be two symmetric matrices. Suppose that* $\mathbf{B} > 0$. *Then the maximum (minimum) of* $\mathbf{x}'\mathbf{A}\mathbf{x}$ *given*

$$\mathbf{x}'\mathbf{B}\mathbf{x} = 1 \tag{A.9.7}$$

is attained when $\mathbf{x}$ *is the eigenvector of* $\mathbf{B}^{-1}\mathbf{A}$ *corresponding to the largest (smallest) eigenvalue of* $\mathbf{B}^{-1}\mathbf{A}$. *Thus if* λ_1 *and* λ_p *are the largest and smallest eigenvalues of* $\mathbf{B}^{-1}\mathbf{A}$, *then, subject to the constraint* (A.9.7),

$$\max_{\mathbf{x}} \mathbf{x}'\mathbf{A}\mathbf{x} = \lambda_1, \qquad \min_{\mathbf{x}} \mathbf{x}'\mathbf{A}\mathbf{x} = \lambda_p. \tag{A.9.8}$$

Proof Let $\mathbf{B}^{1/2}$ denote the symmetric square root of $\mathbf{B}$, and let $\mathbf{y} = \mathbf{B}^{1/2}\mathbf{x}$.

Then the maximum of $\mathbf{x}'\mathbf{A}\mathbf{x}$ subject to (A.9.7) can be written as

$$\max_{\mathbf{y}} \mathbf{y}'\mathbf{B}^{-1/2}\mathbf{A}\mathbf{B}^{-1/2}\mathbf{y} \quad \text{subject to} \quad \mathbf{y}'\mathbf{y}=1. \tag{A.9.9}$$

Let $\mathbf{B}^{-1/2}\mathbf{A}\mathbf{B}^{-1/2}=\mathbf{\Gamma}\mathbf{\Lambda}\mathbf{\Gamma}'$ be a spectral decomposition of the symmetric matrix $\mathbf{B}^{-1/2}\mathbf{A}\mathbf{B}^{-1/2}$. Let $\mathbf{z}=\mathbf{\Gamma}'\mathbf{y}$. Then $\mathbf{z}'\mathbf{z}=\mathbf{y}'\mathbf{\Gamma}\mathbf{\Gamma}'\mathbf{y}=\mathbf{y}'\mathbf{y}$ so that (A.9.9) can be written

$$\max_{\mathbf{z}} \mathbf{z}'\mathbf{\Lambda}\mathbf{z}=\max_{\mathbf{z}} \sum \lambda_i z_i^2 \quad \text{subject to} \quad \mathbf{z}'\mathbf{z}=1. \tag{A.9.10}$$

If the eigenvalues are written in descending order then (A.9.10) satisfies

$$\max \sum \lambda_i z_i^2 \leq \lambda_1 \max \sum z_i^2 = \lambda_1.$$

Further this bound is attained for $\mathbf{z}=(1, 0, \ldots, 0)'$, that is for $\mathbf{y}=\boldsymbol{\gamma}_{(1)}$, and for $\mathbf{x}=\mathbf{B}^{-1/2}\boldsymbol{\gamma}_{(1)}$. By Theorem A.6.2, $\mathbf{B}^{-1}\mathbf{A}$ and $\mathbf{B}^{-1/2}\mathbf{A}\mathbf{B}^{-1/2}$ have the same eigenvalues and $\mathbf{x}=\mathbf{B}^{-1/2}\boldsymbol{\gamma}_{(1)}$ is an eigenvector of $\mathbf{B}^{-1}\mathbf{A}$ corresponding to λ_1. Thus the theorem is proved for maximization.

The same technique can be applied to prove the minimization result. ■

Corollary A.9.2.1 *If $R(\mathbf{x})=\mathbf{x}'\mathbf{A}\mathbf{x}/\mathbf{x}'\mathbf{B}\mathbf{x}$ then, for $\mathbf{x}\neq\mathbf{0}$,*

$$\lambda_p \leq R(\mathbf{x}) \leq \lambda_1. \tag{A.9.11}$$

Proof Since $R(\mathbf{x})$ is invariant under changes of scale of $\mathbf{x}$, we can regard the problem as maximizing (minimizing) $\mathbf{x}'\mathbf{A}\mathbf{x}$ given (A.9.7). ■

Corollary A.9.2.2 *The maximum of $\mathbf{a}'\mathbf{x}$ subject to* (A.9.7) *is*

$$(\mathbf{a}'\mathbf{B}^{-1}\mathbf{a})^{1/2}. \tag{A.9.12}$$

Further

$$\max_{\mathbf{x}} \{(\mathbf{a}'\mathbf{x})^2/(\mathbf{x}'\mathbf{B}\mathbf{x})\}=\mathbf{a}'\mathbf{B}^{-1}\mathbf{a} \tag{A.9.13}$$

and the maximum is attained at $\mathbf{x}=\mathbf{B}^{-1}\mathbf{a}/(\mathbf{a}'\mathbf{B}^{-1}\mathbf{a})^{1/2}$.

Proof Apply Theorem A.9.2 with $\mathbf{x}'\mathbf{A}\mathbf{x}=(\mathbf{a}'\mathbf{x})^2=\mathbf{x}'(\mathbf{a}\mathbf{a}')\mathbf{x}$. ■

Remarks (1) A direct method is sometimes instructive. Consider the problem of maximizing the squared distance from the origin

$$x^2+y^2$$

of a point (x, y) on the ellipse

$$\frac{x^2}{a^2}+\frac{y^2}{b^2}=1. \tag{A.9.14}$$

When y^2 is eliminated, the problem reduces to finding the maximum of

$$x^2+b^2(x^2/a^2-1), \qquad x\in[-a, a].$$

Setting the derivative equal to 0 yields the stationary point $x=0$ which, from (A.9.14), gives $y=\pm b$. Also, at the endpoints of the interval ($x=\pm a$), we get $y=0$. Hence

$$\max(x^2+y^2)=\max(a^2, b^2).$$

This solution is not as elegant as the proof of Theorem A.9.2, and does not generalize neatly to more complicated quadratic forms.

(2) The results (A.9.1)–(A.9.2) follow by direct substitution, e.g.

$$\frac{\partial}{\partial x_1}\mathbf{a}'\mathbf{x}=\frac{\partial}{\partial x_1}(a_1x_1+\ldots+a_px_p)=a_1$$

proves (A.9.1). For (A.9.3) use (A.2.3d).

A.10 Geometrical Ideas

A.10.1 *n*-dimensional geometry

Let $\mathbf{e}_i$ denote the vector in R^n with 1 in the ith place and zeros elsewhere so that $(\mathbf{e}_1, \ldots, \mathbf{e}_n)$ forms an orthonormal basis of R^n. In terms of this basis, vectors $\mathbf{x}$ can be represented as $\mathbf{x}=\sum x_i\mathbf{e}_i$, and x_i is called the ith *coordinate axis.* A point $\mathbf{a}$ in R^n is represented in terms of these coordinates by $x_1=a_1, \ldots, x_n=a_n$. The point $\mathbf{a}$ can also be interpreted as a directed line segment from $\mathbf{0}$ to $\mathbf{a}$. Some generalizations of various basic concepts of two- and three-dimensional analytic Euclidean geometry are summarized in Table A.10.1.

A.10.2 Orthogonal transformations

Let $\mathbf{\Gamma}$ be an orthogonal matrix. Then $\mathbf{\Gamma e}_i=\boldsymbol{\gamma}_{(i)}$, $i=1,\ldots, n$, also form an orthonormal basis and points $\mathbf{x}$ can be represented in terms of this new basis as

$$\mathbf{x}=\sum x_i\mathbf{e}_i=\sum y_i\boldsymbol{\gamma}_{(i)},$$

where $y_i=\boldsymbol{\gamma}'_{(i)}\mathbf{x}$ are new coordinates. If $\mathbf{x}^{(1)}$ and $\mathbf{x}^{(2)}$ are two points with new coordinates $\mathbf{y}^{(1)}$ and $\mathbf{y}^{(2)}$ note that

$$\begin{aligned}(\mathbf{y}^{(1)}-\mathbf{y}^{(2)})'(\mathbf{y}^{(1)}-\mathbf{y}^{(2)})&=(\mathbf{x}^{(1)}-\mathbf{x}^{(2)})'\mathbf{\Gamma\Gamma}'(\mathbf{x}^{(1)}-\mathbf{x}^{(2)})\\&=(\mathbf{x}^{(1)}-\mathbf{x}^{(2)})'(\mathbf{x}^{(1)}-\mathbf{x}^{(2)}),\end{aligned}$$

Table A.10.1 Basic concepts in n-dimensional geometry

Concept	Description $\left(\|\mathbf{x}\| = \left(\sum x_i^2\right)^{1/2}\right)$
Point $\mathbf{a}$	$x_1 = a_1, \ldots, x_n = a_n$
Distance between $\mathbf{a}$ and $\mathbf{b}$	$\|\mathbf{a}-\mathbf{b}\| = \left\{\sum (a_i - b_i)^2\right\}^{1/2}$
Line passing through $\mathbf{a}$, $\mathbf{b}$	$\mathbf{x} = \lambda\mathbf{a} + (1-\lambda)\mathbf{b}$ is the equation
Line passing through $\mathbf{0}$, $\mathbf{a}$	$\mathbf{x} = \lambda\mathbf{a}$
Angle between lines from $\mathbf{0}$ to $\mathbf{a}$ and $\mathbf{0}$ to $\mathbf{b}$	θ where $\cos\theta = \mathbf{a}'\mathbf{b}/\{\|\mathbf{a}\|\,\|\mathbf{b}\|\}^{1/2}$, $0 \leq \theta \leq \pi$
Direction cosine vector of a line from $\mathbf{0}$ to $\mathbf{a}$	$(\cos\gamma_1, \ldots, \cos\gamma_n)$, $\cos\gamma_i = a_i/\|\mathbf{a}\|$; γ_i = angle between line and ith axis
Plane P	$\mathbf{a}'\mathbf{x} = c$ is general equation
Plane through $\mathbf{b}_1, \ldots, \mathbf{b}_k$	$\mathbf{x} = \sum \lambda_i \mathbf{b}_i$, $\sum \lambda_i = 1$
Plane through $\mathbf{0}, \mathbf{b}_1, \ldots, \mathbf{b}_k$	$\mathbf{x} = \sum \lambda_i \mathbf{b}_i$
Hypersphere with centre $\mathbf{a}$ and radius r	$(\mathbf{x}-\mathbf{a})'(\mathbf{x}-\mathbf{a}) = r^2$
Ellipsoid	$(\mathbf{x}-\mathbf{a})'\mathbf{A}^{-1}(\mathbf{x}-\mathbf{a}) = c^2$, $\mathbf{A} > 0$

so that orthogonal transformations preserve distances. An orthogonal transformation represents a rotation of the coordinate axes (plus a reflection if $|\mathbf{\Gamma}| = -1$). When $n = 2$ and $|\mathbf{\Gamma}| = 1$, $\mathbf{\Gamma}$ can be represented as

$$\begin{pmatrix} \cos\theta & -\sin\theta \\ \sin\theta & \cos\theta \end{pmatrix}$$

and represents a rotation of the coordinate axes counterclockwise through an angle θ.

A.10.3 Projections

Consider a point $\mathbf{a}$, in n dimensions (see Figure A.10.1). Its projection onto a plane P (or onto a line) through the origin is the point $\hat{\mathbf{a}}$ at the foot of the perpendicular from $\mathbf{a}$ to P. The vector $\hat{\mathbf{a}}$ is called the *orthogonal projection* of the vector $\mathbf{a}$ onto the plane.

Let the plane P pass through points $\mathbf{0}, \mathbf{b}_1, \ldots, \mathbf{b}_k$ so that its equation from Table A.10.1 is

$$\mathbf{x} = \sum \lambda_i \mathbf{b}_i, \qquad \mathbf{B} = (\mathbf{b}_1, \ldots, \mathbf{b}_k).$$

Suppose rank $(\mathbf{B}) = k$ so that the plane is a k-dimensional subspace. The

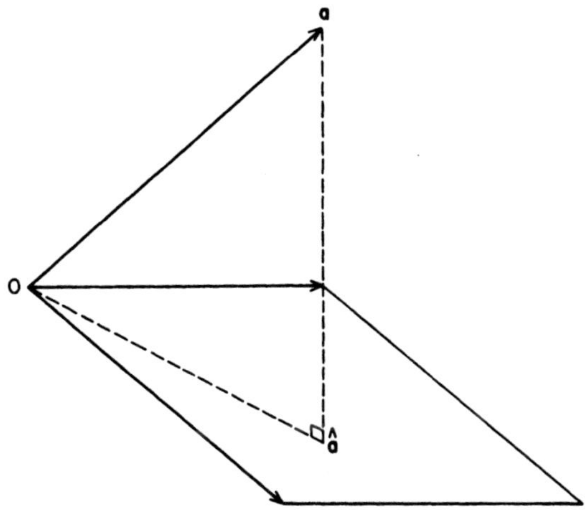

Figure A.10.1 *$\hat{\mathbf{a}}$ is the projection of $\mathbf{a}$ onto the plane P.*

point $\hat{\mathbf{a}}$ is defined by $\mathbf{x}=\sum \hat{\lambda}_i \mathbf{b}_i$, where $\hat{\lambda}_1, \ldots, \hat{\lambda}_k$ minimize

$$\left\| \mathbf{a} - \sum \lambda_i \mathbf{b}_i \right\|$$

since $\hat{\mathbf{a}}$ is the point on the plane closest to $\mathbf{a}$. Using Theorem A.9.1, we deduce the following result.

Theorem A.10.1 *The point $\hat{\mathbf{a}}$ is given by*

$$\hat{\mathbf{a}} = \mathbf{B}(\mathbf{B}'\mathbf{B})^{-1}\mathbf{B}'\mathbf{a}. \quad \blacksquare \tag{A.10.3a}$$

Note that $\mathbf{B}(\mathbf{B}'\mathbf{B})^{-1}\mathbf{B}'$ is a symmetric idempotent matrix. In fact, any symmetric idempotent matrix can be used to represent a projection.

A.10.4 Ellipsoids

Let $\mathbf{A}$ be a p.d. matrix. Then

$$(\mathbf{x}-\boldsymbol{\alpha})'\mathbf{A}^{-1}(\mathbf{x}-\boldsymbol{\alpha}) = c^2 \tag{A.10.4a}$$

represents an ellipsoid in n dimensions. We note that the centre of the ellipsoid is at $\mathbf{x}=\boldsymbol{\alpha}$. On shifting the centre to $\mathbf{x}=\mathbf{0}$, the equation becomes

$$\mathbf{x}'\mathbf{A}^{-1}\mathbf{x} = c^2. \tag{A.10.4b}$$

Definition *Let $\mathbf{x}$ be a point on the ellipsoid defined by* (A.10.4a) *and let $f(\mathbf{x}) = \|\mathbf{x}-\boldsymbol{\alpha}\|^2$ denote the squared distance between $\boldsymbol{\alpha}$ and $\mathbf{x}$. A line through $\boldsymbol{\alpha}$ and $\mathbf{x}$ for which $\mathbf{x}$ is a stationary point of $f(\mathbf{x})$ is called a* principal axis *of* *the ellipsoid. The distance* $\|\mathbf{x}-\boldsymbol{\alpha}\|$ *is called the length of the principal semi-axis.*

Theorem A.10.2 *Let* $\lambda_1, \ldots, \lambda_n$ *be the eigenvalues of* $\mathbf{A}$ *satisfying* $\lambda_1 > \lambda_2 > \ldots > \lambda_n$. *Suppose that* $\boldsymbol{\gamma}_{(1)}, \ldots, \boldsymbol{\gamma}_{(n)}$ *are the corresponding eigenvectors. For the ellipsoids* (A.10.4a) *and* (A.10.4b), *we have*

(1) *The direction cosine vector of the ith principal axis is* $\boldsymbol{\gamma}_{(i)}$.
(2) *The length of the ith principal semi-axis is* $c\lambda_i^{1/2}$.

Proof It is sufficient to prove the result for (A.10.4b). The problem reduces to finding the stationary points of $f(\mathbf{x})=\mathbf{x}'\mathbf{x}$ subject to $\mathbf{x}$ lying on the ellipsoid $\mathbf{x}'\mathbf{A}^{-1}\mathbf{x}=c^2$. The derivative of $\mathbf{x}'\mathbf{A}^{-1}\mathbf{x}$ is $2\mathbf{A}^{-1}\mathbf{x}$. Thus a point $\mathbf{y}$ represents a direction tangent to the ellipsoid at $\mathbf{x}$ if $2\mathbf{y}'\mathbf{A}^{-1}\mathbf{x}=0$.

The derivative of $f(\mathbf{x})$ is $2\mathbf{x}$ so the directional derivative of $f(\mathbf{x})$ in the direction $\mathbf{y}$ is $2\mathbf{y}'\mathbf{x}$. Then $\mathbf{x}$ is a stationary point if and only if for all points $\mathbf{y}$ representing tangent directions to the ellipsoid at $\mathbf{x}$, we have $2\mathbf{y}'\mathbf{x}=0$; that is if

$$\mathbf{y}'\mathbf{A}^{-1}\mathbf{x}=0 \Rightarrow \mathbf{y}'\mathbf{x}=0.$$

This condition is satisfied if and only if $\mathbf{A}^{-1}\mathbf{x}$ is proportional to $\mathbf{x}$; that is if and only if $\mathbf{x}$ is an eigenvector of $\mathbf{A}^{-1}$.

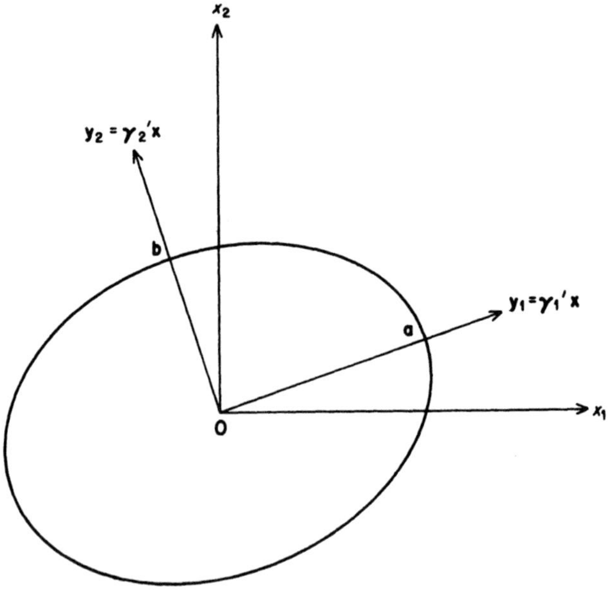

Figure A.10.2 Ellipsoid $\mathbf{x}'\mathbf{A}^{-1}\mathbf{x}=1$. *Lines defined by* $\mathbf{y}_1$ *and* $\mathbf{y}_2$ *are the first and second principal axes,* $\|\mathbf{a}\|=\lambda_1^{1/2}$, $\|\mathbf{b}\|=\lambda_2^{1/2}$.

Setting $\mathbf{x} = \beta \boldsymbol{\gamma}_{(i)}$ in (A.10.4b) gives $\beta^2/\lambda_i = c^2$, so $\beta = c_i\lambda_i^{1/2}$. Thus, the theorem is proved. ■

If we rotate the coordinate axes with the transformation $\mathbf{y} = \boldsymbol{\Gamma}'\mathbf{x}$, we find that (A.10.4b) reduces to

$$\sum y_i^2/\lambda_i = c^2.$$

Figure A.10.2 gives a pictorial representation.

With $\mathbf{A} = \mathbf{I}$, (A.10.4b) reduces to a hypersphere with $\lambda_1 = \ldots = \lambda_n = 1$ so that the λs are not distinct and the above theorem fails; that is, the position of $\boldsymbol{\gamma}_{(i)}$, $i = 1, \ldots, n$, through the sphere is not unique and any rotation will suffice; that is, all the n components are isometric.

In general, if $\lambda_i = \lambda_{i+1}$, the section of the ellipsoid is *circular* in the plane generated by $\boldsymbol{\gamma}_{(i)}$, $\boldsymbol{\gamma}_{(i+1)}$. Although we can construct two perpendicular axes for the common root, their position through the circle is not unique. If $\mathbf{A}$ equals the equicorrelation matrix, there are $p-1$ isotropic principal axes corresponding to the last $p-1$ eigenvalues.

Appendix B
Univariate Statistics

B.1 Introduction

In this appendix, we summarize the univariate distributions needed in the text, together with some of their simple properties. Note that we do not distinguish notationally between a random variable x and its p.d.f. $f(x)$.

B.2 Normal Distribution

If a random variable x has p.d.f.

$$f(x) = (2\pi)^{-1/2} e^{-x^2/2}, \qquad -\infty < x < \infty, \tag{B.2.1}$$

then x is said to have a standardized *normal distribution*, and we write $x \sim N(0, 1)$. Its characteristic function (c.f.) is given by

$$\phi(t) = E(e^{itx}) = e^{-t^2/2}. \tag{B.2.2}$$

Using the formula

$$\mu_r = E(x^r) = (-i)^r \frac{d\phi(t)}{dt}\Big|_{t=0}$$

for the rth moment of x, where r is a positive integer, it is found that

$$\mu_1 = \mu_3 = 0, \qquad \mu_2 = 1, \qquad \mu_4 = 3. \tag{B.2.3}$$

The p.d.f. of $N(\mu, \sigma^2)$ is given by

$$(2\pi\sigma^2)^{-1/2} \exp\{-\tfrac{1}{2}(x-\mu)^2/\sigma^2\}, \qquad -\infty < x < \infty,$$

where $\sigma^2 > 0$, $-\infty < \mu < \infty$. Its characteristic function is given by

$$\exp(i\mu t - \tfrac{1}{2}t^2\sigma^2).$$

B.3 Chi-squared Distribution

Let $x_1, \ldots, x_p$ be independent $N(0, 1)$ variables and set

$$y = x_1^2 + \cdots + x_p^2. \tag{B.3.1}$$

Then y is said to have a *chi-squared distribution* with p degrees of freedom and we write $y \sim \chi_p^2$.

The density of y is given by

$$f(y) = \{2^{p/2}\Gamma(\tfrac{1}{2}p)\}^{-1} y^{p/2-1} e^{-y/2}, \qquad 0 < y < \infty. \tag{B.3.2}$$

Here $\Gamma(x)$ is the gamma function defined by

$$\Gamma(x) = \int_0^\infty x^{u-1} e^{-u}\, du, \qquad x > 0, \tag{B.3.3}$$

which satisfies the relations $\Gamma(x+1) = x\Gamma(x)$ for $x > 0$ and $\Gamma(n+1) = n!$ for integer $n \geq 0$.

Note that (B.3.2) is a density for any real $p > 0$; p need not be an integer. Using (B.3.3), the moments of y are easily calculated as

$$E(y^r) = 2^r\Gamma(\tfrac{1}{2}p + r)/\Gamma(\tfrac{1}{2}p), \tag{B.3.4}$$

for any real number r for which $r > -\tfrac{1}{2}p$. Using $r = 1, 2$, we get

$$E(y) = p, \qquad V(y) = 2p. \tag{B.3.5}$$

Also, if $p > 2$, taking $r = -1$ in (B.3.4), we get

$$E(1/y) = 1/(p-2). \tag{B.3.6}$$

B.4 *F* and Beta Variables

Let $u \sim \chi_p^2$ independently of $v \sim \chi_q^2$, and set

$$x = \frac{u/p}{v/q}. \tag{B.4.1}$$

Then x is said to have an *F distribution* with degrees of freedom p and q. We write $x \sim F_{p,q}$.

Using the same notation, set

$$y = u/(u+v). \tag{B.4.2}$$

Then y is said to have a *beta distribution* (sometimes called the *beta type I distribution*) with parameters $\tfrac{1}{2}p$ and $\tfrac{1}{2}q$. We write $y \sim B(\tfrac{1}{2}p, \tfrac{1}{2}q)$. Its density is given by

$$f(y) = [\Gamma(\tfrac{1}{2}p + \tfrac{1}{2}q)/\{\Gamma(\tfrac{1}{2}p)\Gamma(\tfrac{1}{2}q)\}]y^{p/2-1}(1-y)^{q/2-1}, \qquad 0 < y < 1. \tag{B.4.3}$$

The *beta type II distribution* is proportional to the F distribution and is defined in the above notation by

$$w = px/q = u/v.$$

The F and beta type I distributions are related by the one-to-one transformation

$$y = px/(q + px), \tag{B.4.4}$$

or, equivalently,

$$x = qy/\{p(1-y)\}. \tag{B.4.5}$$

The following theorem shows an important relationship between the beta and chi-squared distributions.

Theorem B.4.1 *Let $u \sim \chi^2_p$ independently of $v \sim \chi^2_q$. Then $z = u + v \sim \chi^2_{p+q}$ independently of $y = u/(u+v) \sim B(\tfrac{1}{2}p, \tfrac{1}{2}q)$.*

Proof Transform the variables (u, v) to (z, y) where $u = zy$ and $v = z(1-y)$. The p.d.f. of (u, v) is known from (B.3.2), and simplifying yields the desired result. ■

B.5 t Distribution

Let $x \sim N(0, 1)$ independently of $u \sim \chi^2_p$. Then

$$t = x/(u/p)^{1/2} \tag{B.5.1}$$

is said to have a *t distribution* with p degrees of freedom. We write $t \sim t_p$. Its p.d.f. is given by

$$[\Gamma(\tfrac{1}{2}p + \tfrac{1}{2})/\{(p\pi)^{1/2}\Gamma(\tfrac{1}{2}p)\}](1 + t^2/p)^{-(p+1)/2}, \qquad -\infty < t < \infty.$$

From (B.3.1) and (B.4.1), note that if $t \sim t_p$, then

$$t^2 \sim F_{1,p}. \tag{B.5.2}$$

The mean and variance of the t distribution can be calculated from (B.2.3) and (B.3.6) to be

$$E(t) = 0, \qquad V(t) = p/(p-2), \tag{B.5.3}$$

provided $p > 1$ and $p > 2$, respectively.

Appendix C
Tables

Table C.1 Upper percentage points of the χ^2_ν-distribution†

ν	$1-\alpha$: 0.90	0.95	0.975	0.99	0.995	0.999
1	2.71	3.84	5.02	6.63	7.88	10.83
2	4.61	5.99	7.38	9.21	10.60	13.81
3	6.25	7.81	9.35	11.34	12.84	16.27
4	7.78	9.49	11.14	13.28	14.86	18.47
5	9.24	11.07	12.83	15.09	16.75	20.52
6	10.64	12.59	14.45	16.81	18.55	22.46
7	12.02	14.07	16.01	18.48	20.28	24.32
8	13.36	15.51	17.53	20.09	21.95	26.12
9	14.68	16.92	19.02	21.67	23.59	27.88
10	15.99	18.31	20.48	23.21	25.19	29.59
11	17.28	19.68	21.92	24.73	26.76	31.26
12	18.55	21.03	23.34	26.22	28.30	32.91
13	19.81	22.36	24.74	27.69	29.82	34.53
14	21.06	23.68	26.12	29.14	31.32	36.12
15	22.31	25.00	27.49	30.58	32.80	37.70
16	23.54	26.30	28.85	32.00	34.27	39.25
17	24.77	27.59	30.19	33.41	35.72	40.79
18	25.99	28.87	31.53	34.81	37.16	42.31
19	27.20	30.14	32.85	36.19	38.58	43.82
20	28.41	31.41	34.17	35.57	40.00	45.31
21	29.62	32.67	35.48	38.93	41.40	46.80
22	30.81	33.92	36.78	40.29	42.80	48.27
23	32.01	35.17	38.08	41.64	44.18	49.73
24	33.20	36.42	39.36	42.98	45.56	51.18
25	34.38	37.65	40.65	44.31	46.93	52.62
26	35.56	38.89	41.92	45.64	48.29	54.05
27	36.74	40.11	43.19	46.96	49.64	55.48
28	37.92	41.34	44.46	48.28	50.99	56.89
29	39.09	42.56	45.72	49.59	52.34	58.30
30	40.26	43.77	46.98	50.89	53.67	59.70
40	51.81	55.76	59.34	63.69	66.77	73.40
50	63.17	67.50	71.42	76.15	79.49	86.66
60	74.40	79.08	83.30	88.38	91.95	99.61
70	85.53	90.53	95.02	100.4	104.2	112.3
80	96.58	101.9	106.6	112.3	116.3	124.8
90	107.6	113.1	118.1	124.1	128.3	137.2
100	118.5	124.3	129.6	135.8	140.2	149.4

For $\nu > 100$, $\sqrt{2\chi^2_\nu} \sim N(\sqrt{2\nu - 1}, 1)$

† Abridged from Catherine M. Thompson: Tables of percentage points of the χ^2 distribution, *Biometrika*, vol. 32 (1941), pp. 187–191, and published here with the kind permission of the author and the editor of *Biometrika*.

Table C.2 Upper percentage points of the t_ν distribution†

ν	$1-\alpha$: 0.90	0.95	0.975	0.99	0.995
1	3.0777	6.3138	12.706	31.821	63.657
2	1.8856	2.9200	4.3027	6.9646	9.9248
3	1.6377	2.3534	3.1824	4.5407	5.8409
4	1.5332	2.1318	2.7764	3.7469	4.6041
5	1.4759	2.0150	2.5706	3.3649	4.0321
6	1.4398	1.9432	2.4469	3.1427	3.7074
7	1.4149	1.8946	2.3646	2.9980	3.4995
8	1.3968	1.8595	2.3060	2.8965	3.3554
9	1.3830	1.8331	2.2622	2.8214	3.2498
10	1.3722	1.8125	2.2281	2.7638	3.1693
11	1.3634	1.7959	2.2010	2.7181	3.1058
12	1.3562	1.7823	2.1788	2.6810	3.0545
13	1.3502	1.7709	2.1604	2.6503	3.0123
14	1.3450	1.7613	2.1448	2.6245	2.9768
15	1.3406	1.7531	2.1314	2.6025	2.9467
16	1.3368	1.7459	2.1199	2.5835	2.9208
17	1.3334	1.7396	2.1098	2.5669	2.8982
18	1.3304	1.7341	2.1009	2.5524	2.8784
19	1.3277	1.7291	2.0930	2.5395	2.8609
20	1.3253	1.7247	2.0860	2.5280	2.8453
21	1.3232	1.7207	2.0796	2.5176	2.8314
22	1.3212	1.7171	2.0739	2.5083	2.8188
23	1.3195	1.7139	2.0687	2.4999	2.8073
24	1.3178	1.7109	2.0639	2.4922	2.7969
25	1.3163	1.7081	2.0595	2.4851	2.7874
26	1.3150	1.7056	2.0555	2.4786	2.7787
27	1.3137	1.7033	2.0518	2.4727	2.7707
28	1.3125	1.7011	2.0484	2.4671	2.7633
29	1.3114	1.6991	2.0452	2.4620	2.7564
30	1.3104	1.6973	2.0423	2.4573	2.7500
31	1.3095	1.6955	2.0395	2.4528	2.7440
32	1.3086	1.6939	2.0369	2.4487	2.7385
33	1.3077	1.6924	2.0345	2.4448	2.7333
34	1.3070	1.6909	2.0322	2.4411	2.7284
35	1.3062	1.6896	2.0301	2.4377	2.7238
36	1.3055	1.6883	2.0281	2.4345	2.7195
37	1.3049	1.6871	2.0262	2.4314	2.7154
38	1.3042	1.6860	2.0244	2.4286	2.7116
39	1.3036	1.6849	2.0227	2.4258	2.7079
40	1.3031	1.6839	2.0211	2.4233	2.7045
60	1.2958	1.6706	2.0003	2.3901	2.6603
120	1.2886	1.6577	1.9799	2.3578	2.6174
∞	1.2816	1.6449	1.9600	2.3263	2.5758

As $\nu \to \infty$, t_ν tends to $N(0, 1)$. Further, $t_\nu^2 = F_{1,\nu}$.

† Taken from Table 2 of K. V. Mardia and P. J. Zemroch (1978).

Table C.3 Upper percentage points of the F_{ν_1,ν_2} distribution†

$1-\alpha$	ν_2	ν_1 = 1	2	3	4	5	6	7	8	9	10	12	15	20	30	60	120	∞
0.90		39.9	49.5	53.6	55.8	57.2	58.2	58.9	59.4	59.9	60.2	60.7	61.2	61.7	62.3	62.8	63.1	63.3
0.95		161	200	216	225	230	234	237	239	241	242	244	246	248	250	252	253	254
0.975	1	648	800	864	900	922	937	948	957	963	969	977	985	993	1,000	1,010	1,010	1,020
0.99		4,050	5,000	5,400	5,620	5,760	5,860	5,930	5,980	6,020	6.060	6,110	6,160	6,210	6,260	6,310	6,340	6,370
0.90		8.53	9.00	9.16	9.24	9.29	9.33	9.35	9.37	9.38	9.39	9.41	9.42	9.44	9.46	9.47	9.48	9.49
0.95		18.5	19.0	19.2	19.2	19.3	19.3	19.4	19.4	19.4	19.4	19.4	19.4	19.5	19.5	19.5	19.5	19.5
0.975	2	38.5	39.0	39.2	39.2	39.3	39.3	39.4	39.4	39.4	39.4	39.4	39.4	39.4	39.5	39.5	39.5	39.5
0.99		98.5	99.0	99.2	99.2	99.3	99.3	99.4	99.4	99.4	99.4	99.4	99.4	99.4	99.5	99.5	99.5	99.5
0.90		5.54	5.46	5.39	5.34	5.31	5.28	5.27	5.25	5.24	5.23	5.22	5.20	5.18	5.17	5.15	5.14	5.13
0.95		10.1	9.55	9.28	9.12	9.01	8.94	8.89	8.85	8.81	8.79	8.74	8.70	8.66	8.62	8.57	8.55	8.53
0.975	3	17.4	16.0	15.4	15.1	14.9	14.7	14.6	14.5	14.5	14.4	14.3	14.3	14.2	14.1	14.0	13.9	13.9
0.99		34.1	30.8	29.5	28.7	28.2	27.9	27.7	27.5	27.3	27.2	27.1	26.9	26.7	26.5	26.3	26.2	26.1
0.90		4.54	4.32	4.19	4.11	4.05	4.01	3.98	3.95	3.93	3.92	3.90	3.87	3.84	3.82	3.79	3.78	3.76
0.95		7.71	6.94	6.59	6.39	6.26	6.16	6.09	6.04	6.00	5.96	5.91	5.86	5.80	5.75	5.69	5.66	5.63
0.975	4	12.2	10.6	9.98	9.60	9.36	9.20	9.07	8.98	8.90	8.84	8.75	8.66	8.56	8.46	8.36	8.31	8.26
0.99		21.2	18.0	16.7	16.0	15.5	15.2	15.0	14.8	14.7	14.5	14.4	14.2	14.0	13.8	13.7	13.6	13.5
0.90		4.06	3.78	3.62	3.52	3.45	3.40	3.37	3.34	3.32	3.30	3.27	3.24	3.21	3.17	3.14	3.12	3.11
0.95		6.61	5.79	5.41	5.19	5.05	4.95	4.88	1.82	4.77	4.74	4.68	4.62	4.56	4.50	4.43	4.40	4.37
0.975	5	10.0	8.43	7.76	7.39	7.15	6.98	6.85	6.76	6.68	6.62	6.52	6.43	6.33	6.23	6.12	6.07	6.02
0.99		16.3	13.3	12.1	11.4	11.0	10.7	10.5	10.3	10.2	10.1	9.89	9.72	9.55	9.38	9.20	9.11	9.02
0.90		3.78	3.46	3.29	3.18	3.11	3.05	3.01	2.98	2.96	2.94	2.90	2.87	2.84	2.80	2.76	2.74	2.72
0.95		5.99	5.14	4.76	4.53	4.39	4.28	4.21	4.15	4.10	4.06	4.00	3.94	3.87	3.81	3.74	3.70	3.67
0.975	6	8.81	7.26	6.60	6.23	5.99	5.82	5.70	5.60	5.52	5.46	5.37	5.27	5.17	5.07	4.96	4.90	4.85
0.99		13.7	10.9	9.78	9.15	8.75	8.47	8.26	8.10	7.98	7.87	7.72	7.56	7.40	7.23	7.06	6.97	6.88
0.90		3.59	3.26	3.07	2.96	2.88	2.83	2.78	2.75	2.72	2.70	2.67	2.63	2.59	2.56	2.51	2.49	2.47
0.95		5.59	4.74	4.35	4.12	3.97	3.87	3.79	3.73	3.68	3.64	3.57	3.51	3.44	3.38	3.30	3.27	3.23
0.975	7	8.07	6.54	5.89	5.52	5.29	5.12	4.99	4.90	4.82	4.76	4.67	4.57	4.47	4.36	4.25	4.20	4.14
0.99		12.2	9.55	8.45	7.85	7.46	7.19	6.99	6.84	6.72	6.62	6.47	6.31	6.16	5.99	5.82	5.74	5.65
0.90		3.46	3.11	2.92	2.81	2.73	2.67	2.62	2.59	2.56	2.54	2.50	2.46	2.42	2.38	2.34	2.31	2.29
0.95		5.32	4.46	4.07	3.84	3.69	3.58	3.50	3.44	3.39	3.35	3.28	3.22	3.15	3.08	3.01	2.97	2.93
0.975	8	7.57	6.06	5.42	5.05	4.82	4.65	4.53	4.43	4.36	4.30	4.20	4.10	4.00	3.89	3.78	3.73	3.67
0.99		11.3	8.65	7.59	7.01	6.63	6.37	6.18	6.03	5.91	5.81	5.67	5.52	5.36	5.20	5.03	4.95	4.86

0.90		3.36	3.01	2.81	2.69	2.61	2.55	2.51	2.47	2.44	2.42	2.38	2.34	2.30	2.25	2.21	2.18	2.16
0.95		5.12	4.26	3.86	3.63	3.48	3.37	3.29	3.23	3.18	3.14	3.07	3.01	2.94	2.86	2.79	2.75	2.71
0.975	9	7.21	5.71	5.08	4.72	4.48	4.32	4.20	4.10	4.03	3.96	3.87	3.77	3.67	3.56	3.45	3.39	3.33
0.99		10.6	8.02	6.99	6.42	6.06	5.80	5.61	5.47	5.35	5.26	5.11	4.96	4.81	4.65	4.48	4.40	4.31
0.90		3.29	2.92	2.73	2.61	2.52	2.46	2.41	2.38	2.35	2.32	2.28	2.24	2.20	2.15	2.11	2.08	2.06
0.95		4.96	4.10	3.71	3.48	3.33	3.22	3.14	3.07	3.02	2.98	2.91	2.84	2.77	2.70	2.62	2.58	2.54
0.975	10	6.94	5.46	4.83	4.47	4.24	4.07	3.95	3.85	3.78	3.72	3.62	3.52	3.42	3.31	3.20	3.14	3.08
0.99		10.0	7.56	6.55	5.99	5.64	5.39	5.20	5.06	4.94	4.85	4.71	4.56	4.41	4.25	4.08	4.00	3.91
0.90		3.18	2.81	2.61	2.48	2.39	2.33	2.28	2.24	2.21	2.19	2.15	2.10	2.06	2.01	1.96	1.93	1.90
0.95		4.75	3.89	3.49	3.26	3.11	3.00	2.91	2.85	2.80	2.75	2.69	2.62	2.54	2.47	2.38	2.34	2.30
0.975	12	6.55	5.10	4.47	4.12	3.89	3.73	3.61	3.51	3.44	3.37	3.28	3.18	3.07	2.96	2.85	2.79	2.72
0.99		9.33	6.93	5.95	5.41	5.06	4.82	4.64	4.50	4.39	4.30	4.16	4.01	3.86	3.70	3.54	3.45	3.36
0.90		3.07	2.70	2.49	2.36	2.27	2.21	2.16	2.12	2.09	2.06	2.02	1.97	1.92	1.87	1.82	1.79	1.76
0.95		4.54	3.68	3.29	3.06	2.90	2.79	2.71	2.64	2.59	2.54	2.48	2.40	2.33	2.25	2.16	2.11	2.07
0.975	15	6.20	4.77	4.15	3.80	3.58	3.41	3.29	3.20	3.12	3.06	2.96	2.86	2.76	2.64	2.52	2.46	2.40
0.99		8.68	6.36	5.42	4.89	4.56	4.32	4.14	4.00	3.89	3.80	3.67	3.52	3.37	3.21	3.05	2.96	2.87
0.90		2.97	2.59	2.38	2.25	2.16	2.09	2.04	2.00	1.96	1.94	1.89	1.84	1.79	1.74	1.68	1.64	1.61
0.95		4.35	3.49	3.10	2.87	2.71	2.60	2.51	2.45	2.39	2.35	2.28	2.20	2.12	2.04	1.95	1.90	1.84
0.975	20	5.87	4.46	3.86	3.51	3.29	3.13	3.01	2.91	2.84	2.77	2.68	2.57	2.46	2.35	2.22	2.16	2.09
0.99		8.10	5.85	4.94	4.43	4.10	3.87	3.70	3.56	3.46	3.37	3.23	3.09	2.94	2.78	2.61	2.52	2.42
0.90		2.88	2.49	2.28	2.14	2.05	1.98	1.93	1.88	1.85	1.82	1.77	1.72	1.67	1.61	1.54	1.50	1.46
0.95		4.17	3.32	2.92	2.69	2.53	2.42	2.33	2.27	2.21	2.16	2.09	2.01	1.93	1.84	1.74	1.68	1.62
0.975	30	5.57	4.18	3.59	3.25	3.03	2.87	2.75	2.65	2.57	2.51	2.41	2.31	2.20	2.07	1.94	1.87	1.79
0.99		7.56	5.39	4.51	4.02	3.70	3.47	3.30	3.17	3.07	2.98	2.84	2.70	2.55	2.39	2.21	2.11	2.01
0.90		2.79	2.39	2.18	2.04	1.95	1.87	1.82	1.77	1.74	1.71	1.66	1.60	1.54	1.48	1.40	1.35	1.29
0.95		4.00	3.15	2.76	2.53	2.37	2.25	2.17	2.10	2.04	1.99	1.92	1.84	1.75	1.65	1.53	1.47	1.39
0.975	60	5.29	3.93	3.34	3.01	2.79	2.63	2.51	2.41	2.33	2.27	2.17	2.06	1.94	1.82	1.67	1.58	1.48
0.99		7.08	4.98	4.13	3.65	3.34	3.12	2.95	2.82	2.72	2.63	2.50	2.35	2.20	2.03	1.84	1.73	1.60
0.90		2.75	2.35	2.13	1.99	1.90	1.82	1.77	1.72	1.68	1.65	1.60	1.54	1.48	1.41	1.32	1.26	1.19
0.95		3.92	3.07	2.68	2.45	2.29	2.18	2.09	2.02	1.96	1.91	1.83	1.75	1.66	1.55	1.43	1.35	1.25
0.975	120	5.15	3.80	3.23	2.89	2.67	2.52	2.39	2.30	2.22	2.16	2.05	1.94	1.82	1.69	1.53	1.43	1.31
0.99		6.85	4.79	3.95	3.48	3.17	2.96	2.79	2.66	2.56	2.47	2.34	2.19	2.03	1.86	1.66	1.53	1.38
0.90		2.71	2.30	2.08	1.94	1.85	1.77	1.72	1.67	1.63	1.60	1.55	1.49	1.42	1.34	1.24	1.17	1.00
0.95		3.84	3.00	2.60	2.37	2.21	2.10	2.01	1.94	1.88	1.83	1.75	1.67	1.57	1.46	1.32	1.22	1.00
0.975	∞	5.02	3.69	3.12	2.79	2.57	2.41	2.29	2.19	2.11	2.05	1.94	1.83	1.71	1.57	1.39	1.27	1.00
0.99		6.63	4.61	3.78	3.32	3.02	2.80	2.64	2.51	2.41	2.32	2.18	2.04	1.88	1.70	1.47	1.32	1.00

† Abridged from Merrington and Thompson (1943) and published here with the kind permission of the author and the editor of *Biometrika*.

Table C.4 Upper percentage points θ_α of $\theta(p, \nu_1, \nu_2)$, the largest eigenvalue of $|\mathbf{B}-\theta(\mathbf{W}+\mathbf{B})|=0$ for $p=2$.† ν_2 = hypothesis degrees of freedom; ν_1 = error degrees of freedom

		ν_2										
ν_1	$1-\alpha$	2	3	5	7	9	11	13	15	17	19	21
	0.90	0.7950	0.8463	0.8968	0.9221	0.9374	0.9476	0.9550	0.9605	0.9649	0.9683	0.9712
5	0.95	0.8577	0.8943	0.9296	0.9471	0.9576	0.9645	0.9696	0.9733	0.9763	0.9787	0.9806
	0.99	0.9377	0.9542	0.9698	0.9774	0.9819	0.9850	0.9872	0.9888	0.9900	0.9910	0.9918
	0.90	0.6628	0.7307	0.8058	0.8474	0.8741	0.8928	0.9066	0.9173	0.9257	0.9326	0.9383
7	0.95	0.7370	0.7919	0.8514	0.8839	0.9045	0.9189	0.9295	0.9376	0.9440	0.9493	0.9536
	0.99	0.8498	0.8826	0.9173	0.9358	0.9475	0.9556	0.9615	0.9660	0.9695	0.9724	0.9748
	0.90	0.5632	0.6366	0.7244	0.7768	0.8120	0.8374	0.8567	0.8720	0.8842	0.8943	0.9027
9	0.95	0.6383	0.7017	0.7761	0.8197	0.8487	0.8696	0.8853	0.8976	0.9076	0.9157	0.9225
	0.99	0.7635	0.8074	0.8575	0.8862	0.9051	0.9185	0.9286	0.9364	0.9427	0.9478	0.9521
	0.90	0.4880	0.5617	0.6551	0.7138	0.7548	0.7854	0.8089	0.8278	0.8433	0.8561	0.8670
11	0.95	0.5603	0.6267	0.7091	0.7600	0.7952	0.8212	0.8413	0.8573	0.8702	0.8810	0.8902
	0.99	0.6878	0.7381	0.7989	0.8357	0.8607	0.8790	0.8929	0.9039	0.9128	0.9202	0.9265
	0.90	0.4298	0.5016	0.5965	0.6587	0.7035	0.7375	0.7644	0.7862	0.8042	0.8194	0.8324
13	0.95	0.4981	0.5646	0.6507	0.7063	0.7459	0.7757	0.7992	0.8181	0.8337	0.8468	0.8580
	0.99	0.6233	0.6770	0.7446	0.7872	0.8171	0.8394	0.8568	0.8706	0.8821	0.8915	0.8997
	0.90	0.3837	0.4527	0.5468	0.6106	0.6577	0.6942	0.7235	0.7475	0.7675	0.7847	0.7993
15	0.95	0.4478	0.5130	0.6003	0.6584	0.7011	0.7338	0.7598	0.7810	0.7989	0.8138	0.8268
	0.99	0.5687	0.6237	0.6954	0.7422	0.7758	0.8013	0.8216	0.8378	0.8512	0.8629	0.8726
	0.90	0.3463	0.4122	0.5043	0.5685	0.6169	0.6550	0.6860	0.7116	0.7334	0.7519	0.7681
17	0.95	0.4065	0.4697	0.5564	0.6160	0.6605	0.6951	0.7232	0.7464	0.7659	0.7825	0.7969
	0.99	0.5222	0.5773	0.6512	0.7008	0.7373	0.7652	0.7875	0.8061	0.8212	0.8346	0.8458
	0.90	0.3155	0.3782	0.4677	0.5315	0.5805	0.6196	0.6517	0.6786	0.7016	0.7214	0.7387
19	0.95	0.3719	0.4327	0.5182	0.5782	0.6238	0.6599	0.6894	0.7139	0.7349	0.7528	0.7684
	0.99	0.4823	0.5369	0.6116	0.6630	0.7014	0.7313	0.7555	0.7756	0.7926	0.8071	0.8198

	0.90	0.2897	0.3493	0.4358	0.4988	0.5479	0.5875	0.6204	0.6482	0.6721	0.6929	0.7112
21	0.95	0.3427	0.4012	0.4847	0.5445	0.5906	0.6277	0.6581	0.6838	0.7058	0.7248	0.7415
	0.99	0.4479	0.5014	0.5762	0.6285	0.6685	0.6997	0.7254	0.7469	0.7652	0.7810	0.7946
	0.90	0.2677	0.3244	0.4080	0.4699	0.5185	0.5584	0.5918	0.6202	0.6448	0.6663	0.6853
23	0.95	0.3177	0.3737	0.4551	0.5143	0.5606	0.5981	0.6294	0.6558	0.6787	0.6986	0.7160
	0.99	0.4179	0.4701	0.5443	0.5971	0.6380	0.6703	0.6970	0.7197	0.7391	0.7558	0.7705
	0.90	0.2488	0.3027	0.3834	0.4439	0.4921	0.5319	0.5655	0.5944	0.6194	0.6415	0.6610
25	0.95	0.2960	0.3498	0.4287	0.4872	0.5333	0.5710	0.6027	0.6298	0.6533	0.6738	0.6920
	0.99	0.3915	0.4424	0.5155	0.5685	0.6096	0.6429	0.6708	0.6941	0.7143	0.7319	0.7474
	0.90	0.2324	0.2839	0.3616	0.4206	0.4682	0.5077	0.5413	0.5704	0.5958	0.6183	0.6383
27	0.95	0.2771	0.3286	0.4052	0.4626	0.5084	0.5462	0.5781	0.6056	0.6296	0.6506	0.6693
	0.99	0.3682	0.4176	0.4895	0.5422	0.5837	0.6175	0.6458	0.6700	0.6909	0.7092	0.7254
	0.90	0.2180	0.2672	0.3420	0.3996	0.4463	0.4855	0.5191	0.5482	0.5738	0.5966	0.6170
29	0.95	0.2604	0.3099	0.3840	0.4404	0.4855	0.5232	0.5553	0.5830	0.6074	0.6288	0.6480
	0.99	0.3475	0.3954	0.4659	0.5182	0.5597	0.5938	0.6225	0.6472	0.6687	0.6876	0.7043
	0.90	0.2053	0.2523	0.3246	0.3805	0.4264	0.4651	0.4985	0.5277	0.5534	0.5763	0.5969
31	0.95	0.2457	0.2931	0.3650	0.4200	0.4647	0.5022	0.5342	0.5620	0.5866	0.6082	0.6278
	0.99	0.3290	0.3754	0.4444	0.4961	0.5374	0.5717	0.6008	0.6258	0.6477	0.6671	0.6843
	0.90	0.1940	0.2389	0.3087	0.3632	0.4081	0.4464	0.4794	0.5084	0.5342	0.5572	0.5781
33	0.95	0.2324	0.2781	0.3478	0.4015	0.4455	0.4825	0.5145	0.5424	0.5671	0.5890	0.6088
	0.99	0.3123	0.3573	0.4247	0.4757	0.5168	0.5511	0.5803	0.6057	0.6279	0.6476	0.6653
	0.90	0.1838	0.2270	0.2943	0.3473	0.3913	0.4290	0.4617	0.4906	0.5163	0.5394	0.5603
35	0.95	0.2206	0.2645	0.3320	0.3845	0.4277	0.4644	0.4962	0.5240	0.5487	0.5708	0.5907
	0.99	0.2972	0.3408	0.4066	0.4569	0.4974	0.5318	0.5612	0.5867	0.6091	0.6292	0.6471
	0.90	0.1747	0.2161	0.2812	0.3328	0.3759	0.4129	0.4453	0.4739	0.4995	0.5226	0.5434
37	0.95	0.2098	0.2521	0.3175	0.3689	0.4113	0.4475	0.4791	0.5068	0.5315	0.5536	0.5737
	0.99	0.2834	0.3257	0.3900	0.4394	0.4797	0.5138	0.5430	0.5688	0.5915	0.6117	0.6299
	0.90	0.1664	0.2062	0.2691	0.3194	0.3616	0.3980	0.4299	0.4583	0.4837	0.5067	0.5277
39	0.95	0.2001	0.2408	0.3044	0.3544	0.3961	0.4319	0.4631	0.4906	0.5152	0.5375	0.5576
	0.99	0.2709	0.3119	0.3747	0.4232	0.4631	0.4969	0.5260	0.5517	0.5747	0.5950	0.6134
	0.90	0.1589	0.1972	0.2582	0.3070	0.3483	0.3840	0.4155	0.4436	0.4689	0.4918	0.5127
41	0.95	0.1912	0.2306	0.2922	0.3411	0.3819	0.4172	0.4480	0.4754	0.4999	0.5221	0.5423
	0.99	0.2594	0.2993	0.3605	0.4079	0.4472	0.4812	0.5100	0.5358	0.5585	0.5792	0.5977

Table C.4 (Continued)

ν_1	$1-\alpha$	ν_2: 2	3	5	7	9	11	13	15	17	19	21
	0.90	0.1295	0.1619	0.2142	0.2571	0.2941	0.3267	0.3559	0.3823	0.4063	0.4284	0.4487
51	0.95	0.1565	0.1900	0.2434	0.2868	0.3239	0.3564	0.3853	0.4114	0.4350	0.4566	0.4765
	0.99	0.2140	0.2486	0.3029	0.3462	0.3828	0.4144	0.4424	0.4675	0.4900	0.5106	0.5295
	0.90	0.1093	0.1373	0.1830	0.2211	0.2544	0.2842	0.3111	0.3357	0.3583	0.3792	0.3987
61	0.95	0.1324	0.1615	0.2086	0.2474	0.2810	0.3109	0.3378	0.3622	0.3847	0.4054	0.4246
	0.99	0.1821	0.2126	0.2610	0.3004	0.3341	0.3637	0.3903	0.4142	0.4361	0.4561	0.4743
	0.90	0.0946	0.1191	0.1597	0.1939	0.2242	0.2514	0.2762	0.2991	0.3203	0.3401	0.3586
71	0.95	0.1148	0.1404	0.1824	0.2175	0.2481	0.2756	0.3006	0.3236	0.3447	0.3644	0.3828
	0.99	0.1584	0.1855	0.2293	0.2651	0.2963	0.3239	0.3488	0.3715	0.3924	0.4118	0.4298
	0.90	0.0833	0.1052	0.1417	0.1727	0.2003	0.2253	0.2483	0.2697	0.2896	0.3082	0.3257
81	0.95	0.1013	0.1242	0.1620	0.1939	0.2221	0.2475	0.2707	0.2922	0.3122	0.3308	0.3483
	0.99	0.1402	0.1646	0.2044	0.2374	0.2662	0.2919	0.3153	0.3368	0.3567	0.3751	0.3924
	0.90	0.0745	0.0942	0.1273	0.1556	0.1810	0.2042	0.2255	0.2455	0.2642	0.2818	0.2984
91	0.95	0.0906	0.1114	0.1457	0.1750	0.2010	0.2245	0.2462	0.2664	0.2852	0.3029	0.3195
	0.99	0.1257	0.1479	0.1844	0.2149	0.2416	0.2657	0.2874	0.3081	0.3270	0.3444	0.3608
	0.90	0.0673	0.0853	0.1155	0.1416	0.1651	0.1866	0.2066	0.2253	0.2428	0.2595	0.2752
101	0.95	0.0820	0.1009	0.1325	0.1594	0.1835	0.2055	0.2258	0.2447	0.2625	0.2792	0.2951
	0.99	0.1140	0.1343	0.1678	0.1961	0.2211	0.2436	0.2644	0.2836	0.3015	0.3184	0.3342
	0.90	0.0565	0.0717	0.0976	0.1200	0.1404	0.1593	0.1768	0.1934	0.2091	0.2240	0.2382
121	0.95	0.0688	0.0849	0.1120	0.1353	0.1563	0.1756	0.1936	0.2105	0.2264	0.2415	0.2559
	0.99	0.0960	0.1133	0.1423	0.1670	0.1891	0.2090	0.2274	0.2447	0.2610	0.2763	0.2907
	0.90	0.0427	0.0544	0.0744	0.0920	0.1081	0.1231	0.1372	0.1507	0.1635	0.1758	0.1876
161	0.95	0.0521	0.0645	0.0856	0.1039	0.1206	0.1361	0.1506	0.1644	0.1775	0.1900	0.2020
	0.99	0.0730	0.0864	0.1092	0.1288	0.1464	0.1627	0.1778	0.1921	0.2056	0.2185	0.2309

† Abridged from F. G. Foster and D. H. Rees: Upper percentage points of the generalized beta distribution I, *Biometrika*, vol. 44 (1957), pp. 237–247, and published here with the kind permission of the authors and the editor of *Biometrika*.

References

Aitchison, J., Habbema, J. D. F., and Kay, J. W. (1977). A critical comparison of two methods of discrimination. *J. Roy. Statist. Soc. C*, **26,** 15–25.

Anderson, E. (1960). A semigraphical method for the analysis of complex problems. *Technometrics*, **2,** 387–391.

Anderson, T. W. (1958). *An Introduction to Multivariate Statistical Analysis.* Wiley, New York.

Anderson, T. W. (1963). Asymptotic theory for principal component analysis. *Ann. Math. Statist.*, **34,** 122–148.

Anderson, T. W. and Bahadur, R. R. (1962). Classification into two multivariate normal distributions with different covariance matrices. *Ann. Math. Statist.*, **33,** 420–431.

Anderson, T. W., Das Gupta, S. P. and Styan, G. P. H. (1972). *A Bibliography of Multivariate Analysis.* Oliver and Boyd, Edinburgh.

Andrews, D. F. (1972). Plots of high dimensional data. *Biometrics*, **28,** 125–136.

Arrow, K. J., Chenery, H. B., Minhas, B. S., and Solow, R. M. (1961). Capital–labor substitution and economic efficiency. *Rev. Econ. Statist.*, **43,** 225–250.

Balakrishnan, V. and Sanghvi, L. D. (1968). Distance between populations on the basis of attribute data. *Biometrics*, **24,** 859–865.

Ball, G. H. and Hall, D. J. (1970). Some implications of inter-active graphic computer systems for data analysis and statistics. *Technometrics*, **12,** 17–31.

Barnard, G. A. (1976). Discussion to "The ordering of multivariate data" by Barnett, V. D., *J. Roy. Statist. Soc. A*, **139,** 318–355.

Barnett, V. D. (1976). The ordering of multivariate data. *J. Roy. Statist. Soc. A*, **139,** 318–355.

Barnett, V. D. and Lewis, T. (1963). A study of the relation between GCE and degree results. *J. Roy. Statist. Soc. A*, **126,** 187–216.

Bartlett, M. S. (1939). The standard errors of discriminant function coefficients. *J. Roy. Statist. Soc.*, Suppl. 6, 169–173.

Bartlett, M. S. (1947). Multivariate analysis. *J. Roy. Statist. Soc. B*, **9,** 176–197.

Bartlett, M. S. (1951). The effect of standardization on a χ^2 approximation in factor analysis (with an appendix by Lederman, W.). *Biometrika*, **38,** 337–344.

Bartlett, M. S. (1954). A note on multiplying factors for various chi-squared approximations. *J. Roy. Statist. Soc. B*, **16,** 296–298.

Bartlett, M. S. (1965). Multivariate statistics. In *Theoretical and Mathematical Biology* (Waterman, T. H. and Morowitz, H. J., eds). Blaisdell, New York.

Beale, E. M. L. (1970). Selecting an optimum subset. In *Integer and Nonlinear Programming* (Abadie, J., ed.). North-Holland, Amsterdam.

Beale, E. M. L., Kendall, M. G., and Mann, D. W. (1967). The discarding of variables in multivariate analysis. *Biometrika*, **54**, 357–366.

Bennett, B. M. (1951). Note on a solution of the generalised Behrens–Fisher problem. *Ann. Inst. Statist. Math.* **2**, 87–90.

Beran, R. J. (1968). Testing for uniformity on a compact homogeneous space. *J. Appl. Prob.*, **5**, 177–195.

Berce, R. and Wilbaux, R. (1935). Recherche statistique des relations existant entre le rendment des plantes de grande cultures et les facteurs meteorologiques en Belgique. *Bull. Inst. Agron. Stn. Rech. Gembloux*, **4**, 32–81.

Bhattacharyya, A. (1946). On a measure of divergence between two multinomial populations. *Sankhya*, **7**, 401–406.

Bingham, C. (1964). Distributions on the sphere and on the projective plane. Ph.D. thesis, Yale University.

Bingham, C. (1972). An asymptotic expansion for the distributions of the eigenvalues of a 3 by 3 Wishart matrix. *Ann. Math. Statist.*, **43**, 1498–1506.

Bingham, C. (1974). An antipodally symmetric distribution on the sphere. *Ann. Statist.*, **2**, 1201–1225.

Bingham, M. S. and Mardia, K. V. (1975). Maximum likelihood characterization of the von Mises distribution. In *A Modern Course on Statistical Distributions in Scientific Work* (Patil, G. P., Kotz, S., and Ord, J. K., eds). D. Reidel, Boston, Vol. 3, pp. 387–398.

Bock, R. D. (1975). *Multivariate Statistical Methods in Behaviourial Research.* McGraw-Hill, New York.

Box, G. E. P. (1949). A general distribution theory for a class of likelihood criteria. *Biometrika*, **36**, 317–346.

Box, G. E. P. and Tiao, G. C. (1973). *Bayesian Inference in Statistical Research.* Addison Wesley, Reading, Mass.

Cavalli-Sforza, L. L. and Edwards, A. W. F. (1967). Phylogenetic analysis: models and estimation procedures. *Evolution*, **21**, 550–570.

Cattell, R. B. (1966). The scree test for the number of factors. *Multivariate Behav. Res.*, **1**, 245–276.

Chernoff, H. (1973). Using faces to represent points in k-dimensional space graphically. *J. Amer. Statist. Assoc.*, **68**, 361–368.

Cochran, W. G. (1934). The distribution of quadratic forms in a normal system, with applications to the analysis of variance. *Proc. Camb. Phil. Soc.*, **30**, 178–191.

Cochran, W. G. (1962). On the performance of the linear discriminant function. *Bull. Inst. Intern. Statist.*, **39**, 435–447.

Cormack, R. M. (1971). A review of classification. *J. Roy. Statist. Soc. A*, **134**, 321–367.

Cornish, E. A. (1954). The multivariate t-distribution associated with a set of normal sample deviates. *Austral. J. Phys.*, **7**, 531–542.

Cox, D. R. (1972). The analysis of multivariate binary data. *Appl. Statist.*, **21**, 113–120.

Cox, D. R. and Brandwood, L. (1959). On a discriminatory problem connected with the works of Plato. *J. Roy. Statist. Soc. B*, **21**, 195–200.

Cox, D. R. and Hinkley, D. V. (1974). *Theoretical Statistics*. Chapman and Hall, London.

Craig, A. T. (1943). A note on the independence of certain quadratic forms. *Ann. Math. Statist.*, **14**, 195–197.

Dagnelie, P. (1975). *Analyse Statistique à Plusieurs Variables*. Vander, Brussels.

Day, N. E. (1969). Estimating the components of a mixture of normal distributions. *Biometrika*, **56**, 301–312.

Deemer, W. L. and Olkin, I. (1951). The Jacobians of certain matrix transformations useful in multivariate analysis. *Biometrika*, **38**, 345–367.

Delany, M. J. and Healy, M. J. R. (1966). Variation in white-toothed shrews in the British Isles. *Proc. Roy. Soc. B*, **164**, 63–74.

Dempster, A. P. (1969). *Elements of Continuous Multivariate Analysis*. Addison-Wesley, Reading, Mass.

Dempster, A. P. (1971). An overview of multivariate data analysis. *J. Multivariate Anal.*, **1**, 316–346.

Dempster, A. P. (1972). Covariance selection. *Biometrics*, **28**, 157–175.

Dhrymes, P. J. (1970). *Econometrics: Statistical Foundations and Applications*. Harper and Row, London.

Dickey, J. M. (1967). Matrix variate generalisations of the multivariate t distribution and the inverted multivariate t distribution. *Ann. Math. Statist.*, **38**, 511–518.

Downs, T. D. (1972). Orientation statistics. *Biometrika*, **59**, 665–675.

Dunnett, C. W. and Sobel, M. (1954). A bivariate generalization of Student's t distribution with tables for certain cases. *Biometrika*, **41**, 153–169.

Eaton, M. L. (1972). *Multivariate Statistical Analysis*. Institute of Mathematics and Statistics, University of Copenhagen.

Edwards, A. W. F. and Cavalli-Sforza, L. L. (1965). A method of cluster analysis. *Biometrics*, **21**, 362–375.

Enis, P. (1973). On the relation $E(Y) = E(E(X \mid Y))$. *Biometrika*, **60**, 432–433.

Everitt, B. (1974). *Cluster Analysis*. Heineman Educational, London.

Fisher, L. and van Ness, J. W. (1971). Admissible clustering procedures. *Biometrika*, **58**, 91–104.

Fisher, L. and van Ness, J. W. (1973). Admissible discriminant analysis. *J. Amer. Statist. Assoc.*, **68**, 603–607.

Fisher, R. A. (1936). The use of multiple measurements in taxonomic problems. *Ann. Eugen.*, **7**, 179–188.

Fisher, R. A. (1947). The analysis of covariance method for the relation between a part and a whole. *Biometrics*, **3**, 65–68.

Fisher, R. A. (1953). Dispersion on a sphere. *Proc. Roy. Soc. A*, **217**, 295–305.

Fisher, R. A. (1970). *Statistical Methods for Research Workers*, 14th ed. Oliver and Boyd, Edinburgh.

Foster, F. G. and Rees, D. H. (1957). Upper percentage points of the generalized beta distribution I. *Biometrika*, 44, 237–247.

Fréchet, M. (1951). Sur les tableaux de correlation dont les mages sont donnees. *Annales de l'Université de Lyon, Section A, Series* 3, **14**, 53–77.

Frets, G. P. (1921). Heredity of head form in man. *Genetica*, **3**, 193–384.

Friedman, H. P. and Rubin, J. (1967). On some invariate criteria for grouping data. *J. Amer. Statist. Assoc.*, **62**, 1159–1178.

Gabriel, K. R. (1970). On the relation between union intersection and likelihood ratio tests. In *Essays in Probability and Statistics*, University of North Carolina, Rayleigh.

Gentleman, W. M. (1965). Robust estimation of multivariate location by minimizing pth power derivatives. Ph.D. thesis, Princeton University.

Giri, N. (1968). On tests of the equality of two covariance matrices. *Ann. Math. Statist.*, **39**, 275–277.

Girshick, M. A. (1939). On the sampling theory of roots of determinantal equations. *Ann. Math. Statist.*, **10**, 203–224.

Girshick, M. A. and Haavelmo, T. (1947). Statistical analysis of the demand for food: examples of simultaneous estimation of structural equations. *Econometrica*, **15**, 79–110.

Glass, D. V. (ed.) (1954). *Social Mobility in Britain*. Routledge and Kegan Paul, London.

Gnanadesikan, R. (1977). *Methods for Statistical Data Analysis of Multivariate Observations*. Wiley, New York.

Gnanadesikan, M. and Gupta, S. S. (1970). A selection procedure for multivariate normal distributions in terms of the generalised variances. *Technometrics*, **12**, 103–117.

Gnanadesikan, R. and Kettenring, J. R. (1972). Robust estimates, residuals and outlier detection with multiresponse data. *Biometrics*, **28**, 81–124.

Goodman, L. A. (1972). Some multiplicative models for the analysis of cross classified data. *Proc. 6th Berk. Symp. Math. Statist. Prob.*, **1**, 649–696.

Goodman, N. R. (1963). Statistical analysis based on a certain multivariate complex Gausian distribution (an introduction). *Ann. Math. Statist.*, **34**, 152–177.

Gower, J. C. (1966). Some distance properties of latent root and vector methods in multivariate analysis. *Biometrika*, **53**, 315–328.

Gower, J. C. (1968). Adding a point to vector diagrams in multivariate analysis. *Biometrika*, **55**, 582–585.

Gower, J. C. (1971*a*). A general coefficient of similarity and some of its properties. *Biometrics*, **27**, 857–874.

Gower, J. C. (1971*b*). Statistical methods of comparing different multivariate analyses of the same data. In *Mathematics in the Archaeological and Historical* Sciences. (Hodson, F. R., Kendall, D. G., and Tautu, P., eds). Edinburgh University Press, Edinburgh, pp. 138–149.

Gower, J. C. and Ross, G. J. S. (1969). Minimum spanning trees and single linkage cluster analysis. *Appl. Statist.*, **18**, 54–64.

Graybill, F. A. (1969). *Introduction to Matrices with Applications in Statistics*. Wadsworth, Belmont, CA.

Green, B. F. (1952). The orthogonal approximation of an oblique structure in factor analysis. *Psychometrika*, **17**, 429–440.

Guttman, L. (1968). A general non-metric technique for finding the smallest co-ordinate space for a configuration of points. *Psychometrika*, **33**, 469–506.

Hartigan, J. A. (1973). Minimum mutation fits to a given tree. *Biometrics*, **29**, 53–65.

Hartigan, J. A. (1975). *Clustering Algorithms*. Wiley, New York.

Hartigan, J. A. (1978). Asymptotic distributions for clustering criteria. *Ann. Statist.*, **6**, 117–131.

Hill, M. O. (1974). Correspondence analysis: a neglected multivariate method. *Appl. Statist.*, **23**, 340–354.

Hill, R. C., Formby, T. B., and Johnson, S. R. (1977). Component selection norms for principal components regression. *Comm. Stat.-Theor. Meth. A*, **6**, 309–334.

Hoerl, A. E. and Kennard, R. W. (1970). Ridge regression. *Technometrics*, **12,** 55–67 and 69–82.

Holgate, P. (1964). Estimation for the bivariate Poisson distribution. *Biometrika*, **51,** 152–177.

Hooper, J. W. (1959). Simultaneous equations and canonical correlation theory. *Econometrica*, **27,** 245–256.

Hopkins, J. W. (1966). Some considerations in multivariate allometry. *Biometrics*, **22,** 747–760.

Horst, P. (1965). *Factor Analysis of Data Matrices*. Holt, Rinehart, and Winston, New York.

Hotelling, H. (1931). The generalization of Student's ratio. *Ann. Math. Statist.*, **2,** 360–378.

Hotelling, H. (1933). Analysis of a complex of statistical variables into principal components. *J. Educ. Psych.*, **24,** 417–441, 498–520.

Hotelling, H. (1936). Relations between two sets of variables. *Biometrika*, **28,** 321–377.

Huber, J. P. (1972). Robust statistics: a review. *Ann. Math. Statist.*, **43,** 1041–1067.

James, A. T. (1964). Distributions of matrix variates and latent roots derived from normal samples. *Ann. Math. Statist.*, **35,** 475–501.

James, G. S. (1954). Tests of linear hypotheses in univariate and multivariate analysis when the ratios of the population variances are unknown. *Biometrika*, **41,** 19–43.

Jardine, N. (1970). Algorithms, methods and models in the simplification of complex data. *Comp. J.*, **13,** 116–117.

Jardine, N. and Sibson, R. (1971). *Mathematical Taxonomy*. Wiley, New York.

Jeffers, J. N. R. (1967). Two case studies in the application of principal component analysis. *Appl. Statist.*, **16,** 225–236.

Jeffreys, H. (1961). *Theory of Probability*. Clarendon Press, Oxford.

Johnson, N. L. and Kotz, S. (1972). *Distributions in Statistics: Continuous Multivariate Distributions*. Wiley, New York.

Jolicoeur, P. and Mosimann, J. (1960). Size and shape variation in the painted turtle: a principal component analysis. *Growth*, **24,** 339–354.

Jolliffe, I. T. (1972). Discarding variables in principal component analysis I: artificial data. *Appl. Statist.*, **21,** 160–173.

Jolliffe, I. T. (1973). Discarding variables in principal component analysis II: real data. *Appl. Statist.*, **22,** 21–31.

Joreskog, K. G. (1967). Some contributions to maximum likelihood factor analysis. *Psychometrika*, **32,** 443–482.

Joreskog, K. G. (1970). A general method for analysis of covariance structures. *Biometrika*, **57,** 239–251.

Joreskog, K. G. and Sorbom, D. (1977). Statistical models and methods for analysis of longtitudinal data. In *Latent Variables in Socio-economic Models* (Aigner, D. J. and Goldberger, A. S., eds). North-Holland, Amsterdam.

Kagan, A. M., Linnik, Y. V., and Rao, C. R. (1973). *Characterization Problems in Mathematical Statistics*. Wiley, New York.

Kaiser, H. F. (1958). The varimax criterion for analytic rotation in factor analysis. *Psychometrika*, **23,** 187–200.

Kelker, D. (1972). Distribution theory of spherical distributions and a location-scale parameter generalization. *Sankhyā A*, **32,** 419–430.

Kendall, D. G. (1971). Seriation from abundance matrices. In *Mathematics in the*

Archaeological and Historical Sciences (Hodson, F. R., Kendall, D. G., and Tautu, P., eds.). Edinburgh University Press, Edinburgh, pp. 215–251.

Kendall, M. G. (1975). *Multivariate Analysis.* Griffin, London.

Kendall, M. G. and Stuart, A. (1958). *The Advanced Theory of Statistics,* Vol. 1, Griffin, London.

Kendall, M. G. and Stuart, A. (1967). *The Advanced Theory of Statistics.* Vol. 2, Griffin, London.

Kendall, M. G. and Stuart, A. (1973). *The Advanced Theory of Statistics.* Vol. 3, Griffin, London.

Kent, J. T. (1978). Limiting behaviour of the von Mises–Fisher distribution. *Math. Proc. Camb. Phil. Soc.,* **84,** 531–536.

Khatri, C. G. (1965). Classical statistical analysis based on a certain multivariate complex Gaussian distribution. *Ann. Math. Statist.,* **36,** 98–114.

Khatri, C. G. and Mardia, K. V. (1977). The von Mises–Fisher matrix distribution in orientation statistics. *J. Roy. Statist. Soc. B,* **39,** 95–106.

Khatri, C. G. and Pillai, K. C. S. (1965). Some results on the non-central multivariate beta distribution and moments of traces of two matrices. *Ann. Math. Statist.,* **36,** 1511–1520.

Kiefer, J. and Schwartz, R. (1965). Admissible Bayes character of T^2, R^2 and other fully invariant tests for classical multivariate normal problems. *Ann. Math. Statist.,* **36,** 747–770.

Kmenta, J. (1971). *Elements of Econometrics.* Collier-Macmillan, London.

Korin, B. P. (1968). On the distribution of a statistic used for testing a covariance matrix. *Biometrika,* **55,** 171–178.

Krishnamoorthy, A. S. (1951). Multivariate binomial and Poisson distributions. *Sankhyā,* **11,** 117–124.

Kruskal, J. B. (1964). Non-metric multidimensional scaling. *Psychometrika,* **29,** 1–27, 115–129.

Kshirsagar, A. M. (1960). Some extensions of multivariate t-distribution and the multivariate generalization of the distribution of the regression coefficient. *Proc. Camb. Phil. Soc.,* **57,** 80–86.

Kshirsagar, A. M. (1961). The non-central multivariate beta distribution. *Ann. Math. Statist.,* **32,** 104–111.

Kshirsagar, A. M. (1972). *Multivariate Analysis.* Marcell Dekker, New York.

Lancaster, H. O. (1969). *The Chi-squared Distribution.* Wiley, New York.

Lance, G. N. and Williams, W. J. (1967). A general theory of classificatory sorting strategies: 1 hierarchial systems. *Comp. J.,* **9,** 373–380.

Lawley, D. N. (1959). Test of significance in canonical analysis. *Biometrika,* **46,** 59–66.

Lawley, D. N., and Maxwell, A. E. (1971). *Factor Analysis as a Statistical Method,* 2nd ed. Butterworths, London.

Lewis, T. (1970). 'The Statistician as a Member of Society'. Inaugural lecture, University of Hull.

Lévy, P. (1937). *Théorie de l'Addition des Variables Aléatoires.* Gauthier-Villars, Paris.

Lingoes, J. C. (1971). Some boundary conditions for a monotone analysis of symmetric matrices. *Psychometrika,* **36,** 195–203.

Lingoes, J. C. and Roskam, E. E. (1973). A mathematical and empirical analysis of two multidimensional scaling algorithms. *Psychometrika,* **38,** Monograph Suppl. No. 19.

Lubischew, A. A. (1962). On the use of discriminant functions in taxonomy. *Biometrics*, **18**, 455–477.

Lyttleton, R. A. (1953). *The Comets and their Origin*. Cambridge University Press, Cambridge.

Lyttleton, R. A. (1961). On the statistical loss of long period comets from the solar system I. *Proc. 4th Berkeley Symp. Math. Statist. Prob.*, **3**, 229–244.

McRae, D. J. (1971). MICKA, a FORTRAN IV iterative K-means cluster analysis program. *Behavioural Sci.* **16**, 423–424.

Mahalanobis, P. C. (1957). The foundations of statistics. *Sankhyā*, **A**, **18**, 183–194.

Mardia, K. V. (1962). Multivariate Pareto distributions. *Ann. Math. Statist.*, **33**, 1008–1015.

Mardia, K. V. (1964*a*). Some results on the order statistics of the multivariate normal and Pareto type I populations. *Ann. Math. Statist.*, **35**, 1815–1818.

Mardia, K. V. (1964*b*). Exact distributions of extremes, ranges and midranges in samples from any multivariate population. *J. Indian Statist. Assoc.*, **2**, 126–130.

Mardia, K. V. (1967*a*). Correlation of the ranges of correlated samples. *Biometrika*, **54**, 529–539.

Mardia, K. V. (1967*b*). A non-parametric test for the bivariate two-sample location problem. *J. Roy. Statist. Soc. B*, **29**, 320–342.

Mardia, K. V. (1967*c*). Some contributions to contingency-type bivariate distributions. *Biometrika*, **54**, 235–249.

Mardia, K. V. (1968). Small sample power of a non-parametric test for the bivariate two-sample location problem. *J. Roy. Statist. Soc. B*, **30**, 83–92.

Mardia, K. V. (1969). On the null distribution of a non-parametric test for the two-sample problem. *J. Roy. Statist. Soc. B*, **31**, 98–102.

Mardia, K. V. (1970*a*). Measures of multivariate skewness and kurtosis with applications *Biometrika*, **57**, 519–530.

Mardia, K. V. (1970*b*). *Families of Bivariate Distributions*. Griffin, London. (No. 27 of Griffin's Statistical Monographs and Courses.)

Mardia, K. V. (1970c). A translation family of bivariate distributions and Fréchet bounds. *Sankhyā*, **A**, **32**, 119–122.

Mardia, K. V. (1971). The effect of non-normality on some multivariate tests and robustness to non-normality in the linear model. *Biometrika*, **58**, 105–121.

Mardia, K. V. (1972*a*). *Statistics of Directional Data*. Academic Press, London.

Mardia, K. V. (1972b). A multisample uniform scores test on a circle and its parametric competitor. *J. Roy. Statist. Soc.*, **B**, **34**, 102–113.

Mardia, K. V. (1974). Applications of some measures of multivariate skewness and kurtosis in testing normality and robustness studies. *Sankhyā B*, **36**, 115–128.

Mardia, K. V. (1975*a*). Assessment of multinormality and the robustness of Hotelling's T^2 test. *J. Roy. Statist. Soc. C*, **24**, 163–171.

Mardia, K. V. (1975*b*). Distribution theory for the von Mises–Fisher distribution and its applications. In *A Modern Course on Statistical Distributions in Scientific Work* (Patil, G. P., Kotz, S., and Ord, J. K., eds). D. Reidel, Boston, pp. 113–130.

Mardia, K. V. (1975*c*). Characterization of directional distributions. In *A Modern Course on Statistical Distributions in Scientific Work* (Patil, G. P., Kotz, S., and Ord, J. K., eds). D. Reidel, Boston, pp. 364–385.

Mardia, K. V. (1975*d*). Statistics of directional data (with discussion). *J. Roy. Statist. Soc. B*, **37,** 349–393.
Mardia, K. V. (1977). Mahalanobis distance and angles. In "Multivariate Analysis—IV". Krishnaiah, P. R. (ed.). North Holland, Amsterdam.
Mardia, K. V. (1978). Some properties of classical multidimensional scaling. *Comm. Statist.-Theor. Meth A*, **7,** 1233–1241.
Mardia, K. V. and Spurr, B. D. (1977). On some tests for the bivariate two-sample location problem. *J. Amer. Statist. Assoc.*, **72,** 994–995.
Mardia, K. V. and Thompson, J. W. (1972). Unified treatment of moment formulae. *Sakhyā A*, **34,** 121–132.
Mardia, K. V. and Zemroch, P. J. (1978). Tables of the F- and Related Distributions with Algorithms. Academic Press, London.
Marriott, F. H. (1952). Tests of significance in canonical analysis. *Biometrika*, **39,** 58–64.
Marriott, F. H. (1971). Practical problems in a method of cluster analysis. *Biometrics*, **27,** 501–514.
Marshall, A. W. and Olkin, I. (1967). A multivariate exponential distribution. *J. Amer. Statist. Assoc.*, **62,** 30–44.
Massy, W. F. (1965). Principal component analysis in exploratory data research. *J. Amer. Statist. Assoc.*, **60,** 234–256.
Maxwell, A. E. (1961). Recent trends in factor analysis. *J. Roy. Statist. Soc. A*, **124,** 49–59.
Merrington, M. and Thompson, C. M. (1943). Tables of percentage points of the inverted beta (F) distribution. *Biometrika*, **33,** 74–88.
Mitra, S. K. (1969). Some characteristic and non-characteristic properties of the Wishart distribution. *Sankhyā A*, **31,** 19–22.
Morgenstern, D. (1956). Einfache Beispiele Zweidimesionaler Verteilungen. *Mill. Math. Statist.*, **8,** 234–235.
Morrison, D. F. (1976). *Multivariate Statistical Methods*, 2nd ed. McGraw-Hill, New York.
Mosimann, J. E. (1970). Size allometry: size and shape variables with characterizations of the lognormal and generalized gamma distributions. *J. Amer. Statist. Assoc.*, **65,** 930–945.
Narain, R. D. (1950). On the completely unbiased character of tests of independence in multivariate normal system. *Ann. Math. Statist.*, **21,** 293–298.
Ogawa, J. (1949). On the independence of linear and quadratic forms of a random sample from a normal population. *Ann. Inst. Math. Statist.*, **1,** 83–108.
Olkin, I. and Tomsky, T. L. (1975). A new class of multivariate tests based on the union-intersection principle. *Bull. Inst. Statist. Inst.*, **46,** 202–204.
Pearce, S. C. (1965). The measurement of a living organism. *Biometrie-praximetrie*, **6,** 143–152.
Pearce, S. C. and Holland, D. A. (1960). Some applications of multivariate methods in botany. *Appl. Statist.*, **9,** 1–7.
Pearson, E. S. (1956). Some aspects of the geometry of statistics. *J. Roy. Statist. Soc. A*, **119,** 125–146.
Pearson, E. S. and Hartley, H. O. (1972). *Biometrika Tables for Statisticians*, Vol. 2. Cambridge University Press, Cambridge.
Pearson, K. (1901). On lines and planes of closest fit to systems of points in space. *Phil. Mag.* (6), **2,** 559–572.
Pillai, K. C. S. (1955). Some new test criteria in multivariate analysis. *Ann. Math. Statist.*, **26,** 117–121.

Pillai, K. C. S. and Jayachandran, K. (1967). Power comparison of tests of two multivariate hypotheses based on four criteria. *Biometrika*, **54**, 195–210.

Press, S. J. (1972). *Applied Multivariate Analysis*. Holt, Rinehart, and Winston, New York.

Puri, M. L. and Sen, P. K. (1971). *Non-parametric Methods in Multivariate Analysis*. Wiley, New York.

Ramabhandran, V. R. (1951). A multivariate gamma-type distribution. *Sankhyā*, **11**, 45–46.

Rao, C. R. (1948). Tests of significance in multivariate analysis. *Biometrika*, **35**, 58–79.

Rao, C. R. (1951). An asymptotic expansion of the distribution of Wilks' criterion. *Bull. Inst. Internal. Statist.*, **33**, 177–180.

Rao, C. R. (1964). The use and interpretation of principal component analysis in applied research. *Sankhyā A*, **26**, 329–358.

Rao, C. R. (1966). Covariance adjustment and related problems in multivariate analysis. In *Multivariate Analysis* (Krishnaiah, P. R., ed.). Academic Press, New York.

Rao, C. R. (1971). Taxonomy in anthropology. In *Mathematics in the Archaeological and Historical Sciences* (Hodson, F. R., Kendall, D. G., and Tautu, P., eds). Edinburgh University Press, Edinburgh.

Rao, C. R. (1973). *Linear Statistical Inference and its Applications*. Wiley, New York.

Rao, C. R. and Mitra, S. K. (1971). *Generalised Inverse of Matrices and its Applications*. Wiley, New York.

Rayleigh, Lord (1919). On the problems of random vibrations and of random flights in one, two or three dimensions. *Phil. Mag.* (6), **37**, 321–347.

Reeve, E. C. R. (1941). A statistical analysis of taxonomic differences within the genus Tamandu Gray (*Xenorthra*). *Proc. Zool. Soc. Lond. A*, **111**, 279–302.

Rosenblatt, M. (1952). Remarks on a multivariate transformation. *Ann. Math. Statist.*, **23**, 470–472.

Ross, G. J. S. (1969). Single linkage cluster analysis (Algorithms AS 13–15). *Appl. Statist.*, **18**, 103–110.

Rothkopf, E. Z. (1957). A measure of stimulus similarity and errors in some paired-associate learning tasks. *J. Exp. Psychol.*, **53**, 94–101.

Roy, S. N. (1957). *Some Aspects of Multivariate Analysis*. Wiley, New York.

Roy, S. N. and Bose, R. C. (1953). Simultaneous confidence interval estimation. *Ann. Math. Statist.*, **24**, 513–536.

Sanghvi, L. D. (1953). Comparison of genetical and morphological methods for a study of biological differences. *Amer. J. Phys. Anthrop.*, **11**, 385–404.

Sanghvi, L. D. and Balakrishnan, V. (1972). Comparison of different measures of genetic distance between human populations. In *The Assessment of Population Affinities in Man* (Weiner, J. A. and Huizinga, H., eds). Clarendon Press, Oxford.

Schatzoff, M. (1966). Exact distributions of Wilks's likelihood ratio criterion. *Biometrika*, **53**, 347–358.

Schmidt-Koenig, K. (1965). Current problems in bird orientation. In *Advances in the Study of Behaviour* (Lehrman, D. S., *et al.*, eds). Academic Press, New York.

Schoenberg, I. J. (1935). Remarks to Maurice Fréchet's article "Sur la definition axiomatique d'une classe d'espace distanciés vectoriellement applicable sur l'espace de Hilbert". *Ann. Math.*, **36**, 724–732.

Scott, A. J. and Symonds, M. J. (1971). Clustering methods based on likelihood ratio criteria. *Biometrics*, **27**, 387–397.

Seal, H. L. (1964). *Multivariate Statistical Analysis for Biologists*. Methuen, London.

Shaver, R. H. (1960). The Pennsylvanian ostracode *Bairdia oklahomaensis* in Indiana. *J. Paleontology*, **34**, 656–670.

Shepard, R. N. (1962*a*). The analysis of proximities: multidimensional scaling with an unknown distance function I. *Psychometrika*, **27**, 125–139.

Shepard, R. N. (1962*b*). The analysis of proximities: multidimensional scaling with an unknown distance function II. *Psychometrika*, **27**, 219–246.

Shepard, R. N. (1963). An analysis of proximities as a technique for the study of information processing in man. *Human Factors*, **5**, 19–34.

Sibson, R. (1978). Studies in the robustness of multidimensional scaling: Procrustes statistics. *J. R. Statist. Soc.*, **B**, **40**, 234–238.

Silvey, S. D. (1970). *Statistical Inference*. Penguin, Baltimore.

Siskind, V. (1972). Second moments of inverse Wishart-matrix elements. *Biometrika*, **59**, 691–692.

Smith, C. A. B. (1947). Some examples of discrimination. *Ann. Eugen.*, **13**, 272–282.

Sneath, P. H. A. and Sokal, R. R. (1973). *Numerical Taxonomy*. W. H. Freeman & Co., San Francisco.

Sokal, R. R. and Michener, C. D. (1958). A statistical method for evaluating systematic relationships. *Univ. Kansas Sci. Bull.*, **38**, 1409–1438.

Spearman, C. (1904). The proof and measurement of association between two things. *Am. J. Psychol.*, **15**, 72 and 202.

Sprent, P. (1969). *Models in Regression and Related Topics*. Methuen, London.

Sprent, P. (1972). The mathematics of size and shape. *Biometrics*, **28**, 23–37.

Stein, C. M. (1956). Inadmissibility of the usual estimator for the mean of a multivariate normal distribution. *Proc. 3rd Berk. Symp. Math. Stat. Prob.*, **1**, 197–206.

Stephens, M. A. (1962). Exact and appropriate tests for directions, I. *Biometrika*, **49**, 463–477.

Subrahmaniam, K. and Subrahmaniam, K. (1973). *Multivariate Analysis. A Selected and Abstracted Bibliography* 1957–1972. Marcel Dekker, New York.

Sugiyama, T. and Tong, H. (1976). On a statistic useful in dimensionality reduction in multivariate linear stochastic system. *Commun. Statist.-Theor. Meth. A*, **5**, 711–721.

Thompson, C. M. (1941). Tables of percentage points of the χ^2 distribution. *Biometrika*, **32**, 187–191.

Thompson, M. L. (1978). Selection of variables in multiple regression I, II. *Int. Statist. Rev.*, **46**, 1–20, 129–146.

Torgerson, W. S. (1958). *Theory and Methods of Scaling*. Wiley, New York.

Tyror, J. G. (1957). The distribution of the directions of perihelion of long-period comets. *Mon. Not. Roy. Astron. Soc.*, **117**, 369–379.

Vinod, H. D. (1976). Canonical ridge and econometrics of joint production. *J. Econometrics*, **4**, 147–166.

Von Mises, R. (1918). Uber die "Ganzahligkeit" der Atomgewicht und verwante Fragen. *Physikal. Z.*, **19**, 490–500.

Watson, G. N. (1944). *A Treatise on the Theory of Bessel Functions*, 2nd ed. Cambridge University Press, Cambridge.

Watson, G. S. (1956). Analysis of dispersion on a sphere. *Mon. Not. Roy. Astron. Soc. Geophys. Suppl.*, **7,** 153–159.

Watson, G. S. and Williams, E. J. (1956). On the construction of significance tests on the circle and the sphere. *Biometrika*, **43,** 344–352.

Weinman, D. G. (1966). A multivariate extension of the exponential distribution. Ph. D. thesis, Arizona State University.

Welch, B. L. (1939). Note on discriminant functions. *Biometrika*, **31,** 218–220.

Welch, B. L. (1947). The generalization of Student's problem when several populations are involved. *Biometrika*, **34,** 28–35.

Wilk, M. B., Gnanadesikan, R., and Huyett, M. J. (1962). Probability plots for the gamma distribution. *Technometrics*, **4,** 1–20.

Wilkinson, J. H. (1965). *The Algebraic Eigenvalue Problem*. Clarendon Press, Oxford.

Wilks, S. S. (1962). *Mathematical Statistics*. Wiley, New York.

Wolfe, J. H. (1971). A Monte Carlo study of the sampling distribution of the likelihood ratio for mixtures of multinormal distributions. *Naval Personnel and Training Res. Lab. Tech. Bull., STB* 72-2.

Wooding, R. A. (1956). The multivariate distribution of complex normal variables. *Biometrika*, **43,** 329–350.

Wright, S. (1954). The interpretation of multivariate systems. In *Statistics and Mathematics in Biology* (Kempthorne, O., *et al.*, eds). Iowa State University Press.

Yao, Y. (1965). An approximate degrees of freedom solution to the multivariate Behrens–Fisher problem. *Biometrika*, **52,** 139–147.

Young, G. and Householder, A. S. (1938). Discussion of a set of points in terms of their mutual distances. *Psychometrika*, **3,** 19–22.

List of Main Notations and Abbreviations

B	between groups SSP matrix
$B(\cdot\,,\cdot)$	beta variable
BLUE	best linear unbiased estimate
$C(\mathbf{x}, \mathbf{y})$	covariance between **x** and **y**
c.f.	characteristic function
D	distance matrix
D^2	Mahalanobis distance
d.f.	distribution function
E	expectation
$F(\cdot)$	distribution function
$f(\cdot)$	probability function
FIML	full information maximum likelihood
GLS	generalized least squares
H	centring matrix
I	identity matrix
i.i.d.	independent and identically distributed
ILS	indirect least squares
IV	instrumental variables
L	likelihood
l	log likelihood
LIML	limited information maximum likelihood
LRT	likelihood ratio test
MANOVA	multivariate analysis of variance
MDS	multidimensional scaling
ML	maximum likelihood
m.l.e.	maximum likelihood estimate
$N_p(\boldsymbol{\mu}, \boldsymbol{\Sigma})$	multinormal distribution
OLS	ordinary least squares
PCA	principal component analysis
p.d.f.	probability density function
S	sample covariance matrix
SLC	standardized linear combination
SSP	sums of squares and products

$\mathbf{S}_u$	unbiased sample covariance matrix
T^2	Hotelling T^2 statistic
UIT	union intersection test
$V(\mathbf{x})$	variance-covariance matrix of $\mathbf{x}$
$\mathbf{W}$	within groups SSP matrix
$W_p(\mathbf{\Sigma}, m)$	Wishart distribution
$\mathbf{X}$	data matrix
Λ	Wilks' Λ statistic
θ	greatest root statistic

Subject Index

LaVergne, TN USA
14 February 2010
173075LV00001B/1/A

9 780124 712522